# FLORA OF WEST TROPICAL AFRICA

# FLORA

OF

# WEST TROPICAL AFRICA

ALL TERRITORIES IN WEST AFRICA SOUTH
OF LATITUDE 18°N. AND TO THE WEST OF
LAKE CHAD, AND FERNANDO PO

FIRST EDITION BY

## J. HUTCHINSON, LL.D., F.R.S., V.M.H., F.L.S.

FORMERLY KEEPER OF THE MUSEUMS OF BOTANY, ROYAL BOTANIC GARDENS, KEW

AND

## J. M. DALZIEL, M.D., B.Sc., F.L.S.

FORMERLY OF THE WEST AFRICAN MEDICAL SERVICE, AND
ASSISTANT FOR WEST AFRICA, ROYAL BOTANIC GARDENS, KEW

SECOND EDITION

REVISION EDITED BY

## F. N. HEPPER, B.Sc., F.L.S.

PRINCIPAL SCIENTIFIC OFFICER, ROYAL BOTANIC GARDENS, KEW

PREPARED AND REVISED AT THE
HERBARIUM, ROYAL BOTANIC GARDENS,
KEW, UNDER THE SUPERVISION OF THE
DIRECTOR

## VOL. III.  PART 1

28TH AUGUST, 1968

PUBLISHED ON BEHALF OF THE GOVERN-
MENTS OF NIGERIA, GHANA, SIERRA LEONE
AND THE GAMBIA

BY THE

CROWN AGENTS FOR OVERSEA GOVERNMENTS      ©
AND ADMINISTRATIONS
MILLBANK, LONDON, S.W.1

*Price 24s.*

*Printed in Great Britain by*
*The Whitefriars Press Ltd., London and Tonbridge.*

# PHYLOGENETIC SEQUENCE OF ORDERS AND FAMILIES CONTAINED IN VOLUME III

(Families Nos. 163–199 in Part 1, Nos. 200–202 in Part 2)
(The cross-lines show breaks in affinity)

KEY TO THE FAMILIES OF MONOCOTYLEDONS (by J. E. Dandy), p. 1.

## ABBREVIATIONS

For a list of abbreviations used in this volume see Vol. II, pp. viii, ix.

# ANGIOSPERMAE

## MONOCOTYLEDONS

## KEY TO THE FAMILIES OF MONOCOTYLEDONS

By J. E. Dandy

Gynoecium composed of 2 or more free or almost free carpels with separate styles and
stigmas  ..    ..    ..    ..    ..    ..    ..    ..    *Group 1* (see below)
Gynoecium composed of 1 carpel or of 2 or more united carpels with free or united
styles and stigmas:
Ovary superior; perianth present or absent:
Perianth present, composed of 4 or more free or united segments, not reduced to
bristles:
Perianth composed of separate calyx and corolla, the calyx often herbaceous, the
corolla petaloid or otherwise different from the calyx, the sepals and petals either
free or united among themselves but not all united into a single perianth-tube
*Group 2* (see p. 2)
Perianth composed of similar or subsimilar segments in 2 series, petaloid or sometimes
herbaceous or dry and glumaceous, either free or all united into a single perianth-
tube  ..    ..    ..    ..    ..    ..    ..    ..    *Group 3* (see p. 2)
Perianth absent or reduced to bristles or to 1–3 free or united scales:
Flowers solitary or few together or arranged in heads or spikes, not enclosed by
scale-like bracts; ovary 1- or more-locular with 1 or more ovules
*Group 4* (see p. 2)
Flowers (florets) arranged in small spikes (spikelets) and enclosed by scale-like bracts
(glumes or lemmas), the spikelets sometimes 1-flowered; ovary 1-locular with
1 ovule; grasses and sedges  ..    ..    ..    ..    *Group 5* (see p. 3)
Ovary inferior or semi-inferior; perianth present:
Androecium composed of 2 or more fertile stamens    ..    *Group 6* (see p. 3)
Androecium composed of 1 fertile stamen sometimes accompanied by 1 or more
petaloid staminodes ..    ..    ..    ..    ..    ..    *Group 7* (see p. 4)

*Group 1*

Saprophytic herbs; leaves reduced to scales  ..    ..    166. **Triuridaceae** (3 : 14)
Plants not saprophytic; leaves well developed, not reduced to scales:
Leaves pinnate; palms ..    ..    ..    ..    ..    ..    193. **Palmae** (3 : 159)
Leaves simple; aquatic or marsh herbs:
Perianth absent; stamens 2, free or united; submerged aquatic plants with narrow
linear leaves:
Flowers bisexual, in 2-flowered spikes; anthers free; carpels 4 or more, with sessile
peltate stigmas; plants of saline water  ..    ..    169. **Ruppiaceae** (3 : 18)
Flowers dioecious, solitary, axillary; anthers united; carpels 2, with 2 filiform
stigmas; plants marine  ..    ..    ..    ..    170. **Zannichelliaceae** (3 : 19)
Perianth present, composed of 2–6 free segments; stamens 3 or more:
Flowers with bracts, whorled in simple or compound inflorescences, the inflorescence
sometimes spike-like or forming an umbel of 1 or more flowers; perianth composed of 3 sepals and 1–3 petals, or sometimes without petals:
Carpels slightly united at the base, in 1 whorl, with numerous ovules scattered over
the wall of the ovary; marsh plants; inflorescence an umbel, sometimes with an
additional whorl of flowers below    ..    ..    .. 163. **Butomaceae** (3 : 5)
Carpels free, in 1 whorl or spirally arranged, with 1 basal ovule; aquatic or marsh
plants  ..    ..    ..    ..    ..    ..    165. **Alismataceae** (3 : 9)
Flowers without bracts, spicate; perianth composed of 2–4 similar segments; plants
aquatic:
Leaves radical; spikes simple or 2-branched on elongated peduncles, at first
enclosed in a spathe; perianth-segments 2–3; anthers borne on elongated
filaments; ovules 2 or more in each carpel    167. **Aponogetonaceae** (3 : 15)
Leaves borne on elongated stems; spikes simple on axillary peduncles, without a
spathe; perianth-segments 4; anthers sessile; ovule 1 in each carpel
168. **Potamogetonaceae** (3 : 16)

1

*Group 2*

Leaves pinnate or flabellate or bifurcate; palms, sometimes climbing

                193. **Palmae** (3: 159)

Leaves simple, not flabellate or bifurcate; herbs:

 Ovary 1-locular with 3 parietal placentas bearing numerous ovules; petals united below into a tube; stamens 3, inserted at the mouth of the corolla-tube and alternating with staminodes; leaves all or mostly radical; flowers in bracteate spikes or heads on elongated peduncles .. .. .. 174. **Xyridaceae** (3: 51)

 Ovary 2–3-locular with axile or apical placentas bearing 1 or more ovules; petals free or united; stamens 6 or fewer, sometimes accompanied by staminodes:

  Style 2–6-branched; flowers monoecious, small, in bracteate heads on elongated peduncles; corolla inconspicuous, not brightly coloured, often minute in the male flowers; leaves narrow, radical or crowded .. 176. **Eriocaulaceae** (3: 57)

  Style simple, unbranched; flowers bisexual or polygamous; corolla conspicuous, often blue or yellow:

   Anthers opening by a terminal pore; flowers actinomorphic, in shortly pedunculate bracteate heads; petals united below into a slender tube; stamens 6, all fertile; ovary with 1 ovule in each loculus .. .. 175. **Rapateaceae** (3: 55)

   Anthers opening by longitudinal slits; flowers actinomorphic or zygomorphic, in lax or dense cymes or panicles often subtended by spathaceous bracts; petals free or united below; stamens 6, all or only 2–5 fertile; ovary with 1 or more ovules in each loculus .. .. .. .. .. .. 172. **Commelinaceae** (3: 22)

*Group 3*

Perianth-segments dry and glumaceous, free; flowers small, clustered in lax or dense (sometimes head-like) terminal panicles; leaves narrow or reduced to sheaths; rushes .. .. .. .. .. .. .. .. ..200. **Juncaceae** (3, Pt. 2)

Perianth-segments petaloid or herbaceous, free or united:

 Aquatic herbs; leaves with a lanceolate to ovate-orbicular lamina, or linear and submerged; flowers solitary or in spikes or racemes; perianth-segments free or united below into a tube .. .. .. .. .. 184. **Pontederiaceae** (3: 108)

 Plants terrestrial:

  Leaves reduced to spines, their function fulfilled by leaf-like acicular or flattened branches (cladodes); herbs or shrubs, often climbing .. 182. **Liliaceae** (3: 90)

  Leaves well developed, sometimes appearing after the flowers:

   Inflorescence a spadix enclosed in a spathe; ovary 1–2-locular; leaves radical, with a cordate or sagittate or hastate lamina .. .. .. 186. **Araceae** (3: 112)

   Inflorescence not a spadix enclosed in a spathe; ovary 3-locular:

    Flowers dioecious, small, in axillary umbels; leaves with reticulate venation and stipular tendrils; climbing shrubs with prickly branches

                185. **Smilacaceae** (3: 111)

    Flowers bisexual; leaves without stipular tendrils:

     Ovary with 2 or more ovules in each loculus; fruit a loculicidal or septicidal capsule; herbs, sometimes climbing .. .. 182. **Liliaceae** (3: 90)

     Ovary with 1 ovule in each loculus; fruit not a capsule:

      Perianth-segments free; fruit a drupe; climbing herbs, the leaves ending in tendrils .. .. .. .. .. .. 173. **Flagellariaceae** (3: 50)

      Perianth-segments united below into a tube; fruit a berry or with a thin pericarp falling away from the berry-like seeds; herbs or shrubs or trees, sometimes climbing, the leaves not ending in tendrils .. ..192. **Agavaceae** (3: 154)

*Group 4*

Leaves absent; minute aquatic herbs, the plant body reduced to a thallus-like " frond " with or without 1 or more pendent rootlets .. .. ..187. **Lemnaceae** (3: 127)

Leaves (or solitary leaf) present, well developed, sometimes appearing after the flowers:

 Leaves (or solitary leaf) petiolate, sometimes 3-partite or much divided, the lamina if undivided often cordate or sagittate or hastate; inflorescence a spadix subtended by or enclosed in a spathe, sometimes with a terminal sterile appendix above the flowers; flowers small, unisexual or sometimes bisexual; herbs or woody climbers, sometimes epiphytic or aquatic .. .. .. 186. **Araceae** (3: 112)

 Leaves sessile, simple and undivided, not cordate or sagittate or hastate; flowers unisexual:

  Leaves opposite, linear, coarsely or minutely toothed, borne on elongated stems; flowers solitary or few together in the axils of the leaves; submerged aquatic herbs

               171. **Najadaceae** (3: 20)

  Leaves alternate; flowers in spikes or heads:

   Trees with aerial roots; leaves broadly linear, armed with spiny teeth on the margin and often also on the midrib beneath .. .. 194. **Pandanaceae** (3: 170)

Herbs, often aquatic; leaves without spiny teeth:
 Floating aquatic plants; leaves obovate or oblong-spathulate, hairy, arranged in a
 cup-like rosette; inflorescence a spadix enclosed in a spathe and adnate to it
 below, with 1 female flower at the base and several male flowers above
186. Araceae (3 : 112)

 Plants not floating; leaves linear or filiform, glabrous:
  Flowers in bracteate heads; ovary 2–3-locular with 1 ovule in each loculus; style
  branched, with 2–3 filiform stigmas .. .. 176. Eriocaulaceae (3 : 57)
  Flowers in dense cylindric spikes, the female below, the male above; ovary
  1-locular with 1 ovule; style simple, unbranched, with 1 stigma
188. Typhaceae (3 : 129)

Group 5

Florets each enclosed by a bract (lemma) on the outside and a bracteole (palea) on the
 inside, the spikelet often having 2 empty bracts (glumes) at its base; perianth absent
 or represented by 2–3 minute scales (lodicules); stems terete or compressed and often
 with hollow internodes; seed adnate to or sometimes free from the pericarp of the
 fruit; grasses, herbaceous or sometimes woody (bamboos) 202 .Gramineae (3, Pt. 2)
Florets each enclosed by a single bract (glume) on the outside, without a bracteole or
 female florets each surrounded by a closed bracteole (utricle); perianth absent or
 represented by bristles or scales; stems often triquetrous and often solid; seed free
 from the pericarp of the fruit; sedges, herbaceous or sometimes arborescent
201. Cyperaceae (3, Pt. 2)

Group 6

Aquatic herbs with submerged leaves:
 Perianth composed of 3 free sepals and 3 free petals, or the petals sometimes vestigial
 or absent; flowers bisexual or dioecious, borne in 1- or more-flowered tubular
 spathes; ovary 1-locular with parietal placentas; styles 3 or more, simple or
 2-branched .. .. .. .. .. 164. Hydrocharitaceae (3 : 5)
 Perianth composed of 6 similar petaloid segments united below into an elongated tube;
 flowers bisexual, borne in few-flowered umbels subtended by 2 free spathaceous
 bracts; ovary 3-locular with axile placentas; style 1, simple
189. Amaryllidaceae (3 : 131)

Plants terrestrial:
 Fertile stamens 5; flowers bisexual or unisexual, zygomorphic, in clusters subtended
 by large spathaceous bracts; perianth 2-lipped, the median inner segment free, the
 others united into a 3–5-toothed sheath; tall herbaceous plants with large leaves
178. Musaceae (3 : 67)

 Fertile stamens 3 or 6:
  Flowers unisexual, small; leaves simple or digitate, with reticulate venation; fruit a
  3-winged capsule; climbing plants .. .. 191. Dioscoreaceae (3 : 144)
  Flowers bisexual:
   Leaves deeply divided with pinnately lobed segments, radical; flowers in bracteate
   umbels, the outer bracts broad, the inner bracts long and filiform; herbs with a
   tuberous rootstock .. .. .. .. .. 197. Taccaceae (3 : 176)
   Leaves (or solitary leaf) simple and undivided, sometimes reduced to scales:
    Stamens 3:
     Stamens opposite to the inner perianth-segments; flowers actinomorphic;
     perianth-segments united below into a tube; saprophytic herbs with a few
     narrow basal leaves or the leaves all reduced to scales
198. Burmanniaceae (3 : 176)

     Stamens opposite to the outer perianth-segments; flowers actinomorphic or
     zygomorphic; perianth-segments free or united below into a tube; narrow-
     leaved herbs with a cormous or rhizomatous rootstock, not saprophytic
190. Iridaceae (3 : 138)

    Stamens 6:
     Saprophytic herbs, the leaves reduced to scales; ovary 1-locular with 3 parietal
     placentas; perianth with an inflated tube .. 198. Burmanniaceae (3 : 176)
     Plants not saprophytic, the leaves (or solitary leaf) well developed; ovary 3-locular:
      Flowers in umbels (sometimes 1-flowered) subtended by 1 or more spathaceous
      bracts and borne on naked peduncles; herbs with a bulbous rootstock and
      radical leaves .. .. .. .. .. 189. Amaryllidaceae (3 : 131)
      Flowers solitary or in racemes or fascicles not subtended by spathaceous bracts;
      rootstock not a bulb:
       Anthers opening by an apical pore; ovary semi-inferior, with 2 ovules in each
       loculus; herbs with a cormous rootstock, each corm bearing a solitary
       petiolate ovate-cordate leaf .. .. .. 183. Tecophilaeaceae (3 : 107)

Anthers opening by longitudinal slits; ovary with numerous ovules in each loculus; leaves not solitary or ovate-cordate:

Perianth composed of 3 free sepals and 3 longer free petals; ovary semi-inferior; herbs with flowers in racemes .. 177. **Bromeliaceae** (3: 67)

Perianth composed of 6 subequal segments free or united below into a slender tube; ovary completely inferior:

Anthers sagittate; perianth (at least the outer segments) hairy outside; ovules 2-seriate on axile placentas; herbs with a cormous or tuberous rootstock .. .. .. .. .. 195. **Hypoxidaceae** (3: 170)

Anthers not sagittate; perianth glabrous outside; ovules multiseriate on axile placentas intruded into the loculi; plants with a woody stem densely covered with the persistent bases of fallen leaves

196. **Velloziaceae** (3: 174)

*Group 7*

Stamen not accompanied by petaloid staminodes; pollen agglutinated into masses (pollinia); ovary 1-locular with numerous ovules on parietal placentas; flowers zygomorphic, the median petal (lip or labellum) more or less different from the lateral petals; terrestrial or epiphytic herbs, sometimes climbing or saprophytic

199. **Orchidaceae** (3: 180)

Stamen accompanied by 1 or more petaloid staminodes; pollen granular, not agglutinated into masses; terrestrial herbs:

Calyx tubular; flowers zygomorphic; stamen with a 2-thecous anther, the filament sometimes petaloid; ovary 3-locular with numerous ovules in each loculus

179. **Zingiberaceae** (3: 69)

Calyx composed of 3 free sepals; flowers asymmetric; stamen semi-staminodial with a 1-thecous anther, the staminodial part petaloid:

Ovary 3-locular with numerous ovules in each loculus .. 180. **Cannaceae** (3: 79)

Ovary 1–3-locular with 1 ovule in each loculus .. 181. **Marantaceae** (3: 79)

# MONOCOTYLEDONS

## 163. BUTOMACEAE

### By F. N. Hepper

Perennial, aquatic or swamp rhizomatous herbs, usually with milky juice. Leaves ensiform to orbicular. Flowers solitary or umbellate. Perianth 2-seriate, the outer 3 usually sepal-like, imbricate, the inner 3 petal-like and usually thin and deciduous. Stamens hypogynous, 8–9 or numerous; anthers basifixed, opening laterally. Carpels free; ovules numerous, scattered on the reticulately branched parietal placentas. Fruits opening by the adaxial suture. Seeds numerous, without endosperm.

Temperate and tropical regions; distinguished from all other Monocotyledons by the peculiar placentation of the ovules, probably a primitive characteristic.

**TENAGOCHARIS** Hochst. in Flora 24: 369 (28 June 1841).
*Butomopsis* Kunth. (July 1841)—F.T.A. 8: 214.

Scapigerous marsh herbs with milky juice; leaves radical, petiolate. Flowers umbellate, with membranous bracts. Sepals 3, persistent. Petals 3, smaller than sepals, fugacious. Stamens 9, in 2 whorls. Carpels about 6; ovules numerous.

Rootstock small, with slender fibrous roots; leaves all radical, long-petiolate, oblanceolate, shortly and subobtusely acuminate, gradually narrowed to the base, about 10 cm. long and 3 cm. broad, glabrous, with 2 pairs of ascending parallel nerves, faintly reticulate between the nerves; umbels long-pedunculate, about 3–8-flowered; pedicels up to 11 cm. long in fruit; sepals ovate, about 1 cm. long in fruit, thin with membranous margins; petals shorter than the sepals; stamens usually 9; carpels usually 6, overtopping the calyx in fruit; seeds minute.. .. .. *latifolia*

**T. latifolia** (*D. Don*) *Buchen.* in Abh. Nat. Ver. Bremen 2: 3 (1869); Van Steenis in Fl. Males. ser. 1, 5: 118 (1954); Carter in F.T.E.A. Butomac.: 1, fig. 1 (1960). *Butomus latifolius* D. Don (1825). *B. lanceolatus* Roxb. (1832). *Butomopsis lanceolata* (Roxb.) Kunth. (1841)—F.T.A. 8: 214; Chev. Bot. 687. *Tenagocharis lanceolata* Dur. & Schinz (1895). Annual herb from a few inches up to more than 1 ft. high (including the inflorescences), with erect leaves; flowers white; in swamps. **Sen.**: *Roger* 69! Bakel *Heudelot* 126! Matam (fl. & fr. Nov.) *Trochain* 983! Carabane, Casamance (fr. Jan.) *Chev.* 2593! **Mali**: Douentza (fr. Dec.) *de Wailly* 5293! Gorinnta, near Dogo (Apr.) *Davey* 607! **Port. G**: Bafata *Esp. Santo* 2350; 2857. Antula, Bissau *Esp. Santo* 2562. **Guin.**: Siguiri (Nov.) *Jac.–Fél.* 520! **Ghana**: Nasia, N.T. (fr. Dec.) *Vigne* FH 4670! Kamba, Lawra (Nov.) *Harris*! Damongo to Yapei (Sept.) *Hall* CC 893! **N. Nig.**: Nupe *Barter* 1509! Dikwa (fl. & fr. Apr.) *H. B. Johnston* 82! Also in Cameroun, C. African Rep., Sudan and Uganda. In Asia it occurs in India, the Malay Islands, and N. Australia.

## 164. HYDROCHARITACEAE

### By F. N. Hepper

Fresh-water or salt-water herbs, partly or wholly submerged; roots sometimes floating. Leaves radical or cauline, alternate to whorled. Flowers hermaphrodite or unisexual, arranged in a tubular spathe or within two opposite bracts, females solitary; peduncle sometimes spirally twisted in fruit. Perianth-segments free, 1–2-seriate, 3 in each series, the outer often green, valvate, the inner petaloid. Stamens numerous to 2. Male flowers with rudimentary ovary. Staminodes usually present in the female flower. Ovary inferior, 1-locular, with parietal placentas sometimes protruding nearly to the middle of the ovary. Ovules numerous. Fruit rupturing irregularly. Seeds numerous, without endosperm.

Warmer regions of the world; advanced relatives of the preceding family, but with the ovary inferior.

Stems very short; leaves all radical; spathes pedunculate:
  Perianth-segments 6, in two series:
    Spathes 2-winged; petals conspicuous and broad; styles 2-fid; leaves broad, long and thin; hermaphrodite .. .. .. .. .. .. .. 1. **Ottelia**

E.M.S.

Fig. 315.—Tenagocharis latifolia (*D. Don*) *Buchen.* (Butomaceae).

1, plant in fruit, × 1. 2, flower with a sepal and petal pulled down, × 3. 3, young carpel opened out, × 9. 4, fruit, × 3. 5, ripe carpel after dehiscence, × 3. 6, seed, × 60. 1, from *Lea* 220. 2–4, from *H. B. Johnston* IV. 82. 5, 6, from *Lind* 260. (Reproduced from F.T.E.A.)

Spathes not winged; petals linear; styles simple; leaves stiff, linear; dioecious
**2. Blyxa**

Perianth-segments 3, in one series; spathes not winged; male flowers minute, very
numerous, freed from the spathe when mature; fruiting peduncle spirally coiled
**3. Vallisneria**

Stems elongated, with numerous small leaves; spathes sessile, male spathes 2–many-
flowered; perianth-segments 6, in two series    ..    ..    **4. Lagarosiphon**

*Hydrocharis chevalieri* (*De Wild.*)*Dandy* occurs from Congo to E. Cameroun and may also
be found within our area.

### 1. OTTELIA Pers., Syn. Pl. 1: 400 (1805); F.T.A. 7: 6.

Roots numerous, slender, yellowish-brown when dry; leaves oblanceolate, subacute,
gradually narrowed into the wing-like petiole, averaging about 30 cm. long and 6 cm.
broad, often much smaller, glabrous, very thin; spathe compressed, narrowly
oblong-elliptic, 2-winged, shortly lobed, 2–5·5 cm. long, faintly nerved; beak of
the ovary not or only shortly exserted from the spathe, narrow; outer perianth-
segments oblong-lanceolate, up to about 2 cm. long, green, inner segments rich
yellow or white, about twice as long as the outer; fruit about as long as and enclosed
in the somewhat expanded spathe    ..    ..    ..    ..    ..    *ulvifolia*

O. **ulvifolia** (*Planch.*) *Walp.* Ann. 3: 510 (1853); Berhaut, Fl. Sén. 185; Obermeyer in Fl. S. Afr. 1: 109,
fig. 35, 1 (1966). *Damasonium ulvaefolium* Planch. in Ann. Sci. Nat. Sér. 3, 11: 81 (1849). *Ottelia lancifolia*
A. Rich. (1851)—F.T.A. 7: 7, incl. var. *fluitans* Ridl. (1886). *O. vesiculata* Ridl. (1886)—F.T.A. 7: 7.
*O. plantaginea* Welw. ex Ridl. (1886)—F.T.A. 7: 7. *O. abyssinica* (Ridl.) Gürke (1904). *Boottia abyssinica*
Ridl. (1886)—F.T.A. 7: 9. A submerged aquatic in muddy pools, leaves often purple-tinged; flowers
just above water-level, yellow or white (dried specimens always appear to be yellow).
**Sen.:** Trochain 1332! Didey *Heudelot* 209! Niokolo-Koba *Berhaut* 182; 4689. **Gam.:** Basse, Upper R.
Div. *Duke* 9! **Mali:** Onario to Kubita (Feb.) *Davey* 40! Ansongo (Sept.) *Hagerup* 420! **Port G.:** S.
Domingos *Esp. Santo* 2251. **S.L.:** Yana (Dec.) *Jordan* 1076! Gbinti, Batkanu to Port Loko (July) *Deighton*
1967! Fadugu (July) *Bakshi* 258! Binkolo (Aug.) *Thomas* 1670! Madina (Sept.) *Adames* 75! **Lib.:**
Nyandamolahun (Feb.) *Bequaert* 79! Gbanga (Sept.) *Linder* 474! Suacoco (July) *Traub* 191! Ganta (May)
*Harley*! **Iv.C.:** Man (May) *Chev.* 21542! Dabou to N'Douci (July) *Aké Assi* 9117! **Ghana:** Folifoli, Afram
Plains (Aug.) *Hall* CC 193! Kpandai (Dec.) *Adams & Akpabla* GC 4035! Tamale (Nov.) *Williams* 413!
**Niger:** Niamey to Gao *Ryff*! Gao (Nov.) *de Wailly* 4890! **N.Nig.:** Nupe *Barter* 910! Zungeru (Sept.)
*Dalz.* 226! Abinsi (June) *Dalz.* 859! Naraguta F. R., Jos, 4,000 ft. (Aug.) *Lawlor & Hall* 174! Gwoza
(Dec.) *McClintock* 75! Vogel Peak, 4,000 ft. (Dec.) *Hepper* 1544! Kakara, Mambila Plateau, 4,900 ft. (Jan.)
*Hepper* 1778! **S.Nig.:** Siluko, Benin Dist. (Nov.) *Onochie* FHI 40421! **W.Cam.:** Bamenda to Banso (Mar.)
*Richards* 5291! **F.Po:** L. Biao, S. shore, 5,700 ft. (Sept.) *Melville* 492! Widespread in tropical Africa,
reaching Transvaal and Madagascar. (See Appendix, p. 464.)

### 2. BLYXA Nor. ex Thou., Gen. Nov. Madag. 4 (1806); F.T.A. 7: 6.

Leaves in a rosette, submerged, narrowly lanceolate, ending in a very sharp point,
about 3 cm. long and 0·5 cm. broad, prominently 3-nerved, shortly setulose on the
nerves below and on the margin; male spathes pedunculate, about 2 cm. long;
peduncle at length elongated and filiform; male flowers on filiform pedicels; female
spathes 3 cm. long in fruit; seeds flattened, lanceolate, acute, with jagged-dentate
wings    ..    ..    ..    ..    ..    ..    ..    ..    ..    *senegalensis*

B. **senegalensis** *Dandy* in J. Bot. 72: 42 (1934); Berhaut, Fl. Sén. 183. A submerged aquatic with a rosette
of about a dozen sharply pointed leaves; the thread-like male peduncle about 5 in. long arises from the
centre of the rosette and bears a spathe with two floating white flowers on pedicels about 2 in. long.
**Sen.:** Tambacounda (Nov.) *Chev.* 34005! 34016! 34018! *Berhaut* 1651! 3256. Kindia (Sept.) *Jac.-Fél.* 1847!
**S.L.:** Kukuna, N. Prov. (Apr.) *Morton & Gledhill* SL 55!

### 3. VALLISNERIA Linn., Sp. Pl. 1015 (1753); F.T.A. 7: 5.

Submerged; leaves elongated, linear, denticulate, up to about 30 cm. long or more
and about 6 mm. broad, glabrous; spathes diœcious; male spathes many-flowered,
female 1-flowered; male flowers becoming detached from the axis and floating in
the water, where pollination is effected with the single female flower, which after
fertilization is drawn down by the spirally coiling peduncle; perianth-segments 3;
stamens 2; ovary 1-celled with 3 parietal placentas; capsule about 5 cm. long
*aethiopica*

V. **aethiopica** *Fenzl* in Sitzb. Akad. Wien, Math.-Nat. 51 Abth. 1: 139 (1865); F.T.A. 7: 5. *V. spiralis* of
F.T.A. 7: 5; F.W.T.A., ed. 1, 2: 301, not of Linn. Submerged aquatic with ribbon-like leaves; in fresh
or brackish water.
**Sen.:** Sedhiou, Casamance *Adam* 3436! *Berhaut* 6417! **Ghana:** Kpong *Irvine* 3010! Ajena, Senchi
*Akabla* 1833! Otisu, Kete-Krachi *Morton* GC 9162! Tefle R., Volta (Apr.) *Pople & Hall* 3627! **S.Nig.:**
Lagos *Barter* 2163! Extending to Ethiopia and Rhodesia.

### 4. LAGAROSIPHON Harv. in Hook., Journ. Bot. 4: 230 (1841); F.T.A. 7: 2.

Leaves more or less alternate, not verticillate, closely inserted, linear-lanceolate, acute,
about 1 cm. long and 1 mm. broad, with about 60 minute teeth on each side; stipules
2, intrafoliaceous, obovate, obtuse; female spathe subentire at the apex, with 2–3
teeth on each side    ..    ..    ..    ..    ..    ..    ..    1. *schweinfurthii*

Fig. 316.—Ottelia ulvifolia (*Planch.*) *Walp.* (Hydrocharitaceae).
A, fruit.  B, cross-section of fruit.

8

Leaves verticillate, linear-lanceolate, very acute, 2–3 cm. long, 3 mm. broad, with numerous minute very acute teeth on each side; female spathe bearing 1 perfect and 1 rudimentary flower; male spathe bilobed bearing 2 perfect and (?) 1 rudimentary flower; outer perianth-segments 3, 3 mm. long; inner perianth-segments 3, 7 mm. long, 4 mm. broad, irregularly trilobed at apex  ..  ..  ..    2. *hydrilloides*

1. **L. schweinfurthii** *Casp.* in Bot. Zeit. 28: 88 (1870); F.T.A. 7: 3. Submerged aquatic with numerous sessile leaves.
   **Mali:** Banani marsh, near Sanga (Dec.) *Monod*! Also in Sudan.
   [The specimen cited above (from IFAN, Dakar) is sterile and the determination is tentative—F.N.H.]
2. **L. hydrilloides** *Rendle* in J. Linn. Soc. 30: 381, t.32, figs. 1–7 (1895); F.T.A. 7: 4. Submerged, densely leafy aquatic rooting in streams; male flowers white, borne on slender pedicels about 2 in. long.
   **Ghana:** Ejisu, Kumasi (May) *Akpabla* 9419! Also in Uganda and Kenya.

## 165. ALISMATACEAE

### By F. N. Hepper

Perennial or annual marsh or aquatic herbs, erect, or rarely with floating leaves; leaves basal, with elongated petioles sheathing but open at the base and linear-lanceolate to ovate-rounded often sagittate blades, the principal nerves parallel with the margins and converging at the apex of the blade, the transverse nerves often close and parallel. Flowers often whorled, racemose or paniculate, bisexual or rarely polygamous, actinomorphic. Torus flat to globose. Perianth 2-seriate, the outer 3 imbricate, persistent, green and sepal-like, the inner 3 petaloid, imbricate and deciduous or rarely absent. Stamens hypogynous, 6 or more, rarely 3, free; anthers 2-locular, extrorse. Carpels free or rarely united at the base, sometimes in a single whorl; style persistent; ovules solitary or several, basal or on the inner angle. Fruit a bunch or whorl of achenes, rarely dehiscing at the base. Seeds curved, with horseshoe-shaped embryo; endosperm none.

Leaves cuneate at base:
  Flowers usually solitary or with up to 3 in a pseudo-umbellate inflorescence of 1 whorl; carpels very numerous, crowded in a subglobose mass, beaked, compressed; receptacle oblong; sepals and petals distinct from each other  ..  1. **Ranalisma**
  Flowers numerous in several whorls or on a panicle:
    Inflorescence unbranched, flowers sessile or subsessile in several whorls; monoecious; bracts united into a membranous sheath; stamens 3; carpels 3–6, 3-ribbed on the back at maturity  ..  ..  ..  ..  ..  ..  ..  2. **Wiesneria**
    Inflorescence branched, paniculate; flowers pedicellate; petals smaller than the sepals; dioecious; stamens 9; carpels 8–20, lateral faces with a flange at maturity
                                                                                6. **Burnatia**
Leaves cordate or sagittate:
  Leaves cordate; inflorescence branched; carpels 10–15; receptacle very small
                                                                                3. **Caldesia**
Leaves sagittate:
  Carpels very numerous, crowded, compressed; receptacle large; flowers few on an unbranched scape-like inflorescence  ..  ..  ..  ..  4. **Sagittaria**
  Carpels 10–20 (–30), ovoid; receptacle very small, inflorescence more or less panicu-late  ..  ..  ..  ..  ..  ..  ..  ..  ..  ..  5. **Limnophyton**

### 1. RANALISMA Stapf in Hook., Ic. Pl. t. 2652 (1900).
*Echinodorus* of F.T.A. 8: 211, partly, not of L. C. Rich.

A small herb with slender roots; leaves all radical, long-petiolate, ovate or oblong-elliptic, subacute, rounded at the base, about 3 cm. long and 1·5 cm. broad, glabrous, with 1 or 2 pairs of ascending nerves and prominent oblique transverse nerves; submerged leaves linear lanceolate, up to 10 cm. long; flowers hermaphrodite, solitary on fairly long pedicels; sepals 3, herbaceous; petals 3, larger than the sepals; stamens 6 or more; carpels numerous in a subglobose head, long-beaked in fruit and compressed  ..  ..  ..  ..  ..  ..  ..  ..  *humile*

**R. humile** (*Kunth*) *Hutch.* in F.W.T.A., ed. 1, 2: 303 (1936); Berhaut, Fl. Sén. 184; Carter in F.T.E.A. Alismatac. 2, fig. 1 (1960). *Alisma humile* Kunth, Enum. Pl. 3: 154 (1841). *Echinodorus humilis* (Kunth) Buchen. (1869)—F.T.A. 8: 211; Chev. Bot. 686; den Hartog in Fl. Males. 5: 325 (1957). A small aquatic 1–2 in. high in mud or up to 6 in. if growing in water; flowers white, head of fruits prickly with persistent styles; in marshy places.
   **Sen.:** *Roger* 58! Kaédi *Chev.* 2592! *Berhaut* 1400. **Mali:** Télé Lake (Apr.) *Leclerq* in *Hb. Chev.* 42467. Nvafemke *Rogeon* 119. Gombomba (fl. & fr. Feb.) *Davey* 49! Ansongo (fl. & fr. Mar.) *de Wailly* 5366! **N.Nig.:** Katagum Dist. *Dalz.* 311! Zaria *Milne-Redhead* 5027! Gashua (fr. Mar.) *Tuley* 1105! Also in C. African Rep., Sudan, Congo, Tanzania and Zambia.

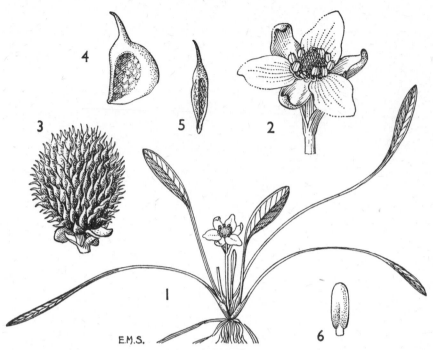

FIG. 317.—RANALISMA HUMILE (*Kunth*) *Hutch.* (ALISMATACEAE).

1, plant, × 1. 2, flower, × 3. 3, fruit, × 3. 4, achene, lateral view, × 6. 5, achene, dorsal view, × 6. 6, seed, × 6. 1, 2 from *Richards* 6380. 3–6 from *Milne-Redhead* 5027. (Reproduced from F.T.E.A.)

**2. WIESNERIA** M. Micheli in DC., Monogr. Phan. 3: 82 (1881); F.T.A. 8: 213; F.W.T.A., ed. 1, 2: 303.

Roots thick, septate; leaves all radical; petiole 11–34 cm. long, 4–6 mm. broad, septate, expanded at base, slightly constricted at apex; lamina elongate-linear—oblanceolate, subobtuse at apex, 8–12 cm. long, 5–14 mm. broad, with an intramarginal nerve and faintly reticulate; flowers monoecious, in an unbranched scape, the females in whorls in the lower part, the males in the upper part; bracts united into a truncate membranous sheath; stamens 3; carpels 3–6, 3-ribbed on the back, beaked, muricate on the margin    ..    ..    ..    ..    ..    ..    ..    *schweinfurthii*

W. **schweinfurthii** *Hook. f.* in Benth. & Hook. f., Gen. Pl. 3: 1007 (1883); F.T.A. 8: 214; Berhaut, Fl. Sén. 184. *W. sparganiifolia* Graebn. (1912). Submerged aquatic with the appearance of a *Potamogeton*; flowers monoecious; in pools on laterite rocks.
     **Sen.:** Tambacounda *Berhaut* 1269; 3257. **Mali:** Sikasso (fl. & fr. Sept.) *Adam* 15304! **S.L.:** Kambia (fr. Aug.) *Jordan* 963! **N.Nig.:** Katagum (fl. & fr. May) *Dalz.* 202! Also in E. Cameroun, C. African Rep., Congo, Sudan, Zambia and Rhodesia.

**3. CALDESIA** Parl., Fl. Ital. 3: 598 (1858). *Alisma* of F.T.A. 8: 207, partly.

Leaves and bracts pellucid-punctate; lamina ovate, 6–10 cm. long along midrib, obtuse at the apex, deeply cordate with obtuse lobes, nerves 9–17 with cross veins 2–5 mm. apart; inflorescences 1–3, well-branched; lower bracts lanceolate, acute, about 3 cm. long; pedicels 1–3·5 cm. long usually arcuate in fruit; sepals 1·5–2 mm. long; petals 2·5 mm. long; achenial fruits kidney-shaped, laterally compressed, 3–4 mm. long, dorsally with 4 longitudinal ribs densely set with blunt spiny warts
                                                1a. *oligococca* var. *echinata*

Leaves and bracts not pellucid-punctate; lamina broadly ovate to suborbicular, 2·5–6 cm. long along midrib, rounded at the apex, deeply cordate at the base with rounded lobes, nerves 13–17 with cross veins about 0·5 mm. apart; inflorescence 1–few, well-branched, rather stout, lower bracts oblong up to 1 cm. long; pedicels 1 cm. long, remaining erect in fruit; sepals 3 mm. long; petals slightly longer; achenial fruits obovate, with persistent style, 2 mm. long, longitudinally ribbed, smooth    ..    ..    ..    ..    ..    ..    ..    *2. reniformis*

1. **C. oligococca** (*F. von Muell.*) *Buchen.* in Engl., Bot. Jahrb. 2: 479 (1882). Vars. *oligococca* and *acanthocarpa* (F. von Muell.) den Hartog occur in tropical Asia and Australia.

1a. **C. oligococca** var. **echinata** *den Hartog* in Fl. Males. 5: 322, fig. 1, 2a–d (1957). An aquatic herb with broad floating leaves and sometimes submerged linear phyllodes; paniculate inflorescences about 2 ft. high; with white flowers and spiny fruits; at the edge of permanent swamps.
    **Sen.:** Niokolo-Koba (fr. Dec.) *Raynal* 6877! **N.Nig.:** Nupe *Barter* 1062! Also in India, Ceylon, Indo-China and Java.

2. **C. reniformis** (*D. Don*) *Mackino* in Bot. Mag. Tokyo 20: 34 (1906); Carter in F.T.E.A., Alismatac. 7, fig. 4 (1960). *Alisma reniforme* D. Don, Fl. Nepal 22 (1825). *A. parnassifolium* of F.T.A. 8: 208; of Chev. Bot. 686; not of Bassi. *Caldesia parnassifolia* var. *nilotica* (M. Micheli) Buchen. (1903). An aquatic with floating leaves; paniculate inflorescences 1–2 ft. high, with white flowers and smooth fruits; in swamps.
    **Dah.:** Khuon lagoon, Abomey Circle *Chev.* 23264! **N.Nig.:** *Thornewill* 96! Matyoro (fr. Oct.) *Thornewill* 143! Widespread in tropical and N. Africa, Madagascar and eastwards to India, Malaysia, China and Australia.

## 4. SAGITTARIA Linn., Sp. Pl. 993 (1753); Bogin in Mem. N.Y. Bot. Gard. 9: 179–233 (1955). *Lophotocarpus* Dur. (1888)—F.T.A. 8: 210; F.W.T.A., ed. 1, 2: 303.

Rootstock small, with numerous slender roots; leaves all radical, long-petiolate, ovate in outline, deeply sagittate at the base, 5–9 cm. long, up to 8 cm. broad, with numerous nerves radiating from the base; scapes few-flowered, unbranched; flowers polygamous, few in each whorl, the lower hermaphrodite; pedicels stout; bracts triangular-ovate, up to 1 cm. long; sepals 3, veined; petals 3, larger; stamens 9–15; carpels numerous, crowded, compressed, with rugose margins
                                        *guayanensis* subsp. *lappula*

**S. guayanensis** *Kunth.* Nov. Gen. & Sp. 1: 250 (1816); Bogin l.c. 192; den Hartog in Fl. Males. 5: 328.
**S. guayanensis** subsp. **lappula** (*D. Don*) *Bogin* l.c. 192, fig. 5c (1955); Aké Assi, Contrib. 2: 11. *S. lappula* D. Don, Prod. Fl. Nep. 22 (1825). *Lophotocarpus guayanensis* (H.B. & K.) Dur. & Schinz (1895)—F.T.A. 8: 210; F.W.T.A., ed. 1, 2: 303; Berhaut, Fl. Sén. 179. *L. gourmacus* A. Chev., Bot. 686 (1920), name only. An aquatic with floating leaves; flowers white; in muddy pools.
    **Sen.:** *Berhaut* 1653; 3258. Richard-Tol (July) *Martine* 100! Port.G.: Piche (Sept.) *Pereira* 3198! **Gam.:** Bansang *Duke* 8! Fulladu West (Nov.) *Frith* 155! **Mali:** Bamako to Ségou (Sept.) *Jaeger* 5118! Sikasso (Sept.) *Adam* 15306! **Iv.C.:** Ouango-Fitini (Dec.) *Aké Assi* 7532! **U.Volta:** Fada-N'Gourma *Chev.* 24486! **Ghana:** Burufo, Lawra (fl. & fr. Sept.) *Hall* CC 733! **N.Nig.:** Katagum Dist. *Dalz.* 198! Also in Sudan, Madagascar, tropical Asia and America.

## 5. LIMNOPHYTON Miq., Fl. Ind. Bat. 3: 242 (1855); F.T.A. 8: 209.

Leaves linear-lanceolate, gradually narrowed to the base, caudate-acuminate, 30 cm. long, 2·5 cm. broad, 3-nerved; flowers verticillate; dry fruits 3 mm. long, 2 mm. broad ..    ..    ..    ..    ..    ..    ..    ..    ..    1. *fluitans*

Leaves sagittate:

Bracts glabrous, triangular, 1–2 cm. long, up to 1 cm. broad; inflorescences with 3–6 bracteoles and 5–10 flowers in each whorl; fruiting pedicels 1–2 mm. diam., 2–4 cm. long; fruits reticulately ridged, pale brown; leaves ovate-sagittate, obtuse at apex, glabrous, midrib 6–9 cm. long, lobes of lamina 7–11 cm. long; petioles 25–50 cm. long ..    ..    ..    ..    ..    ..    2. *obtusifolium*

Bracts pubescent outside, lanceolate, 2–4 cm. long, about 5 mm. broad; inflorescences with at least 10 bracteoles and 15 flowers in each whorl; fruiting pedicels 0·5 mm. diam., 2–4 cm. long; fruits longitudinally ridged otherwise smooth and glossy; leaves markedly sagittate, more or less acute or shortly apiculate at apex, usually slightly pubescent beneath, midrib 10–15 cm. long, lobes of lamina 13–20 cm. long; petioles 10–60 cm. long    ..    ..    ..    ..    ..    3. *angolense*

1. **L. fluitans** *Graebn.* in Engl., Bot. Jahrb. 41: 274 (1908). Submerged aquatic with long, bronze-green leaves; flowers with 3 white petals, anthers cream.
    **S. Nig.:** Oban *Talbot* 708! Owai-Ifunkpa (Apr.) *Talbot* 495 (drawing)! Also in E. Cameroun.

2. **L. obtusifolium** (*Linn.*) *Miq.* Fl. Ind. Bat. 3: 242 (1855); F.T.A. 8: 209; Buchenau in Engl., Pflanzenr. 4, 15: 22, fig. 10 (1903); Berhaut, Fl. Sén. 180; Den Hartog in Fl. Males. 5: 324 (1957); Carter in F.T.E.A. Alismatac.: 9fig. 5, 1–5 (1960); Aké Assi, Contrib. 2: 10, partly; Obermeyer in Fl. S. Afr. 1: 109, fig. 35, 1 (1966). *Sagittaria obtusifolia* Linn., Sp. Pl. 993 (1753). *Lophotocarpus guayanensis* of Chev. Bot. 686, partly (Chev. 2595), not of H.B. & K. Erect tufted herb with long-stalked leaves and inflorescences with spongy peduncles about 2 ft. high; flowers white; at the margin of ponds in mud.
    **Sen.:** (Dec.-Jan.) *Roger* 60! *Heudelot* 231! Dakar *Chev.* 75788. Tambacounda *Adam* 2445! Séléki, Casamance *Chev.* 2595! **Mali:** Banani, near Sanga *Monod*! **Gam.:** Kuntaur *Ruxton* 108! **Lib.:** Nimba Reserve (fr. Dec.) *Adam* 20202! **Ghana:** Accra to Oda (June) *Morton* GC 9233! Savelugu (fr. Dec.) *Morton* GC 9894! **Togo Rep.:** Lomé *Warnecke* 214! *Davidson* 14! **Dah.:**?Whydah *Isert* (photo!) **Niger Rep:** Zinder (fr. Oct.) *Hagerup* 586! **N.Nig.:** Katagum *Dalz.* 199! Sokoto *Lely* 161! Dikwa (fl. & fr. Nov.) *McClintock* 23! Widespread in tropical Africa; also in Madagascar, India, Ceylon and the Malay Islands.

3. **L. angolense** *Buchen.* in Engl., Pflanzenr. 4, 15: 23 (1903); Carter in F.T.E.A. Alismatac.: 11, fig. 5, 6–13. *Lophotocarpus guayanensis* of Chev. Bot. 686, partly (*Chev.* 13327, 14888), not of H.B. & K. *Limnophyton obtusifolium* of F.W.T.A., ed. 1, 2: 303, fig. 281, partly (*Sc. Elliot* 4972; *Deighton* 357; *Thomas* 2520; *Linder* 285 (sheet 2); *Barter* 1532); of Aké Assi, Contrib. 2: 10, partly, not of (Linn.) Miq. *Caldesia reniformis* of F.W.T.A., ed. 1, 2: 304, partly (*Linder* 385, sheet 1). An aquatic herb similar to the last, up to 4 ft. high.
    **Guin.:** Timbo to Farana *Chev.* 13327! Dindia *Caille* in *Hb. Chev.* 14888! Nzérékoré (May) *Schnell* 2765! **S.L.:** Likuru (Feb.) *Sc. Elliot* 4972! Mafinta, near Roruks (fr. July) *Deighton* 4681! Kolia, Mano to Kailahun (Apr.) *Deighton* 3162! Gegbwema (Nov.) *Deighton* 357! Jigaya (fr. Sept.) *Thomas* 2520! Kenema (Jan.) *Thomas* 7631! **Lib.:** Dinklage 3348! Sarbo (July) *Baldwin* 6396! Jabroke (July) *Baldwin* 6486! Gbanga (Sept.) *Linder* 385! **Iv. C.:** Danané (Aug.) *Boughey* GC 18139! **Ghana:** Bimbila (Mar.) *Hepper & Morton* A3106! Amedi Lagoon, Amedikan (Nov.) *Hall* 2740! **N. Nig.:** Nupe *Barter* 1532! **S. Nig.:** Awka Dist. (Feb.) *Jones* FHI 628! Also in E. Africa and Zambia. (See Appendix, p. 464.)

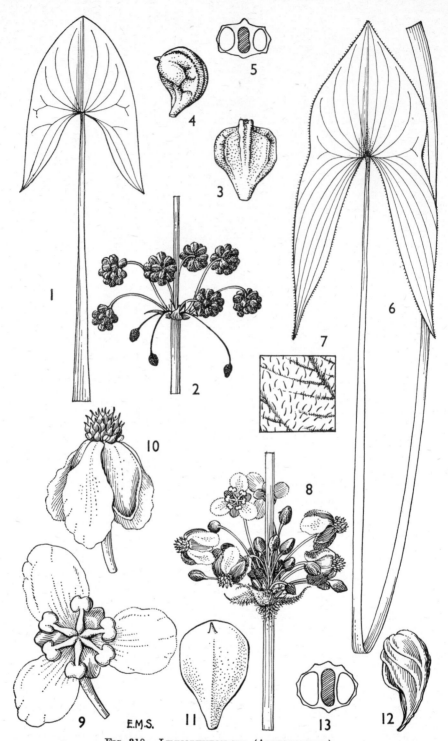

Fig. 318.—Limnophyton spp. (Alismataceae).

*Limnophyton obtusifolium* (Linn.) Miq. 1, leaf, × ¼. 2, lower whorl of inflorescence with bisexual flowers in fruit, × 1. 3, achene, dorsal view, × 4. 4, achene, lateral view, × 4. 5, transverse section of achene showing air chambers. *L. angolense* Buchen. 6, leaf, × ¼. 7, part of undersurface of leaf, × 2. 8, lower whorl of inflorescence, × 1. 9, male flower, × 3. 10, bisexual flower, × 3. 11, dorsal view of achene in fresh state, × 4. 12, lateral view of achene when dried, × 4. 13, transverse section of achene, × 4 . 1, 3–5 from *de Wailly* 5387. 2, from *Welch* 375. 6–8 from *Brown* 136. 9, 11 from *Milne-Redhead* 4082. 10 from *Eggeling* 865. 12, 13 from *Michelmore* 315. (Reproduced from F.T.E.A.)

FIG. 319.—BURNATIA ENNEANDRA *M. Micheli* (ALISMATACEAE).

1, male plant, × ⅓. 2, part of male inflorescence, × 2. 3, male flower, × 9. 4, part of female inflorescence, × 2. 5, female flower, × 9. 6, achene, lateral view, × 9. 7, achene, dorsal view, × 9. 8, seed, × 9. 1 from *Dalziel* 260. 2, 3 from *Rayner* 476. 4, 5 from *A. S. Thomas* 3558. 6, 7 from *Bally* 5237. (Reproduced from F.T.E.A.)

**6. BURNATIA** M. Micheli in DC., Monogr. Phan. **3**: 81 (1881); F.T.A. **8**: 212.

Rootstock shortly rhizomatous and/or with ovoid-globose corms about 1 cm. diam. with numerous slender roots; leaves all radical, long-petiolate, oblong-lanceolate, acute at each end, about 16 cm. long and 3–4 cm. broad, glabrous, with 2–3 lateral nerves on each side, not reticulate; male inflorescence a much-branched panicle longer than the leaves; bracts linear-lanceolate, up to 2 cm. long; pedicels slender, 6–8 mm. long; sepals rounded-elliptic, about 2 mm. long; petals small; female inflorescence much smaller and more compact than the male; carpels about 12, crowded, 1-seeded

*enneandra*

B. **enneandra** *M. Micheli* in DC., Monogr. Phan. **3**: 81 (1881); F.T.A. **8**: 213; Chev. Bot. 686; Aké Assi, Contrib. **2**: 9; Obermeyer in Fl. S. Afr. **1**: 99, fig. 29, 3 (1966). An aquatic with tufted leaves about 1 ft. high and slender branched inflorescences (at least the male) about 2 ft. high; flowers small, white or pink; at margins of muddy pools.
**Sen.**: *Trochain* 1556! **U.Volta**: Fada–N'Gourma *Chev.* 24487! Banfora (Sept.) *Adam* 15383! **Iv.C.**: Sémien Lake, Man road *Aké Assi* IA 5770. **Ghana**: Folifoli, Afram Plains (Aug.) *Hall* CC 190! Mirigu (Oct.) *Vigne* FH 4617! Navrongo (June) *Vigne* FH 4712! W. of Bawku (Aug.) *Hall* CC 544! Yendi (July) *Akpabla* GC 1912! **N.Nig.**: Zungeru (July) *Dalz.* 260! Bida to Badeggi (Aug.) *Onochie* FHI 35418! **S.Nig.**: Awi to Akamkpon, Calabar Prov. (fr. Feb.) *Daramola* FHI 55532! Scattered throughout much of tropical Africa.

## 166. TRIURIDACEAE

### By F. N. Hepper

Leafless saprophytic herbs with simple or subsimple stems furnished with a few pale scales. Flowers very small, racemose or subcorymbose, with decurved bracteate pedicels, actinomorphic, monoecious, dioecious or rarely polygamous. Perianth-segments 3–8, 1-seriate, valvate, sometimes appendaged at the apex, at length reflexed. Male flowers: stamens 2–6, sometimes only half of them fertile; anthers free or immersed in the mass of the receptacle, 2-locular, dehiscing mostly transversely; connective sometimes produced into long subulate appendages. Female flowers rarely with staminodes. Carpels several, free, 1-locular; style terminal to almost basal; ovule solitary, basal. Fruits crowded, opening by a slit. Seed erect, with a fleshy white oily undifferentiated nucleus.

Tropics; rare in Africa.

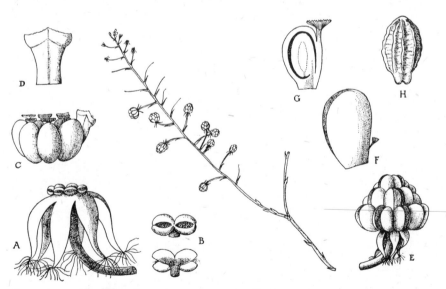

FIG. 320.—SCIAPHILA LEDERMANNII *Engl.* (TRIURIDACEAE).

A, male flower. B, front and back view of stamen. C, carpels and staminodes. D, staminode. E, female flower in fruit. F, carpel. G, vertical section of carpel. H ,seed. After Engl. Bot. Jahrb. 43: 305 (1909). Habit natural size.

**SCIAPHILA** Blume, Bijdr. 514 (1825).

Perianth 3–8-partite or deeply lobed. Anthers sessile or subsessile at the base of the perianth. Style ventral or basal in fruit.

Racemes about 9 cm. long, about 12-flowered; lower flowers female, upper male; leaves reduced to scales, ovate-triangular; bracts ovate-lanceolate, about 2 mm. long; pedicels curved, 7–8 mm. long, glabrous; perianth-segments reflexed, lanceolate, 1·25 mm. long, hairy at the tips; fruits ellipsoid; seeds ribbed, subacute at one end .. .. .. .. .. .. .. .. 1. *ledermannii*
Racemes 15–20 cm. long, with very numerous flowers; leaves linear-lanceolate, 3–4 mm. long; bracts linear, reflexed, persistent; pedicels spreading-recurved, slender, about 1 cm. long, pale glaucous-purple; perianth-segments 6, triangular, ending in a hair, not reflexed; fruits ellipsoid, minutely pitted when dry .. .. 2. *africana*

1. **S. ledermannii** *Engl.* Bot. Jahrb. 43: 304 (1909). A rigid saprophytic herb 3–4 in. high among dead leaves in damp sandy places on the forest floor; aerial parts deep coral pink, perianth paler, subterranean parts whitish.
 **S. Nig.:** Oban (Mar.) *Talbot* 710! 1473! *Richards, Onochie & Coombe* 5141! Also in E. Cameroun.
2. **S. africana** *A. Chev.* in Mém. Soc. Bot. Fr. 2, 8: 96 (1908); Chev. Bot 686. A slender saprophyte about 1 ft. high, with numerous small nodding unisexual flowers on slender pedicels.
 **Iv. C.:** *Aké Assi* 9476! Mt. Copé, Tepos country, Cavally Basin (fr. July.) *Chev.* 19661! **Ghana:** Ankasa F.R. (Jan.) *Enti & Hall* GC 36286!

# 167. APONOGETONACEAE

## By F. N. Hepper.

Fresh-water aquatic herbs with submerged or floating leaves; rhizome tuberous, with fibrous roots. Leaves long-petiolate, or sessile, oblong-elliptic to linear, with few principal parallel nerves and numerous transverse secondary nerves. Flowers hermaphrodite or rarely unisexual, spicate-scapose, spike simple or usually 2- (rarely up to 8-) forked, without bracts. Perianth-segments 1–3, or absent, sometimes petaloid and bract-like, equal or unequal, usually persistent. Stamens 6 or more, free, hypogynous, persistent; anthers extrorse, 2-locular. Carpels free, 3-6, sessile; style short; ovules 2 or more, ascending. Fruits opening on the adaxial side. Seeds without endosperm.

Warm regions from India and South China through Malaya to Australia, but most numerous in Tropical and S. Africa and in Madagascar.

**APONOGETON** Linn. f., Suppl. 32 (1781); F.T.A. 8: 216; K. Krause & Engl. in Engl., Pflanzenr. 4, 13: 9 (1906).

Aquatics. Scapes simple or bifid, with unilateral flowers. Perianth-segments 1–3, white or coloured. Styles distinct.

Leaves long-petiolate, oblong, cordate at the base, obtuse at the apex, 16–20 cm. long, 3–4·5 cm. broad, with a broad midrib and 3 longitudinally parallel nerves on each side, and numerous closely parallel spreading transverse nerves between; spikes long-pedunculate, 2-forked, not secund, arms about 3 cm. long in flower, about 6 cm. long in fruit; fruiting carpels ovoid-ellipsoid, beaked; perianth small and bract-like .. .. .. .. .. .. .. .. 1. *subconjugatus*
Leaves sessile, linear, up to 15 cm. long, subacute, with no distinct midrib but with several distinct longitudinally parallel nerves; peduncles very slender; spikes unbranched, secund, about 2 cm. long; basal bract oblong-lanceolate, apiculate, 1 cm. long; perianth-segments conspicuous, white, veined .. .. 2. *vallisnerioides*

1. **A. subconjugatus** *Schum. & Thonn.* in Beskr. Guin. Pl. 183 (1827); F.T.A. 8: 217; Berhaut, Fl. Sén. 179, 185. *A. heudelotii* Engl. (1887). An aquatic with floating leaves in pools; flowers yellowish.
 **Sen.:** Walo *Heudelot* 433! *Berhaut* 1752. **Mali:** Dogo (Dec.) *Davey* 1! Ouario to Koubita (Feb.) *Davey* 41! Berra, near Gao (Nov.) *de Wailly* 4731! **Ghana:** *Thonning*! Achimota (June) *Irvine* 1468! Accra to Ada (fl. & fr. Dec.) *Morton* GC 8245! Kpong (June) *Irvine* 4917! **N. Nig.:** Katagum Dist. *Dalz.* 226! Also in E. Cameroun.
2. **A. vallisnerioides** *Bak.* in Trans. Linn. Soc. 29: 158 (1875); F.T.A. 8: 218; Troupin in Fl. Sperm. Parc. Nat. Garamba 1: 12, fig. 1 (1953). Aquatic in pools on laterite; flowers white or pale violet.
 **Sen.:** Tambacounda *Chev.* 34017. **S.L.:** Moyamba (Aug.) *Dawe* 566! Kambia (Aug.) *Jordan* 962! *Morton & Gledhill* SL 76! **Iv. C.:** Bassawa (Nov.) *Aké Assi* 9258! **N. Nig.:** " Niger " *Baikie*! Anara F.R., Zaria (July) *Keay* FHI 25961! Also Congo, Sudan, Uganda and Kenya.

*Zostera nana* Roth has been reported from Mauritanian coasts and it may also occur within our area. The genus *Zostera*, in its own family Zosteraceae, is placed by Hutchinson in the Aponogetales. The Engler System (ed. 12) places the genus in Potamogetonaceae.

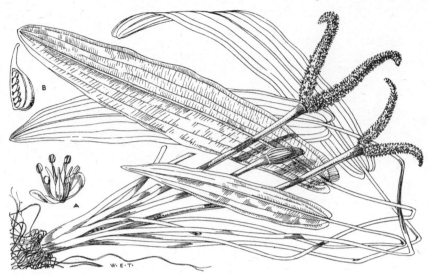

FIG. 321.—APONOGETON SUBCONJUGATUS *Schum. & Thonn.* (APONOGETONACEAE).
A, flower. B, vertical section of carpel.

## 168. POTAMOGETONACEAE

### By F. N. Hepper

Aquatic herbs of fresh water. Leaves alternate or opposite, those immersed thin, those above water often leathery, sheathing at the base, sheath free or partially adnate to the petiole. Flowers hermaphrodite, small, arranged in pedunculate axillary spikes; peduncle surrounded by a sheath at the base; bracts absent. Perianth of 4 free rounded shortly clawed valvate segments. Stamens 4, inserted in the claws of the segments; anthers extrorse, 2-locular, sessile. Carpels 4, free, 1-locular; stigmas sessile or on short styles; ovule solitary, on the adaxial angle. Fruits free, 1-seeded, indehiscent. Seeds without endosperm.

One genus, Potamogeton, widely distributed.

**POTAMOGETON** Linn., Sp. Pl. 126 (1753); F.T.A. 8: 219; Dandy in J. Linn. Soc. 50: 507–540 (1937). Characters of the family.

Submersed leaves about 20 mm. broad, broadly linear, narrowed to each end, 15–25 cm. long, very thin, with about 7 longitudinal nerves, rarely upper leaves floating; spikes many-flowered, about 4 cm. long; fruits obliquely ovoid, beaked, 2·5 mm. long
1. *schweinfurthii*

Submersed lower leaves 1–2 mm. broad, linear, 5–6 cm. long, upper leaves sometimes expanded into elliptic lamina 1–2 cm. long, 3–6 mm. broad, more or less floating; spikes small and few-flowered; peduncle 3·5 cm. long, stout; flowering portion about 1 cm. long; fruits about 2 mm. long .. .. .. .. 2. *octandrus*

1. **P. schweinfurthii** *A. Benn.* in F.T.A. 8: 220 (1901); Dandy l.c. 526, fig. 6 (map); Berhaut, Fl. Sén. 81; Obermeyer in Fl. S. Afr. 1: 66, fig. 18, 2 (1966). *P. lucens* var. *fluitans* of F.T.A. 8: 221, not of Coss. & Germ. Submerged aquatic with very thin rather large leaves.
   **Sen.:** Lake Guier (fr. Jan.) *Leprieur! Roger!* **Mali:** Gao (Feb., Mar.) *de Wailly* 4827! 4976! **Iv. C.:** Banfora to Tiafora (Apr.) *Aké Assi* 8617! **N.Nig.:** Mallam Fatori, Chad Dist. *G. Jackson* FHI 59167! Also widespread in E. and S. Africa.
   [A sterile specimen (*Melville* 488) from L. Biao, F. Po, may also be this species.]
2. **P. octandrus** *Poir.* in Lam., Encycl. Méth., Bot., Suppl. 4: 534 (1816); Dandy l.c. 517, fig. 2 (map); Obermeyer l.c. 65, fig. 16, 1 (1966). *P. javanicus* Hassk. (1856)—F.T.A. 8: 202, partly. *P. preussii* A. Benn. in F.T.A. 8: 222 (1901), partly; Graebner in Engl., Pflanzenr. Potamogetonac. 106, partly. An aquatic with fine submerged leaves and often broader floating ones.
   **Sen.:** Niokolo-Koba (fr. Feb.) *Adam* 17576! **Mali:** Gao (fr. Feb.) *de Wailly* 4834! Gorinnta, Dago (fr. Apr.) *Davey* 608! **S.L.:** Rhombe (fr. Mar.) *Adames* 144! Gbap (fr. Apr.) *Deighton* 4299! **Lib.:** Gbaishelo (fr. Dec.) *H. Vogel* in *Hb. Harley* 2199! **Ghana:** R. Volta, Battor (Apr.) *Pople & Hall* 3625! **N.Nig.:** Jebba

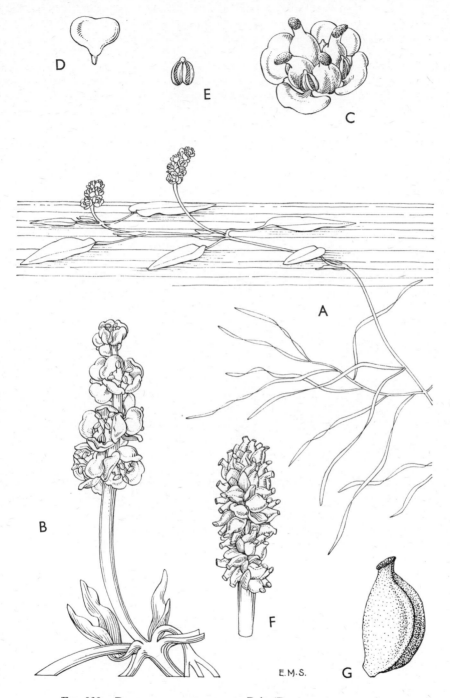

FIG. 322.—POTAMOGETON OCTANDRUS *Poir*. (POTAMOGETONACEAE).

A, habit, × 1.  B, inflorescence, × 3.  C, flower, × 8.  D, tepal, × 8.  E, stamen, × 8.  F, fruiting head, × 3.  G, fruit, × 6.  A, from *Hepper* 1849.  B–E, from *Wood* 461.  F, G, from *Harley* 2199.

17

*Barter* 1069! Near Gembu, Mambila Dist. (Feb.) *Hepper* 1849! **W. Cam.**: Johann-Albrechtshöhe (= Kumba) *Staudt* 462! Barombi-ba-Mbu *Preuss* 451! Throughout the Old World Tropics.

Several other species occur in suitable places near the northern limit of the Flora: *P. hoggarensis* Dandy, *P. panormitanus* Biv. and *P. perfoliatus* Linn.

## 169. RUPPIACEAE

### By F. N. Hepper

Aquatic herbs of saline marshes. Leaves opposite or alternate, linear or setaceous, sheathing at the base. Flowers hermaphrodite, small, in terminal spikes at first enclosed by the sheathing base of the leaves, at length much elongated; bracts and perianth absent. Stamens 2; filaments short, broad; anthers extrorse, loculi reniform and separated by the connective. Carpels 4 or more, free, stigmas peltate or umbonate. Ovule solitary, pendulous. Fruits long-stipitate with spirally twisted stalks, indehiscent. Seeds pendulous, without endosperm.

Salt marshes throughout temperate and tropical regions.

**RUPPIA** Linn., Sp. Pl. 127 (1753); F.T.A. 8: 224.

A very slender profusely branched herb with long linear leaves about 0·5 mm. broad arising from hyaline sheaths; fruit drupaceous, about 3 mm. long    ..  *maritima*

**R. maritima** *Linn.* Sp. Pl. 127 (1753); Obermeyer in Fl. S. Afr. 1: 61, fig. 15, 1 (1966). *R. rostellata* Koch in Reichb., Ic. Pl. Crit. 2: 66, t. 174 (1823); F.T.A. 8: 224; F.W.T.A. .ed. 1, 2: 307. *R. spiralis* of F.T.A. 8: 224, not of Hartm. *Potamogeton pusillus* of Fl. Nigrit. 528, not of Linn. *P. pectinatus* of F.T.A. 8: 223, partly, not of Linn. A submerged aquatic.
**Sen.**: Cayor *Leprieur*! Gandide *Trochain* 2778! **Ghana**: Cape Coast *Don*! Teshi (fr. June) *Irvine* 2777! Labadi Lagoon (fr. Jan.) *de Wit & Morton* A3002! Prampram (fr. Feb.) *Morton* GC 6419! **N.Nig.**? : (fr. Nov.) *E. Vogel*! Widespread in the tropics.

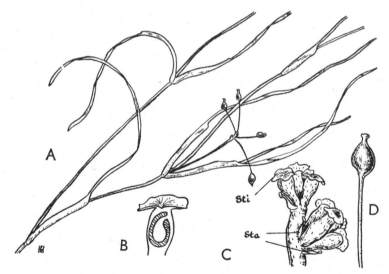

FIG. 323.—RUPPIA MARITIMA *Linn.* (RUPPIACEAE).

A, habit, × 2.  B, section through carpel showing pendulous ovule (diagrammatic).  C, inflorescence, × 6.  D, fruit, × 6.  Sta = stamen.  sti = stigma.

## 170. ZANNICHELLIACEAE

### By F. N. Hepper

Submerged aquatic herbs in fresh, brackish, or salt water; rhizome creeping, slender. Leaves alternate or opposite, or crowded at the nodes, linear, sheathing

at the base, sheaths mostly ligulate at the apex; flowering leaves sometimes reduced to sheaths.  Flowers minute, monoecious or dioecious, axillary, solitary
or in cymes.  Perianth of 3 small free scales, or absent.  Stamens 3, 2 or 1; anthers
2–1-locular, opening lengthwise; pollen globose or thread-like.  Carpels 1–9, free;
style simple or 2–4-lobed.  Ovule solitary, pendulous.  Fruits sessile or stipitate,
indehiscent.  Seed pendulous, without endosperm.

Widely distributed, mainly in salt water.

Styles 2-fid; anthers at an equal height; leaf-apex rounded, toothed      **1. Cymodocea**
Styles simple; one anther attached higher than the other; leaf-apex bicuspidate
                                                                          **2. Halodule**

**1. CYMODOCEA** Konig in Konig & Sims, Ann. Bot. 2: 96 (1805); F.T.A. 8: 228;
                Graebner in Engl., Pflanzenr. 4, 11: 146 (1907).

Perianth absent from the male flowers.  Stamen 1, on a slender filament.  Carpels
curved.

Rhizome creeping, marked with circular scars at the nodes; leaf-sheaths auriculate,
    up to 5 cm. long, deciduous; leaf-blades linear, toothed near the apex, up to 1 m.
    long and 3·5 mm. broad, about 7-nerved; male flowers long-pedunculate; carpels 2,
    collateral, each with 2 long stigmas; fruit sessile, obliquely ovoid, 1 cm. long, keeled,
    shortly beaked  ..     ..     ..    ..     ..     ..     ..      ..      ..        *nodosa*

C. **nodosa** (*Ucria*) *Aschers.* in Sitzungsber. Ges. Naturf. Fr., Berlin 1867: 4 (1867); F.T.A. 8: 229; Graebner
    l.c. 147; Bossier, Fl. Orient. 5: 21; Feldmann in Bull. Soc. Nat. Afr. Nord 29: 111 (1938). *Zostera nodosa*
    Ucria (1790).  A marine aquatic.
    **Maur.**: *Murat*! (*vide* Feldmann *l.c.*).  **Sen.**: Joual, and mouths of the river (*vide* Boissier *l.c.*).  Ponto
    *Raynal* 7240!  Widely distributed on warm shores from Europe and the Canary Is. to the eastern
    Mediterranean.

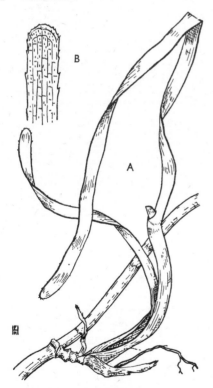

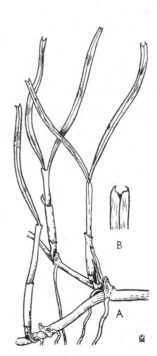

FIG. 324.—CYMODOCEA NODOSA (*Ucria*)          FIG. 325.—HALODULE WRIGHTII *Aschers.*
    *Aschers.* (ZANNICHELLIACEAE).                 (ZANNICHELLIACEAE).

A, habit, × 2.  B, leaf apex, × 4.  From *J. &*      A, habit, × 2.  B, leaf apex, × 4.  From *P.*
        *A. Raynal* 7240.                                *Wilson* 7539.

**2. HALODULE** Endl., Gen. Pl. Suppl. 1: 1368 (1841); den Hartog in Blumea 12: 289–312 (1964). *Diplanthera* Thou. (1806).

Rhizome creeping, with 2–5 roots and a short erect stem at each node; scales elliptic, 4–9 mm. long; sheaths 1·5–4 cm. long; 5–18 cm. long, about 0·5 mm. broad, with a conspicuous midrib sometimes ending in a very small tooth, intra-marginal veins inconspicuous both ending in a narrow triangular lateral tooth giving the leaf tip a bicuspidate appearance; male flowers pedunculate, anthers slender, one higher than the other, 4 mm. long; female flower with compressed elliptic ovary 1·5–2 mm. long, style 11–12 mm. long; fruit ovate, slightly compressed 2 mm. long with a very short rostrum .. .. .. .. .. .. .. .. .. *wrightii*

**H. wrightii** *Aschers.* in Sitzungsber. Ges. Naturf. Fr., Berlin 1868: 19, 24 (1869); den Hartog l.c. 304, figs. 6–8. *Diplanthera wrightii* (Aschers.) Aschers. (1897)—Feldmann in Bull. Soc. Hist. Nat. Afr. Nord 29: 107–112, fig. 1 (1938); Trochain in Mém. I.F.A.N. 2: 108 (1940). A slender marine aquatic herb growing in sand near low tide mark.
**Maur.:** *Murat*! **Sen.:** Nianing (Aug.) *Trochain*! Ponto *Raynal* 7242! Also in Mauritania (Tanoudort) around the African coasts to the Persian Gulf, and in the Carribean area.

# 171. NAJADACEAE

## By F. N. Hepper

Small submerged annual water-plants; stem slender, much branched. Leaves small, subopposite or verticillate, sessile, with a sheathing base and linear entire or toothed blade. Flowers unisexual, monoecious or rarely dioecious, very small axillary, often enveloped in a bract (spathe) with 2-lobed apex. Male flower with 1 stamen; anther sessile, 1–4-locular. Bract (spathe) of female flowers sometimes very thin and adhering to the carpel. Ovary of 1 carpel, 1-locular, with 2–4 stigmas. Ovule solitary erect. Fruit usually embraced by the leaf-sheath, indehiscent. Seed without endosperm.

One genus, widely distributed in temperate and warm regions.

**NAJAS** Linn., Sp. Pl. 1015 (1753); F.T.A. 8: 225, as *Naias*; Horn af Rantzien in Kew Bull. 7: 29–40 (1952). Characters of the family.

Leaves conspicuously toothed (see also No. 2), strongly falcate, the teeth as long as or greater than the width of the leaf; leaf sheath truncate; only male flowers provided with bract (spathe); anther 4-locular.. .. .. .. .. 1. *pectinata*
Leaves obscurely toothed:
Leaf sheaths not auricled but irregularly fimbriate with at least 10 spines; leaf margin with 12–20 rather conspicuous teeth; both male and female flowers enclosed in spathes .. .. .. .. .. .. .. .. 2. *affinis*
Leaf sheaths auricled; leaf margin teeth minute:
Bract (spathe) present, enclosing either male or female flowers or both:
Both male and female flowers enclosed in a bract (spathe), at anthesis male flower exserted from bract on 2 mm.-long pedicel, anthers 4-locular; leaf margin with (12–)16(–21) teeth on each side; fruits 2–2·5 mm. long .. .. .. 3. *meiklei*
Only male flowers enclosed in a bract (spathe), anthers 1-locular; leaf margin with 35–45 teeth on each side .. .. .. .. .. .. 4. *hagerupii*
Bract (spathe) absent from all flowers:
Leaves with about 10 minute teeth; auricles generally bearing 3–5 inconspicuous spines; fruits 1·6–2·1 mm. long, with 20–25 rows of squarish markings 5. *liberiensis*
Leaves with about 20 distinct teeth, often with leaf margin undulate; auricles with about 2 brownish spines at apex; fruits 2–2·5 mm. long, with about 50 rows of elongated markings .. .. .. .. .. .. .. 6. *baldwinii*

1. **N. pectinata** (*Parl.*) *Magnus* in Aschers. & Schweinf., Illustr. Fl. Egypte 145 (1889); H. af Rantz. l.c. 38; Berhaut, Fl. Sén. 53; Obermeyer in Fl. S. Afr. 1: 83, fig. 24, 1 (1966). *Caulina pectinata* Parl. (1858). *Najas horrida* A. Br. ex Magnus (1870)—Rendle in Trans. Linn. Soc., ser. 2, 5: 422, t. 42, fig. 183–191 (1899); F.T.A. 8: 228. Aquatic herb.
**Sen.:** Lake Guier *Roger*. Walo, near Keurmbaye *Leprieur*. St. Louis *Trochain* 2292! **Mali:** Diaka, Macina (fr. Nov.) *Monod*! **Guin.:** Koba (fr. Nov.) *Jac.–Fél.* 7375! **S.L.:** Rokupr (Sept., Oct.) *Jordan* 115! 128! **Ghana:** Lawra (fr. Sept.) *Hall* CC 725! Pong Tamale *Hall* CC 420! **N.Nig.:** Nupe *Barter* 1065! **W.Cam.:** Barombi-Ba-Mbu (Aug.) *Preuss* 452! Kumba (=Johann-Albrechtshöhe) *Staudt* 488! Widespread in Africa and tropical Asia.
2. **N. affinis** *Rendle* in Trans. Linn. Soc., ser. 2, 5: 440 (1899); H. af Rantz. l.c. 32 (1960); Berhaut, Fl. Sén. 53.
**Sen.:** *Leprieur* ! N'Ghar to Namari *Trochain* 3729 *bis*!
3. **N. meiklei** *H. af Rantz.* in Kew Bull. 7: 34, figs. 1, 2 (1962).
**Ghana:** Lawra (Sept.) *Hall* CC 723! Ada (May) *Hall* 3463! **N. Nig.:** near Gwari Hill, Minna (fl. & fr. Dec.) *Meikle* 710! near Gembu, Mambila Dist. (fr. Feb.) *Hepper* 2829!

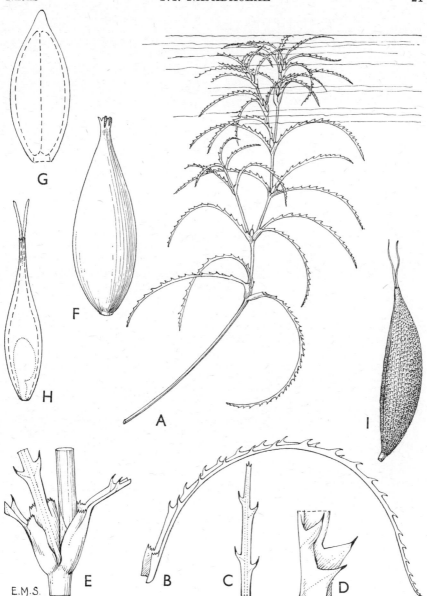

FIG. 326.—NAJAS PECTINATA (*Parl.*) *Magnus* (NAJADACEAE).

A, habit, × 1½. B, leaf, × 3. C, apex of leaf, × 8. D, marginal teeth, × 16. E, section of stem showing sheaths, × 6. F, male flower enclosed by spathe, × 20. G, stamen inside perianth, × 20. H, female flower, × 20. I, fruit, × 18. All from *Milne-Redhead & Taylor* 10910.

4. **N. hagerupii** *H. af Rantz.* in Kew Bull. 7: 35, fig. 3 (1952).
    **Mali:** Gao (Sept.) *Hagerup* 368a! **Ghana:** Baghari, near Lawra (Sept.) *Hall* CC 768! Kamba, near Lawra (Nov.) *T. M. Harris*!
5. **N. liberiensis** *H. af Rantz.* in Medd. Göteb. Bot. Trädg. 18: 190 (1950); Kew Bull. 7: 39 (1952); Aké Assi, Contrib. 2: 12 [222].
    **Mali:** Kita Massif (July) *Jaeger* K1! **Lib.:** Vonjama (Oct.) *Baldwin* 9925! **Iv.C.:** Danané to Nzo *Mangenot & Aké Assi* IA 3276. Also in Chad and C. African Republic.
6. **N. baldwinii** *H. af Rantz.* in Medd. Göteb. Bot. Trädg. 18: 187 (1950); Bot. Notiser 1951: 382; Kew Bull. 7: 40 (1952).
    **S.L.:** Njala (Sept.) *Deighton* 2797! Kambia (July) *Jordan* 471! **Lib.:** Gondolahun, Kolahun Dist. (Nov.) *Baldwin* 10116! Salala (Aug.) *Baldwin* 13206! Gbeidin (Sept.) *Harley* 1656! **Ghana:** Busunu, Damongo to Yapei (fr. Sept.) *Hall* CC 892!

The species are not as clear-cut as the key would suggest and in spite of Horn af Rantzien's work much remains to be done to define the specific limits—if, indeed, this is possible in the traditional sense. Other aquatic plants have been shown to exhibit remarkable morphological variation according to the environmental conditions (moving or still water etc.) and in Najas even the floral characters appear to be of dubious taxonomic value.—F.N.H.

# 172. COMMELINACEAE

## By J. P. M. Brenan

Perennial or annual herbs, often more or less succulent, mostly terrestrial, sometimes aquatic, frequently producing adventitious roots at the nodes ; stems erect to prostrate, rarely somewhat climbing. Leaves alternate (falsely whorled in *Palisota*), with a basal membranous often nervose and closed sheath. Inflorescence composed of single or aggregated cincinni, terminal, lateral or axillary ; sometimes each cincinnus may be reduced to single or (apparently) fascicled flowers. Flowers actinomorphic or zygomorphic, often surrounded by mucilage. Sepals 3, free, usually green or membranous. Petals 3 (one often smaller than the other 2), white or coloured, free or sometimes united below into a tube. Stamens hypogynous, basically 6 in two whorls, but variously modified or suppressed ; fertile stamens 2, 3 or 6 ; staminodes 0, 3 (or rarely 4) ; filaments (with us) free, glabrous or with moniliform hairs ; anthers dorsifixed or basifixed, 2-locular, dehiscing by longitudinal slits (or by basal pores in *Cyanotis*). Ovary superior, 2–3-locular, with a simple terminal style and a small more or less capitate stigma ; ovules 1–6(–10) per loculus, axile. Fruit usually a loculicidal capsule, sometimes partly or wholly indehiscent, or (in *Palisota*) a berry. Seeds usually crowded, with the contiguous faces flat, often muricate, ridged or reticulate, relatively large ; the testa characteristically marked on the outside with a circular or elliptic callosity called the embryostega (or embryotega), under which the embryo is situated ; hilum punctiform or linear ; endosperm abundant, mealy.

A medium-sized family of 500–600 species, widely distributed mainly in tropical and subtropical regions.

The leading features of the family are the leaf-sheaths, the inflorescence of cincinni, the separate calyx and corolla, and the embryostega on the seed. According to Mr. D. P. Stanfield, to whose observations on the family as a whole this account owes much, the family is " readily recognized by the involute margins of the young leaves, a characteristic feature of all the genera in West Tropical Africa except *Cyanotis* ".

The frail flowers are usually fully open for a short time only—sometimes for but a few hours during the day—and closely related species may differ in the times at which their flowers are expanded.

Inflorescences each emerging through a hole at the base of the leaf-sheath, thus always borne laterally :
   Inflorescence on a peduncle 2–8 cm. long ; flowers laxly arranged ; seeds 4–8 per loculus    ..    ..    ..    ..    ..    ..    10. **Buforrestia**
   Inflorescence sessile or nearly so ; peduncle up to about 0·5 cm. long ; seeds 1–2 per loculus :
      Bracts subtending flowers in inflorescence, small, up to about 0·5 cm. long, not strongly pointed ; inflorescence bifurcate, with the branches elongating to about 3 cm. in length as they develop ; petals pink, free    ..    2. **Forrestia**
      Bracts subtending flowers in inflorescence, large, green, 1·5–2·5 cm. long, pointed ; inflorescences appearing as dense leafy clusters at stem nodes ; petals white, united below into a tube    ..    ..    ..    ..    8. **Coleotrype**
Inflorescences not perforating leaf-sheaths, axillary or terminal :
   Inflorescences not enclosed in or closely subtended by a spathaceous bract, usually paniculate and terminal, sometimes axillary and then occasionally each inflorescence with only one or a few flowers :
   Fruit a loculicidal capsule :
      Petals 3, equal :
         Fertile stamens 6 ; staminodes 0 ; inflorescence usually more or less hairy, hairs frequently glandular :
            Capsule 3-locular ; seeds 2–10 per loculus ; bracts subtending pedicels amplexicaul or perfoliate and more or less cupular    ..    ..    1. **Stanfieldiella**
            Capsule 2-locular ; seed 1 per loculus ; bracts more or less narrowed at base, not amplexicaul or perfoliate    ..    ..    ..    ..    4. **Floscopa**

Fertile stamens 3, alternating with 3 yellow hastate or trilobed staminodes; inflorescence glabrous  ..  ..  ..  ..  ..  ..  **3. Murdannia**
Petals 3, unequal, the anticous (lower one) smaller than the others; fertile stamens 3, all anticous (lower); posticous stamens transformed into yellow bilobed staminodes; capsule usually 2-locular, rarely and inconstantly 3-locular
                                               **5. Aneilema**
Fruit indehiscent:
Leaf-margins glabrous or nearly so; fruits hard, brittle, metallic blue-black or purplish, with many seeds  ..  ..  ..  ..  ..  ..  **6. Pollia**
Leaf-margins silky-hairy; fruit a more or less fleshy berry with 3 loculi and 2–3 seeds per loculus, never metallic blue  ..  ..  ..  ..  **7. Palisota**
Inflorescence enclosed in or closely subtended by an often leafy spathaceous bract:
Fertile stamens 5–6, with densely and conspicuously bearded filaments; petals connate into a tube; bracts subtending individual flowers conspicuous and usually more or less falcate; inflorescences usually clustered, terminal axillary or lateral, rarely solitary  ..  ..  ..  ..  ..  ..  **9. Cyanotis**
Fertile stamens 3, with glabrous filaments; petals free; bracts subtending individual flowers 0 or very small and inconspicuous:
Spathes sessile, arranged alternately along more or less elongate terminal and lateral axes which are covered with hooked hairs; spathes small, up to 15 mm. long; petals white  ..  ..  ..  ..  ..  ..  ..  **11. Polyspatha**
Spathes usually pedunculate, sometimes almost sessile, usually solitary and leaf-opposed, sometimes appearing clustered or on a short lateral shoot; spathes often more than 15 mm. long; flowers usually blue or yellow, rarely white or another colour  ..  ..  ..  ..  ..  ..  ..  **12. Commelina**

*Dichorisandra thyrsiflora* Mikan from South America is cultivated in Sierra Leone; *Rhoeo spathacea* (Sw.) Stearn (*R. discolor* (L'Hérit.) Hance) from the West Indies, with sword-shaped, usually green and purple leaves and boat-shaped lateral inflorescences of white flowers subtended by paired leafy bracts, is often cultivated; *Zebrina pendula* Schnizl. from Tropical America, a prostrate herb with ovate or elliptic usually more or less variegated or purple leaves and small pink flowers in close stalked heads, each head subtended by 2 unequal leafy bracts, is also often grown for ornament.

## 1. STANFIELDIELLA Brenan in Kew Bull. 14: 283 (1960). *Buforrestia* C. B. Cl. (1881), (partly)—F.T.A. 8: 76; F.W.T.A., ed. 1, 2: 309.

Inflorescences terminal or more or less aggregated near tops of the erect or suberect flowering stems (lateral inflorescences may be present in addition); hairs on pedicels and outside of sepals (when present) 0·25–1 mm. long:
Seeds 4–10 per loculus, smooth or (in *S. oligantha*) verrucose; capsule much exceeding sepals:
Terminal inflorescence or cluster of inflorescences usually 3–8 cm. wide; leaves not or scarcely aggregated into an apical rosette, usually up to about 13 cm. long and 4 cm. broad (rarely more), elliptic or narrowly elliptic, usually more or less long-ciliate on " petiole " and margin of lamina near base; outside of sepals and pedicels usually pubescent with gland-tipped hairs 0·5–1 mm. long; capsule 1·5–2 mm broad  ..  ..  ..  ..  ..  ..  ..  **1. *imperforata***
Terminal inflorescence or cluster of inflorescences small and compact, about 2–3 cm. long and wide; upper leaves mostly more or less aggregated into apical rosettes, similar to *S. imperforata* but sometimes (not always) larger, 10–16 cm. long; capsule 2·25–3 mm. broad  ..  ..  ..  ..  ..  ..  **2. *oligantha***
Seeds 2–4 per loculus, verrucose; capsule not or scarcely exceeding sepals; leaf-sheaths densely spreading-villous outside; lamina rather densely appressed-hairy on both surfaces; inflorescence pubescent with gland-tipped hairs 0·25–0·5 mm. long  ..  ..  ..  ..  ..  ..  ..  **3a. *brachycarpa* var. *hirsuta***
Inflorescences all lateral and scattered, not terminal or aggregated near top of flowering stem; stems creeping; pedicels and outside of sepals pubescent with short hairs less than 0·25 mm. long; seeds with close low warts or short ridges  ..  **4. *axillaris***

1. **S. imperforata** (*C. B. Cl.*) Brenan in Kew Bull. 14: 284 (1960); J. K. Morton in J. Linn. Soc. 60: 207 (1967). *Buforrestia imperforata* C. B. Cl. in DC., Monogr. Phan. 3: 234, t. 7 (1881); F.T.A. 8: 76; F.W.T.A., ed. 1, 2: 309. A herb with stems 5 in. to 2 ft. high, decumbent and stoloniferous below and rooting from lower nodes; leaves often more or less purplish beneath; flowers with small, white petals, open from about 11.30 a.m. until noon; in lowland rain-forest.
**S.L.:** Kukuna (Jan.) *Sc. Elliot* 4678! Yonibana (Oct.) *Thomas* 4116! Kasewe F.R. (Nov.) *Morton* SL 1484! Tingi Hills above Konelo (fr. Dec.) *Morton & Gledhill* SL 3203! **Lib.:** Du River (July) *Linder* 174! Tenneh (Apr.) *Dinklage* 2562! Jaurazon (Mar.) *Baldwin* 11456! Bahtown, Tchien Dist. (Aug.) *Baldwin* 8031! Gbanga (Sept.) *Linder* 732! Nimba Mt. (July) *Leeuwenberg & Voorhoeve* 4738! **Iv.C.:** Angedédou Forest (Nov.) *Leeuwenberg* 1870! Morénou (Dec.) *Chev.* 22466! Audouin Forest (Aug.) *de Wit* 5766! Oroumba Boca (Aug.) *de Wit* 8274 (partly)! **Ghana:** Prestea (July) *Vigne* FH 1317! Benso (Nov.) *Andoh* FH 5402! Chairaiso, Kumasi Dist. (Nov.) *Morton* A2814! Ejuanema Mt., Mpraeso (Aug.) *Hall* CC 113! Tafo (Sept.) *Deighton* 3420! **S.Nig.:** Idanre F.R., Ondo Dist. (July) *Onochie* FHI 33389! Akure F.R. (Nov.) *Keay* FHI 25534! Owam F.R., Benin (Nov.) *Meikle* 512! Okomu F.R., Benin (Jan.) *Brenan & Richards* 8832! Sapoba F.R., Benin (Nov.) *Keay* FHI 28068! **W.Cam.:** *Preuss* 1297 (partly)! Likomba, Victoria (Oct.) *Mildbr.* 10562! Cam. Mt., 2,500 ft. *Mann* 1340! Bopo to Pelé (Feb.) *Binuyo*

& *Daramola* FHI 35578! Buea, 3,700 ft. (Dec.) *Migeod* 252! Barombi *Preuss* 287! Extends to Uganda, Tanzania and Angola.

[Typical var. *imperforata* is prevalent in our area. Var. *glabrisepala* (De Wild.) Brenan l.c. 285 (1960) with glabrous sepals and pedicels is not certainly recorded, but *Morton* A4129 from near Gbadzeme, Amedzofe, Ghana, and *Morton* GC 9382 from Ghana, Vane may be it. More material is needed.]

2. **S. oligantha** (*Mildbr.*) *Brenan* l.c. 285 (1960); J. K. Morton l.c. (1967). *Buforrestia oligantha* Mildbr. in Notizbl. Bot. Gart. Berl. 9: 258 (1925). A herb very similar in habitat and appearance to *S. imperforata* but with shorter more bunched inflorescences, verrucose seeds, a tendency for the lamina and " petiole " to be less long-ciliate than var. *imperforata*, and the flowers opening much later, about 3–6 p.m.
**Lib.:** Ganta (Aug.) *Harley* 977! Bilimu (Dec.) *Harley* 1546! **Iv.C.:** Oroumba Boca (Aug.) *de Wit* 8274 (partly)! **Ghana:** Amedzofe, 1,800 ft. (Mar., Nov.) *Hepper & Morton* A3028! *Morton* A101! **S.Nig.:** Akure F.R., Owena (Dec.) *Keay* FHI 37271! Oluwa F.R., Ondo Dist. (July) *Onochie* FHI 33410! Orosun Peak, Idanre Hills (Dec.) *Stanfield* 182! Imeri, Owo Dist. (Oct.) *Stanfield* 58! Usonigbe F.R., Benin (Nov.) *Meikle* 567! **W.Cam.:** Victoria to Man O'War Bay (Jan.) *Morton* K1199! *Preuss* 1297 (partly)! Also in E. Cameroun.
[There is some doubt about the identity of the above-cited specimens with *S. oligantha* until more Cameroun material is available.]

3. **S. brachycarpa** (*Gilg & Lederm. ex Mildbr.*) *Brenan* l.c. 285 (1960); J. K. Morton l.c. (1967). *Buforrestia brachycarpa* Gilg & Lederm. ex Mildbr. in Notizbl. Bot. Gart. Berl. 9: 258 (1925).

3a. **S. brachycarpa** var. **hirsuta** (*Brenan*) *Brenan* l.c. 286 (1960). *Buforrestia brachycarpa* Gilg & Lederm. ex Mildbr. var. *hirsuta* Brenan in Kew Bull. 7: 455 (1953). A trailing herb rooting at nodes, with ascending-erect leafy stems 2–10 in. high; leaves long-hairy on both surfaces; flowers greenish-white; in swamp-forest or damp places in lowland rain forest.
**S.Nig.:** Okomu F.R., Benin (Dec.–Feb.) *Brenan* 8486! 9090! Also in E. Cameroun.
[A specimen at Kew collected by Mann on Fernando Po may be this, but is inadequate for certainty.]

4. **S. axillaris** *J. K. Morton in J. Linn. Soc.* 60: 207, t. 5 (1967). *S. sp.* Brenan in Kew Bull. 14: 286 (1960). A herb with creeping leafy stems rooting at nodes; leaves elliptic, about 2–7 cm. long and 1·5–2·8 cm. wide; inflorescences 2–3 cm. long, borne laterally at nodes; petals white or pale pink; fruits not seen, apparently not often set; in forest and plantations.
**Ghana:** Apedwa to Kibbi (Jan.) *Morton* A111! Begoro (Jan.) *Morton* A2671! **S.Nig.:** Usonigbe F.R., Benin (Nov.) *Meikle* 566! Shasha F.R. (Mar.–Apr.) *Richards* 203! 3177! 3366! Ikeja to Ikorodu (July) *Killick* 244!

## 2. FORRESTIA A. Rich., Sert. Astrol. 1 (1834); F.T.A. 8: 77.

Stems glabrous, with long internodes below becoming shorter above; leaf-sheaths 1–2 cm. long, ciliate on margin; lamina obovate to obovate-elliptic, 5–15 cm. long, (2–) 3–6 or more cm. wide, often rather shortly acuminate at apex, narrowed at base and abruptly passing into a distinct false petiole 1–2 cm. long; parallel nerves about 4 (–6) on each side of midrib; inflorescence perforating the base of the leaf-sheath, sessile or subsessile, dividing near base into two branches, at first short, later elongating to 3 cm.; bracts ovate, 3·5–5 mm. long; flowers mauve to reddish-mauve          *tenuis*

**F. tenuis** (*C. B. Cl.*) *Benth.* in Benth. & Hook. f., Gen. Pl. 3: 851 (1883); F.T.A. 8: 77; Brenan in Kew Bull. 7: 456 (1962); J. K. Morton in J. Linn. Soc. 60: 202 (1967). *Buforrestia ? tenuis* C. B. Cl. in DC., Monogr. Phan. 3: 234 (1881). *Forrestia africana* K. Schum. ex C. B. Cl. in F.T.A. 8: 77 (1901). *F. preussii* K. Schum. in Engl., Bot. Jahrb. 24: 344 (1897); F.W.T.A., ed. 2, 2: 309. An erect or ascending herb 1–2 ft. high (? more) with rather broad distinctly " petiolate " leaves and lateral subsessile bifurcate inflorescences of small mauve flowers; in undergrowth, sometimes swampy, of rain forest.
**S.Nig.:** Abe (Jan.) *Morton* K1211! Sapoba (Aug.) *Olorunfemi* FHI 34168! Sapoba F.R. (fl. & fr. Nov.) *Keay* FHI 28070! **W.Cam.:** Banga (Mar.) *Brenan* 9315! 9446! 9472! Bimbia road, Victoria (Apr.) *Maitland* 669! Also in E. Cameroun, Gabon and the Congo.

## 3. MURDANNIA Royle, Illustr. Bot. Himal. 403, t. 95, fig. 3 (1839); Brenan in Kew Bull. 7: 179 (1952). *Baoulia* A. Chev. in Mém. Soc. Bot. Fr. 2, 8: 217 (1912). *Phaeneilema* Brückn. in Engl., Bot. Jahrb. 61, Beibl. 137: 63 (1926). *Aneilema* R. Br. (partly)—F.T.A. 8: 62; F.W.T.A., ed. 1, 2: 312.

Inflorescence normally with distinct lateral branches, cymose or paniculate; rarely (in No. 2) occasional inflorescences may be without lateral branches but are then subumbellately racemose with 3 or more flowers; staminodes normally 3 (–4):
Erect, relatively robust herb about 15–60 cm. high; leaves long, the lower about 7–40 cm. long; axes of partial inflorescences (cincinni) in fruit elongating to about 0·5–2 (–3) cm. with usually close-set conspicuous scars left by the fallen pedicels; petals 5·5–7 mm. long, 4·5–7 mm. broad  ..     ..     ..     ..     .. 1. *simplex*
Weak, prostrate or decumbent herb; leaves relatively short, the lower to 7 (–10) cm. long; axes of partial inflorescences (cincinni) in fruit 0·4–1 (–1·3) cm. long, with less closely spaced and less conspicuous scarring; petals about 3–3·5 mm. long, 2·5 mm. broad ..     ..     ..     ..     ..     ..     ..     .. 2. *nudiflora*
Inflorescence unbranched, consisting of solitary or sometimes paired flowers arising from the axils of 1–3 of the uppermost leaves (bracts); plant slender, grass-like, with usually simple or subsimple stems, glabrous except for margins of sheaths and lower part of leaves; leaves linear-lanceolate, 2–5(–6) mm. wide; fertile stamens 3, with hairy filaments; staminodes 0  ..     ..     ..     ..     .. 3. *tenuissima*

1. **M. simplex** (*Vahl*) *Brenan* in Kew Bull. 7: 186 (1952); Saunders, Handb. W. Afr. Fl. 96 (1958); J. K. Morton in J. Linn. Soc. 60: 202 (1967). *Commelina simplex* Vahl, Enum. 2: 177 (1806). *Aneilema sinicum* Ker-Gawl. in Bot. Reg. t. 659 (1822); C. B. Cl. in DC., Monogr. Phan. 3: 212 (1881); F.T.A. 8: 63; F.W.T.A., ed. 1, 2: 312, t. 286; Sousa in Anais Junta Inv. Col. 5: 47 (1950). *Murdannia sinica* (Ker-Gawl.) Brückn. in E. & P., Pflanzenfam., ed. 2, 15a: 173 (1930). A rather robust, erect or sprawling herb 6 in. to 4 ft. high with long lower leaves and panicles of blue to mauve flowers opening in late afternoon and evening, about 4–6.30 p.m.
**Sen.:** Djembering (Sept.) *Broadbent* 131! Sédhiou, Casamance (July) *Berhaut* 6146! Niokolo-Koba *Adam*

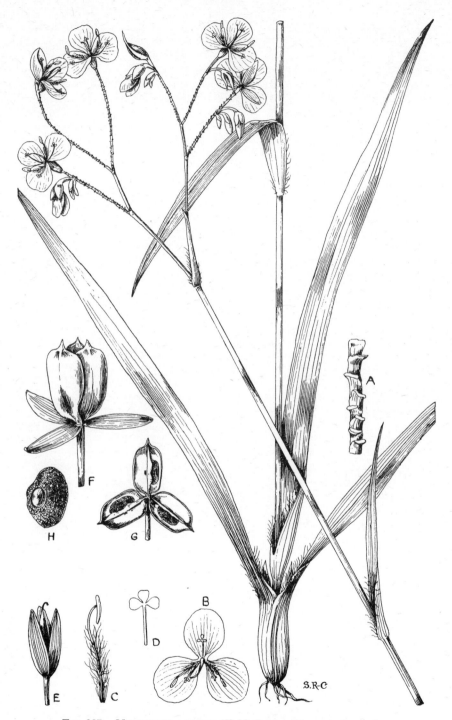

Fig. 327.—MURDANNIA SIMPLEX (*Vahl*) *Brenan* (COMMELINACEAE).

A, portion of rhachis of inflorescence showing scars of fallen pedicels. B, open flower. C, stamen. D, staminode. E, young fruit. F, ripe fruit. G, open fruit. H, seed.

25

17225! Kafountine (Sept.) *Adam* 18179! **Mali**: Sikasso (Sept.) *Adam* 15055! **Guin.**: Kinkou (Apr.) *Pitot*! **Port.G.**: Bissau, Pussubé (July) *Esp. Santo* 1328! **S.L.**: Bumban to Port Loko (Apr.) *Sc. Elliot* 5742! **Iv.C.**: *Scaëtta* 3061! Marabadiassa, Mankono Dist. (July) *Chev.* 22011! Bouaké, Boualé N. Dist. (July) *Chev.* 22089! Rocher d'Issia (fl. & fr. May) *Leeuwenberg* 4132! Assakra, Oroumba Boca (Oct.) *de Wilde* 599! **Ghana**: Tamale to Damongo (Aug.) *Irvine* 4576! Nyamkpale (Dec.) *Adams & Akpabla* GC 4190! Ejura Scarp (Dec.) *Adams & Akpabla* GC 4524! Efiduase (Apr.) *Coz* 26! Nungua (June) *Ankrah* GC 20177! Wenchi to Techiman (Mar.) *Morton* GC 8555! **N.Nig.**: Samaru, Giwa Dist. (June) *Keay* FHI 25859! Anara F.R., Zaria (May) *Keay* FHI 22963! Rimjim Mukur to Bununu, 2,500 ft., Bauchi Dist. (Aug.) *Lawlor & Hall* 611! Abinsi (Mar.) *Dalz.* 802! Near Yola (Aug.) *Dalz.* 258! **S.Nig.**: Oban *Talbot* 752! Ibenó, Eket Dist. (May) *Onochie* FHI 32914! New Ndebiji, Calabar (May) *Ujor* FHI 30162! Abak (June) *Nwauzoee* in *Hb. Tuley* 787! Widespread in tropical Africa from Sierra Leone and the Sudan to Angola, Natal and Zululand; also in Madagascar and Asia.

2. **M, nudiflora** (*Linn.*) Brenan l.c. 189 (1952); J. K. Morton l.c. (1967). *Commelina nudiflora* Linn., Sp. Pl. 41 (1753) partly, as to lectotype, see Merr. in J. Arn. Arb. 18: 64–66 (1937). *Aneilema nudiflorum* (Linn.) Wall., List 182, No. 5224 (1830). *Murdannia malabarica* (Linn.) Brückn. in E. & P., Pflanzenfam. ed. 2, 15a : 173 (1930). A weak, prostrate or decumbent herb with stem often rooting at nodes; flowers small, white to purple, open from 11 a.m. to 1 p.m.
    **S.L.**: Roboli, Magbema (Oct.) *Jordan* 1054! Rokupr (Oct.) *Jordan* 1078! Njala (Aug.–Oct.) *Deighton* 5135! *Katta* 14! Romene, Kasse (Oct.) *Jordan* 360!
    [Probably an introduction in West Africa, as it is otherwise unknown from the continent; widespread in tropical Asia, extending eastwards to Polynesia (Guam); an aggressive weed in Trinidad.]
    [*Deighton* 5135 has been identified as *Murdannia loriformis* (Hassk.) Rolla & Kammathy in Bull. Bot. Surv. India 3: 393 (1961) (*Aneilema nudiflorum* (Linn.) Wall. var. *terminale* (Wight) C. B. Cl. in DC., Monogr. Phan. 3: 211 (1881)), but it does not seem clearly separable from *M. nudiflora*.]

3. **M. tenuissima** (*A. Chev.*) Brenan l.c. 7: 189 (1952); J. K. Morton l.c., fig. 8 (1967). *Baoulia tenuissima* A. Chev. in Mém. Soc. Bot. Fr. 2, 8: 217 (1912). *Aneilema tenuissimum* (A. Chev.) A. Chev. ex Hutch., F.W.T.A., ed. 1, 2: 314 (1936). A slender grass-like herb 8–28 in. high with few inconspicuous pale blue or mauve flowers opening from 10.30 a.m. until noon; in swampy or marshy grassland.
    **Guin.**: Dabola to Faranah (Apr.) *Pitot*! **Iv.C.**: Manikro to Tiegoriakro, N. Baoulé (Aug.) *Chev.* 22318! **Ghana**: Ejura (Oct.–Dec.) *Morton* A3757! GC 9489! **N.Nig.**: Olle F.R., Bunu, Kabba Prov. (fl. & fr. Oct.) *Daramola & Adebusuyi* FHI 38419! Likitaba to Gembu, Mambila Plateau (June) *Chapman* 28! **S.Nig.**: Ilora to Oyo (July–Sept.) *Keay* FHI 28037! *Keay & Stanfield* FHI 46319! *J. B. Gillett* 15186! Owo (Oct.) *Stanfield* 33! Also in the Congo, Uganda, Tanzania, Zambia and Angola.

## 4. **FLOSCOPA** Lour., Fl. Cochinch. 192 (1790); F.T.A. 8: 84.

Inflorescence quite glabrous; seeds smooth, without ribs or conspicuous tubercles; leaves linear-lanceolate, 1·5–8 mm. broad  ..    ..    ..    ..      1. *leiothyrsa*
Inflorescence more or less pubescent or tomentose:
  Petals yellow; stems short, up to 20 cm. high, often branching from near base, rooting at base only; leaves lanceolate to linear-lanceolate, 2–8 mm. broad; inflorescence very shortly and inconspicuously pubescent with hairs up to 0·25 mm. long; seeds tuberculate  ..    ..    ..    ..    ..    ..    ..    ..      2. *flavida*
  Petals blue or lilac or sometimes white; stems straggling or erect, rooting at base or from lower nodes as well, often exceeding 15 cm.; seeds with radial ridges or smooth, without conspicuous tubercles:
    Capsule 1·25–1·75 mm. broad; seeds 0·5–0·8 mm. long, 0·5–0·75 mm. broad, 12–14 (rarely –17)-ribbed; cincinni rather narrow and elongate; annual, 10–30 cm. high, rooting only at or near base; leaves not narrowed or falsely petiolate at base
                     3. *axillaris*
    Capsule 2–3 mm. broad; seeds 1–1·5 mm. long, 0·75–1 mm. broad, 13–21-ribbed:
      Leaf-sheaths densely villous-pubescent all over outside; lamina elliptic, subacute or acute at apex, not at all acuminate; inflorescence lax, open, with main axis only about 1·5–3·5 cm. long, and relatively few branches    ..    4. *mannii*
      Leaf-sheaths usually ciliate on margins, otherwise glabrous or with an inconspicuous line of pubescence running longitudinally along the fused margins; lamina lanceolate or elliptic-lanceolate, more or less acutely acuminate at apex; inflorescence usually dense and many-flowered, with main axis usually exceeding 3·5 cm. in length, and usually with numerous branches:
        Leaves much narrowed at base into a more or less marked false petiole; inflorescence shortly and inconspicuously pubescent or puberulous; stems ascending, decumbent or creeping, rooting from lower nodes:
          Pedicels of fruit 0·5–1·5 mm. long; sepals mostly 2–2·75 mm. long; style not or scarcely exserted beyond the sepals:
            Stems erect, ascending or straggling, about 15–90 cm. high; plant with reddish-purple pigment particularly in stems, undersides of leaves and inflorescences; leaves with lamina usually four or more times as long as broad, 4–15 cm. long
                             5a. *africana* subsp. *africana*
            Stems wholly low and creeping; flowering stems up to about 15–20 cm. high; plant lacking reddish-purple pigment; leaves with lamina up to about three times as long as broad, about 2·5–6 cm. long      5b. *africana* subsp.*petrophila*
          Pedicels of fruit 1·5–2 mm. long; sepals 3–4 mm. long; style exserted beyond the sepals, becoming reflexed; robust plant up to 1·2 m. high with a panicle 7–10 cm. long    ..    ..    ..    ..    5c. *africana* subsp. *majuscula*
        Leaves not at all narrowed at base, or if somewhat narrowed then without any false petiole:
          Capsule with stipe 0·5–1 mm. long, almost half as long as the body of the capsule; seeds with 20–23 clear radial ribs; plant erect, rooting only near base; lamina

8–15 cm. long, not narrowed at base; leaf-sheaths with a line of pubescence down the outside along the fused margins .. .. .. 6. *polypleura*

Capsule with usually much shorter stipe up to 0·5(–0·75) mm. long, at most ¼–⅓ as long as the body of the capsule; seeds usually with 11–18 (rarely to 21) radial ribs, rarely ribs obscure or absent:

    Leaves with lamina gradually narrowed towards the sheath, not at all petiolate nor amplexicaul, mostly 6·5–13 cm. long, 0·5–1·7 cm. broad; inflorescence usually dense and many-flowered, densely and shortly pubescent; seeds with 11–15 (–16) faint ribs .. .. .. .. .. .. .. 7. *confusa*

    Leaves with lamina not narrowed towards the sheath, rounded and more or less amplexicaul at base; seeds usually with well-marked ribs, ribs rarely obscure or absent:

        Terminal inflorescence with few (up to about 5) branches or simple; indumentum on inflorescence sparse (if densely villous-pubescent, compare *F. glomerata*); plant creeping or straggling, often floating in water, rooting freely from lower nodes; lamina short, up to 7 cm. long; seeds with 16–21 radial ribs

                                                8. *aquatica*

        Terminal inflorescence usually with numerous (5 or more) branches and more or less densely villous-pubescent; seeds with 13–18(–20) radial ribs, rarely ribs obscure or absent; lamina up to 11 cm. long:

            Sepals more or less densely hairy with hairs of varying length, among which are rather numerous long flexuous hairs 0·75–2·5 mm. long; terminal cell of hairs usually enlarged; leaves mostly 10–17 mm. broad, but sometimes only 5 mm. broad .. .. .. .. 9a. *glomerata* subsp. *glomerata*

            Sepals more or less densely hairy with short hairs 0·2–0·5 mm. long:

                Hairs on calyx with terminal cell not or scarcely enlarged in comparison with those below it; leaves mostly very narrow, 4–8 mm. broad, rarely more

                                9b. *glomerata* subsp. *pauciflora*

                Hairs on calyx with terminal cell clearly enlarged in comparison with those below it, the hairs thus appearing glandular-capitate; leaves mostly 8–18 mm. broad .. .. .. .. 9c. *glomerata* subsp. *lelyi*

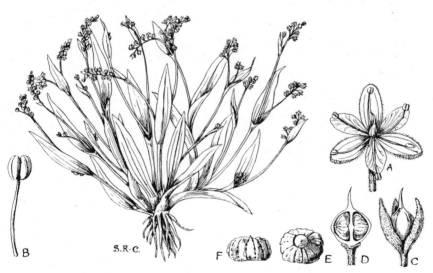

FIG. 328.—FLOSCOPA FLAVIDA *C.B. Cl.* (COMMELINACEAE).

A, open flower. B, stamen. C, fruit. D, vertical section of fruit. E, top and F, side view of seeds.

1. **F. leiothyrsa** *Brenan* in Kew Bull. 7: 206 (1952). A probably annual herb, apparently erect, 8–10 in. high, subsimple to much branched; flowers pale violet, in few- to many-branched inflorescences; in or near water or marshes.
   **Mali:** Fetodie, nr. Dogo (Jan.) *Davey* 107! Also in Chad, Tanzania, Zambia and Botswana.
   [Immediately separable from all other species of *Floscopa* by the glabrous inflorescences.]
2. **F. flavida** *C. B. Cl.* in DC., Monogr. Phan. 3: 269 (1881); F.T.A. 8: 87; Berhaut, Fl. Sén. 210; Nielson, Introd. Fl. Pl. W. Trop. Afr. 184 (1965). Small annual herb usually much branched from base 1¾–8 in. high; stem-leaves mostly short and less than 2 in. long, sometimes to 3 in.; sepals dark reddish-purple; petals yellow, differing thus from those of all other *Floscopa* species in our area.
   **Sen.:** Elinkine, Bignona (Oct.) *Adam* 18275! Niokolo-Koba (Oct.) *Adam* 15891! **Port.G.:** Buruntuma (Dec.) *Pereira* 2160! **Guin.:** Youkounkoun (Oct.) *Pitot*! **Ghana:** Yendi to Zabzugu (Dec.) *Morton* A1464! **N.Nig.:** Takum, Borgu Dist. *Barter* 760! Kabba road, Lokoja Dist. *Parsons* 61! Kontagora (Nov.) *Dalz.* 271! Naraguta, Jos Plateau (Aug.) *Lely* 554! Also in Chad, Sudan, Tanzania, Malawi, Zambia and Angola.

3

3. **F. axillaris** (*Poir.*) *C. B. Cl.* in DC., Monogr. Phan. 3: 268 (1881); F.T.A. 8: 87; J. K. Morton in J. Linn.
Soc. 60: 201 (1967). *Polygala axillaris* Poir., Encycl. 5: 489 (1804). *Floscopa elliotii* C. B. Cl. in F.T.A. 8:
88 (1901). Annual herb about 3–12 in. high; leaves ovate-lanceolate to lanceolate, up to about 2 in. long,
with short sheaths which are non-ciliate on margins; flowers blue or dull purple; in marshes and swamps.
**Sen.:** Oussouye (Oct.) *Adam* 18301! Bissine (Dec.) *Berhaut* 6695! **Port. G.:** Orango Grande, Bijagos Is.
(Jan.) *Raimundo & Guerra* 933! Prabis, Bissau (Nov.) *Raimundo & Guerra* 172! Pussubé, Bissau (Nov.)
*Esp. Santo* 1000! **Guin.:** Tristao Is. (Nov.) *Jac.-Fél.* 7326! Friguiagbé (Nov., Dec.) *Chillou* 907! 2362!
Filicoundji (Oct.) *Adam* 12587! Kounounkan (Oct.) *Schnell* 7622! **S:L.:** Tisana, Bonthe Is. (Nov.)
*Deighton* 2304! Bonthe (Oct.) *Adames* 87! York to No. 2 River (Nov.) *Morton* SL 1455! Kitchom,
Scarcies (Dec.) *Sc. Elliot* 4340! Kasewe F.R. (Nov.) *Morton* SL 1536!
4. **F. mannii** *C. B. Cl.* in DC., Monogr. Phan. 3: 268 (1881); F.T.A. 8: 86; J. K. Morton l.c. (1967). A low
herb about 4–6 in. high; leaves with elliptic lamina 3–6·5 cm. long and 1·5–3 cm. wide. A very distinctive
but little-known species, readily known by the combination of falsely petiolate leaves and densely pubescent leaf-sheaths; probably in wet places in rain-forest.
**S.Nig.:** Oban *Talbot* 756! Also in Rio Muni.
5. **F. africana** (*P. Beauv.*) *C. B. Cl.* l.c. 267 (1881); F.T.A. 8: 85; J. K. Morton l.c. 199 (1967). *Aneilema
africanum* P. Beauv., Fl. Oware 2: 57, t. 93 (1818).
5a. **F. africana** (*P. Beauv.*) *C. B. Cl.* subsp. **africana.** An erect or straggling herb up to 3 ft. high, variable in
robustness, often rooting at the nodes; leaves lanceolate, with a false petiole; flowers whitish or pale
lilac; in or near water, usually by rivers and streams.
**Gam.:** Genieri (Feb.) *Fox* 73! **Port.G.:** Fá (Oct.) *Guerra* 3878! **Guin.:** Kounounkan valley (Oct.) *Schnell*
7619! **S.L.:** Musaia (Dec.) *Deighton* 4495! Rogbin, N. Prov. (Oct.) *Glanville* 49! Kameron to Kurubonla
(Nov.) *Morton* SL 2509! Mamaka (Nov.) *Thomas* 4634! Gebwema (Nov.) *Deighton* 399! **Lib.:** Wohmen,
Vonjama Dist. (Oct.) *Baldwin* 10065! 10096! Banga (Oct.) *Linder* 1176! Fayapulu (Oct.) *Linder* 1139!
Loma F.R. (Dec.) *Voorhoeve* 750! **Iv.C.:** 56 Km. N. of Sassandra, E. of Béyo (Jan.) *Leeuwenberg* 2553!
Akabossué to Ebrinahoué, Mid. Comoé (Dec.) *Chev.* 22606! **Ghana:** Kumasi (Oct.) *Darko* 426! Shiare,
Togo Hills (Dec.) *Jenik & Hall*! Birrim R., 18 mi. N. of Anyinam (Mar.) *Irvine* 1758! Dawa Male Kole
(Dec.) *A. S. Thomas* D2! **S.Nig.:** Ibadan (Oct., Nov.) *Stanfield* 43! *Keay* FHI 37189! Okomu F.R.,
Benin Dist. (fr. Dec., Feb.) *Brenan* 8485! 8528! Orem, Calabar to Mamfe (Jan.) *Onochie* FHI 36055x!
**W.Cam.:** Victoria (Jan.) *Maitland* 1304! Also from E. Cameroun to Mayombe and eastwards to Uganda.
[An odd variant occurs in Ghana with dense indumentum all over, the leaf-sheaths outside and on the
lower surface of the leaves. *Hall* 2754, from Kakum F.R. is an example.—J.P.M.B.]
5b. **F. africana** subsp. **petrophila** *J. K. Morton* in J. Linn. Soc. 60: 200 (1967). Similar to subsp. *africana*, but
stems much lower and leaves smaller. Some perplexing intermediates occur.
**Lib.:** Vonjama (Oct.) *Baldwin* 9869! **Iv.C.:** W. of Béyo (fr. Nov.) *Leeuwenberg* 2735! **Ghana:** Apiso (Dec.)
*Morton* A3849! Kpene, Togo Plateau (Nov.) *Morton* A3444! **W.Cam.:** Likomba Plantation, Victoria
(Dec.) *Mildbraed* 10775! Apparently also in E. Cameroun, Congo and Uganda.
5c. **F. africana** subsp. **majuscula** (*C. B. Cl.*) *Brenan* in Kew Bull. 22: 387 (1968). *F. africana* var. *majuscula*
C. B. Cl. in F.T.A. 8: 85 (1901).
**Port.G.:** Antula, Bissau (Nov.) *Esp. Santo* 1413! **Lib.:** Vonjama (Oct.) *Baldwin* 9886! Ganta (Dec.)
*Barker* 1121! **Dah.:** *Le Testu* 90! **S.Nig.:** Akure (Dec.) *Adams* 1460! Akure F.R., Aponmu (Oct.) *Keay*
FHI 25487! Owo (Oct.) *Stanfield* 5! Idanre F.R. (Dec.) *Keay* FHI 37265! Ibadan (Dec.) *Meikle* 859!
Lagos *Punch*! *Millen* 97! Okomu F.R., Benin (Jan.) *Brenan* 8872! Similar plants in E. Cameroun,
Rio Muni and the Congo.
[From Dahomey to E. Cameroun this is a very distinct taxon, but elsewhere it is not always so marked
and the characteristic exserted style less obvious.—J.P.M.B.]
6. **F. polypleura** *Brenan* in Kew Bull. 7: 207 (1952); J. K. Morton in J. Linn. Soc. 60: 201 (1967). Erect or
suberect, annual herb 8 in.—2 ft. 6 in. high; inflorescence glandular-pubescent; flowers lilac-purple or
sometimes white; seeds with remarkably numerous (20–23) radial ribs; in savanna.
**Ghana:** Tamale to Lumbungu (Nov.) *Morton* A3787! Yendi (Nov.) *Akpabla* 476! Also in Tanzania and
Zambia.
7. **F. confusa** *Brenan* in Kew Bull. 15: 225 (1961); J. K. Morton l.c. 200 (1967). A coarse straggling herb
about 2–5 ft. high, with stems rooting at lower nodes; leaves linear-lanceolate, leaf-sheaths ciliate at
mouth, otherwise glabrous or sparsely pubescent along the connate margins; flowers mauve, in densely
and shortly pubescent panicles; seeds with very faint ribs; in upland marshes or swamps.
**N.Nig.:** Mambila Plateau, Sardauna Prov. (Jan.) *Hepper* 1719! **S.Nig.:** Ikwette Plateau, Obudu Div.
(Dec.) *Savory & Keay* FHI 25246! Also in the E. Cameroun, Congo and Uganda.
[Related to *F. glomerata*.]
8. **F. aquatica** *Hua* in Bull. Mus. Hist. Nat. Paris 1: 122 (1895); Brenan in Kew Bull. 15: 228 (1961); J. K.
Morton in J. Linn. Soc. 60: 201 (1967). *F. myosotoides* Hutch., F.W.T.A., ed. 1, 2: 311 (1936), and in
Kew Bull. 1939: 241; Brenan in Kew Bull. 7: 457 (1952). Herb with creeping or floating stems 2½–16 in.
long, rooting freely from the nodes; leaves ovate to lanceolate, mostly 2–5 cm. long, thin, often wavy-
margined, not or scarcely narrowed at base; sheaths short, glabrous; flowers whitish to lilac or blue, in
rather small inflorescences; often floating in river and rice-fields, sometimes creeping in swamps.
**Sen.:** Santiaba, Oussouye (Nov.) *Berhaut* 6503! Bissine (Nov.) *Berhaut* 6450! Sédhiou (Mar.) *Berhaut*
7175! Niokolo-Koba (Nov.) *Jac.-Fél.* 7238! **Port.G.:** Nova Lamego, Cabuca (Dec.) *Pereira* 2452! **Guin.:**
Rio Nunez *Heudelot* 668! Friguiagbé (Sept.) *Chillou* 1064! **S.L.:** Masactaba (Oct.) *Glanville* 50! Subu,
Nongoba Bullom (Nov.) *Jordan* 595! Yonibana (Oct.) *Thomas* 4134! Mabouto (Oct.) *Thomas* 3614!
Gegbwema (Oct.) *Pyne* 20! **Lib.:** Monrovia (Nov.) *Linder* 1412! **Iv.C.:** Mossou (Aug.) *Schnell* 6562!
**S.Nig.:** Sapoba, Jamieson R. (Feb.) *Onochie* FHI 31943! *Richards* 3928! Ologbo, Ossiomo R. (Feb.)
*Stanfield* 158! Also in Middle Congo.
9. **F. glomerata** (*Willd. ex J. A. & J. H. Schult.*) *Hassk.*, Commel. Ind. 166 (1870); C. B. Cl. in DC., Monogr.
Phan. 3: 267 (1881); F.T.A. 8: 86; J. K. Morton in J. Linn. Soc. 60: 200 (1967). *Tradescantia glomerata*
Willd. ex J. A. & J. H. Schult., Syst. 7: 1175 (1830).
9a. **F. glomerata** (*Willd. ex J. A. & J. H. Schult.*) *Hassk.* subsp. **glomerata.** *F. rivularis* (A. Rich.) C. B. Cl. l.c.
(1881); F.T.A. 8: 86; F.W.T.A., ed. 1, 2: 311; Sousa in Anais Junta Inv. Col. 5: 147 (1950); Berhaut,
Fl. Sén. 210; Nielsen, Introd. Fl. Pl. W. Trop. Afr. 184 (1965). *Aneilema rivulare* A. Rich., Tent. Fl.
Abyss. 2: 342 (1850–51). Annual or perennial herb, erect or straggling, sometimes rhizomatous at base;
stems about 6 in.–5 ft. long, rooting at base or from lower nodes; flowers mauve, in usually much-branched
but generally rather compact hairy inflorescences; in marshes and swamps.
**Sen.:** Bissine, Casamance (Nov.) *Berhaut* 6451! Niokolo-Koba (Nov.) *Adam* 15949! **Port.G.:** Farim,
Bejene to Barro (Nov.) *Esp. Santo* 3658! **Guin.:** Mankoutan (Nov.) *Jac.-Fél.* 7346! **S.L.:** Morea to
Fintonia (Jan.) *Morton & Gledhill* SL 583! Juring (Dec.) *Deighton* 308! Kainkordu, Kono (Dec.) *Morton
& Jarr* SL 3283! **Ghana:** Burufo (Dec.) *Adams* 4379! **N.Nig.:** Mada, Plateau Dist. (Nov.) *Hepburn* 161!
Kaduna, Zaria Dist. (Nov.) *J. A. D. Jackson* 164! Jos Plateau (Oct.) *Lely* P842! Panyam, Bauchi Dist.
(Sept.) *Lely* 729! Widespread in tropical Africa (although rare in our area) from the Sudan southwards
to S. Africa, also in Madagascar.
9b. **F. glomerata** subsp. **pauciflora** (*C. B. Cl.*) *J. K. Morton* l.c. (1967). *F. pauciflora* C. B. Cl. in F.T.A. 8: 88
(1901); F.W.T.A., ed. 1, 2: 311. Similar in habit and habitat to subsp. *glomerata* but generally narrower-
leaved.
**Port.G.:** Barro to Ingoré (Nov.) *Esp. Santo* 3652! **Guin.:** Dalaba (Nov.) *Adames* 419! Sombalako to
Boulivel (Sept.) *Chev.* 18255! Mamou (Sept.) *Schnell* 6787! Kouroussa (Oct.) *Pobéguin* 578! 579! **S.L.:**
Gorahun (Nov.) *Deighton* 361! Yetaya (Sept.) *Thomas* 2319! Seli R. (Oct.) *Thomas* 3112! Kaballa (Sept.)

*Thomas* 2259! Kamasigi (Oct.) *Glanville* 47! Saiama, Kono (Dec.) *Morton* SL 3278! **Lib.:** Gondolahun (Nov.) *Baldwin* 10120! Sodu (Jan.) *Bequaert* 46! Extending to Chad.

9c. **F. glomerata** subsp. **lelyi** (*Hutch.*) *Brenan* in Kew Bull. 22: 387 (1968). *F. lelyi* Hutch., F.W.T.A., ed. 1, 2: 311 (1936), and in Kew Bull. 1939: 242 (1939). Similar in habit and habitat to subsp. *glomerata* but indumentum on inflorescence shorter.

**S.L.:** Ronietta (Nov.) *Thomas* 5528! **N.Nig.:** Kontagora, Niger Dist. (Dec.) *Dalz.* 270! Minna (Dec.) *Meikle* 730! Anara F.R., Zaria (Dec.) *Meikle* 75! Naraguta, Plateau Prov. (Nov.) *Lely* 704! Vogel Peak, Sardauna Prov. (Nov.) *Hepper* 1454! **S.Nig.:** Owo (Oct.) *Stanfield* 32! 138! Igarra (Dec.) *Stanfield* 74! Also apparently in Brazil and Paraguay, the specimens cited by C. B. Cl. in DC., Monogr. Phan. 3: 270 (1881) under typical *F. glabrata* Hassk. not seeming separable. The other varieties of *F. glabrata* may well represent different taxa.

*Imperfectly known species.*

**F.** sp. Annual herb 4–12 in. high. Rather similar to *F. axillaris*, with similar elongate cincinni up to 5 cm. long, but flowers and cymules larger, the latter 2–2·5 mm. across.
**Sen.:** Oussouye (Apr.) *Berhaut* 5844! Ziguinchor (Dec.) *Berhaut* 6698! Bignona (Dec.) *Berhaut* 6738! **Port.G.:** Nhampassaré, Gabu (Oct.) *Esp. Santo* 3526!

## 5. **ANEILEMA** R. Br., Prod. 290 (1810); F.T.A. 8: 33, excluding *Tricarpellaria*; Brückner in E. & P., Pflanzenfam., ed. 2, 15A: 174 (1930); J. K. Morton in J. Linn. Soc. 60: 168 (1967).*

Petals yellow; capsule truncate at apex with two rather projecting subacute corners like little horns; cells on surface of capsule isodiametric and projecting, giving a minutely papillose surface easily seen under a × 10 lens; leaves ovate-lanceolate, ovate-oblong or ovate-elliptic, mostly 2–6·5 cm. broad; plant with hooked clinging hairs; inflorescence lax     ..     ..     ..     ..     ..     ..     .. 1. *aequinoctiale*
Petals white, pink, lilac or mauve; capsule rounded and often emarginate at apex, without projecting corners; cells on surface of capsule usually transversely elongate, wider than tall (except in *A. paludosum* where they are more or less isodiametric but scarcely projecting, so that the capsule-surface is shiny); leaves variable:
Capsule with a single seed in each loculus; capsule usually about as wide as or wider than long:
Leaves mostly 2–4·5 cm. broad, ovate to elliptic-lanceolate, mostly narrowed into a false petiole above the sheath; inflorescence dense, about 3–5 cm. long and 2–4 cm. broad, its axis puberulous, rarely with some longer hairs as well
4. *dispermum*
Leaves mostly less than 2 cm. broad, lanceolate to linear-lanceolate, mostly sessile on the sheath; inflorescence about 1–2·5 cm. broad:
Capsule broader than long, usually emarginate; stems slender; plant annual with fibrous roots; seeds globose:
Bracts on main inflorescence-axis elongate, linear, 5–7 mm. long, exceeding the flowers at least in bud; inflorescence all or at least the better-developed ones densely cylindrical, 2–4 cm. long, 1–1·5 cm. broad; sepals with projecting glands at apex; capsule very sparsely and shortly pubescent     .. 10. *mortonii*
Bracts on main inflorescence-axis short, ovate, 1–3 mm. long, not exceeding the flowers or flower-buds; inflorescence capitate, rarely slightly elongate and 2–3 cm. long; sepals without projecting glands at apex; capsule more densely and longly pubescent:
Inflorescences single or 1–2 together at ends of flowering shoots
9a. *paludosum* subsp. *paludosum*
Inflorescences several together near ends of flowering shoots which are thus much branched distally, small, few-flowered 9b. *paludosum* subsp. *pauciflorum*
Capsule longer than broad, not emarginate; stems slender to stout; plant perennial with fusiform tuberous roots clustered at the rootstock; seeds longer than wide:
Stems slender, their internodes mostly exposed to near the inflorescence; no red bristly hairs; inflorescence 1·5–3 cm. long and about 1·5 cm. diam.; seeds more or less ellipsoid, about 2·5 mm. long and 1·25 mm. broad
9c. *paludosum* subsp. *pseudolanceolatum*
Stems stout, their internodes usually totally concealed in the upper part of the stem by the imbricate leaf-sheaths, occasionally partly exposed but then stems and sheaths with spreading red bristly hairs; inflorescence about 2–4 cm. long and 1·5–2·5 cm. broad; seeds ellipsoid, about 4 mm. long and 2 mm. broad:
Leaf-sheaths with dark rusty-red spreading bristly hairs outside and on margins
8a. *setiferum* var. *setiferum*
Leaf-sheaths with colourless bristly hairs usually on margins only, sometimes outside also     ..     ..     ..     .. 8b. *setiferum* var. *pallidiciliatum*
Capsule with two or more seeds in each loculus; capsule longer than wide:
Stems erect or suberect from a basal rootstock with fusiform tubers; leaves lanceolate to linear; savanna species:
Leaves increasing markedly in length towards the base of the stem where there is a

* I am greatly indebted to Prof. J. K. Morton, who kindly made available prior to publication his valuable papers on the West African species of *Commelinaceae*, which have been of the utmost use in the present account— J.P.M.B.

rosette of linear-lanceolate leaves (6–)15–25(–30) cm. long which are tapered
below into false petioles; leaf-margins ciliate with more or less spreading conical
hairs; flowers opening after mid-day    ..    ..    ..    ..    7. *pomeridianum*
Leaves on stem more or less uniform in length except that the lamina becomes reduced or absent towards base of stem; basal rosette thus absent; leaf-margins
closely and minutely cartilaginous-serrate; flowers opening in the morning:
Flowers usually produced with the leaves which are then mostly more than 5 cm.
long and 8 mm. broad; inflorescence mostly 1·5–5 cm. diam., its axis usually
not deep purple or blue    ..    ..    ..    6a. *lanceolatum* subsp. *lanceolatum*
Flowers usually produced before the leaves develop fully (at flowering time leaves
up to about 5 cm. long and 6–8 mm. broad, later elongating to about 13 cm. but
still as narrow); inflorescence often narrower than in 6a, its axis often deep
purple or blue    ..    ..    ..    ..    6b. *lanceolatum* subsp. *subnudum*
Stems straggling to creeping, or with erect branches, producing fibrous roots from the
lower nodes, but without fusiform tubers at the rootstock; leaves lanceolate to
elliptic or ovate; mostly forest species, rarely in savanna:
Inflorescence a more or less lax open panicle; leaves ovate to elliptic, usually with
distinct false petioles:
Leaf-sheaths usually with some rusty-red (sometimes colourless) spreading bristly
hairs on margins and outside surface; inflorescence-branches (4–)8–31; leaves
up to 13 cm. long and 4 cm. broad    ..    2a. *umbrosum* subsp. *umbrosum*
Leaf-sheaths without rusty-red hairs; sometimes a few colourless cilia on margins
but not on outside surface which is puberulous; inflorescence-branches 2–8;
leaves mostly up to 8 cm. long, 3 cm. broad (rarely 10 cm. long and 4 cm. broad)
2b. *umbrosum* subsp. *ovato-oblongum*
Inflorescence dense, subcapitate to ovoid:
Inflorescence small, 1–2 cm. long and wide; capsule with distinct acumen at
apex; leaves elliptic to lanceolate, up to about 7 cm. long and 2·5 cm. broad;
ciliate on margins, with a distinct false petiole    ..    ..    5. *silvaticum*
Inflorescence larger, more than 2 cm. long; capsule rounded, subtruncate or
minutely apiculate at apex; leaves not ciliate on margins, falsely petiolate or not:
Leaves linear to lanceolate, usually without a petiole; fusiform tubers always
present at rootstock, but the stems may straggle and root at lower nodes;
capsule with short spreading pubescence outside; a savanna species (see also
above)    ..    ..    ..    ..    ..    6a. *lanceolatum* subsp. *lanceolatum*
Leaves lanceolate to elliptic or ovate, more or less clearly falsely petiolate, up to
about 24 cm. long and 6 cm. broad; no fusiform tubers; roots all fibrous;
capsule glabrous; a forest species    ..    ..    ..    ..    3. *beniniense*

1. **A. aequinoctiale** (*P. Beauv.*) *Kunth* Enum. 4: 72 (1843); C. B. Cl. in DC., Monogr. Phan. 3: 221 (1881),
partly, excl. vars. *minor* and *kirkii*; F.T.A. 8: 65; Chev. Bot. 666, at least partly; Irvine, Pl. Gold
Coast 30 (1930); Nielsen, Introd. Fl. Pl. W. Afr. 183 (1965); J. K. Morton in J. Linn. Soc. 59: 443,
fig. 4 (1966). *Commelina aequinoctialis* P. Beauv., Fl. Oware 1: 65, t. 38 (1806–9). A scrambling or sometimes erect robust herb up to 6 ft. high or more, climbing with clinging hairs; flowers bright or orange-
yellow, opening about 7–10 a.m.; in forest, often secondary; said sometimes to be a weed in farmland.
**Lib.:** *Harley* 1205! Peahtah (Oct.) *Linder* 1010! Ganta (Feb.) *Harley* 905! Nimba (June) *Adam* 21580!
**Ghana:** Kumasi (Oct.) *Darko* 349! Agogo (Jan.) *Irvine* 877! Ampam, Kwahu (Dec.) *Morton* 5031!
Koforidua (Apr.) *Morton* A822! Kpeme (Nov.) *Morton* A3447! **S.Nig.:** Ughoton, Iyekuselu Dist. (Jan.)
*Daramola & Emwiogbon* FHI 32782! Aweba, Idanre (Jan.) *Brenan & Keay* 8686! Uyere to Owo (Nov.)
*Meikle* 647! Owo (Nov.) *Stanfield* 311! Calabar (Feb.) *Mann* 2338! Widespread in the forest regions of
tropical and S. Africa.
2. **A. umbrosum** (*Vahl*) *Kunth* Enum. 4: 71 (1843); Berhaut, Fl. Sén. 210; J. K. Morton in J. Linn. Soc. 59:
459 (1966), 60: 168 (1967). *Commelina umbrosa* Vahl, Enum. 2: 179 (1836).
2a. **A. umbrosum** (*Vahl*) *Kunth* subsp. **umbrosum**—J. K. Morton in J. Linn. Soc. 59: 461, fig. 10 (1966).
*A. ovato-oblongum* var. *nigritanum* C. B. Cl. in DC., Monogr. Phan. 3: 227 (1881); F.T.A. 8: 69; Chev.
Bot. 667; C. B. Cl.) Hutch., F.W.T.A., ed. 1, 2: 312 (1936). A herb variable in size and
indumentum, with stems straggling or decumbent or with erect branches, up to 3 ft. high; flowers white
to mauve or purple (particularly in Bamenda), opening from 2.30–6 p.m.; in forest or in damp shady
places (e.g. *Raphia* groves) in savanna.
**S.L.:** Limbaia, Malaba (Oct.) *Small* 424! Segbwema (July) *Deighton* 3746! **Iv.C.:** Tonkoui (Sept.)
*Aké Assi* 6265! Danané (Mar.) *Leeuwenberg* 2991! Gbénou Forest, Bouaké (Apr.) *Leeuwenberg* 3300!
**Ghana:** Atonso, Ashanti (Oct.) *Baldwin* 13510! Oti R., Kete Krache (May) *Morton* GC 7289! Komosu
to Koforidua (Jan.) *Morton* GC 8291! Ajena (fl. & fr. Nov.) *Morton* GC 8049! Bisa (Jan.) *Morton* GC
8236! **Togo Rep.:** Kpandu (June) *Andoh* FH 5288! **Dah.:** Porto Novo (Oct.) *Newton* 14! **N.Nig.:**
Jebba *Barter*! Vogel Peak, Sardauna Prov. (fl. & fr. Nov.) *Hepper* 1494! **S.Nig.:** Oyo F.R., Afo R.
(Sept.) *Latilo* FHI 23544! Ibadan (Aug.) *Gillett* 15340! Lagos (Oct.) *Baldwin* 13630! Owo (Oct.) *Stan-
field* 7! Oban *Talbot* 755! **W.Cam.:** Batibo (Apr.) *Brunt* 1059! Wum (June) *Daramola* FHI 41092!
Bali-Ngemba F.R. (May) *Ujor* FHI 30359! Bamenda, 4,800 ft. (Jan.–Feb.) *Daramola* FHI 40584!
*Migeod* 474! **F.Po:** (Jan., Feb.) *Milne*! Guinea 1811! Also in E. Cameroun, Gabon, Cabinda and the
Congo.
2b. **A. umbrosum** subsp. **ovato-oblongum** (*P. Beauv.*) *J. K. Morton* in J. Linn. Soc. 59: 461, fig. 11 (1966).
*A. ovato-oblongum* P. Beauv., Fl. Oware 2: 72, t. 104 (1818); Benth. in Fl. Nigrit. 545 (1849); C. B. Cl.
in DC., Monogr. Phan. 3: 226 (1881); F.T.A. 8: 69; Chev. Bot. 667; Irvine, Pl. Gold Coast 31 (1930).
A straggling or ascending herb, usually smaller and more slender than in subsp. *umbrosum*; flowers
white to pale lilac; habitat similar to that of subsp. *umbrosum*.
**S.L.:** Bonjema (Oct.) *Deighton* 5969! Rokupr (Sept.) *Jordan* 95! Makump (Aug.) *Deighton* 1204!
Binkolo (Sept.) *Harvey* 119! Bumbuna (Oct.) *Thomas* 3714! **Lib.:** Du River (Aug.) *Baldwin* 6109!
**Iv.C.:** Davo R., E. of Béyo (Feb.) *Leeuwenberg* 1370! **Ghana:** Pra Suhien F.R.
(July) *Morton*! Kibbi (May, Dec.) *Morton* P1259! GC 8096 (cult.)! Begoro (Jan.) *Morton* P1333!
Asiakwa to Bomo (Dec.) *Morton* GC 8170! **N.Nig.:** Patti Lokoja (Nov.) *Dalz.* 255! **S.Nig.:** Ikoyi

Plains, Lagos (May) *Dalz.* 1289! Ibadan (Apr.) *Meikle* 1472! Olokemeji F. R. (Oct.) *Hepper & Keay*
928! Orosun Mt., Idanre (Aug.) *J. B. Gillett* 15314! **W.Cam.**: Mamfe (Apr.) *Morton* GC 7135! Also in
E. Cameroun, Uganda and S. America; possibly in Kenya.
[Intermediates between subsp. *ovato-oblongum* and subsp. *umbrosum* occur.]
3. **A. beniniense** (*P. Beauv.*) *Kunth* Enum. 4: 73 (1843); Benth. in Fl. Nigrit. 546 (1849); C. B. Cl. in DC.,
Monogr. Phan. 3: 224 (1881); F.T.A. 8: 68; Chev. Bot. 666; Irvine, Pl. Gold Coast 30 (1930); Sousa
in Anais Junta Inv. Col. 5: 48 (1950); Mangenot in Ic. Pl. Afr. 4: 76 (1957); Saunders, Handb. W. Afr.
Pl. 96 (1958); Nielsen, Introd. Fl. Pl. W. Afr. 183 (1965); J. K. Morton in J. Linn. Soc. 59: 464, fig. 12,
13 (1966) incl. subsp. *sessilifolium*. A robust straggling herb up to 2–3 ft. high, producing fibrous roots
from the lower nodes, usually subglabrous, rarely rather densely pubescent; leaves mostly elliptic to
ovate, rarely all narrow and lanceolate; inflorescence dense, about 1–2½ in. across in fruit; flowers open
from 7 a.m. to noon; petals white to lilac; in rain forest.
**Sen.**: Koulikan to Palmeraie, Casamance (Feb.) *Berhaut* 6822! Boudié Forest, Sédhiou (Mar.) *Berhaut*
7159! **Port.G.**: Catio (Aug., Dec.) *Raimundo & Guerra* 674! *Esp. Santo* 2157! **Guin.**: Kala, Dalaba
(Sept.) *Schnell* 7355! N'Zérékoré (Sept.) *Baldwin* 13277! 13279! **S.L.**: Dambaea to Wania, Kuru Hills
(May) *Morton* SL 555! Morea, Kuru Hills *Morton* SL 1471! Musaia (Dec., Apr.) *Deighton* 4572! 5409!
Ronietta (Nov.) *Thomas* 5562! Baoma (Dec.) *Deighton* 3482! **Lib.**: Bushrod I. (Aug.) *Baldwin* 13055!
Grand Bassa (July) *T. Vogel* 40! Kulo, Sinoe Dist. (Mar.) *Baldwin* 11434! Zwedru, Tchien Dist. (Mar.,
Aug.) *Baldwin* 7039! *de Wilde* 3713! Genne Loffa, Kolahun Dist. (Nov.) *Baldwin* 10075! **Iv.C.**: Banco
(May, Oct.) *Bégué* 33! *Gruys* L7! Niapidou (Apr.) *Leeuwenberg* 3205! Angedédou Forest, Abidjan
(Nov.) *Leeuwenberg* 1871! Lamé (Oct.) *Leeuwenberg* 1733! **Ghana**: Detana Wiairso (Dec.) *Morton* A3606!
Asiakwa to Pusupusu (Dec.) *Morton* GC 8208! Kade (Jan.) *Morton* GC 8331! Kibi (Dec.) *Morton* GC
8094! Apedwa to Kukurantumi (Aug.) *Morton* A931! Volta River F.R. (Nov.) *Morton* GC 6056! Ajena
(Nov.) *Morton* P1238! Shiare (Apr.) *Hall* 1376! **S.Nig.**: Lagos (Mar.) *Millen* 22! Ibadan (Nov.–Jan.)
*Meikle* 986! *Hambler* 362! Gambari F.R. (Mar.) *Hepper* 2269! Idanre (Jan.) *Brenan & E. W. Jones*
8734! Usonigbe F.R. (Nov.) *Meikle* 570! Owo (Oct.) *Stanfield* 9! **W.Cam.**: Victoria (Dec.) *Morton*
K902! *Maitland* 187! S. Bakundu F.R. (fl. & fr. Apr.) *Daramola* FHI 41005! Barombi *Preuss* 1195!
**F.Po**: (Nov., Dec., June) *T. Vogel*! *Barter*! *Mann* 92! Widespread in the rain-forest regions of tropical
Africa. (See Appendix, p. 464.)
[Morton in J. Linn. Soc. 59: 466 (1966) distinguishes 1: subsp. *beniniense*, with more or less pubescent
or ciliate leaf-sheaths, rather small inflorescences of pinkish-mauve flowers, and truncate to slightly
emarginate capsules with pale brown seeds; and 2: subsp. *sessilifolium* (Benth.) J. K. Morton, with
puberulous rarely slightly ciliate leaf-sheaths, broader leaves, larger inflorescences of white flowers, and
apiculate capsules with whitish-scaly seeds. The latter subspecies is said to be common from Ghana to
Cameroun. I have not found it possible to distinguish clearly between these two taxa in the herbarium.
The more recently described subsp. *leonense* J. K. Morton in J. Linn. Soc. 60: 169, t. 1 (1967) from
Sierra Leone, with rather large leaves inflorescences and capsules, and an indumentum of short hooked
hairs, seems to need further consideration before it is accepted—J.P.M.B.]
4. **A dispermum** *Brenan* in Kew Bull. 7: 198 (1952); J. K. Morton in J. Linn. Soc. 59: 445 (1966). A straggling herb up to about 2½ ft. high with dense ovoid-oblong inflorescences; flowers white in W. Africa but
said to be bluish or pinkish elsewhere; in montane forest.
**W.Cam.**: Cam. Mt., 4,500–6,500 ft. (Nov.–Mar.) *Morton* K694! GC 6726! *Migeod* 31! *Breteler et al.*
MC 286! **F.Po**: *Monod* 10422! Also in Tanzania and Malawi.
5. **A. silvaticum** *Brenan* in Kew Bull. 7: 203 (1952) (var. *silvaticum*); J. K. Morton l.c. 459 (1966). *Aneilema
affine* De Wild. in Bull. Jard. Bot. Brux. 5: 83 (1915), not of R. Br. (1810). A slender, straggling herb
with fibrous roots at the lower nodes and ascending flowering stems up to about 1 ft. high; flowers white
or very pale blue.
**S.Nig.**: *Kennedy* 2674! Sapoba *Kennedy* 1758a! Sapoba to Agbadi (Nov.) *Meikle* 637! Also in E.
Cameroun and the Congo.
[The variety in our area is typical var. *silvaticum* resembling a miniature *A. beniniense* but with
ciliate-margined leaves and acuminate capsules.]
6. **A. lanceolatum** *Benth.* in Fl. Nigrit. 546 (1849); C. B. Cl. in DC., Monogr. Phan. 3: 227 (1881); F.T.A. 8:
72, partly; Irvine, Pl. Gold Coast 30 (1930); F.W.T.A., ed. 1, 2: 314; Nielsen, Introd. Fl. Pl. W. Afr.
183 (1965); J. K. Morton l.c. 453 (1966). (See Appendix, p. 464.)
6a. **A. lanceolatum** *Benth.* subsp. **lanceolatum**—J. K. Morton in J. Linn. Soc. 59: 453, fig. 8 (1966). *A.
lanceolatum* var. *evolutius* (as "evolutior") C. B. Cl. l.c. (1881), partly, excl. lectotype. *A. schweinfurthii*
C. B. Cl. l.c. (1881); F.T.A. 8: 71; Chev. Bot. 667. *A. soudanicum* C. B. Cl. in F.T.A. 8: 72 (1901).
*A. lanceolatum* Benth. var. *lanceolatum*—Brenan in Kew Bull. 7: 202 (1952), partly, excl. syn. *Lam-
prodithyros gracilis, Aneilema gracile, A. subnudum. A. lanceolatum* var. *pilosum* Brenan in Kew Bull.
15: 219 (1961). A variable perennial herb with fusiform tuberous roots at the rootstock and erect
ascending or prostrate stems up to about 2 ft. long; flowers mauve, opening in the morning, from 7.30
a.m. to noon; in savanna-woodland and grassland.
**Mali**: Gourma (July) *Chev.* 24469! Bamako (Sept.) *Adam* 15012! **U.Volta**: Sindou to Banfora (June)
*Leeuwenberg* 4312! Po (Nov.) *Darko* 459! **Ghana**: Gambaga (May) *Morton* GC 7320! Manga, Bawku
(Aug.) *Irvine* 4651! Bawku (Aug.) *Irvine* 4607! Nakpanduri (Aug.) *Hall* CC 490! **Togo Rep.**: Defalé
(Sept.) *Morton* A4387! **N.Nig.**: Kufena Rock, Zaria Dist. (July) *Keay* FHI 25919! Liruwen-Kano Hills
*Carpenter*! Adari, Sara Hills, Jos Plateau (Apr., Oct.) *Lely* P203! *Hepper* 1123! Biu, Bornu (Aug.)
*Noble* 11! Gurum, Sardauna Prov. (Nov.) *Hepper* 1307! Mambila Plateau (July) *Chapman* 33! **S.Nig.**:
Alagbee F.R., Ilorin Dist. (Feb.) *Ejiofor* FHI 19808! Idogun, Owo Dist. (Mar.–May) *Stanfield* 64! 64a!
Extending eastwards to the Sudan, Uganda and Kenya, also in E. Cameroun and the Congo.
6b. **A. lanceolatum** subsp. **subnudum** (*A. Chev.*) *J. K. Morton* in J. Linn. Soc. 59: 455, fig. 9 (1966). *Lam-
prodithyros gracilis* Kotschy & Peyr., Pl. Tinn. 47, t. 23 fig. A (1867). *Aneilema gracile* (Kotschy & Peyr.)
C. B. Cl. in DC., Monogr. Phan. 3: 228 (1881); F.T.A. 8: 73. *A. subnudum* A. Chev. in Mém. Soc. Bot.
Fr. 2, 8: 216 (1912); Chev. Bot. 668; F.W.T.A., ed. 1, 2: 314. Similar in habit to subsp. *lanceolatum*
and linked by intermediates particularly in Nigeria but in general with reduced narrow foliage, a greater
tendency to precocious flowering, and often narrower inflorescences.
**Guin.**: Farana (Apr.) *Pitot*! **Iv.C.**: Buandougou to Marabadiassa (July) *Chev.* 22006! Bouna (Feb.)
*de Wilde* 3486! **Ghana**: Ejura Scarp (Dec.) *Morton* GC 9743! Banda (Mar.) *Morton* GC 8603! Kampaha
(June) *Adams* 719! Walemboi (Apr.) *Kitson* 820! Tamale to Yendi (Mar.) *Adams* 3867! Anyabani to
Juketi (Jan.) *Morton* GC 6464! **N.Nig.**: Ilorin to Jebba (Feb.) *Meikle* 1173! 1199! Nupe *Barter* 1474!
7. **A. pomeridianum** *Stanfield & Brenan* in Kew Bull. 15: 217 (1961); J. K. Morton l.c. 451, fig. 7 (1966).
*A. lanceolatum* var. *evolutius* (as "evolutior") C. B. Cl. in DC., Monogr. Phan. 3: 227 (1881), partly,
as to lectotype; Brenan in Kew Bull. 7: 203 (1952), partly. *A. lanceolatum* of F.W.T.A., ed. 1, 2: 314
(1936), partly; Brenan in Kew Bull. 7: 202 (1952), partly, not of Benth. A perennial herb 4 in.–3 ft.
high, similar in appearance to *A. lanceolatum* but with the lower and basal leaves elongate and falsely
petiolate; flowers mauve to bluish-purple, open from about 1 p.m. until dusk; in grassland or woodland,
sometimes in cultivated ground.
**Ghana**: Sutawa (Mar.) *A. S. Thomas* D158! Kintampo (Mar.) *Goodall* GC 16371! Tamale to Yendi
(Apr.) *Morton* GC 9076! Yendi (Apr.) *Lloyd Williams* 163! Volta R. *Anderson* 17! **N.Nig.**: Nupe *Barter*
1117! Zungeru (July) *Dalz.* 267! Sokoto (June) *Dalz.* 449! Abinsi (Mar., July) *Dalz.* 806! 807! **S.Nig.**:
Eruwa, W. Lagos *Rowland*! Igboora, Ibadan (Apr.) *Stanfield* 192! Idogun, Owo (Mar., Oct.) *Stanfield*
65! 157!

8. **A. setiferum** *A. Chev.* in Mém. Soc. Bot. Fr. 2, 8: 215 (1912); Chev. Bot. 667; F.W.T.A., ed. 1, 2: 314; Sousa in Anais Junta Inv. Col. 7: 63 (1952); J. K. Morton in J. Linn. Soc. 59: 449 (1966).

8a. **A. setiferum** *A. Chev.* var. **setiferum**—J. K. Morton l.c. 451, fig. 6 (1966). A perennial herb with fusiform tuberous roots at the rootstock and erect or ascending stems up to about 2 ft. high; leaves linear, sometimes oblong-lanceolate, up to about 6 in. long and ⅔ in. broad; petals white; in savanna and grassland. **Mali:** Sikasso (Aug.) *Chev.* 801! **Port.G.:** Pitche to Buruntuma, Gabu (Aug.) *Esp. Santo* 2738! **Guin.:** Gali to Hinde, Fouta Djalon (July) *Adames* 317! **Iv.C.:** *Chev.* 21805! Oroumba-Boka (Oct.) *de Wilde* 660! **U.Volta:** Banfora (June) *Leeuwenberg* 4361! **Ghana:** Folifoli, Afram Plains (Aug.) *Hall* CC 185! Kete Krache (Apr.) *Morton* GC 9129! N. of Banda (Dec.) *Morton* GC 25082! **N.Nig.:** Abinsi (May) *Dalz.* 810! Gawu to Abuja (Aug.) *Onochie* FHI 35450!

[*Adames* 317 is aberrant in having only the bristles round the sheath-margins slightly rusty, the rest colourless.]

8b. **A. setiferum** var. **pallidiciliatum** *J. K. Morton* in J. Linn. Soc. 59: 451 (1966) (as " pallidociliatum "). Similar to last, but stem and leaf-sheaths without red bristles; stem often more completely concealed by the leaf-sheaths and appearing stouter. **Ghana:** Damongo (Mar.) *Morton* GC 8740! Pong Tamale (Nov.) *Akpabla* 385! Tamale *Anderson* 10! Tamale to Nyankpala (May) *Darko* 5073! Ejura (Aug.) *Andoh* 5053!

9. **A. paludosum** *A. Chev.* in Mém. Soc. Bot. Fr. 2, 8: 215 (1912); Chev. Bot. 667; Berhaut, Fl. Sén. 210; J. K. Morton in J. Linn. Soc. 59: 446 (1966), partly.

9a. **A. paludosum** *A. Chev.* subsp. **paludosum**—J. K. Morton l.c. 447 (1966) partly, excl. cited Ghana specimens and t. 1. A more or less slender-stemmed straggling annual herb, often rooting at the lower nodes; leaves 3–8(–10) cm. long, 3–20(–25) mm. broad; inflorescences usually capitate, sometimes to 2 cm. long; bracts on main axis short, ovate, 1–2 mm. long; cincinni usually sessile or up to about 1 mm. pedunculate; petals white or sometimes mauve, open at 2.30 p.m.; in damp or marshy savanna and grassland, sometimes in cultivated ground. **Sen.:** Bougnadou, Casamance (Oct.) *Adam* 18488! Bignona, Casamance (Sept.) *Adam* 18132! **Port.G.:** Piche, Buruntuma (Nov.) *Pereira* 3522! Fá (Sept.) *Guerra* 3815! **S.L.:** Mange, Bure Makonte (Oct.) *Jordan* 581! Mange Ferry (Nov.) *Morton* SL 93! Musaia, Koinadugu Dist. (Aug.) *Haswell* 42! Sasa, Tonko Limba (Sept.) *Jordan* 331! **Iv.C.:** N'Douci (Aug.) *de Wit* 6045! Alangouassou to Mboyakro, Nzi valley (Aug.) *Chev.* 22246! **Ghana:** *Lloyd Williams* 392! Nakwaly, Bole (Nov.) *T. M. Harris*! Lumbaga, Tamale (Dec.) *Morton* GC 6257! Ejura (Dec.) *Morton* GC 9719! 9729! Folifoli, Afram Plains (Aug.) *Hall* CC 172! **N.Nig.:** Kumu (Oct.) *Lely* 670! **S.Nig.:** Oyo to Iseyin (July) *J. B. Gillett* 15233! *Stanfield* FHI 45646!

[The intermediates between subsp. *pauciflorum* and subsp. *pseudolanceolatum* referred to by Morton (l.c. 449) appear to me to be states or variants of subsp. *paludosum* rather than genuine intermediates. *Lely* 670 is aberrant in having the cincinni on peduncles up to 5 mm. long and the flowers on longer arcuate pedicels to 5 mm. long. More material may prove this to be a separate taxon.]

9b. **A. paludosum** subsp. **pauciflorum** *J. K. Morton* l.c. 447 t. 2 (1966). Similar to subsp. *paludosum* but with the flowering shoots each branched distally into several small inflorescences. Habitat similar to that of subsp. *paludosum* but said to be more usually in cultivated or disturbed ground. **Sen.:** Sedhiou, Casamance (Nov.) *Berhaut* 6365! **S.L.:** Mateboi, Sanda Tenraran (Nov.) *Jordan* 825! **Ghana:** Burufo (Dec.) *Adams* 4401! Bugari (Nov.) *Leather* 2536! Pong Tamale (Dec.) *Morton* GC 9867! Kete Krache to Otisu (Jan.) *Morton* A1503!

[Intermediates with subsp. *paludosum* occur, e.g. *Adam* 18379 from Senegal, Djibelor-Figuin-Chor, *T. M. Harris* from Ghana, Wa, and *Cudjoe* from Ghana, Accra—Winneba road.]

9c. **A. paludosum** subsp. **pseudolanceolatum** *J. K. Morton* l.c. 448, t. 3 (1966). A perennial herb up to about 8 in. high with tuberous roots; leaves up to 10 cm. long and 16 mm. wide, usually smaller; inflorescences ovoid to cylindrical, dense; bracts on main axis lanceolate, 1–3 mm. long; cincinni on peduncles 2–3·5 mm. long; flowers white to mauve. **Ghana:** Adamsu (Dec., May) *Adams* 3090! *Morton* A4250! Berekum to Sampa (Apr., Dec.) *Morton* A1118! A3229! **Togo Rep.:** Defalé (Sept.) *Morton* A4435!

[As stated by Morton (l.c.) this distinctive plant may prove either a hybrid of *A. paludosum* and *A. lanceolatum*, or even a separate species.]

10. **A. mortonii** *Brenan* in Kew Bull. 22: 387 (1968). *A. paludosum* of J. K. Morton in J. Linn. Soc. 59: 447 (1966) partly (as to cited Ghana specimens and t. 1). An apparently annual herb with fibrous roots similar in habit to *A. paludosum* subsp. *paludosum*; leaves up to about 12 cm. long and 12 mm. broad; flowers white; in marshy savanna. **Ghana:** Walewale (Dec.) *Adams & Akpabla* 4206! Lumbaga, Tamale (Dec.) *Morton* GC 6229! Pong Tamale (Dec.) *Morton* GC 9836! Yendi to Zabsugu (Dec.) *Morton* A1455! Attebubu (Nov.) *Morton* A2737!

## 6. POLLIA Thunb., Nov. Gen. Pl. 1: 11 (1781).

Inflorescence lax, with well-spaced, distinct lateral branches (0·5–)1–2·5(–3) cm. long; leaves with lamina mostly 7–14 cm. long and 2–3(–3·5) cm. broad, with about 5 parallel nerves on either side of midrib, rather quickly tapering at base into a distinct slender false petiole about 1–2 cm. long and 1–2 mm. broad; fertile stamens 6, equal; fruits ellipsoid, shining (though less so than in the next species), pinkish to purplish, 5–6 mm. long  ..    ..    ..    ..    ..    ..    ..    ..    ..    .. 1. *mannii*

Inflorescence very dense, with close, very short, indistinct lateral branches to about 5 mm. long; leaves with lamina mostly 15–30 cm. long and 5–9 cm. broad, with about 8–9 parallel nerves on either side of midrib, gradually tapering at base into an indistinct rather stout false petiole 3–6 mm. broad; fertile stamens 3, with 3 very small staminodes; fruits globose or subglobose, with a strong metallic lustre, bluish-purple to blue or blue-black  ..    ..    ..    ..    ..    .. 2. *condensata*

1. **P. mannii** *C. B. Cl.* in DC., Monogr. Phan. 3: 124 (1881); F.T.A. 8: 26; J. K. Morton in J. Linn. Soc. 60: 206 (1967). A straggling herb about 1–2 ft. high, sometimes to 5 ft., with stems decumbent below or stolons rooting freely at the lower nodes, oblong-elliptic or oblong-lanceolate leaves, and lax terminal panicles of white flowers, open about 10 a.m.; in the herb-layer of primary and secondary rain-forest, sometimes in extended colonies over a wide area. **Iv.C.:** Yabouakrou to Tingouéla (fr. Dec.) *Chev.* 22750! Divo Forest (fr. Sept.) *Aké Assi* 5699! **Ghana:** Akokoasa (fr. Nov.) *Fishlock* 71! Apedwa (fl. & fr. Oct.) *Deighton* 3424! Bosumkese F.R. (Dec.) *Adams* 2903! Mampong (Dec.) *Morton* GC 9650! Bobini F.R. (Jan.) *Foggie* 515! **S.Nig.:** Aponmu (fl. & fr. Sept.) *Stanfield* 124! Aponmu F.R. (Aug.) *J. B. Gillett* 15321! Akure F.R. (July–Aug.) *Jones* FHI 20205! *Onochie* FHI 33372! Shasha F.R. (fl. & fr. Nov.) *Emwiogbon* FHI 43529! **W.Cam.:** Nkom-Wum (July) *Ujor* FHI 30458! Bova to Kuke Bova (Sept.) *Olorunfemi* FHI 30779! Also in E. Cameroun, S. Tomé, Uganda and N.W. Tanzania.

2. **P. condensata** *C. B. Cl.* in DC., Monogr. Phan. 3: 125 (1881); F.T.A. 8: 27; J. K. Morton l.c. (1967).
A stout herb 9 in. to 4 ft. high, with stems often decumbent at base or stolons and these rooting freely from the nodes; leaves rather large, with a tendency to terminal rosetting, so that the plant rather resembles *Palisota*, but the margins of leaves and leaf-sheaths lack long hairs; flowers white, in dense clusters, open from 10.30 a.m. till 12.30 p.m.; in primary and secondary rain-forest, flowering mainly in gaps.
**S.L.:** Duunia (fr. Feb.) *Sc. Elliot* 4863! Talla (Feb.) *Sc. Elliot* 4851! Bumbuna (fr. Oct.) *Thomas* 3757! 3865! Kougohun (Jan.) *Deighton* 3865! Tingi Hills above Konelo (Dec.) *Morton & Gledhill* SL 3207! **Lib.:** Gbanga (fr. Sept.) *Linder* 673! Bilimu Mt. (May) *Harley* 1356! Monrovia town (Aug.) *Baldwin* 8013! Zwedru (Aug.) *Baldwin* 7034! Webo (July) *Baldwin* 6253! **Iv.C.:** 61Km. N. of Sassandra, W. of Niapidou (fr. Jan.) *Leeuwenberg* 2494! Béyo (fr. Jan.) *Leeuwenberg* 2454! **Ghana:** Kumasi (fr. Feb.) *Irvine* 90! Assin-Yan-Kumasi *Cummins* 82! S. Fomang Su F.R. (fr. Oct.) *Brown* FH 2069! Amentia (fr. Mar.) *Vigne* FH 1887! Accra to Kibi (fr. Jan.) *Morton* GC 25342! **N.Nig.:** Mayo Ndaga, Sardauna Prov. (fr. Jan.) *Latilo & Daramola* FHI 28987! **S.Nig.:** Gambari F.R. (Aug.) *Onochie* FHI 19141! Apomu F.R. (Aug.) *J. B. Gillett* 15320! Owo (fl. & fr. Oct.) *Stanfield* 22! Okomu F.R. (fr. Dec.) *E. W. Jones* in *Hb. Brenan* 8607! Oban *Talbot* 44! **W.Cam.:** Likomba, Victoria (fr. Oct.) *Mildbraed* 10528! **F.Po:** (Dec.) *Mann* s.n. ! 93! (fr. June) *Barter* 1518! Also in E. Cameroun, S. Tomé, Principe, Congo, Cabinda, Sudan, Ethiopia, Uganda, Tanzania and Angola. (See Appendix, p. 467.)

## 7. PALISOTA Reichb. ex Endl., Gen. Pl. 125 (1836); F.T.A. 8: 27. *Nom. cons.*

Plants with a distinct aerial main stem, sometimes short and 15–100 cm. high, sometimes branched and up to 5(–6) m. high:
Inflorescence lax, elongate, 10–30 cm. long, with rather slender lateral peduncles usually about 1–2 cm. long; stems up to about 6 m. high, somewhat branched, with leaves mostly in terminal rosettes; leaves oblanceolate to obovate, 15–30(–40) cm. long, 4–11·5 cm. wide, more or less long-hairy near base, particularly on margins; inflorescences often several per rosette .. .. .. .. .. 1. *hirsuta*
Inflorescence either dense, up to 15(–20) cm. long, or lax, but then only 4–10 cm. long and normally 1 per rosette:
Inflorescence rather lax and narrow, 3–10 cm. long, with lateral peduncles very short, 0–3 mm. long; flower-bearing part of each partial inflorescence at first very short, becoming elongate in fruit, 5–30 mm. long, and thickened; inflorescence puberulous; stems up to about 2 m. high, often branched, with leaves mostly in rosettes; leaves obovate to oblanceolate or elliptic, 12–23 cm. long, 4–10 cm. broad; petiole short, up to 2·5 cm.; inflorescence normally 1 per rosette
2. *ambigua*
Inflorescence dense:
Leaves scattered along the upper part of the stem, not whorled; petiole 1–2·5 cm. long; lamina elliptic to oblong-elliptic, 13–27 cm. long, 4·5–7 cm. broad; stems slender, about 3–5 mm. diam.; inflorescence solitary, terminal, about 5–9 cm. long and 1·5–2 cm. broad .. .. .. .. .. .. 3. *preussiana*
Leaves mostly whorled in rosettes; petiole 2–30 cm. long; stems stouter, mostly 5 mm. or more in diameter; inflorescences solitary to several per rosette:
Inflorescence short, about 2–2½ times as long as broad; stem about 15–100 cm. tall; leaf-lamina elliptic to somewhat ovate or obovate, 12–50 cm. long, 5–18 cm. broad, narrowed at base to a distinct petiole about 2·5–30 cm. long　8. *barteri*
Inflorescence elongate, about 4–7 times as long as broad:
Petioles short, 1–8 cm. long; peduncles 2–6 cm. long; leaf-lamina usually elliptic to oblong-elliptic, 15–40 cm. long, 4–14·5 cm. broad; inflorescence about 6–15 cm. long and 1·5–3·5 cm. broad .. .. .. 4. *schweinfurthii*
Petioles long, 10–33 cm. long; peduncles about 8–15 cm. long; leaf-lamina mostly about 40–100 cm. long, 13–21 cm. broad; inflorescence about 7–19 cm. long and 1·5–4 cm. broad .. .. .. .. .. .. 5. *mannii*
Plants without a distinct aerial main stem; rosettes of leaves at ground-level:
Peduncle of inflorescence and inflorescence itself (except for flowers) completely covered with very long dense silky-woolly hairs; lamina 15–36 cm. long, 8–12 cm. broad, obovate-oblanceolate, tomentose beneath; petiole about 5–7 cm. long; inflorescence about 2·5–5 cm. long and 2–2·5 cm. broad .. .. 6. *lagopus*
Peduncle of inflorescence not or scarcely tomentose; inflorescence (except sometimes for bract-margins) not tomentose; lamina not or scarcely tomentose beneath:
Inflorescence with conspicuous ovate bracts about 5–10 mm. broad subtending each lateral branch and projecting among the flowers; bracts long-ciliate; leaf-lamina elliptic to oblanceolate or narrowly obovate, about 25–40 cm. long and 5·5–14 cm. broad, often with a pale stripe along the midrib, gradually cuneate at base, almost sessile or with a petiole up to 25 cm. long; inflorescence about 4–8(–18) cm. long and 1·5–4 cm. broad; fruits pilose, beaked　7. *bracteosa*
Inflorescence with inconspicuous lanceolate to linear bracts mostly 2–3 mm. broad or less:
Inflorescence about 4–8 times as long as broad, about 10–23 cm. long and 2–4 cm. broad, in fruit up to 50 cm. long and 10 cm. broad; leaf-lamina about 60–110 cm. long and 11–34 cm. broad, tapering at base (sometimes rather abruptly) into a petiole 20–30 cm. long; fruits glabrous or nearly so (in our area) .. 5. *mannii*
Inflorescence about 1½–2 times as long as broad, about 2–9 cm. long and 1·5–4 cm. broad; leaf-lamina about 20–60 cm. long and 6–12 cm. broad, gradually narrow-

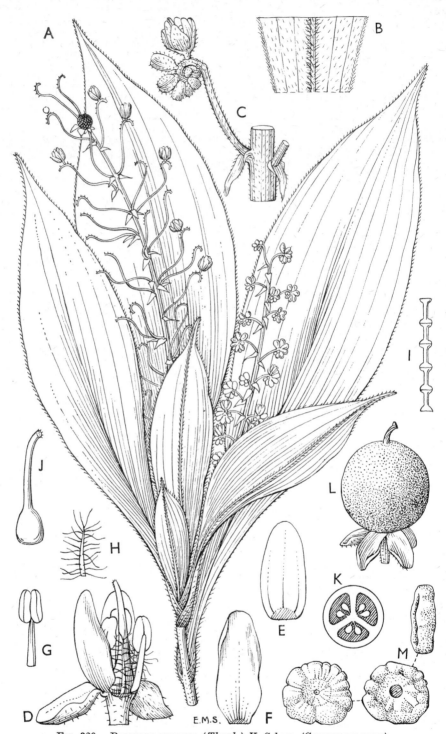

FIG. 329.—PALISOTA HIRSUTA (*Thunb.*) *K. Schum.* (COMMELINACEAE).

A, habit, × ¼.  B, undersurface of leaf, near base, × 2.  C, section of inflorescence, × 2.  D, flower with sepal and one petal removed, × 6.  E, sepal, × 6.  F, petal, × 6.  G, stamen, × 6. H, staminode, × 6.  I, section of staminode filament, × 60.  J, gynoecium, × 6.  K, T.S. ovary, diagram.  L, fruit, × 4.  M, seed, 3 different aspects, × 6.  A, B, from *Dalziel* 922. C–K from *Brenan* 8505.  L, M from *Dalziel* 922.

ing to the petiole which is sometimes short and indistinct sometimes up to 25 cm.
long; fruits hairy, beaked        ..        ..        ..        ..        ..        .. 8. *barteri*

1. **P. hirsuta** (*Thunb.*) *K. Schum.* in Engl., Bot. Jahrb. 24: 347 (1897); Mangenot in Ic. Pl. Afr. 4: t. 91 (1957);
J. K. Morton in J. Linn. Soc. 60: 204 (1967). *Dracaena hirsuta* Thunb., Diss. Dracaena 7 (1808). *Dianella
triandra* Afzel., Stirp. Guin. Med. Sp. Nov. 6 (1818). *Palisota thyrsiflora* Benth. in Fl. Nigrit.: 544 (1849),
illegit. name, excl. syn. *Commelina ambigua*; F.T.A. 8: 31. *P. prionostachys* C. B. Cl. in DC., Monogr.
Phan. 3: 134 (1881). *P. maclaudii* Cornu in Bull. Soc. Bot. Fr. 43: 30 (1896). *P. cf. preussiana* of Sousa
in Anais Junta Inv. Col. 6: 46 (1957). Robust herb 4–9(–12) ft. high, with lax inflorescences with whitish
to pinkish lateral branches, white to purplish flowers open from 4 p.m. until dusk, and blackish glossy
fruits; in lowland rain-forest.
**Sen.:** *Heudelot* 845! Boudié forest, Sédhiou (fl. Mar.) *Berhaut* 7158! **Port.G.:** Junta to Buba (fr. Feb.)
*Juan d'Orey* 301! Bissalanca, Bissau (Jan.) *Esp. Santo* 1666! Cumura to Bór, Bissau (fl. & fr. Mar.)
*Esp. Santo* 1877! Xitole (Mar.) *Pereira* 1663! **S.L.:** Newton, Peninsula (fl. & fr. Apr.) *Deighton* 5922!
Gbinti, Dibia (Jan.) *Deighton* 5899! Bo (Jan.) *Thomas* 7441! Bumban (Dec.) *Sc. Elliot* 4234! Gola Forest
*Small* 542! **Lib.:** *Linder* 60! Monrovia *Whyte*! Kakatown *Whyte*! Dukwia R. (fr. June) *Cooper* 359!
Sinoe Basin *Whyte*! Grand Bassa (July) *T. Vogel* 64! Nimba Reserve *Adames* 782! **Iv.C.:** Adiopodoumé
(fl. & fr. Oct.) *Leeuwenberg* 1709! *Gruys* 6! Banco (July) *de Wit* 78721 Audouin Forest (Aug.) *de Wit*
6017! **Ghana:** Axim (fl. & fr. Feb.) *Irvine* 2367! Swedru (Feb.) *Dalz.* 8293! Aburi Scarp (Jan.) *de Wit
& Morton* A2864! New Tafo *Lovi* WACRI 3949! Kibbi (fr. Jan.) *Morton* GC 6395! **Dah.:** Pobé
*Adjanohoun* 122! **N.Nig.:** Patti Lokoja (Feb.) *Dalziel* 922! Wamba Div. (fr. Mar.) *Thornewill* 40! Lafia,
Benue Prov. (Feb.) *Jones* FHI 462! Jemaa (Feb.) *McClintock* 193! **S.Nig.:** Epe (fr. Aug.) *J. B. Gillett*
15261! Sapoba F.R. (Jan.) *Keay* FHI 37738! Okomu F.R. (Dec.) *Brenan* 8505! *Brenan, Jones &
Richards* 8617! Calabar (Feb.) *Mann* 2339! **W.Cam.:** *Preuss* 1133! Su, Bamenda, 3,500 ft. (Apr.)
*Maitland* 1735! Kumba (Apr.) *Daramola* FHI 29836! Banga, Kumba Dist. (Mar.) *Binuyo & Daramola*
FHI 35628! Buea Rd., S. Bakundu (June) *Dundas* FHI 8351! **F.Po:** (fl. & fr. June, Nov., Dec.)
*T. Vogel* 77! *Barter*! *Mann* 95! Balacha to Ureka (Mar.) *Guinea* 2287! Also in E. Cameroun, Rio Muni,
Cabinda and the Congo. (See Appendix, p. 466.)
[Leaves variable in hairiness: upper surface subglabrous to rather densely hairy; hairiness of leaf-base
also variable.]

2. **P. ambigua** (*P. Beauv.*) *C. B. Cl.* in DC., Monogr. Phan. 3: 131 (1881); F.T.A. 8: 31; J. K. Morton l.c.
(1967). *Commelina ambigua* P. Beauv., Fl. Oware 1: 26, t. 15 (1805). Robust herb 1–7½ ft. high, with
inflorescences at first lax and narrow, becoming broader as the lateral branches elongate; sepals and petals
violet; flowers open from 4 p.m. until dusk; fruits green then indigo or purple; in lowland rain-forest.
**S.Nig.:** Usonigbe F.R., Benin (Nov.) *Keay* FHI 25622! Okomu F.R., Benin (Dec.) *Brenan* 8402! Benin
(Aug.) *Stanfield* 109! Port Harcourt (Jan.) *Jones* FHI 616! Oban *Talbot*! **W.Cam.:** Barombi F.R.,
Kumba (fl. & fr. Apr.) *Daramola* FHI 41013! Bopo to Banga, S. Bakundu (fr. Mar.) *Binuyo & Daramola*
FHI 35636! Bimbia road, Victoria (fr. Feb.) *Maitland* 420! Victoria (Nov.) *Mildbraed* 10730! Also in E.
Cameroun, Rio Muni, Gabon and the Congo.
[Lower surface of leaves varying from almost glabrous to rather densely appressed-grey-silky.]

3. **P. preussiana** *K. Schum. ex C. B. Cl.* in F.T.A. 8: 30 (1901); J. K. Morton l.c. 205 (1967). Perennial herb
with a rather slender erect stem with leaves scattered in the upper part and a single terminal inflorescence
of mauve flowers; in upland forest.
**W.Cam.:** Buea (Dec.) *Migeod* 87! *Preuss* 996! **F.Po:** Moka (Dec.) *Boughey* 51!

4. **P. schweinfurthii** *C. B. Cl.* in DC., Monogr. Phan. 3: 132 (1881) partly, as to *Schweinfurth* 3054 (lectotype),
3647; F.T.A. 8: 29 partly; J. K. Morton l.c. (1967). *P. mannii* C. B. Cl. in DC., Monogr. Phan. 3:
132 (1881) partly, as to *Mann* 2139, *Kalbreyer* 163, but excl. lectotype; F.T.A. 8: 29, partly, as to β;
F.W.T.A., ed. 1, 2: 315, partly. Robust herb with stems up to 6 ft. high, leaves mostly in terminal rosettes
and dense spikes of dirty white to pink flowers; in rain-forest.
**W.Cam.:** Cam. Mt., 2,000–3,000 ft. (Apr.) *Maitland* 1071! *Kalbreyer* 163! (Dec.) *Mann* 2139! Buea
(Feb.) *Breteler et al.* MC 297! Ndop Plain (Apr.) *Brunt* 369! Also in E. Cameroun, Rio Muni, the Congo,
Sudan, Uganda, Tanzania, Zambia and Angola.

5. **P. mannii** *C. B. Cl.* in DC., Monogr. Phan. 3: 132 (1881) partly, as to *Mann* 2340 (lectotype); F.T.A. 8:
29 partly, as to α; J. K. Morton l.c. 205 (1967). *P. megalophylla* Mildbr. in Notizbl. Bot. Gart. Berl. 9:
246 (1925), incl. var. *robusta*. Herb usually stemless, sometimes with stems 1–6 ft. high; leaves up to
4½ ft. long; peduncles 8–24 in. long; flowers white, going pale pink or lilac, open from 1–2 p.m.; fruits
(with us) white turning orange (elsewhere said to be green when young, turning yellow, orange, red, or
deep purple-brown when mature); in rain-forest.
**S.Nig.:** Orosun Peak, Idanre (fl. Mar., Aug., Oct., fl. & fr. Dec.) *Stanfield* 63! 149! *Richards* 5115! *Keay*
FHI 25518! *J. B. Gillett* 15301! Afi River F.R., Ogoja (May) *Jones & Onochie* FHI 18613! **W.Cam.:**
Bafut-Ngemba F.R., 6,500 ft. (July) *Brunt* 839! Buea, 4,000–5,000 ft. (Feb., May) *Dalz.* 8349! *Preuss*
894! Mimbia (Mar.) *Brenan & Jones* 9501! Bamenda (Jan.) *Daramola* FHI 40634! Likomba Plantation
(Nov.) *Mildbraed* 10618! Ada to Oshie (fl. & fr. Aug.) *Brunt* 1236! **F.Po:** 3,000 ft. (Mar.) *Mann* 2340!
Also in E. Cameroun, Cabinda, Congo, Sudan, and Uganda.

6. **P. lagopus** *Mildbr.* in Notizbl. Bot. Gart. Berl. 9: 248 (1925); J. K. Morton l.c. (1967). Herb with leaves
arising at or near ground-level, with remarkably woolly peduncles and inflorescences; flowers white to
pink or violet; in rain-forest.
**W.Cam.:** Mamfe (Mar.) *Richards* 5218! Kumba to Victoria road, 15 miles from Kumba *Morton* K922!
Also in E. Cameroun.

7. **P. bracteosa** *C. B. Cl.* in DC., Monogr. Phan. 3: 133 (1881); F.T.A. 8: 28; J. K. Morton l.c. (1967). Stemless herb (or merely rhizomatous) with a rosette of leaves arising at about ground-level; lamina sometimes
with a greenish-white stripe along midrib above; flowers pinkish-white or white; fruits red; in rain-
forest.
**Guin.:** Kinidougou (fr. Feb.) *Martine* 257! **S.L.:** Battiema, Kowa (fr. Dec.) *Pyne* 77! Makump (Oct.)
*Thomas* 3940! Yonibana (Nov.) *Thomas* 4992! Sirabu, near Blama (fr. Dec.) *Deighton* 3452! Gegbwema
(fr. Mar.) *Deighton* 4122! **Lib.:** *Straub* 143! *Harley* 496! Dinamu to St. John River (May) *Harley* 496a!
Toroke, Tchien District (July) *Baldwin* 6731! Diala (Oct.) *Voorhoeve* 521! Nimba (Nov.) *Adames* 679!
**Iv.C.:** 43 km. E. of Soubré (Dec.) *Leeuwenberg* 2190! **Ghana:** Sampa, W. Prov. (Feb.) *Vigne* 2917!
Banka, Ashanti *Irvine* 496! Chairaiso, Ashanti (Nov.) *Morton* A2761! **S.Nig.:** Akarara, Calabar (May)
*Ujor* FHI 30844! Oban *Talbot* 741! **W.Cam.:** Banga, S. Bakundu F.R. (Mar., Aug.) *Richards* in *Hb.
Brenan* 7945! *Dundas* FHI 8387! Also in E. Cameroun and S. Tomé. (See Appendix, p. 466.)

8. **P. barteri** *Hook.* in Bot. Mag. t. 5318 (1862); C. B. Cl. in DC., Monogr. Phan. 3: 132 (1881); Troll in Beitr.
Biol. Pfl. 36: 359, figs. 20 I, 21, 22 (1961); J. K. Morton l.c. 204 (1967). Usually stemless herb; flowers
white, open from 4–6 p.m.; fruits green to purplish or red; in rain-forest.
**S.L.:** Tingi Hills (Dec.) *Morton & Gledhill* SL 3020! **Lib.:** Gbanga (Aug.–Sept.) *Linder* 638! *Baldwin*
13236! Sanokwele (Sept.) *Baldwin* 9543! Kulo (fr. Mar.) *Baldwin* 11400! S. Nimba (fr. June) *Adam*
21523! **Iv.C.:** Banco Forest (Jan.) *Leeuwenberg* 2328! **Ghana:** Ejura Scarp (fr. Dec.) *Morton* GC 9756!
Geaso (fr. Dec.) *Andoh* FHI 5110! **S.Nig.:** Okomu F.R., Benin (fl. & fr. Dec.–Feb.) *Lyaninko* in *Hb.
Brenan* 8884! Ifon (fr. Dec.) *Stanfield* 52! (Aug.) *Stanfield* 118! Omo F.R. (fl. & fr. Feb.) *Jones & Onochie*
FHI 17515! Stubbs Creek F.R., Eket Dist. (May) *Onochie* FHI 32919! **W.Cam.:** Buea (Apr.) *Hambler*
182! Victoria (Dec.) *Ndi* FHI 50345! S. Bakundu F.R., Kumba (Jan.) *Binuyo & Daramola* FHI 35054!

Above Buea (Mar.) *Morton* GC 6714! **F.Po:** *Mann* 94! (June) *Barter*! (fr. Sept.) *Melville* 454! Balea Mt. (Dec.) *Guinea* 384! Also in E. Cameroun and Congo. (See Appendix, p. 466.)
　　The above-cited specimens are all typical stemless *P. barteri*. Occasionally stems up to c. 120 cm. long are developed. The taxonomic status of such plants is doubtful. Examples are: **S.L.:** Tingi Hills (Dec.) *Morton & Gledhill* SL 3020! **S.Nig.:** Usonigbe F.R., Benin (Nov.) *Keay* FHI 125580! 25625! Okomu F.R., Benin (Dec., Feb.) *E. W. Jones* in *Hb. Brenan* 8606! *Brenan* 9019!
　　[The Liberia and Ghana specimens, with larger leaves more rounded at base than usual, may prove to be taxonomically separable.
　　*P. staudtii* K. Schum. in Engl., Bot. Jahrb. 24: 346 (1897). **W.Cam.:** Victoria to Bimbia (fl. Apr., May), *Preuss* 1157; 1229 may well be a synonym of *P. barteri* as suggested in Ed. 1. The very slender pedicels up to 1 cm. long seem unusual, however—J.P.M.B.]

*Uncertain species.*

**P.** bicolor *Mast.* in Gard. Chron., n.s. 9: 527 (1878); C. B. Cl. in DC., Monogr. Phan. 3: 134 (1881); F.T.A. 8: 32. Stem abbreviated; petiole 7·5–10 cm. long; lamina 30–35 cm. long; midrib brown-woolly-pubescent beneath; flowers and fruits not described.
　　Described from a plant cultivated by Messrs. Veitch, originally from Fernando Po.

*Imperfectly known species.*

**P. sp.** Stems 6 ft. long, straggling; infructescence lax, about 14 cm. long; in dense forest. **Ghana:** Pusopuso ravine, Asiakwa (fr. Feb.) *Morton* A3864!
　　[Resembling *P. hirsuta* (Thunb.) K. Schum., but fruiting pedicels 5–6 mm. long (1–3 mm. in *P. hirsuta*). Near to and possibly conspecific with *P. pedicellata* K. Schum. from S. Tomé. More material needed.]

## 8. COLEOTRYPE C. B. Cl. in DC., Monogr. Phan. 3: 238 (1881).

A rather robust straggling herb, often producing aerial roots from the lower nodes, with vegetative parts almost glabrous except for short hairs on margins of leaves and sheaths; leaves elliptic, 9–20 cm. long, 3–6·5 cm. broad, not at all rosetted, very acuminate at apex, gradually tapering to a short false petiole at base; inflorescences piercing the leaf-sheaths, appearing as dense sessile clusters of conspicuous green leafy bracts mixed with white flowers; the clusters about 1·5–3 cm. long and broad, mainly at the lower nodes or even at ground-level .. 　　.. 　　.. 　　*laurentii*

**C.** laurentii *K. Schum.* in Engl., Bot. Jahrb. 33: 377 (1903); Brenan in Kew Bull. 7: 456 (1952); Aké Assi Contrib. 2: 222, t. 19 (1964); J. K. Morton in J. Linn. Soc. 60: 171, fig. 1 (1967). A straggling herb with stems up to about 6 ft. long and 4 ft. high, often stoloniferous, with compact clusters of white flowers and green bracts sessile on the stem; flowers open from about 11 a.m. to 3 p.m.; in rain-forest, sometimes abundant where clearing has taken place.
　　**Iv.C.:** Divo Forest (Oct.) *Aké Assi* 5410! **Ghana:** Mampong (Dec., Jan.) *Morton* GC 9676! A1543! Mpraeso Scarp (Oct.) *Hall* CC 65! **S.Nig.:** Akure F.R. (Aug.) *Tamajong* FHI 19516! Owo (fl. & fr. Nov.) *Stanfield* 46! Okomu F.R. (Feb.) *Brenan & Onochie* 9001! Usonigbe F.R. (Nov.) *Keay* FHI 25630! Oban *Talbot* 742! Also in the Congo and Uganda.

## 9. CYANOTIS D. Don, Prod. Fl. Nepal. 45 (1825); F.T.A. 8: 78. *Nom. cons.*

Plant with a bulb-like thickened basal stock, or with tuberously thickened roots, or with the stem abruptly thickened and corm-like (but then also producing fibrous roots from stem-nodes above the corm); plants always perennial, with usually more or less erect stems:
Outer bract of each cincinnus (partial inflorescence) or cluster of cincinni not or only slightly exceeding them; base of stem covered by the persistent leaf-bases which are themselves clothed with long silky hairs; tuberously thickened roots radiating from stem-base; leaves in a radical rosette, up to 21 cm. long and 3·5 cm. broad; flowering stems often produced before the radical leaves expand, the stems themselves without developed leaves but with scales or small bracts, ending in an irregular inflorescence of several small mostly pedunculate cincinni, mostly borne separately (i.e. singly) .. 　　.. 　　.. 　　.. 　　.. 　　.. 　　5. *caespitosa*
Outer bract of each cincinnus (if borne singly) or cluster of cincinni clearly exceeding them:
Plant with a conspicuously enlarged bulb-like base 1·5–3 cm. across whose scales are the persistent papery leaf-bases more or less clothed with long straight silky hairs; leaves mostly radical, linear, up to 20 cm. long, 2–7 mm. broad; stems usually with 1–2 reduced leaves and irregular inflorescence of pedunculate cincinni borne separately and singly .. 　　.. 　　.. 　　.. 　　.. 　　6. *angusta*
Plant without a bulb-like base, either with a small corm (tuberously thickened stem) below ground-level or with tuberously thickened roots only:
Plant with a pea-like corm about 1–1·5 cm. diam., below ground-level; fibrous roots produced from nodes above the corm; no radical rosette of leaves; leaves linear, up to about 8 cm. long; stems more or less spreading-pilose above; individual cincinni sessile or subsessile, normally in terminal or axillary clusters of 2 or more together, with more or less falcate outer leafy bracts .. 　　7. *barbata*
Plant without a corm but with tuberously thickened roots radiating from stem-base; a radical rosette of leaves usually present, their sheaths often with long straight silky hairs:
Stems very short, 5–10 cm. high; cincinni 1–5 per stem, borne singly and pedunculate, each cincinnus with 1 outer leafy bract; radical leaves up to 5 cm. long and 7 mm. broad; stem leaves 0 or very reduced:

Cincinni (including bract) up to about 1·1 cm. long; bracteoles and sepals
  glabrous .. .. .. .. .. .. .. .. 3. *scaberula*
Cincinni (including bract) about 1·5–2·7 cm. long; bracteoles and sepals hairy
  4. *lourensis*
Stems 15–100 cm. high; cincinni normally sessile and clustered 2 or more together
  in terminal or axillary heads; heads mostly 1·5–3 cm. diam., usually with 2 or
  more leafy bracts; leaves 5–22 cm. long, 0·3–2·5 cm. broad; cauline leaves well
  developed:
Leaves (0·1–)0·4–1·8 cm. broad, the lowest leaves mostly 10–20 cm. long, the
  upper ones decreasing in length; lamina not or scarcely amplexicaul at base;
  radical rosette usually present .. .. .. .. .. 1. *longifolia*
Leaves mostly 3–20 mm. broad; plant usually tall, 30–100 cm. high:
Plant robust; leaves mostly 8–20 mm. broad; sepals generally with a dorsal
  longitudinal line of pubescence .. .. 1a. *longifolia* var. *longifolia*
Plant more slender; leaves 3–8 mm. broad; sepals pubescent all over back
  1b. *longifolia* var. *gracilis*
Leaves 1–3 mm. broad; plant slender, 7–50 cm. high
  1c. *longifolia* var. *rupicola*
Leaves 2–2·5 cm. broad (the bracts slightly narrower), the lowest leaves 7–9 cm.
  long, the upper scarcely shorter; lamina amplexicaul at junction with sheath;
  no radical rosette.. .. .. .. .. .. .. 2. *ake-assii*
Plant producing only fibrous roots (bulbs, corms and tuberously thickened roots absent,
  but beware specimens of *C. barbata* showing only fibrous roots, whose corm has been
  left behind in the ground!; these are accounted for in the key):
Bracts on stems subtending cincinni ovate, 1–1·5 cm. broad above base, about 1·5–3 cm.
  long; plant densely spreading-hairy; cincinni all sessile, single, or clustered near
  ends; plant apparently perennial .. .. .. .. .. .. 9. *hepperi*
Bracts on stems subtending cincinni linear to lanceolate, usually less than 1 cm.
  broad (if, very rarely, as broad as above, then bracts much longer than 3 cm., plant
  without spreading hairs and annual):
Cincinni in clusters of 2 or more together, mostly sessile or nearly so; clusters usually
  with 2 or more leafy bracts; plant annual (*C. lanata*):
Heads of cincinni (other than terminal one) some at least usually pedunculate;
  bracteoles mostly subacutely pointed; occurring between 6,000 and 12,000 ft.
  altitude .. .. .. .. .. .. .. .. 7. *barbata*
Heads of cincinni (other than terminal one) almost always sessile; bracteoles acute;
  below 5,000 ft. altitude .. .. .. .. .. .. 11. *lanata*
Cincinni mostly single, sometimes paired at end of stem; plant apparently perennial
  (certainly so in *C. arachnoidea*):
Cincinni (including bract) 1·5–2·5 cm. long, on peduncles 1·5–5·5 cm. long; leaves
  linear, 2–4·5 mm. broad .. .. .. .. .. 8. *ganganensis*
Cincinni (including bract) 0·5–1·5 cm. long, those other than the falsely pedunculate
  terminal ones mostly sessile, occasionally pedunculate; leaves mostly more than
  4·5 mm. broad:
Plant perennial:
Plant more or less silky or cottony, but not spreading-hairy
  10a. *arachnoidea* var. *arachnoidea*
Plant densely spreading-hairy .. .. .. 10b. *arachnoidea* var. *pilosa*
Plant annual .. .. .. .. .. .. .. .. .. 11. *lanata*

1. **C. longifolia** *Benth.* in Fl. Nigrit. 543 (1849); F.T.A. 8: 81; Brenan in Kew Bull. 7: 205 (1952); Schnell
  in Bull. I.F.A.N. 19, Sér. A: 729 (1957); J. K. Morton in J. Linn. Soc. 60: 195 (1967).
1a. **C. longifolia** *Benth.* var. **longifolia.** *C. djurensis* C. B. Cl. in DC., Monogr. Phan. 3: 256 (1881); F.T.A. 8:
  82; F.W.T.A., ed. 1, 2: 317. *C. deightonii* Hutch., F.W.T.A., ed. 1, 2: 317 (1936); in Kew Bull. **1939**:
  243. *C. lanata* of Sousa in Anais Junta Inv. Col. 5: 48 (1950); l.c. 6: 47 (1951); not of Benth. *C.
  longifolia* var. *albo-lanescens* Schnell in Bull. I.F.A.N. 19, Sér. A: 731 (1957); J. K. Morton l.c. 197
  (1967). *C. longifolia* var. *deightonii* (Hutch.) Schnell in Bull. I.F.A.N. 19, Sér. A: 732 (1957). J. K.
  Morton l.c. (1967). *C. longifolia* subsp. *deightonii* (Hutch.) J. K. Morton l.c. 196 (1967). *C. longifolia*
  subsp. *longifolia*—J. K. Morton l.c. 197 (1967). A perennial herb up to about 3 ft. high, with thickened
  roots and dense clusters of blue (occasionally white or pink) flowers; in savanna or grassland.
  A very variable plant particularly in its indumentum: sometimes it is almost glabrous, usually more
  or less grey-cottony, sometimes with spreading hairs, or rarely densely velvety-hairy all over (*Adames*
  309, Guineé, Fouta Djalon, near Labé airport; *Morton* SL 2643, Sierra Leone, Bintimani).
  **Gam.:** Jollofin to Genieri, E. Kiang (July) *Fox* 180! 181! **Port.G.:** Farim to Begene (Aug.) *Esp. Santo*
  2401! Fulacunda (Aug.) *Esp. Santo* 2160! **Guin.:** Macenta (Oct.) *Baldwin* 9763! Foot of Fon Massif
  (Aug.) *Schnell* 6583! Koiré Mt., N'Zérékoré, 2,500 ft. (Sept.) *Baldwin* 13275! 13276! 13286! **S.L.:**
  Freetown (Oct.) *Deighton* 43! Wellington (Aug.) *Harvey* 62! Moyamba (Aug.) *Dawe* 558! Kabala to
  Kamabai (Sept.) *Deighton* 3977! Gberia Fotumbu (Sept.) *Small* 290! **Iv.C.:** Nimba, 2,000 ft. (Aug.)
  *Boughey* GC 18154! Beyla to Touba, 1,750 ft. (July) *Collenette* 65! Orodougou, Touna to Siné (June)
  *Chev.* 21806! **U.Volta:** Mossi (Aug.) *Chev.* 24773! Boromo, at edge of Black Volta (Sept.) *Aké Assi* 6459!
  **Ghana:** Banda (Aug.) *Hall* 1967! Kwahu (Nov.) *Irvine* 1714! Folifoli, Afram Plains (Aug.) *Hall* CC 206!
  **Togo Rep.:** Aledjo, Atacora Mts. (June) *Villiers*! **N.Nig.:** Mt. Elphinstone Fleming, nr. Badeggi, *Barter*
  492! Anara F.R., Zaria (July) *Keay* FHI 25972! Dogon Kurmi, Jemaa Dist. (Aug.) *Killick* 29! Abinsi
  (July) *Dalz.* 808! Mambila Plateau, 5,500 ft. (July) *Chapman* 44! **S.Nig.:** Obodu Plateau (Nov.) *Tuley*
  1065! **W.Cam.:** Kishong, 6,650 ft. (June) *Brunt* 723! Widespread in the savanna regions of tropical
  Africa.

1b. **C. longifolia** var. **gracilis** (*Schnell*) *Schnell* in Bull. I.F.A.N. 19, Sér. A: 733 (1957); J. K. Morton in J. Linn. Soc. 60: 197 (1967). *C. lanata* Benth. var. *gracilis* Schnell in Rev. Gén. Bot. 57: 287 (1950). A perennial herb similar to var. *longifolia* but more slender, up to 2½ ft. high, more or less densely grey-cottony; in savanna, particularly where damp.
**Mali:** Bougouni *Roberty* 3133 (225)! **Guin.:** *Scaëtta* 3247! Pita, Fouta Djalon (Sept.) *Adames* 366! **S.L.:** Loma Mts. (Sept., Nov.) *Morton* SL 2608! SL 2642! **Iv.C.:** Nimba, 2,000 ft. (Aug.) *Boughey* GC 18149! *Schnell* 6212! *Gruys* 35! Koualé, Dourou Mt., 3,369 ft. (May) *Chev.* 21712! Séguéla (Aug.) *Schnell* 6419! **Ghana:** Banda Hills (Dec.) *Morton* GC 25268! Shiare to Chilinga, Nkwanta Hills (May) *Morton* A4007! **Togo Rep.:** Mt. Sotto, Lama-Kara (Sept.) *Morton* A4495! Defale (Sept.) *Morton* A4388! **Dah.:** Birni to Kouandé (Sept.) *Morton* A4544!

1c. **C. longifolia** var. **rupicola** (*Schnell*) *Schnell* in Bull. I.F.A.N. 19, Sér. A: 734 (1957); J. K. Morton l.c.198 (1967). *C. rupicola* Schnell in Rev. Gén. Bot. 57: 289, fig. 6c (1950). *C. longifolia* Benth. var. *pseudo-rupicola* Schnell in Bull. I.F.A.N. 19, Sér. A: 735 (1957). *C. longifolia* Benth. var. *maliensis* Schnell in Bull. I.F.A.N. 19, Sér. A: 736 (1957); J. K. Morton l.c. (1967). *C. longifolia* Benth. var. *fonensis* Schnell in Bull. I.F.A.N. 19, Sér. A: 737 (1957). A perennial herb similar to var. *gracilis* but even more slender and narrower-leaved, and not so tall.
**Guin.:** Fouta Djalon (Sept.) *Schnell* 7069! Nimba Mts. (Aug.) *Schnell* 6213! Massif de Fon (Aug.) *Schnell* 6601! **S.L.:** Kponkporto to Waia, Koinadugn Dist. (July) *Marmo* 299! Sakasakala, Kambia Dist. (Sept.) *Deighton* 5158!

2. **C. ake-assii** *Brenan* in Kew Bull. 22: 389 (1968). A perennial herb with rather stout hairy stems about 2½ ft. long and densely clustered thickened roots; leaves rather broad, with more or less grey-cottony hair on both surfaces.
**Mali:** Bamako, Baguineda (Aug.) *Roberty* 2663! 64 km. W. of Ségou (Aug.) *Roberty* 3769! **Iv.C.:** Boundiali to Korhogo (July) *Aké Assi* 5449!

3. **C. scaberula** *Hutch.* in Kew Bull. 1939: 243; F.W.T.A., ed. 1, 2: 317 (1936), English descr. only; Schnell in Bull. I.F.A.N. 19, Sér. A: 749 (1957). A perennial herb about 4 in. high with tuberously thickened roots.
**Guin.:** Kouroussa *Pobéguin* 258! 368!

4. **C. lourensis** *Schnell* in Bull. I.F.A.N. 19, Sér. A: 744 (1957); J. K. Morton in J. Linn. Soc. 60: 194 (1967). A perennial herb 1–4 in. high, with tuberously thickened roots, rose-purple flowers, blue or violet staminal hairs and yellow anthers; similar to the last but with larger hairier cincinni and probably wider leaves; the leaves are all radical and appear to be purplish, at least when dry; in shallow poor gravelly soil, seasonally moist, near bare rocks.
**Guin.:** Mt. Loura, Fouta Djalon, 4,800 ft. (Sept.) *Schnell* 7126! Misside Banga, Pita (July) *Adames* 298!

5. **C. caespitosa** *Kotschy & Peyr.* Pl. Tinn. 48 (1867); F.T.A. 8: 82; Schnell in Bull. I.F.A.N. 19, Sér. A: 748 (1957); J. K. Morton l.c. 192, fig. 7 (1967). A perennial herb with a stout hairy base and purple flowering stems 1½–12 in. high; flowers normally blue (occasionally lilac, purple or even white) with blue staminal hairs and yellow anthers; in grassland, often flowering after fires.
**Iv.C.:** Bouna (Jan.) *de Wilde* 3493! **Ghana:** Banda Ravine (Mar.) *Morton* GC 8640! Banda Watershed F.R. (Dec.) *Morton* GC 25194! Pong Tamale to Walewale (Mar.) *Morton* GC 8826! Wapuli to Sapoba (Apr.) *Morton* GC 9088! **N.Nig.:** Sokoto Prov. (June) *Dalz.* 542! Katagum Dist. *Dalz.* 384! Kwudoga, Birnin Gwari (Feb.) *Daggash* FHI 31415! Samaru to Shika, Zaria (May) *Keay* FHI 25746! Kaduna (Dec.) *Meikle* 796! Vogel Peak, 4,900 ft. (Dec.) *Hepper* 1570! Nguroje to Kakara, Mambila Plateau, 5,400 ft. (Jan.) *Hepper* 1762! **S.Nig.:** Onitsha *Talbot* 1920! **W.Cam.:** Bafut-Ngemba F.R., 6,000 ft. (Feb., Mar.) *Richards* 5312! Daramola FHI 40512! Bum, Bamenda, 4,000 ft. (May) *Maitland* 1518! Also in E. Cameroun, Congo, Sudan, Uganda, Kenya, Tanzania, Zambia and Angola. (See Appendix, p. 466.)

6. **C. angusta** *C. B. Cl.* in DC., Monogr. Phan. 3: 260 (1881); F.T.A. 8: 79; F.W.T.A., ed. 1, 2: 317; Schnell in Bull. I.F.A.N. 19, Sér. A: 749 (1957). *C. bulbifera* Hutch. in Kew Bull. 1939: 242; F.W.T.A., ed. 1, 2: 317 (1936), English descr. only; Schnell l.c. (1957); J. K. Morton l.c. (1967). A perennial bulbous herb about 4–10 in. high, with blue or purple flowers; cincinni 1–5 per stem.
**Mali:** Sikasso (Apr.) *Chev.* 747! Kinkou (Apr.) *Pitot*! **S.L.:** Bintimani, *Morton & Gledhill* SL 1099! Loma Mts., above Yifin (Mar.) *Morton & Gledhill* SL 1042! **U.Volta:** 4 km. N. of Banfora (June) *Leeuwenberg* 4338! **Ghana:** Paga, Navrongo Dist. (June) *Andoh* FH 5894! Navrongo (June) *Vigne* FH 4532! Tumu (May) *Morton* GC 7579! Gambaga (May) *Morton* GC 7367! **N.Nig.:** Sokoto Prov. *Dalz.* 543! Nupe *Barter* 1476! Anara F.R., Zaria (May) *Keay* FHI 22979! Hoss, Jos Plateau (July, Aug.) *Keay* FHI 12724! *Lely* P377! **S.Nig.:** Ipapo, N. of Iseyin (May) *Hambler* 1091! Udi (Mar.–Apr.) *Kitson*! Onitsha to Enugu (Mar.) *Morton* K3! Enugu (Mar.) *Morton* GC 6680!

7. **C. barbata** *D. Don* Prodr. Fl. Nepal. 46 (1825); J. K. Morton in J. Linn. Soc. 60: 198 (1967). *C. vaga* of Merr. in Trans. Amer. Phil. Soc., n.s. 24, 2: 102 (1935), partly; Rao in Notes R.B.G. Edinb. 25: 186 (1964), partly, as to syn. *C. barbata* Don, prob. not of (Lour.) Schult. & Schult. f. (1830). *Cyanotis hirsuta* Fisch. & Mey., Ind. Hort. Sem. Petrop. 8: 57 (1842), nomen subnudum; F.T.A. 8: 78. *C. mannii* C. B. Cl. in DC., Monogr. Phan. 3: 258 (1881); F.T.A. 8: 83; F.W.T.A., ed. 1, 2: 317. A perennial herb 2–18 in. high, variable in appearance, with normally blue, rarely mauve or white flowers; in our area the stems and sheaths are more or less spreading-pilose. Differs from *C. angusta* in its slender stem at ground-level, only swollen below the surface of the soil. In grassland at high altitudes.
**Ghana:** Wudidi, Togo Plateau (Nov.) *Morton* A3448! **W.Cam.:** Cam. Mt., 7,000–12,000 ft. (Nov.–Jan.) *Mann* 1310! N2140! *Johnston* 34! *Deistel* 79! *Morton* K776! *Maitland* 1296! Bamenda, 6,500 ft. (Jan.) *Migeod* 417! **F.Po:** Clarence Peak, 9,000 ft. (Dec.) *Mann* 616! Also in Congo, Sudan, Ethiopia, Uganda, Kenya, Tanzania, Malawi and Rhodesia; extending also along the Himalayas from India to China.
[Merrill's adoption, followed by Rao, of *C. vaga* as the correct name of this species seems unjustified because no type-specimen appears extant, the original description does not agree very well with *C. barbata*, and it seems unlikely that Loureiro could have found true *C. barbata*, only known at high altitudes in China, near Canton, the type-locality of *C. vaga*.]

8. **C. ganganensis** *Schnell* in Bull. I.F.A.N. 19, Sér. A: 746 (1957); J. K. Morton l.c. 199 (1967). A herb with stems 8–16 in. long, with narrow leaves scattered along the stems and not in a basal rosette, and blue to pink flowers in falcate cincinni, the latter borne singly on peduncles.
**Guin.:** Gangan Mt., Kindia (Oct.) *Schnell* 7487! Pita, Fouta Djalon (Sept.) *Adames* 367! Kinkou to Pita (Oct.) *Pitot*!

9. **C. hepperi** *Brenan* in Kew Bull. 22: 390 (1968). A densely spreading-hairy, apparently perennial herb with leaves up to 9 cm. long and 1·2 cm. wide in basal clusters and erect or straggling stems up to 18 in. long; hairs on stamen-filaments white; in rock crevices in upland savanna.
**N.Nig.:** Vogel Peak Massif, above Gurum, 4,700 ft. (Nov.) *Hepper* 1360!

10. **C. arachnoidea** *C. B. Cl.* in DC., Monogr. Phan. 3: 250 (1881); J. K. Morton in J. Linn. Soc. 60: 194, t. 2, 1 (1967).

10a. **C. arachnoidea** *C. B. Cl.* var. **arachnoidea.** A perennial, succulent, much branched, prostrate herb, with stems up to about 1½ ft. long forming loose mats on rocks, more or less clothed with white indumentum like silk or cotton-wool; flowers pink to mauve, open from 7–10 a.m. Similar in appearance to *C. lanata* but perennial, with the cincinni frequently borne singly, and with the leaf-sheaths usually larger and more overlapping.
**Lib.:** Sinoe *Dinklage* 2138! **Ghana:** Aboma F.R., Ashanti, 1,000 ft. (Nov.) *Vigne* 3440! Atonso, Ashanti (Oct.) *Baldwin* 13496! **S.Nig.:** Mt. Orosun, Idanre (Oct.) *Keay* FHI 22584! Akure W.R., Ondo (Dec.) *Morton* K956! Okelife, Ondo, 1,200 ft. (Nov.) *Onochie* FHI 34217! Igarra, Owo (Oct.) *Stanfield* 30!

Fig. 330.—Cyanotis angusta *C.B. Cl.* (Commelinaceae).

A, showing habit with bulb-like base (in section). B, flower. C, stamen. D, pistil. E, open capsule. F, seeds.

Oba, Owo (Nov.) *Stanfield* 48! Owo (June) *Stanfield* 97! Also in Kenya, Tanzania, India and? Ceylon.
10b. **C. arachnoidea** var. **pilosa** *Brenan* in Kew Bull. 22: 389 (1968). Similar to var. *arachnoidea* but stems, leaves bracts and peduncles densely spreading-hairy.
**S.Nig.:** Oba, Owo (June) *Stanfield* 98! Owo F.R. (July) *Onochie* FHI 33351, partly!
11. **C. lanata** *Benth.* in Fl. Nigrit. 542 (1849); F.T.A. 8: 80; Sousa in Anais Junta Inv. Col. 7: 63 (1952); Schnell in Bull. I.F.A.N. 19, Sér. A: 739 (1957); Berhaut, Fl. Sén. 209; J. K. Morton in J. Linn. Soc. 60: 194 (1967). *C. lanata* var. *schweinfurthii* C. B. Cl. in DC., Monogr. Phan. 3: 258 (1881); Sousa l.c. 5: 48 (1950). *C. rubescens* A. Chev. in Mém. Soc. Bot. Fr. 2, 8: 216 (1912); F.W.T.A., ed. 1, 2: 317. *C. lanata* var. *sublanata* C. B. Cl. in DC., Monogr. Phan. 3: 258 (1881); Schnell, Vég. Fl. Nimba 522 (1952). *C. lanata* var. *rubescens* (A. Chev.) Schnell in Bull. I.F.A.N. 19, Sér. A: 741 (1957). *C. longifolia* of Berhaut, Fl. Sén. 209, partly. An annual herb with fibrous roots and stems about 2–12 in. high often branched from base; plant usually more or less cottony-hairy, sometimes subglabrous or glabrous; flowers with petals purple, pink, or white with pink border and hairs on stamen-filaments blue, pink or pinkish-purple; in terminal and axillary clusters, in the top 1–4 axils of each stem; flowers open from 7–10 a.m.; often in seasonally wet places on and near rocks, also sometimes in cultivated ground, usually at low elevations (4,500 ft. or less).
**Sen.:** Forêt de Vélor, Kaolak (July) *Berhaut* 1548, partly! Velingara, Gouloumbou (Oct.) *Adam* 18618! **Mali:** Koulikoro (Oct.) *Chev.* 2587! Massif de Kita (Sept.) *Jaeger*! S. Macina, Kiri to Koru (Aug.) *Chev.* 24825! **Port.G.:** Pirada, Gabú (June) *Esp. Santo* 3047! Bafatá (July, Aug.) *Esp. Santo* 2699! 3287! Bissau, Pussubé (Nov.) *Esp. Santo* 911! Piche, Buruntuma (Dec.) *Pereira* 2158! **Guin.:** Kouria to Trébéléya (Sept.) *Chev.* 18237! Mali (Sept.) *Schnell* 7050! Dalaba (Sept.) *Schnell* 6929! *Adames* 382! Macenta (Oct.) *Baldwin* 9804! **S.L.:** Rokupr (Aug.) *Jordan* 79! Musala (Sept.) *Small* 210! Kamabai, nr. Makeni (Sept.) *Harvey* 92! Baima (Sept.) *Deighton* 3053! Kulafaga to Bendugu (Sept.) *Deighton* 5160! **Lib.:** Monrovia (Aug.) *Baldwin* 13010! Sanokwele (Sept.) *Baldwin* 9555! **Iv.C.:** Nimba (Aug.) *Boughey* GC 18153! Séguéla (Aug.) *Boughey* GC 18452! Orumbo-Boka Mt. (Dec.) *Boughey* 14431! Brafouédi (Dec.) *Leeuwenberg* 2292! Near Assahra S.W. of Dimbokro (Nov.) *de Wilde* 3215! **U.Volta:** Banfora (Sept.) *Adam* 15221! Koumbia (Sept.) *Jangoux* 873! 886! 1137! Pissi (Sept.) *Irvine* 4692! Mossi to Yako (Aug.) *Chev.* 24769! **Ghana:** Burufu, Lawra (Sept.) *Irvine* 4738! Tamale (Oct.) *Baldwin* 13574! Ntonso (Oct.) *Darko* 403! Aburi *Anderson* 16! Mankrong, R. Afram (Aug.) *Hall* CC 214! **Togo Rep.:** Mt. Kakaroua, Defalé (Sept.) *Morton* A4359! **Dah.:** Tenaka F.R., N. of Kopargo (Sept.) *Morton* A4514! Kouandé (Sept.) *Morton* A4570! **N.Nig.:** Yola (Sept.) *Dalz.* 256a! Hepham to Ropp (July) *Lely* 382! Dogon Kurmi, Jemaa Div. (Aug.) *Killick* 43! Naraguta Hills, Jos Plateau (July) *Lawlor & Hall* FHI 46505! Mando F.R., Birnin Gwari (Aug.) *Keay* FHI 26000! **S.Nig.:** Ibadan (Oct.) *Oseni* FHI 5200! Lagos (Oct.) *Dalz.* 1286! *Baldwin* 13629! Owo (Oct.) *Stanfield* 19! 66! Afikpo, Ondo Prov. (Sept.) *Stone* 92! Widely distributed in tropical Africa, reaching as far S. as the Transvaal. (See Appendix, p. 466.)
[The plant is variable in its indumentum from more or less densely grey-cottony to subglabrous or glabrous. *Chev.* 22196 (Iv.C., Nzi Valley, Tiébissou to Languira) appears to be an unusual variant with spreading hairs. Certain Ghana gatherings (e.g. Agomenda to Somanya *Adams* 4386; Bole to Wa *Rose Innes* in Herb. Dept. Agric. 345; Tumu *Irvine* 4705), also Iv.C., Séguéla *Aké Assi* 6578, have few-flowered often single cincinni and may represent a distinct taxon.]

## 10. BUFORRESTIA C. B. Cl. in DC., Monogr. Phan. 3: 233 (1881); F.T.A. 8: 76 partly; Brenan in Kew Bull. 14: 280–281 (1960).

Stems up to about 1–1·2 m. high; leaf-sheaths about 2–3 cm. long, glabrous to very shortly ciliate on margins; leaf-lamina elliptic, broadest about middle, 12–30 cm. long, 5–11 cm. broad, tapering above into a conspicuous narrow acumen about 1·5–3 cm. long; inflorescence-axes glabrous, about 6–15 cm. long, bearing 2–4 clusters of flowers (more if bifurcate); sepals in fruit about 1·5 cm. long; petals small; seeds rather closely and coarsely warted .. .. .. .. .. .. 1. *mannii*
Stems about 10–35 cm. high; leaf-sheaths about 1·5 cm. long, more densely and conspicuously ciliate on margins than above; leaf-lamina obovate, broadest in upper part, rather abruptly narrowed above to a very short point about 2–7 mm. long, not or scarcely acuminate; inflorescence-axes pubescent, about 2–4 cm. long, bearing 1 cluster of flowers and 1(–2) lower bracts without flowers; otherwise similar to above
2. *obovata*

1. **B. mannii** *C. B. Cl.* in DC., Monogr. Phan. 3: 233, t. 6 (1881); F.T.A. 8: 76; Brenan in Kew Bull. 14: 283, fig. 1, 1–10 (1960), partly, excluding *Linder* 561; Aké Assi, Contrib. 2: 222 (1964), partly, excluding *Linder* 561; J. K. Morton in J. Linn. Soc. 60: 169 (1967). A stout herb about 1½–5 ft. high with acuminate leaves and lateral inflorescence-axes each emerging through a hole at the base of the leaf-sheath; flowers small, white, open from 10 a.m. until 3 p.m.; in lowland rain-forest.
**S.Nig.:** Aponmu, Akure Dist. (July) *Olorunfemi* FHI 34155! Akure F.R. (Mar.) *Richards* 5134! Owo (July, Nov.) *Stanfield* 45! 103! Ifon (Dec.) *Stanfield* 51! Oban F.R., Calabar (Jan.) *Onochie & Okafor* FHI 36133x! **W.Cam.:** Barombi (fr. Dec., Mar., May) *Preuss* 294! *Morton* K822! *Brenan & Richards* 9469! **F.Po:** (fr. Dec.) *Mann* 96! Also in E. Cameroun.
2. **B. obovata** *Brenan* in Kew Bull. 22: 390 (1968). *B. mannii* of Brenan in Kew Bull. 14: 283 (1960), partly, as to *Linder* 461 and note; Aké Assi, Contrib. 2: 222 (1964) partly; *B. sp.* J. K. Morton l.c. (1967). A shorter, less stout herb than above with not or scarcely acuminate leaves and shorter inflorescence-axes emerging, however, as above; flowers small, white; in lowland rain-forest.
**Guin.:** Foot of Nimba Mts. (Sept.) *Schnell* 3598! **S.L.:** Fourah Bay (Oct.) *Morton* SL 2395! **Lib.:** Gbanga (Sept.) *Linder* 561! Sanokwele (Sept.) *Baldwin* 9533! Siaple (Sept.) *Baldwin* 9462! Bilimu Mt. (Dec.) *Harley* 1283! Nimba Mts. (Oct.) *Adames* 625! **Iv.C.:** Yapo Forest (Sept.) *de Wilde* 567a! *Giovannetti* 360! **Ghana:** Ankasa F.R. (Dec.) *Hall & Enti* GC 36223!

## 11. POLYSPATHA Benth. in Fl. Nigrit. 543 (1849); F.T.A. 8: 61.

Inflorescence-axes (i.e. those bearing spathes) elongate, at least the longer ones on each stem 6–20 cm. long; leaves glabrous or almost so, rarely shortly pubescent on surface but not long-pilose, elliptic or subrhombic-elliptic, cuneate at base, acutely more or less acuminate at apex, the larger ones 7–20 cm. long, 3–7 cm. broad; inflorescence usually of several flexuose axes apparently terminally clustered but sometimes single, leafless but bearing small recurved spathes about their own distance apart; vigorous plants with some axillary inflorescence-axes in addition which perforate the leaf-

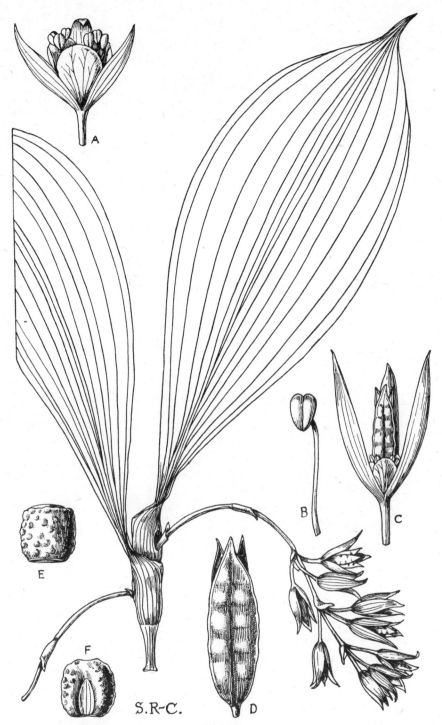

Fig. 331.—Buforrestia mannii *C.B. Cl.* (Commelinaceae).
A, flower. B, stamen. C, young fruit and persistent floral parts. D, fruit. E and F, seeds.

41

sheaths; capsule 4–5 mm. long; seeds 2·5–3 mm. long, with ribs radiating from
embryostega   ..        ..      ..    ..     ..     ..     ..     **1.** *paniculata*
Inflorescence-axes (i.e. those bearing spathes) all short, about 1·5–4 cm. long; leaves
more or less scattered-long-pilose on upper surface, glabrous or sparsely pilose below,
elliptic or oblong-elliptic, cuneate at base, gradually and acutely acuminate at apex,
the larger ones 6–13 cm. long, 2–4(–4·5) cm. broad; inflorescence a short dense
apparently terminal cluster of axes bearing spathes more or less as above but about
half to three-quarters their own distance apart; capsule about 3 mm. long; seeds
2–2·5 mm. long, more or less faintly ribbed ..    ..    ..    ..    ..  **2.** *hirsuta*

1. **P. paniculata** *Benth.* in Fl. Nigrit. 543 (1849); C. B. Cl. in DC., Monogr. Phan. 3: 194 (1881), incl. var.
*glaucescens* C. B. Cl. l.c. 195; F.T.A. 8: 61; J. K. Morton in J. Linn. Soc. 60: 206 (1967). *P. glaucescens* (C. B. Cl.) Hutch. in Kew Bull. 1939: 244; F.W.T.A., ed. 1, 2: 320 (1936), English descr. only.
Herb with decumbent stems or horizontal stolons rooting at nodes and erect leafy and flowering stems
6 in. to 2 ft. high, with usually clustered more or less zig-zag inflorescence-axes and small white flowers,
open from 2.30–5 p.m.; in lowland rain-forest.
  **Guin.:** Bafing (Oct.) *Adam* 12657! Kouria *Caille* in *Hb. Chev.* 14894! **S.L.:** York Pass, 1,300 ft. (Nov.)
*T. S. Jones* 184! Kumrabai-Mamila (Nov.) *Deighton* 4949! Mabonto (Oct.) *Thomas* 3515! Kambui F.R.,
Kenema (Apr.) *Jordan* 2025! Gola North F.R. (Feb.) *Bakshi* 44! **Lib.:** Zuie, Boporo Dist. (Nov.) *Baldwin*
10262! Nekabozu, Vonjama Dist. (Oct.) *Baldwin* 9976! Bilimu Mt. (Nov.–Dec.) *Harley* 1284! 2070!
Bobei Mt., Sanokwele Dist. (Sept.) *Baldwin* 9569! **Iv.C.:** Guédéyo to Soubré (Dec.) *Leeuwenberg* 2185!
Sassandra to Fresco (Mar.) *Leeuwenberg* 3082! S. of Koungramba (Dec.) *de Wit* 7426! **Ghana:** Atonso,
Ashanti Dist. (Oct.) *Baldwin* 13511! Fume (Nov.) *Morton* A112! Akim (Dec.) *Johnson* 250! Tafo (Sept.)
*Deighton* 3423! Suhun (Nov.) *Morton* 8009! Dawa Malo Kole (Dec.) *A. S. Thomas* D34! **S.Nig.:** Gambari
North F.R., Ibadan (Nov.) *Idahosa* FHI 23881! Owo (Nov.) *Stanfield* 41! Okomu F.R., Benin (Dec.)
*Brenan* 8566! Oban (Mar.) *Richards* 5205! Calabar to Mamfe (Jan.) *Onochie* FHI 36066! **W.Cam.:**
*Preuss* 1186! Likomba, Victoria (Oct.) *Mildbraed* 10530! Cam. Mt., 3,000 ft. *Mann* 2138! Mombo to
Ebonji, Kumba (May) *Olorunfemi* FHI 30584! **F.Po:** *Barter* 1474! 2055! (Dec.) *T. Vogel*! 2,000 ft.
*Mann*! Musola (Jan.) *Guinea* 939! Also in E. Cameroun, Rio Muni, Congo, Cabinda and Uganda.

2. **P. hirsuta** *Mildbr.* in Notizbl. Bot. Gart. Berl. 9: 256 (1925), incl. var. *togoensis* Mildbr. l.c. 257; J. K.
Morton l.c. (1967). Herb similar to last, with white flowers, opening from about 2.30 p.m. until evening;
in lowland rain-forest, sometimes locally abundant and covering extensive areas.
  **S.L.:** Tingi Hills above Konelo (fr. Dec.) *Morton & Gledhill* SL 3206! **Iv.C.:** Koroukoro to Touna (May)
*Chev.* 21796! **Ghana:** Akokoaso, Birrim Dist. (Nov.) *Fishlock* 70! Amentia (Mar.) *Vigne* FH 1886!
Kwahu Tafo (Aug.) *Hall* CC 261! Pawmpawm R., Anyaboni, Afram Plains (Dec.) *Morton* GC 6102!
Anum (fr. Nov.) *Morton* GC 7944! **Togo Rep.:** Atakpame (Nov.) *Mildbraed* 7452! **N.Nig.:** Patti Lokoja
(Nov.) *Dalz.* 254! Near Jagindi, Jemaa Div. (Oct.) *Hepper* 1034! **S.Nig.:** Olokemeji F.R. (Nov.) *Nchami*
FHI 14339! Gambari F.R., Ibadan (Jan.) *Onochie & Okafor* FHI 31522! Igbara Oke, Akure (Nov.)
*Meikle* 509! Okomu F.R. (fr. Feb.) *Brenan & Onochie* 8998! *Brenan* 9103! Sapoba (Nov.) *Baldwin*
13728! Also in E. Cameroun and Uganda.

## 12. COMMELINA Linn., Sp. Pl. 40 (1753); F.T.A. 8: 33.

Spathe-margins free to base:
  Leaves linear, 1–5(–7) mm. broad, mostly 3–15 cm. long; plant annual, subsimple to
    much branched, 10–60 cm. high, producing fibrous roots from lowest nodes; spathes
    small, 5–15 mm. long, sessile or subsessile (peduncle to 5 mm.), deflexed; flowers
    apricot to buff, rarely blue; seeds with 2–5 coarse, deep, irregular furrows radiating
    from the embryostega on the side opposite the hilum   ..     ..     **6.** *subulata*
Leaves ovate to lanceolate, rarely linear but then spathes more than 2 cm. long, on
    peduncles 2–5 cm. long, and flowers blue:
  Flowers yellow:
    Spathes borne singly, not clustered; plant glabrous to hairy, very rarely with
      rusty hairs; transverse nerves on leaves and spathes obscure or absent in dried
      specimens; capsule with dorsal loculus developed but indehiscent and with one
      seed; ventral loculi each with 2 ovules, one or both of which usually abort;
      seeds smooth or minutely reticulate; lamina lanceolate, sessile or subsessile on the
      sheath (*C. africana*):
      Leaves glabrous or almost so (some long marginal hairs normally present near
        mouth of leaf-sheath):
        Leaves normally up to 6(–7) cm. long or less; spathes mostly about 1·5–3·5 times
          as long as broad:
          Base of leaves rounded to rounded-cuneate; if cordate then leaves more than
            4 cm. long    ..      ..      ..     ..      ..       **1a.** *africana* var. *africana*
          Base of leaves cordate; leaves up to 4 cm. long, ovate, sessile
                                                       **1b.** *africana* var. *mannii*
        Leaves (at least larger ones) normally 7–13 cm. long; spathes about 1·5–5 times
          as long as broad, often (by no means always) more acuminate than above
                                                      **1c.** *africana* var. *lancispatha*
      Leaves more or less densely pubescent:
        Leaves normally up to 7 cm. long or less   ..    **1d.** *africana* var. *villosior*
        Leaves (at least larger ones) normally 7–13 cm. long
                                                      **1e.** *africana* var. *boehmiana*
    Spathes more or less clustered at ends of stems; rusty marginal hairs usually present
      round spathes and/or mouths of leaf-sheaths, occasionally absent; transverse
      nerves on leaves and spathes usually conspicuous in dried specimens; capsule
      with dorsal loculus absent; ventral loculi each developing 2 seeds; seeds large,

pale, elongate, conspicuously and deeply pitted; lamina elliptic-lanceolate or oblanceolate, asymmetric at base, often with a short distinct petiole          7. *capitata*
Flowers blue, violet or white:
Plant, particularly stems, sheaths and spathes, more or less densely pubescent with
  colourless hairs hooked at tip, scandent; spathes on divergent peduncles on
  special mostly lateral shoots bearing mostly leafless sheaths; spathes not or
  scarcely acuminate; flowers blue; capsule with dorsal loculus developed, 1-
  seeded, indehiscent, ventral loculi with 2 ovules of which 1 or both often abort;
  seeds elongate, coarsely ridged    ..    ..    ..    ..    ..    4. *ascendens*
Plant without hooked hairs, erect, straggling or prostrate; spathes not on special
  more or less leafless lateral shoots, usually acuminate:
Spathes 4–9 times as long as broad, elongate, 2·5–6·3 cm. long, narrow, 4–8 mm.
  broad (folded), gradually long-acuminate; stems not rooting from lower nodes
  above base; leaves linear-lanceolate, 5–16 cm. long, 5–11(–15) mm. broad,
  glabrous or subglabrous; seeds large, elongate, reticulate, covered with white
  powder    ..    ..    ..    ..    ..    ..    ..    5. *macrospatha*
Spathes 2–4 times as long as broad, often more than 10 mm. broad (folded); stems
  usually rooting freely from lower nodes above base; leaves often (not always)
  more than 11 mm. broad and shorter in proportion to width; seeds not covered
  with white powder:
Seeds smooth:
Leaves lanceolate to linear-lanceolate, gradually attenuate-acute at apex,
  mostly 4–11 cm. long, 1–1·8 cm. broad; spathes acuminate, mostly 2·5–5 cm.
  long; flowers whitish to very pale violet; leaf-sheaths often more or less
  spreading-hairy all over outside    ..    ..    ..    ..    3. *thomasii*
Leaves ovate-elliptic, obtuse to acute, not attenuate at apex, 1·5–4·5(–5) cm.
  long, 1–1·6(–2) cm. broad; spathes not or scarcely acuminate, 1·2–2·3 cm.
  long; flowers pale or bright blue; leaf-sheaths pubescent on margins and on a
  longitudinal line opposite insertion of lamina, sometimes all over
                                          2b. *diffusa* subsp. *montana*
Seeds reticulate; prostrate or straggling plant up to 1·2 m. high; lamina lanceolate
  to elliptic, rarely more than 8 cm. long; spathes 1·5–3 cm. long; flowers blue
  to violet; capsule 5-seeded    ..    ..    ..    2a. *diffusa* subsp. *diffusa*
Spathe-margins more or less connate at base above insertion of peduncle; spathe sometimes forming a wide funnel-shaped tube:
Leaves about 2–6 times as long as broad, lanceolate, elliptic, oblong-elliptic, ovate or
  ovate-elliptic:
Flowers white* to yellow; species of high forest except for *C. congesta* and *C. bracteosa*,
  (very rare albino variants):
Flowers yellow; spathes more or less clustered at ends of stems; rusty marginal
  hairs usually present round spathes and/or mouths of leaf-sheaths, occasionally
  absent; capsule with dorsal loculus absent; ventral loculi each developing
  2 large, pale, elongate, conspicuously and deeply pitted seeds          7. *capitata*
Flowers white:
Leaves obovate, distinctly ((0·5–)1–3 cm.) petiolate, up to about 18 cm. long and
  7 cm. broad; stems decumbent, rooting at nodes; spathes clustered and imbricate at ends of stems ..    ..    ..    ..    ..    ..    8. *longicapsa*
Leaves of various shapes but never obovate, always broadest at or below middle,
  sessile or with a very short petiole up to about 5 mm. long:
Leaf-sheaths rusty-hairy; lamina elliptic-lanceolate to ovate-lanceolate, acuminate
  at apex, 4–12 cm. long, 1·7–4 cm. broad; spathes sessile or subsessile, clustered
  1–3 together at or near ends of stems; capsule with 2 loculi each containing 1
  smooth ellipsoid seed    ..    ..    ..    ..    ..    19. *cameroonensis*
Leaf-sheaths without rusty hairs:
Seeds reticulate-pitted, large, about 4 mm. long and 2·5 mm. broad; stems stiff,
  slender, not rooting at nodes; leaves ovate-elliptic, acute at apex, asymmetric
  at base with one side rounded and the other cuneate; spathes mostly single,
  sometimes more than one together, on short bracteate shoots among upper
  leaves; spathes semicircular or semicordate, about 10–17 mm. long, not or
  scarcely acuminate; capsule with 2 single-seeded loculi    ..    18. *macrosperma*
Seeds smooth or nearly so; capsule normally with 3 single-seeded loculi; stems
  often rooting at nodes:
Spathes sessile and imbricate in terminal clusters, each spathe with margins
  connate for some distance, thus more or less obliquely funnel-shaped or
  triangular in outline; leaf-sheaths often spreading-hairy; flowers very pale
  blue or violet, almost white ..    ..    ..    ..    ..    20. *congesta*

---

* Occasional casual white-flowered individuals may occur in certain normally blue-flowered species, notably
*C. benghalensis* Linn., *C. lagosensis* C. B. Cl. and *C. erecta.* These are not accounted for in this part of the
key.

Spathes more or less pedunculate, solitary or several together, more or less semicircular in outline; flowers normally bright blue, very rarely and casually white    ..    ..    ..    ..    ..    ..    ..    ..    16. *bracteosa*
Flowers blue:
Leaf-margins rather strongly undulate, at least when dry, cartilaginous; lamina oblong-elliptic, rarely exceeding 7 cm. long and 1·7 cm. broad, usually obtuse to subacute, occasionally acute but never much tapering at apex; spathes small, about 8–13 mm. long, not or scarcely acuminate, borne singly and shortly pedunculate; cleistogamous flowers on basal underground shoots present, but rusty hairs on leaf-sheaths quite absent    ..    ..    ..    ..    10. *forskalaei*
Leaf-margins not or scarcely undulate (occasionally more or less undulate in *C. petersii*, but then leaves normally longer than 7 cm. and spathes longer than 2 cm.; also in *C. erecta* subsp. *maritima* where spathes are usually more than 1·5 cm. long and seeds are laterally ridged); leaves (other than those towards base of stem) acute to very acute at apex; cleistogamous flowers absent except in *C. benghalensis* which has rusty hairs at least on the margins of the leaf-sheaths:
Seeds rough with coarse more or less reticulate ridges or transverse wrinkles on the side opposite the hilum; leaf-sheaths (at least on margins) usually with more or less elongate red or rusty hairs; cleistogamous flowers often present on short underground stems; leaves ovate to ovate-lanceolate; spathes usually more or less clustered:
Spathes on peduncles 1·5–3·5 cm. long, the spathes themselves 2–3 cm. long; rusty hairs on leaf-sheaths and cleistogamous flowers absent    13. *petersii*
Spathes sessile or subsessile (peduncles to about 7 mm. long); spathes usually less than 2 cm. long; rusty hairs usually present on leaf-sheaths; cleistogamous flowers often present:
Leaves narrowed but not or scarcely attenuate to the acute to subacute apex; leaf-sheaths rusty-ciliate on margins, with sometimes a few rusty hairs on surface outside    ..    ..    ..    9a. *benghalensis* var. *benghalensis*
Leaves more or less gradually attenuate-acute at apex; leaf-sheaths more or less densely rusty-hairy all over outside    9b. *benghalensis* var. *hirsuta*
Seeds smooth, at most faintly and shallowly reticulate, or sometimes with a low ridge on either side but otherwise smooth; red or rusty hairs and cleistogamous flowers absent:
Spathes pedunculate:
Laminae (at least on some of the upper leaves) amplexicaul, sessile on the leaf-sheath and clasping the stem by a rounded base; spathes shortly pubescent outside, 2–3 cm. long, on peduncles 1·5–4 cm. long; peduncle of lower cincinnus normally projecting 1–2 cm. beyond spathe; capsule bilocular, each loculus with 2 ellipsoid-oblong seeds ..    ..    ..    12. *imberbis*
Laminae not amplexicaul, usually very shortly petiolate above the sheath, which itself may bear small green auricles distinct from the lamina; loculi 1–2-seeded:
Capsule (4–)5-seeded, ventral loculi each 2-seeded; seeds ellipsoid to subglobose, usually very shallowly reticulate, without lateral ridges; plant robust with more or less erect stems; leaves often 10–16 cm. long and 2·5–4·5(–5·5) cm. broad; spathes 1·7–3·4 cm. long, not or scarcely acuminate, on peduncles (1–)1·5–3·5(–5) cm. long    ..    ..    ..    11. *zambesica*
Capsule 2–3-seeded, ventral loculi each 1-seeded; seeds not reticulate, often with lateral ridges; peduncle of spathe 0·5–1·2 cm. long:
Seeds globose or nearly so, without lateral ridges:
Leaves small, up to about 6 cm. long, ovate to ovate-lanceolate, often rounded at base; stems prostrate, up to 1 m. long, rooting at nodes; sheaths spreading-hairy or pubescent outside; peduncles 5–10 mm. long
17. *lagosensis*
Leaves much larger, 5–13 cm. long, ovate-lanceolate to lanceolate, often cuneate at base; stems erect or straggling, normally not rooting freely at nodes; indumentum of leaf-sheaths variable; peduncles 8–22 mm. long
16. *bracteosa*
Seeds ellipsoid to subglobose, with a lateral ridge on each side below the embryostega; ridge sometimes continuous round the seed; spathes often (not always) more or less acuminate at apex    21. *erecta* aggregate
(see note on *Berhaut* 897)
Spathes sessile or subsessile, usually more or less congested:
Seeds smooth, without 2 lateral or 1 peripheral thickened ridge, ellipsoid; stems rooting at nodes; leaf-sheaths often more or less clothed with long soft spreading hairs, without spreading green auricles at mouth; flowers very pale violet, almost white    ..    ..    ..    ..    ..    20. *congesta*
Seeds with 2 lateral or 1 peripheral thickened ridge, ellipsoid to subglobose; stems often not rooting at nodes; leaf-sheaths without long soft spreading

hairs except a few on margins, often with small spreading green auricles at
mouth; flowers bright blue:
Leaves mostly about 4–8 times as long as broad, not markedly fleshy, often
cuneate at base into a short petiole; plant straggling to erect
21a. *erecta* subsp. *erecta*
Leaves mostly about 2–3 times as long as broad, fleshy, mostly sessile or almost
so on the sheath; plant prostrate, almost glabrous
21b. *erecta* subsp. *maritima*
Leaves about 9–10 times or more as long as broad, linear to linear-lanceolate:
Flowers white to yellow or orange or deep purple, not blue:
Spathes large, 2·8–3·8 cm. long, 1·5–2 cm. deep (folded), on peduncles 4–7 cm. long;
stems to 1 m. long, straggling and rooting at nodes; leaves to about 12 cm. long
and 1·2 cm. broad; flowers deep purple ..    ..    ..    ..    .. 14. *sp.* B
Spathes smaller, not more than 1·5 cm. long and 1 cm. deep, on peduncles not
exceeding 1·5 cm. in length; flowers white, yellow or orange:
Spathes scattered singly at nodes, not more than 1–3 on each stem or branch;
peduncles usually short but distinct, 0·5–1·5 cm. long; spathes usually abruptly
narrowed at base into the peduncle:
Seeds ellipsoid, smooth except for 2 prominent pits on the side opposite the hilum,
and usually some tiny scattered paler tubercles; loculi of capsule single-seeded
22a. *nigritana* var. *nigritana*
Seeds subglobose, reticulate; loculi (partly at least) 2-seeded
22b. *nigritana* var. *gambiae*
Spathes in terminal clusters of usually more than 3; peduncles up to about 5 mm.
long; spathes narrowed at base into a pale funnel-shaped part tapering into the
peduncle; capsule with 3 single-seeded loculi; seeds subglobose, smooth
23. *aspera*

Flowers blue:
Spathes single on each stem (rarely 2 at separate nodes), terminal, (1·5–)2·3–3 cm.
long, on a peduncle (1–)2–4 cm. long; leaves more or less amplexicaul at base
and with rather revolute margins ..    ..    ..    15. *schweinfurthii*
Spathes usually clustered several together, sometimes single, subsessile or shortly
(to 1 cm.) pedunculate; leaves not markedly amplexicaul at base (though sometimes auriculate at sheath-mouth) and with involute not revolute margins
21c. *erecta* subsp. *livingstonii*

1. **C. africana** *Linn.* Sp. Pl. 1: 41 (1753); C. B. Cl. in DC., Monogr. Phan. 3: 164 (1881); F.T.A. 8: 45;
J. K. Morton in J. Linn. Soc. 55: 515, fig. 11 (1956), 60: 174 (1967); Brenan in Mitteil. Bot. Staatssamml.
München 5: 199–203 (1964). *C. bakueana* A. Chev., Bot. 663, name only, fide F.W.T.A. ed. 1.
1a. **C. africana** *Linn.* var. **africana**—Brenan l.c. 203 (1964). Herb with stems up to about 3 ft. long, often
prostrate and rooting at nodes, sometimes suberect, with yellow flowers open from 7 to 10 a.m.
**Port.G.:** Bambadinca (July) *Pereira* 3047! **S.L.:** Roboli, nr. Rokupr (July) *Jordan* 287! Warantamba,
Gberia Fotombu (Oct.) *Small* 372! Miligi, Tonko Limba (July) *Jordan* 44! Kagberi, Sela-Limba (Sept.)
*Jordan* 546! Makuta, 3,400 ft. (June) *Thomas* 477! **Ghana:** Mangoase (July) *Deighton* 3403! Aburi
Scarp (Sept.) *Morton* GC 9327! Aburi to Nsawam (Dec.) *Morton* GC 8058! **N.Nig.:** Jos Plateau (June)
*Lely* P323! *Batten-Poole* 305! Hepham to Ropp, 4,600 ft. (July) *Lely* 372! Neill's Valley, Naraguta
(June) *Lely* 263! **S.Nig.:** Obudu Plateau, Ogoja, 5,000 ft. (May) *Head* 79! *Tuley* 630! **W.Cam.:** Jakiri,
5,000 ft. (June) *Brunt* 553! Bafut-Ngemba F.R., 7,000 ft. (Feb.) *Hepper* 2149! *Daramola* FHI 41160!
Widespread in tropical and S. Africa.
1b. **C. africana** var. **mannii** (*C. B. Cl.*) Brenan l.c. 206 (1964). *C. mannii* C. B. Cl. l.c. 167 (1881); F.T.A.
8: 48; F.W.T.A., ed. 1, 2: 318; J. K. Morton in J. Linn. Soc. 55: 318, fig. 1 (1956), 60: 188 (1967).
Small prostrate herb with stems rooting at lower nodes and small, ovate-elliptic leaves; clearings
in mist-forest and montane grassland.
**W.Cam.:** Cam. Mt., 7,000 ft. (Dec.) *Mann* 2136! Mann's Spring, 7,500 ft. (Apr., Dec.) *Morton* K844!
7093! Also in Ethiopia.
1c. **C. africana** var. **lancispatha** *C. B. Cl.* in Th.-Dyer, Fl. Cap. 7: 10 (1897); Brenan l.c. 211 (1964); J. K.
Morton in J. Linn. Soc. 60: 175 (1967). *C. elliotii* C. B. Cl. & Rendle in J. Linn. Soc. 30: 98 (1894).
Herb similar to var. *africana* but more robust.
**S.L.:** Port Loko (May) *Deighton* 4782! Yengema (Dec.) *Deighton* 3581! Kahreni to Lokko, Limba
*Sc. Elliot* 5749! Kurubonla to Seredu, foot of Loma Mts. (Nov.) *Morton* SL 2529! Seria to Kameron
(Nov.) *Morton* SL 2453! Widespread in eastern and central Africa from about 3° S. latitude southwards;
rare and local in W. Africa.
1d. **C. africana** var. **villosior** (*C. B. Cl.*) Brenan l.c. 207 (1964). *C. barbata* Lam. var. *villosior* C. B. Cl. in DC.,
Monogr. Phan. 3: 167 (1881). *C. africana* of Schnell, Vég. Fl. Minka 522 (1952). *C. africana* var. *kreb-
siana* of J. K. Morton in J. Linn. Soc. 55: 515 (1956), 60: 175 (1967), not of (Kunth) C. B. Cl. Herb
similar to var. *africana*, but hairy on leaves.
**Sen.:** Gorom (Sept.) *Berhaut* 2070, partly! **S.L.:** Roboli, nr. Rokupr (Sept.) *Jordan* 287a! Musaia (July)
*Deighton* 4794! Sasa, Tonko Limba (Sept.) *Jordan* 330! Loma Mts., above Yifin (Mar.) *Morton &
Gledhill*! Kambia to Kukuna (Nov.) *Morton & Gledhill* SL 39! **Lib.:** Bushrod Is. (Aug.) *Baldwin*
13097! **Iv.C.:** Adiopodoumé (Nov.) *Leeuwenberg* 2104! Nimba Mts. (Mar.–Apr.) *Schnell* 878! 932!
Brafouedi (Dec.) *de Wit* 7595! Yapo Forest, nr. Abidjan (July) *Boughey* GC 14548! **Ghana:** Nsemre
F.R., Borku to Anka (Dec.) *Adams* 2837! Mlabo, Togo Plateau (May) *Morton* A3934! Pepiasi, Abetifi
(Apr.) *Morton* A788! Lartey Scarp, Akwapim (Dec.) *Morton* A2839! Akatsi, nr. Ohawu (July) *Irvine*
4950! **S.Nig.:** Owerri (Sept.) *English*! Owo (Oct.) *Stanfield* 15! **W.Cam.:** Bamenda, 6,000 ft. (Jan.)
*Migeod* 355! Bafut-Ngemba F.R., Bamenda, 6,000 ft. (fl. Mar., fl. & fr. May) *Richards* 5304! *Ujor* FHI
30308! Bamenda, 5,000 ft. (Mar.) *Morton* K19! Widespread in tropical and S. Africa; also in Madagascar.
1e. **C. africana** var. **boehmiana** (*K. Schum.*) Brenan l.c. 213 (1964). *C. boehmiana* K. Schum. in Engl., Pflanzenw.
Ost-Afr. C: 135 (1895). Herb similar to var. *lancispatha*, but hairy on leaves.
**S.L.:** Port Loko (May) *Deighton* 4781! Kamabai, nr. Makeni (Sept.) *Harvey* 90! Distribution similar to
that of var. *lancispatha* (see above).

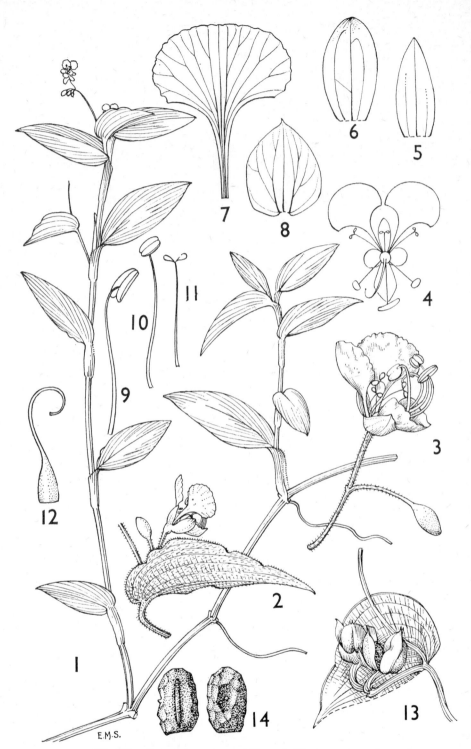

FIG. 332.—Commelina diffusa *Burm. f.* (Commelinaceae).

1, habit, × ⅔. 2, inflorescence, × 2. 3, flower and bud, × 6. 4, floral diagram. 5, dorsal sepal, × 9. 6, sepal, × 9. 7, petal, × 9. 8, keeled petal, × 9. 9, stamen, × 9. 10, stamen (one of pair), × 9. 11, staminode, × 9. 12, gynoecium, × 9. 13, fruit, × 2. 14, seeds, × 6. 1 from *Gillett* 15348. 2–12 from *Gillett* 15295a. 13, 14 from *Okeke* FHI 18228.

2. **C. diffusa** *Burm. f.* Fl. Ind. 18, t. 7, fig. 2 (1768); Merrill in J. Arn. Arb. 18: 64 (1937); J. K. Morton in J. Linn. Soc. 55: 521, fig. 18 (1956), 60: 181 (1967). *C. nudiflora* Linn., Sp. Pl.: 41 (1753), partly (excl. lectotype, see Merrill l.c.); C. B. Cl. in DC., Monogr. Phan. 3: 144 (1881), including var. *werneana* (Hassk.) C. B. Cl.; F.T.A. 8: 36; F.W.T.A., ed. 1, 2: 318; Dalz., Usef. Pl. W. Trop. Afr. 465 (1937); Sousa in Anais Junta Inv. Col. 5: 48 (1950); Berhaut, Fl. Sén. 208. *C. agraria* Kunth, Enum. 4: 38 (1843), illegit. name; Benth. in Fl. Nigrit. 541 (1849). *C. aquatica* J. K. Morton in J. Linn. Soc. 55: 515, fig. 12 (1956). *C. communis* of Benth. in Fl. Nigrit. 541 (1849), fide C. B. Cl., not of Linn.

2a. **C. diffusa** *Burm.f.* subsp. **diffusa**—J. K. Morton in J. Linn. Soc. 60: 181, fig. 3 (1967). Annual or perennial herb; stems rooting freely at nodes; leaves 0·8–2·5 cm. broad; flowers deep blue to violet or lilac-blue, open from about 8.30 a.m. until noon. Variable in habit and pubescence, probably with various local races, of which *C. aquatica* is a robust perennial one. Open wet places.

    **Sen.:** Djembering (Sept.) *Broadbent* 167! Kafountine, Oussouye (Sept.) *Adam* 18181! Ile du Diable, Sedhiou (Oct.) *Adam* 18447! **Gam.:** Kuntaur *Ruxton* 93! **Mali:** Sarédina (Mar.) *Davey* 55! Fafa rapids, Gao (Mar.) *de Wailly* 5359! **Port.G.:** Teixeira Pinto, Cacheu, Pijame (Dec.) *Raimundo & Guerra* 520! Bissau, Pussubé (Apr.) *Esp. Santo* 1174! Bafatá, Geba (Jan.) *Raimundo & Guerra* 888! **S.L.:** (June) *T. Vogel* 11! *Hart*! Freetown (May) *Deighton* 1197! Makene (Sept.) *Deighton* 4034! Magbile (Dec.) *Thomas* 6099! Rokupr (Apr.) *Hepper* 2667! **Lib.:** Firestone Plantation No. 3, Du River (Aug.) *Linder* 324! Peahtah (Oct.) *Linder* 1059! Gbanga (Oct.) *Okeke* 21! Nana Kru (Mar.) *Baldwin* 11589! Webo, Cavally R. (Apr.) *Dinklage* 2645! **Iv.C.:** Adiopodoumé (Oct.) *Leeuwenberg* 1787! **Ghana:** Gambaga (Mar.) *Hepper & Morton* A3138! Gonokrom to Dormaa (Dec.) *Adams* 2986! Afram Mankrong F.R. (Dec.) *Adams* 4973! Adaiso to Kade (Jan.) *Morton* GC 8340! Ada to Accra (Oct.) *Morton* A30! **N.Nig.:** (July) *Lely* P549! Jos Plateau (May) *Lely* P261! Gwoza, Dikwa Div. (Dec.) *McClintock* 73! **S.Nig.:** Ikoyi Plains, Lagos (Oct.) *Dalz.* 1287! Orosun Peak, Idanre Hills (Dec.) *Stanfield* 181! Owo (July, Oct.) *Stanfield* 18! 101! Calabar (Aug.) *Holland* 75! **W.Cam.:** *Preuss* 1320! Babungo, Ndop Plain (Mar.) *Brunt* 117! Buea (Jan.) *Morton* K945! **F.Po:** (Nov.) *T. Vogel* 67! Widespread in the tropics.

    [A colony of apparent hybrids between *C. diffusa* and *C. thomasii* is recorded from S. of Kade in Ghana (*Morton* GC 8350!) in J. Linn. Soc. 55: 510 (1956).]

2b. **C. diffusa** subsp. **montana** *J. K. Morton* in J. Linn. Soc. 60: 181 (1967). Annual, prostrate herb with stems up to 16 in. long rooting at nodes. Very close to typical *C. diffusa*, differing in the lamina normally sessile and cordate or subcordate at base, and (especially) in the smooth seeds.

    **N.Nig.:** Nuguroje, Mambila Plateau (Jan.) *Hepper* 1745! **W.Cam.:** Buea (Dec., Mar.) *Morton* K811 *Morton* GC 7039! Bamenda, 6,000 ft. (Jan.) *Migeod* 437! L. Bambelue to Santa (Apr.) *Morton* K281! Cheddar Gorge road, Bamenda (Mar.) *Morton* K20! **F.Po:** Moka, 3,500–4,000 ft. (Dec.) *Boughey* 67!

3. **C. thomasii** *Hutch.* in Kew Bull. 1939: 243, and F.W.T.A., ed. 1, 2: 318 (1936), English descr. only; J. K. Morton in J. Linn. Soc. 55: 528, fig. 26 (1956), and 60: 189 (1967). A straggling or sprawling perennial herb, with stems up to 3 ft. high, rooting at the lower nodes, and white to pale lilac flowers opening throughout the morning; in secondary lowland rain-forest and persisting in farms and plantations.

    **S.L.:** Masimo (June) *Marmo* 109! Yonibana (Oct., Nov.) *Thomas* 3962! 5205! Kurubonla to Seredu (Nov.) *Morton* SL 2531! **Lib.:** Gbanga (Sept.) *Linder* 476! Nimba Mts. (Sept.) *Adames* 545! **Ghana:** Mankrong, Kwahu (Dec.) *Adams* 5082! Kade (Jan.) *Morton* GC 8332! Amanase, Nsawam to Suhum (Nov.) *Morton* GC 8003! Bisa (Jan.) *Morton* GC 8235! Ekumfi Swedru, Mankessin (Sept.) *Hall* 2375! **Togo Rep.:** Baumann Peak, Palime (Aug.) *Morton* A4276! **S.Nig.:** Ibadan (Nov.) *Stanfield* 180! Owerri (July) *English*! Owo (Oct.–Nov.) *Stanfield* 4! 35! Idogun (Oct.) *Stanfield* 155!

4. **C. ascendens** *J. K. Morton* in J. Linn. Soc. 55: 517, fig. 13 (1956), 60: 176, fig. 2 (1967). Scandent herb to 8 ft. or more with stems rooting at lower nodes, lanceolate leaves up to 11 cm. long and 3 cm. broad, and pale blue flowers opening early in the day and fading within 3 hours of dawn (about 7–9 a.m.); in primary and secondary lowland rain-forest, often by rivers or streams.

    **Ghana:** Afram Mankrong F.R. (Dec., Apr.) *Morton* 4963! A551! Mampong, Akwapim *Morton* GC 8424! Nsawam to Aburi (Dec., Apr.) *Morton* A3653! GC 8061! **S.Nig.:** Owo (May–Dec.) *Stanfield* 14! 100! Idogun (Oct.) *Stanfield* 154!

5. **C. macrospatha** *Gilg & Lederm. ex Mildbr.* in Notizbl. Bot. Gart. Berl. 9: 253 (1925); Berhaut, Fl. Sén. 208; J. K. Morton in J. Linn. Soc. 55: 527, fig. 24 (1956), 60: 188, fig. 6 (1967). Perennial herb with erect or suberect stems 10–32 in. high arising from a rhizome; lowest sheaths brown, with laminae reduced or absent; flowers bright pure blue, with 3 well-developed petals; in grassland or savanna woodland.

    **S.L.:** Bintumane Peak, 6,000 ft. (May) *Deighton* 5098! Loma Mts. (Nov., Feb., Mar.) *Morton* SL 2586! *Gledhill* 355! *Morton & Gledhill* SL 1050! **Ghana:** Ejura (Dec.) *Morton* GC 9730! 9824! Mankrong, Afram Plains (Apr.) *Morton* A721! Folifoli, Afram Plains (Aug.) *Hall* CC 153! Yeji (Aug.) *Hall* CC 367! **N.Nig.:** Naraguta (June) *Lely* 328! Jos Plateau (Oct., May) *Lely* P777! P290! Mbakon F.R., Benue Prov. (July) *Keay* FHI 37128! Abinsi (June) *Dalz.* 803! Also in E. Cameroun, the Congo, Uganda and Kenya.

6. **C. subulata** *Roth* Nov. Pl. Sp. 23 (1821); C. B. Cl. in DC., Monogr. Phan. 3: 148 (1881); F.T.A. 8: 38; Berhaut, Fl. Sén. 208; Chikkanaiah in Phytomorphology 13: 174–184 (1963); J. K. Morton in J. Linn. Soc. 60: 189 (1967). An annual herb, usually slender, with linear leaves and small subsessile deflexed spathes at the nodes; flowers usually yellowish, rarely blue; in wet places.

    **Sen.:** Kaolak (Sept.) *Berhaut* 1243! M'Bambey (Oct.) *Monod*! Djembering, Oussouye (Oct.) *Adam* 18347! **Ghana:** Manga, Bawku (Aug.) *Irvine* 4650! Yendi (Oct.) *Morton* A3743! N. of Nandom (Sept.) *Hall* CC 796! Widespread in eastern and southern Africa; also in southern India.

7. **C. capitata** *Benth.* in Fl. Nigrit. 541 (1840); C. B. Cl. in DC., Monogr. Phan. 3: 176 (1881); F.T.A. 8: 54; Sousa in Anais Junta Inv. Col. 5: 48 (1950); J. K. Morton in J. Linn. Soc. 55: 519, fig. 16 (1956), 60: 179 (1967). Straggling, rather robust herb up to 4 ft. high, often rooting from lower nodes, characterized by yellow flowers (open from 8–11 a.m.) and the usual presence of reddish hairs round margins of leaf-sheaths and spathes; in high forest.

    **Sen.:** Casamance (Apr.) *Perrottet*! Tendimaim, Bignona (Dec.) *Berhaut* 6752! **Port.G.:** Bissau, Cumura (Nov.) *Esp. Santo* 2227! **Guin.:** Dalaba–Diaguissa Plateau (Oct.) *Chev.* 18777! Dyeke (Oct.) *Baldwin* 9666! **S.L.:** Luseniya (Dec.) *Sc. Elliot* 4080! Gberia Fotombu (Oct.) *Small* 411! Njala (Dec.) *Pyne* 111! Ronietta (Nov.) *Thomas* 5667! Kenema (Nov.) *Deighton* 3867! **Lib.:** Soplima, Vonjama Dist. (Nov.) *Baldwin* 10041! Cape Mount *Dinklage* 2145! Belefanai, Gbanga Dist. (Dec.) *Baldwin* 10548! Maisi forest, Ganta (Dec.) *Harley* 1465a! Ganta (Dec.) *Barker* 1116! **Iv.C.:** High Forest, nr. Nigbi II (Nov.) *de Wilde* 3278! nr. Guédéyo (Dec.) *Leeuwenberg* 2171! Banco forest, nr. Abidjan (Dec.) *de Wit* 7352! Orumba-Boka Mt., nr. Toumodi (Dec.) *Boughey* A18! A26! **Ghana:** Banda (Dec.) *Adams & Akpabla* GC 4048! Axim (Feb.) *Morton* GC 8444! Benso, Tarkwa Dist. (Nov.) *Andoh* FH 5401! Assin West-Skinn 274! Amedzofe (Oct.) *Morton* A966! Togo Plateau, W. side, 1,700 ft. (Apr.) *Morton* GC 9200! **Dah.:** Sakété to Pedjilé (Feb.) *Chev.* 22900! **N.Nig.:** Gangumi, Gashaka Dist. (Dec.) *Latilo & Daramola* FHI 28788! **S.Nig.:** Omo & Shasha F.R. (Mar.) *Jones & Onochie* FHI 17200! Idanre Hills, Orosun (Oct.) *Keay* FHI 22682! Usonigbe F.R., Benin (Nov.) *Meikle* 576! Okomu F.R., Benin (Dec.) *Brenan* 8555! Sapoba (Nov.) *Ejiofor* FHI 24627! Lagos (Apr.) *Moloney* 9! **W.Cam.:** Bafut-Ngemba F.R., Bamenda (Feb.) *Tiku* FHI 22162! Buea, 3,300–4,000 ft. (Nov.–Dec.) *Maitland* 1193! *Migeod* 86! Victoria (Jan.) *Kalbreyer* 17! **F.Po:** (Dec.) *Mann* N91! Moka, 3,500–4,000 ft. (Dec.) *Boughey* 74! Extending also eastwards to Uganda and southwards to Angola.

8. **C. longicapsa** *C. B. Cl.* in DC., Monogr. Phan. 3: 176 (1881); F.T.A. 8: 55; Brenan in Kew Bull. 7: 455 (1952); J. K. Morton in J. Linn. Soc. 60: 188 (1967). A decumbent herb rooting at nodes, on the floor of lowland rain-forest.

**Lib.:** Kulo, Sinoe Co. (Mar.) *Baldwin* 11412! **S.Nig.:** Okomu F.R., Benin (Dec.–Mar.) *Onochie* FHI 38304! *Brenan* 8525! 8525a! **W.Cam.:** Victoria to Kumba (Jan.) *Morton* K923! Also in Rio Muni, Gabon, Cabinda and Congo.

9. **C. benghalensis** *Linn.* Sp. Pl. 41 (1753); C. B. Cl. in DC., Monogr. Phan. 3: 159 (1881); F.T.A. 8: 41; Sousa in Anais Junta Inv. Col. 5: 49 (1950); Berhaut, Fl. Sén. 208; J. K. Morton in J. Linn. Soc. 55: 519 (1956), 60: 176 (1967); Saunders, Handb. W. Afr. Fl. 96 (1958).

9a. **C. benghalensis** *Linn.* var. **benghalensis**—J. K. Morton l.c. 178 (1967) as to diploid form and fig. 15f. *C. benghalensis* subsp. *benghalensis*—J. K. Morton l.c. (1967). Prostrate, ascending or sprawling herb with mostly ovate to elliptic leaves, pale leaf-sheaths, and blue flowers opening from 8 a.m. until noon; whitish, cleistogamous, subterranean flowers produced on basal shoots; a weedy plant of open cultivated and waste ground, also in savanna.

**Gam.:** Kuntaur *Ruxton* 151! **Port.G.:** Bissau, Pussubé (Apr.) *Esp. Santo* 1513! Fá (Oct.) *Guerra* 3872! **S.L.:** Wallia, Scarcies (Apr.) *Sc. Elliot* 4250! **Iv.C.:** Nambonkaha (Nov.) *Leeuwenberg* 1976! Tonkoui, Nimba Mts. (Aug.) *Boughey* GC 18320b! **Ghana:** Larabanga, Damongo to Sawla (Oct.) *Rose Innes* GC 30994! Pong Tamale (Aug.–Nov.) *Williams* 850! *Irvine* 4598! Kwadaso, Kumasi (Dec.) *Darko* 647! Achimota (Aug.) *Irvine* 849! **Togo Rep.:** Lomé *Warnecke* 276! **N.Nig.:** Kontagora (Oct.) *Dalz.* 265! Katsina (Sept.) *Dalz.* 809! Jos Plateau (Sept.) *Lely* P791! Sara Hills, Jos Plateau, 4,700 ft. (Oct.) *Hepper* 1124! Katagum Dist. *Dalz.* 231! Vogel Peak, Sardauna Prov. (Nov.) *Hepper* 1301! **S.Nig.:** Lagos *Millen* 31! Owo (Oct.) *Stanfield* 12! Widespread in tropical Africa and Asia. (See Appendix, p. 465.)

9b. **C. benghalensis** var. **hirsuta** C. B. Cl. in DC., Monogr. Phan. 3: 160 (1881); F.T.A. 8: 42; J. K. Morton l.c. (1967), as to tetraploid form. *C. benghalensis* subsp. *hirsuta* (C. B. Cl.) J. K. Morton in J. Linn. Soc. 60: 178 (1967). Herb similar to var. *benghalensis*, but with often longer and more straggling stems, mostly ovate to lanceolate leaves often more narrowed at base, leaf-sheaths less pale and (apparently, but more observation needed) no subterranean cleistogamous flowers; often in forest, but also in montane grassland.

**Guin.:** Guéaso to Moussadougou (Mar.) *Chev.* 20956! **S.L.:** Bumbuna (Oct.) *Thomas* 3913! 3915! Jaiama (Dec.) *Deighton* 3481! Musaia (Dec.) *Deighton* 4535! Sekorella, foot of Loma Mts. (Nov.) *Morton* SL 2814! **Lib.:** Gbanga (Oct.) *Linder* 1159! Vonjama (Oct.) *Baldwin* 9926! **Ghana:** Bosumkese F.R., Ashanti (Dec.) *Adams* 5250! Asuboi (Nov.) *Morton* 8001! Vane (Nov.) *Morton* A102! **Togo Rep.:** Misahöhe *Baumann* 413! **N.Nig.:** Mambila Plateau, 5,600 ft. (Jan.) *Hepper* 1746! **S.Nig.:** Ogwashi (Dec.) *Thomas* 2057! **W.Cam.:** Buea, 3,000 ft. (Nov.) *Migeod* 2! Keshong, Bamenda, 7,000 ft. (Oct.) *Gregory* 222! Bamenda (Jan.) *Daramola* FHI 40570! Bambui, 5,000 ft. (June) *Brunt* 495! Also in E. Cameroun, Congo, Rwanda, Ethiopia, Uganda,? Kenya, Tanzania, Malawi and Zambia.

10. **C. forskalaei** *Vahl* Enum. 2: 172 (1806); C. B. Cl. in DC., Monogr. Phan. 3: 168 (1881), including var. *hirsutula* C. B. Cl.; F.T.A. 8: 44; Berhaut, Fl. Sén. 208; J. K. Morton in J. Linn. Soc. 55: 522, fig. 19 (1956), 60: 185 (1967); Maheshwari & Baldev in Phytomorphology 8: 277–298 (1958); Aké Assi, Contrib. 2: 223 (1964); *C. zenkeri* of Chev. Bot. 665, not of C. B. Cl., fide F.W.T.A., ed. 1. Annual, prostrate or widely straggling weedy plant with stems rooting at nodes; flowers rich pure blue; grassland, open waste and cultivated ground, roadsides, sand-dunes, etc.

**Sen.:** *Roger!* *Heudelot* 276! Malika (Aug.) *Broadbent* 40! Roland Mt., Thiès (Aug.) *Broadbent* 49! Richard Toll (Oct.) *Roger* 93! **Gam.:** Kuntaur *Ruxton* 36! 144! 153! Fulladu West (Nov.) *Frith* 152! **Mali:** Toguerè of Saredina (May) *Davey* 93! Timbuktu (Aug.) *Hagerup* 265! Gao to Korogoussou (Aug.) *de Wailly* 4730! **Ghana:** N. Reg. (June) *Saunders* 3! Manga, Bawku (Aug.) *Irvine* 4652! Tema (May) *Morton* A889! Savelugu (Aug.) *Hall* CC 402! **Dah.:** Tanguieta (Sept.) *Morton* A4631! **N.Nig.:** Ilorin (Sept.) *Stanfield* FHI 45640! Nupe *Barter* 1477! Mongu, 4,300 ft. (July) *Lely* 390! Katagum Dist. *Dalz.* 232! Maiduguri (Oct.) *Noble* A6! Widely distributed in tropical Africa except in the forest areas; also in Madagascar, Arabia and India. (See Appendix, p. 465.)

11. **C. zambesica** C. B. Cl. in DC., Monogr. Phan. 3: 161 (1881); F.T.A. 8: 43; Brenan in Kew Bull. 15: 209 (1961); J. K. Morton in J. Linn. Soc. 60: 190 (1967). Robust herb with weakly upright stems up to 4 ft. high or sometimes widely straggling and rooting at nodes, and blue flowers; often all 3 petals well-developed; in woodland and grassland.

**N.Nig.:** Jos Plateau (July) *Lely* P551! Cheche Boku F.R., Gwari *E. W. Jones* FHI 42340! Sanga River F.R., Jemaa Dist. (July) *Keay* FHI 37119! 37119a (cult.)! *Stanfield* 191 (cult.)! Also in Uganda, Tanzania, Zanzibar, Mozambique, Malawi, Zambia and Rhodesia.

12. **C. imberbis** *Ehrenb. ex Hassk.* in Schweinf. Beitr. Fl. Aethiop. 209 (1867); F.T.A. 8: 49; J. K. Morton l.c. 185 (1967). Prostrate or straggling herb with stems sometimes up to 6 ft. long, leaves lanceolate-attenuate, dark green, more or less amplexicaul, glabrous or only slightly pubescent, leaf-sheaths glabrous to only slightly pubescent outside, and blue flowers open from 9–11 a.m.; in cultivated ground, probably introduced.

**S.Nig.:** Moor Plantation, Ibadan (Oct.–Nov.) *Stanfield* FHI 45709! 45718! Also in eastern Africa from the Sudan to Rhodesia.

13. **C. petersii** *Hassk.* in Flora 46: 385 (1863); C. B. Cl. in DC., Monogr. Phan. 3: 169 (1881); F.T.A. 8: 50; J. K. Morton l.c. 189 (1967). Straggling or erect herb 1–2½ ft. high; leaves lanceolate to ovate-lanceolate, mostly 7–15 cm. long, 1–4 cm. wide; flowers blue. Similar to *C. imberbis* but leaves not or scarcely amplexicaul and seeds furrowed not smooth.

**N.Nig.:** Biu, Bornu Prov., 2,500 ft. (Aug.) *Noble* N4! Gwoza, 3,500 ft. (Jan.) *McClintock* 154! Also in Kenya, Tanzania, Mozambique, Malawi, Rhodesia, Botswana, Angola, and S.W. Africa.

14. **C. sp. B.** Robust herb with long stems rooting at nodes and deep purple flowers.

**N.Nig.:** Vogel Peak, 4,700 ft. (Nov.) *Hepper* 1351!

15. **C. schweinfurthii** C. B. Cl. in DC., Monogr. Phan. 3: 158 (1881); J. K. Morton in J. Linn. Soc. 60: 189 (1967). *C. lateriticola* A. Chev. in Mém. Soc. Bot. Fr. 2, 8: 214 (1911). Herb with linear leaves, spathes usually one per stem, and blue or brownish-orange flowers. Apparent variation in flower-colour needs further investigation.

**Guin.:** Dalaba (Sept.–Oct.) *Chev.* 18821! **S.L.:** Bumban, nr. Makeni (Sept.) *Harvey* 82! **Iv.C.:** Nimba Mts., 2,500–5,000 ft. (Aug.) *Boughey* GC 18059! **Togo Rep.:** Oti Sansame Mango (Sept.) *Morton* A4640! **N.Nig.:** Jos Plateau (Aug.) *Lely* P666!

16. **C. bracteosa** *Hassk.* in Flora 46: 385 (1863); C. B. Cl. in DC., Monogr. Phan. 3: 180 (1881); F.T.A. 8: 55. *C. bainesii* C. B. Cl. l.c. 184 (1881); F.T.A. 8: 57. *C. aspera* of Berhaut, Fl. Sén. 209, not of Benth. Perennial herb with erect to straggling or even almost prostrate stems and blue flowers in spathes which are borne singly or laxly aggregated; peduncle of spathe 8–22 mm. long. (See note under *C. lagosensis*.)

**Sen.:** *Heudelot* 476! Kaolak (Sept.) *Berhaut* 234! Sangalkam (Sept.) *Berhaut* 2074! Kalounayes, Bignona (Sept.) *Adam* 18079! Diégoune, Bignona (Sept.) *Adam* 18049! **Gam.:** Genieri to Kaiaaf (July) *Fox* 117! Kuntaur *Ruxton* 119! **Mali:** Kanikombolé ravine, Macina (Sept.) *Chev.* 24850! **Port.G.:** S. Domingos (Aug.) *Pereira* 3122! **N.Nig.:** Zungeru (Aug.) *Dalz.* 266! Panshanu Pass, 2,900 ft. (Aug.) *Lawlor & Hall* 248! 274! Abinsi *Dalz.* 804! Also in E. Cameroun and eastern Africa from Sudan southwards to Mozambique and Malawi;? in S.W. Africa.

17. **C. lagosensis** C. B. Cl. in F.T.A. 8: 57 (1901); F.W.T.A., ed. 1, 2: 320; J. K. Morton in J. Linn. Soc. 55: 526, fig. 22 (1956), 60: 185, fig. 5 (1967); Saunders, Handb. W. Afr. Fl.: 98, fig 339 (1958). *C. lagosensis* var. *subglabra* A. Chev. in Mém. Soc. Bot. Fr. 2, 8: 214 (1911). Herb with prostrate stems up to 3 ft. long, radiating from centre and rooting at nodes; leaves comparatively small; flowers blue, rarely and casually white, open from dawn until 9.30 a.m.

**Sen.:** Djembering (Sept.) *Broadbent* 160! **Port.G.:** Fá (Sept.) *Guerra* 3833! **S.L.:** Rokupr (Aug.) *Harvey* 16! Rowala (July) *Thomas* 1127! Pendembu (July) *Thomas* 844! New England (May) *Marmo* 106!

Fourah Bay College (Mar.) *Morton* SL 947! **Iv.C.:** Diahbo to Bouaké (July) *Chev.* 22072! **Ghana:** Gambaga to Yendi (Apr.) *Morton* GC 9045! Achimota (Oct.) *Morton* GC 7838! Bantrosie to Senya Beraku (Oct.) *Morton* GC 7896! Accra to Nsawam (Oct.) *Morton* GC 7789! Tonogo, Togo Plateau (Nov.) *Morton* A3461! **Togo Rep.:** Lomé *Warnecke* 281! **S.Nig.:** Lagos *Millen* 21! (June) *Dalz.* 1418! Ibadan (Aug.) *Keay* FHI 37139! Owo (Oct.) *Stanfield* 8! Benin (Dec.) *Brenan & E. W. Jones* 8633a! Sapoba (Nov.) *Meikle* 583! Also in Kenya, Tanzania and Zanzibar; ? in Mozambique.

[Very near *C. bracteosa* Hassk., and may prove to be better placed as a subspecies of it.]

18. **C. macrosperma** *J. K. Morton* in J. Linn. Soc. 55: 528, fig. 25 (1956), 60: 188 (1967). Perennial herb with stiff slender stems up to about 2 ft. high; flowers open from dawn until 9 a.m.; in wet areas of lowland rain-forest, also under cocoa.
**S.L.:** Kasiri, Scarcies R. (Aug.) *Deighton* 3022! **Ghana:** Bobiri F.R. (May) *Morton* A3358! Pawtroasi, Kibi Rd. (Jan.) *Morton* A109! Brenase, Akim (Apr.) *Irvine* 553! **S.Nig.:** Boje to Katabong, Afi River F.R., Ikom (May) *Jones & Onochie* FHI 18722! Owo (June–Oct.) *Stanfield* 21! 102!

19. **C. cameroonensis** *J. K. Morton* in J. Linn. Soc. 55: 318, fig. 1 (1956), 60: 179 (1967). Straggling herb with stems up to 2½ ft. high, rooting at lower nodes; in and on edges of montane rain-forest.
**S.Nig.:** Obudu Plateau, Ogoja Prov. (Nov., Dec.) *Savory & Keay* FHI 25210! *Tuley* 1040! 1066! **W.Cam.:** Buea (Feb.) *Breteler et al.* MC 229! Bafut-Ngemba F.R., 7,000–7,900 ft. (Feb., Mar.) *Hepper* 2088! *Onochie* FHI 34886! Daramola FHI 40527! L. Oku, 7,500 ft. (Jan.) *Keay* FHI 28475! *Onochie* FHI 34886! **F.Po:** *Monod*!

[Near *C. capitata*, but differs in white flowers, 2-seeded capsule and smooth not pitted seeds.]

20. **C. congesta** *C. B. Cl.* in DC., Monogr. Phan. 3: 160 (1881); F.T.A. 8: 43; Berhaut, Fl. Sén. 208; J. K. Morton in J. Linn. Soc. 55: 521, fig. 17 (1956), and 60: 179 (1967). *C. condensata* C. B. Cl. l.c. (1881); F.T.A. 8: 43; F.W.T.A., ed. 1, 2: 320. *C. heudelotii* C. B. Cl. l.c. 184 (1881). *C. amphibia* A. Chev. in Mém. Soc. Bot. Fr. 2, 8: 214 (1911), including var. *hirsuta* A. Chev. *C. congesta* C. B. Cl. var. *hirsuta* (A. Chev.),Hutch., F.W.T.A., ed. 1, 2: 320 (1936). Creeping, perennial herb with rather stout stems rooting at nodes, elliptic to lanceolate leaves, spathes crowded 2 or more together at ends of stems, flowers very pale blue or violet almost white, open from dawn until nearly noon, and capsules with 3 single-seeded loculi; usually in forest, sometimes in open.
**Guin.:** *Heudelot* 788! **Port.G.:** Catió (Dec.) *Raimundo & Guerra* 680! Susana (Dec.) *Raimundo & Guerra* 268! Nova Lamego, Cabuca (Dec.) *Pereira* 2458! **S.L.:** Mano Salija (Nov.) *Deighton* 279! Njala (Oct.) *Deighton* 1344! Yonibana (Nov.) *Thomas* 4800! Robola, Rokupr (Oct.) *Jordan* 135! Magbema, Rokupr (Dec.) *Jordan* 710! Mussaia (Dec.) *Deighton* 4496! **Lib.:** Banga (Oct.) *Linder* 1177! Webo, Cavally R. (Apr.) *Dinklage* 2619! **Iv.C.:** Angouakoukro to Toumodi (Aug.) *Chev.* 22411! Coula to Nékaougnié (July) *Chev.* 19589! Nékaougnié to Grabo (July) *Chev.* 19597! Azuretti (Sept.) *de Wit* 7597! Yapo (July) *Boughey* 14549! **Ghana:** Nsawam, Cape Coast (Jan.) *Morton* GC 8337! Dunkwa to Pusupusu, Kibbi Dist. (Dec.) *Morton* GC 8204! Abure to Nsawam (Dec.) *Morton* 8057! **S.Nig.:** Sapoba *Kennedy* 2681! 2696! Ilobi, Egbado Dist. (Dec.) *Onochie* FHI 31882! Owo (Oct.) *Stanfield* 2! 3! Owerri (Sept.) *English*! Oban *Talbot* 759! **F.Po:** *Mann* 91 (partly)! Also in E. Cameroun, Gabon, Congo and C. African Republic.

[Variable in indumentum. *C. condensata* represents a variant with the leaf-sheaths clothed outside with long soft spreading hairs.]

21. **C. erecta** *Linn.* Sp. Pl. 41 (1753); Fernald in Rhodora 42: 435–441 (1940); J. K. Morton in J. Linn. Soc. 60: 183 (1967).

21a. **C. erecta** *Linn.* subsp. **erecta.** *C. undulata* R. Br., Prodr. Fl. Nov. Holl. 1: 270 (1810); C. B. Cl. in DC., Monogr. Phan. 3: 179 (1881). *C. umbellata* Thonn., Beskr. Guin. Pl. 21 (1827). *C. sulcata* of Benth. in Fl. Nigrit. 541 (1849),? of Link. *C. gerrardii* C. B. Cl. l.c. 183 (1881); J. K. Morton in J. Linn. Soc. 55: 523, fig. 20 (1956). *C. kurzii* C. B. Cl. l.c. 185 (1881); Rao Rolla & Kammathy in J. Bombay N.H.S. 59: 61–65, t. 1A, D (1962). *C. vogelii* C. B. Cl. l.c. 189 (1881), incl. var. *angustior* C. B. Cl.; F.T.A. 8: 56; Dalz., Usef. Pl. W. Trop. Afr. 465 (1937); Berhaut, Fl. Sén. 209; Saunders, Handb. W. Afr. Fl. 98, fig. 338 (1958). *C. guineensis* Hua in Bull. Mus. Hist. Paris 1: 119 (1895), partly, as to *Paroisse* 195; Brenan in Kew Bull. 15: 211 (1961). A prostrate, decumbent or sometimes weakly erect to erect herb with stems 4 in.–6 ft. long, lanceolate leaves and usually clustered sometimes single spathes with blue flowers opening from about an hour after dawn until nearly noon; plant usually more or less pubescent; a weedy plant of roadsides and waste or cultivated ground, particularly in forest and woodland areas.
**Iv.C.:** Séguéla (Apr.) *Leeuwenberg* 3278! Diahbo to Bouaké (July) *Chev.* 22062! 22064! **U.Volta:** Koudougou to Boromo (June) *Leeuwenberg* 4397! **Ghana:** Nchenenchena to Kwahu Tafo (May) *Kitson* 1149! Kibi (Nov.) *Morton* GC 8013! Shai Hills (Dec.) *Morton* GC 8080! Senya Beraku (Oct.) *Morton* GC 7913! Labadi Lakes (Oct.) *Morton* GC 7839! Teshie (Oct.) *Morton* GC 7845! **S.Nig.:** Benin (Dec.) *Brenan & E. W. Jones* 8634! Port Harcourt (Jan.) *Stubbings* 108! Lagos (Jan.) *Dalz.* 1288! Ibadan (Sept.) *Keay* FHI 37174! Owo (Oct.) *Stanfield* 16! **F.Po:** (Dec.) *T. Vogel* 261! Widespread in the tropics and subtropics of the Old and New Worlds.

[*Berhaut* 897 from Senegal, Tambacounda, cited as *C. sp.* in Berhaut, Fl. Sén. 209, and *Berhaut* 4528 from the same locality have densely pubescent ovate-lanceolate leaves up to about 15 cm. long and 4 cm. broad and small shortly pedunculate non-acuminate spathes, but the seeds resemble those of *C. erecta*, of which they may be a variety, though in appearance the plants are more like *C. bracteosa*. More material and observation needed.]

21b. **C. erecta** subsp. **maritima** (*J. K. Morton*) *J. K. Morton* in J. Linn. Soc. 60: 184 (1967). *C. gerrardii* C. B. Cl. subsp. *maritima* J. K. Morton op. cit. 55: 525, fig. 21 (1956). A fleshy, creeping, almost glabrous herb with blue or purple flowers; in coastal sandbar vegetation under coconut-palms.
**S.L.:** Hamilton, Peninsula (Feb.) *Morton* SL 2361! John Obey (Nov.) *Morton* SL 143! **Lib.:** Monrovia (Aug.) *Baldwin* 13001! **Iv.C.:** Grand Bassam (Mar.) *Leeuwenberg* 3141! **Ghana:** Axim (Nov.) *Boughey* GC 10238! Princestown (Mar., May) *Morton* A329! P1388! GC 6609! GC 8452! **Dah.:** Cotonou (Jan.) *Chev.* 22698!

[Due to introgression, intermediates with subsp. *erecta* occur in Ghana, see Morton l.c. 526. *Morton* GC 8448 from Princestown castle walls and *Morton* A171 from New Ningo are examples.]

21c. **C. erecta** subsp. **livingstonii** (*C. B. Cl.*) *J. K. Morton* in J. Linn. Soc. 60: 184 (1967). *C. livingstonii* C. B. Cl. in DC., Monogr. Phan. 3: 190 (1881); F.T.A. 8: 59; J. K. Morton op. cit. 55: 526, fig. 23 (1956). *C. albescens* Hassk. var. *occidentalis* C. B. Cl. l.c. 185 (1881). *C. albescens* of F.T.A. 8: 58, partly (as to W. Afr. specimens); Aké Assi, Contrib. 2: 223 (1964), partly, as to W. Afr. specimens; not of Hassk. *C. schweinfurthii* of F.T.A. 8: 41, partly, as to *Sc. Elliot* 5164, not of C. B. Cl. *C. sphaerosperma* C. B. Cl. in F.T.A. 8: 58 (1901). *C. umbellata* of Schnell, Vég. Fl. Nimba 522 (1952), not of Thonn. *C. subalbescens* Berhaut, Fl. Sén. 208. A perennial herb up to about 3 ft. high, usually erect, rarely straggling, usually not rooting at nodes. Usually in drier areas than subsp. *erecta*, in savanna and woodland N. of the forest belt.
**Sen.:** Almadies, Ngor (Aug.) *Broadbent* 20! Bargny (Aug., Oct.) *Berhaut* 5442! *Adam* 14919! Niembato, Gambia R. (July) *Berhaut* 929! Medina Gouasse (July) *Adam* 14849! **Gam.:** Kuntaur *Ruxton* 37! 120! **Port.G.:** Bambadinca (July) *Pereira* 3044! **Guin.:** *Paroisse* 195! **S.L.:** York (May) *Deighton* 5527! Mange, Bure Dist. (May) *Deighton* 5941! Sendugu (June) *Thomas* 634! Falaba (Mar.) *Sc. Elliot* 5164! Sulimania (Mar.) *Sc. Elliot* 5385! **Iv.C.:** Nzérékoré (May) *Schnell* 2786! **Ghana:** Weija, Accra (Feb.) *Morton* GC 8417! Tema Harbour (May) *Morton* A889! Shai Hills (Dec.) *Morton* 8084! Damongo Scarp (Mar.) *Morton* GC 8654! Damongo to Kpirri Lake (Mar.) *Adams* 3964! Gambaga Cliff (June) *Harris*! **Dah.:** *Burton*! **N.Nig.:** Munchi, Abinsi *Dalz.* 805! Cheche F.R., Gwari Dist. (Apr.–June) *E. W. Jones* 150! Bindawa (Aug.) *Grove* 11! Jos (June) *Lely* 313! Jos to Bauchi (June) *Onochie* FHI 47714! Katagum Dist. *Dalz.* 230! Widespread in tropical Africa.

[Intermediates with subsp. *erecta* occur which may be hard to identify exactly.]

22. **C. nigritana** *Benth.* in Fl. Nigrit. 541 (1849). *C. umbellata* of C. B. Cl. in DC., Monogr. Phan. 3: 179 (1881); F.T.A. 8: 55; Berhaut, Fl. Sén. 208; J. K. Morton in J. Linn. Soc. 55: 529 (1956), 60: 190 (1967), not of Thonn. *C. gourmaensis* A. Chev. in Mém. Soc. Bot. Fr. 2, 8: 213 (1912), fide F.W.T.A., ed. 1. *C. gourmaca* A. Chev. Bot. 664, fide F.W.T.A., ed. 1.

22a. **C. nigritana** *Benth.* var. **nigritana.** Annual, erect or decumbent herb 4–16 in. high, rather slender, with linear leaves and yellowish to orange flowers open from 9–11 a.m.; in grassland or savanna.

**Sen.:** Kafountine, Oussouye (Sept.) *Adam* 18157! **Guin.:** Kindia (Oct.) *Schnell* 7694! **S.L.:** Falaba (Sept.) *Small* 325! (Nov.) *Morton* SL 2853! Mt. Babaana, Limbaia (Oct.) *Small* 438! **Ghana:** Mampong (Aug.) *Irvine* 5142! Tamale (Oct.) *Baldwin* 13593! Damongo (Dec.) *Morton* GC 9967! Dutukpene, Krachi (Oct.) *Morton* A3725! Vane (Nov.) *Morton* A105! **N.Nig.:** Anara F.R., Zaria (Sept.) *Olorunfemi* FHI 24370! Nupe *Barter* 1473! Igarra (Dec.) *Stanfield* 79! **S.Nig.:** Owo (Oct.) *Stanfield* 1! Enugu F.R., Udi (Sept.) *Onochie* FHI 34118! Jesse, Warri Prov. (Sept.) *Butler-Cole* 16! Also in Uganda, Tanzania, Zambia and Rhodesia.

[Spathes are somewhat variable in size and indumentum. A specimen from Ivory Coast (Diahbo to Bouaké (July) *Chev.* 22063!) is *C. umbellata* but it is uncertain which variety.]

22b. **C. nigritana** var. **gambiae** (*C. B. Cl.*) *Brenan* in Kew Bull. 22: 392 (1968). *C. gambiae* C. B. Cl. in DC., Monogr. Phan. 3: 146 (1881); F.T.A. 8: 38; F.W.T.A., ed. 1, 2: 318; Berhaut, Fl. Sén. 208. *C. umbellata* var. *gambiae* (C.B.Cl.) J. K. Morton l.c. 531, fig. 28 (1956). Herb similar to last and only safely separable by seeds. Intermediates between the two varieties are said to occur frequently in Ghana (Morton in J. Linn. Soc. 55: 531 (1955)).

**Sen.:** Djembering (Sept.) *Broadbent* 130! Ziguinchor (Nov.) *Roberty* 6408! Casamance *Heudelot* 577! Bailá (Sept.) *Adam* 18114! Sedhiou (Oct.) *Adam* 18426! **Gam.:** Ingram! *Hayes* 578! **Guin.:** Pita, Fouta Djalon (Sept.) *Adames* 350! **Port.G.:** Orango Grande, Bijagos (Jan.) *Raimundo & Guerra* 941! Mansoa-Uague (Noy.) *Pereira* 1955! Fá (Sept.–Oct.) *Guerra* 3862! 3908! **S.L.:** Bonthe I. (Nov.) *Deighton* 2286! 2316! Balamuya, Gbinle (Oct.) *Jordan* 793! Kontobe, Malal (Nov.) *Pyne* 182! Mateboi, Sanda Tenraran (Nov.) *Jordan* 824! Gbenge Hills (Nov.) *Morton* A 2911! **Lib.:** Monrovia (Nov.) *Linder* 1523! **Ghana:** Gambaga Scarp (Nov.) *Morton* A2723! Demon to Yendi (Oct.) *Morton* A3773! Banda, Wenchi Dist. (Dec.) *Morton* GC 25290! Ejura (Dec.) *Morton* GC 9829! **N.Nig.:** Zaria F.R. (Aug.) *Keay* FHI 28028! Anara F.R., Zaria (July) *Keay* FHI 25975! Panshanu, Bauchi Prov., 2,800 ft. (Aug.) *Lawlor & Hall* 71! Igarra (Oct.–Dec.) *Stanfield* 29! 80!

23. **C. aspera** *Benth.* in Fl. Nigrit. 542 (1849); C. B. Cl. in DC., Monogr. Phan. 3: 180 (1881); F.T.A. 8: 56; J. K. Morton in J. Linn. Soc. 60: 176 (1967). Annual, erect, simple or branched herb 3–14 in. high with narrow leaves, clusters of small deflexed sharply pointed spathes, and yellowish to white flowers open from about 9–11 a.m.; in grassland and savanna, and sometimes in farmland.

**Port.G.:** Piche (Sept.) *Pereira* 3216! 3229! **S.L.:** Between Kurubonla, Seredu, Sekorella and foot of Loma Mts. (Nov.) *Morton* SL 2551! **Iv.C.:** Namboukaha to Ouorossantiakara (Nov.) *Leeuwenberg* 1991! **Ghana:** Damongo (Dec.) *Morton* A104! Tamale (Aug.) *Irvine* 4564! Yendi to Demon (Oct.) *Morton* A3749! Dutukpene to Kete Krachi (Oct.) *Morton* A3729! Pawpaw Mt., Nkwanta Krache (Oct.) *Morton* A3721! **N.Nig.:** Jos (Aug.) *Keay* FHI 20190! Lokoja *T. Vogel*! Benue R. *Talbot* 762! **S.Nig.:** Owo (May, Oct.) *Stanfield* 76! 156! Also in E. Cameroun, Chad, Congo, Uganda, Tanzania, Zambia, Rhodesia, Angola, and S.W. Africa.

[Broader-leaved variants occur in south tropical Africa but not (so far) in our area.]

*Imperfectly known species.*

1. **C. sp.** A perennial, apparently erect herb about 20 cm. high; leaves linear, 8–17 cm. long, 7–10 mm. broad, more or less spreading-hairy at least on margins; sheaths more or less hairy; spathes spreading-hairy, conspicuously purple-veined, 1·5–4·5 cm. long, 1–1·5 cm. deep, on peduncles 4–12 cm. long; capsule probably 4-seeded.

**Guin.:** Kinkou (June) *Adam* 14677!

[Very close to *C. carsonii* C. B. Cl. from southern tropical Africa, to which it is referred by Morton in J. Linn. Soc. 60: 179 (1967), but more robust; also close to the east African *C. elgonensis* Bullock, which may not be specifically distinct from *C. carsonii*.]

2. **C. pilosissima** *Hutch.*, F.W.T.A., ed. 1, 2: 320 (1936), described only in English. A perennial herb with stems about 50 cm. long and linear leaves up to about 15 cm. long and 1 cm. broad; the whole plant densely hairy, very like *C. erecta* subsp. *livingstonii* of which it may well be only a variant. It also somewhat resembles *C. schweinfurthii* except for being densely hairy and having peduncles c. 5–7 mm. long. It has not been refound.

**Iv.C.:** Sassandra valley, Orodougou, between Sifié and Séguéla (June) *Chev.* 21826!

# 173. FLAGELLARIACEAE

## By F. N. Hepper

Erect or climbing. Leaves sometimes ending in a tendril; leaf-sheath embracing the stem, closed. Flowers hermaphrodite or dioecious, in terminal panicles. Perianth persistent, segments 6, 2-seriately imbricate, dry or somewhat petaloid. Stamens 6; anthers 2-locular, introrse, opening lengthwise by slits. Ovary superior, 3-locular; style 3-lobed. Ovules solitary in each loculus, spreading or pendulous from the central axis. Fruit indehiscent, fleshy or drupaceous. Seeds with copious endosperm.

Tropics and subtropics of Old World.

**FLAGELLARIA** Linn., Sp. Pl. 333 (1753); F.T.A. 8: 90.

Flowers hermaphrodite; perianth subpetaloid, 2-seriate, the outer shorter; stamens 6; ovary 3-celled, with a solitary ovule in each cell; fruit a small berry.

A tall herbaceous climber; branches covered by the encircling leaf-sheaths, the latter deeply split on one side, glabrous, 1·5–3 cm. long; leaves linear-lanceolate, sessile on the sheath, with a slender tendriliform tip, 10–20 cm. long, 1·5–3 cm. broad, many- and closely-nerved, glabrous; flowers numerous, in a terminal panicle 6–10 cm. long; perianth-segments in 2 series, the outer half as long as the inner, subpetaloid; stamens 6, exserted; ovary glabrous, trigonous; fruit subglobose, mucronate, about 6 mm. diam., bright red    ..    ..    ..    ..    ..    .. *guineensis*

**F. guineensis** *Schumach.* in Schum. & Thonn., Beskr. Guin. Pl. 181 (1827); F.T.A. 8: 90; Chev. Bot. 672. A tough herbaceous forest climber, usually near rivers; flowers small, yellow or white; fruits red in dense clusters.
    **Iv.C.**: Bingerville *Chev.* 15206! Bouroukrou (Jan.) *Chev.* 16754! Grand Bassam (fl. Oct., fr. Jan.) *Roberty* 12268! *Leeuwenberg* 2358! Port Bouët (fl. & fr. Apr.) *Aké Assi* 8314! **Ghana**: Cape Coast (July) *T. Vogel* 14! Axim (fl. & fr. Mar.) *Chipp* 393! Assuantsi (fl. & fr. Jan.) *Baldwin* 14029! Amentia, Ashanti *Irvine* 453! **Togo Rep.**: Lomé *Warnecke* 102! **Dah.**: nr. Cotonou (Feb.) *Raynal* 13507! **S.Nig.**: Lagos *Barter* 2172! Ikoyi Plains (Mar.) *Dalz.* 1416! Ogun River F.R. (fr. Mar.) *Hepper* 2258! Majidun Ilaje (Feb.) *Hambler* 405! Calabar *Robb*! In tropical Africa generally, especially in coastal forest, to Natal.

FIG. 333.—FLAGELLARIA GUINEENSIS *Schumach.* (FLAGELLARIACEAE).
A, flower. B, anther. C, pistil. D, fruit. E, cross-section of fruit. F, fruits. G, portion of leaf showing lower surface.

## 174. XYRIDACEAE

### By F. N. Hepper

Perennial or annual herbs. Leaves mostly radical, tufted, linear, terete or filiform, sheathing at the base. Flowers hermaphrodite, slightly zygomorphic, arranged in pedunculate terminal globose or cylindrical heads; bracts imbricate, leathery or rigid, the lower sometimes forming an involucre. Sepals 3 or rarely 2, the lateral 2 exterior, boat-shaped, keeled, glumaceous, the third interior, membranous, forming a hood over the corolla and pushed aside as the latter develops. Corolla with a short or long tube and 3 equal spreading lobes. Stamens 3, opposite the corolla-lobes, and 3 alternate staminodes or the latter absent; anthers 2-locular, opening by slits. Ovary superior, 1-locular, with 3 parietal placentas or imperfectly 3-locular at the base; style simple or 3-lobed. Ovules numerous to few. Fruit a capsule enclosed in the persistent corolla-tube. Seeds numerous, with copious endosperm and small embryo.

Warm regions; in saline and freshwater marshes and seepage areas.

**XYRIS** Linn., Sp. Pl. 42 (1753); F.T.A. 8: 7.

Bracts acutely mucronate, scabrid; lateral sepals with ciliate keel and spinous apex; perennials:

Outer (lower) bracts larger than the inner bracts, up to 2 cm. long, broadly lanceolate, laciniate on margins, median nerve extending as a spine, reflexed on fruiting; inflorescence many-flowered, at first spherical, about 1 cm. diam. extending to nearly 2 cm. in fruit; leaves up to 60 cm. long, 3–6 mm. broad, stiffly acute at apex (Fig. 334/F) .. .. .. .. .. .. .. .. .. 1. *leonensis*

Outer bracts not markedly different from inner ones, nearly orbicular, entire; keel of lateral sepals ciliolate:

Inflorescence spherical, less than 1 cm. diam.; flowers about 1 cm. diam.:

Leaf-bases abruptly broadened, half-width about 1 cm., shining dark brown forming a thick tuft, lamina up to 60 cm. long, 2–3 mm. broad, finely acute at apex; inflorescence nearly 1 cm. diam., not elongating markedly in fruit; bracts 4 mm. diam., with a median keel extending as a short (0·5 mm.) spine    2. *rehmannii*

Leaf-bases narrow, half-width 2–3 mm., pale brown or reddish with hyaline margin forming a light tuft, lamina up to 30 cm. long, 1·5–2 mm. broad, curved, acute; inflorescence 4–7 mm. diam.; bracts 3 mm. diam., outer bracts finely mucronate, inner bracts with a horny apex and more or less mucronate ..    .. 3. *barteri*

Inflorescence elongated (see also below)    ..    ..    ..    .. 4. *hildebrandtii*

Bracts at most acute, usually obtuse or emarginate:

Inflorescences elongated up to 15 mm. long and 7 mm. broad (rather larger in fruit), closely imbricated with numerous closely imbricated bracts; bracts all similar, sometimes minutely mucronate, nearly orbicular, 3 mm. diam., margin entire; keel of lateral sepals ciliolate the whole length; peduncle terete, up to 80 cm. long; leaves 2 mm. broad, in the fresh state flat on one side and convex on the other (Fig. 334/D) ..    ..    ..    ..    ..    ..    ..    .. 4. *hildebrandtii*

Inflorescences more or less spherical, at least not elongated:

Peduncle flattened and 2-edged or almost winged, without ribs between the edges; annuals; (see also note after No. 7):

Leaves and margins of peduncles not rugulose; bracts pale greenish yellow (at length pale brown), suborbicular, with membranous margins and a hardened greenish keel or triangle towards the apex; lateral sepals hyaline, broadly oblanceolate, with a sharp hyaline entire glabrous keel; inflorescence 5–8 mm. long; leaves 2–4 mm. broad and up to 24 cm. long (Fig. 334/C)    ..    ..    ..    .. 5. *anceps*

Leaves and usually margins of peduncles markedly rugulose; bracts reddish brown, occasionally pale brown, suborbicular, entire margins, green patch at apex very small and linear or absent; keel of lateral sepals smooth, entire; inflorescence 3–5 cm. long; leaves 1–2 mm. broad, 1–8 cm. long    ..    ..    .. 6. *rubella*

Peduncle terete (sometimes more or less ridged or angular):

Keel of lateral sepals quite smooth and entire:

Bracts chestnut shining brown, almost opaque, orbicular-obovate, obtuse, scarcely keeled; inflorescences 4–8 (–10) mm. long, (3–)5–8 mm. broad (up to 14 mm. in fruit); peduncle up to 70 cm. long, usually stout 1–2 mm. diam. and terete (sometimes compressed and 1 mm. diam., see note); leaves up to 20 cm. long, 2–4 mm. broad, broadly sheathing at the base .. ..    ..    .. 7. *capensis*

Bracts transparent or translucent and tinged pink or brown, elliptic-oblong, acute, not keeled; inflorescences 3–6 mm. long, 2–3 mm. broad; peduncle up to 30 cm. long, usually about 15 cm., very slender 0·5 mm. or less diam.; leaves 3–8(–12) cm. long, 1–2·5 mm. broad, sometimes minutely transversely rugulose when dry (Fig. 334/A)    ..    ..    ..    ..    ..    ..    .. 8. *straminea*

Keel of lateral sepals ciliate or serrulate:

The keel of lateral sepals ciliate or ciliolate:

Lowland annual; leaves 6–11 cm. long, 2–3 mm. broad, minutely transversely rugulose when dry; peduncles very slender, 10–42 cm. long; inflorescences obovate, 3–6 mm. long, bracts chestnut brown, horny with a linear rough area towards the apex; lateral sepals hyaline, ciliolate on the keel at least towards the apex ..    ..    ..    ..    ..    ..    ..    .. 9. *filiformis*

Highland perennials with bulbous base; inflorescences more or less spherical, 6–7 mm. diam.:

Bracts dark brown, thin with recurved margins, rather prominently keeled, apex usually emarginate; keel of lateral sepals ciliate at least in the lower ⅔; peduncles 17–50 cm. long; leaves 13–32 cm. long, 2–3 mm. broad    10. sp. near *obscura*

Bracts pale brown to straw-coloured, thin with recurved margins, not keeled, apex rounded; keel of lateral sepals long-ciliate; peduncles about 45 cm. long, minutely rugulose; leaves about 30 cm. long and 2 mm. broad .. 11. sp. *A.*

The keel of lateral sepals serrulate:

Leaves (2)3–6 mm. broad, up to 50 cm. long, not shiny at the base; peduncles stout, strongly angular-ribbed, up to 75 cm. long; inflorescences about 1 cm. diam.; bracts broadly obovate-orbicular, about 5 mm. long, brown with an elliptic median green patch, margins slightly jagged (Fig. 334/B) 12. *decipiens*

Leaves 1 mm. broad, 5–35 cm. long, greatly expanded at the base and shiny

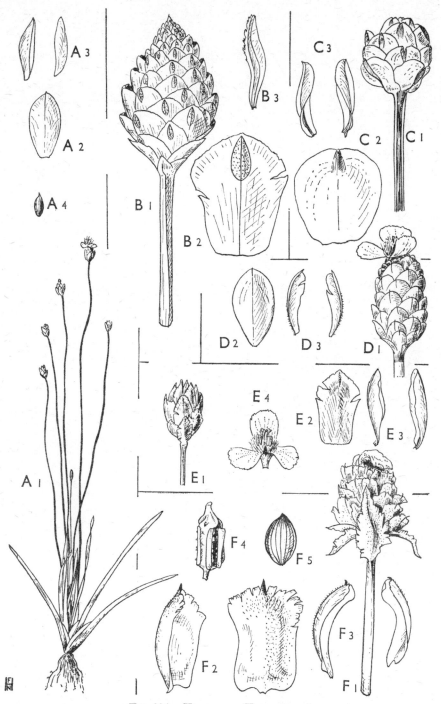

**Fig. 334.—Xyris spp. (Xyridaceae).**

*X. straminea* Nilss.—A1, habit, × ⅔. A2, bract, × 4. A3, lateral sepals, × 4. A4, seed, × 16. *X. decipiens* N.E. Br.—B1, inflorescence, × 2. B2, bract, × 4. B3, lateral sepal, × 4. *X. anceps* Lam.—C1, inflorescence, × 2. C2, bract, × 4. C3, lateral sepals, × 4. *X. hildebrandtii* Nilss.—D1, inflorescence, × 2. D2, bract, × 4. D3, lateral sepals, × 4. *X. festucifolia* Hepper—E1, inflorescence, × 2. E2, bract, × 4. E3, lateral sepals, × 4. E4, flower, × 4. *X. leonensis* Hepper—F1, inflorescence, × 2. F2, bract, × 4. F3, lateral sepals, × 4. F4, dehisced capsule, × 4. F5, seed, × 16. A from *Hepper* 1488. B from *Sc. Elliot* 4220. C from *Sc. Elliot* 3931, *Deighton* 908. D from *Hepper* 1771. E from *Jaeger* 7884. F from *Jones* 90.

brown; peduncles slender, more or less angular; inflorescences 9 mm. long, 3–5 mm. broad; bracts about 5 mm. long, brownish, with small triangular patch towards apex, margins entire becoming broken (Fig. 334/E)    13. *festucifolia*

1. **X. leonensis** *Hepper* in Kew Bull. 21: 422, fig. 2, 1–8 (1968). Stout tufted herb with large aristate outer bracts.
    **Guin.:** Ditinn (Feb.) *des Abbayes* 858! **S.L.:** Bintumane Peak, 5,400 ft. (Dec., Jan.) *T. S. Jones* 90! *Morton* SL 408! **Lib.:** Nimba Reserve (fr. Feb.) *Adam* 20882! **Iv.C.:** Sifié (Aug., Sept.) *Aké Assi* 6173! 7022! Séguéla (Oct.) *Aké Assi* 6180! **Ghana:** Ejura (Dec.) *Morton* GC 9705!

2. **X. rehmannii** *Nilss.* in Kongl. Svensk Vet. Akad. Handl. 24, No. 14: 28 (1892); Hepper in Kew Bull. 21: 424 (1968). *X. dispar* N. E. Br. in F.T.A. 8: 12 (1901). Perennial herb forming large tufts with peduncles 1½–2 ft. high; flowers bright yellow; bracts very dark brown; sometimes abundant in upland wet places.
    **N.Nig.:** Nguroje, Mambila Plateau (Jan.) *Hepper* 1758! **W.Cam.:** Oku to Kumbo (= Banso) (Feb.) *Hepper* 2018! Also in Rhodesia and Transvaal.

3. **X. barteri** *N. E. Br.* in F.T.A. 8: 22 (1901). *X. ledermannii* Malme (1912). Erect perennial herb 1–2 ft. high, the terete peduncles with only one ridge; flowers rather large, yellow, bracts chestnut brown; in permanently wet places.
    **Guin.:** Kala Plateau, Fouta Djalon (Nov.) *des Abbayes* 860! **N.Nig.:** Bida (Jan.) *Meikle* 1014! Nupe *Barter*! Kontagora (Nov.) *Dalz.* 255! Vom, 4,000 ft. (Oct.) *Hepper* 1154!

4. **X. hildebrandtii** *Nilss.* in Öfvers. Vet. Akad. Förhandl. Stockh. 1891: 155; F.T.A. 8: 24; Perrier de la Bathie, Fl. Madag. 35: 10, fig. 2, 8–11 (1946); Hepper in Kew Bull. 21: 421 (1968). A massively tufted perennial herb with peduncles 2½–3½ ft. high; flowers bright yellow, bracts dark brown in an elongated head; dominant in open wet depressions in upland.
    **N.Nig.:** Kakara, Mambila Plateau, 4,900 ft. (fl. & fr. Jan.) *Hepper* 1771! **W.Cam.:** Oku to Kumbo (= Banso), 6,000 ft. *Hepper* 2839! Known in eastern and southern tropical Africa and in Madagascar.

5. **X. anceps** *Lam.* Tabl. Encycl. 1: 132 (1791); F.T.A. 8: 12; Meikle & Baldwin in Am. J. Bot. 39: 50 (1952); Berhaut, Fl. Sén. 184; Hepper l.c. 419 (1968). *X. perroteti* (sic) Steud. (1855). *X. minima* Steud. (1855). Herb with a fan of leaves reddish at the base; inflorescence pale yellowish, 1 ft. or more high; in wet places, rice fields and mangrove swamps.
    **Sen.:** *Perrottet* 840! Kaolak *Berhaut* 642! 2353. Cayor *Leprieur*! Casamance *Heudelot* 562, partly! **Port G.:** Catió (Dec.) *Raimundo & Guerra* 716! **Guin.:** Koba (Nov.) *Jac.–Fél.* 7229! **S.L.:** Aberdeen (Nov.) *Small* 785! 787! *Gledhill* 4! Mahela (Dec.) *Sc. Elliot* 3931! Rokupr (Dec., Apr.) *Deighton* 908! *Hepper* 2631! Pujehun (Dec.) *Deighton* 263! **Lib.:** Monrovia (Nov., Dec.) *Linder* 1474! *Barker* 1471! *Baldwin* 10511! Duport (Nov.) *Linder* 1513! **Iv.C.:** Moussou (Aug.) *de Wilde* 227! **Ghana:** Atwabo *Fishlock* 2a! Beyin (fr. Feb.) *Hall* 2931! **S.Nig.:** Lagos *Barter* 2200! *Millen* 152! *Dalz.* 1278! 1279! 1428! *Baldwin* 14013! Ogoyo to Igbosere (Aug.) *Savory* UCI 271! R. Nyaba, near Enugu (May) *Onochie* FHI 35814! Extending to Angola and Zanzibar and Natal; apparently absent from some intermediate countries.

6. **X. rubella** *Malme* in Engl., Bot. Jahrb. 48: 303 (1912); Hepper l.c. 424 (1968). *X. subrubella* Malme ex Hutch., F.W.T.A., ed. 1, 2: 322 (1936), English descr. only. Small annual herb about 6 in. high with small reddish or pale brownish heads and yellow flowers; in seasonal flushes on rock outcrops and sand-dunes.
    **Sen.:** Basséré, Casamance (Nov.) *Berhaut* 6666! **Mali:** Sikasso (Sept.) *Jaeger* 5161! Bamako (Sept.) *Waterlot* 1346! Loutana (Sept.) *Demange* 2751! **Guin.:** Madina Tossekré (Oct.) *Adam* 12551a! **Ghana:** Gambaga Scarp (Sept.) *Hall* 799b! Nakpanduri to Yendi (Sept.) *Rose Innes* GC 32157! Atwabo *Fishlock* 2 of 1931! **N.Nig.:** Lokoja (Nov.) *Dalz.* 240! Guduma, near Minna (Oct.) *Hepper* 972! Anara F.R., Zaria (Sept., Oct.) *Olorunfemi* FHI 24396! *Hepper* 990! 991! Kogigiri, Jos Plateau (Oct.) *Hepper* 1057! **S.Nig.:** Badagry (Aug.) *Onochie* FHI 33496! Ogoyo to Igbosere (Aug.) *Savory* UCI 271a! Widespread in tropical Africa.

7. **X. capensis** *Thunb.* Prod. 12 (1794); F.T.A. 8: 13. A herb with several fan-like tufts of leaves arising from a perennial rhizome; flower yellow on 1–2 ft. high peduncles; in wet places.
    **Sen.:** *Leprieur*! **Guin.:** Kollangui (Mar.) *Chev.* 12207bis! Dalaba-Diaguissa Plateau, Fouta Djalon (Mar.) *Chev.* 18713! **S.L.:** Rokon (May) *Marmo* 296! **N.Nig.:** Jos Plateau (Jan.) *Lely* 716! Rafin Bauna North F.R., Jos Plateau (Oct.) *Hepper* 1184! Miango, Jos (Dec.) *B. J. Harris*! Vogel Peak, 4,000 ft. (Dec.) *Hepper* 1529! Kakara, Mambila Plateau, 4,900 ft. (Jan.) *Hepper* 1768! **S.Nig.:** Koloishe, Obudu Plateau, 5,500 ft. (Dec.) *Savory & Keay* FHI 25085! **W.Cam.:** Jakiri, Bamenda, 5,800 ft. (Feb.) *Hepper* 1955! 1959! Bafut-Ngemba to Bamenda Nkwe (Mar.) *Daramola* FHI 40534! Bum, Bamenda, 4,000 ft. (May) *Maitland* 1525! Mbenguri to Mankon, 4,000 ft. (Apr.) *Brunt* 1108! Throughout tropical and S. Africa.
    [A variable species without clearly defined limits and several varieties have been described. Var. *medullosa* N. E. Br. in F.T.A. 8: 14 (1901) (syn. var. *microcephala* Malme in Svensk Bot. Tidsk. 6: 558 (1912)) appears to occur in our area (N. Nigeria: Naraguta *Lely* 282—cited as *X. straminea* in Ed. 1). It has pithy peduncles, rather than hollow ones, and smaller leaves and heads. The peduncles are also compressed and 2-ridged: a feature sometimes found in hollow peduncle plants.]

8. **X. straminea** *Nilss.* in Öfvers. Vet. Akad. Förhandl. Stockh. 1891: 153; F.T.A. 8: 19; Meikle & Baldwin in Am. J. Bot. 39: 50 (1952). A slender annual herb a few inches to 1 ft. high; flowers yellow, bracts almost transparent; in seasonal wet flushes on rock outcrops.
    **Mali:** Sicoro (Jan.) *Chev.* 231! Bamako (Nov.) *Waterlot*! **Guin.:** Dalaba, Fouta Djalon (Oct.) *des Abbayes* 677! Madina Tossekré (Oct.) *Adam* 12551b! **Lib.:** Tawato, Boporo Dist. (Nov.) *Baldwin* 10330! Gbau, Sanokwele Dist. (Sept.) *Baldwin* 9445! **Iv.C.:** Singrobo (Dec.) *Aké Assi* 7241! Mankono (June) *Chev.* 21880! **Ghana:** Kwahu Tafo (Aug.) *Hall* CC 137b! Banda, Wenchi (Dec.) *Morton* GC 25098! Aboma F.R., Ashanti (Nov.) *Vigne* FH 3439! Gambaga to Nakpanduri (Mar.) *Hepper & Morton* A3141! **N.Nig.:** Nupe *Barter* 764! Anara F.R., Zaria Dist. (Oct.) *Keay* FHI 20101! Naraguta F.R., Jos Plateau (Oct.) *Hepper* 1012! Vogel Peak, 1,900–3,500 ft. (Nov.) *Hepper* 1245! 1488! **S.Nig.:** Ado Rock, Olokemeji to Iseyin (June) *Hambler* 301! Mt. Orosun, Idanre Hills (Oct.) *Keay* FHI 22592! Okelife, Ondo Dist. (Nov.) *Onochie* FHI 34216! Oba, Owo Dist. (Aug.) *Stanfield* 111! Igarra, NE. of Owo (Dec.) *Stanfield* 70! 71! 72! Widespread in tropical and S. Africa.

9. **X. filiformis** *Lam.* Tabl. Encycl. 1: 132 (1791); F.T.A. 8: 21; Meikle & Baldwin in Am. J. Bot. 39: 50, figs. 28–41 (1952). Slender annual, leaves often reddish at the base, flowers yellow (exceptionally purple), bracts rich brown, horny; in seasonal wet flushes.
    **Sen.:** Basséré, Casamance (Nov.) *Berhaut* 6665! **Port. G.:** Bijagos (Jan.) *Raimundo & Guerra* 934! **Mali:** Sikasso (Oct.) *Demange* 2761! **Guin.:** Dalaba, Fouta Djalon (Nov.) *des Abbayes* 848! **S.L.:** *Smeathman*! Kambia, Magbema (Nov.) *Jordan* 660! Port Loko, Mforki (Oct.) *Jordan* 633! Brookfields, Freetown (Oct.) *Deighton* 2139! Yoni to Ngepe, Bonthe I. (Nov.) *Deighton* 2288! Mafare (Nov.) *Marmo* 209! **Lib.:** Genne-Loffa, Kolahun Dist. (Nov.) *Baldwin* 10087! Paynesville (Apr.) *Barker* 1251! Duport (Nov.) *Bequaert* in Hb. *Linder* 1462! Grand Bassa (Nov.) *Dinklage* 2294 ! Nimba (Oct.) *Adames* 701! **Ghana:** Kwahu Tafo (Aug.) *Hall* CC 137a! **S.Nig.:** Oba (Nov.) *Stanfield* 127!

10. **X. sp. near obscura** *N. E. Br.* in F.T.A. 8: 16 (1901). Perennial herb with fine leaves dying back to a stout fleshy bulbous base; flowers yellow, bracts very dark brown; in highlands.
    **W.Cam.:** Bafut-Ngemba F.R., Bamenda, 6,000 ft. (fl. Sept., Oct., fr. Feb.) *Savory* UCI 481! *Lightbody* FHI 26274! *Ujor* FHI 30203! *Hepper* 2850!

[Typical *X. obscura* has mucronate bracts and *X. makuensis* N. E. Br. has a glabrous keel to the lateral sepals.]

11. **X. sp. A.** Tufted perennial about 1½ ft. high; flowers yellow; in wet ground.
**Guin.:** Labé (July, Nov.) *Chev.* 34450! *Adames* 307!
[This plant shows certain affinities with several species described from Angola, e.g. *X. welwitschii* Rendle, but the strongly perennial habit, pale scape, and round head give it a characteristic appearance. —F.N.H.]

12. **X. decipiens** *N. E. Br.* in Dyer, Fl. Cap. 7: 3 (1897); F.T.A. 8: 22; Chev. Bot. 660, partly; Meikle & Baldwin in Am. J. Bot. 39: 50 (1952); Hepper in Kew Bull. 21: 419 (1968). *X. angularis* N. E. Br. in F.T.A. 8: 22 (1901); F.W.T.A., ed. 1, 2: 322; Chev. Bot. 660. A perennial herb about 2 ft. high in sandy moist places; flowers yellow, rather large, in hard heads at first round and becoming ovoid, cone-like; bracts firm with large grey dorsal patch.
**Sen.:** Basséré, Casamance (Nov.) *Berhaut* 6654! **Port.G.:** Bijagos (Jan.) *Raimundo & Guerra* 927! **Mali:** Sikasso (Oct.) *Demange* 2762! **Guin.:** Conakry (Oct.) *Adam* 12620! Dalaba to Boulivel *Chev.* 12656! Ditinn (Feb.) *des Abbayes* 859! Pita (Feb.) *des Abbayes* 896! Coyah (fr. Feb.) *des Abbayes* 955! Macenta (Oct.) *Baldwin* 9814! **S.L.:** *Smeathman!* Samu country (fr. Dec.) *Sc. Elliot* 4220! Mayumbo (Oct.) *Jordan* 634! Matamba to Manyakoi (Oct.) *Glanville* 29! Mano Salija (Nov.) *Deighton* 344! Bintumane Peak (Jan.) *T. S. Jones* 68! **Lib.:** Tawata (Nov.) *Baldwin* 10332! Duport (Nov.) *Linder* 1457! Paynesville (Oct.) *Voorhoeve* 94! Monrovia (June) *Baldwin* 5847! **Iv.C.:** Adiopodoumé (Dec.) *Aké Assi* 7169! Abouabou (Oct.) *de Wilde* 3167! Toumodi (Aug.) *Chev.* 22398! Moussou (Jan.) *Aké Assi* 6880! **Dah.:** Natitingou to Bocorona (June) *Chev.* 24180! **N.Nig.:** Nupe *Barter!* **S.Nig.:** Igarra, NE. of Owo *Stanfield* 27! Udi F.R., Onitsha Prov. (Nov.) *Keay* FHI 22279! Enugu (Dec.) *Killick* 285! Calabar *Robb!* **W.Cam.:** Basu, Mamfe Dist. (Nov.) *Tamajong* FHI 22106! Also in E. Cameroun, Congo, Angola and Uganda.

13. **X. festucifolia** *Hepper* in Kew Bull. 21: 419, fig. 1, (1968). Perennial herb with very fine leaves from stout congested rhizomatous bases; flowers yellow; in highlands.
**Guin.:** Macenta, 2,000–2,500 ft. (Oct.) *Baldwin* 9823! Nimba Mt. *Jac.-Fél* 1924! *Schnell* 3720! **S.L.:** Bintumane Peak, 5,400–6,390 ft. (Jan.) *T.S. Jones* 80! 98! *Nichols* 5! *Jaeger* 396! 7622! 7884! Tingi Mts. (Dec.) *Morton & Gledhill* SL 3047!

# 175. RAPATEACEAE

## By F. N. Hepper

Perennial herbs with a thick rhizome. Leaves radical, narrow, with parallel lateral nerves. Inflorescence scapose, capitate or unilaterally spicate. Flowers hermaphrodite, actinomorphic. Perianth double, the outer hyaline, lobes chaffy, rigid, imbricate, inner tubular, hyaline, lobes ovate, spreading, broadly imbricate. Stamens 6, inserted in the tube; anthers basifixed, 4-locular, loculi confluent at the top and opening by 1 or 2 pores or by a terminal cleft. Ovary superior, perfectly or imperfectly 3-locular; style simple. Ovules few to solitary, basal or axile. Fruit a capsule, opening by 3 valves septate in the middle. Seeds with copious mealy endosperm.

In West Africa (Sierra Leone, Liberia, Iv.Coast), otherwise entirely South American.

**MASCHALOCEPHALUS** Gilg & K. Schum. in Engl., Bot. Jahrb. 28: 148 (1900); F.T.A. 8: 89.

A tufted herb with fibrous roots; leaves in a basal rosette, all in one plane, elongate linear-lanceolate, narrowed to a subobtuse apex, gradually contracted into the sheath at the base, up to 45 cm. long and 3 cm. broad, with a prominent midrib and several less prominent lateral parallel nerves, with distant faint cross-nerves between; sheath folded on each side of the sheath of the next leaf, winged on the back, about 12 cm. long; flower-head sessile in axils of the leaves; bracts 2, almost concealing the flowers; bracteoles several; perianth-tube slender, about 4 cm. long; lobes 6; stamens 6; anthers 5 mm. long, with a produced connective; capsule triangular
*dinklagei*

**M. dinklagei** *Gilg & K. Schum.* l.c. (1900); F.T.A. 8: 89; Baldwin in Amer. J. Bot. 37: 402 (1950); Mangenot in Ic. Pl. Afr. I.F.A.N. 7: 61 (1965). A tufted sedge-like herb with an inflorescence amongst the leaf-bases; flowers yellow; in wet forested areas in sandy savanna.
**S.L.:** Bagroo R. (Apr.) *Mann* 900! **Lib.:** Grand Bassa (Oct.) *Dinklage* 2088! E. of Monrovia (Aug.) *Baldwin* 13054! Kolobanu to Fayapulu, St. Paul R. (Oct.) *Bequaert!* Gbanga *Linder* 1002c! Kotoma, Sanokwele Dist. *Baldwin* 14062! **Iv.C.:** Yapo Forest (Jan., Oct.) *Aké Assi* 6902! *Leeuwenberg* 3148! *Raynal* 13617!

## Excluded genus

*Langevina* Jac.-Fél. in Bull. Mus. Hist. Nat. Paris, sér. 2, 19: 88 (1947). Although this was described as an additional genus in *Rapateaceae*, its true identity is *Mapania amplivaginata* K. Schum. (*M. oblonga* C.B.Cl.) in *Cyperaceae*.

Fig. 335.—Maschalocephalus dinklagei *Gilg & K. Schum.* (Rapateaceae).
A, flower with bract and bracteoles. B, stamen. C, same, from the back.

# 176. ERIOCAULACEAE

## By R. D. MEIKLE

Annuals or perennials; stems often very short, occasionally elongate; leaves narrow, often crowded or rosulate, frequently with large, conspicuous cells. Inflorescence an involucrate capitulum comprising numerous small, densely crowded flowers often subtended by floral bracts; peduncle leafless, unbranched, usually arising from a well-developed basal sheath; flowers unisexual, the males and females mixed in the same capitulum, or the males in the middle and the females around, or rarely males and females in separate capitula; perianth membranous or scarious, the segments usually in 2 distinct series, the outer segments ("sepals") generally free in female flowers, often connate in male, the inner ("petals") free or rarely connate in female flowers, connate and often very reduced in male flowers. Stamens equal in number to, or twice as many as the inner perianth segments; anthers 2-thecous, or less often 1-thecous, introrse. Ovary 2–3-locular, ovules solitary, pendulous in each loculus; style usually distinct, divided above into 2–3 elongate, simple (or occasionally bifid) stigmas, sometimes with alternating appendages. Fruit a membranous loculicidal capsule; seeds relatively large, testa generally scabrid or papillose; endosperm copious.

A small homogeneous family of 13 genera, widely distributed in the tropics of the Old and New Worlds, with a few outlying representatives in temperate areas of America, Europe (Skye, Coll and western Ireland) and eastern Asia; many of the species are restricted to wet habitats, a few are strictly aquatic.

The subdivision of the family into genera with stamens equal in number to the inner perianth segments (conveniently named "petals") and those with twice as many stamens as inner perianth segments, is generally valid, but subject to exception in some African *Eriocaulon* species; it should not be employed as the sole criterion for separating *Eriocaulon* and *Mesanthemum* from *Paepalanthus* and *Syngonanthus*, though usually a reliable pointer to generic identity.

Stamens 6 or 4, twice as many as the petals, in 2 series; petals often glandular within
   near apex:
  Petals of female flowers free; leaves, peduncles and basal sheaths generally glabrous
                                                     **1. Eriocaulon**
  Petals of female flowers connate except at apex and base; leaves, peduncles and basal
    sheaths generally pubescent . .    . .  . .    . .    . .  **2. Mesanthemum**
Stamens 3, as many as the petals; petals rarely glandular within near apex:
  Petals of female flowers free:
    Leaves, peduncles and basal sheaths glabrous; petals glandular near apex; male
      flowers with much reduced, but distinct, petals (spp. 15 & 25). .  **1. Eriocaulon**
    Leaves, peduncles and basal sheaths pubescent; petals eglandular; male flowers
      without any apparent petals    . .    . .    . .    . .  **3. Paepalanthus**
  Petals of female flowers connate except at apex and base; leaves, peduncles and basal
    sheaths usually pubescent and often glandular  . .    . .    **4. Syngonanthus**

**1. ERIOCAULON** Linn., Sp. Pl. 87 (1753); F.T.A. 8: 231; Ruhland in Engl., Pflanzenr.
              Eriocaulac. 30 (1903).

Capitula sessile; stamens 3 (Fig. 337/27)  . .  . .    . .    . .    25. *sessile*
Capitula pedunculate; stamens 4 or 6, very rarely 3:
  Involucral bracts radiating beyond the head of flowers:
    Sepals of female flowers strongly keeled:
      Floral bracts cuspidate-acuminate; sepals of female flowers 2, not overlapping;
        petals of female flowers 2 or 3; ovary 2–3-locular (Fig. 337/23)    21. *irregulare*
      Floral bracts blunt and rounded at apex; sepals of female flowers 2, overlapping;
        petals of female flowers 2; ovary 2-locular (Fig. 338/24)  . .    22. *remotum*
    Sepals of female flowers not keeled:
      Capitula 5–9 mm. diam.; leaves 2–6 mm. wide at base; receptacle pilose (Fig.
        338/22)  . .    . .    . .    . .    . .    . .    . .    20. *togoënse*
      Capitula 3–5 mm. diam.; leaves 1–2 mm. wide at base; receptacle glabrous (Fig.
        338/25)  . .    . .    . .    . .    . .    . .    . .    23. *pulchellum*

Involucral bracts not radiating beyond the head of flowers:
Ovary 2-locular; male flowers with 4 stamens; female flowers with 2 sepals and
    2 petals (Fig. 338/26)    ..    ..    ..    ..    24. *adamesii*
Ovary 3-locular; male flowers with 6 (or very rarely 3) stamens:
  Female flowers without petals and usually with 2 (rarely without) sepals; receptacle
    glabrous; anthers pallid (Fig. 337/21) ..    ..    ..    19. *cinereum*
  Female flowers with 3 petals:
    Female flowers with 2 sepals; male flowers with 2 free sepals:
      Peduncles 15–35 (–50) cm. long; capitula subglobose, 5–8 mm. diam. (Fig. 337/6)
                        6a. *plumale* subsp. *plumale*
      Peduncles 5–15 (–20) cm. long; capitula 3–5 (–7) mm. diam.:
        Involucral bracts dark brown, rather rigid; capitula subglobose (Fig. 337/7)
                        6b. *plumale* subsp. *jaegeri*
        Involucral bracts pallid or fuscescent, membranous; capitula hemispherical
        (Fig. 337/8)    ..  ..    ..    ..    6c. *plumale* subsp. *kindiae*
    Female flowers with 3 sepals and 3 petals; male flowers with 2–3 free or united
      sepals:
      Leaves and basal sheaths distinctly reddish; peduncles commonly more than
        30 cm. long, much longer than the relatively short (1·5–3·5 cm. long) leaves;
        capitula globose; flowers densely white–papillose (Fig. 336/5)  ..  5. *rufum*
      Leaves and basal sheaths not reddish:
        Leaves large, either exceeding 15 cm. in length or 1 cm. in width at the base:
          Anthers pallid; leaves rather flaccid, 15–40 (–50) cm. long; sepals of female
            flowers not distinctly winged-keeled; capitula rarely viviparous (Fig. 336/4)
                            4. *latifolium*
          Anthers fuscous; leaves not flaccid, rather succulent, broad and obtuse at
            apex; sepals of female flowers strongly winged-keeled; capitula commonly
            viviparous (Fig. 336/1)    ..    ..    ..    1. *zambesiense*
      Leaves generally less than 10 cm. long or 1 cm. wide at base:
        Male flowers with well-developed plumose, glanduliferous petal-lobes; capitula
        generally more than 5 mm. diam., flowers conspicuously white-papillose:
          Leaves flaccid, long-acuminate, usually exceeding basal sheaths; capitula
            5–8 mm. diam., peduncles slender, about 5-sulcate (Fig. 336/3)
                            3. *mamfeënse*
          Leaves rigid, excavate at apex, much shorter than basal sheaths; capitula
            about 8 mm. diam., peduncles stout, 7–8-sulcate (Fig. 337/2)  2. *intrusum*
        Male flowers with minute (generally eglandular) petal-lobes; flowers shortly
        and sparsely papillose or glabrous:
          Floral bracts acuminate-cuspidate, exceeding the flowers and giving the
          capitula an echinate appearance:
            Stamens 3, anthers pallid; petals linear; capitula subglobose, pale,
            stramineous (Fig. 338/17) ..    ..    ..    ..    15. *jordanii*
            Stamens 6, anthers fuscous; petals oblanceolate; capitula hemispherical,
            pallid or fuscescent towards centre (Fig. 337/16)    ..  ..14. *meiklei*
          Floral bracts not acuminate-cuspidate; capitula not echinate in appearance:
            Stem elongate; leaves filiform, scattered, not in a rosette; capitula blackish;
            aquatic plant (Fig. 337/9) ..    ..    ..    ..    7. *setaceum*
            Stem very short or almost wanting; leaves in a dense tuft or rosette; plants
            generally terrestrial:
              Involucral bracts minute or apparently wanting; capitula small, globose,
              black or more or less densely white-papillose (Fig. 336/14)
                            12. *elegantulum*
             Involucral bracts well developed and easily visible, especially in the young
             capitula:
              All the sepals of the female flowers concave, but without a spongy or winged
              keel, not markedly unequal:
                Sepals of male flowers 3, free to base; sepals and petals of female flowers
                linear or filiform (Fig. 338/18)..    ..    ..    16. *bongense*
                Sepals of male flowers connate into a spathe:
                  Receptacle glabrous or very thinly pilose; leaves subulate with a long
                  slender acumen; petals of female flowers narrowly linear (Fig. 338/19)
                              17. *abyssinicum*
                Receptacle pilose; leaves linear or strap-shaped, acute or shortly
                  acuminate; petals of female flowers narrowly oblanceolate:
                    Involucral bracts erect, appressed to capitulum, rather rigid; capitula
                    greyish, shortly papillose; floral bracts rather regularly imbricated
                    (Fig. 336/10) ..    ..    ..    ..    ..    8. *afzelianum*
                  Involucral bracts spreading or reflexed; capitula blackish, papillose;
                  floral bracts not regularly imbricated (Fig. 336/11)
                    9a. *transvaalicum* var. *hanningtonii*

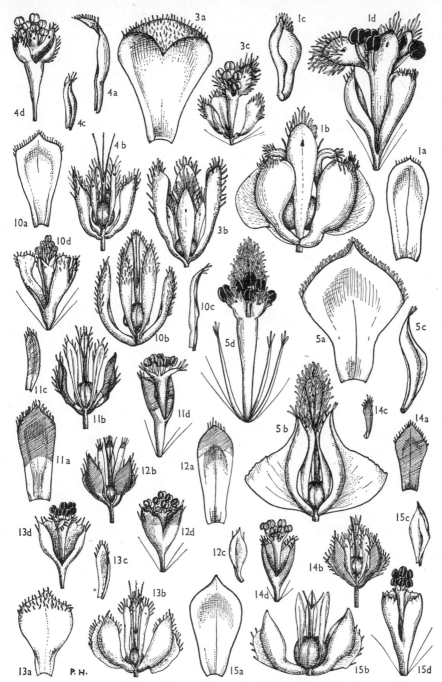

FIG. 336.—ERIOCAULON SPP. (ERIOCAULACEAE).

1, *E. zambesiense* Ruhl—a, floral bract. b, female flower. c, 3rd sepal of female flower. d, male flower. All approx. × 13. 3, *E. mamfeënse* Meikle—a, floral bract, × 13. b, female flower, ×13. c, male flower, approx. × 6. 4, *E. latifolium* Sm.—a, floral bract. b, female flower. c, 3rd sepal from female flower. d, male flower. All approx. × 13. 5, *E. rufum* Lecomte—a, floral bract. b, female flower. c, 3rd sepal from female flower. d, male flower. All approx. × 13. 10, *E. afzelianum* Wikstr. ex Koern.—a, floral bract. b, female flower. c, 3rd sepal from female flower. d, male flower. All approx. × 13. 11, *E. transvaalicum var. hanningtonii* (N.E. Br.) Meikle—a, floral bract. b, female flower. c, 3rd sepal from female flower. d, male flower. All approx. × 13. 12, *E. nigericum* Meikle—a, floral bract. b, female flower. c, 3rd sepal from female flower. d, male flower. All approx. × 13. 13, *E. inundatum* Moldenke—a, floral bract. b, male flower. c, 3rd sepal from male flower. d, female flower. All approx. × 13. 14, *E. elegantulum* Engl.—a, floral bract. b, female flower. c, 3rd sepal from female flower. d, male flower. All approx. × 13. 15, *E. deightonii* Meikle—a, floral bract. b, female flower. c, 3rd sepal from female flower. d, male flower. All approx. × 13. 1, from *Maitland* 1400. 3, from *FHI* 22107. 4, from *Jordan* 427. 5, from *Arrieu* 230 in *Hb. Chillou* 3139. 10, from *Raynal* 6795. 12, from *Jones FHI* 20718. 13, from *Monod* s.n. (28 Oct. 1943). 14, from *Greenway* 6613. 15, from *Jordan* 946.

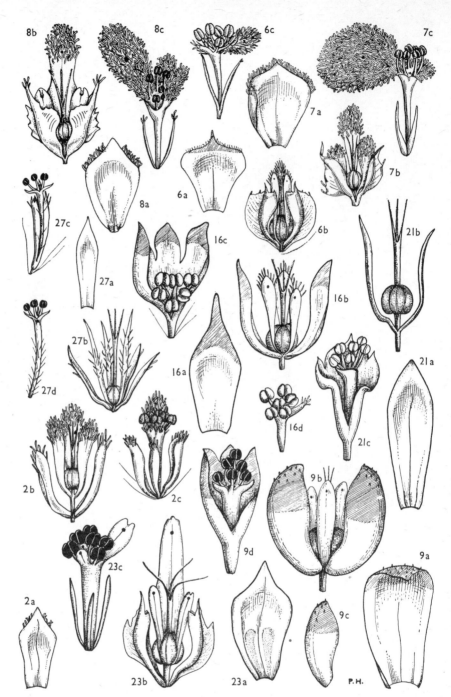

FIG. 337.—ERIOCAULON SPP. (ERIOCAULACEAE).

2. *E. intrusum* Meikle—a, floral bract. b, female flower. c, male flower. All approx. × 6.
6, *E. plumale* N.E. Br. subsp. *plumale*— a, floral bract. b, female flower. c, male flower.
All approx. × 6. 7, *E. plumale* subsp. *jaegeri* (Moldenke) Meikle— a, floral bract. b, female
flower. c, male flower. All approx. × 13. 8, *E. plumale* subsp. *kindiae* (Lecomte) Meikle—
a, floral bract. b, female flower. c, male flower. All approx. × 13. 9, *E. setaceum* Linn.— a,
floral bract. b, female flower. c, third sepal from female flower. d, male flower. All approx.
× 26. 16, *E. meiklei* Moldenke—a, floral bract, approx. × 13. b, female flower, approx. × 26.
c, male flower, approx. × 26. d, male flower with sepals removed, approx. × 26. 21,
*E. cinereum* R. Br.—a, floral bract. b, female flower. c, male flower. All approx. × 26.
23, *E. irregulare* Meikle—a, floral bract. b, female flower. c, male flower. All × 20. 27,
*E. sessile* Meikle—a, floral bract. b, female flower. c, male flower. d, male flower with sepals
removed. All approx. × 6. 2, from *Lely* 283. 6, from *Jordan* 658. 7, from *Chillou* 906. 8,
from *Boismaré* 385 in *Hb. Chillou* 3903. 9, from *Barter* 1021. 16, from *Meikle* 1043. 21, from
*de Wailly* 5002. 23, from *Adames* 353. 27, from *Chev.* 5841 bis.

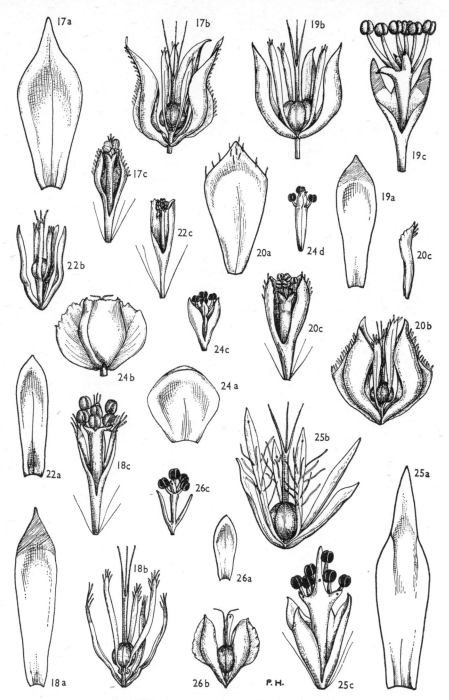

FIG. 338.—ERIOCAULON SPP. (ERIOCAULACEAE).

17, *E. jordanii* (Moldenke) Meikle—a, floral bract. b, female flower. c, male flower. All approx. × 13. 18, *E. bongense* Engl. & Ruhland ex Ruhland—a, floral bract. b, female flower. c, male flower. All approx. × 13. 19, *E. abyssinicum* Hochst.—a, floral bract. b, female flower. c, male flower. All approx. × 26. 20, *E. fulvum* N.E. Br.—a, floral bract. b, female flower. c, 3rd sepal from female flower. d, male flower. All approx. × 13. 22, *E. togoense* Moldenke—a, floral bract. b, female flower. c, male flower. All approx. × 13. 24, *E. remotum* Lecomte—a, floral bract. b, female flower. c, male flower. d, male flower with sepals removed. All approx. × 13. 25, *E. pulchellum* Koern.—a, floral bract. b, female flower. c, male flower. All approx. × 26. 26, *E. adamesii* Meikle—a, floral bract. b, female flower. c, male flower. All approx. × 13. 17, from *Jordan* 1051. 18, from *Meikle* 1015. 19, from *Lely* P786. 20, from *Roberty* 13305. 22, from *Raynal* 5306 quinto. 24, from *Pitot*. 25, from *Boismaré* 408 in *Hb. Chillou* 3926. 26, from *Adames* 97.

Two of the larger sepals (or all the sepals) of the female flowers with a
distinct spongy or winged keel, the third sepal usually much smaller
than the others:

Leaves subulate, with a long slender acumen, usually very numerous;
capitula greyish-fuscescent; involucral and floral bracts concolorous
(Fig. 336/15)   ..    ..    ..    ..    ..    ..    13. *deightonii*

Leaves linear, oblong or strap-shaped, obtuse, acute or shortly acuminate,
usually fewer than 12 in each rosette:

Floral bracts rounded or blunt at apex, regularly imbricated, fulvous or
bronze-tinged, shining; sepals of female flowers often with dentate
keels (Fig. 338/20)   ..    ..    ..    ..    ..    .. 18. *fulvum*

Floral bracts acute or subacute; not regularly imbricated nor shining;
sepals of female flowers usually with entire or subentire keels:

Involucral bracts usually conspicuous, larger and paler than floral
bracts; capitula hemispherical; petals of female flowers eglandu-
liferous (Fig. 336/12)..    ..    ..    ..    ..    10. *nigericum*

Involucral bracts not conspicuous, concolorous with floral bracts;
capitula globose; petals of female flowers glanduliferous (Fig. 336/13)
11. *inundatum*

1. **E. zambesiense** *Ruhland* in Engl., Bot. Jahrb. 27: 75 (1899); F.T.A. 8: 252. Peduncles up to 2 ft. 8 in.
high; capitula 5–8 mm. diam. covered with whitish papillae, generally viviparous.
  **W.Cam.:** Pinyin, 5,500 ft. (Apr.) *Brunt* 1092! Kumbo to Oku, Bamenda, 6,000 ft. (Feb.) *Hepper* 2021!
  Lakom, Bamenda, 6,000 ft. (June) *Maitland* 1400! **F.Po:** (Dec.) *Monod* 10358!
  [The specimen from F.Po closely resembles typical *E. zambesiense*, but those from W.Cam. are alto-
  gether more robust, with very broad and blunt leaves, up to 1.5 cm. wide at the base.]
2. **E. intrusum** *Meikle* in Kew Bull. 22: 141 (1968). *E. lacteum* of F.W.T.A., ed. 1, 2: 327, not of Rendle.
  Each plant generally with only one scape; peduncle up to 16 in. long; capitulum white-papillose with
  conspicuous pale brown, blunt involucral bracts.
  **N.Nig.:** Naraguta, Plateau Prov. (June) *Lely* 283!
3. **E. mamfeënse** *Meikle* l.c. 141 (1968). Each plant with several scapes; peduncle up to 1 ft. long, shining;
  capitula subglobose, densely white-papillose.
  **W.Cam.:** Mamfe (Nov., Mar.) *Morton* K676! *Migeod* 276! *Tamajong* FHI 22107! *Richards* 5245!
4. **E. latifolium** *Sm.* in Rees, Cyclop. 13 (1819); F.T.A. 8: 243; Ruhland in Engl., Pflanzenr. 4, 30: 78 (1903).
  *E. rivulare* G. Don in Hook., Fl. Nigrit. 547 (1849). *E. thunbergii* Wickstr. ex Koern. in Linnaea 27: 677
  (1856). *E. banani* Lecomte in Bull. Soc. Bot. Fr. 55: 645 (1909). *E. latifolium* Sm. form a *proliferum*
  Moldenke in Phytologia 8: 387 (1962). *Mesanthemum radicans* Stapf in Johnston, Liberia 2: 662 (1906),
  not of (Benth.) Koern. Robust aquatic perennial; capitula densely white-papillose, to 14 mm. diam.
  **Mali:** Banankoro (Mar.) *Chev.* 524! Sikasso (May) *Chev.* 803! **Port G.:** Boé, Madina (Feb.) *Pereira* 2999!
  **Guin.:** Conakry (Jan.) *Dalz.* 8247! Dalaba (Oct., Nov., Mar.) *dés Abbayes* 699! *Adames* 416! Langdale-
  Brown 2638! **S.L.:** *Afzelius*! York (Apr.) *Adames* 206! Tawia *Deighton* 1031! Gbangabatok (Jan.)
  *Capstick* in *Hb. Deighton* 5301! Magbema (Mar.) *Jordan* 427! Farangbaya *Crisp* A!! Bandakafaia (Oct.)
  *Jaeger* 8096! **Lib.:** Mouth of Loffa R., Grand Cape Mount (Dec.) *Baldwin* 10946! Genne Tanyehun,
  Grand Cape Mount (Dec.) *Baldwin* 10091! Kakatown *Whyte*! Nimba (Dec.) *Adames* 808! Also in Congo
  and Angola.
5. **E. rufum** *Lecomte* in Bull. Soc. Bot. Fr. 55: 644 (1909). *E. plumale* of F.W.T.A., ed. 1, 2: 327 partly,
  not of N.E. Br. Strictly erect; scapes usually numerous; capitula globose to 10 mm. diam., florets
  generally concealing involucral bracts in fully developed capitula.
  **Guin.:** Conakry *Maclaud*! Ouassou (Nov.) *Jac.-Fél.* 7256! Kindia (Oct.) *Pobéguin* 1312! Bena, Fouta
  Djalon (Feb.) *Arrieu* 230 in *Hb. Chillou* 3139! **S.L.:** Kasawe F.R., Yoni Mamila Chiefdom (Dec.) *King*
  55b!
6. **E. plumale** *N.E. Br.* in F.T.A. 8: 251 (1901); Ruhland in Engl., Pflanzenr. 4, 30: 106 (1903); F.W.T.A.,
  ed. 1, 2: 327 partly. *E. senegalense* N.E. Br. in F.T.A. 8: 251 (1901).
6a. **E. plumale** N.E. Br. subsp. **plumale.** Each plant with numerous leaves and scapes; leaves narrow,
  linear-subulate; capitula subglobose, white, plumose; involucral bracts pale shining brown.
  **Sen.:** Badi (Dec.) *Berhaut* 1121! Bala (Nov.) *Berhaut* 1258! **Guin.:** Rio Nunez *Heudelot* 680! **S.L.:**
  Wellington (Nov.) *Deighton* 1867! Port Loko (Oct.) *Jordan* 632! Kambia (Oct.) *Jordan* 658! 944!
  **Iv.C.:** Cissédougou, Boundiali to Mankono (Nov.) *Aké Assi* 8294!
6b. **E. plumale** subsp. **jaegeri** (*Moldenke*) *Meikle* in Kew Bull. 22: 142 (1968). *E. jaegeri* Moldenke in Phytologia 5: 338 (1956). *Mesanthemum necopinatum* Moldenke l.c. 8: 389 (1962). Scapes and leaves numerous;
  leaves subulate-filiform; capitula white, plumose, 5–7 mm. diam., subglobose; involucral bracts rather
  rigid, dark brown.
  **Guin.:** Friguiagbé *Chillou* 906! Fouta Djalon (Sept., Oct.) *Adames* 342! *Schnell* 7393! Loursa Mt.,
  Dabola (Sept.) *Jaeger* 4917!
6c. **E. plumale** subsp. **kindiae** (*Lecomte*) *Meikle* l.c. 142 (1968). *E. kindiae* Lecomte in Bull. Soc. Bot. Fr. 55:
  646 (1909). *E. pumilum* of F.W.T.A., ed. 1, 2: 326 partly, not of Afzel. ex Koern. *Mesanthemum*
  *chilloui* Moldenke in Phytologia 8: 389 (1962). Scapes and leaves numerous; leaves subulate-filiform;
  capitula hemispherical, 3–5 mm. diam., white, plumose, said to be fragrant; involucral bracts pale
  brown or greenish, not rigid.
  **Guin.:** Friguiagbé (Sept.) *Chillou* 716! 717! *Boismaré* 385 in *Hb. Chillou* 3903! Kindia *Pobéguin* 1359
  bis! **S.L.:** Waterloo (Aug.) *Melville & Hooker* 276! Kambia (Aug., Sept.) *Jordan* 303! *Adames* in *Hb.*
  *Jordan* 554! *Harvey* 33! Also in Chad.
7. **E. setaceum** *Linn.* Sp. Pl. 87 (1753). *E. melanocephalum* Kunth, Enum. Pl. 3: 549 (1841). *E. bifistulosum*
  van Heurck & Muell. Arg. in van Heurck, Obs. Bot. 105 (1870); F.T.A. 8: 239 (1901); Ruhland in
  Engl., Pflanzenr. 4, 30: 90 (1903); F.W.T.A., ed. 1, 2: 326. *E. limosum* Engl. & Ruhl. in Engl., Bot.
  Jahrb., 27: 74 (1899). Floating aquatic; stems elongate, clothed with numerous filiform leaves; scapes
  generally numerous; capitula blackish, small, seldom exceeding 4 mm. diam.
  **Sen.:** Oussouye (Nov.) *Berhaut* 6502! **Mali:** Macina (Nov.) *Monod*! Sarekoye (Feb.) *Davey* 22! **Port.G.:**
  Saucunda to Buba (Oct.) *Esp. Santo* 2195! **Guin.:** Boffa (Dec.) *Jac.-Fél.* 7361! Fetoré Plain, Fouta Djalon
  (Sept.) *Adames* 374! Mali (Sept.) *Schnell* 7056! **S.L.:** Kasanko (Oct.) *Jordan* 815! Malako (Nov.) *Marmo*
  226! Njala (Sept.) *Deighton* 2794! Bonthe (Oct.) *Adames* 94! **Lib.:** Nimba (Sept.) *Adames* 566! **Ghana:**
  Tamale to Bolgatanga (Aug.) *Hall* CC 431! **N.Nig.:** Nupe *Barter* 1021! Amo, Jos Plateau (Oct.) *Hepper*
  1030! Panshanu Pass (Aug.) *Lawlor & Hall* 441! Jalo, Gashaka Dist. (Dec.) *Latilo & Daramola* FHI
  28884! Widespread in tropical Africa; also in tropical Asia, Australia and America.
8. **E. afzelianum** *Wikstr. ex Koern.* in Linnaea 27: 680 (1856); F.T.A. 8: 250; Ruhland in Engl., Pflanzenr.
  4, 30: 83 (1903). *E. kouroussense* Lecomte in Bull. Soc. Bot. Fr. 55: 644 (1909). *E. heterochiton* Lecomte

l.c. 647 (1909), partly, not of Koern. Scapes erect, few to numerous, peduncles shining, pale brown, capitula depressed-globose, to about 5 mm. diam., involucral bracts pale brownish.
**Sen.:** Oussouye (Nov.) *Berhaut* 6633! Bignona (Dec.) *Berhaut* 6769! Niokolo-Koba, Tambacounda (Dec.) *Raynal* 6795! **Guin.:** Pobéguin 615! Iles Tristao (Nov.) *Jac.-Fél.* 7327! Dalaba to Diaguissa (Oct.) *Chev.* 18810 *bis*! Kala, nr. Dalaba (Oct.) *Schnell* 7448! **S.L.:** *Afzelius*! Peninsula coast (Dec.) *Dawe* 417! Waterloo (Aug.) *Melville & Hooker* 283! Mapotolon (Sept.) *Adames* 246! Roboli, nr. Rokupr (Nov.) *Jordan* 163! **Iv.C.:** Séguéla to Mankono (Nov.) *de Wilde* 937! **Ghana:** Ejura (Dec.) *Morton* GC 9823b! Kintampo (Sept.) *Hall* CC 905! Kamba marsh, Lawra Dist. (Nov.) *Harris*! Navrongo (Oct.) *Vigne* FH 4607! **N.Nig.:** Nupe *Barter* 1019! Gurum, Vogel Peak (Nov.) *Hepper* 1236! **S.Nig.:** Igarra, NE. of Owo (Dec.) *Stanfield* 56! Also in Chad.

9. **E. transvaalicum** *N.E. Br.* in Fl. Cap. 7: 54 (1897); Ruhland in Engl., Pflanzenr. 4, 30: 81 (1903).

9a. **E. transvaalicum var. hanningtonii** (*N.E. Br.*) *Meikle* in Kew Bull. 22: 142 (1968). *E. hanningtonii* N.E. Br. in F.T.A. 8: 253 (1901); Ruhland l.c. 74. *E. monodii* Moldenke in Phytologia 3: 165 (1949), without Latin descr. Scapes numerous, capitula subglobose, about 4 mm. diam., involucral bracts pale brown.
**Mali:** Koulouba to Bamako (Dec.) *Monod*! Also in Tanzania and Mozambique; typical *E. transvaalicum* is widespread in tropical Africa outside our area.

10. **E. nigericum** *Meikle* in Kew Bull. 5: 231 (1950). Scapes numerous, usually less than 4 in. high; leaves often rather blunt; capitula 4–5 mm. diam., involucral bracts conspicuous, pale, shining.
**Sen.:** Niokolo-Koba, Tambacounda (Dec.) *Raynal* 6879! **S.L.:** Sefadu (Sept.) *Jordan* 522! **Lib.:** Tawata, Boporo Dist. (Nov.) *Baldwin* 10336! Genne Loffa, Kolahun Dist. (Nov.) *Baldwin* 10088! Palilah, Gbanga Dist. (Aug.) *Baldwin* 9145! Gbau, Sanokwele Dist. (Sept.) *Baldwin* 9456! **S.Nig.:** Shaki, Oyo Dist. (Nov.) *Stanfield* 189! Shabe Rock, Oyo Dist. (Nov.) *Hambler* 739! Akure, Owo Dist. (Oct.) *Stanfield* 140! Akure F. R. (Oct.) *Keay* FHI 25471! Orosun Peak, Idanre Hills (Oct.) *Keay* FHI 22595! Idanre, Ondo Prov. (Aug.) *Jones* FHI 20718!

11. **E. inundatum** *Moldenke* in Phytologia 3: 413 (1951). Scapes erect, about 4 in. high; leaves acuminate; capitula globose, pale brown, about 4 mm. diam.
**Sen.:** Fremarin (Oct.) *Monod*!

12. **E. elegantulum** *Engl.* Pflanzenr. Ost.-Afr. C: 133 (1895); Ruhland in Engl., Bot. Jahrb. 27: 83 (1899); F.T.A. 8: 254. Scapes generally numerous, erect; capitula quite globose, about 4 mm. diam., white-papillose, without any visible involucral bracts.
**Ghana:** Lumbaga (Dec.) *Morton* GC 6250b! Nungua (July) *Hall* 3729! **S. Nig.:** Badagry (Aug.). Onochie (FHI)! Also in Sudan, Tanzania, Mozambique, Rhodesia.

13. **E. deightonii** *Meikle* in Kew Bull. 22: 143 (1968). Scapes very numerous; leaves numerous, narrow, subulate; capitula subglobose, about 4 mm. diam., involucral bracts inconspicuous.
**Guin.:** Friguiagbé (Oct.) *Chillou* 726! Pita (Oct.) *Pitot*! **S.L.:** Mapotolon (Sept.) *Adames* 245! Rokon (Oct.) *Jordan* 574! Manye (Sept.) *Jordan* 533! Njala (Sept.) *Deighton* 2795! Bonthe (Oct.) *Adames* 88!

14. **E. meiklei** *Moldenke* in Phytologia 3: 164 (1949). Scapes usually numerous, less than 6 in. high; leaves rather thick and opaque, acuminate; capitula hemispherical, pallid or fuscescent, about 4–5 mm. diam., distinctly echinate with protruding bracts.
**Sen.:** Oussouye (Feb.) *Berhaut* 6983! Bignona (Dec.) *Berhaut* 6739 in *Hb. Bambey*! Bissine (Dec.) *Berhaut* 6691! **Mali:** Valley of Balasoko, Bamako to Koulouba (Dec.) *Monod*! Sicoro (Jan.) *Chev.* 218! **Guin.:** Dabola to Faranah (Apr.) *Pitot*! **N.Nig.:** Auna to Kontagora (Jan.) *Meikle* 1043! Vogel Peak, Sardauna Prov. (Nov.) *Hepper* 1450!

15. **E. jordanii** (*Moldenke*) *Meikle* in Kew Bull. 22: 143 (1968). *Syngonanthus jordanii* Moldenke in Phytologia 5: 91 (1954). Scapes numerous, about 6 in. high; capitula subglobose, 5–6 mm. diam., glossy, stramineous, conspicuously echinate with protruding bracts.
**S.L.:** Kontabana (Jan.) *Jordan* 1051! Mange (Dec.) *Jordan* 721!

16. **E. bongense** *Engl. & Ruhland ex Ruhland* in Engl., Bot. Jahrb. 27: 75 (1899); Ruhland in Engl., Pflanzenr. 4, 30: 100 (1903); F.T.A. 8: 246. Plants robust, sometimes 1 ft. high; leaves rather few; capitula depressed-globose or sometimes subconical when fully mature, 5–8 mm. diam.; involucral bracts stramineous, flowers and floral bracts shining silvery-grey.
**Sen.:** Niassia (Nov.) *Berhaut* 6662! **Gam.:** Kombo *Dawe* 23! **Ghana:** Burufo (Dec.) *Adams & Akpabla* 4398! 4399! Kintampo (Sept.) *Hall* CC 905a! Tamale (Dec.) *Adams & Akpabla* 4155! Bimbila (Dec.) *Morton* GC 6272! **N.Nig.:** Nupe *Barter* 1019a! Lokoja (Nov.) *Dalz.* 239! Bida to Zungeru (Jan.) *Meikle* 1015! **S.Nig.:** Olokemeji to Eruwa (Nov.) *Daramola* FHI 43846! Owo (Oct., Nov.) *Stanfield* 132! 133! Igarra (Dec.) *Stanfield* 55! Also in C. African Rep. and Sudan.

17. **E. abyssinicum** *Hochst.* in Flora 28: 341 (1845); F.T.A. 8: 257; Ruhland in Engl., Pflanzenr. 4, 30: 282 (1903). Small and inconspicuous; scapes few, 1½–3½ in. high; leaves narrowly subulate; capitula greyish-fuscous, subglobose, 2–3 mm. diam.
**N.Nig.:** Jos Plateau (Oct.) *Lely* P786! Sara Hills, Jos Plateau (Oct.) *Hepper* 1126! Also in Ethiopia, Kenya, Uganda, Tanzania, Rhodesia and S. Africa.

18. **E. fulvum** *N.E. Br.* in F.T.A. 8: 248 (1901); Ruhland in Engl., Pflanzenr. 4, 30: 101 (1903). Scapes usually numerous, rather rigidly erect, to about 5 in. high; leaves few, often conspicuously short; capitula subglobose, 2·5–4 mm. diam.
**Sen.:** Badi (Dec.) *Berhaut* 1175! **Mali:** Sotuba (Dec.) *Raynal* 5202 *bis*! Bougouni to Sikasso (Dec.) *Roberty* 13305! **Guin.:** Dinguiraye (Dec.) *Roberty* 16336a! **Ghana:** Lungbunga to Tamale (Dec.) *Morton* GC 6248! Sakogu (Aug.) *Hall* CC 454! Foot of Gambaga Scarp (Sept.) *Hall* 747! **N.Nig.:** Nupe *Barter*! Fuka, nr. Minna, Niger Prov. (Oct.) *Hepper* 980! Anara F.R., Zaria Prov. (Oct.) *Hepper* 985! Kogigiri, Jos Plateau (Oct.) *Hepper* 1060! Gurum, Vogel Peak area (Nov.) *Hepper* 1236a! 1237! **S.Nig.:** Oba, Owo Dist. (Oct., Nov.) *Stanfield* 36! 143! 144! Igarra, Benin Prov. (Dec.) *Stanfield* 54!
[Closely akin to *E. maculatum* Schinz (S. Africa) and *E. strictum* Milne-Redhead (Tanzania).]

19. **E. cinereum** *R. Br.* Prodr. 254 (1810). *E. sieboldianum* Sieb. & Zucc. ex Steud., Syn. Pl. Cyp. 2: 272 (1855); Ruhland in Engl., Pflanzenr. 4, 30: 111 (1903). *E. heudelotii* N.E. Br. in F.T.A. 8: 258 (1901); F.W.T.A., ed. 1, 2: 326. Scapes slender, usually less than 4 in. high; leaves numerous, setaceous; capitula small, about 2·4–3 mm. diam., brownish or fuscescent.
**Sen.:** *Heudelot* 677! **Mali:** Kita Mts. (Nov.) *Jaeger* 5581! Gao to Berra (Mar., May) *de Wailly* 5002! 5006a! **Guin.:** Port. G. frontier *Maclaud* 03.9.105! Seriba (Oct.) *Pitot*! **S.L.:** Kambia (Oct.) *Jordan* 943! **Ghana:** Burufo, nr. Lawra (Sept., Oct.) *Hall* CC 742! *Hinds* 5005! Busuno, Damongo to Yapei (Sept.) *Hall* CC 885! Also in Chad, Tanzania, tropical Asia, China, Japan and Australia.

20. **E. togoënse** *Moldenke* in Known Distrib. Eriocaul. 42 (1946). *E. xeranthemoides* Van Heurck & Muell. Arg. in Van Heurck, Obs. Bot. 105 (1870); F.T.A. 8: 237; Ruhland in Engl., Pflanzenr. 4, 30: 96 (1903); F.W.T.A., ed. 1, 2: 326, not of Bong. Neat tufted annual, usually less than 2½ in. high; scapes numerous; capitula 5–8 mm. diam. with conspicuous whitish involucral bracts.
**Mali:** Sotuba (Dec.) *Raynal* 5306 *quinto*! **Iv.C.:** Bemouni (Jan.) *de Wit* 7901 sub. no. 545! Singrobo (Dec.) *Aké Assi* 7243! **Ghana:** Burufo (Dec.) *Adames* 4400! Ejura Scarp (Dec.) *Morton* GC 9582! Kwahu Tafo (Aug.) *Hall* CC 92! **Togo Rep.:** Sokode (Nov.) *Schroeder* 155! 162! **N.Nig.:** Fakun, Borgu *Barter* 778! Kabba (Oct.) *Parsons* L102! Dakwakuku, Bonu F.R., Niger Prov. (Dec.) *Daley* FHI 32294! Gimba, Maigana Dist. (Oct.) *Philcox* 167! Gwari Hill, nr. Minna (Dec.) *Meikle* 703! **S.Nig.:** Oba, Owo Dist. (Aug.) *Stanfield* 117! Igarra, NE. of Owo (Dec.) *Stanfield* 57!

21. **E. irregulare** *Meikle* in Kew Bull. 22: 143 (1968). *E. heterochiton* Lecomte in Bull. Soc. Bot. Fr. 55: 647 (1909), partly, incl. *Chev.* 18488, not of Koern. Scapes numerous, erect, less than 1¼ in. high; leaves

narrowly subulate or setaceous; capitula generally less than 5 mm. diam. with whitish, glossy involucral bracts.
**Guin.:** Fetoré Plain, Koubi, Fouta Djalon (Sept.) *Adames* 353! Timbo to Ditinn (Sept.) *Chev.* 18488! Kala, nr. Dalaba (Oct.) *Schnell* 7379! **S.L.:** *Jaeger* 184!

22. **E. remotum** *Lecomte* in Bull. Soc. Bot. Fr. 55: 644 (1909). *E. heterochiton* Lecomte in Bull. Soc. Bot. Fr. 55: 647 (1909) partly, incl. *Chev.* 18770, not of Koern. Scapes rather numerous, usually less than 1¼ in. high; leaves relatively broad and opaque; capitula 3–4 mm. diam., fuscous, subglobose at maturity; involucral bracts conspicuous, stramineous, blunt.
**Guin.:** Kandiara R. (Oct.) *Pitot*! Ditinn (Sept.) *Schnell* 7372! Dalaba to Diaguissa (Oct.) *Chev.* 18770! 18810! **S.L.:** Brookfields (Oct.) *Deighton* 2178!

23. **E. pulchellum** *Koern.* in Linnaea 27: 622 (1856); Ruhland in Engl., Pflanzenr. 4, 30: 97 (1903). *E. pumilum* Afzel. ex Koern. in Linnaea 27: 621 (1856); F.T.A. 8: 237; Ruhland in Engl., Pflanzenr. 4, 30: 97 (1903); F.W.T.A., ed. 1, 2: 326 not of Raf. *E. sierraleonense* Moldenke in Phytologia 3: 398 (1950). *E. heterochiton* Lecomte in Bull. Soc. Bot. Fr. 55: 647 (1909), partly, incl. *Chev.* 18770a, not of Koern. Very slender, usually less than 2½ in. high; leaves numerous, narrowly subulate; capitula to about 5 mm. diam. with conspicuous, white, radiating involucral bracts.
**Mali:** Bamako *Garnier*! **Guin.:** Friguiagbé (Sept.) *Boismaré* 376 in *Hb. Chillou* 3894! Friguiagbé to Toumoukouré (Sept.) *Boismaré* 408 in *Hb. Chillou* 3926! Kala, nr. Dalaba (Sept.) *Schnell* 7373! Dalaba (Aug., Sept.) *Schnell* 6814! *Adames* 329! Dalaba to Diaguissa (Oct.) *Chev.* 18770a! Macenta (Oct.) *Baldwin* 9800! **S.L.:** *Afzelius*! Lumley Beach, Freetown (Sept.) *Harvey* 135! Waterloo (Aug.) *Melville & Hooker* 277! Kissy Flat (Aug.) *Jordan* 2160! Mafare (Nov.) *Marmo* 31! Mateboi (Oct.) *Glanville* 21! Binkolo (Aug.) *Jordan* 504! Bumban (Aug.) *Deighton* 1306! Kanya *Thomas* 2976! Loma Mts. (Sept.) *Jaeger* 7600! Bintumane Peak, 6,390 ft. (Jan.) *T. S. Jones* 99! **Lib.:** Monrovia (Aug.) *Baldwin* 9176! Paynesville (Oct.) *Harley* 1679! Mount Bele, Nimba Mts. (Oct.) *Adames* 708!

24. **E. adamesii** *Meikle* in Kew Bull. 3: 472 (1948). Slender annual; scapes erect, to about 4 in. high; leaves numerous, narrowly subulate; capitula globose, 2–3·5 mm. diam., greenish, without noticeable involucral bracts.
**S.L.:** Wellington (Sept.) *Harvey* 112! Kent (Nov.) *Deighton* 5630! Sembehun (Jan.) *Adames* in *Hb. Deighton* 4151! Bonthe (Oct.) *Adames* 97! *Jordan* 621! Taigbe (Oct.) *Adames* 91! **Lib.:** Paynesville (Sept.) *Dinklage* 3009!

25. **E. sessile** *Meikle* in Kew Bull. 9: 275 (1954). *Paepalanthus sessilis* Lecomte in Bull. Soc. Bot. Fr. 55: 596 (1909). *Eriocaulon diaguissense* Bourdu in Bull. Soc. Bot. Fr. 104: 156 (1957). A remarkable dwarf, less than 1 cm. high; the capitula sessile and forming a central "disk" surrounded by numerous, narrowly subulate, often reddish, leaves.
**Guin.:** Dalaba (Oct.) *des Abbayes* 777! Also in C. African Republic.

2. **MESANTHEMUM** Koern. in Linnaea 27: 572 (1856); F.T.A. 8: 260; Ruhland in Engl., Pflanzenr. 4, 30: 117 (1903).

Innermost series of involucral bracts greatly exceeding the florets, and resembling the ray-florets of a member of the Compositae:
  Leaves narrow, grass-like, less than 5 mm. wide; peduncles slender; involucral bracts shining, white, thinly adpressed-pilose .. .. .. .. 1. *prescottianum*
  Leaves broad, ligulate, often more than 10 mm. wide; peduncles stout; involucral bracts dirty white or pale brown, not shining, rather densely adpressed-pilose
      3. *bennae*
Innermost series of involucral bracts not much exceeding the florets, or shorter than the florets, not resembling the ray-florets of a member of the Compositae:
  Robust perennials; scapes 30–60 cm. high; leaves generally more than 15 cm. long:
    Florets immersed in a cushion of black receptacular hairs; stem often swollen, but not elongate; plants usually terrestrial .. .. .. .. 2. *jaegeri*
    Florets immersed in a cushion of whitish or pale grey receptacular hairs; stem often elongate; plants commonly aquatic .. .. .. .. 4. *radicans*
  Slender annuals, scapes less than 25 cm. high; leaves usually less than 4 cm. long:
    Capitula golden-yellow, orange or red; floral bracts plumose, involucral bracts narrow, fuscous .. .. .. .. .. .. 5. *auratum*
    Capitula whitish; floral bracts wanting, involucral bracts broad, blunt, pale brown, glossy .. .. .. .. .. .. .. .. 6. *albidum*

1. **M. prescottianum** (*Bong.*) *Koern.* in Mart., Fl. Brasil. 3, 1: 472, t. 60 (1863); F.T.A. 8: 261; Ruhland in Engl., Pflanzenr. 4, 30: 118 (1903). *Eriocaulon prescottianum* Bong. in Act. Petrop. Sci. Math., ser. 6, 1: 635, t. 36 (1831). *Mesanthemum tuberosum* Lecomte in Bull. Soc. Bot. Fr. 55: 598 (1909). Slender perennial; leaves and peduncles thinly clothed with spreading hairs; scapes up to 18 in. high; leaves narrow, grass-like; capitulum with conspicuous radiating involucral bracts, like a white daisy.
**Guin.:** Dalaba (June) *Adames* 273! Mamou (Sept.) *Schnell* 6789! Timbo (June) *Miquel* 64! Timbo to Kouroussa (June) *Pobéguin* 734! Macenta (Oct.) *Baldwin* 9772! Nimba Mts., 3,900 ft. (Aug.) *Schnell* 6218! **S.L.:** *Hb. Lindley*! Sugar Loaf Mt. (Aug.) *Melville & Hooker* 358! Picket Hill (Nov.) *T. S. Jones* 231! Binkolo (Aug.) *Deighton* 1277! Kponkponto to Waia (July) *Marmo* 298! Kulafaga to Bendugu (Sept.) *Deighton* 5154! Loma Mt. (Sept.) *Jaeger* 7595! Koya Mt., Bafodia, Koinadugu Dist. (Aug.) *Haswell* 80! **Lib.:** Sanokwele (Sept.) *Baldwin* 9516! **Iv.C.:** Nimba Mts., 2500–5000 ft. (June, Aug.) *Boughey* GC 18053! *Gruys* 56! Tonkoui Mt. (Aug.) *Schnell* 6363!

2. **M. jaegeri** *Jac-Fél.* in Bull. Soc. Bot. Fr. 94: 146 (1947). Robust perennial with a swollen, bulbous base; leaves and peduncles pilose with spreading hairs; scapes to 16 in. high, capitula 1–1·8 cm. diam.; involucral bracts dirty whitish; florets whitish, immersed in a cushion of long black receptacular hairs.
**S.L.:** Loma Mt. (Sept.) *Jaeger* 1625! 7655! Sankan Biriwa, 6,080 ft. (Jan.) *Cole* 155! **S.Nig.:** Koloishe Mt., 5,000–5,400 ft., Obudu Div. (Dec.) *Savory & Keay* FHI 25079!

3. **M. bennae** *Jac.-Fél.* in Bull. Soc. Bot. Fr. 94: 145 (1947). Tufted perennial; leaves pilose or pubescent; peduncles thinly pilose with spreading hairs, 16–18 in. high; capitula 2–2·5 cm. diam.; involucral bracts whitish or pale brown; florets pallid, immersed in a cushion of long whitish receptacular hairs.
**Guin.:** Kindia (Oct.) *Jac.-Fél.* 2091!

4. **M. radicans** (*Benth.*) *Koern.* in Linnaea 27: 573 (1856); F.T.A. 8: 260; Ruhland in Engl., Pflanzenr. 4, 30: 119 (1903). *Eriocaulon radicans* Benth. in Hook., Fl. Nigrit. 547 (1849). Robust perennial, often forming loose mats in swamps or streams; stems often elongate, leaves and peduncles thinly pilose or glabrescent; scapes up to 2 ft. high; capitula 1–1·3 cm. diam.; involucral bracts whitish; florets pallid, immersed in a cushion of long, whitish or pale grey receptacular hairs.

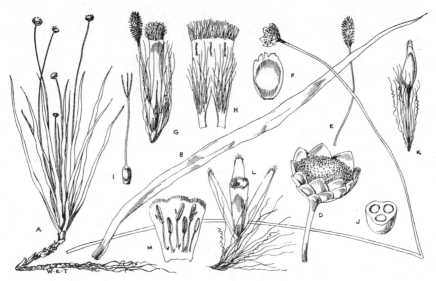

FIG. 339.—MESANTHEMUM RADICANS (*Benth.*) *Koern.* (ERIOCAULACEAE).

A, whole plant. B, leaf. C, flower head and peduncle. D, flower-head. E, bracteole. F, bract. G, female flower and bracteole. H, part of female perianth from inside. I, pistil. J, cross-section of ovary. K, bracteole. L, male flower. M, male perianth with stamens.

**Sen.:** *Perrottet*! Santiaba, Casamance (Apr.) *Berhaut* 5822! **S.L.:** Lumley Beach, Freetown (Sept.) *Harvey* 143! Kabala to Bandakarfaia (Sept.) *Jaeger* 2074! **Lib.:** Monrovia (Jan.) *H. C. D. de Wit* 9122! Duport (Sept.) *Voorhoeve* 1217! Paynesville (Dec.) *Harley* 1958! Nimba Mts. (Dec.) *Adam* 20332! Mt. Bele Rd. (Aug.) *Adames* 461! **Iv.C.:** Moossou (Aug., Nov.) *Schnell* 6542! *de Wilde* 219! 809! Grand Bassam (Sept.) *de Wit* 7902! **Ghana:** Atwabo *Fishlock* 34/1931! **Dah.:** Porto Novo (Sept.) *Adjanohoun* 248! Sémé (Feb.) *Raynal* 13543! **S.Nig.:** Lagos *Barter* 2201! Ogoya, Lagos (Feb.) *Richards* 5086! Sapoba *Kennedy* 2798! Mnobi, Onitsha Dist. (Dec.) *Onyeagocha* FHI 7790! Ibeno, Eket Dist. (May) *Onochie* FHI 32099! **F.Po:** *Milne*! Also in Congo and Angola and eastwards to Uganda and Tanzania.

5. **M. auratum** *Lecomte* in Bull. Soc. Bot. Fr. 55: 599 (1909). *M. rubrum* Moldenke in Phytologia 8: 390 (1962). Slender annual, plants often in rather close tufts; leaves thinly pilose or subglabrous, usually less than 3·5 cm. long; peduncles to about 10 in. high, slender, shining, thinly clothed with spreading hairs; involucral bracts narrow, fuscous, pilose; floral bracts filiform, plumose, orange or reddish; florets yellow or red, concealed by the floral bracts and pale receptacular hairs.
　　**Port G.:** Orango Grande, Bijagos (Jan.) *Raimundo & Guerra* 931! **Guin.:** Mt. Gangan (Oct.) *Schnell* 7479! Friguiagbé (Oct., Jan.) *Chillou* 675! 676! 679! *Pobéguin* 30! Kindia (Dec.) *Jac.-Fél.* 7451! **S.L.:** Loma Mt. Plateau, 3,900 ft. (Nov.) *Jaeger* 576!

6. **M. albidum** *Lecomte* in Bull. Soc. Bot. Fr. 55: 60 (1909). *Eriocaulon guineense* Moldenke in Phytologia 8: 386 (1962). *E. hirsutulum* Moldenke l.c. 387 (1962). Slender annual, plants scattered in open colonies; leaves subglabrous, generally less than 1½ in. long; peduncles slender, shining, to about 6 in. high, clothed with spreading hairs; involucral bracts broad, pallid; floral bracts wanting; florets whitish, pilose, receptacle pilose with pallid hairs.
　　**Sen.:** Kabrousse (Oct.) *Adam* 18370! **Guin.:** Friguiagbé (Sept.) *Chillou* 3904! Kindia (Aug.) *Pobéguin* 1359! Pita, Fouta Djalon (Sept.) *Adames* 363! Kouroussa (Aug.) *Pobéguin* 1153! **S.L.:** Gbap, Nongoba, Bullom (Oct.) *Jordan* 588!

## 3. PAEPALANTHUS Kunth., Enum. 3: 498 (1841); F.T.A. 8: 262; Ruhland in Engl., Pflanzenr. 4, 30: 121–223 (1903). *Nom. cons.*

Leaves linear, flattened, 2–4 mm. wide; stem usually somewhat elongated; scapes numerous, arising from the centre of the leaf-rosette　　..　　..　　1. *lamarckii*
Leaves filiform subterete, less than 1 mm. wide, densely crowded; stem wanting or almost wanting, scapes rather few, arising laterally from the rosettes　2. *pulvinatus*

1. **P. lamarckii** *Kunth* Enum. Pl. 3: 506 (1841); Ruhland in Engl., Pflanzenr. 4, 30: 160 (1903); Lecomte in Bull. Soc. Bot. Fr. 55: 595 (1909); Milne-Redhead in Kew Bull. 4: 472 (1949). Plants usually 3–7 mm. high; capitula subglobose, greyish, pilose, 2–3·5 mm. diam.; sepals of female flowers hardening and recurving at maturity and throwing out the ripe seeds.
　　**Guin.:** Friguiagbé (Oct.) *Chillou* 776! Friguiagbé to Toumoukouré (Dec.) *Boismaré* 52 in *Hb. Chillou* 3482! Grandes Chutes, Fouta Djalon (Dec.) *Chev.* 20307! Koba (Nov.) *Jac.-Fél.* 7210! **S.L.:** Roboli, nr. Rokupr (Nov.) *Jordan* 161! Newton *Deighton* 1440! Bomatok (Oct.) *Adames* in *Hb. Deighton* 4128! Taigbe (Oct.) *Adames* 90! **Lib.:** Duport (Nov.) *Bequaert* in *Hb. Linder* 1451! Also in Congo and Tanzania (Mafia Island) and S. America.

2. **P. pulvinatus** *N.E. Br.* in F.T.A. 8: 263 (1901); Ruhland in Engl., Pflanzenr. 4, 30: 221 (1903). Plants forming dense cushions; leaves greyish, strongly recurved; capitula greyish, pilose, hemispherical, about 5 mm. diam.; sepals of female flowers not recurving at maturity.
　　**S.L.:** Kissy *Bockstatt*!
　　[A very rare plant, only once collected, and unsatisfactorily localized; its rediscovery would be very welcome.]

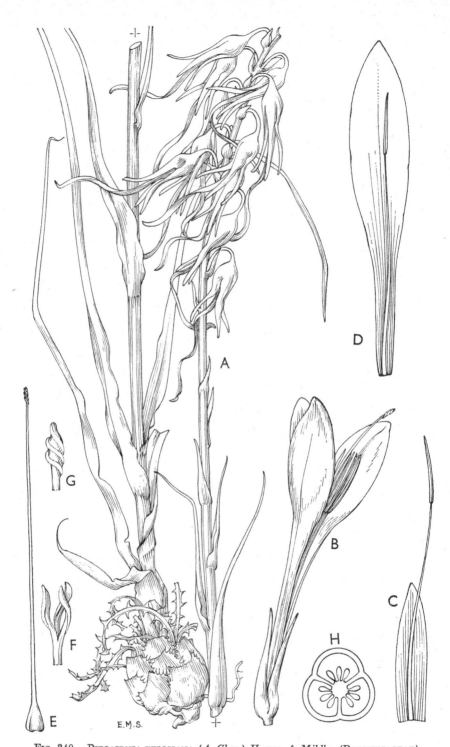

Fig. 340.—Pitcairnia feliciana (*A. Chev.*) *Harms & Mildbr.* (Bromeliaceae).

A, habit, × 1. B, flower, × 1½. C, sepal and stamen, × 1½. D, petal and stamen, × 1½. E, gynoecium, × 1½. F, stigma, with style arms free, × 6. G, stigma with style-arms curved, × 6. H, T.S. ovary, diagram. A from *Schnell* 3126. B–H from *Jacques-Félix* 1856.

**4. SYNGONANTHUS** Ruhland in Urban, Symb. Antill. 1: 487 (1900), and in Engl., Pflanzenr. 4, 30: 242–282 (1903).

Leaves thinly pubescent or nearly glabrous above, lanuginose at base, densely tufted, up to 3 cm. long; scapes often numerous, arising from the centre of the leaf-tuft, up to 25 cm. long, sheaths and peduncles subglabrous or more often with patent glandular hairs; capitula hemispherical, about 5–7 mm. diam., involucral bracts brownish; sepals thinly pilose towards apex .. .. .. .. *wahlbergii*

S. **wahlbergii** (*Koern.*) *Ruhland* in Engl., Pflanzenr. 4, 30: 247 (1903). *Paepalanthus wahlbergii* Koern. in Mart., Fl. Brasil. 3, 1: 459 (1863); F.T.A. 8: 263. *Syngonanthus chevalieri* Lecomte in Bull. Soc. Bot. Fr. 55: 597 (1909); F.W.T.A., ed. 1, 2: 328. Plants usually forming crowded tufts; leaves narrowly linear-subulate, often distinctly recurved, and white-lanuginose at base; peduncles slender, wiry, minutely glandular (especially just below capitulum); flowers brownish, sometimes dark brown.
**N.Nig.:** Nupe *Barter* 1539! Widespread in tropical Africa and south to the Transvaal.

# 177. BROMELIACEAE

## By F. N. Hepper

Mostly short-stemmed epiphytes or growing on rocks. Leaves usually in a dense cluster, long and strap-shaped, rigid and spinulose-toothed or rarely flaccid, often coloured towards the base. Flowers in a terminal head, spike, or panicle often with highly coloured bracts, actinomorphic, bisexual or rarely unisexual. Perianth hypogynous to epigynous, segments in two series, the outer calyx-like, imbricate, the inner corolla-like and free or variously connate, imbricate. Stamens 6, mostly inserted at the base of the segments, free or partially adnate to them; anthers free or rarely connate in a ring, linear, usually versatile, 2-locular, opening by longitudinal slits. Ovary superior to inferior, 3-locular; style slender, elongated, stigmas 3. Ovules numerous in each loculus, the axile placentas sometimes divided. Fruit fleshy and indehiscent or rarely opening unevenly, or rarely a septicidal or loculicidal capsule. Seeds with abundant mealy endosperm and a small embryo, sometimes winged.

A tropical American family with only the following species represented in Africa. The pineapple, *Ananas comosus* (Linn.) Merrill (syn. *A. sativus* Schult. f.), is widely cultivated and often naturalized.

### PITCAIRNIA L'Hérit., Sert. Angl. 7, t.11 (1789), *nom. cons.*

Rhizomatous, glabrous herb; leaves linear, 30–50 cm. long, 5–9 mm. broad, distantly serrate with spiny curved teeth; inflorescence racemose, 8–15 cm. long, borne on an erect leafy scape about 30 cm. long; flowers 5 cm. long; sepals coriaceous, lanceolate, acute 15–22 cm. long, petals 5 cm. long; stamens ¾-length of petals; ovary 3-lobed, style longer than the stamens; style capitate and contorted; seeds numerous, linear
*feliciana*

P. **feliciana** (*A. Chev.*) *Harms & Mildbr.* in Notizbl. Bot. Gart. Berl. 14: 118 (1938); Hepper in Webbia 19: 601 (1965). *Willrussellia feliciana* A. Chev. in Bull. Soc. Bot. Fr. 84: 502, fig. 1 (1937). A tufted herb with stout rhizomes and narrow spiny leaves; flowers orange-yellow; in crevices on rock out-crops.
**Guin.:** Kindia (fl. July, fr. Sept.) *Jac.-Fél.* 1856! *Schnell* 2201! 3126! Télimélé (fr. Jan.) *Roberty* 16453!

# 178. MUSACEAE

## By F. N. Hepper

Stems formed by the imbricate bases of the petioles, erect. Leaves spirally arranged, very large, with a thick midrib and numerous pinnately parallel nerves. Flowers mostly unisexual, clustered and subtended by large green spathaceous bracts, the male flowers within the upper bracts, the female within the lower. Calyx elongated, at first narrowly tubular, soon splitting on one side, variously toothed at apex. Corolla more or less 2-lipped. Stamens 5 perfect, with a rudimentary sixth, or 6 perfect; filaments filiform; anthers 2-locular, the loculi parallel and contiguous. Ovary inferior, 3-locular; ovules numerous, axile; style filiform, with a lobulate stigma. Fruit fleshy, indehiscent, 3-locular. Seeds with a thick hard testa and straight embryo in copious endosperm.

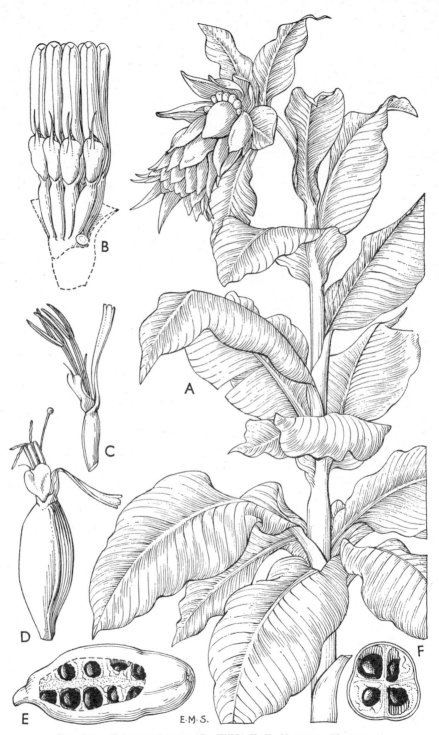

FIG. 341.—ENSETE GILLETII (*De Wild.*) *E. E. Cheesman* (MUSACEAE).

A, habit, × ⅛. B, young male flowers, × 1. C, male flower, × 1. D, female flower, × 1. E,
fruit, × ⅔. F, T.S. fruit, × ⅔. All from *Keay* FHI 28003.

Many varieties of bananas and plantains (*Musa* spp.) are cultivated in our area. The wild African bananas are now distinguished as a separate genus *Ensete*, which is monocarpic. *Musa* differs from *Ensete* in being perennial and producing suckers; bracts and flowers are inserted independently on the axis; seeds are small and rarely produced. See Simmonds, Bananas (1959).

*Ravenala madagascariensis* Sonn., the fan-like Travellers Tree from Madagascar, is occasionally cultivated, and *Strelitzia* spp. from S. Africa may be cultivated as ornamentals; both genera are now usually placed in the family Strelitziaceae.

**ENSETE** Horan., Prod. Monogr. Scit. 40, t.4 (1862); Cheesman in Kew Bull. 2: 97 (1947); Baker & Simmonds in Kew Bull. 8: 405 (1953).

Plant 1·5–3 m. high; leaves spread along the stem and not aggregated towards apex, lower leaves up to 1·5 m. long reducing in size upwards so that the upper leaves become bracts; male bracts 4·5–9 cm. long, 17–25 cm. broad; stamens 6, 12 mm. long; female flower with rudimentary stamens; fruit squat angular, rather obconic, about 5 cm. long; seeds 7–9 mm. diam., hard . . . . . . . . . . *gilletii*

E. **gilletii** (*De Wild.*) *E. E. Cheesman* l.c. 103 (1947); Baker & Simmonds l.c. 407 (1953); Koechlin in Fl. Cam. 4: 11, t.3. *Musa gilletii* De Wild. in Rev. Cult. Colon. 8: 102 (1901); Not. Pl. Ut. Intér. Fl. Congo 1: 73, t.5, 6, 7(1–2) (1903). *M. chevalieri* Gagnepain (1908). *M. schweinfurthii* of F.W.T.A., ed. 1, 2: 328 not of K. Schum. & Warb. ex K. Schum. A wild banana of hilly grassland savanna. The leafy young stem dies back to a hard round corm about 1 ft. across remaining on the soil surface; fertile stem nodding, bearing a mass of small fruits about 2 in. long with yellow or orange pulp and large dark brown seeds which are later exposed by longitudinal splitting of the fruit; male bracts mauve.
**Guin.**: Yambéring (Sept.) *Schnell* 7319! **S.L.**: Falaba (fr. Noy.) *Deighton* 2582! **N.Nig.**: *B. D. Burtt!* Mando F.R., Birnin Gwari, Zaria Prov. (Aug.) *Keay* FHI 28003! Gawn to Abuja, Niger Prov. (fr. Aug.) Onochie FHI 35747! Vogel Peak, Sardauna Prov. (fr. Nov.) *Hepper* 1546! Bellel, Mambila Dist. *Hepper* (photo)! **S.Nig.**: *Freeman* 49! **W.Cam.**: Bambui (June) *Russell* FHI 14987! Jakiri *Nicholas* (photo)! Also in E. Cameroun, Congo, Malawi and Angola.

*Musa martretiana* A. Chev. and *M. riperti* A. Chev., Bot. 632, names only, cited in Ed. 1 as synonyms of *M. schweinfurthii*, are rejected by Simmonds (l.c. 409) as being *nomina nuda* and having no standing.


# 179. ZINGIBERACEAE

## By F. N. Hepper

Perennial herbs, usually aromatic, with horizontal tuberous rhizomes; stems sometimes very short, leafy or bearing only flowers. Leaves in two rows or spirally arranged, with an open or closed sheath, sessile or stalked on the sheath, the blade usually large with numerous closely parallel pinnate nerves diverging obliquely from the midrib. Flowers solitary or in a distinct inflorescence accompanying or separate from the leaves, mostly bisexual, symmetric or asymmetric. Perianth 6-merous, 2-seriate, the outer calyx-like, the inner corolla-like and often very showy and delicate; outer segments united into a tube, inner more or less united, the posterior segment usually the largest. Stamen 1, with a 2-locular anther, sometimes accompanied by petaloid staminodes which may form the conspicuous part of the flower (see Fig. 341). Ovary inferior, (2–)3-celled with axile placentas, or 1-locular with parietal or rarely basal placentas; style sometimes enveloped in a groove of the fertile stamen. Ovules mostly numerous. Fruit fleshy and indehiscent or a capsule. Seeds mostly arillate; endosperm abundant, white.

Tropics and subtropics.

Flowers paniculate, inflorescences terminating leafy stems or arising separately; lateral staminodes rudimentary or suppressed; filament and connective of stamen not petaloid . . . . . . . . . . . . . . . . . . . . 1. **Renealmia**
Flowers spicate often in a strobiliform inflorescence, or solitary:
Flowers at ground level appearing before the leaves (or when the leafy shoot is beginning to grow); lateral staminodes large and petaloid, forming the conspicuous part of the flower; stamen connective not petaloid . . . . . . . 2. **Kaempferia**
Flowers appearing with the mature leaves either on leafy shoots (sometimes at ground-level) or on separate basal inflorescences; lateral staminodes rudimentary or suppressed; conspicuous part of the flower usually a large labellum (= petaloid staminode):
Leaves distichously arranged on tall stems; inflorescences always separate with 1-several flowers; filament and connective not petaloid, but surmounted by 2 lateral horns and usually an apical process between them . . 3. **Aframomum**

Leaves spirally arranged on tall or short stems or only 1 or 4 leaves present; inflorescences terminal or separate; filament and connective petaloid bearing the anther
in the middle ..    ..    ..    ..    ..    ..    ..    ..    ..    **4. Costus**

Besides the above indigenous genera the following from Asia have been introduced into our area: *Alpinia
speciosa* (Wendl.) K. Schum. (ornamental); *Curcuma domestica* Valeton (in Bull Jard. Bot. Buit. ser. 2,
27: 31 (1918), syn. *C. longa* of F.W.T.A. ed. 1, 2: 334 and Appendix, p. 473, not of Linn.), the turmeric;
*Hedychium coronarium* Koenig (ornamental); *Zingiber officinale* Rosc. (ginger: widely cultivated for the
rhizomes—see Appendix, p. 474).

**1. RENEALMIA** Linn. f., Suppl. 7 (1781); F.T.A. 7: 311; K. Schum. in Engl., Pflanzenr.
4, 46, Zingib.: 282 (1904). *Nom. cons.*

Inflorescence borne at the top of the leafy shoot:
Leaves broadly oblanceolate, acutely acuminate, 20 cm. long, 5–5·5 cm. broad, with
very numerous ascending parallel lateral nerves; sheath broadened in contact with
the stem, not ligulate; flowers not seen; bracts lanceolate, 2 cm. long, ribbed,
glabrous; fruits oblong, 2 cm. long ..    ..    ..    ..    ..    1. *battenbergiana*
Leaves elongate-lanceolate, acutely acuminate, about 35 cm. long and 4–5 cm. broad,
with numerous very closely parallel lateral ascending nerves; sheath abruptly
broadened in contact with the stem; bracts densely pubescent; fruits ellipsoid,
1–1·3 cm. long ..    ..    ..    ..    ..    ..    ..    2. *longifolia*
Inflorescence borne directly on the rhizome and apart from the leafy shoot:
Inflorescence about 1 m. long with the erect cincinni almost appressed (even in fruit)
to the axis rather widely spaced from one another, axis shortly and distinctly
pubescent; calyx-tube at apex of fruit about 5 mm. long, narrowly funnel-shaped;
lamina elongate-oblanceolate, acutely acuminate, about 30 cm. long, 3–6 cm.
broad, with numerous close ascending lateral nerves    ..    ..    3. *cincinnata*
Inflorescence 8–35 cm. long with the cincinni spreading especially in fruit, usually
glabrous:
Calyx in fruit campanulate funnel-shaped, about 5 mm. long and broad; flowers
sessile; bracts nearly orbicular, 1 cm. long; inflorescence about 13 cm. long;
lamina oblong-elliptic, caudate-acuminate, rather shortly cuneate at the base,
about 20 cm. long and 6–7 cm. broad, with very numerous ascending lateral
nerves ..    ..    ..    ..    ..    ..    ..    ..    ..    4. *mannii*
Calyx in fruit narrowly funnel-shaped, up to 10 mm. long and much longer than
broad; flowers pedicellate; bracts oblong-elliptic to broadly lanceolate:
Lamina obovate-elliptic, shortly acuminate, rather shortly cuneate at the base,
20–30 cm. long, 6–10 cm. broad, with maroon patches on surfaces; inflorescence
about 10 cm. long ..    ..    ..    ..    ..    ..    ..    5. *maculata*
Lamina oblanceolate, caudate-acuminate, long-cuneate at the base, 30–50 cm. long,
4–10 cm. broad; inflorescence 10–35 cm. long ..    ..    ..    6. *africana*

1. **R. battenbergiana** *Cummins ex Bak.* in F.T.A. 7: 313 (1898); K. Schum. in Engl., Pflanzenr. 4, 46, Zingib.:
289 (1904). A herb about 1 ft. high with red subtorulose fruits.
    **Iv.C.:** Alépé, Attie *Chev.* 17480! 17900! Malamalasso *Chev.* 17504! Maféré (fr. Feb.) *Aké Assi* 7330!
    Yapo Forest (fr. Oct.) *Aké Assi* 72! **Ghana:** Assin-Yan-Kumasi *Cummins* 197!
    [This may not be really distinct from the next species—F.N.H.]
2. **R. longifolia** *K. Schum.* l.c. (1904); Stapf in Johnston, Liberia 656. *R. ivorensis* A. Chev. in Mém. Soc. Bot.
    Fr. 2, 8: 305 (1917); *Chev. Bot.* 628. A herb with leafy flowering stems 4–6 ft. high; flowers pink, fruits
    red, turning black; in forest.
    **Lib.:** Grand Bassa (fr. May) *Dinklage* 1945! Dukwai R., Monrovia (fr. Oct., Nov.) *Cooper* 47! Zwedru,
    29 Km. NW. of Chien, Putu Dist. (Mar.) *de Wilde* 3727! Mnanulu, Webo Dist. (fr. June) *Baldwin* 6046!
    **Iv.C.:** Grabo, Cavally Basin (July) *Chev.* 19608!
3. **R. cincinnata** (*K. Schum.*) *Bak.* in F.T.A. 7: 312 (1898); K. Schum. l.c. 295; Koechlin in Fl. Gabon 9: 23,
    t. 4, and in Fl. Cam. 4: 30, t. 5. *Ethanium cincinnatum* K. Schum. (1892). Herb with leafy stems about
    5 ft. high, leaves purplish on the lower surface; inflorescences about 3 ft. high; fruits red; in high forest.
    **S.Nig.:** Oban *Talbot* 876! *Robb*! *Onochie* FHI 34819! Afi River F.R., Ogoja Prov. (fr. May) *Jones &
    Onochie* FHI 5837! **W.Cam.:** Victoria to Bimbia *Preuss* 1348! Also in E. Cameroun and Gabon.
4. **R. mannii** *Hook. f.* in Hook., Ic. Pl. sub. t. 1430 (1883); F.T.A. 7: 312; K. Schum. l.c. 291. Several in-
    florescences about 5 in. long with rather crowded bracts.
    **F.Po:** *Mann* 1172!
5. **R. maculata** *Stapf* in Johnston, Liberia 656 (1906). Leafy stems 5–6 ft. high, leaves spotted maroon; flowers
    white or orange.
    **Lib.:** Kakatown *Whyte*! Gbanga (Sept.) *Linder* 619! Nimba Mt. (Oct.) *Adames* 616! 668! **Iv.C.:** Binger-
    ville *Chev.* 15400! 15501! Upper Nuon (fr. Apr.) *Chev.* 21137! Teke Forest *de Wit* 7851! Mopri Forest
    (fr. Jan.) *Aké Assi* 8483!
6. **R. africana** *Benth. ex Hook. f.* in Hook., Ic. Pl. t. 1430 (1883); K. Schum. l.c. 294; Koechlin in Fl. Gabon 9:
    30, t. 5, and in Fl. Cam. 38, t. 8. *R. talbotii* Hutch., F.W.T.A., ed. 1, 2: 332 (1936), English descr. only.
    Perennial rhizomatous herb with leafy stems 3–5 ft. high; inflorescences with reddish axis up to 1 ft. long,
    arising from the rhizome; flowers white, fruits red turning black; in high forest in hilly districts.
    **S.Nig.:** Idanre F.R., Ondo (fr. July) *Onochie* FHI 33404! Umudike (fl. & fr. June) *Ariwaodo* ARS 1255!
    Oban *Talbot* 1652! Oron Dist. (July) *Kitson*! **W.Cam.:** Cam. Mt., 5,000 ft. *Johnston* 106! N. Korup F.R.
    (fr. June) *Olorunfemi* FHI 30656! Bambuko F.R., Kumba (fl. & fr. Jan.) *Keay* FHI 37478! Bu F.R.,
    Bamenda (Apr.) *Ujor* FHI 30088! **F.Po:** *Mann* 323! El Pico (fl. & fr. Dec.) *Boughey* 174! Moka (fl. & fr.
    Sept.) *Wrigley* 533! Also in E. Cameroun, Gabon and Congo.

*Imperfectly known species*

    Although the types of the following species were seen for F.W.T.A., ed. 1, and the species were maintained
they seem to be doubtfully distinct from *R. africana*:

1. **R. albo-rosea** *K. Schum.* l.c. 293 (1904).
 **W.Cam.:** Ikassa *Rudatis* 6. Victoria *Winkler* 357. [Both destroyed at Berlin.]
2. **R. macrocolea** *K. Schum.* l.c. 294 (1904).
 **Togo Rep.:** Misahöhe, in primary moist forest *Baumann* 315. *Midlbr.* 7299. [Both destroyed at Berlin.

2. **KAEMPFERIA** Linn., Sp. Pl. 2 (1753); F.T.A. 7: 294 (1898); K. Schum. in Engl.,
 Pflanzenr. 4, 46, Zingib.: 64 (1904); F.W.T.A., ed. 1, 2: 334.

Calyx 3–4 cm. long, shortly 3-lobed, lobes rounded; petals lanceolate, 4 cm. long, united
into a slender tube; petaloid staminodes about 10 cm. long, limb nearly as wide, the
tip broad and emarginate; flowers arising in a separate cluster from the rhizome;
leafy stems arising separately from the rhizomes; pseudostems of petioles well
developed, about 50 cm. long; basal leaves reduced to sheaths, mature upper leaves
oblong-lanceolate to oblanceolate, up to 38 cm. long and 4·5(–9) cm. broad, very
acute at apex; ligule conspicuous, 1–1·5(–3) cm. long; tubers oblong-fusiform, about
5 cm. long .. .. .. .. .. .. .. .. 1. *aethiopica*
Calyx about 1·5 cm. long; ligule inconspicuous, 2–5 mm. long; mature upper leaves
almost elliptic about 40 cm. long and 15 cm. broad; pseudostem of petioles not well
developed, about 20 cm. long when fully grown .. .. .. .. 2. *nigerica*

The Asiatic species *K. galanga*, Linn. is cultivated in southern Sierra Leone.

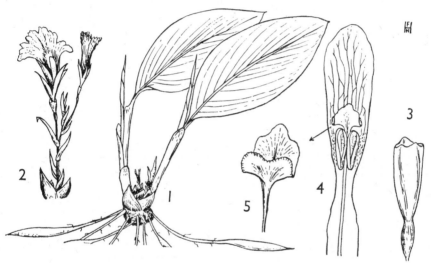

Fig. 342.—Kaempferia nigerica *Hutch. ex Hepper* (Zingiberaceae).

1, habit with young leaves after flowering, × ⅓. 2, inflorescence, × ⅓. 3, calyx and ovary, × 2
4, fertile stamen, with style and stigma, × 2. 5, stigma (viewed from opposite side to that
shown in 4), × 4. All from *Dalziel* 276.

1. **K. aethiopica** (*Schweinf.*) *Solms-Laub.* in Schweinf. l.c. 198 (1867); F.T.A. 7: 294; K. Schum. l.c. 66,
fig. 10 A–C; F.W.T.A., ed. 1, 2: 334; Troupin in Expl. Parc Nat. Garamba 1: 225, fig. 41 (1956); Berhaut,
Fl. Sén. 4, 207; Morton, W. Afr. Lilies and Orchids fig. 40; Hambler in Bot. Notes Nig. Coll. Tech. Ibadan
1: 33 (1960); Koechlin in Fl. Cam. 4: 24, t. 4. *Cienkowskya aethiopica* Schweinf., Beitr. Fl. Aethiop. 197,
t. 1 (1867). Tuberous rooted perennial with a leafy shoot about 1 ft. high appearing towards the end of
or after the flowering season; flowers about 3 in. high, often appearing in considerable numbers in savanna
and sometimes mistaken for ground orchids; the apparent corolla is the united petaloid purple or blue
staminodes with yellow markings.
 **Sen.:** *Berhaut* 256. *Heudelot* 876! **Gam.:** (Mar.) *Brooks* 62! N. bank, Gambia R. (July) *Ozanne!* Genieri
to Kaiaaf (July) *Fox* 120! **Mali:** Bamako (July) *Waterlot* 1550! **Guin.:** Kadé *Pobéguin!* **S.L.:** Kasikeri
(Apr.) *Thomas* 174! **Iv.C.:** Séguéla to Siana *Chev.* 21845! **Ghana:** Tumu (May) *Morton* GC 7492! Lawra
(May) *Vigne* FH 3810! Tamale to Ejura (Apr.) *Bally* 174a! Salaga Apr.) *Krause!* Bole (Mar.) *Hepper &
Morton* A3184! Kete-Krachi (Apr.) *Morton* GC 7151! **Dah.:** Djougou *Chev.* 23902! **Niger:** Fada to
Koupéla, Gourma *Chev.* 24531! **N.Nig.:** Liruwen–Kano Hills *Carpenter!* Zaria (June) *Lamb* 60! Birnin
Gwari (June) *Keay* FHI 25852! Biu, Bornu Prov. (May) *Noble* 48! Musgu (May) *E. Vogel* 99! Vom, Jos
Plateau *Dent Young* 238! Abinsi (May) *Dalz.* 829! Lokoja (Apr.) *Parsons* L118! **S.Nig.:** Oyo to Ilorin
*Barter* 3427! Iseyin to Shaki (Mar.) *Hambler* 875 (FHI 50704)! Widespread in the savanna regions of
tropical Africa. (See Appendix, p. 473.)
2. **K. nigerica** *Hutch. ex Hepper* in Kew Bull. 22: 465 fig. 8 (1968); Hutch., F.W.T.A., ed. 1, 2: 334 (1936),
English descr. only; Hambler l.c. A tuberous rooted herb with precocious flowers.
 **Ghana:** Tamale to Ejura (Apr.) *Bally* 174! Kete-Krachi (Apr.) *Morton* GC 2101! Dutukpene (Mar.)
*Hepper & Morton* A3065! **N.Nig.:** *Hambler* 415! *Meikle* 1300! Birnin Gwari (June) *Keay* FHI 25830!
Afikpo, Ogoja Prov. (Apr.) *Stone* 1! Zungeru (Apr., May) *Dalz.* 276! Abinsi (June) *Dalz.* 830! Lokoja
(Mar.) *Elliott* 42! **S.Nig.:** Ileshe to Benin (Mar.) *Morton* K451! Lagos (Feb.) *Cons. of For.* 449! (See
Appendix, p. 473.)

**3. AFRAMOMUM** K. Schum. in Engl., Pflanzenr. 4, 46, Zingib.: 201 (1904). *Amomum* Linn., partly—F.T.A. 7: 302.

Leaves hairy on both surfaces or only beneath:
Leaves loosely pilose with rather long hairs; lateral parallel nerves very numerous:
Inflorescences very short (about 4 cm.) and clustered, 1-flowered; labellum and hood very narrow; fruit oblong, 2·5 cm. long; lamina subsessile on the pilose sheath, unequal-sided at base, long-acuminate, nearly 30 cm. long and 5–6 cm. broad, laxly pilose on both surfaces, with a conspicuous striated midrib, densely ciliate (Fig. 344D)   ..   ..   ..   ..   ..   ..   ..   1. *pilosum*
Inflorescences nearly as long as or longer than the lamina, several-flowered; lamina unequal-sided at the base, rounded into a distinct short stalk on the sheaths, long-acuminate, 20–35 cm. long, 5–10 cm. broad, glabrous above, sheath pilose or more or less glabrous:
Labellum about 4 cm. long; peduncle covered with stiff imbricated bracts about 4 cm. long; flowers rather few; fruits ovoid, 2–4 enveloped in upper bracts
2. elliotii

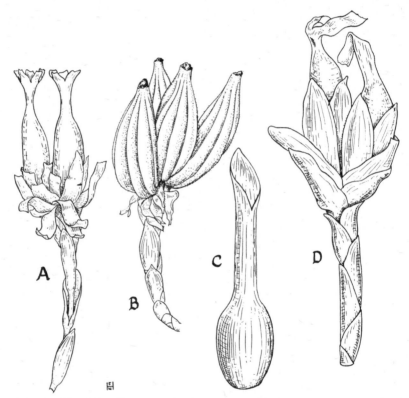

FIG. 343.—FRUITS OF AFRAMOMUM SPP. (ZINGIBERACEAE).

A, *Aframomum sceptrum* (Oliv. & Hanb.) K. Schum. B, *A. exscapum* (Sims) Hepper. C, *A. citratum* (Pereira) K. Schum. D, *A. longiscapum* (Hook.f.) K. Schum. A after Hook. Ic. Pl. t. 2477. B from *Pyne* 87. C after Hook. Ic. Pl. t. 2478. D after Hook. Ic. Pl. t. 2481.

Labellum about 2 cm. long; peduncle with few papery bracts hardly imbricated, about 3 cm. long; flowers up to 20; fruits longitudinally ribbed   4. *polyanthum*
Leaves shortly felted-tomentose beneath, glabrous above, elongate-oblong, subobtuse at base, about 35 cm. long and 6 cm. broad, with very numerous crowded ascending parallel lateral nerves; ligule 6 mm. long, ribbed; leaf-sheath with conspicuous transverse veinlets; inflorescences short (but see note after sp.), about 10 cm. long, 1–2-flowered, with closely imbricated bracts; fruits narrowly obovoid, about 7 cm. long (Fig. 344G)   ..   ..   ..   ..   ..   ..   3. *subsericeum*
Leaves glabrous beneath except sometimes on the midrib and along the margins:
Leaves auriculate-cordate and sessile at the base at the junction with the sheath,

elongate-oblong, acuminate, about 30 cm. long and 5 cm. broad, with numerous spaced ascending parallel nerves; sheath ribbed but without transverse veinlets; inflorescences much elongated, simple or branched at the top; bracts shortly overlapping; fruits with a long, persistent, tubular beak (Figs. 343D, 344E)

6. longiscapum

Leaves not auricled, at most rounded-subcordate at the junction with the sheath and false petiole:
Fruit grooved or corrugated (usually 9 regular longitudinal lines), pubescent or glabrous; labellum broad:
Fruit (and ovary) pubescent, corrugated when fresh, about 6 cm. long, several clustered at apex of 8–18 cm.-long inflorescence; bracts becoming narrower and thinner towards apex; leaves oblong-lanceolate, almost 20 cm. long and 6 cm. broad, rounded at each end and tailed at apex, lateral veins more than twice their own thickness apart* (Fig. 343B)  ..    ..    ..    ..        7. exscapum
Fruit glabrous, deeply grooved or ribbed:
Lateral veins of leaves very close and less than twice their thickness apart*; fruits ovoid, deeply grooved, about 5 cm. long, crowned by long persistent calyx:
Flowers mauve, well exserted from bracts, labellum obovate 3·5 cm. broad; anther connective with an oblong-deltoid appendage at the apex; inflorescences simple or sometimes branched; leaves oblong-lanceolate, acuminate, 10–30 cm. long, 3–7 cm. broad, hardly petiolate      ..      ..      ..      8. sulcatum
Flowers orange-red, crowded by bracts, labellum oblong, 2 cm. broad; anther connective with a rounded appendage at apex; inflorescences branched; leaves 30–50 cm. long, 7–10 cm. broad (Fig. 344 B) ..      ..      9. daniellii
Lateral veins of leaves more than twice their own thickness apart*, lamina oblong-elliptic, 18–28 cm. long, 5–10 cm. broad, midrib pubescent, false petiole 1 cm. long; ligules short and rounded; inflorescence simple, with numerous flowers, labellum 2 cm. long, less than 1 cm. broad; fruits strongly ribbed, ovoid, about 4 cm. long, with short calyx      ..      ..      ..      ..    .. 5. chlamydanthum
Fruit smooth, not grooved or corrugated (apart from perhaps a few folds towards the apex and shrinkage wrinkles in dried specimens); ovary glabrous:
Lateral nerves of the leaves more than twice their thickness apart*:
Leaf-margins the whole length pubescent with short brown hairs; lamina linear-oblong, 20–31 cm. long, 3–6 cm. broad midrib tomentose or glabrous; ligule rounded, up to 3 mm. long; inflorescences very short; labellum broad (Fig. 344F)      ..      ..      ..      ..      ..      ..      ..      10. limbatum
Leaf-margins glabrous or only upper part pubescent; labellum very broad and conspicuous:
Ligules up to 4 mm. long; inflorescences at base of stem; ligule rigid, not divided:
Flowers white:
Tufted herb lacking slender rhizomes; leaves oblong to oblong-elliptic, 13–38 cm. long, 2–8 cm. broad, cuspidate at apex, rounded at base, shortly petiolate; fruits long beaked; seeds angular, grey; bracts closely imbricated, stiff, striate      ..      ..      ..      ..      ..      ..      11. strobilaceum
Rhizomatous herb; leaves elliptic, 10–27 cm. long, 4–7·5 cm. broad, long-tailed at apex, cuneate to rounded at base, distinctly petiolate; fruits ovoid, not beaked; seeds rounded, glossy brown; bracts papery 12. stanfieldii
Flowers mauve, inflorescences several-flowered, short up to 10 cm. long, with broad imbricated bracts; leaves oblong, 20–30 cm. long, rounded to a short petiole at the base, abruptly acuminate at apex, 4–5 cm. broad ..   13. sp. A
Ligules 1–3 cm. long or if less then papery and paired on each sheath:
Inflorescences arising at base of the stem; plant tufted or rhizomatous:
Inflorescence slender, 10–30 cm. long, distinctly pedunculate and covered with rather stiff bracts, few-flowered:
Rhizomatous herb; inflorescences at ground level, 10–30 cm. long, covered with papery bracts; labellum very broad, mauve; fruits narrowly ovoid; leaves oblong, 12–20 (–28) cm. long, acuminate, cuneate to rounded at base, with more or less distinct false petiole; ligule very variable, more or less triangular apex, easily breaking off (Figs. 343A, 344C ..      14. sceptrum
Tufted herb; inflorescences about 12 cm. long, closely imbricated by stiff bracts; labellum broadly elliptic, 5·5 cm. long, 4·5 cm. broad

15. erythrostachyum

Inflorescences short and massive with numerous broad imbricated bracts 6 cm. broad; flowers numerous; ligules obtuse or acute, 1–3 cm. long; fruits ovoid, about 4·5 cm. long with persistent calyx at least the same length (Figs. 343C, 344A)  ..      ..      ..      ..      ..      16. citratum

---

\* Lateral veins referred to in this key should be observed on the underside of the leaf (preferably on dried specimens) as only the more widely spaced principal nerves are usually discernable on the upper surface.

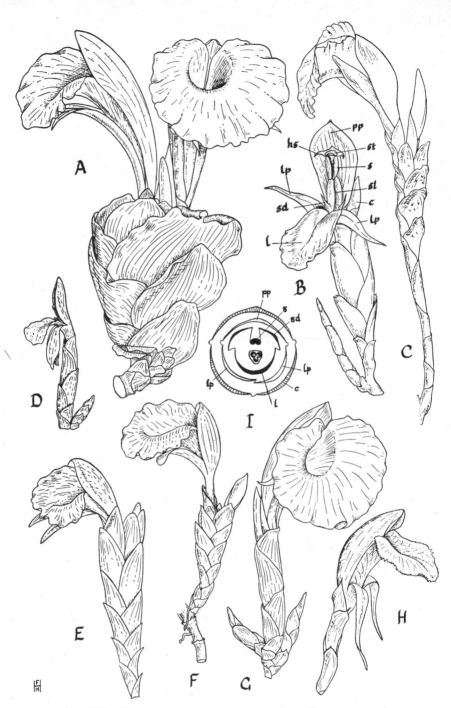

FIG. 344.—INFLORESCENCES OF AFRAMOMUM SPP. (ZINGIBERACEAE).

A, *Aframomum citratum* (Pereira) K. Schum. B, *A. daniellii* (Hook.f.) K. Schum. s, fertile stamen. st, stigma. sl, style. sd, staminode. lp, lateral petal. hs, horns of staminal connective. p.p., posterior petal. c, calyx. l, labellum. C, *A. sceptrum* (Oliv. & Hanb.) K. Schum. D, *A. pilosum* (Oliv. & Hanb.) K. Schum. E, *A. longiscapum* (Hook.f.) K. Schum. F, *A. limbatum* (Oliv. & Hanb.) K. Schum. G, *A. subsericeum* (Oliv. & Hanb.) K. Schum. H, *A. melegueta* K. Schum. I, floral diagram of *Aframomum*. A after Hook. Ic. Pl. t. 2478. B after Hook. J. Bot. vol. 4, t. 5. C after Bot. Mag. t. 5761. D after Hook. Ic. Pl. t. 2483. E after Hook. Ic. Pl. t. 2481. F after Hook. Ic. Pl. t. 2480. G after Hook. Ic. Pl. t. 2484. H after Bot. Mag. t. 5987.

Inflorescences arising from slender rhizomes some distance from the stem, very short with pale papery bracts 1 cm. broad, flowers usually paired
                                             17. *leptolepis*
Lateral veins of leaves very close, twice their own thickness apart or less; margins glabrous; ligules short:
  Midrib beneath pubescent or pilose its whole length at least along the side of the midrib; lamina oblong-oblanceolate, cuneate at base, gradually but acutely acuminate, 20–33 cm. long, 4–7 cm. broad; inflorescences short at ground level, labellum oblong and similar in size to the hood; fruits broadly ovoid (see also No. 2.)    ..    ..    ..    ..    ..    18. *baumannii*
  Midrib glabrous; labellum about 4 cm. long, rather or very broad and much larger than the hood:
    Leaves narrow (broadly-linear), about 3 cm. broad, 20 cm. long; numerous lateral nerves very prominent and all similar on both sides of the lamina, tailed-acuminate, ligule short and rounded; inflorescences very short and narrow, 1-flowered; bracts rounded and mucronate; fruits ovoid, 4 cm. long, beaked by the persistent calyx; tufted plant (Fig. 344H) ..    19. *melegueta*
    Leaves lanceolate to oblong-lanceolate, the veins often rather indistinct, about 20 cm. long, 3–10 cm. broad, acute to shortly tailed at apex, rounded at base, false petiole wanting; ligules short; inflorescences several flowered, about 10 cm. long; fruits broadly ovoid, 7 cm. long, shortly beaked; creeping rhizomatous plant    ..    ..    ..    ..    ..    ..    20. *latifolium*

1. **A. pilosum** (*Oliv. & Hanb.*) *K. Schum.* in Engl., Pflanzenr. 4, 46, Zingib.: 206 (1904); Koechlin in Fl. Cam. 4: 45, t. 9 (1965). *Amomum pilosum* Oliv. & Hanb. in J. Linn. Soc. 7: 110 (1863); Baker in Hook., Ic. Pl. t. 2483 (1892); F.T.A. 7: 307. A herb 6–8 ft. high, with very short bracteate 1-flowered inflorescences at the base of the stem; labellum yellow.
  **S.Nig.**: near Umudike (fl. & fr. June) *Ariwaodo* ARS 1227! Oban *Talbot* 94! **F.Po.**: *Mann* 1415! Also in E. Cameroun.

2. **A. elliotii** (*Bak.*) *K. Schum.* l.c. 209 (1904); Chev. Bot. 625; Aké Assi, Contrib. 2: 16 [225]. *Amomum elliotii* Bak. in F.T.A. 7: 309 (1898). A herb in shady woods; inflorescences from the base of the stem, stout, about 1 ft. long, with a cluster of vanilla-scented flowers at the top; bracts on the peduncle about 2 in. long; petaloid staminode white, yellowish in the throat, fruits red and shining; in swamp forest, brackish lagoon bank, or wet grassland.
  **Sen.**: *Perrottet*! Yambalba, Cape Verde (May) *Leprieur*! **Guin.**: Erimakuna (= Hérémakuna) (fr. Mar. *Sc. Elliot* 5249! source of the Sassandra R., near Sahadougou (Mar.) *Chev.* 20872! **S.L.**: Njala (fr. Apr.) *Deighton* 4290! **Lib.**: Monserrado Co.: Marshall (fr. Feb.) *Baldwin* 11048! Bushrod I. (Feb.) *Baldwin* 11064! **Iv.C.**: Tabou to Bériby (fl. & fr. Aug.) *Chev.* 19954! Port Bouét (Feb.) *Leeuwenberg* 2711! Abouabou Forest (fr. Jan.) *Leeuwenberg* 2371! **Ghana**: Essiama (fl. & fr. Feb.) *Irvine* 2329! *Morton* A1642!
  [The Liberian specimens cited have only the midrib pubescent, but the long peduncle distinguishes them from *A. baumannii*.]

3. **A. subsericeum** (*Oliv. & Hanb.*) *K. Schum.* l.c. 217 (1904); Aké Assi, Contrib. 2: 16 [226]; Koechlin in Fl. Gabon 9: 36 (1964), and in Fl. Cam. 4: 48 (1965). *Amomum subsericeum* Oliv. & Hanb. in J. Linn. Soc. 7: 110 (1864); Baker in Hook., Ic. Pl. t. 2484 (1896); F.T.A. 7: 307. *Amomum glaucophyllum* K. Schum. in Engl., Bot. Jahrb. 15: 415 (1892); F.T.A. 7: 306; (type not seen for Ed. 2). *Aframomum glaucophyllum* (K. Schum.) K. Schum. in Engl., Pflanzenr. 4, 46: 216 (1904). A herb with a creeping rhizome; leaves bluish white beneath; inflorescences very short, in a cluster at the base of the stem or disposed on a creeping shoot, about 2-flowered; bracts closely imbricate, gradually increasing from the base upwards; petaloid staminode reddish or magenta, 2½ in. diam.; fruits narrowly pear-shaped 3 in. long.
  **Iv.C.**: Mamba Forest *Mangenot & Aké Assi* IA 543. **S.Nig.**: Calabar *Milne*. **W.Cam.**: Mbaw Plain (May) *Brunt* 464! Bopo to Banga, S. Bakundu F.R. (Mar.) *Binuyo & Daramola* FHI 35633! Barombi *Preuss* 254. Also in E. Cameroun, Gabon, Congo and Cabinda.
  [Plants with felted leaves from Nimba Mt. (*Adam* 20381, *Adames* 604) have inflorescences up to about 40 cm. long and pure white flowers; otherwise they are similar to *A. subsericeum*.]

4. **A. polyanthum** (*K. Schum.*) *K. Schum.* l.c. 207 (1904); F. W. Andrews, Fl. Sudan 3: 253; Koechlin in Fl. Cam. 4: 49, t. 9; N. Hallé in Adansonia, sér. 2, 7: 73, tt. 1, 2 (1967). *Amomum polyanthum* K. Schum. (1892)—F.T.A. 7: 309. Tall herb with rather large leaves softly pilose beneath; flowers rather small and numerous in a pedunculate head, labellum yellowish-purple.
  **W.Cam.**: Ndop Plain (Mar.) *Brunt* 202! Also in Gabon and Sudan.

5. **A. chlamydanthum** *Loes. & Mildbr.* in Notizbl. Bot. Gard. Berl. 10: 706 (1929). A herb 3 ft. or more high; flowers small, dirty white or yellowish in a congested inflorescence about 6 in. high.
  **S.Nig.**: Obudu Plateau, 5,000 ft. (fr. Nov.) *Tuley* 1041! **W. Cam.**: Likomba, Victoria Dist. (Nov.) *Mildbr.* 10712. Bamenda (Mar.) *Daramola* FHI 40549! **F.Po**: L. Biao (Sept.) *Melville* 487!
  [*Mildbr.* 10712, the type, was seen for Ed. 1, but it has since been destroyed in Berlin. The specimens cited are closely related to *A. zambesiacum* (Bak.) K. Schum. and *A. keniense* R. E. Fries. *Melville* 487 has much narrower leaves than the others.]

6. **A. longiscapum** (*Hook. f.*) *K. Schum.* l.c. 212 (1904). Erect herb with leafy stem 5–9 ft. high; flowers mauve, fruits dark red; in secondary forest.
  **Guin**: Mt. Nimba (fr. June) *Schnell* 3054! **S.L.**: Regent (Dec.) *Daniell* 3! Jagwema, Kouno (July, Aug. *Dawe* 540! Koyema, Gola of Tingi Hills (fr. Dec.) *Morton & Gledhill* SL 3226! **Lib.**: Gbanga (Sept.) *Linder* 724! Bilimu Mt. (Dec.) *Harley* 1319! Yratoke, Webo Dist. (July) *Baldwin* 6240! Mt. Bele, Nimba (Sept.) *Adames* 579! **Iv.C.**: Béyo (Jan.) *Leeuwenberg* 2457! Baléko (fr. Sept.) *Nozerau*! (See Appendix, p. 471.)

7. **A. exscapum** (*Sims*) *Hepper* in Kew Bull. 21: 133, figs. 1 & 2 (1967). *Amomum exscapum* Sims in Ann. Bot. 1: 548, t. 13 (1805). *Amomum grana-paradisi* of Hook. in Bot. Mag. t. 4603 (1851); F.T.A. 7: 304, partly (descr. & *Daniell, Afzelius*). *A. cuspidatum* (Gagnep.) K. Schum. l.c. 209 (1904); F.W.T.A., ed. 1, 2: 331 *A. leonense* K. Schum. l.c. 214 (1904). *A. oleraceum* A. Chev., Bot. 626, name only. A forest herb with leafy stems 4–6 ft. high, the leaves borne on the upper part; flowers white, flushed with pink, throat yellow; fruits red, pubescent and deeply corrugated longitudinally when fresh, corrugations become indistinct on drying and are often difficult to distinguish; seeds shiny black, pungent.
  **Guin.**: Conakry *Maclaud*! Sankaran, Kankan (fl. & fr. Mar.) *Pobéguin* 943! Ditinn (Apr.) *Chev.* 13461! **S.L.**: *Afzelius*! *Daniell*! Njala (fr. Jan., Apr.) *Deighton* 4287! 4289! *Pyne* 87! Kambia (Jan.) *Sc. Elliott* 4511! **Lib.**: Bilimu (Dec.) *Harley* 1553b! **Iv.C.**: Mt. Niénokué, Mid. Cavally *Chev.* 19468! Toupa, Lower Cavally *Chev.* 19851!

8. **A. sulcatum** (*Oliv. & Hanb.*) *K. Schum.* l.c. 205, fig. 26 A–D (1904); Chev. Bot. 626. *Amomum sulcatum*

Oliv. & Hanb. in F.T.A. 7: 304 (1898). Slender creeping rhizome, leafy stems up to 13 ft. high, tinged red together with midrib; flowers pale mauve or orange-yellow; fruits deeply sulcate, seeds (as in other species) like grape stones and very shiny; in forest and derived farmland.
**Guin.:** N'Zo (fr. Mar.) *Chev.* 21043! **S.L.:** Musaia (fr. Dec.) *Deighton* 4528! Njala (Apr.) *Deighton* 4288! Port Loko (July) *Deighton* 4344! **Lib.:** Cape Palmas (Sept.) *Hoffman* 14! *Gibson!* Bilimu Mt. (Dec.) *Harley* 1553a! Ganta (Feb.) *Baldwin* 11002! 11006! Nimba (Aug.) *Adames* 422! E. of Béyo (fr. Jan.) *Leeuwenberg* 2467! Doubtfully recorded by Koechlin from Cameroun and Gabon. (See Appendix, p. 472.)

9. **A. daniellii** (*Hook. f.*) *K. Schum.* l.c. 218 (1904); Aké Assi, Contrib. 2: 15 [225]; Koechlin in Fl. Gabon 9: 55, t. 10, and in Fl. Cam. 4: 64, t. 12. *Amomum daniellii* Hook. f. in Hook. Kew Journ. Bot. 4: 129, t. 5 (as *A. afzelii*) (1852); Baker in Bot. Mag. t. 4764. *A daniellii* var. *purpureum* Hook. f. l.c. 6: 294 (1854). *A. angustifolium* of F.T.A. 7: 308, partly (syn. & W. Afr. specimens). *Aframomum hanburyi* K. Schum. l.c. 217 (1904); Koechlin in Fl. Gabon 9: 56, and in Fl. Cam. 4: 65. A tall herb in moist shady places; flowers clustered at the base or on short peduncles, decorative with waxy red perianth and yellowish labellum; fruits red, in erect bunches.
**Guin.:** Conakry *Maclaud* 40! **S.L.:** Loma Mts. (fl. & fr. June) *Morton* SL 3663! **Iv.C.:** Grabo (July) *Chev.* 19639! Mamba *Mangenot & Aké Assi* IA 547. Yapo *Aké Assi* IA 4032. **S.Nig.:** Akilla F.R., Ijebu Dist. (fr. Mar.) *Emwiogbon* FHI 47172! Owo (fl. & fr. July) *Stanfield* 105! Oban F.R., Calabar (Feb.) *Onochie* FHI 36343x! Afi River F.R., Ogoja Prov. (May) *Jones & Onochie* FHI 17338! **F.Po.:** *Daniell! Barter! Mann* 1170! Mioko, 4,500 ft. (Feb.) *Exell* 855! Also in S. Tomé, Principe, E. Cameroun and Gabon. (See Appendix, p. 470.)

10. **A. limbatum** (*Oliv. & Hanb.*) *K. Schum.* l.c. 215 (1904); Koechlin in Fl. Gabon 9: 37, and in Fl. Cam. 4: 50 *Amomum limbatum* Oliv. & Hanb. in J. Linn. Soc. 7: 110 (1864); Baker in Hook., Ic. Pl. t. 2482; F.T.A. 7: 305. *A. lycobasis* K. Schum. l.c. (1904); F.W.T.A., ed. 1, 2: 331. A tall forest herb with leafy shoots up to 15 ft. high, with creeping rhizomes and short bracteate inflorescences, labellum pale purple.
**S.Nig.:** Usonigbe F.R., Benin (Feb.) *Onochie* FHI 31932! *Meikle* 577! Nyaje, Oban Dist. (Mar.) *Onochie* 36446x! Oban *Talbot* 95! *Richards* 3902! **W.Cam.:** Johann-Albrechtshöhe (= Kumba) (fl. & fr. Mar.) *Staudt.* 680! **F.Po:** *Mann* 99! 1171! Also in Gabon.
[*A. biauriculatum* K. Schum., tentatively included as a synonym of this species in Ed. 1, should be deleted, even though the type is apparently destroyed, as there is no evidence from the description that it is likely to be synonymous—F.N.H.]

11. **A. strobilaceum** (*Sm.*) *Hepper* in Kew Bull. 21: 136, fig. 3 (1967). *Amomum strobilaceum* Sm. in Rees, Cyclop. 39: Amomum (1819). *A. afzelii* Rosc. (1807). *A. palustre* Afzel. (1829.) *A. citratum* of Chev. Bot. 625. A forest herb with leafy stems 3–5 ft. high; flowers white; fruits scarlet.
**Guin.:** Kindia (Mar.) *Chev.* 12785! Boola to Manonkoro *Chev.* 20943! Nimba Mt. (fr. Aug.) *Schnell* 3387! **S.L.:** *Afzelius* 4! Guma, Freetown (fr. Mar.) *Hepper* 2500! **Lib.:** Mt. Bele road, Nimba (fr. Sept.) *Adames* 605! Nimba Mt. (Apr.) *Harley* 2216! *Adam* 21151! **Iv.C.:** Mt. Tonkoui, near Man (Mar.) *Breteler* 2979! E. of Béyo (fl. & fr. Jan.) *Leeuwenberg* 2539! Mid. Cavally, Keéta (fr. July) *Chev.* 19346! (See Appendix, p. 470.)

12. **A. stanfieldii** *Hepper* in Kew Bull. 22: 462, fig. 7 (1968). A rhizomatous herb about 3 ft. high; flowers white; fruits glossy scarlet, 3-angled.
**S.Nig.:** Owo (fr. July) *Stanfield* 87! Gambari F.R., Ibadan (fl. Mar., fr. Apr.) *Emwiogbon* FHI 45514! *Adebuwyi* FHI 45907! *Stanfield* 47087!

13. **A. sp.** *A. A. dalzielii* Hutch, F.W.T.A., ed. 1, 2: 331 (1936), English descr. only.
**W.Cam.:** Buea (Feb.) *Dalz.* 8235!
[Although this appears to be a distinct species, validation of the name is deferred until adequate material is available for a full description to be made.—F.N.H.]

14. **A. sceptrum** (*Oliv. & Hanb.*) *K. Schum.* l.c. 214 (1904); Morton, Lilies & Orch. fig. 38; Koechlin in Fl. Gabon 9: 40, t. 7, and in Fl. Cam. 4: 52, t. 10; Chev. Bot. 626. *Amomum sceptrum* Oliv. & Hanb. (1864) —Baker in Bot. Mag. t. 5761; F.T.A. 7: 306. *Aframomum granum-paradisi* of K. Schum. l.c. 213 (1904) partly (excl. *Daniell*); F.T.A. 7: 304; F.W.T.A., ed. 1, 2: 331. *A cereum* (Hook. f.) K. Schum. l.c. 210 (1904). *Amomum cereum* Hook. f. (1854)—Baker in Hook. Ic. Pl. t. 2477; F.T.A. 7: 309. ? *Aframomum rostratum* K. Schum. l.c. 215 (1904). Forest herb with pale lilac flowers; inflorescences may be about 1 ft. long at the foot of the leafy shoot or at ground level from the rhizome at some distance from the leaves.
**S.L.:** Regent *Daniell!* **Lib.:** Suacoco, Gbanga (Sept.) *Daniel & Okeke* 16! Jaurazon (Mar.) *Baldwin* 11475! Nimba Reserve (Sept.) *Adames* 596! **Iv.C.:** Grabo to Taté (Aug.) *Chev.* 19768! Noé, Bas Cavally (Aug.) *Chev.* 19839! **Ghana:** Pokoase (Apr.) *Plumptre* 102! Aburi (Mar.) *Johnson* 865! Amedzofe (Mar.) *Morton* GC 6494! **Dah.:** Ouidah (fr. Apr.) *Chev.* 23444! Port-Novo (Mar.) *Chev.* 23345! **N.Nig.:** Jemaa Dist. (Aug.) *Killick* 8! Nupe *Barter* 1543! **S.Nig.:** Akilla Plantations (Apr.) *Jones & Onochie* FHI 17401! Ikoyi Plains (Jan.) *Dalz.* 1280! Onitsha *Barter* 1787! Brass *Barter* 38! Port Harcourt (Sept.) *Stubbings* 58! Also in E. Cameroun, Gabon, Middle Congo and Angola. (See Appendix, p. 472.)

15. **A. erythrostachyum** *Gagnep.* in Bull. Soc. Bot. Fr. 51: 444 (1904); Hepper in Kew Bull. 21: 137 (1967).
**Guin.:** Sankaran (Mar.) *Pobéguin* 941!
[The status of this species is not clear: further material is required.]

16. **A. citratum** (*Pereira*) *K. Schum.* in Engl., Pflanzenr. 4, 46: 214 (1904); Koechlin in Fl. Gabon 9: 42, t. 8, and in Fl. Cam. 4: 54, t. 11. *Amomum citratum* Pereira in Pharm. Journ. 9: 313 (1850)—F.T.A. 7: 308; Baker in Hook., Ic. Pl. t. 2478. *A. macrolepis* K. Schum. (1892)—F.T.A. 7: 306. A herb to 10 ft. high; inflorescence short and almost globose with large imbricate bracts; labellum pink, about 3 in. high; in forest.
**S.Nig.:** Oban *Talbot* 32! 33! 90! **W.Cam.:** Barombi *Preuss* 5. Also in E. Cameroun and Gabon.

17. **A. leptolepis** (*K. Schum.*) *K. Schum.* l.c. 216 (1904). *Amomum leptolepis* K. Schum. (1892)—F.T.A. 7: 307. *A. sceptrum* of F.W.T.A., ed. 1, 2: 221, partly (*Chipp* 129, *Johnson* 866), not of (Oliv. & Hanb.) K. Schum. A forest herb 5–6 ft. high; flowers pink, at ground level.
**Iv.C.:** Montézo, Altié (Feb.) *Chev.* 17384! Danané *Chev.* 21257! **Ghana:** Adeambra (Mar.) *Chipp* 129! L. Bosumtwe (Apr.) *Adams* 2456! Aburi (Mar.) *Johnson* 866! Agogo (Apr.) *Adams* 2590! Hohoe (Apr.) *Morton* GC 9172! **W.Cam.:** Barombi *Preuss* 555! Also in E. Cameroun.

18. **A. baumannii** *K. Schum.* in Engl., Pflanzenr, 4, 46: 220, fig. 26 F–H (1904); Aké Assi, Contrib. 2: 15 [224]. *A. sceleratum* A. Chev. (1917). Leafy stems about 6 ft. high; flowers narrow in short inflorescences, pink or carmine, fruits smooth, glossy scarlet; in marshy shady places in hilly savanna.
**Guin.:** Farana *Chev.* 20666! Bérégia *Chev.* 20782! **Iv.C.:** Yapo *Aké Assi* IA 5776! Grabo *Chev.* 19639. **Ghana:** Amedzofe (Mar.) *Hepper & Morton* A3041! **Togo Rep.:** Misahöhe (June) *Baumann* 106 (seen for Ed. 1). **Dah.:** Atacora Mts. *Chev.* 24210! **N.Nig.:** Bonu F.R., Gwari Dist., Niger Prov. (fl. & fr. June) *Onochie* FHI 38499! Gidan Tukura, Gawun Dist., Niger Prov. (Dec.) *Daley* FHI 32276! Mambila Plateau (fl. & fr. Jan.) *Latilo & Daramola* FHI 34406! *Hepper* 1751! **S.Nig.:** Awba Hills F.R., Ibadan (fr. Oct.) *Jones* FHI 5995! Afi River F.R., Ikom (May) *Jones & Onochie* FHI 17338! (See Appendix, p. 470.)

19. **A. melegueta** *K. Schum.* l.c. 204 (1904); Hepper in Kew Bull. 21: 130, t. 7 (1967); Koechlin in Fl. Gabon 9: 58, t. 12, and in Fl. Cam. 4: 68, t. 14. *Amomum melegueta* Rosc., Monandr. Pl. t. 98 (1828); F.T.A. 7: 303, non Griseke (1792). *Aframomum meleguetella* K. Schum. l.c. 205 (1904). Tufted with leafy stems up to 3 ft. high; inflorescences at the base of the leafy stems, 1-flowered, very short; labellum pink or lilac; in shade, often cultivated. Melegueta or (by corruption) Alligator pepper, Grains of Paradise.
**Guin.:** Sampouyara to Bérézia (cult., Feb.) *Chev.* 20773! **S.L.:** Regent, (cult.) *Daniell!* Makene (cult., May) *Deighton* 4144! Njala (cult., Feb., May) *Deighton* 5014! Bayabaya *Sc. Elliot* 4577! **Lib.:** Nimba (cult. fl. Dec., fr. Mar.) *Adam* 21105! *Adames* 838! **Ghana:** Aburi Gardens (cult.) *Irvine* 2049! **Iv.C.:**

Alépé (Mar.) *Chev.* 17458! Kassignié (fr. Jan.) *Chev.* 17205! **Dah.**: Allada (Mar.) *Chev.* 23418! **N.Nig.**: Lapai (Mar.) *Yates* 38! **S.Nig.**: Yoruba *Daniell!* Ikure (Jan.) *Holland* 254! Aguku Dist. *Thomas* 845! Ibadan (cult. fr. Sept.) *Binuyo & Onyeachusim* FHI 56543! **F.Po**: *Daniell!* Apparently native in W. Africa but also widely cultivated in various parts of tropical Africa, S. America and elsewhere in the tropics. (See Appendix, p. 471.)

20. **A. latifolium** (*Afzel.*) *K. Schum.* l.c. 209 (1904); Aké Assi, Contrib. 2: 16 [225]. *Amomum latifolium* Afzel. (1813)—F.T.A. 7: 305. ? *A. kayseranum* K. Schum. in Engl., Bot. Jahrb. 25: 415 (1892); F.T.A. 7: 305; Koechlin in Fl. Cam. 4: 66, t. 14. Herb with leafy shoots 5–10 ft. high and white or pinkish flowers from short inflorescences borne separately on long creeping rhizome; in savanna.
**Guin.**: Nimba Mts. *Schnell* 555! **S.L.**: *Afzelius!* Kissy *Daniell* 2! Sugar Loaf Mt. *Daniell* 5! **Lib.**: Du R. (July) *Linder* 99! Ganta (Sept.) *Baldwin* 12499! Bilimu (Sept.) *Harley* 15266! Nimba (Aug.) *Adames* 436! **Iv.C.**: Kinou, near Boundiali *Aké Assi* IA 3060. **Ghana**: Ejura (Aug.) *Andoh* FH 5034! **N.Nig.**: Sanga F.R., Jemaa Div. (May) *E. W. Jones* 80! Acharane F.R., Ankpa Dist. (June) *Daramola* FHI 38031! **S.Nig.**: Olokemeji F.R. (July) *Hopkins* FHI 52220! Owo (July) *Stanfield* 104! Ogoja Prov. *Rosevear* 24/30a! **W.Cam.**: Buea (Feb.) *Preuss* 826! Ndop Plain (Apr.) *Brunt* 353! Also in E. Cameroun. (See Appendix, p. 471.)
[The type of *A. kayseranum* was collected at Buea in forest: it may not, therefore, be conspecific with *A. latifolium* a savanna species—F.N.H.]

# 4. COSTUS Linn., Sp. Pl. 2 (1753); F.T.A. 7: 297. *Cadalvena* Fenzl (1865)—F.T.A. 7: 296.

Leaves in a basal rosette or solitary; inflorescence amongst leaves:
  Leaves several in a basal rosette flat on the ground, obovate-orbicular, rounded to a mucronate apex, broadly cuneate at the base, up to 12 cm. long and 10 cm. broad, with fairly numerous ascending nerves and fine parallel transverse veins, pubescent beneath; flowers arising from the centre of the rosette; calyx spathaceous, tube 4 cm. long; savanna herb .. .. .. .. .. .. 1. *spectabilis*
  Leaves solitary, broadly obovate-elliptic, not acuminate, shortly narrowed at the base, 8–12 cm. long, 6–8 cm. broad, with numerous ascending nerves, glabrous; flower-spike small, sessile at the apex of the stem within the leaf-sheath, 2–3-flowered
                                          2. *engleranus*
Leaves numerous borne on tall stems; inflorescences may be separate, terminating leafy stems or lateral:
  Inflorescences lateral with few flowers on leafy stems; leaves glabrous; ligules tubular and very long, reaching nearly to the next leaf:
    Inflorescences surrounded by a few imbricated ovate bracts up to 2 cm. long forming a small compact subsessile head about 2 cm. across; floral bracts lanceolate, pubescent; calyx 1·5 cm. long, 3-toothed, glabrous; leaves oblong, long-acuminate, narrowed and shortly petiolate, 12–20 cm. long, 4·5–5 cm. broad .. .. 3. *talbotii*
    Inflorescences with small bracts at the base about 5 mm. long, head shortly stalked; floral bracts oblong, truncate, pubescent; calyx 1–1·5 cm. long, 3-toothed, shortly pubescent; leaves ovate, acuminate, cuneate at the base, shortly petiolate (8–)10–18 cm. long, (3–)4–8 cm. broad .. .. .. .. 4. *lateriflorus*
  Inflorescences terminal or on separate shoots, with numerous flowers; leaves obovate-elliptic, cuneate at the base, long-acuminate, 10–31 cm. long, 3–7 cm. broad, glabrous to pubescent, shortly petiolate:
    Each bract covering one flower:
      Bracts all leafy at apex with broadly ovate appendage up to 3 cm. long, and 2 cm. broad; inflorescences apparently always terminal; plant densely pilose
                                          5. *fissiligulatus*
      Bracts (except sometimes outer ones) not leafy at apex; leaf-sheaths without a rim below the apex; plant glabrous or pubescent:
        Leaves and stems densely setose and with fine short pubescence beneath i.e. two kinds of hairs; edge of leaf-sheath lanate with long matted hairs; inflorescences always basal; flowers uniform pale pink .. .. .. 6. *schlechteri*
        Leaves pubescent (at the most sparsely setose on the stem), or glabrous:
          Flowers white with yellow throat; inflorescences basal and terminal; ochrea as long as broad; leaves shortly appressed-pubescent .. .. .. 7. *dubius*
          Flowers white with rich pink tip; inflorescences always terminal; ochrea very short; leaves pilose beneath .. .. .. .. .. .. 8. sp. *A*
    Each bract covering two flowers (at different stages of development); flowers pink, yellow or white; large cone-like inflorescences usually terminating leafy shoots and sometimes short leafless stems:
      Leaf-sheaths usually without a rim below the apex or if present then lacking long bristles:
        Calyx-tube included behind the bract, more or less glabrous:
          Leaf-sheaths expanded 2–5 mm. beyond insertion of petiole; leaves glabrous or slightly pubescent; stems usually stout .. .. .. .. 9. *afer*
          Leaf-sheaths prolonged into a cylindrical ochrea 1 cm. or more long; leaves glabrous; stems slender; inflorescence cylindrical .. .. 10. *deistelii*
        Calyx-tube longer than the bract, usually densely puberulous .. 11. *lucanusianus*
      Leaf-sheaths with a long bristly rim below the apex; calyx-tube usually much longer than the bract, puberulous; inflorescence globose .. 11. *lucanusianus*

1. **C. spectabilis** (*Fenzl.*) *K. Schum.* in Engl., Bot. Jahrb. 15: 422 (1892), and in Engl., Pflanzenr. 4, 46: 421; F.W.T.A., ed. 1, 2: 334; Berhaut, Fl. Sén. 62; Koechlin in Fl. Gabon 9: 65, t. 13, and in Fl. Cam. 4: 75, t. 15. *Cadalvena spectabilis* Fenzl. (1865)—F.T.A. 7: 297; Bot. Mag. t. 7992; Chev. Bot. 624. *C. dalzielii* C. H. Wright in Hook., Ic. Pl. t. 3013 (1915). Herb with a rosette of about 4 suborbicular fleshy leaves spread flat on the ground; flowers bright orange or yellow; in rocky savanna.
   **Sen.:** Tambacounda *Berhaut* 3150! Patako *Trochain* 3900! **Mali:** Kita *Paroisse* 34! **Guin.:** Botala, near Kankan (July) *Collenette* 80! **S.L.:** Gberia Timbako, Sulima (June) *Bakshi* 223! Kafuru *Dawe* 498! **Iv.C.:** Gouéhouma, Haute Sassandra *Chev.* 21659. **Ghana:** Gambaga (June) *Akpabla* 663! Shiare to Chilinga Nkwanta Hills (May) *Morton* A3965! **N.Nig.:** Sokoto Prov. (June) *Dalz.* 560! Damau, Anchau Dist. (June) *Ogua* FHI 7842! Zaranda Mt., 5,000 ft. (May) *Lely* 198! Vom *Dent Young* 239! Keana *Hepburn* 48! Kolba, Yola (July) *Dalz.* 229! **S.Nig.:** Obudu Plateau, 3,000–5,000 ft. *Stone* 88! **W.Cam.:** Ndop Plain (Mar.) *Brunt* 280! *Mildbr.* 8954! Extends to Sudan, E. Africa, Rhodesia and Angola. (See Appendix, p. 473.)

2. **C. engleranus** *K. Schum.* in Engl., Bot. Jahrb. 15: 419, t. 13 (1892), and in Engl., Pflanzenr. 4, 46: 424; F.T.A. 7: 300; Koechlin in Fl. Gabon 9: 66, t. 13, and in Fl. Cam. 4: 75, t. 15. A creeping unifoliolate fleshy herb forming dense ground cover in moist forest; flowers white with yellow throat.
   **Iv.C.:** Alépé (Mar.) *Chev.* 17407! Bianoua (Sept.) *Aké Assi* 6181! **Ghana:** Tarkwa (June) *Enti* FH 6439! S. Scarp F.R., Obomeng (Dec.) *Adams* 5150! Kibi *Morton*! Begoro, Akim (Apr.) *Irvine* 1357! **S. Nig.:** Idanre F.R. (July) *Onochie* FHI 33385! Omo F.R., Ijebu Dist. (Apr.) *Onochie* FHI 15512! Afl River F.R., Obudu Dist. (June) *Onochie & Jones* FHI 17341! Oban *Talbot* 885! **F.Po:** Ureka (Mar.) *Guinea* 2293! Also in E. Cameroun.

3. **C. talbotii** *Ridl.* in Cat. Talb. 111 (1913). Herb with wiry stem covered with long cylindrical ligules; flowers in apparently lateral heads.
   **S.Nig.:** Oban *Talbot* 1521!
   [*Talbot* 839 included as *C. dinklagei* K. Schum. in Ed. 1 appears to be a mixture. A leafy shoot with densely hairy sheaths may be *C. dinklagei* but it is sterile and the type from E. Cameroun is not available for comparison. The remainder of *Talbot* 839 included glabrous detached leaves, a leafless stem with long sheaths and a few flowers resembling those of *C. talbotii* or *C. lateriflorus*—F.N.H.]

4. **C. lateriflorus** *Bak.* in F.T.A. 7: 301 (1898); Koechlin in Fl. Gabon 9: 76, t. 16. An epiphyte in forest; flowers pale yellow with a red patch inside.
   **S. Nig.:** Orukim to Unyene, near Stubbs Creek, Eket Dist. (May) *Onochie* FHI 33177! Ikwette, Obudu *Savory & Keay* FHI 25202! Extending to Gabon.

5. **C. fissiligulatus** *Gagnep.* in Bull. Soc. Bot. Fr., sér. 4, 2: 93 (1902); Koechlin in Fl. Gabon 9: 78, t. 17, and in Fl. Cam. 4: 86, t. 18. A hairy herb with foliaceous bracts in a cone-like inflorescence.
   **S.Nig.:** Calabar to Mamfe (Feb.) *Onyeachusim & Latilo* FHI 54099! Ubiaja to Ihishi, Benin (Dec.) *Lowe* FHI 47973! Extending to Gabon.
   [Several similar species have been described and the specimens cited are tentatively referred to *C. fissiligulatus* Gagnep. They are possibly the robust var. *major* Gagnep. l.c. 94 (1902), which approaches *C. dewevrei* De Wild. & Th. Dur.—F.N.H.]

6. **C. schlechteri** *Winkler* in Engl., Bot. Jahrb. 41: 275 (1908). A herb with hairy leaves and stem, up to 6 ft. high, with terminal inflorescence.
   **Lib.:** Sarbo, Webo Dist. (July) *Baldwin* 6410! **Iv.C.:** Dyolas country, Upper Sassandra (Aug.) *Chev.* 21517! **S.Nig.:** Afi River F.R. (May) *Jones & Onochie* FHI 18913! Calabar to Mamfe, Ikpai Dist. (Feb.) *Onyeachusim & Latilo* FHI 54099! **W.Cam.:** Victoria *Winkler* 25a.

7. **C. dubius** (*Afzel.*) *K. Schum.* in Engl., Pflanzenr. 4, 46: 395 (1904); Hepper in Kew Bull. 22: 464 (1968). *Zingiber dubium* Afzel., Remed. Guin. 3: 9 (1810). *Costus zechii* K. Schum. l.c. (1904). *C. albus* A. Chev. ex J. Koechl., Fl. Gabon 9: 68, t. 14 (1964), and in Fl. Cam. 4: 78, t. 16; Chev. Bot. 607 (1920), name only; F.W.T.A., ed. 1, 2: 334. Tall perennial herb up to 6 ft. high; flowers white in succulent inflorescences either terminating the leafy shoots or short leafless stems at their foot; in rain forest.
   **S.L.:** Gberia Fotombu (Sept.) *Small* 286! Rokel R. Bridge (Apr.) *Hepper* 2565! Mokebi, Kori (May) *Jordan* 900! Bonjema, Kori (Dec.) *Pyne* 88! Bwedu (Apr.) *Deighton* 3172! **Lib.:** Brewersville (Nov.) *Barker* (1085! Dukwai R. (Oct., Nov.) *Cooper* 14! **Iv.C.:** Bingerville *Chev.* 15217! Bettié (Mar.) *Chev.* 17579! Lamé (Nov.) *Leeuwenberg* 1897! **Ghana:** Bunso (Apr.) *Morton* A523! Bubrum (Dec.) *Thomas* D124! **N.Nig.:** Sanga River F.R., Jemaa Div. (Apr., May) *E. W. Jones* 70! 212! **S.Nig.:** Usonigbe F.R. (Feb.) *Onochie* FHI 27696! Benin (Aug.) *Stanfield* 115! Calabar (Aug.) *Holland* 77! Also in E. Cameroun, C. African Rep., Middle Congo and Gabon.
   [*C. littoralis* K. Schum. (l.c. 1904) in Liberia have a small terminal inflorescence and glabrous leaves like *C. deistelii*, but with one flower per bract it must be close to or part of *C. dubius*.]

8. **C. sp. A.** Tall herb 6–7 ft. high with terminal inflorescences; flowers pink tipped; in savana forest.
   **Ghana:** Sorebella, Turner (Mar.) *Morton* GC 8883! Nsawkaw (Dec.) *Adams* GC 5404!
   [When better known this plant may prove to be taxonomically distinct.]

9. **C. afer** *Ker-Gawl.* in Bot. Reg. t. 683 (1823); Hook. in Bot. Mag. t. 4979; F.T.A. 7: 299; K. Schum. l.c. 392; Chev. Bot. 627; F.W.T.A., ed. 1, 2: 334; Koechlin in Fl. Gabon 9: 84, t. 3, and in Fl. Cam. 4: 92, t. 20. *C. oblitterans* K. Schum. l.c. 393 (1904). *C. anomocalyx* K. Schum. l.c. 396 (1904). *C. bingervillensis* A. Chev., partly; *C. insularis* A. Chev.; *C. luteus* A. Chev., Bot. 627–628, names only. Tall perennial herb with leafy stems up to 10 ft. high; flowers white and yellow with pink tip in succulent terminal inflorescences.
   **Sen.:** Sedhiou, Casamance *Chev.*! **Port.G.:** (Jan.) *Esp. Santo* 1111! Teixeira Pinto (Jan.) *d'Orey* 160! **Guin.:** Los Is. *Selwyn!* Conakry *Maclaud*! Kindia (Mar.) *Chev.* 12786! **S.L.:** *Don*! Musaia (Dec.) *Deighton* 4570! Batkanu (Apr.) *Thomas* 80! **Lib.:** Monrovia *Whyte*! Baiima *Bunting* 46! Du R. (Aug.) *Linder* 264! Dukwai R. (Oct., Nov.) *Cooper* 14! Kakatown *Whyte*! Nyaake, Webo Dist. (June) *Baldwin* 6126! **Iv.C.:** Tonkoui Mt. (Mar.) *Leeuwenberg* 2963! Anguédodou Forest, Adiopodoumé (May, Nov., Dec.) *de Wit* 7850! *Leeuwenberg* 1874! *Gruys* 26! **Ghana:** Anwhiaso F.R. (Sept.) *Andoh* FH 5241! Kwahu-Tafo to Mpraeso (Apr.) *Morton* A737! Nsawkaw (Dec.) *Adams* 3168! Fumso (Mar.) *Darko* 545! **Dah.:** Pira to Cabolé (May) *Chev.* 23765! **N.Nig.:** Zungeru (Aug.) *Dalz.* 277! Anara F.R., Zaria (Oct.) *Keay* FHI 21106! Tep, Mambila Dist. (Feb.) *Hepper* 1853! **S.Nig.:** Epe (Aug.) *Gillett* 15258! Abo *Barter* 417! Ibu *T. Vogel* 1! **F.Po:** Moka (Aug.) *Wrigley* 505! Widespread in the forest region of tropical Africa. (See Appendix, p. 472.)

10. **C. deistelii** *K. Schum.* in Engl., Pflanzenr. 4, 46: 393 (1904). *C. pulcherrimus* A. Chev., Bot. 628, name only. A herb with several leafy stems 3–8 ft. high; flowers in terminal heads, yellow or orange; in forest.
   **S.L.:** Guma (Mar.) *Hepper & Pyne* 2501! York Pass (Sept., Dec.) *Deighton* 3352! *Melville & Hooker* 625! Kenema (Oct.) *Deighton* 5236! **Lib.:** Gbanga (Sept.) *Linder* 387! Reputa (Jan.) *Harley* 1103! Jabroke (July) *Baldwin* 6427! **Iv.C.:** Toula to Nékaougnié *Chev.* 19568! Grabo *Chev.* 19717! 19732! **Ghana:** Bunso (Aug.) *Darko* WACRI 953! Atewa Range F.R. (Jan.) *de Wit & Morton* A2931! Ajaka (Dec.) *Adams* 2202!

11. **C. lucanusianus** *J. Braun & K. Schum.* in Mitteil. Deut. Schutzgeb. 2: 151 (1889); F.T.A. 7: 299; Schlechter, Kautschuk-Exp. 64, with fig. (1901); K. Schum. l.c. 392, incl. var. *major* K. Schum. (1904); Koechlin in Fl. Gabon 9: 83, t. 19. *C. bingervillensis* A. Chev., Bot. 627, partly (*Chev.* 17280), name only. A tall herb, stems 6–10 ft. high with leaves spirally arranged and a terminal inflorescence; flowers with a red lobe; in forest.
   **Guin.:** Conakry (June) *Debeaux!* Beyla (Mar.) *Chev.* 20866! **S.L.:** Alikalia, N. Prov. (Apr.) *Glanville* 197! **Lib.:** Gbanga (Sept.) *Linder* 390! Belefanai, Gbanga (Dec.) *Baldwin* 10554! Ganta (Oct.) *Harley*! **Iv.C.:** Haut Sassandra (May) *Chev.* 21516! Adiopodoumé (Nov.) *Roberty* 15406! Bingerville *Chev.* 17280!

Prolo to Bliéron *Chev.* 19883! **S.Nig.**: Benin *Dennett* 17! Sapoba (Oct.) *Kennedy* 1750! *Onyeachusim & Emwiogbon* FHI 43220! Oban *Talbot* 840! Aboh *T. Vogel!* Port Harcourt (Sept.) *Stubbings* 40! **W. Cam.**: Buea, 3,000 ft. (Nov.) *Migeod* 105! **F.Po**: (June) *Mann* 439! Extends to Gabon. (See Appendix, p. 473.)

[Ker-Gawler (in Reg. Bot. t. 683 (1823)) pointed out that he obtained from seeds of his *C. afer* a number of plants with a great range of variation including some with a ring of hairs at the base of the ligule. Such plants were later separated as *C. lucanusianus* and it may be that Ker-Gawler used mixed seed, or population studies will prove that *C. lucanusianus* cannot be maintained as a species distinct from *C. afer*—F.N.H.]

## 180. CANNACEAE

### By F. N. Hepper

Tall leafy perennial rhizomatous herbs. Leaves large, broad, pinnately nerved, with a distinct midrib. Flowers racemose or paniculate, bracteate, zygomorphic, bisexual, mostly large and brightly coloured. Perianth double, the outer calyx-like, the inner corolla-like. Sepals 3, imbricate, free, herbaceous. Petals 3, connate at the base and adnate to the staminal column. Stamens petaloid, 3 outer sterile, imbricate, 2 inner more or less connate, 1 free; anther solitary, 1-locular, adnate to the side of the petaloid portion. Ovary inferior, 3-locular; ovules numerous, axile. Fruit a capsule, pericarp often warted. Seeds many, rounded, with very hard endosperm.

Mainly in tropical and subtropical America.

**CANNA** Linn., Sp. Pl. 1 (1753); F.T.A. 7: 327; Kränzlin in Engl., Pflanzenr. 4, 47 Cannac.: (1912).

Stems glabrous, erect; leaves ovate-elliptic, broadly acuminate, abruptly cuneate at the base, up to 40 cm. long and 25 cm. broad, glabrous, with numerous fine close parallel nerves diverging from the midrib at an angle of 45°; racemes terminal, few-flowered; pedicels very short; bracts ovate, about 1·3 cm. long; sepals lanceolate, 1·3 cm. long; petals linear-lanceolate, acute, about 4 cm. long; staminodes 3, one of them bidentate at the apex, spathulate-oblanceolate, about 4 cm. long; style as long as the staminodes, flattened; fruit closely muricate .. .. .. *indica*

**C. indica** *Linn.* Sp. Pl. 1 (1753); Morton, W. Afr. Lilies & Orchids 52, fig. 62. *C. indica* subsp. *orientalis* (Rosc.) Bak. in F.T.A. 7: 328 (1898). *C. orientalis* Rosc. (1827). *C. bidentata* Bertol. (1859)—Kränzlin l.c. 46; F.W.T.A., ed. 1, 2: 335. Herb 3–4 ft. high with scarlet or orange-red flowers. The Indian Shot plant. (See Appendix, p. 474.)

[Kränzlin (l.c.) recognised the tropical African plants as a distinct native species under the name of *C. bidentata* Bertol. But it is found in and near towns and villages, mainly in the forest belt, and it was presumably originally an escape from cultivation. It would seem better to regard the wild African plants as naturalised *C. indica*, a native of tropical America, and the cultivated plant as a stouter, finer form. D. P. Stanfield reports two forms sometimes occurring in mixed populations in Nigeria: *form* 1 with purplish leaf margins and red fruit, *form* 2 with leaves uniformly green and green fruits. L. Aké Assi also records two forms growing side by side in Ivory Coast.]

## 181. MARANTACEAE

### By F. N. Hepper

Perennial herbs. Leaves in two rows, differentiated into an open sheath, stalk and blade, the stalk often winged, but terete and pulviniform towards the apex, this being referred to as the calloused portion (see Fig. 345), the blade sometimes with one straight and one curved side, with closely parallel numerous nerves diverging obliquely from the midrib. Flowers bisexual, asymmetric, in a terminal bracteate spike or panicle, or the inflorescence arising from the rhizome. Perianth mostly differentiated into calyx and corolla; outer segments free, inner more or less tubular, divided into 3 mostly unequal parts. Fertile stamen 1; anther 1-locular; staminodes variously petaloid. Ovary inferior, 3–1-locular, sometimes 2 of the loculi infertile; style stout, often dilated at the apex. Ovule solitary, erect. Fruit fleshy or a loculicidal capsule. Seeds with abundant endosperm and much incurved or folded embryo, and often with an aril.

Tropics and subtropics, mostly in moist or swampy primary forest.

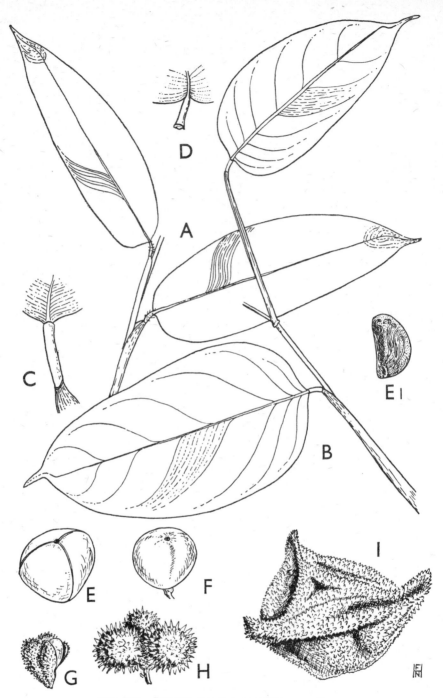

FIG. 345.—LEAVES AND FRUITS OF MARANTACEAE

A, *Marantochloa congensis* (K. Schum.) Léonard & Mullenders—leafy shoot, × ⅔. B, *M. filipes* (Benth.) Hutch.—leafy shoot, × ⅔. C, *Hypselodelphys violacea* (Ridl.) Milne-Redh.— calloused portion of petiole, × 2. D, *Trachyphrynium braunianum* (K. Schum.) Bak.— calloused portion of petiole, × 2. E, *Megaphrynium macrostachyum* (Benth.) Milne-Redh.— fruit, × 1. El, seed, × 1. F, *Sarcophrynium prionogonium* (K. Schum.) K. Schum.— fruit, × 1. G, *Trachyphrynium braunianum* (K. Schum.) Bak.— fruit, × 1. H, *Hypselodelphys poggeana* (K. Schum.) Milne-Redh.—fruit, × 1. I, *H. violacea* (Ridl.) Milne-Redh.— fruit, × 1.

Stems simple, usually bearing simple leaf and inflorescence, or stem absent with petiole
arising from rhizome; petiole with no distinct circular groove at junction of petiole
with the calloused portion; ovary and fruit smooth, not muricate:
Inflorescence arising at ground-level, 6–15 cm. long:
  Inflorescence arising from the lowest node of the leaf, simple or forked; fruits 3-winged,
    3 cm. diam.; leaves very large, 60 cm. or more long, 40 cm. broad
                                                      1. **Thaumatococcus**
  Inflorescence arising separate from the leaves, simple; leaves up to 35 cm. long and
    10 cm. broad ..    ..    ..    ..    ..    ..    ..    ..    2. **Afrocalathea**
Inflorescence aerial; fruits more or less rounded or cylindrical, not winged:
  Leaves strongly asymmetrical with one side nearly straight and the other rounded:
    Fruit 3-seeded, more or less rounded; (see also below) ..    ..    3. **Marantochloa**
    Fruit 1-seeded, cylindrical; inflorescence with closely imbricated bracts and erect
      branches    ..    ..    ..    ..    ..    ..    ..    ..    4. **Halopegia**
  Leaves not strongly asymmetrical, both sides rounded:
    Ovary 1-locular with 1 erect ovule; fruit 1-seeded, the seed filling the cavity,
      arillate; inflorescences lax panicles; in open swamps ..    ..    ..    5. **Thalia**
    Ovary 3-locular; inflorescence borne on a simple stem terminating in a single leaf;
      in shady places:
      Nodes of inflorescences each with 1 cymule; each bract enclosing the part of the
        inflorescence above it, caducous at anthesis; each cymule with 1 fleshy bracteole;
        fruit with well marked sutures; seed arillate ..    ..    ..    6. **Megaphrynium**
      Nodes of inflorescences each with 2–4 cymules; bracts not enclosing the part of
        the inflorescence above it, persistent; each cymule with 2 fleshy bracteoles;
        fruit without obvious sutures; seeds without an aril ..    .. 7. **Sarcophrynium**
Stems branched, sometimes repeatedly so to form a scrambling plant, leaves numerous;
ovary and fruit 3-locular, smooth or muricate:
  Junction of the petiole with the calloused portion marked by a distinct circular groove
  with thickening (see Fig. 345); ovary and fruit muricate:
    Calloused portion of the petiole separated from the midrib on the lower surface by a
    slight but distinct thickening and on the upper surface by a prominent beak;
    fruits indehiscent; seeds not arillate    ..    ..    ..    .. 8. **Hypselodelphys**
    Calloused portion of petiole continuous with the midrib on the lower surface and not
    marked by any disjunction, and on the upper surface the junction with the midrib
    marked by an interruption but not beaked; fruits dehiscent; seeds arillate
                                          9. **Trachyphrynium**
  Junction of the petiole with the calloused portion not marked by a circular groove,
  only marked by a change in texture; calloused portion uninterrupted with midrib on
  lower surface (see Fig. 345); fruits not muricate:
    Bracts suborbicular; stems often prickly; leaves hardly asymmetric, both sides
    curved ..    ..    ..    ..    ..    ..    ..    ..    ..    10. **Haumania**
    Bracts oblong; stems not prickly; leaves strongly asymmetric with one side curved
    and the other side almost straight; cymules without fleshy bracteoles; seeds
    arillate:
      Bracts acute; flowers at different levels; fruit ultimately dehiscent; inflorescence
      erect ..    ..    ..    ..    ..    ..    ..    ..    ..    3. **Marantochloa**
      Bracts obtuse; flowers more or less side by side; fruits indehiscent; inflorescence
      more or less reflexed    ..    ..    ..    ..    ..    ..    11. **Ataenidia**

The arrow-root, *Maranta arundinacea* Linn., a native of tropical S. America, is widely cultivated in
gardens in the forest zone of W. Africa.

## 1. THAUMATOCOCCUS Benth. in Benth. & Hook. f., Gen. Pl. 3: 652 (1883); F.T.A. 7: 320.

Leaves ovate-elliptic, rounded-truncate at the base, shortly acuminate, up to 46 cm.
long and 30 cm. broad, papery, with very numerous parallel nerves diverging from
the midrib at an angle of about 45°; petiole subterete; spikes simple or forked,
about 10 cm. long; bracts imbricate, 4 cm. long; flowers as long as the bracts;
sepals broadly linear, 1 cm. long; corolla-tube very short, lobes oblong, 2·5 cm. long;
ovary silky; fruit 3-winged, hard, 3 cm. diam.    ..    ..    ..    ..    *daniellii*

T. **daniellii** (*Benn.*) *Benth.* l.c. (1883); F.T.A. 7: 321; K. Schum. in Engl., Pflanzenr. 4, 48: 40, fig. 8; Chev.
Bot. 630; Mangenot in Ic. Pl. Afr., I.F.A.N. 4: 95 (1957); Schnell in Bull. I.F.A.N. 19: 1133 (1957);
Koechlin in Fl. Gabon 9: 141, t. 30 (1964), and in Fl. Cam. 4: 144, t. 28 (1965). *Phrynium daniellii*
Benn. (1855). *Monostiche daniellii* (Benn.) Horan. (1862). *Donax daniellii* (Benn.) Roberty (1955), as
(Horan.) Roberty. A herb up to 10 ft. high; rhizome slender, creeping; spikes arising from the base, rough
with the scars of fallen flowers, the latter pale purple; fruit crimson, at or below ground-level; seeds
black, hard, shining; in large clumps in forest.
**S.L.:** (cult.) *Daniell!* *Johnston!* *Bockstatt!* **Lib.:** Gbanga (Sept.) *Linder* 668! Nimba Mt. (fr. Feb.) *Adam*
20266! **Iv.C.:** Mt. Nuon (fr. Apr.) *Chev.* 21135! Bouroukrou (fr. Jan.) *Chev.* 16940! Banco (Mar.) *Aké
Assi* 9523! Niapidou *Leeuwenberg* 2504! **Ghana:** basin of Volta R. *Krause!* Assin-Yan-Kumasi *Cummins*
135! Akwawa, nr. Begoro *Plumptre* 97! **N.Nig.:** R. Benue *Talbot* 879! **S.Nig.:** Oba, Owo (July) *Stanfield*
106! Mamu F.R. (Sept.) *Onochie* FHI 7654! Onitsha *Barter* 263! 1546! Sapoba *Kennedy* 2716! L. Odule,

Degema Div. (fr. Aug.) *Talbot* 3808! **W.Cam.**: Bambuko F.R., Kumba Dist. (Jan.) *Keay* FHI 37462! Ambas Bay (Dec.) *Mann* 2145! Also in E. Cameroun, Gabon, Congo, Cabinda, Principe and S. Tomé. See Appendix, p. 476.)

## 2. AFROCALATHEA K. Schum. in Engl., Pflanzenr. 4, 48: 51 (1902).

Perennial rhizomatous herb; leaves in pairs or solitary from the rhizome, petiole about 12–60 cm. long, the calloused portion about 2 cm. long, continuous with the petiole and midrib beneath, interrupted above at junction with lamina, pubescent; lamina oblong-lanceolate, slightly asymmetrical, shortly acuminate at apex, deltoid at the base, 10–35 cm. long, 4–10 cm. broad, glabrous; inflorescences up to 15 cm. long, arising directly from the rhizome, 3–4-flowered, spicate with imbricate bracts; sepals linear, 1 cm. long; ovary densely pubescent    ..    ..    ..   *rhizantha*

**A. rhizantha** (*K. Schum.*) *K. Schum.* l.c. (1902); Koechlin in Fl. Gabon 9: 134, t. 30 (1964), and in Fl. Cam. 4: 137, t. 28 (1965). *Calathea rhizantha* K. Schum. (1892)—F.T.A. 7: 327. Rhizomatous herb; flowers whitish; in forest.
**S.Nig.**: Unyene to Stubbs Creek F.R., Eket Dist. (Oct.) *Daramola* FHI 55320! Also in E. Cameroun and Gabon.

## 3. MARANTOCHLOA Brongn. ex Gris in Bull. Soc. Bot. Fr. 7: 321 (1860). *Clinogyne* of Benth. in small part, not of Salisb. *Donax* of K. Schum. in Engl. Bot. Jahrb. 15: 434, and of Baker in F.T.A. 7: 315, not of Lour. *Phrynium* of K. Schum. (1902), and of F.W.T.A., ed. 1, 2: 337, partly.

Inflorescence congested, about 5 cm. long and nearly as broad composed of numerous short branches, erect, sessile; bracts about 3 cm. long with several flowers to each bract, finely pubescent; leaves oblong-elliptic, rounded to a very slightly acute base, acumen eccentric, 14–31 cm. long, 6–14 cm. broad, with very numerous parallel lateral nerves, petiole up to 28 cm. long, the calloused portion 1·5–3·5 cm. long and shortly pubescent on the upper side, portion above sheath up to 8 cm. long  1. *mannii*
Inflorescence lax or spiciform:
  Leaves dissimilarly asymmetric, mirror image of the next above, i.e. asymmetry of the leaves on alternate sides in relation to the midrib (see Fig. 345A), 7–15(–25) cm. long; calloused portion of petiole 2–7(–10) mm. long, pilose, sheath contiguous or almost so with callous portion; branchlets and nodes normally glabrous, sometimes villous; inflorescence lax and delicate, simple or with few branches, bracts and pedicels usually spreading at right angles to the glabrous rhachis; bracts 12–20 mm. long  ..  ..  ..  ..  ..  ..  ..  ..  ..  ..  2. *congensis*
  Leaves all similar in their asymmetry (see Fig. 345B), or plant monophyllous:
    Aerial stems absent, leaf arising from rhizome:
      Inflorescences about 5 cm. long, very narrow, unbranched; ovary and fruit thinly pubescent; sepals broadly lanceolate, 3 mm. long; bracts about 1·5 cm. long, shortly pubescent; rhizomes bearing only one leaf; leaf oblong-lanceolate, slightly cordate or broadly and very shortly cuneate at the base, 21–30 cm. long, 7–12 cm. broad, with distinct transverse nerves between the numerous lateral nerves; calloused portion of petiole about 1 cm. long, sheathing part of petiole much shorter than the unsheathed portion  ..  ..  ..  3. *holostachya*
      Inflorescences 15–18 cm. or more long, apparently unbranched, upper bracts imbricate, 3 cm. long, folded half-width about 5 mm.; leaves solitary or paired from rhizome, petioles about 30–50 cm. long, slightly pubescent, calloused portion about 3 cm. long; lamina acuminate, rounded to truncate at base with cuneate portion at top of petiole, about 20 cm. long and 7 cm. broad, coriaceous, glossy above  4. sp. *A*
    Aerial stems bearing leaves:
      Nodes and branchlets glabrous:
        Leaves 7–16 cm. long, 3·5–7(–11) cm. broad; calloused portion of petiole 3–7 mm. long, shortly pubescent above, contiguous with sheath; inflorescence with slender rhachis simple or once branched; bracts 1·5–3·5 cm. long, caducous; sepals 2·5–6 mm. long; seeds 3–4 mm. long  ..  ..  ..  5. *filipes*
        Leaves 10–25 cm. long, 6·5–12 cm. broad; calloused portion of petiole 5–14 mm. long, shortly pubescent above, contiguous with sheath; inflorescence 1–3 times branched; sepals 5 mm. long; fruits spherical, 1 cm. diam. with 3 suture grooves, sparsely pubescent; seeds 5–6 mm. long  ..  ..  ..  6. *leucantha*
      Nodes and branches pubescent:
        Inflorescences lax and well-branched, up to 20 cm. long, bracts about 3·5 cm. long, persistent:
          Flowers white (see also above)  ..  ..  ..  ..  ..  6. *leucantha*
          Flowers deep purple; leaves broadly ovate-elliptic, truncate at the base, abruptly apiculate at apex, up to 40 cm. long and 20 cm. broad, calloused portion of petiole 1–6 cm. long, portion above sheath up to 7 cm. long (may be absent or very short in upper leaves only); fruits spherical, 8 mm. diam., pubescent
                                     7. *purpurea*

Inflorescences narrow, 10–25 cm. long, simple or with erect narrow branches; seeds
   with laminate aril:
Inflorescence unbranched or usually only divided at the base, 3–4 cm. long, half
      width 4 mm., enclosing rather large white flowers; corolla-tube about 1 cm.
      long; stem much-branched; leaves oblong-elliptic, rounded to a very shortly
      cuneate or slightly cordate base, abruptly tailed-acuminate, up to 26 cm. long
      and 14 cm. broad..    ..    ..    ..    ..    ..    ..    8. *ramosissima*
Inflorescence with several parallel branches; bracts more or less imbricate 3 cm.
      long, half width 3 mm.; corolla-tube about 5 mm. long; stems sparingly
      branched; leaves ovate to ovate-lanceolate or oblong, rounded to a very shortly
      cuneate base, abruptly tailed-acuminate at apex, up to 40 cm. long and 26 cm.
      broad    ..    ..    ..    ..    ..    ..    ..    ..    9. *cuspidata*

1. **M. mannii** (*Benth.*) *Milne-Redh.* in Kew Bull. 7: 167 (1952), 454 (1953), and in F.T.E.A. Marantac.: 8
   (1952); Koechlin in Fl. Gabon 9: 128, t. 25 (1964), and in Fl. Cam. 4: 130, t. 24 (1965). *Calathea mannii*
   Benth. (1883). *Phrynium hensii* Bak. in F.T.A. 7: 323 (1898); Léonard & Mullenders in Bull. Soc. Roy.
   Bot. Belg. 83: 22. *P. mannii* (Benth.) K. Schum. in Engl. Pflanzenr. 4, 48: 56 (1902); F.W.T.A., ed. 1,
   2: 337. *Clinogyne hensii* (Bak.) K. Schum. l.c. 62 (1902). A herb up to 8 ft. high with several leaves and
   clustered pink flowers; in moist forest.
   **Iv.C.:** Maféré (Feb.) *Aké Assi* 7331! **Ghana:** Koforidua (Oct.) *Baldwin* 13475! Kumasi (Dec.) *Darko* 486!
   **S.Nig.:** Idanre F.R. (Dec.) *Keay* FHI 37268! Okomu F.R., Benin (Dec.) *Brenan* 8470! **F.Po:** *Mann*
   1173! Also in Gabon, Congo, Sudan, Uganda and Tanzania.
2. **M. congensis** (*K. Schum.*) *Léonard & Mullend.* in Bull. Soc. Roy. Bot. Belg. 83: 17 (1950), Contrib.
   2: 28[236]; Koechlin in Fl. Gabon 9: 120, t. 26, and in Fl. Cam. 4: 123, t. 25. *Donax congensis* K. Schum.
   (1892)—F.T.A. 7: 317. *Clinogyne congensis* (K. Schum.) K. Schum. in Engl., Pflanzenr. 4, 48: 67
   (1902). A slender rhizomatous branched herb, erect, usually only a few feet high; flowers white or
   yellowish; in small colonies in forest.
   **S.L.:** Vama (Apr.) *Deighton* 3668! Kamasu, Tunkia (Dec.) *Deighton* 5220! Rowala (July) *Thomas* 1181!
   Gondama, Maje (May) *Deighton* 5783! Pampanu (June) *Marmo* 200! **Lib.:** Peahtah (fl. & fr. Oct.) *Linder*
   924! Toroke, Webo Dist. (July) *Baldwin* 6734! Bilimu Mt. (Dec.) *Harley* 1311! Mt. Nelo road (Sept.)
   *Adames* 502! **Iv.C.:** Guitry (Dec.) *Boughey* GC 13529! Bianouan to frontier (Apr.) *Leeuwenberg* 3935!
   **Ghana:** Essem F.R. (Jan.) *Morton* GC 8354! Awaso, W. Prov. (Dec.) *Morton* A3622! Fumso, Ashanti
   (Nov.) *Darko* 672! Kwabeng, Akim (Apr.) *Adams* GC 2700! Kwahu (Apr.) *Johnson* 691! **S.Nig.:** Usonigbe
   F.R., Benin (Nov.) *Keay* FHI 25568! Sapoba (Mar., Nov.) *Kennedy* 2315! *Jones* FHI 1089! *Meikle* 557!
   Orlu (Apr.) *Thompson*! Oban *Talbot* 881! Extends to Cabinda and Congo.
   [The plants with pilose branches and nodes have been distinguished as var. *pubescens* Léonard & Mullend.
   l.c. 18 (1950).]
3. **M. holostachya** (*Bak.*) *Hutch.* F.W.T.A., ed. 1, 2: 338 (1936); Koechlin in Fl. Gabon 9: 116, t. 25, and in
   Fl. Cam. 4: 120, t. 24. *Phrynium holostachyum* Bak. in F.T.A. 7: 322 (1898). *Clinogyne holostachya*
   (Bak.) K. Schum. in Engl., Pflanzenr. 4, 48: 65 (1902). A herb apparently bearing a single leaf, with
   yellow flowers.
   **S.Nig.:** Oban *Talbot* 875! Also in E. Cameroun, Gabon and Congo.
4. **M. sp. A.** A herb with 1 or 2 leaves from the rhizome and a simple inflorescence; further material is required.
   **S.Nig.:** Oban *Talbot* 93! 1262!
5. **M. filipes** (*Benth.*) *Hutch.* F.W.T.A., ed. 1, 2: 338 (1936); Koechlin in Fl. Gabon 9: 122, t. 26, and in Fl.
   Cam. 4: 126, t. 25. *Phrynium filipes* Benth. (1849). *Donax filipes* (Benth.) K. Schum. (1892)—F.T.A. 7:
   316. *Clinogyne filipes* (Benth.) K. Schum. (1902). *C. eburnea* A. Chev., Bot. 629 (1920) partly, name only.
   *Marantochloa oligantha* (K. Schum.) Milne-Redh. in Bull. Soc. Roy. Bot. Belg. 83: 20 (1950). *Donax*
   *oligantha* K. Schum. (1892)—F.T.A. 7: 316. *Clinogyne oligantha* (K. Schum.) K. Schum. (1902). Branched
   herb more or less climbing, up to 8 ft. high, locally gregarious in high forest; flowers white; fruits red,
   splitting and exposing 3 black seeds.
   **Guin.:** N'Zérékoré (Sept.) *Baldwin* 13278! **S.L.:** Benikoro (fr. Oct.) *Thomas* 2912! **Lib.:** Zwedru (Aug.)
   *Baldwin* 12495! **Iv.C.:** Bingerville *Chev.* 15397! Yapo Forest (Oct.) *Leeuwenberg* 1833! Adiopodoumé
   (fr. Jan.) *Aké Assi* 7293! **Ghana:** Juaso, Ashanti (fr. Sept.) *Lovi* WACRI 3921! Wankye, Birrim Dist.
   (fr. Aug.) *Chipp* 565! Assuantsi (July) *Irvine* 5081! 5082! **S.Nig.:** Okomu F.R., Benin (fr. Dec.) *Brenan*
   8502! Oban *Talbot* 881!(?) **F.Po:** (Dec.) *T. Vogel* 163! Musola to Concepcion (fr. Sept.) *Wrigley* 553!
   Also in E. Cameroun, Gabon and Congo.
   [F. R. Irvine noted under his No. 5081 that the stem had no nodes whereas No. 5082 had knee-like
   nodes: otherwise the two gatherings appear to be similar—F.N.H.]
6. **M. leucantha** (*K. Schum.*) *Milne-Redh.* l.c. 19 (1950), and in F.T.E.A. Marantac.: 6 (1952). *Donax leucantha*
   K. Schum. (1892)—F.T.A. 7: 317. *Clinogyne leucantha* (K. Schum.) K. Schum. in Engl., Pflanzenr. 4,
   48: 66 (1902). *C. eburnea* A. Chev., Bot. 629, partly, name only. *Donax cuspidata* of F.T.A. 7: 315 partly,
   not of K. Schum. *Marantochloa flexuosa* of F.W.T.A., ed. 1, 2: 338, partly (syn. *M. leucantha*). More or less
   scrambling herb up to 12 ft. high, often much less, with rather large leaves, swollen nodes and branched
   inflorescences; flowers whitish, fruits whitish or reddish; seeds black; in forest.
   **Guin.:** Macenta (fl. & fr. May) *Collenette* 9! **S.L.:** Duunia *Sc. Elliot* 4820! Gorahun, Kpola Mende (June)
   *Small* 727! Bobobu, Tunkia (fr. Dec.) *Deighton* 3807! Njala (fr. Feb.) *Deighton* 2458! Taiama (fr. Dec.)
   *Deighton* 2933! **Lib.:** Gletown, Tchien Dist. (fr. July) *Baldwin* 6736! Yratoke, Webo Dist. (fr. July)
   *Baldwin* 6259! Bilimu Mt. (fr. Aug.) *Harley* 1450a! Montézo, Attié (fr. Feb.) *Chev.* 17439! Bingerville
   *Chev.* 17310! **Iv.C.:** Adiopodoumé (fl. & fr. Oct.) *Leeuwenberg* 1772! Abengourou (fl. & fr. Feb.)
   *Baldwin* 13476! Juaso, Ashanti (fr. Sept.) *Lovi* WACRI 3922! Mpraeso Scarp (Apr.) *Morton* A524!
   Aburi swamps (fl. & fr. July) *Johnson* 757! Sibiri, Akim (fr. Apr.) *Irvine* 1185! **Togo Rep.:** Takpla (fr.
   Feb.) *Schröder* 213! **N.Nig.:** Dogon Kurmi, Jemaa Dist. (fl. & fr. Sept.) *Killick* 69! **S.Nig.:** Ibadan South
   F.R. (Apr.) *Keay* FHI 22841! Aponmu F.R., Akure (fl. & fr. Aug.) *J. B. Gillett* 15281! Okomu F.R.,
   Benin (fl. & fr. Jan.) *Brenan* 8764! Sapoba *Kennedy* 2657! Owo (fl. & fr June) *Stanfield* 107! **W.Cam.:**
   Buea *Deistel* 170! Barombi *Preuss* 354! S. Bakundu F.R., Kumba Dist. (fr. Feb.) *Binuyo & Daramola*
   FHI 35527! Also in E. Cameroun, Congo, Sudan, Uganda and Tanzania.
   [*Clinogyne lasiocolea* K. Schum. in Engl., Pflanzenr. 4, 48: 66 (1902), (W. Cam.: Bali *Conrau* 240) is
   said to differ from *M. leucantha* only in having the leaf sheaths golden hairy towards the base. The type
   has not been seen, but it may be a synonym of the above; Koechlin in Fl. Cam. 4: 127 regards it as a
   variety—F.N.H.]
7. **M. purpurea** (*Ridl.*) *Milne-Redh.* l.c. 21 (1950); and in F.T.E.A. Marantac. 6, fig. 2; Schnell in Bull. I.F.A.N.
   19: 1128, 1131; Mangenot in Ic. Pl. Afr., I.F.A.N. 7: 160 (1965); Koechlin in Fl. Gabon 9: 125, t. 27,
   and in Fl. Cam. 4: 129, t. 26. *Clinogyne purpurea* Ridl. in J. Bot. 25: 132 (1887). *C. baumannii* K. Schum.
   in Engl., Pflanzenr. 4, 48: 63 (1902). *Donax cuspidata* of F.T.A. 7: 315, partly, not of K Schum.
   *Marantochloa flexuosa* of F.W.T.A., ed. 1, 2: 338, partly. A swamp forest herb 4–8 ft. high; bracts pale
   pink; petals and outer staminodes deep purple (pink in Congo and Uganda plants), with stamen appendage
   and inner staminode yellow; fruits scarlet.
   **S.L.:** Heddle's Farm, Freetown (Dec.) *Sc. Elliot* 3900! Faiama, Nomo (Mar.) *Deighton* 4110! Nienia

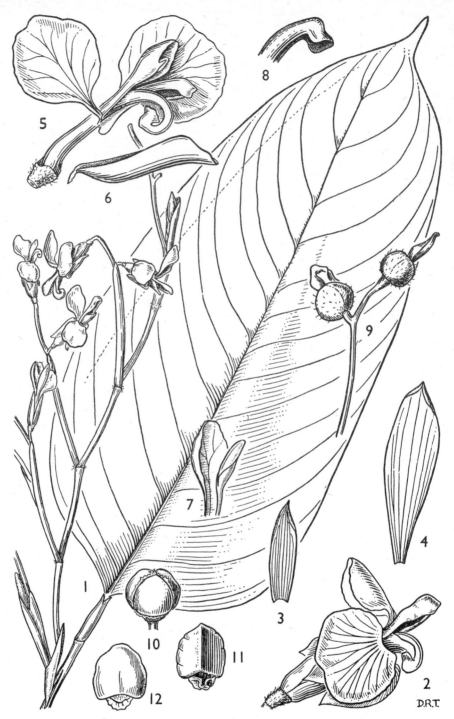

FIG. 346.—MARANTOCHLOA PURPUREA (*Ridl.*) *Milne-Redh.* (MARANTACEAE).

1, part of inflorescence and upper leaf, × 1. 2, flower with style reflexed, × 3. 3, sepal, × 3. 4, corolla lobe, × 3. 5, flower with perianth and one inner staminode removed to show style, × 3. 6, the inner staminode removed from 5, × 3. 7, stamen, × 4. 8, upper part of style and stigma, × 4. 9, fruiting cymule, × 1. 10, fruit with valves removed to show seeds, × 2. 11, 12, dorsal and ventral view of seed, × 3. Mainly from *Thomas* 2774. (Reproduced from F.T.E.A.)

(Feb.) *Sc. Elliot* 4902! Bagroo R. (Apr.) *Mann* 901! **Iv.C.:** Mt. Tonkoui, Man (Feb., Mar.) *Aké Assi* 7357! *Leeuwenberg* 2938! **Ghana:** Apapam, Kibbi (Dec.) *Morton* GC 8151! Offinso, Kumasi (Nov.) *Darko* 643! Bame *Krause*! Mpraeso (May) *Robertson* K7! Mt. Equanema, Southern Scarp F.R. (Nov.) *Foggie* FH 4934! Swedru (Feb.) *Dalz.* 8290! **Togo Rep.:** Lomé *Warnecke* 453! **N.Nig.:** Lokoja (Mar.) *Shaw* 7! Dogon Kurmi, Jemaa Dist. (Nov.) *Keay* FHI 21043! Matyoro Lakes, Gombe Div. (Apr.) *Thornewill* 61! Gangumi, Gashaka Dist. (Dec.) *Latilo & Daramola* FHI 28860! **S.Nig.:** Lagos *Millen* 141! Owo (fl. & fr. May) *Stanfield* 108! Angiama *Barter* 97! Bashi-Okwango F.R., Obudu (Apr.) *Latilo* FHI 30913! Oban Rock (Mar.) *Coombe* 184! Abakaliki Dist. (Mar.) *Kitson*! **W.Cam:** Johann-Albrechtshöhe (= Kumba) *Staudt* 485! Bambuko F.R., Kumba Dist. (fl. & fr. Jan.) *Keay* FHI 37433! Nkambe to Wum (Feb.) *Hepper* 1917! Also in E. Cameroun, Congo, Sudan, Uganda and Tanzania.

8. **M. ramosissima** (*Benth.*) Hutch. F.W.T.A., ed. 1, 2: 338 (1936); Koechlin in Fl. Cam 4: 33. *Phrynium ramosissimum* Benth. (1849)—F.T.A. 7: 326. *Clinogyne ramosissima* (Benth.) K. Schum. l.c. 64 (1902). *Donax arillata* K. Schum. (1892)—F.T.A. 7: 316. *Clinogyne arillata* (K. Schum.) K. Schum. in Engl. Pflanzenr. 4, 48: 62 (1902), partly. Stems much branched, 4–8 ft. high; calyx rosy white, flowers white tinged pink with yellow centre.

> **Iv.C.:** Nigbi Soubré (Nov.) *de Wilde* 3264! **S.Nig.:** Oban (Mar.) *Onochie* FHI 36464x! *Talbot* 1262! Ikom (Dec.) *Rosevear* 27/30a! *Keay* FHI 28269! Ukpon F.R., Obubra Dist. (July) *Latilo* FHI 31869! **W.Cam.:** Kake, Barombi (Aug.) *Preuss* 379! Cam. R. *Mann* 2141! **F.Po:** (fl. & fr. Dec.) *T. Vogel* 165! 178! *Mann* 100! 1174! Clarence *Barter* 1544! Also in E. Cameroun. (See Appendix, p. 475.)

9. **M. cuspidata** (*Rosc.*) Milne-Redh. in Proc. Linn. Soc. 165: 30 (1954); Schnell in Bull. I.F.A.N. 19: 1128, 1131 (1957); Mangenot in Ic. Pl. Afr., I.F.A.N. 7: 159 (1965). *Maranta cuspidata* Rosc., Scitam. t. 31 (1828). *Phrynium flexuosum* Benth. (1849). *Clinogyne flexuosa* (Benth.) K. Schum. in Engl., Pflanzenr. 4, 48: 63 (1902). *C. chrysantha* Gagnep. (1904). *C. arcta* Stapf (1906). *Marantochloa flexuosa* (Benth.) Hutch., F.W.T.A., ed. 1, 2: 338 (1936). Little-branched herb 2–6 ft. high; flowers yellow; in forest undergrowth, often secondary or by rivers in savanna.

> **Guin.:** *Heudelot* 729! Conakry (Dec.) *Maclaud* 138! **S.L.:** *Don*! Port Lokko (Apr.) *Sc. Elliot* 5877! Kambia (fr. Dec.) *Sc. Elliot* 4196! Njala (fl. & fr. Dec.) *Deighton* 2934! Kambui Hills (Mar.) *Small* 503! Bonkababa, near Kasawa (fl. & fr. Dec.) *King* 61b! **Lib.:** Bilimu Mt. (Dec.) *Harley* 1310! Du R. (July) *Linder* 98! Peahtah (Oct.) *Linder* 1023! Ganta (Jan.) *Baldwin* 14052! Kakatown *Whyte*! **Iv.C.:** Guédéyo (Dec.) *Leeuwenberg* 2161! Fresco, Sassandra (Mar.) *Leeuwenberg* 3064! **Ghana:** Aboma F.R. (Oct.) *Foggie* FH 4932! Pamu-Berekum F.R. (Sept.) *Vigne* FH 2497! Mampong (Aug.) *Darko* 710! Ejura (Nov.) *Vigne* FH 3473! Dutukpene, Buem-Krachi Dist. (Mar.) *Hepper & Morton* A3056! (See Appendix, p. 475.)

## 4. HALOPEGIA K. Schum. in Engl., Pflanzenr. 4, 48, Marantac.: 49 (1902).

Leaves elongate-oblong, very broadly and shortly cuneate at the base, abruptly caudate-acuminate, 20–44 cm. long, 6–14 cm. broad, very thin, with very numerous parallel nerves, a few slightly more prominent than the others; petiole very long, sheathing in the lower half, quite terete towards the apex; inflorescence erect-branched and narrow; bracts persistent, imbricate, narrowly lanceolate, about 5 cm. long; corolla mottled; ovary silky  ..      ..       ..      ..      ..        ..       ..      ..      *azurea*

**H. azurea** (*K. Schum.*) K. Schum. l.c. 50 (1902); Léonard & Mullenders in Bull. Soc. Roy. Bot. Belg. 83: 30 (1950); Schnell in Bull. I.F.A.N. 19: 1129 (1957); Koechlin in Fl. Gabon 9: 136, t. 29, and in Fl. Cam. 140, t. 29. *Donax azurea* K. Schum. in Engl., Bot. Jahrb. 15: 434 (1892). *Calathea vaginata* A. Chev., Bot. 631 (1920), name only. Several long-petiolate leaves 3–4 ft. high arising from the rhizome, leaves sometimes reddish beneath; inflorescences shorter than the leaves, narrow, 1 or 2 flowers appearing at a time, decorative purple with yellow staminode; in marshy places by streams in forest.

> **S.L.:** Bumbuna (Oct.) *Thomas* 3312! Dodo, Dea (Oct.) *Deighton* 3769! Vaama, Koya (Dec.) *Deighton* 3842! Tingi Mts., Kono (Mar.) *Morton & Gledhill* SL 3199! **Lib.:** Wanau Forest (Nov.) *Harley* 2059! Gbanga (Sept.) *Linder* 782! Peahtah (Oct.) *Linder* 782a! Suen (Nov.) *Linder* 1394! **Iv.C.:** Bouroukrou *Chev.* 16520! 16686! 16714! Mid. Cavally *Chev.* 19266bis! 19408! Yapo Forest (Oct.) *de Wilde* 700! Guéyo (July) *Leeuwenberg* 4546! **Ghana:** Amentia, Ashanti *Irvine* 519! Akwapim (Dec.) *F. W. Brown* 788! Cadbury Hall, Kumasi (Nov.) *Darko* 471! Krokusu Hills F.R. South (Nov.) *Foggie* FH 4935! Puso Puso Ravine (Oct.) *Morton* A145! **S.Nig.:** Okomu F.R. (Dec.) *Brenan* 8398! Boje to Aboabam, Ikom Dist. (Dec.) *Keay* FHI 28243! Ikom *Rosevear* 26/30A! Oban *Talbot* 56! 57! 92! Also in E. Cameroun, Gabon, Congo and Cabinda. (See Appendix, p. 475.)

## 5. THALIA Linn., Sp. Pl. 1193 (1753); F.T.A. 7: 313.

A straggling herb; leaves ovate-lanceolate, rounded or subcordate at base, gradually and shortly acuminate, 15–40 cm. long, 8–20 cm. broad, with numerous closely parallel lateral nerves forming a cartilaginous margin; petiole sheathing in the lower part or nearly up to the calloused portion; flowers in a lax panicle with slender branches; bracts 2-flowered, deciduous, lanceolate, 1·5–2 cm. long, boat-shaped; capsule 1 cm. long; seed filling the capsule, broadly oblong, with a basal aril
*welwitschii*

**T. welwitschii** *Ridl.* in J. Bot. 25: 132 (1887); F.T.A. 7: 314; Koechlin in Fl. Gabon 9: 139, t. 29, and in Fl. Cam. 4: 142, t. 29. *T. coerulea* Ridl. l.c. (1887); F.T.A. 7: 314. *T. geniculata* of F.T.A. 7: 314, and of K. Schum. in Engl., Pflanzenr. 4, 48: 173, Afr. distrib. only; not of Linn. Erect herb with several stems arising from each plant, stems and petioles spongy in the lower part, leaves usually about 3 ft. high; lax inflorescence up to 8 ft. high bearing purple flowers opening in the morning; in rice-fields and wet places.

> **Sen.:** *Heudelot* 665! **Guin.:** N'Zérékoré (Sept.) *Baldwin* 13309! **S.L.:** Roruks (Feb.) *Jordan* 977! Musaia (Sept.) *Small* 260! Falaba (Mar., Sept.) *Deighton* 5182! *Sc. Elliot* 5098! Binkolo (Aug.) *Deighton* 1268! Njala (Oct.) *Deighton* 3224! **Lib.:** Palilah, Gbanga Dist. (Aug.) *Baldwin* 9161! Ganta (May, Sept.) *Harley* s.n.! 312! **Iv.C.:** Ferkéssedougou (Nov.) *Leeuwenberg* 2025! Oroumba-Boka (Oct.) *de Wilde* 663! **Ghana:** Babile, Burufu (Nov.) *T. M. Harris*! Navrongo (Nov.) *Darko* 448! Pong-Tamale (Dec.) *Morton* GC 9833! Berekum to Wamfie (Dec.) *Adams* 2926! Gidan Bazamfari (May) *Krause*! Dzolokpuita, Vane, (Aug.) *de Wit & Morton* A2916! **Dah.:** Porto-Novo *Chev.* 22720! Godomé (Sept.) *Newton* 16! **N.Nig.:** Odeke, Igala Dist. (Feb.) *Jones* 2857! Lokoja (Oct.) *Dalz.* 228! Nupe *Barter* 1024! Nabardo (Oct.) *Lely* 678! Yola (Nov.) *Hepper* 1206! **S.Nig.:** Abeokuta *Irving* 158! Lagos *Millen*! *MacGregor* 348! Ikoyi Plains (Mar.) *Dalz.* 1282! Ubiaja to Illushi, Ishan Dist. (Aug.) *Onochie* FHI 33295! Widespread in tropical Africa. (See Appendix, p. 476.)

> [The African plants have glabrous bracts and calyx and they are therefore regarded as a species distinct from the American *T. geniculata* Linn. in which they were formerly placed.]

> *T. dealbata* Fraser is cultivated in the Botanic Garden, Victoria, W. Cameroun.

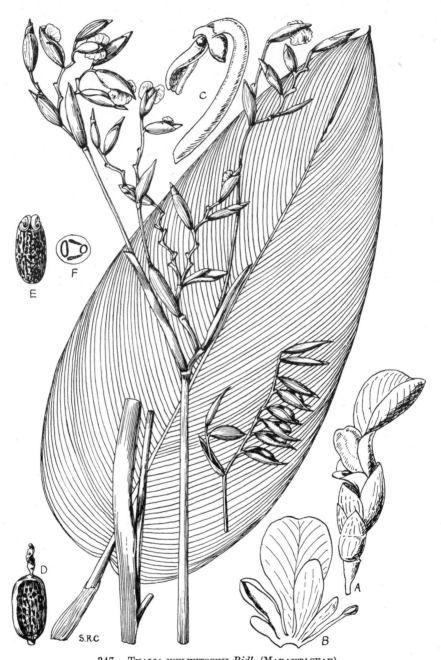

347.—THALIA WELWITSCHII *Ridl.* (MARANTACEAE).

A, flower. B, androecium. C, style. D, fruit with part of pericarp removed. E, seed. F, cross-section of seed.

**6. MEGAPHRYNIUM** Milne-Redh. in Kew Bull. 7: 169 (1952). *Sarcophrynium* K. Schum. (1902), partly—F.W.T.A., ed. 1, 2: 336, partly.

Inflorescence arising from a sheath up to 10 cm. below the calloused portion of the apical leaf petiole, borne on a simple stem 2–3 m. high; inflorescences up to 20 cm. long with a few lax branches, the branches spike-like and articulated with numerous nodes, cymules solitary at the nodes; bract enclosing the part of the inflorescence above it and caducous at anthesis; flowers side by side; bracteole solitary per cymule; outer staminodes lanceolate, linear, subulate or one absent; calyx-lobes free or more or less connate; fruit 3-lobed, 2 cm. diam. with conspicuous sutures, glabrous, tardily dehiscent; seeds arillate, not keeled, 16 by 10 mm.; leaves arising from the base with erect petioles 1–2 m. high, calloused portion up to 10 cm. long, lamina broadly ovate, rounded at base, shortly acuminate, up to 40 cm. long and 25 cm. broad  ..  ..  ..  ..  ..  ..  ..  ..  .. 1. *macrostachyum*

Inflorescence arising from a sheath 50–80 cm. below the calloused portion of the apical leaf petiole, borne on a simple stem 1 m. high; inflorescences about 8 cm. long, compact with few short branches; young fruits ciliate; seeds keeled, 12 by 8 mm.

2. *distans*

1. **M. macrostachyum** (*Benth.*) *Milne-Redh.* l.c. 170 (1952), and in F.T.E.A. Marantac.: 4; Koechlin in Fl. Gabon 9: 156, and in Fl. Cam. 4: 154, t. 31. *Phrynium macrostachyum* Benth. in Benth. & Hook. f., Gen. Pl. 3: 653 (1883). *Sarcophrynium macrostachyum* (Benth.) K. Schum. in Engl., Pflanzenr. 4, 48: 37 (1902); F.W.T.A., ed. 1, 2: 336. *S. macrophyllum* (K. Schum.) Hutch. F.W.T.A. ed. 1, 2: 336 (1936), as to name. *Phyllodes macrophyllum* K. Schum. (1892). *Phrynium macrophyllum* (K. Schum.) Bak. in F.T.A. 7: 323 (1898). *Sarcophrynium adenocarpum* (K. Schum.) K. Schum. (1902). *S. oxycarpum* (K. Schum.) K. Schum. (1902). *Phrynium adenocarpum* (K. Schum.) Bak. in F.T.A. 7: 324 (1898). *P. oxycarpum* (K. Schum.) Bak. l.c. (1898). *P. benthamii* Bak. l.c. (1898). A perennial herb forming extensive colonies in open high forest, the large solitary leaves forming a dense blanket 3 ft. or more high and separate taller inflorescences with one leaf on top, up to 10 ft. high; flowers whitish with red or purplish calyx; fruits glossy red; seeds black; leaves sometimes purple on the under surface.
**S.L.:** Kowama, Peri (fr. Nov.) *Deighton* 5270! **Lib.:** Ganta (fr. Dec.) *Harley* 1621a! Nimba Mt. (fr. Jan.) *Adam* 20660! **Iv.C.:** Adiopodoumé, (fl. & fr. Oct.) *Leeuwenberg* 1752! Bacanda, Tiassalé (fr. Jan.) *Aké Assi* 8423! **Ghana:** Begoro *Plumptre* 101! Obomeng, Kwahu (fl. & fr. Dec.) *Adams* 5169! **Togo Rep.:** *Baumann* 174! **S.Nig.:** Ibadan (July) *Ahmed & Chizea* FHI 20012! Lagos (fr. Mar.) *Millen* 19! *Dawodu* 120! Benin (Feb.) *Maggs* 164! Oban *Talbot* 880! Owo (June) *Stanfield* 86! **W.Cam.:** Ambas Bay *Mann* 1335! Barombi to Ninga *Preuss* 343! 381! Mamfe (Dec.) *Baldwin* 13822! Also in E. Cameroun, Rio Muni, Gabon, Cabinda, Congo, Sudan and Uganda. (See Appendix, p. 475.)
2. **M. distans** *Hepper* in Kew Bull. 22:461 (1968). *Sarcophrynium macrophyllum* of F.W.T.A., ed. 1, 2: 336 (as to descr. and cited specimens. Leaves and inflorescences 3 or 4 ft. high; flowers yellow with a red spot.
**Lib.:** Monrovia (fr. June) *Baldwin* 5904! Ganta (fr. Sept.) *Baldwin* 12492! **Iv.C.:** Amitioro Forest, Tiassalé Reg. (Dec.) *Aké Assi* 8390! **Ghana:** W. Afao Hills F.R., Awaso (fl. & fr. Feb.) *Darko* 762! S. Scarp F.R., Kwahu (fr. Dec.) *Adams* 5168! Asamankese (fr. Dec.) *Plumptre* 48! Assin-Yan-Kumasi *Cummins* 206!

*Imperfectly known species.*

**Sarcophrynium spicatum** *K. Schum.* in Engl., Pflanzenr. 4, 48: 40 (1902); F.W.T.A., ed. 1, 2: 336; Milne-Redhead in Kew Bull. 7: 170 (1952).
**Lib.:** Grand Bassa, near the coast in primary forest on the R. Cestos (May) *Dinklage* 1946. This specimen was not seen for Ed. 1 and it is not now in the Berlin Herbarium. Milne-Redhead stated that although it belongs to *Megaphrynium*, he was not prepared to make a new combination or reduce it to synonymy until material was available for study.—F.N.H.

**7. SARCOPHRYNIUM** K. Schum. in Engl., Pflanzenr. 4, 48 Marantac.: 35 (1902); Milne-Redhead in Kew Bull. 7: 168 (1952).

Inflorescence sessile, branched from the base, up to 7 cm. long; bracts overlapping, boat-shaped, 1·5–2 cm. long; leaves ovate-elliptic, rounded at the base 20–30 cm. long, 10–15 cm. broad, with very numerous closely parallel lateral nerves, glabrous or puberulous beneath; calloused portion of petiole 4–6 cm. long; fruits subglobose, scarcely bilobed, shining and wrinkled when dry, about 1·5 cm. diam.

1. *brachystachys*

Inflorescence pedunculate, paniculate; bracts very lax, containing 2 to several flowers, 4 cm. long; peduncle thinly villous; large leaves elliptic, acuminate, rounded at the base, 30–47 cm. long, 12–19 cm. broad; calloused portion of petiole 5–9 cm. long:
Leaves glabrous or more or less softly puberulous 2a. *prionogonium* var. *prionogonium*
Leaves pilose beneath, hairs arising from tubercles     2b. *prionogonium* var. *ivorense*

1. **S. brachystachys** (*Benth.*) *K. Schum.* in Engl., Pflanzenr. 4, 48: 36 (1902); Koechlin in Fl. Gabon 9: 145, t. 31 (1964), and in Fl. Gabon 4: 146, t. 30 (1965). *Maranta* (?) *brachystachys* Benth. in Fl. Nigrit. 531 (1849). *Phrynium brachystachys* (Benth.) Koernicke (1862)—F.T.A. 7: 322; Chev. Bot. 630. *P. molle* A. Chev., Bot. 631 (1920), name only. *Sarcophrynium strictifolium* Schnell in Bull. I.F.A.N. 15: 1390, fig. 1 (1953) & 19: 1132 (1958). Herb in forest 1½–5 ft. high with single long-petioled leaf arising in groups from the rhizome; flowers white; fruits red.
**Sen.:** Boukitingo Forest, Casamance (fr. Nov.) *Berhaut* 6544! **Guin.:** Ya Valley, Nimba Mts. (Apr.) *Schnell* 5217! **S.L.:** Tawia, Scarcies (fr. Jan.) *Sc. Elliot* 4473a! Kamahi (fr. May) *Thomas* 315! Njala (fl. & fr. Feb.) *Deighton* 4703! Kambui F.R., Kenema (fr. Mar.) *Jordan* 2007! Gola Forest (fr. Mar.) *Small* 568! **Lib.:** Monroviatown (fr. Aug.) *Baldwin* 7098! Grand Bassa *T. Vogel* 77! Ganta (fr. Sept.) *Baldwin* 12490! Nana Kru (Mar.) *Baldwin* 11575! Jabroke, Webo Dist. (fr. July) *Baldwin* 6450! Cape Palmas (July) *T. Vogel* 29! **Iv.C.:** Oubi, Mid. Cavally (fr. July) *Chev.* 19350! Adiopodoumé (fr. Jan.) *Aké Assi* 7294! Béyo (fl. & fr. Jan.) *Guédéyo* (fl. & fr. Dec.) *Leeuwenberg* 2165! Yapo (Oct.) *Roberty* 15351! **Ghana:** Mampong Scarp (fr. Dec.) *Morton*! Jimira F.R. (fl. & fr. Feb.) *Foggie* FH 4965! Fumso (fl. & fr. Feb.) *Darko* 671! Juaso (fr. June) *Irvine* 321! Kwabeng, Akim (fr. Apr.) *Adams* 2698! **N.Nig.:** Lapai (Mar.–Apr.) *Yates* 39! Lokoja (Mar.) *Shaw* 29! 30! Mayo Ndaga, Mambila Plat.

(fr. Jan.) *Latilo & Daramola* FHI 28986! **S.Nig.**: Ibadan South F.R. (fr. Apr.) *Keay* FHI 22835! Owo (fl. & fr. June) *Stanfield* 96! Okomu F.R. (Dec.) *Brenan* 8490! Onitsha *Barter* 1545! Nun R. (fl. & fr. Sept.) *Mann* 517! Cross R. North F.R. (fr. Dec.) *Keay* FHI 28311! **W.Cam.**: Kumba (Apr.) *Daramola* FHI 29842! Extends to Congo. (See Appendix, p. 475.)

2. **S. prionogonium** (*K. Schum.*) *K. Schum.* in Engl., Pflanzenr. 4, 48: 39 (1902); Schnell in Bull. I.F.A.N. 19: 1132 (1955); Koechlin in Fl. Gabon 9: 147, and in Fl. Cam. 4: 149. *Phyllodes prionogonium* K. Schum. (1892). *Phrynium prionogonium* (K. Schum.) Bak. in F.T.A. 7: 325 (1898).

2a. **S. prionogonium** (*K. Schum.*) *K. Schum.* var. **prionogonium**—*S. prionogonium* var. *puberulifolium* Schnell in Bull. I.F.A.N. 15: 1392, fig. 2 (1953). *Phrynium cerasiferum* A. Chev., Bot. 630 (1920), name only. *Sarcophrynium leiogonium* (K. Schum.) K. Schum. (1902). Erect herb about 6 ft. high; inflorescence branched, flowers whitish, fruits red; in forest.
　　**Guin.**: Nimba Mt. (fr. Feb.) *Schnell* 4346! **S.L.**: Kenema (fl. & fr. Oct.) *Deighton* 5240! **Lib.**: Bilimu (fr. Dec.) *Harley* 1307! Kitoma, Sanokwele Dist. (fr. Jan.) *Baldwin* 14061! Banga (fl. & fr. Oct.) *Linder* 1256! Ganta (fr. Apr.) *Harley* 1130! Jabroke, Webo Dist. (fr. July) *Baldwin* 6604! **Iv.C.**: Malamalasso (fr. Mar.) *Chev.* 17500! Teke Forest (Aug.) *de Wit* 7237! Bocanda to Tiépo (fr. Jan.) *Aké Assi* 8447! **Ghana**: Mt. Ejuanema, Kwahu (fl. & fr. Dec.) *Adams* 5113! **S.Nig.**: Okomu F.R., Benin (Dec.) *Brenan* 8583! Oban *Talbot* 877! **W.Cam.**: Barombi-Ba-Mbu *Preuss* 458! Extends to Gabon and Congo.

2b. **S. prionogonium** var. **ivorense** *Schnell* in Bull. I.F.A.N. 19: 1133 (1957).
　　**Iv.C.**: Bouroukrou *Chev.* 16746! Agnéby Valley *Chev.* 17032! Lamé (fr. Nov.) *Leeuwenberg* 1906! Bocanda to Tiépo (fr. Jan.) *Aké Assi* 8448! **S.Nig.**: Okomu F.R., Benin (fl. Mar., June, fr. Jan.) *Onochie* FHI 34613! 38309! *Brenan* 8869! Afl River F.R., Ogoja Prov. (fl. & fr. May) *Jones & Onochie* FHI 5815!

## 8. HYPSELODELPHYS (K. Schum.) Milne-Redh. in Kew Bull. 5: 158 (1950). *Trachyphrynium* Benth. (1883), partly; F.W.T.A., ed. 1, 2: 337, partly. *T.* subgen. *Hypselodelphys* K. Schum. in Engl., Pflanzenr. 4, 48: 42 (1902).

Inflorescences congested with several (often many) lateral branches arising from about the same point; bracts persistent, closely overlapping, acutely acuminate, about 3 cm. long; leaves elliptic, rounded at base, gradually acuminate, 12–14 cm. long, about 6 cm. broad, with numerous closely parallel lateral nerves; petiole sheathing in the lower ¾, the upper quarter articulated in the middle with the calloused portion about 2 cm. long　..　..　..　..　..　..　..　1. *zenkerana*

Inflorescences loosely arranged, little-branched:
Fruits acutely triangular, each side 4 cm. long, flattened, densely and shortly muricate; inflorescences usually zig-zag, (5–)8–16(–24) cm. long from lowest node, with internodes about 1 cm. apart; bracts 2–3·5 cm. long; flowers 2·5 cm. diam.; leaves oblong-elliptic, rounded-truncate at base, abruptly acuminate, 9–21 cm. long, 5–9 cm. broad, with numerous close parallel lateral nerves; petiole sheathing up to the junction with the calloused portion　..　..　..　..　..　2. *violacea*
Fruits 3-lobed with rounded corners, muricate:
Tubercles of fruit short and dense, about 2 mm. long; fruit up to 4–5 cm. diam.; inflorescences nearly straight, 5–9 cm. long from the lowest node, with internodes about 5 mm. apart; bracts 2–3·5 cm. long; leaves oblong-elliptic to ovate-oblong, rounded-truncate at base, abruptly acuminate, 8–15 cm. long, 3–9 cm. broad, with numerous close parallel lateral nerves; petiole as the last　..　3. *poggeana*
Tubercles of fruit long and apparently less numerous, up to 5 mm. long, often curved; fruit up to 5 cm. diam.; inflorescence branches usually zig-zag, about 20 cm. long, with internodes about 1 cm. apart; bracts 3·5–4 cm. long; leaves elliptic, oblong-elliptic or ovate-elliptic, rounded-truncate at base, shortly acuminate, 12–35 cm. long, 5–17 cm. broad with numerous close parallel lateral nerves; calloused portion of petiole above articulation up to 3 cm. long; midrib pubescent or glabrous
　　　　　　　　　　　　　　　　　　4. *scandens*

1. **H. zenkerana** (*K. Schum.*) *Milne-Redh.* in Kew Bull. 5: 161 (1950); Koechlin in Fl. Cam. 4: 109, t. 22. *Trachyphrynium zenkeranum* K. Schum. in Engl., Pflanzenr. 4, 48: 45 (1902); F.W.T.A., ed. 1, 2: 337. A bamboo-like plant in forest.
　　**W.Cam.**: Barombi *Preuss* 873. Buea *Deistel* 206; 504; 519. *Preuss* 873b. *Dusen* 415. Also in E. Cameroun.
　　[Although Schumann described the fruit as having globose cocci, *Zenker* 274 (BR) has a detached fruit with pointed cocci. Field observations and further careful collections are required to settle the question.]

2. **H. violacea** (*Ridl.*) *Milne-Redh.* l.c. 160 (1950); Koechlin in Fl. Gabon 9: 104, t. 23 (1964), and in Fl. Cam. 4: 112, t. 22 (1965). *Trachyphrynium violaceum* Ridl.(1887)—F.T.A. 7: 320; K. Schum. l.c. 44; F.W.T.A., ed. 1, 2: 337. *T. preussianum* K. Schum. (1892)—F.T.A. 7: 320. *Donax violacea* (Ridl.) Roberty (1955); Straggling climber with slender stems branching at all angles forming a tangle in forest about 12 ft. high; flowers violet and white.
　　**S.L.**: Potoru (Dec.) *Deighton* 3796! Sefadu, Bari (Nov.) *Deighton* 5272! Pujehun (Nov.) *Deighton* 5273! **Lib.**: Bushrod I. (Dec.) *Barker* 1100! Kakatown *Whyte*! Vahon, Kolahun Dist. (Nov.) *Baldwin* 10242! Gbanga (Sept.) *Linder* 629! Yratoke, Webo Dist. (July) *Baldwin* 6261! **Iv.C.**: Man to Touba (fl. & fr. Feb.) *de Wit* 9188! Grabo to Olodio (fr. Feb.) *Aké Assi* 9432! **S.Nig.**: Olokemeji F.R. (Mar.) *Hepper* 2298! Osse R., Idogun (Apr.) *Stanfield* 163! Babiri, Obudu Dist. (May) *Latilo* FHI 30921! **W.Cam.**: Barombi *Preuss* 321! Extends to Cabinda and Angola.
　　[The Nigerian specimens present some difficulty in that they have the typical angular fruit of *H. violacea*, but the straighter inflorescence of *H. poggeana*. Further field observations are desirable.—F.N.H.]

3. **H. poggeana** (*K. Schum.*) *Milne-Redh.* l.c. 160 (1950); Léonard & Mullenders in Bull. Soc. Roy. Bot. Belg. 83: 12, t. 1 C, D, 2 G; Koechlin in Fl. Gabon 9: 104, t. 23, and in Fl. Cam. 110, t. 22. *Trachyphrynium poggeanum* K. Schum. (1892)—F.T.A. 7: 320; K. Schum. l.c. 44, fig. 9 G–K. *Hybophrynium braunianum* of F.W.T.A., ed. 1, 2: 337, partly (*Sc. Elliot, Thomas* etc.). Widely branching, often about 6 ft. high but sometimes 20 ft. or more; flowers violet and white.
　　**S.L.**: Mt. Leicester (May) *Barter*! Kambia, Scarcies (Jan.) *Sc. Elliot* 4508! Taiama (Mar.) *Deighton* 3372! Kumrabai (Dec.–Jan.) *Thomas* 6776! Bumpe (Dec.) *Deighton* 4968! **Lib.**: *Linder* 347! Brewersville (June) *Barker* 1323! Diebla, Webo Dist. (July) *Baldwin* 6338! Nimba (Fr.) *Baldwin* 10242! **Ghana**: Techiman road, Ejura (Dec.) *Morton* GC 9559! Kumasi (Feb.) *Darko* 516! Nsawam (Feb.) *Dalz.* 8291! Benso, Tarkwa Dist. (Dec.) *Andoh* FH 5433! Kpokoase (Mar.) *Irvine* 193! **S.Nig.**: Gambari F.R., Ibadan (Feb.) *Meikle* 1195! Idanre, Akure Div.(Jan.) *Brenan & Keay* 8689! Orlu (Apr.) *Thompson* 6! Iso-

Bendiga to Bendiga Afi, Ikom Dist. (Dec.) *Keay* FHI 28268! Oban *Talbot* 874! Extends to Congo and Angola.

4. **H. scandens** *Louis & Mullend.* in Bull. Soc. Roy. Bot. Belg. 83: 14, t. 1 E, F, 2 F (1950); Koechlin in Fl. Gabon 9: 106, tt. 21, 23, and in Fl. Cam. 113, t. 22. A bamboo-like climber 10–30 ft. high in moist forest; bracts and pedicels dull purplish-brown; flowers pale violet, white and brown in arching inflorescences. **Iv.C.:** Mt. Tonkoui, Man (fr. Feb.) *de Wit* 9150! **S.Nig.:** Cross River North F.R., Ikom Dist. (June) *Latilo* FHI 31835! *Eze* FHI 23647! Etomi to Ikom (May) *Jones* FHI 18873! Oban *Talbot* 874! Also in E. Cameroun, Gabon and Congo.

## 9. TRACHYPHRYNIUM Benth. in Benth. & Hook. f., Gen. Pl. 3: 651 (1883); Milne-Redhead in Kew Bull. 5: 157 (1950). *Hybophrynium* K. Schum. in Engl., Pflanzenr. 4, 48: 41 (1902).

Leaves oblong-elliptic, subcordate, acuminate, 7–14 cm. long, 3·5–6·5 cm. broad, glabrous, with numerous closely parallel lateral nerves diverging from the midrib at an angle of 45°; petiolar sheath encircling the shoot; calloused portion 5–17 mm. long, continuous and no disjunction with midrib below; inflorescence terminal, simple or slightly branched, shorter than the leaves; axis pubescent to glabrous; bracts caducous, 2·5 cm. long; capsule dehiscent, deeply 3-lobed, lobes rounded, densely muricate; seeds ellipsoid, with a large lamellate basal aril             ..           *braunianum*

**T. braunianum** (*K. Schum.*) *Bak.* in F.T.A. 7: 319 (1898); Léonard & Mullenders in Bull. Soc. Roy. Bot. Belge 83: 11, t. 1 A, B (1950); Milne-Redhead in F.T.E.A. Marantac.: 2 (1952); Koechlin in Fl. Gabon 9: 98, t. 22, and in Fl. Cam. 4: 106, t. 21. *Hybophrynium braunianum* K. Schum. in Engl., Bot. Jahrb. 15: 428, fig. A, F, H, K (1892); Chev. Bot. 629; F.W.T.A., ed. 1, 2: 336. Straggling herb forming a tangle in forest, up to 10 ft. high; flowers white; fruits orange-yellow.
**Guin.:** Kindia *Chev.* 12783. **S.L.:** Makump (July) *Thomas* 902! Kafogo (Apr.) *Sc. Elliot* 4464! **Lib.:** Ganta (Feb.) *Baldwin* 11021! **Iv.C.:** Sassandra (Dec.) *Leeuwenberg* 2258! Soubré to Grabia *Chev.* 19160! Tiassalé (Dec.) *Leeuwenberg* 2136! **Ghana:** *Burton & Cameron*! Lomanya, Akuse (fr. Oct.) *Morton* A146! **N.Nig.:** Katsina Ala (Aug.) *Dalz.* 832! Tonti to Beku, S. Muri Dist. (Nov.) *Latilo & Daramola* FHI 28746! **Dah.:** Porto-Novo (fr. Jan.) *Chev.* 22799! **S.Nig.:** Ibadan (Feb., Mar.) *Meikle* 1139! 1264! Ikoyi Plains, Lagos (June) *Dalz.* 1281! Benin (Aug.) *Stanfield* 164! Onitsha *Barter* 1784! Calabar *Thomson* 108! **W.Cam.:** Bambuko F.R., Kumba (Jan.) *Keay* FHI 37469! Banga, Kumba Dist. (Nov.) *Dundas* FHI 8492! **F.Po:** *Mann* 1175! Also in E. Cameroun, Cabinda, Congo, Uganda and Sudan. (See Appendix, p. 475.)

## 10. HAUMANIA J. Léonard in Bull. Jard. Bot. Brux. 19: 453 (1949).

Climbing slender-stemmed shrub bearing numerous reflexed prickles; leaves ovate to ovate-lanceolate, truncate at base, acuminate, 8–22(–27) cm. long, 3–8(–13) cm. broad, with numerous closely parallel lateral nerves, glabrous; petiole 6–23 cm. long, sheathing for much of its length; spikes pendulous, up to 10 cm. long; basal bracts sheathing the pubescent rhachis, sterile, lanceolate, 3–7 cm. long; fertile abaxial bracts spreading at right angles to the rhachis, 2 cm. long, broad and folded, the half-width 1–1·5 cm., with very numerous nerves; sepals about 4 mm. long; ovary spiny    ..        ..        ..        ..        ..        ..        ..     *danckelmaniana*

**H. danckelmaniana** (*J. Braun & K. Schum.*) *Milne-Redh.* in Kew Bull. 5: 162 (1950); Koechlin in Fl. Gabon 9: 110, t. 24, and in Fl. Cam. 116, t. 23. *Trachyphrynium danckelmaniana* J. Braun & K. Schum. in Mitth. Deutsch Schutzg. 2: 153 (1889); F.T.A. 7: 319; K. Schum. in Engl., Pflanzenr. 4, 48: 42; F.W.T.A., ed. 1, 2: 337. A straggling or climbing bamboo-like shrub with wiry prickly stems; bracts and flowers white; in forest clearings in high rainfall districts.
**W.Cam.:** Barombi *Preuss* 255! Dengding *Mildbr.* 8825! Also in E. Cameroun, Rio Muni and Gabon.

## 11. ATAENIDIA Gagnep. in Bull. Soc. Bot. Fr. 55: xli (1908).

Straggling, branching herb, 1–2 m. high; leaves rarely arising from rhizome, usually inserted beneath the inflorescence; petiole 11–20 cm. long, with narrow membranous sheath along half its length; lamina oblong, shortly cuneate at base, shortly and eccentrically acuminate, 20–30(–40) cm. long, 7–12(–15) cm. broad, glabrous above, finely pubescent beneath; inflorescence a sessile spike, congested, deflexed, about 4 cm. long and broad; bracts sheathing, ovate, glabrous; hooded staminode without a spur-like appendage; fruit membranous, indehiscent   ..       ..      ..     *conferta*

**A. conferta** (*Benth.*) *Milne-Redh.* in Kew Bull. 7: 168 (1952); and in F.T.E.A. Marantac.: 9; Schnell in Bull. I.F.A.N. 19: 1129 (1957); Koechlin in Fl. Gabon 9: 131, t. 28, and in Fl. Cam. 4: 136, t. 27. *Calathea conferta* Benth. in Benth. & Hook. f., Gen. Pl. 3: 653 (1883); F.T.A. 7: 327. *Phrynium confertum* (Benth.) K. Schum. (1902)—F.W.T.A., ed. 1, 2: 337; Léonard & Mullenders in Bull. Soc. Roy. Bot. Belg. 83: 24, t. 1 K (1950). *P. crista-galli* A. Chev., Bot. 631 (1920), name only. *Ataenidia gabonensis* Gagnep. (1908). *Donax conferta* (Benth.) Roberty (1955). A herb 4 ft. high, forming tangles in gaps in moist high forest; undersurfaces of the leaves sometimes purplish; bracts of the congested inflorescences red or purplish, flowers pale mauve or pinkish; fruits red.
**Iv.C.:** Malamalasso, Bas-Comoé *Chev.* 17499! Niapidou *Leeuwenberg* 2563! **Ghana:** Koechlin in Fl. Gabon. Amentia, Ashanti *Irvine* 437! Assin-Yan-Kumasi *Cummins* 64! Begoro (Nov.) *Morton* A149! Kibbi to Akim (Dec.) *Johnson* 246! Pra Anum F.R. (Sept.) *Andoh* FH 5892! **S.Nig.:** Omo F.R. (Feb.) *Jones & Onochie* FHI 14729! Oluwa F.R., Ondo Dist. (July) *Onochie* FHI 33406! Bendiga Ayuk to Aboabam, Ikom Div. (May) *Jones & Onochie* FHI 14132! Oban *Talbot* 884! **W.Cam.:** Cam. Mt., 3,000 ft. (Dec.) *Mann* 2144! Johann-Albrechtshöhe (= Kumba) (Apr.) *Staudt* 215! Also in E. Cameroun, Gabon, C. African Rep., Rio Muni, the Congos, Cabinda, Sudan and Uganda. (See Appendix, p. 475.)

## 182. LILIACEAE

### By F. N. Hepper

Herbs, mostly perennial, or rarely soft-wooded shrubs; roots from a rhizome, corm or bulb, or tuberous; stem erect or climbing, leafy or scapose. Flowers bisexual or rarely unisexual, actinomorphic or slightly zygomorphic, sometimes large and showy, never in umbels. Perianth mostly corolla-like, with or without a tube; segments usually 6, rarely 4 or more, in 2 similar series, imbricate or the outer valvate. Stamens usually 6, hypogynous or adnate to and always opposite to the perianth segments; filaments usually free; anthers 2-celled, usually opening by a slit lengthwise. Ovary superior, mostly 3-locular with axile placentas, or rarely 1-locular with parietal placentas; style entire or divided, rarely styles free. Ovules usually numerous and mostly 2-seriate in each loculus. Fruit a capsule or berry. Seeds with copious endosperm.

World-wide distribution, more abundant in temperate and subtropical regions.

Leaves large and fleshy, spiny on the margin; inflorescence up to 1 m. high    1. **Aloë**
Leaves not spiny on the margin, usually not fleshy:
  Leaves much reduced and scale-like, their function fulfilled by linear and acicular cladodes; fruit a berry; flowers very small; rootstock rhizome with root tubers; stems wiry, often armed..    ..    ..    ..    ..    2. **Asparagus**
  Leaves not reduced to scales; branches not modified into cladodes:
    Stems more or less climbing, leafy; leaves often tendriliform at the apex; style spreading at a right angle from the top of the ovary    ..    ..    13. **Gloriosa**
    Stems erect or much reduced; leaves not tendriliform at the apex:
      Rootstock a rhizome or tuber (neither a bulb nor a corm); roots fibrous or sometimes thick and tuberous:
        Rootstock a bulb-like tuber, usually surmounted by numerous fibres; seeds villous; flowers on long pedicels ..    ..    ..    ..    4. **Eriospermum**
        Rootstock not thick and bulb-like, but sometimes the roots ending in tubers; seeds glabrous; pedicels not elongated:
          Fruits globose or shallowly lobed; seeds not flat:
            Perianth segments united into a tube; inflorescence a stout spike; highland
                                                       3. **Kniphofia**
            Perianth segments free at the base; inflorescence slender, more or less branched; base of leaves more or less reddish mottled; widespread .. 5. **Anthericum**
          Fruits deeply lobed; seeds compressed; inflorescence branched or subspicate; reddish mottling absent from the base of the leaves; widespread
                                                   6. **Chlorophytum**
      Rootstock a bulb or corm:
        Perianth segments free to the base:
          Capsule more or less deeply lobed; corolla rotate    ..    ..    7. **Urginea**
          Capsule shallowly lobed:
            Stamens hypogynous:
              Raceme not spike-like, upper flowers fertile:
                Leaves basal; corolla more or less campanulate; bulb    8. **Albuca**
                Leaves cauline; flowers long-pedicellate; corm    ..    14. **Iphigenia**
              Raceme spike-like, upper flowers infertile; bulb    ..    9. **Drimiopsis**
            Stamens perigynous; flowers bluish or mauve; bulb ..    10. **Scilla**
        Perianth segments united at the base:
          Outer perianth-segments markedly caudate-acuminate; bulb with white coloration; lowland and upland savanna    ..    ..    ..    11. **Dipcadi**
          Outer perianth-segments not caudate-acuminate:
            Bulb with purplish coloration; seeds flattened; inflorescence many-flowered; upland savanna ..    ..    ..    ..    ..    ..    12. **Drimia**
            Tunicate corm; seeds subglobose; inflorescence 2-flowered; montane grassland
                                                         15. **Wurmbea**

### 1. ALOË Linn., Sp. Pl. 319 (1753); F.T.A. 7: 454; Keay in KewBull. 17: 65 (1963).

#### By R. W. J. Keay

Leaf-bases not forming a bulb-like swelling; perianth pink or red; fruits, where known, not exceeding 2·5 cm. long:
  Plants never suckering; lower leaves rosulate and spreading or lying on surface of ground; perianth with pronounced basal swelling truncate at base, abruptly con-

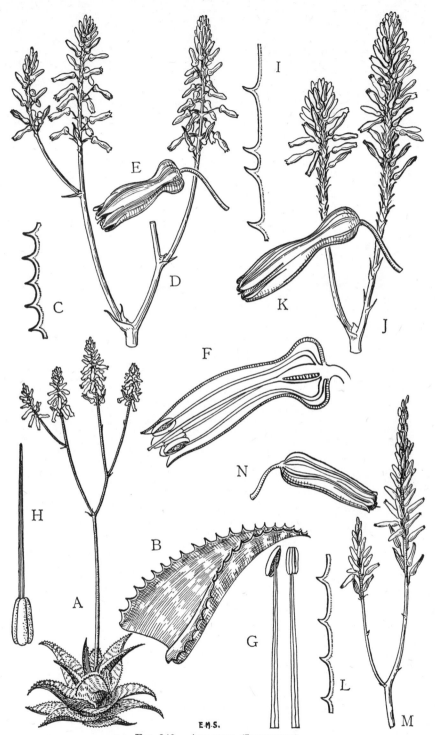

E.M.S.

Fig. 348.—Aloe spp. (Liliaceae).

*Aloe macrocarpa* var. *major* A. Berger—A, flowering plant, × $\frac{1}{15}$. B, leaf, × $\frac{2}{3}$. C, margin of leaf, × 1. D, part of inflorescence, × $\frac{1}{4}$. E, flower, × 1. F, flower in longitudinal section, × 2. G, stamens, × 2. H, pistil, × 2. *A. buettneri* A. Berger—I, margin of leaf, × 1. J, inflorescence, × $\frac{1}{4}$. K, flower, × 1. *A. schweinfurthii* Bak.—L, margin of leaf, × 1. M, inflorescence, × $\frac{1}{4}$. N, flower, × 1. A (partly), B, C from *FHI* 21031. A (partly), D, from *FHI* 22725. E–H from *FHI* 25326. I, J from *FHI* 25329. K from *FHI* 25328. L, M from *FHI* 22582.

stricted above the ovary with narrowest part about ¼ length from base and enlarging
   to the throat (see Fig. 348 E, F); on rocky hills  .. 1a. *macrocarpa* var. *major*
Plants often suckering; perianth rounded at base, cylindrical or very slightly constricted (see Fig. 348 N):
Leaves up to 8 cm. broad, grey-green; stems sometimes long; perianth about
   2·8 cm. long, cylindrical; fruits 2·5 cm. long; on rocky hills  2. *schweinfurthii*
Leaves 10–12 cm. broad; stems short; perianth about 3 cm. long, very slightly
   constricted above ovary; Accra Plains  .. .. .. .. 3. *keayi*
Leaf-bases forming an underground bulb-like swelling; plants never suckering;
   perianth green, yellow or orange or dull red, rounded at base, gradually constricted
   with the narrowest part about ¼ length from base (see Fig. 348 K); fruits 3·7–4 cm.
   long; in grassy places in savanna woodland .. .. .. 4. *buettneri*

1. **A. macrocarpa** *Todaro* Hort. Bot. Panorm. 1: 36, t. 9 (1875); F.T.A. 7: 462; Reynolds, Aloes Trop. Afr.
   91 (1966). In Eritrea and Ethiopia.
1a.**A. macrocarpa** var. **major** *A. Berger* in Engl., Pflanzenr. 4, 38, 3, 2: 210 (1908); Keay in Kew Bull. 17:
   67 (1963); Reynolds l.c. fig. 98. *A. edulis* A. Chev. in Rev. Bot. Appliq. 31: 592, t. 32A (1952), French
   descr. only. *A. barteri* of Schnell in Ic. Pl. Afr. t. 23 (1953), not of Bak. A perennial herb with fleshy
   leaves up to about 1½ ft. long and 4 in. broad, margins sharply toothed, leaf-surfaces strongly mottled
   and becoming reddish in the dry season; perianth dull scarlet; among rocks.
   **Dah.:** Somba, Atacora Mts. (cult., June) *Chev.* 24157! **N.Nig.:** Pankshin, 5,100 ft. (July) *Lely* 433!
   Jos (May–Oct.) *Batten-Poole* FHI 12874a! *Keay* FHI 22725! *Latilo* FHI 25326!
2. **A. schweinfurthii** *Bak.* in J. Linn. Soc. 18: 175 (1880); F.T.A. 7: 467; Reynolds in J.S. Afr. Bot. 20:
   165, t. 10, 11 (1954), and Aloes Trop. Afr. 288 fig. 293, 294, t. 63; Keay l.c. *A. barteri* Bak. in J. Linn.
   Soc. 18: 168 (1880), partly; F.T.A. 7: 464, partly; F.W.T.A., ed. 1, 2: 345, partly. *A. trivialis* A. Chev.
   l.c. 594, t. 32B (1952), French descr. only. A perennial herb with fleshy grey-green leaves up to about
   2 ft. long and about 3 in. broad (narrower than in the other W. African species), with toothed margins;
   leaf-surface mottled, turning red in dry season; stem up to 2¼ in. diam. and 2 ft. long, usually falling over
   and giving rise to several suckers; perianth dull pink or red; on rocky hills and sometimes planted in
   villages.
   **Ghana:** Accra Plains (Feb.) *Brown* 928! Nakpanduri (Aug.) *Hall* CC 495! **N.Nig.:** Sara Hills, Dutsen
   Adari, Jos Plateau (Oct.) *Hepper* 1131! Alantika Mts., Adamawa (Dec.) *Hepper* 1603! **S.Nig.:** Lagos
   (Feb.) *Millen* 172! Ado Rock, Iseyin (Sept.) *Keay* FHI 22578! Mt. Orosun, Idanre (Oct.) *Keay* FHI
   22582! Also in C. African Rep., Sudan and the Congos.
3. **A. keayi** *Reynolds* in J.S. Afr. Bot. 29: 43, pl. 6, 7 (1963), and Aloes Trop. Afr. 286, figs. 290–292. A
   perennial herb with fleshy leaves up to 2 ft. long and 5 in. broad; stem short, stout, surrounded by remains
   of dead leaves, forming suckers from the base; perianth orange-pink; flowering during rainy season;
   grassy places.
   **Ghana:** Nyanyanu to Oduponkpeake, nr. Accra *Keay & Adams* FHI 37757a! Mouree, near Cape Coast
   (Apr.) *Hall* 2585!
4. **A. buettneri** *A. Berger* in Engl., Bot. Jahrb. 36: 60 (1905), and in Engl., Pflanzenr. 4, 38, 3: 201 (1908);
   Keay l.c.; Reynolds, Aloes Trop. Afr. 41, figs. 37–43, t. 8–9. *A. barteri* Bak. l.c. 168 (1880), partly; F.T.A.
   7: 464, partly; F.W.T.A., ed. 1, 2: 345, partly. *A. paedogona* A. Berger (1906). *A. bulbicaulis* Christian
   in Fl. Pl. S. Afr. 16: t. 630 (1936). *A. paludicola* A. Chev. in Rev. Bot. Appliq. 31: 597, t. 33 C (1952),
   French descr. only. A perennial herb with a rosette of fleshy leaves up to 2 ft. long and 5 in. broad at base,
   margins sharply toothed; in grassy places in the moister savanna areas; after the bush fires destroy the
   mature leaves the plant passes the dry season with the leaf-bases partly buried and with only the central
   leaf-tips exposed; perianth varied in colour.
   **Mali:** Kita to Badinko (Jan.) *Chev.* 115! **Iv.C.:** Assakra (Aug.) *de Wit* 5615! Singobro forest (Oct.)
   *de Wilde* 606! **U.Volta:** Diondiou, nr. Bobo-Dioulasso (June) *Chev.* 1014! **Ghana:** Kpeve (Dec.) *Howes*
   1073! Kintampo (June) *Vigne* FH 3046! **Togo Rep.:** Bismarckburg (July) *Büttner* 24! **Dah.:** Natitingou
   to Bokorona, Atacora Mts. (June) *Chev.* 24217! **N.Nig.:** Randa (Sept.) *Hepburn* 126! Anara F.R., Zaria
   (Oct.) *Keay* FHI 5488! Mando F.R., Zaria (June) *Keay* FHI 25819! Zungeru to Birnin Gwari (June)
   *Latilo* FHI 25329! **S.Nig.:** Iseyin (May) *Hambler* 455! Olokemeji (June) *Jones & Keay* FHI 18816!
   Ado Rock, Oyo (Feb.–July) *Keay & Brenan* FHI 22403! *Latilo* FHI 25328! 32970! 32971! 32972!
   Extending to C. African Rep., Congo, Angola, Zambia and Malawi.

   Besides the indigenous species, two forms of what appears to be *A. barbadensis* Mill. are cultivated in
Sierra Leone (Freetown and Njala) and at Victoria, W. Cameroun.


## 2. **ASPARAGUS** Linn., Sp. Pl. 313 (1753); F.T.A. 7: 425.

Flowers in racemes; cladodes flattened and more or less falcate; branches smooth and
   glabrous; spines short and recurved, only on the main shoots:
Pedicels jointed about or above the middle, about 5 mm. long in flower; cladodes
   about 2 cm. long and 1 mm. broad; racemes about 3 cm. long usually paired or
   solitary .. .. .. .. .. .. .. .. .. 1. *racemosus*
Pedicels jointed nearly at the base, about 5 mm. long in flower; cladodes about 4 cm.
   long and 2 mm. broad; racemes about 5 cm. long, often clustered  2. *warneckei*
Flowers fasciculate or solitary; cladodes subulate or filiform (flattened in *A. schroederi*),
   straight or nearly so:
Pedicels filiform 3–15 mm. long; branches smooth, terete or angular, glabrous or
   pubescent:
Pedicels jointed above the middle, about 15 mm. long; perianth 5 mm. long; branchlets very slender; cladodes filiform, over 2 cm. long  .. .. 3. *longipes*
Pedicels jointed below the middle:
   Leaf spines mostly limited to the nodes from which the principal branches arise,
   strongly reflexed; younger branches densely velvety pubescent; cladodes
   1–2 cm. long, subulate, rather numerous at each node; pedicels 5 mm. long;
   perianth 4 mm. long .. .. .. .. .. .. .. 4. *pubescens*

Leaf spines at most nodes, acute:
Cladodes 2–4 cm. long, filiform; branchlets glabrous; pedicels 7 mm. long;
perianth 2·5 mm. long  ..  ..  ..  ..    5. *schweinfurthii*
Cladodes about 1–1½ cm. long, subulate, stiff; branchlets glabrous or pubescent,
smooth and rounded; pedicels 4 mm. long in flower; perianth 2 mm. long
        6. *flagellaris*
Pedicels short and stout, about 3 mm. long, jointed about the middle; stems glabrous,
angular and minutely scabrid on the angles:
Cladodes rigidly subulate, sharply angular 1(–2) cm. long; spines very short and
curved from a broad base  ..  ..  ..  ..  ..  7. *africanus*
Cladodes flattened, more or less triangular in section, about 2 cm. long; spines fairly
long, slightly curved  ..  ..  ..  ..  ..  ..  8. *schroederi*

Besides the above indigenous species, *A. sprengeri* Regel, from Natal, and *A. plumosus* Bak., from E. Africa, are cultivated as ornamental plants.

1. **A. racemosus** *Willd.* Sp. Pl. 2: 152 (1799); Chev. Bot. 645. Climbing plant up to 6 ft. high in thickets; flowers green or white; berries red.
**Port.G.:** Sonaco (Oct.) *Pereira* 3333! **Guin.:** Kaba to Haut-Mamou *Chev.* 12752! Farana *Chev.* 13175! Diaguissa *Chev.* 13472! **S.L.:** Konta, Kambia Dist. (Apr.) *Hepper* 2647! Konnoh country *Burbridge* 488! Musaia (Dec.) *Deighton* 4471! Kabala (fl. & fr. Feb.) *Deighton* 1873! Loma Mts. (Nov.) *Jaeger* 8209! **Ghana:** Akroful *Cummins* 31! Pawn-pawna (Jan.) *A. S. Thomas* D103! Dawa Mate Kole (Jan.) *A. S. Thomas* D79! Achimota & Aburi (cult.) *Irvine* 2028! **S.Nig.:** Yoruba *Millson* 25! Widespread in the drier parts of tropical Africa and in tropical Asia.

2. **A. warneckei** (*Engl.*) *Hutch.* in Kew Bull. 1939: 245; Morton, Lilies & Orch. fig. 18. *A. drepanophyllus* var. *warneckei* Engl., Bot. Jahrb. 32: 97 (1902). Climber on shrubs in thickets; leafless at flowering time; flowers white or creamy with strong sickly odour.
**Guin.:** Pita (Nov.) *Pobéguin.* **Ghana:** Accra-Dodowah Plains (Feb.) *Irvine* 1502! Achimota (Dec.) *Milne-Redh.* 5176! *Akpabla* 1265! Accra to Aburi (Apr.) *Baldwin* 11989! **Togo Rep.:** Lomé *Warnecke* 28!

3. **A. longipes** *Bak.* in Kew Bull. 1901: 134. Slender climbing unarmed plant with pale green-brown flowers. **W.Cam.?:** "Cameroon Mountains." Imported by Messrs. Sanders of St. Albans, and flowered at Kew in Nov. 1898. Also cultivated at Achimota, Ghana, in 1936 (*Akpabla* 485!). Further information is required about the origin and occurrence of this species.

4. **A. pubescens** *Bak.* in Trans. Linn. Soc. ser. 2, Bot. 1: 254 (1878); F.T.A. 7: 430. Trailing or climbing much-branched plant 4–6 ft. high, only the main branches bearing acutely reflexed spines; flowers white. **N.Nig.:** Jos Plateau (Apr.) *Lely* P222! **S.Nig.:** Shaki, Oyo Prov. *Keay* FHI 37574! Also in E. Cameroun, Zambia, S. Rhodesia, Bechuanaland and Angola.

5. **A. schweinfurthii** *Bak.* in J. Linn. Soc. 14: 616 (1875); F.T.A. 7: 429. Erect armed shrub about 4 ft. high branched in the upper part; in upland grassland among rocks.
**W.Cam.:** Bum, Bamenda Div., 4,000 ft. (fr. May) *Maitland* 1702! Kimbi R. bridge, 3,000 ft. *Brunt* 1016! Also in E. Cameroun, C. African Rep. and Sudan.

6. **A. flagellaris** (*Kunth*) *Bak.* in J. Linn. Soc. 14: 614 (1875); F.T.A. 7: 430; Chev. Bot. 644. *Asparagopsis flagellaris* Kunth, Enum. Pl. 5: 103 (1850). *Asparagus pauli-guilelmi* Solms-Laub. (1867)—F.T.A. 7: 428, incl. var. *daltoni* Bak. (1875); F.W.T.A., ed. 1, 2: 352; Berhaut, Fl. Sén. 53. *A. africanus* of Chev. Bot. 644, not of Lam. Climbing or more or less erect plant with arching spiny branchlets; flowers white, fragrant; fruits orange when ripe.
**Sen.:** Sousoun *Chev.* 15754! Kaolack 55 *Kaichinger*! **Mali:** Bamako (Apr.) *Hagerup* 31! San *Chev.* 2576! **Guin.:** Dalaba (Mar.) *Langdale-Brown* 2625! Kouroussa (Apr.) *Pobéguin* 217! 701! **S.L.:** *Glanville* 382! Falaba to Berria (fr. Mar.) *Sc. Elliot* 5215! **U.Volta:** Sindou (fr. June) *Leeuwenberg* 4323! Diapaga to Fada *Chev.* 24483! **Ghana:** Sissu, Ashanti (fr. July) *Chipp* 495! Okroso to Otiso Ferry, Buem-Krachi Dist. (fl. & fr. Mar.) *Hepper & Morton* A3036! Kete Krache *Krause*! Anyaboni (Jan.) *A. S. Thomas* D109! Krepi Plains (Jan.) *Johnson* 534! Ho (Mar.) *Lloyd Williams* 59! **Togo Rep.:** Lomé *Warnecke* 479! **Dah.:** Abomey (Feb.) *Chev.* 23157! **N.Nig.:** Nupe *Barter* 1125! 1519! Jebba *Meikle* 1187! Lokoja (Feb.)

FIG. 349.—Asparagus africanus *Lam.* (Liliaceae).
A, cladode. B, pedicel. C, flowers and cladodes. D, open flower.

*Dalz.* 921! Vom, *Dent Young* 253! Katagum Dist. *Dalz.* 245! L. Chad (Jan.–Feb.) *Talbot* 733! **S.Nig.:** Abeokuta (Feb.) *Irving* 70! Iseyin to Igbetti *Hambler* 137! Ijaiye F.R., Oyo Prov. (Mar.) *Keay* FHI 21178! Milliken Hill, Enugu (Mar.) *Hepper* 2218! Widespread in tropical Africa.

7. **A. africanus** *Lam.* Encycl. Méth. Bot. 1: 295 (1783); F.T.A. 7: 433; Berhaut, Fl. Sén. 53; Morton, Lilies & Orch. fig. 20. *A. gourmacus* A. Chev., Bot. 645, name only. Erect herb 3–5 ft. high with numerous spiny branches; flowers greenish white; occurring in clumps in moist sandy savanna.
    **Sen.:** Thiès (fr. June) *Berhaut* 382! Cayor *Heudelot* 462! **Gam.:** Kuntaur *Ruxton* 46! **U.Volta:** Fada to Diapaga, Gourma Dist. (July) *Chev.* 24484! **Ghana:** Bamboi to Kintampo (Mar.) *Hepper & Morton* A3194! Sissu, Ashanti (June) *Chipp* 494! **N.Nig.:** Nupe *Barter* 1516! Samaru to Shika, Giwa Dist. (May) *Keay* FHI 25794! Nabardo (May) *Lely* 204! Rimjim Mukur to Bununu, Jarawa R. (fl. & fr. Aug.) *Lawlor & Hall* 613! Widespread in the drier parts of tropical Africa, Socotra and Arabia.

8. **A. schroederi** *Engl.* Bot. Jahrb. 32: 97 (1902); Aké Assi, Contrib. 2: 20 [229]. *A. falcatus* of Chev. Bot. 644, not of Linn. Erect spiny herb 1–3 ft. high from tuberous roots; flowers creamy with orange anthers, almost leafless in full flower; in savanna.
    **Iv. C.:** Ferkéssedougou *Serv. For.* 2283! Touba to Odienné *Mangenot & Aké Assi* 4245. **Ghana:** Banda ravine, Wenchi Dist. (fr. Apr.) *Morton* A3270! Yendi, Dagomba Dist. (Mar.) *Hepper & Morton* A3074! **Togo Rep.:** Sokode *Schroeder* 20. **Dah.:** Natitengou to Bokorona, Atacora Mts. (June) *Chev.* 24167! Pelebina to Djougou *Chev.* 23847! **N.Nig.:** Kontagora Prov. (Apr.) *Dalz.* 444! Samaru, Giwa Dist. (May) *Keay* FHI 25728! Kabama F.R., Zaria Dist. (May) *Keay* FHI 25739! Jos (Mar.) *Hill* 23! Katagum Dist. *Dalz.* 245!

**3. KNIPHOFIA** Moench., Meth. 631 (1794); F.T.A. 7: 450. *Notosceptrum* Benth. (1883)—F.T.A. 7: 454; F.W.T.A., ed. 1, 2: 342.

Leaves in a basal tuft and forming a cylindrical "tube" at the base, linear, up to 60 cm. long and 1·5 cm. broad, glabrous except the minutely scabrid margin, closely nerved; spike stout, about 60 cm. long, the upper two-thirds bearing subsessile flowers, the latter rather dense; bracts boat-shaped, soon reflexed, 6 mm. long, scabrid on the margin; perianth campanulate, 1 cm. long, 6-lobed, lobes half as long as the tube, obtusely triangular, 1-nerved to the base of the tube; ovary broadly ovoid, glabrous; style entire, a little longer than the ovary    ..    *reflexum*

**K. reflexum** *Hutch. ex Codd* in Bot. Notis. 120: 44 (1967). *Notosceptrum reflexum* Hutch., F.W.T.A. ed. 1, 2: 342 (1936), English descr. only. A tufted perennial herb with a single sterile inflorescence about 2 ft. high; flowers small, campanulate, yellow; "scattered in considerable numbers in the grass on the plateau" (Maitland).
    **W.Cam.:** Lakom, Bamenda, 6,000 ft. (Apr.) *Maitland* 1624!

**4. ERIOSPERMUM** Jacq., Collect. Suppl. 72 (1796); F.T.A. 7: 470.

Tuber bulb-like or somewhat irregular in shape, crowned by the fibrous remains of the leaf-bases; leaves 1–3, long-petiolate, linear-lanceolate, acute at each end, about 15 cm. long and 1–1·3 cm. broad, markedly nerved, glabrous; raceme about 25 cm. long, slender; pedicels ascending, the lower up to 10 cm. long in fruit; perianth-segments about 8 mm. long, narrowly oblanceolate; capsule obovoid, 8 mm. long; seeds densely villous    ..    ..    ..    ..    ..    ..    *abyssinicum*

**E. abyssinicum** *Bak.* in J. Linn. Soc. 15: 263 (1876); F.T.A. 7: 471; Troupin in Expl. Parc Nat. Garamba 4: 196; Morton, Lilies & Orch. t. 8. *E. elatum* Bak. in F.T.A. 7: 471 (1898). *E. togoense* Dammer (1905). *Hypoxis villosa* of Chev. Bot. 635, partly (*Chev.* 937, 978), not of Linn. f. Tuber giving rise to an inflorescence about 1 ft. high towards the end of the dry season with long-pedicellate yellow flowers which open in the morning or late afternoon: the erect leaf grows later; in rocky savanna.
    **Sen.:** Tambacounda *Berhaut* 1658! Ouassadou *Berhaut* 2154! **Guin.:** Sambailo *Adam* 14837! **Iv.C.:** Bouna Reserve (Apr.) *Aké Assi* 8666! **U.Volta:** Banankalidoro to Bama *Chev.* 937! Simona to Toro *Chev.* 978! **Ghana:** Damongo Scarp (Mar.) *Morton* GC 8699! Tamale to Yendi (Mar.) *Adams* 3866! Tumu (May) *Morton* A3329! **Togo Rep.:** Aledzo, Sokode-Basari *Kersting* 310. **Dah.:** Pobéyon to Birni *Chev.* 23954! Kouandé *Chev.* 23998! **N.Nig.:** Nupe *Barter* 1538! Anara F.R., Zaria (May) *Keay* FHI 25773! Vom *Dent Young* 255! Yola (fr. May) *Dalz.* 233! Mambila Plateau, 4,900 ft. (bud Jan.) *Hepper* 2824! Widespread in the savanna of tropical Africa and as far south as Transvaal.

**5. ANTHERICUM** Linn., Sp. Pl. 310 (1753); F.T.A. 7: 477.

Rhachis of the inflorescence pubescent or puberulous:
  Leaves linear, expanding to the base, about 12–40 cm. long and 8–14 mm. broad in the middle, often arcuate, minutely serrulate-scabrid on the margins, closely nerved; inflorescence shorter than the leaves, axillary; lower bracts up to 1 cm. long; pedicels 5–10 mm. long ..    ..    ..    ..    ..    ..    1. *pubirhachis*
  Leaves lanceolate, narrowed towards the base, 8–15(–30) cm. long and about 2 cm. broad, thin and with cross veins visible between the rather widely spaced nerves, minutely serrulate-scabrid on the margins and on some nerves beneath; inflorescence shorter than the leaves, terminal; lower bracts about 2 cm. long; pedicels up to 4 mm. long    ..    ..    ..    ..    ..    ..    2. *subpetiolatum*
Rhachis of the inflorescence glabrous:
  Peduncle winged:
    Inflorescence unbranched; pedicel jointed near the base; fruits 7 mm. long, closely ribbed:
      Perianth about 1 cm. long; flowers congested near the apex of the inflorescence; bracts about 10 mm. long, decreasing in size upwards to about 5 mm., acutely acuminate; leaves narrowly linear-oblanceolate, about 30 cm. long and 2(–3) cm.

FIG. 350.—ERIOSPERMUM ABYSSINICUM *Bak.* (LILIACEAE).

A, flower. B, perianth-segment and stamen. C, stamens. D, pistil. E, opened capsule showing seeds. F, seed, with G, hairy covering removed.

broad, closely nerved with about 25 nerves on each side of the midrib, not transversely barred with purple towards the base      ..    ..    ..  3. *pterocaulon*
Perianth about 1·5 cm. long; flowers close together towards apex of the inflorescence;
bracts 1·5–4 mm. long, gradually pointed; leaves linear, about 40(–70) cm. long,
6–16 mm. broad, very closely nerved, leaf sheaths transversely barred with purple
towards the base   ..    ..    ..    ..    ..    ..  4. *uyuiense*
Inflorescence branched; fruits 5 mm. long, coarsely ribbed:
Inflorescence with several branches more than 3 cm. long; flowers usually paired,
pairs about 1 cm. apart; pedicels jointed near the base; perianth 6 mm. long;
leaves as long as or shorter than the panicle, linear, 1–1·5(–2·5) cm. broad, minutely
scabrid on the margins, with about 20 lateral nerves on each side of the midrib
                                                               5. *limosum*
Inflorescence shortly branched, branches up to 2·5 cm. long; flowers congested on
branches, several flowers together; pedicels jointed in the middle; perianth 8 mm.
long; leaves narrowly linear, about 5 mm. broad, with about 6 lateral nerves on
each side of the midrib    ..    ..    ..    ..    ..  6. *zenkeri*
Peduncle not winged:
Pedicels jointed near the base or at least below the middle:
Leaf sheaths mottled towards the base; lower part of lamina and upper part of
leaf sheaths ciliate-bristly; leaves linear up to 5 mm. broad; inflorescence
paniculate to more or less simple, slender; pedicels very short; perianth
8–10 mm. long; fruit up to 6 mm. long, transversely ribbed  ..    7. *dalzielii*
Leaf-sheaths un-mottled:
Inflorescences ascending, usually with a few equal branches from near the base
spreading in opposite directions, very slender, about 20 cm. long usually less,
flowers few, evenly spaced; perianth 3–4 mm. long; pedicels up to 5 mm. long
in fruit; fruits about 4 mm. broad; leaves as long as the inflorescences, 2–3 mm.
broad ..    ..    ..    ..    ..    ..    ..    ..  8. *warneckei*
Inflorescences erect, 23–52 cm. high, flowers few towards apex; bracts broadly
ovate hyaline; pedicels about 6 mm. long; perianth-segments 7 mm. long; fruits
5 mm. diam., strongly ribbed; leaves linear 3 mm. broad, shorter than the
inflorescence ..    ..    ..    ..    ..    ..  9. *immaculatum*
Pedicels jointed near the top or at least above the middle:
Lamina conspicuously setose on the margins    ..    ..    ..  10. *sp. A.*
Lamina glabrous at most scabrid-ciliolate on the margin (in *A. caulescens*), leaf
sheaths glabrous or setose (in *A. nigericum*); pedicels jointed above the middle:
Inflorescence erect, branched (or sometimes simple in *A. nubicum*):
Leaves present at flowering time, linear, 2–2·5 cm. broad; inflorescences well-
branched, slender, up to 60 cm. high; bracts narrow, small; pedicels 1–1·5 cm.
long in fruit; perianth 1 cm. long, segments narrow; fruits 4 mm. long, 6 mm.
broad, transversely ribbed    ..    ..    ..    ..  11. *caulescens*
Leaves absent at flowering time; leaves linear up to 1 cm. broad; inflorescences
simple or branched, up to about 35 cm. long, with scarious leaf sheaths and
bracts; pedicels 1 cm. long; perianth 1·2 cm. long, segments rather broad;
fruits 7 mm. diam. with a few transverse oblique ribs ..    ..  12. *nubicum*
Inflorescences deflexed and curved, simple, appearing with and shorter than the
leaves:
Margins of leaf sheaths glabrous, lamina linear, about 20 cm. long, 3–4 mm.
broad, long-tailed at apex; inflorescence-rhachis angular; bracts about 5 mm.
long, flowers usually paired; fruits nearly oblong, 9 mm. long, 6 mm. broad,
surfaces lightly veined, sutures thickened    ..    ..  13. *inconspicuum*
Margins of leaf-sheaths setose, lamina linear, about 20 cm. long, 2 mm. broad;
inflorescence-rhachis cylindrical; bracts 2 mm. long  ..    ..  14. *nigericum*

1. **A. pubirhachis** *Bak.* in J. Linn. Soc. 15: 302 (1876); F.T.A. 7: 481; Berhaut, Fl. Sén. 182, 186. *Dasy-stachys graminea* A. Chev., Bot. 652 (1920), name only. Small rhizomatous herb with arching leaves and sometimes recurved inflorescences; flowers white.
    **Sen.:** Kounpentoum (June) *Trochain* 3675! Niokolo-Koba (June) *Adam* 14162! **Mali:** Baguineda (fr. Aug.) *Roberty* 294! **Guin.:** Kouroussa (June) *Pobéguin* 293! **Ghana:** Yendi (Apr.) *Morton* GC 9081! **Dah.:** Nioro to Kouandé, Atacora Mts. *Chev.* 24010! **N.Nig.:** Nupe *Barter*! Zaria to Jos (fl. & fr. June) *Stanfield* FHI 54823! **S.Nig.:** Upper Ogun, N. of Iseyin (May) *Stanfield* 59874! Extends to E. Cameroun, C. African Rep. and ? Congo.
2. **A. subpetiolatum** *Bak.* in J. Linn. Soc. 15: 302 (1876); Troupin, Fl. Sperm. Parc Nat. Garamba 1: 184. *A. limbamenense* Engl. & K. Krause in Engl., Bot. Jahrb. 45: 126 (1910); F.W.T.A., ed. 1, 2: 341. Small herb with a short inflorescence of white flowers.
    **N.Nig.:** Abuja, Niger Prov. (July) *Latilo* FHI 54857! Jos Plateau (July) *Lely* P354! Bukure to Hepham *Lely* 350! Birni, Kwaya Dist., Bornu (Aug.) *Daggash* FHI 24856! Widespread in the savanna of eastern tropical Africa.
3. **A. pterocaulon** *Welw. ex Bak.* in Trans. Linn. Soc., ser 2, 1: 258 (1878). *A. korrowalense* Engl. & K. Krause in Engl., Bot. Jahrb. 45: 125 (1910); F.W.T.A., ed. 1, 2: 341; Troupin, Fl. Sperm. Parc Nat. Garamba 1: 184. *A. atacorense* A. Chev., Bot. 648 (1920), name only. A rather stout herb with erect inflorescence about 1 ft. high; flowers white.
    **Dah.:** Bokorono to Kouandi, Atacora Mts. (June) *Chev.* 24219! **N.Nig.:** Anara F.R., Zaria Prov. (May) *Keay & Mulch* FHI 22927! Zaria to Samaru (fl. & fr. July) *Keay* FHI 25923! Wana (fl. & fr. June, Aug.) *Hepburn* 111a! Keana, Benue Dist. *Hepburn* 56! Yola (Oct.) *Shaw* 97!

4. **A. uyuiense** *Rendle* in J. Linn. Soc. 30: 415 (1894); F.T.A. 7: 485; Troupin in Bull. Jard. Bot. Brux. 25: 233 (1955), and in Fl. Sperm. Parc Nat. Garamba 1: 185, fig. 30. *A. speciosum* of F.W.T.A., ed. 1, 2: 341, not of Rendle. *A. koutiense* A. Chev. in Mém. Soc. Bot. Fr. 2, 8: 89 (1908). Leaves and inflorescence appearing together, 1–2½ ft. high; flowers white.
**U.Volta:** Zorgongo to Ouagnan, Mossi *Chev.* 24635! **Ghana:** Djowany, Gambaga Dist. (June) *Akpabla* 699! **Dah.:** Koussokonigou, Atacora Mts. (June) *Villiers*! Farfara to Toukountouna (June) *Chev.* 24058! **N.Nig.:** Liruwen-Kano Hills *Carpenter*! Samaru, Giwa Dist. (July) *Keay* FHI 25948! Neill's Valley, Jos (June) *Lely* 258! Naraguta F.R. (June) *Kennedy* FHI 7257! Tangale-Waja, Bauchi Prov. (July) *G. V. Summerhayes* 72! Extending to Uganda, Congo, Tanzania and Zambia.

5. **A. limosum** *Bak.* in Trans. Linn. Soc., ser. 2, 1: 257 (1878); F.T.A. 7: 482. *A. usseramense* var. *occidentalis* A. Chev. in Mém. Soc. Bot. Fr. 2, 8: 90 (1908); Bot. 648. *Chlorophytum nigericum* A. Chev., Bot. 651 (1920), name only. A slender herb with leaves and flowers together; inflorescence rhachis flattened, branched above, about 2 ft. high; in moist rocky savanna.
**Gam.:** *Saunders* 27! Dipa Kunda (July) *Fox* 146! **Mali:** Kita (July) *Duong* 454! **Guin.:** Ditinn *Chev.* 12180! Timbo to Niger *Chev.* 13271! Kindia *Chev.* 13542! Kolenté (June) *Chillou* 1574! **S.L.:** Matamba (Oct.) *Glanville* 28! Samaia (May) *Deighton* 4789! Musaia (Apr., Aug.) *Haswell* 3! *Deighton* 4733! Kundita, Falaba (Mar.) *Sc. Elliot* 5203! **Iv.C.:** Mt. Dourou, near Koualé *Chev.* 21744! Issia Rock (fl. & fr. Oct.) *Aké Assi* 5681! 8841! **U.Volta:** Diapaga to Fada *Chev.* 24440. **Dah.:** Konkobiri *Chev.* 24329! **N.Nig.:** Kargi Hill, Birnin Gwari Dist. (fl. & fr. June) *Keay* FHI 25883! Vom *Dent Young* 250! Naraguta F.R. (Aug.) *Keay* FHI 12731! Yola (fl. & fr. Aug.) *Dalz.* 235! Also in E. Cameroun and Angola.

6 **A. zenkeri** *Engl.* Bot. Jahrb. 32: 91 (1902). Slender herb with erect inflorescences 1–2 ft. high and grass-like leaves; flowers pure white.
**N.Nig.:** Vogel Peak, Sardauna Prov., 5,000 ft. (fl. & fr. Nov.) *Hepper* 1510! Also in E. Cameroun.

7. **A. dalzielii** *Hutch. ex Hepper* in Kew Bull. 22: 456, fig. 5, 1–3 (1968); F.W.T.A., ed. 1, 2: 342 (1936), English descr. only. Leaves about 1 ft. high with the inflorescence taller, sometimes densely fibrous at the base; leaf sheaths conspicuously red-mottled; flowers white; in sandy places.
**N.Nig.:** Bukuru, Jos Plateau (fl. & fr. July) *Lely* 459! Tangale-Waja, Bauchi Prov. (May) *G. V. Summerhayes* 64! S. of Keana *Hepburn* 54! Abinsi (June) *Dalz.* 854!

8. **A. warneckei** *Engl.* Bot. Jahrb. 32: 91 (1902); Morton, W. Afr. Lilies & Orch. fig. 11. *A. deightonii* Hutch. ex Berhaut, Fl. Sén. 184 (1954), French descr. only. Small herb with grass-like leaves; inflorescence arching near the ground; flowers small, white; amongst grass.
**Sen.:** Tambacounda *Berhaut* 1644! **Guin.:** Kouroussa (fl. & fr. June) *Pobéguin* 1116! **Ghana:** Elmina (fl. & fr. July) *Hall* 1523! Accra (Mar.) *Deighton* 584! Achimota (fl. & fr. May) *Irvine* 1771! Adidome to Ho (fl. & fr. Apr.) *Morton* A3218! **Togo Rep.:** Schlickboden, near Lomé (May) *Warnecke* 304. Also in Gabon.

9. **A. immaculatum** *Hepper* in Kew Bull. 22: 458, fig. 6, 1–3 (1968). Slender herb about 1 ft. high with white flowers.
**Sen.:** Oussouye, Casamance (fl. & fr. July) *Berhaut* 6178! Diouloulou (fl. & fr. Sept.) *Adam* 18213! Kédougou (June) *Adam* 14422! **Port.G.:** Dandum to Saber Capeidje (June) *Esp. Santo* 2922! **Guin.:** Sambailo (July) *Adam* 14843!

10. **A. sp. A.** *A. deflexum* A. Chev., Bot. 648 (*Chev.* 938); Hutch., F.W.T.A., ed. 1, 2: 342 (1936), English descr. only, not of A. Chev. (1913). Small herb with narrow conspicuously ciliate leaves.
**Mali:** Banankalidoro to Bama (June) *Chev.* 938!
[Further material is required to establish the identity of this plant.]

11. **A. caulescens** *Bak.* in J. Linn. Soc. 15: 303 (1876). Herb with rather broad leaves and tuberous roots; flowers inconspicuous dull green on a branched inflorescence; in dry rocky ground.
**N.Nig.:** Nupe *Barter* 1515! Ilorin to Jebba (fr. June) *Onochie* FHI 40253!

12. **A. nubicum** *Bak.* l.c. 301 (1876); F.T.A. 7: 484; Milne-Redhead in Kew Bull. 3: 471 (note under *Urginea glaucescens*). *A. fibrosum* Hutch., F.W.T.A., ed. 1, 2: 342 (1936), English descr. only; Hepper in Bull. I.F.A.N. 27: 504 (1965). *A. articulatum* Hutch. l.c. (1936), English descr. only, partly (*Lely* 690). Perennial herb with a fibrous corm flowering leafless or nearly so with slender usually unbranched inflorescences a few inches to 1½ ft. high bearing scale leaves; flowers white with a pink median stripe on each tepal; flowering in the dry season in savanna grassland.
**Guin.:** Kouroussa (Jan.) *Pobéguin* 629! 635! Labé (Apr.) *Chev.* 12306! **N.Nig.:** Kontagora (Jan.) *Dalz.* 261! Minna (Jan.) *Onochie* FHI 40852! Vom (Feb.) *McClintock* 204! Naraguta (Nov.) *Lely* 690! Wana (Apr.) *Hepburn* 111b! Yola (fl. & fr. Feb.) *Dalz.* 230! Vogel Peak (Nov.) *Hepper* 1388! **S.Nig.:** Kishi to Igbeti, Oyo Dist. (Feb.) *Keay* FHI 22497! Extending to Sudan and Uganda.

13. **A. inconspicuum** *Bak.* in J. Bot. 15: 71 (1877); F.T.A. 7: 480. Small herb with arching grass-like leaves about 8 in. high; inflorescence arching and bearing small flowers.
**Sen.:** Ngor (fl. & fr. July) *Berhaut* 4768! Extending to Somalia.

14. **A. nigericum** *Hepper* in Kew Bull. 22: 459, fig. 5, 4–6 (1968). A slender perennial herb with short fat root tubers and very narrow leaves; inflorescence angled near the base 6–8 in. long, about the same length as the leaves; flowers white; in rocky savanna.
**N.Nig.:** Anara F.R., Zaria Prov. (July) *Keay* FHI 37081! Jos Plateau (Aug.) *Lely* P575!

## 6. CHLOROPHYTUM Ker-Gawl. in Bot. Mag. t. 1071 (1807); F.T.A. 7: 493.

Pedicels jointed near the top or at least above the middle:
  Inflorescence stout, subspicate or congested lateral racemes or rarely with one branch, not paniculate:
    Leaves linear, 1–2 cm. broad, up to 45 cm. long, ciliate; flowering part of inflorescence 20–35 cm. long:
      Bracts pectinate-ciliate, linear-filiform from a broader base, about 1 cm. long; leaves up to 2 cm. broad .. .. .. .. .. .. .. 1. *senegalensis*
      Bracts not ciliate, triangular-ovate, shortly acuminate; perianth 1 cm. long; leaves up to 1·5 cm. broad with about 20 closely parallel nerves .. 2. *aureum*
    Leaves lanceolate or linear-lanceolate, 3–8 cm. broad:
      Leaves lanceolate to oblanceolate, erect, long-acuminate, attenuate towards the base into an indistinct petiole and broadening to the point of insertion, up to 60 cm. long from base to apex, up to 8 cm. broad; peduncle 12–30 cm. long, apparently elongating in fruit up to 40(–60) cm. long, flowers congested in upper third; fruits 7–10 cm. long .. .. .. .. .. .. 3. *macrophyllum*
      Leaves linear-lanceolate, close to ground, acute at apex, broadly petiolate, up to 40 cm. long from base to apex, about 4 cm. broad; inflorescence about ⅛ as long

as the leaves, a continuous dense raceme about 8 cm. long; bracts almost imbricate, ribbed, 1 cm. long; fruits 5 mm. long    ..    ..    ..    11. *stenopetalum*
Inflorescence slender, paniculate or if simple then leaves distinctly petiolate :*
  Leaves linear or nearly so, 1–2 cm. broad, up to 80 cm. long; inflorescence a well-branched panicle longer than the leaves, rhachis not scabrid :
    Lobes of fruit divergent, cuneate to the base, whole fruit 4 mm. long, 6 mm. broad, strongly nerved; perianth 4 mm. long    ..    ..    ..    ..    4. *polystachys*
    Lobes of fruit rounded, lobed at base, whole fruit 3 mm. long, 5 mm. broad, more or less nerved; perianth 6 mm. long:
      Inflorescences much-branched panicles; pedicels up to 5 mm. long    5. *altum*
      Inflorescences sparingly branched usually with 3 equal branches up to 20 mm. long; pedicels 10–12 mm. long    ..    ..    ..    ..    ..    6. *nzii*
  Leaves not linear, 2–5 cm. broad:
    Rhachis of inflorescence smooth, more or less branched; fruits broader than long:
      Leaves indistinctly petiolate* linear-lanceolate, 30–45 cm. long, 3–5 cm. broad, with undulate margins, very thin and with numerous slender transverse veins; inflorescences paniculate or with a few slender branches; perianth 5 mm. long; fruit 4–5 mm. long, 5–6 mm. broad (see also *C. laxum* below) .. 7. *gallabatense*
      Leaves with a distinct slender petiole about as long as the lanceolate lamina, leaf variable in length, 10–40 cm. long, 2–4 cm. broad, very acute with about 6 lateral nerves on each side of the midrib; inflorescences slender, usually much longer than the leaves with a few slender branches or more or less simple; perianth 4 mm. long; fruit 4–6 mm. long, 5–7 mm. broad    ..    8. *togoense*
    Rhachis of inflorescence minutely scabrid, simple, about as long as the leaves; perianth about 8 mm. long; fruits as long as or longer than broad, longitudinally broadly elliptic, 8 mm. long; leaves lanceolate, very acute, up to 30 cm. long and 4 cm. broad, with about 9 lateral nerves on each side of the midrib 9. *inornatum*
Pedicels jointed at or below the middle:
  Leaves ciliate, at least towards the base:
    Inflorescence racemose or shortly branched, axis 10–40 cm. long, smooth; bracts subulate-lanceolate, about 1 cm. long, the lower ones much longer, caducous; perianth 8 mm. long; fruits suborbicular, about 1 cm. diam.; leaves linear-lanceolate, erect, the base nearly as broad as the lamina in young plants, older ones narrowed to a petiole and up to 40 cm. long with about 15 nerves on each side of the midrib    ..    ..    ..    ..    ..    ..    ..    10. *blepharophyllum*
    Inflorescence subcapitate, up to 4 cm. long, with short congested branches; bracts lanceolate less than 1 cm. long, ribbed; perianth 5 mm. long; leaves forming a rosette, lanceolate-elliptic, 12–21 cm. long, 2·5–6 cm. broad, usually distinctly petiolate, petiole 3–6 cm. long    ..    ..    ..    ..    ..    12. *geophilum*
  Leaves not ciliate:
    Leaves appressed to ground, ovate or ovate-lanceolate, sessile, 5–10 cm. long, 2–4 cm. broad, with about 12 lateral nerves on each side of the midrib; inflorescence very short and subsessile in the cluster of leaves; fruits suborbicular, 5 mm. diam.
                                        13. *pusillum*
    Leaves erect, not as above:
      Petiole long and distinct:
        Axis of inflorescence minutely scabrid, usually longer than the leaves, few-flowered with the flowers widely spaced, simple, often arcuate and viviparous; fruits about 5 mm. diam.; leaves obovate to oblanceolate, very acutely acuminate, thin, about 15 cm. long, 3–7 cm. broad; petiole about as long as the lamina
                                       14. *sparsiflorum*
      Axis of the inflorescence smooth:
        Leaves ovate to ovate-lanceolate, 15–35 cm. long, 5–10 cm. broad, with about 20 strong nerves; inflorescence branched, as long as or longer than the leaves; perianth 6 mm. long, segments linear; fruits suborbicular, 7 mm. diam.; pedicels jointed in the middle    ..    ..    ..    ..    15. *orchidastrum*
        Leaves smaller, linear to oblanceolate; pedicels jointed below the middle:
          Leaves linear or narrowly linear-oblanceolate, very gradually narrowed to the base, 6–23 cm. long, 1–2 cm. broad, thin, with about 5 parallel nerves on each side of the midrib; inflorescences very slender, shorter than the leaves and with a few short branches; bracts all as long as or longer than the pedicels; fruits suborbicular, 5 mm. diam.    ..    ..    ..    ..    16. *laxum*
         Leaves lanceolate, rather abruptly narrowed into the petiole, the lamina (4–)6–14 cm. long, 1·5–3·5 cm. broad, with about 8 parallel nerves on each side of the midrib; inflorescences slender, as long as the leaves, more or less branched; bracts (except the lowermost) much shorter than the pedicels; fruit obovate, the lobes appearing divergent, about 5 mm. long and 9 mm. broad
                                       17. *alismifolium*

---

\* The term petiole is used in this key for what is strictly a false petiole.

Petiole not clearly developed:
  Segments of perianth nearly 2 cm. long, 4 mm. or more broad; pedicels jointed
    in the middle; inflorescence usually simple, about as long as the leaves; bracts
    ovate, long-acuminate, as long as or longer than the flowers; fruits oblong-
    elliptic, 1·3 cm. long; leaves linear, 1–2·5 cm. broad  ..    .. 18. *tuberosum*
  Segments of perianth up to 1 cm. long, linear:
    Pedicels 1–1·5 cm. long:
      Fruit broader than long, 5 mm. broad, 3 mm. long; pedicels about 1 cm. long
        in flower, slightly longer in fruit; perianth segments 6 mm. long; inflorescence
        up to 57 cm. long, usually with 3 equal branches up to 20 cm. long; leafy at
        flowering time, leaves linear lanceolate, about 1·5 cm. broad, acute   6. *nzii*
      Fruits about as broad as long, nearly square in outline, 7–8 mm. diam.; perianth
        segments linear-filiform, 1 cm. long; inflorescences up to 80 cm. long with
        several long branches; bracts 2 mm. long; leaves linear about 10 cm. long
        and 5 mm. broad, not developed at flowering time or only towards the base of
        the rhachis, upper ones reduced to sheaths   ..    ..    .. 19. *andongense*
    Pedicels a few mm. long:
      Pedicels jointed below the middle; inflorescences slender, about 10 cm. long;
        fruits suborbicular, 5 mm. diam. (for other characters see above)  16. *laxum*
      Pedicels jointed in the middle:
        Leaves linear, about 1 cm. broad; inflorescence a slender panicle up to 30 cm.
          high; perianth segments linear, 5 mm. long, bracts subulate from a broader
          base, 3 mm. long    ..    ..    ..    ..    .. 20. *bequaertii*
        Leaves lanceolate, 2–3 cm. broad; inflorescence an interrupted raceme up to
          30 cm. high; perianth segments lanceolate, 4 mm. long; bracts ovate-
          lanceolate, about 1 cm. long   ..    ..    ..    ..    ..21. *deistelianum*

1. **C. senegalense** (*Bak.*) *Hepper* in Kew Bull. 21: 496 (1968). *Dasystachys senegalensis* Bak. in Bull. Herb. Boiss., sér. 2, 1: 782 (1901); F.W.T.A., ed. 1, 2: 3. *D. macensis* A. Chev., Bot. 652 (1920), name only. Stout, unbranched inflorescence 3–4 ft. high arising from a thick tuber or rhizome; flowers white; in savanna.
  **Sen.:** Kaolak (July) *Berhaut* 2157! Tambacounda to Sambailo (fr. Oct.) *Pitot* 6! **Mali:** Thou to Kiri, S. Macina (Aug.) *Chev.* 24854! Bamako (July) *Waterlot* 1218! **Port.G.:** Geba to Banjara, Bafata (Aug.) *Esp. Santo* 3060! **Guin.:** Kouroussa (Aug.) *Pobéguin* 374! Farabana, near Kankan (July) *Collenette* 78!
2. **C. aureum** *Engl.* Bot. Jahrb. 15: 469 (1892). *Dasystachys aurea* (Engl.) Bak. in F.T.A. 7: 512 (1898). *D. atacorensis* A. Chev., Bot. 652 (1920), name only; F.W.T.A., ed. 1, 2: 345 (1936), English descr. only; Aké Assi, Contrib. 2: 19[228]. *D. sombae* A. Chev., Bot. 652 (1920), name only. Inflorescence about 3 ft. high; flowers white; in savanna woodland.
  **Iv.C.:** Foro-Foro, Bouaké *Aké Assi* IA 4316. **Dah.:** Bokorona to Kouandé, Atacora Mts. (June) *Chev.* 24224! Farfa to Toukountouna (June) *Chev.* 24057! **S.Nig.:** Zomi. N. of Iseyin (May) (Apr.) *Stanfield* FHI 59873! Extends to Sudan.
3. **C. macrophyllum** (*A. Rich.*) *Aschers.* in Schweinf., Beitr. Fl. Aethiop. 294 (1867); F.T.A. 7: 498; Chev. Bot. 651; Berhaut, Fl. Sén. 186; Morton, W. Afr. Lilies & Orch. fig. 15. *C. macrophyllum* var. *albiflorum* A. Chev. in Mém. Soc. Bot. Fr. 2, 8: 91 (1908). Tuberous-rooted herb with rather large leaves in a rosette; flowers white, fragrant; in riverine forest.
  **Sen.:** Dougar (Sept.) *Berhaut* 5438! **Guin.:** Sériba to Gaoual (fr. Oct.) *Pitot*! **S.L.:** Sankan Biriwa Massif (fr. Jan.) *Cole* 168! **Lib.:** Nimba Mts. (July, Sept.) *Leeuwenberg & Voorhoeve* 4754! *Adames* 586! **Iv.C.:** Bouroukrou *Chev.* 16812! Assikasso, Mid. Comoe *Chev.* 22591! **Ghana:** Abetifi, Kwahu (fr. June) *Irvine* 1724! **N.Nig.:** Bonu, Birnin Gwari Dist. (fl. & fr. Aug.) *Onochie* FHI 18260! Acharane, Ankpa Dist. (fl. & fr. June) *Daramola* FHI 38034! Nupe *Barter* 1513! Patti Lokoja (fl. & fr. Nov.) *Dalz.* 23! **S.Nig.:** Olokemeji F.R. (fr. Nov.) *Jones, Keay & Onochie* FHI 14543! Ibadan F.R. (June) *Tamajong* FHI 23282! Atikiriji, Omo F.R. (fr. Dec.) *Tamajong* FHI 20282! Afi River F.R. (June) *Jones & Onochie* FHI 18952! Also in E. Cameroun, C. African Rep., Congo, Sudan, Ethiopia, Uganda, Kenya, Tanzania and Rhodesia
  [*Berhaut* 1246, from Senegal, cited as this sp. by Berhaut (l.c.) has much smaller fruits than others cited here.]
4. **C. polystachys** *Bak.* in J. Bot. 16: 326 (1878); F.T.A. 7: 509; Hepper in Kew Bull. 21: 496 (1968). *C. palustre* Engl. & K. Krause (1910). Herb with narrow leaves; inflorescence with erect branches; damp places in savanna.
  **U.Volta:** Diapaga, Gourma (July) *Chev.* 24379! **Ghana:** Bahare, Gambaga (June) *Akpabla* 664! Also in E. Cameroun, Sudan, Kenya, Tanzania, Zambia and Rhodesia.
5. **C. altum** *Engl. & K. Krause* in Engl., Bot. Jahrb. 45: 131 (1910). Perennial tuberous rooted herb with a large panicle about 2 ft. high.
  **Ghana:** White Volta R., nr. Bawku (fl. & fr. Aug.) *Hall* CC 543! Also in E. Cameroun.
6. **C. nzii** *A. Chev. ex Hepper* in Kew Bull. 22: 459, fig. 6, 4–6 (1967); Chev. Bot. 651 (1920), name only; F.W.T.A., ed. 1, 2: 345 (1936), English descr. only.
  **Iv.C.:** Tiébissou to Languira, Nzi valley, Baoulé Nord (fr. July) *Chev.* 22195! Danané (fr. Sept.) *Nozerau*!
7. **C. gallabatense** *Schweinf. ex Bak.* in F.T.A. 7: 504 (1898); Chev. Bot. 650. *C. andongense* of Chev. Bot. 649, not of Bak. Herb with a lax branched inflorescence nearly 2 ft. high; flowers blue-green with white filaments; moist places in savanna.
  **Sen.:** Tambacounda (fr. Sept.) *Berhaut* 1652! **Guin.:** Kouroussa *Chev.* 15641! Farana (May) *Chev.* 13398! **U.Volta:** Gourma (July) *Chev.* 24375! **N.Nig.:** Gwalor Hill, Gwari Dist. (fl. & fr. June) *Onochie* FHI 40170! Sanga F.R., Jemaa Div. (May) *E. W. Jones* 46! Also in Chad, C. African Rep., Sudan, Uganda, Kenya, Tanzania, Zambia, Malawi, Rhodesia and Mozambique. (See Appendix, p. 477.)
  [*Onochie* FHI 40253 from Jebba, N. Nigeria, may also be this species.]
8. **C. togoense** *Engl.* Bot. Jahrb. 32: 92 (1902); Morton, W. Afr. Lilies & Orch. fig. 16. *C. fosteri* A. Chev. in Mém. Soc. Bot. Fr. 2, 8: 90 (1907). *C. talbotii* Rendle, Cat. Talb. 113 (1913). *C. toumodiense* A. Chev. Bot. 652 (1920), name only. A slender herb about 1 ft. high with small greenish flowers; in shady places in high forest.
  **Lib.:** (fl. & fr. June) *Harley* 962! **Iv.C.:** Abouabou (Oct.) *Adam* 6552! Toumodi to Dimbroko *Chev.* 22252b! **Ghana:** Banda, N.W. Ashanti (Apr.) *Morton* A3281! Asafo to Kwanyaku, Cape Coast (Mar.) *Morton* GC 8527! **Togo Rep.:** Jägge stream, near Misahöhe (fr. Sept.) *Büttner* 172. Fasugu (bud May) *Büttner* 665. **Dah.:** Pénésoulon to Pélébina (May) *Chev.* 23835! Kouba to Farfa (June) *Chev.* 24037! Koussohoingou, Atacora Mts. (June) *Villiers*! **N.Nig.:** Wana (June) *Hepburn* 117! Gindiri, Jos Plateau (fl. & fr. Oct.) *Hepper* 1111! Abinsi (June) *Dalz.* 849! **S.Nig.:** Ikoyi woods, Lagos (May) *Dalz.* 1285!

Ibadan North F.R. (fr. June) *Tamajong* FHI 23293! Ewokini to Idumuje, Ishan Dist. (Aug.) *Onochie* FHI 33272! Olokemeji (fl. & fr. July) *Foster & Chev.* 14099! Ogun R., Oyo to Iseyin (July) *J. B. Gillett* 15221! Awka, Onitsha Prov. (June) *Jones* FHI 6601! Oban *Talbot* 731!

9. **C. inornatum** *Ker-Gawl.* in Bot. Mag. t. 1071 (1807); F.T.A. 7: 499; Chev. Bot. 650. *C. afzelii* Bak. (1876)—F.T.A. 7: 496; Chev. Bot. 649. *C. baoulense* A. Chev. and *C. cavalliense* A. Chev., Bot. 649 (1920), names only. Stemless herb about 1 ft. high, with the leaves as long as the inflorescence, pink towards the base; flowers green; in forest often by streams.
**Guin.:** Kissidougou (July) *Martine* 364! Trébél (fr. Sept.) *Chev.* 18615! **S.L.:** Wallia (fr. Jan.) *Sc. Elliot* 4642! Kambia (Aug.) *Harvey* 20! Magbema (Aug.) *Jordan* 965! Makump (fr. Sept.) *Deighton* 1287! Njala (fr. Dec.) *Deighton* 5976! **Lib.:** Mission Hill (Aug.) *Harley* 1653! **Iv.C.:** Tos forest, Bouaflé to Sinfra (Dec.) *Aké Assi* 7228! Lamé (fr. Nov.) *Leeuwenberg* 1899! Mbayakro, Nzi (fr. Aug.) *Chev.* 22264! **Ghana:** Amosima, Cape Coast (Dec.) *Hall* 2543! **S.Nig.:** Idanre F.R., Ondo Dist. (fr. July) *Onochie* FHI 33382! Ibadan to Oyo (fr. Jan.) *Meikle* 982! Sapoba, Benin Dist. (fr. Sept.) *Jones* FHI 4970! **W.Cam.:** Bamenda *Daramola* FHI 40543!

10. **C. blepharophyllum** *Schweinf. ex Bak.* in J. Linn. Soc. 15: 327 (1876); F.T.A. 7: 501; Berhaut, Fl. Sén. 186. *C. ciliatum* Bak. (1878)—F.T.A. 7: 505; Chev. Bot. 649. *C. kerstingii* Dammer (1912). *C. fibrosum* Engl. & K. Krause (1910). *C. stipitatum* v. Poelln. in Portug. Acta Biol., sér. B, 1: 225 (1945). Herb about 1 ft. high; flowers pale brown; amongst stones in moist savanna.
**Sen.:** Tambacounda (fl. June, fr. Sept.) *Trochain* 3631! *Berhaut* 1632! **Mali:** Sikasso (May) *Chev.* 799! Bamako (June) *Waterlot* 1158! **Guin.:** Kaba to Mamou (Apr.) *Chev.* 12753! Timbo (July) *Pobéguin* 1590! Kouroussa (May) *Pobéguin* 314! **S.L.:** Pendembu (fr. July) *Thomas* 779! Makunde (fr. Apr.) *Sc. Elliot* 5703! Musaia (fl. & fr. Mar.) *Deighton* 5482! Sulimania (Mar.) *Sc. Elliot* 5306! Loma Mt., 2,800 ft. (fr. June) *Morton* SL 3538! **Iv.C.:** Séguéla to Siana (June) *Chev.* 21842! Niangbo (fr. Apr.) *Aké Assi* 8592! **Ghana:** Japei (Mar.) *Adams* 3932! Banda (fr. Aug.) *Hall* 1997! Gambaga (fr. June) *Akpabla* 665! **Togo Rep.:** Aledyoi Sokodé-Basari (fl. & fr. Feb.) *Kersting* 314 (seen for Ed. 1). **Dah.:** Kouandé to Konkobiri (fl. & fr. June) *Chev.* 24295! **N.Nig.:** Nupe *Barter*! Mokwa, Bida Dist. (Apr.) *Keay* FHI 25715! Anara F.R., Zaria (fl. & fr. May) *Keay* FHI 22976! Afaka F.R., Jos (fr. July) *Okafor & Binuyo* FHI 47549! Zelau, Bauchi Dist. (Apr.) *Lely* 113! Abinsi (May) *Dalz.* 851! Also in E. Cameroun, C. African Rep., Congo, Sudan, Uganda, Kenya, Tanzania, Zambia, Rhodesia and Mozambique.
[The young leaves expand at the time of flowering and they are then short and broad. They reach their maximum length in the fruiting state when they appear to be narrowed towards the base.]

11. **C. stenopetalum** *Bak.* in J. Linn. Soc. 15: 331 (1876); Aké Assi, Contrib. 2: 18 [228]. *C. deistelianum* of F.W.T.A., ed. 1, 2: 345, partly (Chev. 24470), not of Engl. & K. Krause. A herb with a rosette of leaves close to the ground; inflorescence short with numerous congested white flowers; in savanna woodland.
**Iv.C.:** Kong to Tafiré *Aké Assi* IA 5680. **U.Volta:** Bobo-Dioulasso (May) *Chev.* 925! **Ghana:** Boro, Wa (June) *Adams* 632! Kpiri Lake, Damongo (Sept.) *Hall* CC 833! **Dah.:** Birni, Kouandé *Chev.*! Konkobiri (July) *Chev.* 24303! **Niger:** Diapaga to Fada (July) *Chev.* 24470! **N.Nig.:** Nupe *Barter*! Zungeru (June) *Dalz.* 274! Samaru, Giwa Dist. (June) *Keay* FHI 25905! Anara F.R., Zaria (May) *Keay* FHI 22941! Naraguta (May) *Lely* 245! Also in E. Cameroun, Congo, Uganda, Zambia, Malawi and Angola.

12. **C. geophilum** *Peter ex v. Poelln.* in Ber. Deutsch. Bot. Ges. 61: 127 (1943); *C. subcapitatum* Hutch. F.W.T.A., ed. 1, 2: 343 (1936), name only. *C. afzelii* of Chev. Bot. 649, not of Bak. *C. gourmacum* A. Chev., Bot.650 (1920), partly (Chev. 24690, 24297). Small plant with tuberous roots and leaves more or less flattened on the ground; inflorescence almost sessile; flowers white; pedicels elongating after flowering and reflexing, burying the fruit.
**Guin.:** Kouroussa (July) *Pobéguin* 305! **Iv.C.:** Koualé to Kouroukoro, Haute Sassandra (May) *Chev.* 21755! **U.Volta:** Ouagadougou to Septemga (Aug.) *Chev.* 24690! **Ghana:** Kampaha (June) *Adams* 717! **Dah.:** Kouandé to Konkobiri (June) *Chev.* 24297! **N.Nig.:** Bonu, Niger Prov. (June) *Onochie* FHI 38456! *E. W. Jones* 175! Naraguta, 4,000 ft. (May) *Lely* 248! Samaru to Shika, Zaria (June) *Keay* FHI 25913! Also in E. Cameroun and Tanzania.

13. **C. pusillum** *Schweinf. ex Bak.* in J. Bot. 16: 325 (1878); F.T.A. 7: 502; Aké Assi, Contrib. 2: 18 (227). *C. mossicum* A. Chev., Bot. 651 (1920), name only. *C. gourmacum* A. Chev., Bot. 650 (1920), partly (Chev. 24108, 24320). Leaves appressed to the ground forming a mat; flowers white; under moist rocks in savanna.
**Mali:** Bamako (July) *Waterlot* 1221! **Iv.C.:** Touba *Aké Assi* IA 5661. **U.Volta:** Yako to Ouahigouya, Mossi (Aug.) *Chev.* 24770! **Dah.:** Konkobiri (July) *Chev.* 24320! Tanguéta, Atacora Mts. (June) *Chev.* 24108! **Ghana:** Gambaga (fl. July, fr. Nov.) *T. M. Harris*! *Morton* A2725! **N.Nig.:** Panshanu (fl. & fr. Aug.) *Lawlor & Hall* 302! Kilba Hills, near Yola (fr. Aug.) *Dalz.* 224! Also in E. Cameroun, Uganda, Tanzania, Zambia and Malawi.

14. **C. sparsiflorum** *Bak.* in J. Linn. Soc. 15: 325 (1876); F.T.A. 7: 498. *C. grewenii* Engl. & K. Krause (1910). *C. petrophilum* K. Krause (1914). *C. viviparum* A. Chev. (incl. var. *maritimum* A. Chev.) Bot. 652 (1920), names only. *C. orchidastrum* of F.W.T.A., ed. 1, 2: 343, partly (*Preuss* 191), not of Lindl. A rather fleshy herb about 1½ ft. high of wet places in forest; flowers white or greenish; sometimes viviparous.
**S.L.:** Kambui Hills South, Kenema Prov. (Nov.) *Small* 827! **Lib.:** Duoh, Sinoe Dist. (Mar.) *Baldwin* 11355! Tawata, Boporo Dist. (Nov.) *Baldwin* 10287! Kondessu, Boporo Dist. (Dec.) *Baldwin* 10678! Gbanga (Sept.) *Linder* 382! Bobei Mt., Sanokwele Dist. (Sept.) *Baldwin* 9573! **Iv.C.:** Nékaougnié to Grabo, Cavally (July) *Chev.* 19602! Assinie, Sanvi (Apr.) *Chev.* 17886! **S.Nig.:** Ikwette, Obudu Dist. (fl. & fr. Dec.) *Savory & Keay* FHI 25262! Oban F.R. (Jan.) *Onochie* FHI 36149x! **W.Cam.:** Victoria *Grewen*! Cam. Mt., 4,000–5,000 ft. (Feb.) *Dalz.* 8345! Barombi (fl. & fr. Jan.) *Preuss* 191! **F.Po:** (Apr.) *Mann* 388! Moka (Sept.) *Wrigley* 526! Also in E. Cameroun, Ethiopia, Uganda, Kenya and Tanzania.

15. **C. orchidastrum** *Lindl.* in Bot. Reg. t. 813 (1824); Trans. Hort. Soc. 6: 78 (1825); F.T.A. 7: 500; Morton, W. Afr. Lilies & Orch. fig. 14. *C. petiolatum* Bak. (1876)—F.T.A. 7: 500. A stemless herb with woolly tomentose roots and broad leaves about 1 ft. long; inflorescences usually branched, about 1½ ft. long, flowers pale green or white; in forest.
**Guin.:** Mt. Boola (fr. Mar.) *Chev.* 20923! **S.L.:** Don! York Pass (Mar.) *Lane-Poole* 462! Mount Gonkwi, Ninia, Talla Hills (fr. Feb.) *Sc. Elliot* 4995! **Lib.:** Gbanga (fr. Sept.) *Linder* 637! Yah R., Sanokwele Dist. (fr. Sept.) *Adames* 540! Bilimu Mt., Sanokwele Dist. (fr. Jan.) *Harley* 1539! Palipo, Webo Dist. *Baldwin* 12187! Filoke, Webo Dist. (fr. July) *Baldwin* 6669! **Iv.C.:** Béyo (fl. & fr. Dec.) *Leeuwenberg* 2220! Bingerville (fr. Feb.) *Chev.* 15503! Guébo (fr. Feb.) *Chev.* 17033! **Ghana:** Atroni, W. Ashanti (fl. & fr. Aug.) *Vigne* FH 2474! Subiri F.R., Tarkwa Dist. (Sept.) *Andoh* FH 4233! Bunso F.R., E. Prov. (fl. & fr. Aug.) *Darko* 955! **N.Nig.:** Dekina, Benue Dist. (fr. June) *Elliot* 248! **S.Nig.:** Olokemeji (Mar.) *Hambler* FHI 50706! Oban *Talbot* 731! Orem, Ikpai Dist. (fl. & fr. Feb.) *Onyeachusim & Latilo* FHI 54075! Owhy, Cross R. (Dec.) *Holland* 168! **W.Cam.:** Cam. Mt., 2,000 ft. (fl. & fr. Dec.) *Mann* 2132! Kumba (fl. & fr. Jan.) *Binuyo & Daramola* FHI 35155! Extends to the Congo and Zambia.

16. **C. laxum** *R. Br.* Prodr. Fl. Nov. Holl. 277 (1810); F.T.A. 503; Berhaut, Fl. Sén. 182. A small herb of shady places; inflorescences usually curved; flowers white.
**Sen.:** Tambacounda (Sept.) *Berhaut* 1677! 3163! **Port.G.:** Fá (Oct.) *Guerra* 3899! **Guin.:** Kindia *Pobéguin*! **S.L.:** Freetown (Aug.) *Deighton* 2062! Fourah Bay (Aug.) *Harvey* 6! Roruks (Aug.) *Deighton* 5338! Mapaki-Bomonto (Aug.) *Deighton* 1300! Pendembu (July) *Thomas* 843! **Lib.:** Monrovia (Aug.–Nov.) *Baldwin* 13008! *Linder* 1571! Grand Bassa (July) *T. Vogel* 41! Tawata, Boporo Dist. (Nov.) *Baldwin* 10329! Soplima, Vonjama Dist. (Nov.) *Baldwin* 10032! **U.Volta:** Banfora (Sept.) *Jaeger* 5265! **Ghana:** Apaaso, Volta Gap (Nov.) *Morton* A157! **Togo Rep.:** Lomé *Warnecke* 304! **N.Nig.:** Nupe *Barter* 1514! Zungeru (Sept.) *Dalz.* 236! *Elliott* 10! Dogon Kurmi, Jemaa Dist. (Sept.) *Killick* 65! Abinsi (July) *Dalz.* 850! **S.Nig.:** Oban *Talbot* 860! Widespread in tropical Africa.

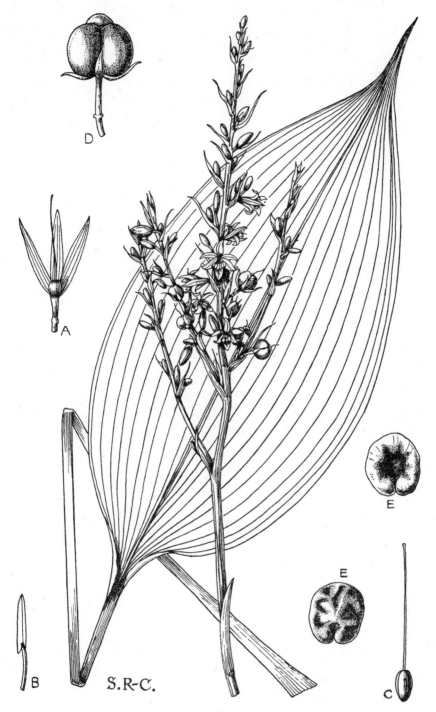

FIG. 351.—CHLOROPHYTUM ORCHIDASTRUM *Lindl.* (LILIACEAE).
A, calyx and pistil. B, stamen. C, pistil. D, capsule. E, seeds.

17. **C. alismifolium** *Bak.* in J. Linn. Soc. 15: 324 (1876); F.T.A. 7: 496; Morton, W. Afr. Lilies & Orch. fig. 13; Aké Assi, Contrib. 2: 18 [227]. *C. baillaudii* A. Chev. in Mém. Soc. Bot. Fr. 2, 8: 91 (1908); F.W.T.A., ed. 1, 2: 343. Small herb with a short stem and petiolate leaves; flowers and young fruits white, seeds black; tubers short; amongst stones in streams in forest.
**S.L.:** Waterloo to York (fl. & fr. Sept.) *Melville & Hooker* 629! Mamunta (Nov.) *Deighton* 4945! Makump *Deighton* 1954! Mabonto (July) *Deighton* 3271! **Lib.:** Kondessu, Boporo Dist. (Dec.) *Baldwin* 10664! Vonjama (Oct.) *Baldwin* 9908! Gbanga (Sept.) *Linder* 647! Sanokwele (Sept.) *Baldwin* 9523! Zeahtown, Tchien Dist. (July) *Baldwin* 6940! **Iv. C.:** Davo (fr. Aug.) *de Wilde* 367! Guidéko to Zozro (June) *Chev.* 19068! Mid. Sassandra to Mid. Cavally *Chev.* 19247! 21789! Tiapleu Forest *Aké Assi* IA 3892. Danané to N'zo *Aké Assi* IA 5652. **Ghana:** Abra (fl. & fr. Feb.) *Hall* 2900! **S.Nig.:** Ekeji-Ipetu F.R., Ilesha Dist. (Dec.) *Onochie* FHI 5241! Omo F.R., Etemi (Dec.) *Onochie, Ntima & Okeke* FHI 36594! Etemi to Atikiriji, Ijebu Ode Dist. (fl. & fr. May) *Tamajong* FHI 16784! Akure F.R., Ondo Dist. (fl. & fr. Oct.) *Keay* FHI 21575! Oban *Talbot* 859! Also in E. Cameroun, Gabon and Congo.

18. **C. tuberosum** (*Roxb.*) *Bak.* in J. Linn. Soc. 15: 332 (1876); F.T.A. 7: 508. *Ornithogalum tuberosum* Roxb. (1798). Inflorescence developing before the leaves reach full size; flowers large and ornamental; in masses in suitable moist places in savanna woodland.
**N.Nig.:** (May) *Thornewill* 160! Katagum *Dalz.* 235! Filiya, Bauchi Prov. (June) *G. V. Summerhayes* 63! Biu, Bornu Prov. (May, June) *Noble* 55! *Fishwick* FHI 42093! Yola (July) *Dalz.* 234a! Also in C. African Rep., Sudan, Ethiopia, Uganda, Kenya and Tanzania.

19. **C. andogense** *Bak.* in Trans. Linn. Soc., ser 2, Bot. 1: 260 (1878); F.T.A. 7: 506. *C. guineense* A. Chev., Bot. 650 (1920), name only; F.W.T.A., ed. 1, 2: 345 (1936), English descr. only; v. Poelln. in Portug. Acta Biol. sér. B, 1: 223 (1945), German descr. only.
**Guin.:** Grandes Chutes (fr. Dec.) *Chev.* 20218! Teliko *Pobéguin* 1120! **S.L.:** Wallia, Scarcies (Feb.) *Sc. Elliot* 4584! Also in E. Cameroun, C. African Rep. and Angola.

20. **C. bequaertii** *De Wild.* Pl. Bequaert. 1: 14 (1921). *C. inundatum* A. Chev., Bot. 651 (1920), name only; F.W.T.A., ed. 1, 2: 345 (1936), English descr. only; Morton, W. Afr. Lilies & Orch. fig. 17. Herb with narrow leaves and panicle about 1 ft. high; flowers greenish brown with white filaments and style; on rock outcrops in savanna.
**Mali:** Macina (July) *Roberty* 2546! **Guin.:** Irébéléya to Timbo, Fouta Djalon (Sept.) *Chev.* 18328! **S.L.:** Koya Mt., near Bafodia (Aug.) *Haswell* 11! 60! **Ghana:** *fide* Morton *l.c.* Also in C. African Rep., Congo and Rwanda.

21. **C. deistelianum** *Engl. & K. Krause* in Engl., Bot. Jahrb. 45: 134 (1910). *C. stenopetalum* of Chev. Bot. 652, not of Bak. Leaves thin with undulating margins; inflorescence about 1 ft. high, longer than the leaves.
**W.Cam.:** Buea *Deistel* 649!
[I am not certain about the status of this species; it is close to *C. sparsiflorum*—F.N.H.]

## 7. **URGINEA** Steinh. in Ann. Sci. Nat. Paris, sér 2, 1: 321 (1834); F.T.A. 8: 536.

Pedicels about 3–5 cm. long, slender, not jointed, arching in flower, erect in fruit; inflorescences few-flowered, 30–70 cm. high; perianth segments 9–15 mm. long, shortly connate at the base; capsule ovate, 2 cm. long, erect; bracts minute, early-caducous; leaves linear, 6(–20) cm. broad [see note after sp.], developing after flowering   ..   ..   ..   ..   ..   ..   ..   ..   ..   ..   1. *indica*
Pedicels less than 2 cm. long:
Pedicels 1–2 cm. long; inflorescences stout, very many-flowered, up to 2 m. high; perianth segments about 6 mm. long; capsule depressed globose in outline, deeply 3-lobed, about 1 cm. broad; bracts small and linear, early caducous; leaves broadly linear, about 3 cm. broad, developing after flowering   ..   ..   2. *altissima*
Pedicels of flower less than 1 cm. long (about 1 cm. in fruit); bracts more or less persistent:
Inflorescence about 60 cm. long with rather numerous small flowers in the upper third; perianth segments about 6 mm. long, 3-nerved with broad hyaline margins; capsule deeply 3-lobed, 1 cm. long and broad, transversely veined, truncate at the base; leaves linear, about 1·5 cm. broad, developing during flowering
3. *ensifolia*
Inflorescence up to 10 cm. long, few-flowered, densely clustered near apex; perianth segments 6 mm. long, with a broad dark median nerve and hyaline margins; leafless at flowering time; leaves and fruit unknown   ..   ..   4. *pauciflora*

1. **U. indica** (*Roxb.*) *Kunth* Enum. Pl. 4: 333 (1843); F.T.A. 7: 540; Chev. Bot. 655; Berhaut, Fl. Sén. 5, 181; Morton, W. Afr. Lilies & Orch. fig. 1. *Scilla indica* Roxb. (1831). Inflorescence 1–4 ft. high appearing before the leaves; flowers dull green; in rocky savanna.
**Maur.:** M'Bout (Mar.) *Roberty* 16883! **Sen.:** Diourbel *Chev.* 4503! St. Louis *Berhaut* 878. **Mali:** Dialiba *Chev.* 258. Sareya *Chev.* 20625. San *Chev.* 1099. **Guin.:** Kindia *Chev.* 13541. Farana *Chev.* Conakry Airfield (May) *Baldwin* 5791! **S.L.:** Regent (Apr.) *Hepper* 2550! Karina (Feb.) *Glanville* 161! Duunia (Feb.) *Sc. Elliot* 4817! Wallia (Feb.) *Sc. Elliot* 4572! Musaia (fr. Apr.) *Deighton* 5371! **Ghana:** Afram Plains, Ashanti *Chev.* Cansdale 9! Damongo Scarp (Mar.) Morton GC 8683! Nangodi *Morton* A3165! **Togo Rep.:** Lomé *Warnecke* 94! **N.Nig.:** Nupe *Barter* 1099! Jebba (Feb.) *Meikle* 1172! 1182! Naraguta Hill 35! Abinsi (Apr.) *Dalz.* 855! Widespread in tropical Africa and extending to India and tropical Asia.
[*U. sebirei* Berhaut (in Mém. Soc. Bot. Fr. 1953–54: 7 (1954), as *sebirii*. Senegal *Berhaut* 461!) has been segregated from *U. indica* on account of its much broader leaves and peltate bracteoles. As the leaves usually develop after the inflorescence and the bracteoles are caducous, it may not be easily distinguished. It is one of a number of forms exhibited by the widely distributed *U. indica* aggregate—F.N.H.]

2. **U. altissima** (*Linn. f.*) *Bak.* in J. Linn. Soc. 13: 221 (1872); Chev. Bot. 655; Berhaut Fl. Sén. 5, 181; Morton, W. Afr. Lilies & Orch. t. 3. *Ornithogalum altissimum* Linn. f., Suppl. 199 (1781). *O. giganteum* Jacq., Hort. Schoen. t. 87 (1797). *Drimia altissima* (Linn. f.) Ker-Gawl. in Bot. Mag. t. 1074 (1808). *D. barteri* (1870)—F.T.A. 7: 526. *Urginea micrantha* (A. Rich.) Solms-Laub. (1867)—Chev. Bot. 656. Inflorescence 1½–7 ft. high appearing without leaves during the dry season from a large bulb; flowers whitish; very common in savanna.
**Sen.:** Kaolak (Feb.) *Berhaut* 898! Bondou (Mar.) *Heudelot* 355! **Gam.:** N. Bank, Gambia R. (July) *Ozanne* 33! **Mali:** Bamako (Jan.) *Chev.* 236! Kati (Jan.) *Chev.* 192! Couroula *Chev.* 735. **Guin.:** Timbo to Farana *Chev.* 13313! 13417! Diaguissa *Chev.* 13615! Labé *Chev.* 12398. Beyla *Chev.* 20859! **S.L.:** Burbridge 541! Kafogo (Apr.) *Sc. Elliot* 5537! Musaia (Apr.) *Deighton* 4736! Loma Mt. (Mar.) *Jaeger* 9566! **Iv.C.:** Mt. Zan, Upper Sassandra *Chev.* 21562! Gouékoura, Upper Sassandra *Chev.* 21656! **U.Volta:** Banankalidoro to Bama *Chev.* 930. **Ghana:** Kintampo Scarp, Wenchi-Sunyani Dist. *Hepper & Morton* A3187! Salaga *Krause*! Dutukpene, Buem-Krachi Dist. (Mar.) *Hepper & Morton* A3062! Afram Plains

(Feb.) *Cansdale* 10! Kpong, Volta R. Dist. (Jan.) *Johnson* 618! **N.Nig.**: Nupe *Barter* 1183! Sokoto (Dec.) *Dalz.* 442! Vom, Jos Plateau (Feb.) *McClintock* 234! Katagum *Dalz.* 385! Lake Chad & Bornu (Jan., Feb.) *Talbot* 736! Nguroje to Kakara, Mambila Plateau, 5,400 ft. (Jan.) *Hepper* 1765! **S.Nig.**: Ado *Rowland*! Enugu (fr. Mar.) *Irvine* 3623! Bendi, Obudu Dist. (fr. June) *Jones & Onochie* FHI 18998! **W.Cam.**: Bamenda, 6,000 ft. (Jan.) *Migeod* 353! Bum, Wum Div. (Feb.) *Hepper* 1922!

[A variable species. Morton (l.c.) draws attention to at least three forms occurring in W. Africa: large, medium and small sized plants which usually grow in uniform populations.]

3. **U. ensifolia** (*Thonning*) *Hepper* in Kew Bull. 21: 497 (1968). *Ornithogalum ensifolium* Thonning in Schum., Beskr. Guin. Pl. 173 (1827). *U. glaucescens* Engl. & K. Krause in Engl. Bot. Jahrb. 45: 146 (1910); Milne-Redhead in Kew Bull. 3: 470 (1949); Morton, W. Afr. Lilies & Orch. fig. 2. *Ornithogalum ndellense* A. Chev. (1913), name only. *Albuca narcissifolia* of Chev. Bot. 654 (Dahomey), not of A. Chev. (1907) (Ubangi-Shari). *Urginea narcissifolia* (A. Chev.) Hutch. in F.W.T.A., ed. 1, 2: 348 (1936), partly. *Anthericum articulatum* Hutch. in F.W.T.A., ed. 1, 2: 341 (1936), partly (*Lely* 127), English descr. only. Inflorescence 1–2 ft. high, flowers white with a broad green median stripe on each tepal, leaves commencing growth at flowering time; in seasonally wet savanna and amongst rocks.

**Ghana**: Coastal areas *Thonning*! Achimota *Irvine* 1005! Labadi (Apr.) *Irvine* 2651! Volta R. *Anderson* 18! **N.Nig.**: Jira (May) *Lely* 127! Anara F.R., Zaria (May) *Keay* FHI 22930! Katagum *Dalz.* 236! Yola (July) *Dalz.* 234! Abinsi (Apr.) *Dalz.* 856! Also in E. Cameroun and C. African Republic.

4. **U. pauciflora** *Bak.* in F.T.A. 7: 539 (1898); Chev. Bot. 656. A small plant with a small cluster of flowers perianth dark-coloured; bulb scales also dark.

**Guin.**: *Jac.-Fél.*! Labé, 3,100 ft. (Apr.) *Chev.* 12304! **S.L.**: Wallia, Scarcies (Feb.) *Sc. Elliot* 4580!

FIG. 352.—URGINEA ALTISSIMA (*Linn.f.*) Bak. (LILIACEAE).
A, flower. B, stamen. C, fruit.

## 8. ALBUCA Linn., Sp. Pl. 308 (1753); F.T.A. 8: 542.

Flowers bright yellow, 2·5 cm. long, perianth segments free to the base, with broad thin margins; bracts persistent, tailed acuminate from an ovate-lanceolate base, 2·5–4·5 cm. long, many-nerved; pedicels about 7 mm. long, persistent and jointed at the top; capsule ovoid, 2 cm. long, transversely nerved; seeds black, angular, 6 mm. long; leaves developing after flowering          .. 1. *abyssinica*

Flowers yellowish green, 1–2 cm. long:

Perianth 1·5–2 cm. long; pedicels about 5 mm. long, elongating considerably in fruit; fruit broadly ovoid, 1·5–2 cm. long; bracts variable in length, 1–2·5 cm. long; leaves glabrous, developing during or after flowering          .. 2. *nigritana*

Perianth 1 cm. long; pedicels about 4 mm. long, not markedly elongating in fruit; fruit ovoid, about 1 cm. long; bracts usually less than 1 cm. long; leaves sometimes ciliate on the margin, developing during or after flowering          .. 3. *sudanica*

1. **A. abyssinica** *Murray* Syst. Veg. ed. 14, 326 (1784); Hepper in Kew Bull. 21: 493 (1968); F.T.A. 7: 533. Herb with a bulb 1–2 in. diam.; inflorescence up to 7 ft. high with large pendulous pale yellow flowers; wet places in hilly savanna.

**N.Nig.**: Jos (Aug.) *Keay* FHI 12705! Naraguta F.R. (July) *Lawlor & Hall* FHI 45573! Hepham to Ropp, 4,600 ft. (July) *Lely* 380! Mada Hills (Aug.) *Hepburn* 77! Fobur, Korom, Jarawa Dist. (July) *Summerhayes* 18! **S.Nig.**: Mt. Orosun, Idanre Hills (Oct.) *Keay* FHI 22590! Extending to Ethiopia and Arabia.

2. **A. nigritana** (*Bak.*) *Troupin* in Bull. Jard. Bot. Brux. 25: 231 (1955), and in Expl. Parc Nat. Garamba 4: 181; Morton, W. Afr. Lilies & Orch. fig. 4. *Urginea nigritana* Bak. (1873)—F.T.A. 7: 542; F.W.T.A., ed. 1, 2: 348; Berhaut, Fl. Sén. 181. *Albuca schweinfurthii* Engl. (1892)—F.T.A. 7: 533. *A. ledermannii* Engl. & K. Krause (1910). *A. sudanica* A. Chev. (1908), partly. Single inflorescence about 2½ ft. high

arising from the leafless bulb (or with young leaves) during the dry season; perianth segments with a median green stripe and pale green or whitish margins; in savanna.

**Sen.:** Cape Verde (Mar.) *Adam* 871! Cambérène (July) *Berhaut* 2155! **Mali:** Kita to Bamako (Jan.) *Chev.* 161! Dio (Jan.) *Chev.* 172! **Guin.:** Mamou to Conakry (fr. May) *Pitot*! **S.L.:** Falaba (Mar.) *Sc. Elliot* 5289! **Iv.C.:** Touba (July) *Collenette* 57! **U.Volta:** Ouagadougou (Sept.) *Irvine* 4678! **Ghana:** Kintampo (Mar.) *Dalz.* 78! Balai Kpandai to Salaga (Apr.) *Morton* A3906! **Dah.:** Kétou to L. Azri (fl. & fr. Feb.) *Chev.* 23041! **N.Nig.:** Ijayo to Ilorin *Barter* 3335! Kontagora (Jan.) *Dalz.* 258! Mando F.R., Birnin Gwari (July) *Keay* FHI 25992! Bukuru, Jos Plateau (July) *Lely* 460! Korom, Jarawa Dist., 3,700 ft. (July) *Summerhayes* 17! Ngel Nyaki, Mambila Dist., 5,400 ft. (Jan.) *Hepper* 1727! **S.Nig.:** Lagos *Rowland*! Abeokuta (fl. & fr. Jan.) *Burtt* 24! **W.Cam.:** Bamessi, Ndop Plain, 3,800 ft. (Mar.) *Brunt* 270! Kumbo to Oku, Bamenda Div., 6,000 ft. (Feb.) *Hepper* 2017! Bamenda, 5,000 ft. (Jan.) *Migeod* 297! Extending to Congo and Tanzania.

3. **A. sudanica** *A. Chev.* in Mém. Soc. Bot. Fr. 2, 8: 93 (1908), partly (*Chev.* 12578), incl. var. *gracilis* A. Chev., Bot. 655, name only; Hepper in Kew Bull. 21: 494 (1968). *A. mankonensis* A. Chev., Bot. 654 and *A. sassandrensis* A. Chev., Bot. 655, names only. *A. narcissifolia* of A. Chev. Bot. 654, not A. Chev. (1907). *Urginea narcissifolia* (A. Chev.) Hutch. in F.W.T.A., ed. 1, 2: 348 (1936), partly. *U. mankonensis* (A. Chev.) Hutch. l.c. (1936). A slender bulbous plant with the inflorescence about 1 ft. high usually appearing before the leaves; perianth greenish; in savanna.

**Port.G.:** Bafata (Jan.) *Raimundo & Guerra* 886! **Guin.:** Dalaba (fr. Apr.) *Chev.* 18121! Boulivel (Apr.) *Chev.* 12932! Timbo *Pobéguin* 120! *Chev.* 12578! Kadé (Apr.) *Pobéguin* 2099! Mamou (Mar.) *Dalz.* 8432! Musaia (Apr.) *Deighton* 4734! 5370! **Iv.C.:** Mt. Dourou (May) *Chev.* 21726! **Ghana:** Nsawkaw, Wenchi Dist. (Jan.) *Vigne* FH 3538! Kete Krache (Mar.) *Hepper & Morton* A3087! **N.Nig.:** Jebba (Feb.) *Meikle* 1171! Samaru, Giwa Dist. (May) *Keay* FHI 25749! Zaria (May) *Keay* FHI 25743! Naraguta *Hill* 34! Wana (Apr.) *Hepburn* 133! **S.Nig.:** Kishi to Igbetti, Old Oyo F.R. (Feb.) *Keay* FHI 22496! Probably extending well to the east in savanna country.

## 9. DRIMIOPSIS Lindl. in Paxt., Fl. Gard. 2: 73, fig. 172 (1851); F.T.A. 7: 542.

Leaves ascending, linear-lanceolate, up to 20 cm. long and 2 cm. broad, with spaced parallel nerves, margins subhyaline; inflorescence long-pedunculate, spicate, the flowers subsessile on the axis; perianth campanulate, about 3·5 mm. long; seeds black   ..   ..   ..   ..   ..   ..   ..   ..   ..   *barteri*

**D. barteri** *Bak.* in Saunders, Ref. Bot. 3, App. 18 (1870); F.T.A. 7: 543; Chev. Bot. 656; Troupin in Expl. Parc Nat. Garamba 4: 195; Morton, W. Afr. Lilies & Orch. fig. 6. *D. aroidastrum* A. Chev. in Mém. Soc. Bot. Fr. 2, 8: 93 (1908). *Drimia barteri* (in error) of Chev. Bot. 653, not of Bak. Bulb about 2 in. long, giving rise to flowers and leaves at the same time; leaves sometimes purple beneath and spotted above; flowers greenish, apparently occasionally cleistogamous with the perianth remaining closed.

**Ghana:** Yendi (Apr.) *Morton* GC 9094! **N.Nig.:** Lokoja (Apr.) *Parsons* L114! Nupe *Barter* 1512! 3449! Sokoto (June) *Dalz.* 541! Anara F.R., Zaria Prov. (May) *Keay* FHI 22924! 25770! Ancho (May) *Hepburn* 144! Mada Hills *Hepburn* 79! Yola (Apr.) *Dalz.* 236! Also in E. Cameroun and C. African Rep.

## 10. SCILLA Linn., Sp. Pl. ed. 2: 438 (1762); F.T.A. 7: 548.

Small bulbous herb; raceme occurring together with the leaves and usually rather longer than them; flowers in the upper ¼ of the peduncle; pedicels 2–4 mm. long; perianth 4 mm. long; leaves about 8 cm. long or up to 15 cm. long, 6–12 mm. broad, closely nerved   ..   ..   ..   ..   ..   ..   ..   ..   *sudanica*

**S. sudanica** *A. Chev.* in Mém. Soc. Bot. Fr. 2, 8: 94 (1908); Hepper in Kew Bull. 21: 497 (1968). *S. picta* A. Chev. ex Hutch. in Kew Bull. 1939: 244; Chev. Bot. 657, name only; F.W.T.A. ed. 1, 2: 350; Morton, W. Afr. Lilies & Orch. fig. 7. *S. camerooniana* of F.W.T.A., ed. 1, 2: 350, partly (*Chev.* 23719) not of Bak. *S. dahomensis* A. Chev., Bot. 657, name only. *S. mankonensis* A. Chev., name only. Leaves often mottled, sometimes almost flat on the ground; inflorescence a few inches to nearly a foot high with inconspicuous greenish flowers tinged dull purple; amongst rocks in savanna.

**Sen.:** Birkelane (fl. & fr. Aug.) *Berhaut* 286! Dalafing Koussan *Trochain* 3571! **Gam.:** (Mar.) *Brooks* 60! Niana Bantang (July) *Rhind* in Hb. *Deighton* 5585! **Guin.:** Timbo (May) *Pobéguin* 1585! Farana (May) *Chev.* 13413! Kankan to Kouroussa *Brossart* in Hb. *Chev.* 15719! **Iv.C.:** Nandala to Mankono (June) *Chev.* 21871 bis! Siana to Nandala (June) *Chev.* 21855! Boundiali (July) *Aké Assi* 5650! **U.Volta:** Karankasso (May) *Chev.* 9021! Linoré to Gampéla, Mossi (Aug.) *Chev.* 24624! **Ghana:** Kwahu Tafo (fl. & fr. May) *Hall* 3021! Afram Plains (Mar.) *Johnson* 702! Sesiamang (Feb.) *A. S. Thomas* D131! Dutukpene, Buem-Krachi Dist. (Mar.) *Hepper & Morton* A3060! Lawra (May) *Vigne* FH 4510! **Dah.:** Savalou to Gouka (May) *Chev.* 23719! Banté to Pira, Savalou *Chev.* 23742! **N.Nig.:** Anara F.R., Zaria (May) *Keay* FHI 22907! Jos Plateau (May) *Lely* P286! Abinsi (Apr.) *Dalz.* 857! Yola *Badderly* (cult. Kew)!

[*Chev.* 23719 recorded as *S. camerooniana* in Ed. 1 seems to be only a luxuriant form of *S. sudanica*. The latter name has priority over *S. picta*, but when the genus is revised for tropical Africa as a whole, a still earlier name may be applicable—F.N.H.]

### Excluded species

**S. camerooniana** *Bak.* in Saund., Ref. Bot. 3, App. 9 (1870); F.T.A. 7: 554; F.W.T.A. ed. 1, 2: 350. Although this species was included in Ed. 1, *Mann* 728, 2230 were collected outside the area of this Flora.

**S. bertholetii** *Webb* Phyt. Canar. 3: 337, t. 232 (1847); Hook. in Bot. Mag. t. 5308; F.T.A. 7: 550. This species was included in F.W.T.A. ed. 1 on the basis of a bulb collected by Mann near Cameroons R., E. Cameroun, and grown at Kew. However, in F.T.A. it was attributed to Barter and so is the specimen in Kew Herbarium. It is unlikely to be the Canary I. species and it seems to be *S. camerooniana* Bak.

## 11. DIPCADI Medic. in Act. Palat., Mannheim 6: 431 (1787); F.T.A. 7: 516.

Inflorescence very slender, few-flowered (10–)20–30 cm. long; leaves up to about 5 mm. broad, linear, usually appearing after the inflorescence; bulb 1–2 cm. diam.; bracts only a little longer than the pedicels; outer perianth segments conspicuously tailed   1. *tacazzeanum*

Inflorescence many-flowered, 30–100 cm. long; leaves 8–15 mm. long, linear, usually appearing with the inflorescence; bulb up to 4 cm. diam.; bracts much longer than the pedicels; outer perianth segments conspicuously tailed, reflexed during anthesis   2. *longifolium*

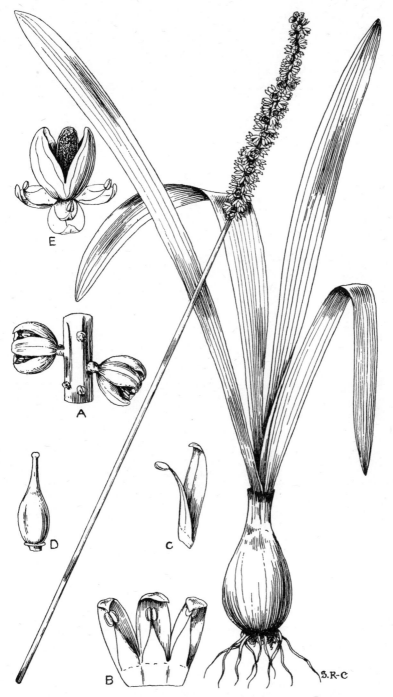

FIG. 353.—DRIMIOPSIS BARTERI *Bak*. (LILIACEAE).

A, two flowers. B and C, perianth-segments and stamens. D, pistil. E, capsule and seed.

1. **D. tacazzeanum** (*Hochst. ex A. Rich.*) *Bak.* in J. Linn. Soc. 11: 400 (1870); F.T.A. 7: 520. *Uropetalum tacazzeanum* Hochst. ex A. Rich., Tent. Fl. Abys. 2: 325 (1851). *D. occidentale* Bak. (1895)—F.T.A. 7: 521; F.W.T.A., ed. 1, 2: 350; Aké Assi, Contrib. 2: 19 [228]. *D. dahomensis* A. Chev., Bot. 653, name only. *D. gourmaensis* A. Chev. ex Hutch. in F.W.T.A., ed. 1, 2: 350 (1936), English descr. only. A slender bulbous plant with small narrow leaves, often leafless at time of flowering; flowers yellowish or purplish-green, honey-scented.
**Mali:** Gao (Aug.) *de Wailly* 5110! **Guin.:** Kolenté (June) *Chillou* 1589! **S.L.:** Hastings to Lumley *Deighton* 2683! Brookfields (May) *Deighton* 5522! 5523! Wallia, Scarcies (Feb.) *Sc. Elliot* 4840! Loma Mts. (fl. Jan., fr. Mar.) *Jaeger* 8794! 9434! **Iv.C.:** Kong *Mangenot & Aké Assi* IA 4283. **U.Volta:** Diapaga to Fada *Chev.* 24422! 24502! **Ghana:** Adamsu (fr. Apr.) *Morton* A3879! White Volta to Lumbunga, Tamale (Apr.) *Morton* A3896! Salaga (Apr.) *Morton* A3909! **Dah.:** Banté to Pira (May) *Chev.* 23738! **N.Nig.:** Kufena Rock, Zaria (May) *Keay* FHI 25736! Anara F.R., Zaria Prov. (May) *Keay* FHI 25775! **S.Nig.:** Ado Rock, nr. Ibadan (Oct.) *Hambler* 943! **W.Cam.:** Bafut-Ngemba F.R. (Feb.) *Hepper* 2120! Apparently widely distributed in the drier parts of tropical Africa.
    [See note after next sp.]

2. **D. longifolium** (*Lindl.*) *Bak.* l.c. 397 (1870); F.T.A. 7: 519. *Uropetalum longifolium* Lindl. in Bot. Reg. t. 974 (1826). *Dipcadi kerstingii* Dammer (1905). *D. tacazzeanum* of Morton, W. Afr. Lilies & Orch. fig. 5, not of (Hochst ex A. Rich.) Bak. A bulbous herb with twisted leaves; inflorescence up to 2 ft. high; flowers green; in savanna.
**Port.G.:** *Unknown Coll.* 1187! **Iv.C.:** Kouroukourounga (fl. & fr. Apr.) *Aké Assi* 8743! **Ghana:** Ejura (fl. & fr. Mar.) *Hepper & Morton* A3203! Yendi to Bawku (fr. Apr.) *Morton* A3905! Gambaga (May) *Morton* A3339! **N.Nig.:** Shuifuri (May) *Thornewill* 187! Ilorin *Barter* 3441! Yola (Apr.) *Dalz.* 232! Apparently widespread in tropical Africa.
    [The division of our specimens into two species on the characters used in the key is arbitrary. It is apparent that the size of the plants in a population can depend on their age and nourishment, but no morphological characters appear to be reliable. A critical study of living material is essential for an appreciation of the range of variation exhibited by *Dipcadi* before a satisfactory taxonomic revision can be attempted—F.N.H.]

## 12. DRIMIA Jacq., Ic. Rar. t. 373–377 (1788); F.T.A. 7: 525.

Racemes slender, about 40 cm. long, flowering in the upper third, not accompanied by leaves; pedicels 3 mm. long; bracts linear-filiform, early caducous; perianth about 1 cm. long, segments connate in the lower half, linear, reflexed and curled; stamens long-exserted; fruit 3-lobed, orbicular in outline, nearly 1 cm. diam., glabrous; leaves linear, acute, pilose-ciliate, imperfectly known and appearing after the inflorescences   ..   ..   ..   ..   ..   ..   ..   ..   *zombensis*

**D. zombensis** *Bak.* in F.T.A. 7: 525 (1898); Hepper in Bull. I.F.A.N. 28: 123 (1966). *D. incerta* A. Chev. ex Hutch. in Kew Bull. 1939: 245; Chev. Bot. 653, name only; F.W.T.A., ed. 1, 2: 351 (1936), English descr. only. Inflorescence about 1 ft. high or more arising from a bulb 3 in. long and 2 in. thick with purplish scales inside and out and a long neck of purple leaf bases; flowers brownish green; leaves appearing separately; in rocky upland savanna.
**Guin.:** Zangama, Dalaba (fr. Apr.) *Chillou* 1235! **S.L.:** Loma Mts. (Feb.) *Jaeger* 9369! *Morton & Gledhil* SL 1057! **Iv.C.:** Gouékouma and Koualé, Toura country, Haute-Sassandra (May) *Chev.* 21656bis. 21733 **N.Nig.:** Ancho (Apr.) *Hepburn* 139! Mambila Plateau (Jan.) *Hepper* 1802! *Wimbush* FHI 48400! **W.Cam.** Bafut-Ngemba F.R. (Feb.) *Hepper* 2114! *Daramola* FHI 40529! Also in Uganda, Kenya, Zambia, Malawi Rhodesia, Angola and Mozambique.

## 13. GLORIOSA Linn., Sp. Pl. 305 (1753); F.T.A. 7: 563.

Perianth-segments with very crisped-wavy margins for the whole length, narrowly linear, 6–9 cm. long, about 1 cm. broad; leaves sessile, in whorls of 3 or opposite or alternate, the lamina ovate-lanceolate, about 8–10 cm. long and 4 cm. broad, ending in a long slender cirrhose apex; style longer than the perianth, acutely reflexed; filaments shorter than the perianth; anthers about 1 cm. long; fruits about 9 cm. long and 2 cm. broad; climber in forest   ..   ..   ..   1. *superba*
Perianth-segments only slightly undulate towards the apex or with flat margins, broadly oblanceolate and about 4 cm. long or narrowly linear and about 9 cm. long; lower leaves sometimes hardly tendriliform; erect or climbing in savanna country

                                        2. *simplex*

1. **G. superba** *Linn.* l.c. (1753); F.T.A. 7: 563; Bot. Reg. t. 77; Andr., Bot. Rep. t. 129; Berhaut, Fl. Sén. 148; Morton, W. Afr. Lilies & Orch. fig. 12. Common forest climber; flowers turning from yellow to red. Widely distributed in tropical Africa to northern Transvaal and in tropical Asia. (See Appendix, p. 478.)
2. **G. simplex** *Linn.* Mant. 62 (1767). *G. caerulea* Mill. (1768). *G. virescens* Lindl. in Bot. Mag. t. 2539 (1825); F.T.A. 7: 563. *G. virescens* var. *grandiflora* (Hook.) Bak. (1879)—Bot. Mag. t. 5216. Mainly in savanna, erect or climbing; typically with smaller flower than the last species, although large flowered plants do occur.
Widespread in tropical Africa. (See Appendix, p. 478.)
    [Owing to great variation the taxonomy of *Gloriosa* remains confused and specimens have not been cited here. *Brunt* 424 resembles highland specimens from E. Africa known as *G. abyssinica* A. Rich., if this is maintainable as a species distinct from *G. simplex* Linn.—F.N.H.]

## 14. IPHIGENIA Kunth, Enum. Pl. 4: 212 (1843); F.T.A. 7: 561.

Stems very flexuous, bearing 4–5 narrowly linear acute leaves about 10 cm. long with about 4 parallel nerves on each side of the midrib; flowers solitary; pedicels about 2 cm. long, at length elongating; perianth-segments free to the base, linear, 5–7 mm. long; fruit 1–2 cm. long, laterally lobed, slightly rugose   ..   ..   *ledermannii*
**I. ledermannii** *Engl. & K. Krause* in Engl., Bot. Jahrb. 45: 123 (1910); Morton, W. Afr. Lilies & Orch. fig 9; Aké Assi, Contrib. 2: 19 [228]; Hepper in Kew Bull. 21: 497 (1968). *Helonias guineensis* Thonn. in Schum., Beskr. Guin. Pl. 182 (1827), not *Iphigenia guineensis* Bak. (1898). *I. sudanica* A. Chev., Bot. 658, name only. Rootstock a small corm with brown coverings; flowering stem 5–15 in. high, flowers deep crimson inside with the 6 thread-like segments spreading in the form of a star; in moist places in savanna.

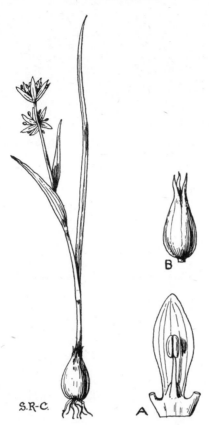

FIG. 354.—WURMBEA TENUIS (*Hook.f.*) *Bak.* (LILIACEAE).
A, perianth-segment and stamen. B, pistil.

**Iv.C.:** Ouango-Fitini (fr. July) *Aké Assi* IA 5659! **U.Volta:** Mossi (Aug.) *Chev.* 24675! Fada, Gourma (July) *Chev.* 24490! Konkobiri to Diapaga, Gourma *Chev.* 24355. **Ghana:** Tema (June) *Hall* 2599! Navrongo (fl. & fr. June) *Vigne* FH 4533! **N.Nig.:** Zaria (June) *Keay* FHI 25863! Samaru (fl. & fr. Aug.) *Keay* FHI 28008! Birni, Bornu (fl. & fr. Aug.) *Daggash* FHI 24854! Also in E. Cameroun.

## 15. **WURMBEA** Thunb., Nov. Gen. 18 (1781); F.T.A. 7: 560.

Leaves 1 or 2 arising from the bulb, linear, acute, 6–16 cm. long, 1·5–4 mm. broad, glabrous; corm ovoid about 1 cm. long, covered with brown tunics appearing bulb-like; membranous sheath around the base of the stem truncate, 1–3 cm. long; stem leaves 2–3 decreasing in size upwards; spike 2–6-flowered; perianth segments shortly connate at the base, narrowly oblanceolate, 6 mm. long; anthers broadly ovate-elliptic, cordate at the base ..    ..    ..    ..    ..    .. *tenuis*

**W. tenuis** (*Hook.f.*) *Bak.* in J. Linn. Soc. 17: 436 (1879); F.T.A. 7: 560. *Melanthium tenue* Hook. f. in J. Linn. Soc. 7: 223 (1864). Tiny tunicated corm giving rise to flowering stem 3–8 in. high; perianth white and purple, fading in older flowers; amongst rocks in montane grassland.
**W.Cam.:** Cam. Mt.? *Preuss* 933! Bafut-Ngemba F.R. (Mar., Apr.) *Richards* 5322! *Ujor* FHI 30007! Lakom (fl. & fr. Apr.) *Maitland* 1407! 1511! **F.Po:** 9,000 ft. (May) *Mann* 1454!

## 183. **TECOPHILAEACEAE**

### By F. N. Hepper

Herbs with fibrous tunicated corms or thick orbicular flattened tubers. Leaves radical or towards the base of the flowering stems, linear to ovate-orbicular and cordate, glabrous. Flowers bisexual, actinomorphic, in simple racemes separately from the tuber or corm, or in panicles; bracts large and membranous to small.

Perianth-tube short or absent; lobes 6, spreading or reflexed, subequal, imbricate Stamens 6, perfect, or 3 and with 3 staminodes, inserted at the throat of the perianth; anthers 2-locular, the connective often produced at both ends, the base then swollen or spur-like, cells opening by a terminal pore, rarely by a slit. Ovary semi-inferior, 3-locular; style subulate or filiform. Ovules numerous, axile, 2-seriate in each loculus. Fruit a loculicidal capsule. Seeds numerous, with fleshy endosperm.

Mainly Southern Hemisphere, especially S. Africa and Chile.

**CYANASTRUM** Oliv. in Hook., Ic. Pl. 20: t. 1965 (1891); S. Carter in Kew Bull. 16: 190 (1962).

Corms solid, depressed-globose, arranged in tiers, with few roots; leaves single from the corm, long-petiolate, very broadly ovate or ovate-orbicular; widely cordate at the base, acutely triangular at the apex, about 15 cm. long and up to 15 cm. broad, with numerous arcuate nerves radiating from the base and faint transverse parallel nerves; petiole up to 30 cm. long; flowering stems short, leafless, arising directly from the corm, clothed with a few membranous oblanceolate sheathing bracts; perianth with 6 equal spreading coralline obovate lobes; stamens 6, opposite the lobes; anthers linear, opening by terminal pores; ovary half inferior; fruit deeply 3-lobed  ..      ..      ..      ..      ..      ..      ..      ..      ..          *cordifolium*

C. **cordifolium** *Oliv.* l.c. (1891); Engl. in Engl. & Prantl, Pflanzenfam. ed. 2, 15a: 189, fig. 72 J-M; S. Carter l.c. 192, fig. 1, 4–9. *Schoenlandia gabonensis* Cornu (1896). Herb of the rain-forest floor, often gregarious; flowers bright blue.
**S. Nig.:** Oru to Ibadan (May) *Millson* 89! Ibadan South F.R. (July) *Ahmed & Chizea* FHI 20009! Okomu F.R., Benin *Brenan* 8462! Sapoba *Kennedy* 206! 2713! Oban *Talbot* 911! **W. Cam.:** Eyo, Mamfe Dist. (Nov.) *Tiku* FHI 29416! Barombi, S. Bakundu (Aug.) *Binuyo & Daramola* FHI 35520! Victoria (Oct., Mar.) *Mildbr.* 10583! *Kalbreyer* 89! Ambas Bay (Feb.) *Mann* 769! Extends to Gabon.

# 184. PONTEDERIACEAE

## By F. N. Hepper

Aquatic erect or floating herbs. Leaves with floating or immersed blades sheathing at the base. Flowers bisexual, actinomorphic, arranged in racemes or panicles subtended by a spathe-like leaf-sheath; bracts minute or absent. Perianth hypogynous, corolline; lobes 6 or 4, sub-biseriate. Stamens 6 or 3, rarely 1, inserted on the perianth, sometimes somewhat unequal in length or 1 the longest of all; filaments free; anthers 2-locular, opening lengthwise by slits or rarely by pores. Ovary superior, 3-locular, with axile placentas or 1-locular with 3 parietal placentas; style entire or shortly lobed. Ovules numerous to solitary and then pendulous. Fruit a capsule opening by 3 valves, or indehiscent. Seeds ribbed, with copious endosperm.

Fresh water aquatics, in the tropics and subtropics.

Stamens 6 in the normal flowers:
Perianth-segments free to the base; leaves all alike      ..      ..    1. **Monochoria**
Perianth-segments partially united into a tube; submerged leaves linear, floating ones ovate or ovate-orbicular      ..      ..      ..      ..      ..      ..    2. **Eichhornia**
Stamens 3 in the normal flowers, sometimes 1 in the cleistogamous flowers; leaves all alike      ..      ..      ..      ..      ..      ..      ..      ..      ..    3. **Heteranthera**

*Scholleropsis lutea* H. Perr. with 4 perianth-segments has been found once in E. Cameroun: it is otherwise only known from Madagascar. Letouzey in Adansonia, sér. 2, 7:33 (1967).
*Pontederia lanceolata* Nutt., native in N. America, is cultivated at Victoria, W. Cameroun.

1. **MONOCHORIA** C. Presl, Rel. Haenk. 1: 127 (1827); F.T.A. 8: 5; Verdcourt in Kirkia 1: 80–83 (1961).

Basal leaves reduced to membranous sheaths; leaves narrowly lanceolate, acutely acuminate, about 6 cm. long and 1 cm. broad, narrowed into a false petiole about as long as the blade and expanded into a membranous sheath at the base; racemes at first embraced by the leaf-sheath, soon exserted, about 8–12-flowered; pedicels stout, 4 mm. long; perianth-segments oblong, about 1 cm. long, 6–7 nerved, with membranous margins; anthers subequal, but one larger, 4–5 mm. long ; fruit enclosed by the persistent perianth ..      ..      ..      ..      ..          *brevipetiolata*

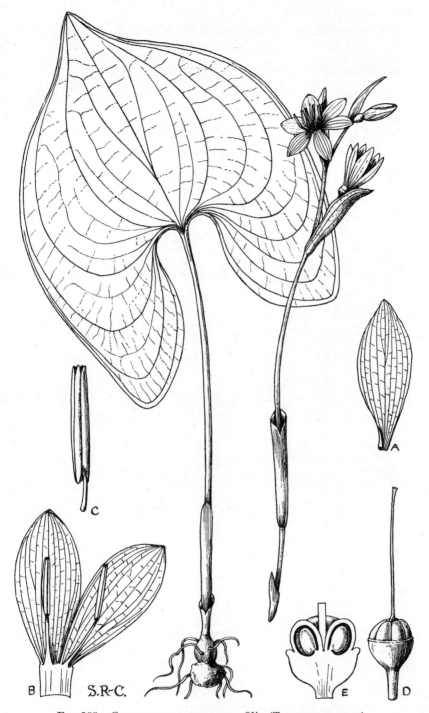

Fig. 355.—Cyanastrum cordifolium *Oliv.* (Tecophilaeaceae).

A, perianth-segment. B, two perianth-segments and stamens. C, stamen. D, pistil. E, vertical section of ovary.

**M. brevipetiolata** *Verdc.* l.c., t. 9, 10 (1961). *M. vaginalis* var. *plantaginea* of F.W.T.A., ed. 1, 2: 354; of Sousa in Anais Junta Inv. Col. 5: 55 (1950), not of Solms-Laub. *M. africana* of Berhaut, Fl. Sén. 185, not of (Solms-Laub.) N. E. Br. An aquatic herb nearly a foot high with blue or violet flowers; in pools on laterite rocks.

    **Sen.**: Tambacounda *Berhaut* 1268; 3136. Galam *Heudelot* 241!   **Gam.**: Fulladu West (Nov.) *Frith* 153! Bansang *Duke* 14!  **Mali**: between R. Senegal and R. Niger *Bellamy* 57!  **Port. G.**: Gabu (Sept.) *Esp. Santo* 2306! 2777!  **Guin.**: Kankan (Sept.) *Pobéguin* 1107! Kindia *Pobéguin*!  **S.L.**: Wonkifu, Tonko Limba (Oct.) *Jordan* 966!  **Iv.C.**: Boloma, Boundiali Reg. (Nov.) *Aké Assi* 8287!

## 2. EICHHORNIA Kunth, Eichhornia, genus novum (1842); F.T.A. 8: 4; Robyns in Bull. Acad. Sci. Col. Brux. 1: 1116 (1956). *Nom cons.*

Herb; submerged leaves linear, 3–6 cm. long, floating ones long-petiolate, ovate to ovate-orbicular, cordate at the base, 1·5–2·5 cm. long, entire, minutely pustulate on the upper surface, with very numerous and close nerves below; flowers solitary; perianth 6-lobed; stamens 6 ..     ..    ..    ..    ..    ..    *natans*

**E. natans** (*P. Beauv.*) *Solms-Laub.* in Abh. naturw. ver. Bremen 7: 254 (1882); F.T.A. 8: 4; Chev. Bot. 659; Sousa in Anais Junta Inv. Col. 5: 56 (1950); Berhaut, Fl. Sén. 179. *Pontederia natans* P. Beauv. (1810). *Eichhornia diversifolia* of Troupin in Fl. Sperm. Parc Nat. Garamba 1: 177, fig. 26 (1956), not of (Vahl) Urb. Aquatic herb with floating upper leaves; flowers purplish with yellow spot; in pools and streams.

    **Sen.**: Diohine (Sept.) *Berhaut* 562! St. Louis *Leprieur*! Galam *Heudelot* 258!  **Gam.**: Bansang *Duke* 2! Wallikunda *Macluskie* 27! Kuntaur *Ruxton* 27!  **Mali**: Bamako *Waterlot* 1052! Gao *de Wailly* 4857!

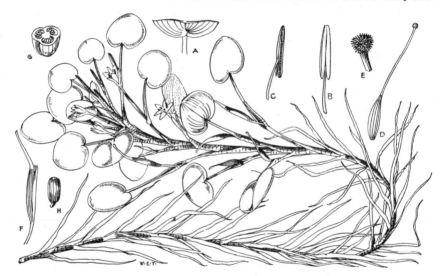

FIG. 356.—EICHHORNIA NATANS (*P. Beauv.*) *Solms-Laub.* (PONTEDERIACEAE).
A, base of leaf. B and C, stamens. D, pistil. E, stigma. F, leaf-sheath. G, cross-section of ovary. H, seed.

    Tabacco *Chev.* 123!  **Port G.**: Pussubé, Bissau *Esp. Santo* 1101.  **Guin.**: Kouria *Caille* in *Hb. Chev.* 14664! 14927!  **S.L.**: Falaba (Feb.) *Sc. Elliot* 5290! Mange (Dec.) *Jones* 57! Kasawa F.R. (Jan.) *King* 211! Njala (Nov.) *Deighton* 2815! Gegbwema to Faiama *Pyne* 30!  **Lib.**: Vonjama (Oct.) *Baldwin* 9913! Yila, St. John R. (Aug.) *Baldwin* 9132! Gbanga *Linder* 379! Salala (Aug.) *Baldwin* 13218! Ganta (June) *Harley* 596!  **Iv.C.**: Alangouassou to Mbayakro *Chev.* 22242! Soukourougban (Apr.) *Aké Assi* 8748!  **U. Volta**: Fada *Chev.* 24491!  **Ghana**: Ada to Dawa (Dec.) *Adams* 3614! Pong-Tamale (Dec.) *Akpabla* FH 452! Yendi (Nov.) *Thorold* 270! Bimbila (Apr.) *Akpabla* FH 1852! Navrongo (Oct.) *Vigne* FH 4606!  **N. Nig.**: Nupe *Barter*! Zungeru (Sept.) *Dalz.* 273! Yola (Nov.) *Hepper* 1199! Gashaka Dist., Sardauna Prov. (Nov.). Latilo & Daramola FHI 28761! Donga R., Mambila Dist. *Hepper* 2830!  **S. Nig.**: Iseyin to Shaki, Oyo Prov. (Nov.) *Hambler* 377! Extends in savanna to Sudan, Uganda, Congo, Zambia, Rhodesia, Angola and S.W. Africa; also in Cuba and tropical S. America.

    *E. crassipes* (Mart.) Solms-Laub., the water hyacinth, is a tropical American floating aquatic with inflated petioles which is now widespread in the Old World Tropics. It is sometimes cultivated in pools as an ornamental and it has become a serious pest as a water weed in parts of tropical Africa, as yet outside our area (see Wild in Kirkia 2: 9, with fig. (1961)).

## 3. HETERANTHERA Ruiz & Pav., Prod. 9, t. 2 (1794); F.T.A. 8: 2 *Nom. cons.*

Aquatic herb; leaves petiolate, ovate to lanceolate, cordate to rounded at the base, 5–7 cm. long, 1–5 cm. broad, thin, glabrous, with numerous slender nerves ascending from the base; petiole sheathing at the base and embracing the capsule; flowers spicate, few; perianth-tube about 5 mm. long, lobes 6, oblong, about 4 mm. long, nervose; stamens 3 or reduced to 1 in the cleistogamous flowers    ..   *callifolia*

**H. callifolia** *Rchb. ex Kunth* Enum. 4: 121 (as 123) (1843); F.T.A. 8: 2; Chev. Bot. 659; Sousa in Anais Junta Inv. Col. 5: 56 (1950); Berhaut, Fl. Sén. 180; Troupin in Fl. Sperm. Parc Nat. Garamba 1: 179, fig. 27, 28 (1956). *H. kotschyana* Fenzl ex Solms-Laub. (1867)—F.T.A. 8: 3. *H.potamogeton* Solms-Laub. (1883)—F.T.A. 8: 3. *H.pubescens* A. Chev., Bot. 659, name only. An aquatic herb about 6 in. high, with hollow petioles raising the lamina above the water; flowers white.

**Sen.:** *Sieber* 58! Pozo Cervalo *Perrottet* 779! Galam *Heudelot* 230! Diohine ( Sept). *Berhaut* 538! **Mali:** Koulikoro (Oct.) *Chev.* 2596! Bentia, S. of Ansongo (Sept.) *Hagerup* 419! **Port G.:** Farim *Esp. Santo* 2296. **S.L.:** Yana, Tambakha (Dec.) *Jordan* 1071! **Lib.:** Ganta (Sept.) *Baldwin* 1096! Sarbo *Baldwin* 6395! Jabroke (July) *Baldwin* 6483! Gletown (July) *Baldwin* 6761! **Iv. C.:** Buandougou to Marabadissa (July) *Chev.* 21986! Grabo *Chev.* 19611! Languira to Bouakro *Chev.* 22215! Ferkéssédougou (Nov.) *Leeuwenberg* 2034! Assalerna (May) *Aké Assi* 8872! **Ghana:** Ohawu (July) *Irvine* 4958! Gonokrom to Dormaa (Dec.) *Adams* 2996! Mampong to Kumasi (June) *Morton* A3404! Kumasi (Oct.) *Darko* 382! Numia F.R. (Nov.) *Enti* FH 6743! **N. Nig.:** S. of Gombe (Oct.) *Lely* 667! Kabama, Zaria (Dec.) *Daggash* FHI 31405! Katagum Dist. *Dalz.* 227! **S. Nig.:** Ibadan *Meikle* 863! Shagamu (Jan.) *Burtt* 18! Ahoada, Degema *Talbot* 3809! Throughout tropical Africa and into the Transvaal and SW. Africa.

## 185. SMILACACEAE

### By F. N. Hepper

Shrubs, climbing or straggling, often with tendril-like petioles and prickly stems and branches; roots from an often stout rhizome; stems leafy. Leaves alternate or opposite, 3-nerved, reticulate-veiny between the nerves. Flowers dioecious or rarely bisexual, small, arranged in axillary umbels, racemes or spikes. Perianth-segments 6, free or rarely united. Stamens 6; filaments free or united; anthers apparently 1-locular by the confluence of the cells, introrse. Ovary superior, 3-locular; ovules 1–2 in each loculus, pendulous. Staminodes present in the female flower. Fruit a berry. Seeds 1–3; embryo small in hard endosperm.

Widely distributed. A small family close to *Liliaceae*.

**SMILAX** Linn., Sp. Pl. 1028 (1753); F.T.A. 7: 423.

Flowers dioecious, in umbels or panicles; perianth-segments free; stamens 6 or more, free.

A climbing shrub with prickly shoots; leaves alternate, ovate-elliptic to broadly elliptic, abruptly and very shortly acuminate (acumen oblique in dried specimens), rounded to subacute at the base, averaging about 12 cm. long and 7 cm. broad, glabrous, prominently 3-nerved from the base, larger leaves widely subcordate; flowers dioecious; umbels axillary, shortly pedunculate, many-flowered; peduncle bracteate at the base; pedicels slender, about 6 mm. long; perianth 5 mm. long; fruit subglobose, nearly 1 cm. diam. .. .. .. .. .. *kraussiana*

FIG. 357.—SMILAX KRAUSSIANA *Meisn.* (SMILACACEAE).
A, flowering shoot. B, male flower. C, stamen. D, female flower. E, fruits. F, fruit. G, seed.

**S. kraussiana** *Meisn.* in Flora 28: 312 (1845); A. DC., Monogr. Phan. 1: 171, incl. vars. *dregei* A. DC., *morsoniana* A. DC., & *senegambiae* A. DC. l.c. 172 (1878); F.T.A. 7: 424; Chev. Bot. 643; Berhaut, Fl. Sén. 148. A climbing or scrambling shrub; flowers greenish white; fruits red; often in secondary forest. **Sen.**: *Berhaut* 431. Samandiniéry *Chev.* 2577. Bignona *Chev.* 2578. **Gam.**: Albreda (May) *Perrottet* 789! **Mali**: Kénlégué *Chev.* 279 *bis*! **Guin.**: Péla (fr. Sept.) *Baldwin* 13319! **S.L.**: *Morson*! Hastings (Sept.) *Melville & Hooker* 430! Njala (fr. Nov.) *Small* 807! Bonthe I. (Mar.) *Deighton* 2466! Baiima (fr. Sept.) *Deighton* 3056! **Lib.**: Moylakwelli (fr. Oct.) *Linder* 1313! Ganta *Harley*! Brewersville (fr. Sept.) *Barker* 1062! **Iv. C.**: Flansobly Forest (July) *Aké Assi* 9062! N. of Ferkéssédougou (fr. Nov.) *Leeuwenberg* 1988! **Ghana**: Benso (Sept.) *Andoh* FH 5885! Tafo (Aug.) *Darko* WACRI 903! Kumasi (June) *Irvine* 4508! Kete-Krachi to Dutukpene (fr. Dec.) *Adams* 4659! **Dah.**: Atacora Mts., Bokorona to Kouandé *Chev.* 24225. **N. Nig.**: Acharane F.R., Ankpa Dist. (June) *Daramola* FHI 38047! Sokoto *Dalz.* 545! Tonti, Muri Dist. (fr. Nov.) *Latilo & Daramola* FHI 28745! **S. Nig.**: Lagos *Batten-Poole* 61! Iguobazuwa to Siluko, Benin (fr. Nov.) *Onochie* FHI 40418! Warri to Sapele (fr. Sept.) *Onochie* FHI 34292! Aguku Dist. *Thomas* 837! Degema *Talbot*! **W. Cam.**: Cam. Mt., 3,000–4,000 ft. (Feb.) *Mann* 1271! *Maitland* 308! Nchan, 5,000 ft., Bamenda (Apr.) *Maitland* 1474! **F. Po**: Moka (fr. Sept.) *Wrigley* 573! Widespread in tropical Africa and in S. Africa. (See Appendix, p. 479.)

# 186. ARACEAE

## By F. N. Hepper

Herbs with watery, bitter or milky juice, with a tuberous or elongated rhizome, rarely woody and climbing. Leaves solitary or few, sometimes appearing after the flowers, mostly radical, when cauline then alternate and distichous or spirally arranged, entire or variously divided, often hastate or sagittate, with a membranous sheath at the base. Flowers small, arranged on a spadix enclosed in a spathe, bisexual or monoecious, the males in the upper part, females below, rarely dioecious. Perianth present in the bisexual flowers or absent from the unisexual flowers. Stamens hypogynous, 2–4–8, opposite the perianth-segments; anthers opening by pores or slits, free or united. Ovary superior or immersed, 1–many-locular; style various or absent. Ovules parietal, axile, basal or apical. Fruit a berry, or coriaceous and rupturing, 1–many-seeded. Seeds mostly with copious endosperm.

Temperate and tropical regions, particularly tropical America and tropical Asia. Usually easily recognised by the characteristic spathe enclosing the inflorescence (spadix).

Floating aquatic herb, with a tuft of fibrous roots and a rosette of flabellately nerved leaves; flowers unisexual, without a perianth, the female part of the spadix adnate to the spathe .. .. .. .. .. .. .. .. .. .. 1. **Pistia**
Terrestrial or epiphytic herbs or climbers, if aquatic not floating; female flowers more than one:
  Flowers hermaphrodite:
    Herb with large radical sagittate leaves; petiole prickly .. .. 2. **Cyrtosperma**
    Climber with cauline leaves cunate at the base; petiole smooth    3. **Rhaphidophora**
  Flowers unisexual:
    Perianth present; male flowers covering the upper part of the spadix; savanna or forest herbs .. .. .. .. .. .. .. .. 4. **Stylochiton**
    Perianth absent:
      Herbs without distinct aerial stems, at most rhizomatous:
        Spadix with a terminal appendage destitute of flowers; rootstock tuberous:
          Leaves divided; leaves and flowers produced at different times:
            Leaves pedate with a few segments; spadix with sterile process above the female portion .. .. .. .. .. .. .. 5. **Sauromatum**
            Leaves 3-branched at the apex of the petiole, branches dichotomously divided, the ultimate segments each with one apex; spadix without sterile process
                               6. **Amorphophallus**
          Leaves peltate; leaves and flowers produced at the same time; introduced and cultivated, more or less naturalized .. .. .. .. 7. **Colocasia**
        Spadix without a terminal sterile appendage:
          Leaves not peltate:
             Leaves sagittate, elliptic etc. not divided (3-lobed, but not to the base, in *Nephthytis constricta*):
              Introduced, cultivated and more or less naturalized in waste places; leaves sagittate .. .. .. .. .. .. .. .. 8. **Xanthosoma**
            Forest plants or herbs of wet places:
              Herbs of streams and wet places:
                Spathe inconspicuous greenish colour; ovules numerous in each of the 2 or 3 cells; lateral nerves of leaves numerous and closely parallel; streams in lowland forest:

Anthers inserted on the sides of the synandria; leaves cordate-sagittate, hastate, or linear-elliptic .. .. .. .. **9. Anubias**
Anthers crowded on top of the synandria; leaves hastate to 3-lobed
          **10. Amauriella**
Spathe conspicuous golden-yellow; leaves hastate; ovules 2–3 in each of several cells; marsh herb in highland .. .. **11. Zantedeschia**
Herbs of drier places in forest; leaves sagittate or 3-lobed; ovules solitary in each cell .. .. .. .. .. .. .. **12. Nephthytis**
Leaves much divided, the upper ultimate segments each with 2 apices
          **13. Anchomanes**
Leaves peltate, cordate at the base:
Flowers and leaves produced in alternate seasons, the uniform-green leaves accompanied by a peduncle bearing burr-like bulbils covered with hooked prickles .. .. .. .. .. .. .. **14. Remusatia**
Flowers and leaves produced together, leaves usually brightly coloured; burr-like bulbils not produced; naturalized introduction .. **15. Caladium**
Herbs with well developed aerial stems, climbing or creeping, rooting at the nodes, or erect; spadix with an appendage:
Ovary 2–3-celled with numerous ovules in each cell; anthers connate; herb of wet places (for other characters see above) .. .. .. **9. Anubias**
Ovary 1–2-celled with solitary ovule; anthers free; herbs of forest, often climbers:
Leaves unequally pinnatisect or perforated .. .. **16. Rhektophyllum**
Leaves neither pinnatisect nor perforated:
Seeds with endosperm; ovule basal; leaves more or less ovate, not hastate or 3-lobed .. .. .. .. .. .. .. .. **17. Culcasia**
Seeds without endosperm; ovule inserted above the base of the cell; leaves cordate, sagittate or hastate or 3-lobed .. .. .. **18. Cercestis**

*Dieffenbachia* sp. from America is cultivated as an ornamental in Sierra Leone and the Asiatic *Typhonium trilobatum* (Linn.) Schott has been introduced into the vicinity of Bingerville, Ivory Coast. The Asiatic elephant's ear, *Alocasia macrorhiza* Schott, is occasionally cultivated in our area. Other genera are probably grown as ornamentals.

### 1. PISTIA Linn., Sp. Pl. 963 (1753); F.T.A. 8: 140.

Leaves sessile in a rosette, oblong-spathulate, rounded, truncate or widely emarginate at the apex, up to about 12 cm. long and 5 cm. broad, with several subparallel flabellate nerves, softly puberulous to tomentellous on both surfaces, tomentose towards the base; spathe axillary, shortly pedunculate, up to 1·2 cm. long, tubular below, villous outside, ovate, with a broad white ciliolate margin; spadix shorter than the spathe, monoecious, the female part adnate to the spathe; male part a stipitate whorl of 3–8 flowers, each flower composed of 2 connate 2-celled anthers; ovary solitary, 1-celled, with a capitate stigma .. .. .. .. *stratiotes*

P. **stratiotes** *Linn.* l.c. (1753); F.T.A. 8: 140; Chev. Bot. 685; Berhaut, Fl. Sén. 184; Morton, W. Afr. Lilies & Orch. fig. 61; Wild in Kirkia 2: 13, t. 2. A floating herb, propagates mainly by stolons and forming immense colonies; roots numerous, fibrous, reaching the bottom in shallow water; flowers obscure beneath the rosette of glandular viscid leaves. Water lettuce.
**Sen.**: *Berhaut* 1008. **Mali:** Djenné *Chev.* 1106; 1142. **S.L.:** Gbundapi (Mar., Sept.) *Adames* 22! 71! Serabu, S. Prov. (Apr.) *Deighton* 1678! **Lib.:** Gbaishela *Harley* 1298! Toroke (July) *Baldwin* 6730! **Iv.C.:** Adiopodoumé *de Wit* 5569! **U. Volta:** Ouagadougou *Chev.* 24669! **Ghana:** Bjury (July) *Chipp* 505! Kumasi *Cummins*! Nungua (Feb.) *Rose Innes* GC 30095! **Niger:** Niamey to Gao *Ryff*! **N. Nig.:** Lom *Barter* 176! Gulumba (Dec.) *McClintock* 44! **S. Nig.:** Lagos *Batten-Poole* 75! Okomu F.R. (fr. Jan.) *Brenan* 8770! Burutu (Sept.) *Parsons* 4! Sapoba (May) *Kennedy* 1421! Aboh *T. Vogel* 12! Isoba, Port Harcourt (Jan.) *Stubbings* 129! Cross R. (Jan.) *Holland* 233! **W. Cam.:** *Preuss* 1361! An aquatic widely distributed throughout the tropics. (See Appendix, p. 482.)

### 2. CYRTOSPERMA Griff., Notul. 3: 149 (1851), and Ic. 3: t. 169 (1851); F.T.A. 8: 197.

Leaves sagittate, variable in size, up to about 80 cm. long, and 40 cm. broad, the basal lobes lanceolate, acuminate, forming a narrow sinus, the terminal lobe ovate-lanceolate, acute, glabrous, laxly reticulate; petiole up to 2 m. long or more, prickly-toothed; spathe long-pedunculate (overtopping the leaves), enveloping and much longer than the spadix, 20–45 cm. long, acuminate; spadix sessile, cylindric; flowers hermaphrodite; berries 1·5 cm. long, oblong; seeds covered with dentate crests
          *senegalense*

C. **senegalense** (*Schott*) *Engl.* in DC., Monogr. Phan. 2: 270 (1879); F.T.A. 8: 198; Bot. Mag. t. 7617; Chev. Bot. 679; Berhaut, Fl. Sén. 180; Morton, W. Afr. Lilies & Orch. fig. 54. *Lasimorpha senegalensis* Schott and *L. afzelii* Schott, Gen. Aroid. t. 85, figs. 1–10 and 1–20 (1858). Terrestrial herb with a rhizome giving rise to large sagittate leaves about 5 ft. high with prickly petioles; spathe green outside, inside cream with long dull red-purple linear marking presenting a longitudinally striped appearance; peduncle prickly, longer than the leaves and up to 12 ft. high; at the margins of forest, and in swamps and ravines in savanna country.
**Sen.**: *Berhaut* 1319. **Port G.:** Teixeira Pinto (Jan.) *d'Orey* 133! **Guin.:** *Heudelot* 639! **S.L.:** Newton (Nov.) *Deighton* 1513! N'tunga (Dec.) *Sc. Elliot* 4333! Mayoso (Aug.) *Thomas* 1470! Pendembu (July) *Thomas*

798! **Lib.**: Monrovia (fr. June) *Baldwin* 5915! Grand Bassa (July) *T. Vogel* 94! Gbanga (Sept.) *Linder* 779! **Iv.C.**: Angedédou Forest, Abidjan (Nov.) *Leeuwenberg* 1873! Grand Bassam (Sept.) *de Wit* 5775! **Ghana:** Aiyinase (July) *Irvine* 5000! Axim (Dec.) *Johnson* 874! Tarkwa (fr. Mar.) *Morton* GC 6569! **N.Nig.:** Ankpa to Oturkpa, Kabba Prov. (June) *Daramola* FHI 38024! Gongoroko, Nupe *Barter* 1467! Munchi (leaf June) *Dalz.* 863! **S.Nig.:** Benin (Nov.) *Meikle* 875! Osho, Omo F.R. (fr. Apr.) *Jones & Onochie* FHI 17236! Nkpoku, Port Harcourt *Stubbings* 120! Calabar (July) *Holland* 64! **F.Po:** *Barter* 9! (Jan.) *Mann* 244! Extends to the Congo.

**3. RHAPHIDOPHORA** Hassk. in Flora 25, 2, Beibl. 11 (1842); F.T.A. 8: 199, as *Raphidophora. Afroraphidophora* Engl. (1906).

A tall climber; stem slender, rooting at the nodes; leaves entire, oblong-oblanceolate, shortly cuneate at the base, gradually acuminate, 15–30 cm. long, 4–8 cm. broad, glabrous, with numerous pinnate parallel nerves spreading from the midrib at an angle of about 45°; petiole up to 20 cm. long, sheathing; peduncle 8–10 cm. long; spathe closely wrapped around the spadix, beaked, 7–10 cm. long; spadix sessile, up to 10 cm. long, 1·4 cm. diam.; flowers hermaphrodite; stigmas discoid

        *africana*

**R. africana** *N. E. Br.* in Kew Bull. 1897: 286; F.T.A. 8: 200; Chev. Bot. 678; Morton, Lilies & Orch. fig. 60. *R. ovoidea* A. Chev. (1909)—Chev. Bot. 678. *R. pusilla* of Chev. Bot. 679, not of N.E. Br. *Afroraphidophora africana* (N. E. Br.) Engl. (1906). Climbing up trees in forest, up to 80–100 ft. high, but apparently often much less; spathe greenish yellow; spadix flowers white.
**S.L.:** *Thomas* 9488! Njala (Mar.) *Deighton* 3123! Sakuru (Feb.) *Sc. Elliot* 4940! Kurusu (Apr.) *Sc. Elliot* 5524! Faiama (Mar.) *Deighton* 4123! **Lib.:** Kitoma *Harley* 1604! Dukwia R. (Feb.) *Cooper* 284! **Iv.C.:** Béyo (Jan.) *Leeuwenberg* 2619! Bouroukrou *Chev.* 16849. Grabo, Cavally *Chev.* 19643; 19669. Malamalasso *Chev.* 17505! **Ghana:** *Burton*! Enchi to Boinso (Dec.) *Adams* 2188! Kade (Feb.) *Morton* GC 8388! Assin-Yan-Kumasi *Cummins* 47! **S.Nig.:** Omo F.R. (Feb.) *Jones & Onochie* FHI 17558! Babatope to Ipetu, Ife-Ilesha Dist. (Apr.) *Onochie* FHI 15501! Okomu F.R. (Jan.) *Brenan* 8787! Eket Dist. *Talbot* 3138! **F.Po:** *Mann* 103! S. Carlos *Guinea* 671!

**4. STYLOCHITON** Lepr. in Ann. Sci. Nat., Sér. 2, 2: 184 (1834); F.T.A. 8: 187.

Leaves cordate or hastate with broad obtuse lobes, or lanceolate:
  Flowers axillary, accompanied by mature leaves; spathe narrowly oblong-lanceolate about 5 cm. long and 5 mm. broad, margins shortly connate below into a tube; spadix 1 cm. long with 2–3 female flowers in a whorl; peduncle 5–25 mm. long; leaves obovate or nearly orbicular, cordate at the base, abruptly acuminate at the apex 8–16 cm. long, 6–12 cm. broad, with about 8 faint lateral nerves on each side of the midrib; petiole 10–20 cm. long; in forest    ..    ..    ..    1. *zenkeri*
  Flowers appearing before the leaves; in savanna:
    Radial flanges absent in between the 5 female flowers at the base of the spadix, male flowers slightly separated from the female portion; spathe lanceolate, mucronate, 2·5–5 cm. long, about 1 cm. broad, tubular in the lower half; leaves hastate to lanceolate, 10–18 cm. long, 5–8 cm. broad, glabrous, with the main nerves radiating from the apex of the puberulous petiole, basal lobes of the lamina very obtuse and rounded, ovate or broadly triangular    ..    ..    2. *lancifolius*
    Radial flanges separating the 3 female flowers; spadix 1·5 cm. long; spathe lanceolate, 2 cm. long; leaves ovate-elliptic, 6–10 cm. long, 4–5 cm. broad, with about 5 pairs of lateral nerves; peduncle 2 cm. long    ..    ..    ..    3. *hostifolius*
Leaves hastate with linear lobes 4–6 cm. long, lamina narrowly lanceolate, when fully expanded about 15 cm. long and 1·5–3 cm. broad; petiole up to 25 cm. long; flowers appearing before the leaves; spathe about 4 cm. long, more or less hooded and obtuse at the apex; spadix nearly as long as the spathe; in savanna    ..    4. *hypogaeus*

1. **S. zenkeri** *Engl.* Bot. Jahrb. 26: 424 (1899); F.T.A. 8: 189. *S. gabonicus* N. E. Br. in F.T.A. 8: 190 (1901). Leaves about 1 ft. high appearing with the axillary flowers; spathe light pink; in forest, especially in wet places.
    **S.L.:** Njala (July) *Deighton* 3960! Robat, Rokupr (Aug.) *Deighton* 3021! **W.Cam.:** Likomba *Mildbr.* 10593! **F.Po:** *Barter* 1470! Extends to Gabon.
2. **S. lancifolius** *Kotschy & Peyr.* Pl. Tinn. 42, t. 20 (1867); F.T.A. 8: 193. *S. warneckei* Engl. (1905)— F.W.T.A., ed. 1, 2: 359; Morton, W. Afr. Lilies & Orch. t. 56. *S. chevalieri* Engl. (1907)—Chev. Bot. 685. *S. dalzielii* N.E. Br. (1910). Rhizome erect, fleshy, yellow; flowers appearing towards the end of the dry season in savanna woodland, spathe apex dull purplish outside, whitish inside; spadix whitish; leaves about 1 ft. high produced during the rainy season.
    **Sen.:** Kaolak *Berhaut* 1631! **Mali:** Thiédiana *Chev.* 994! **Guin.:** Farana, Sangara Dist., 3,300 ft. (Mar.) *Sc. Elliot* 5373! **S.L.:** Waterloo (May) *Deighton* 2763! Kambia Magbema (fl. Apr., fr. May) *Jordan* 440! *Hepper* 2635! Mange Bure (May) *Deighton* 4784! **Ghana:** Bolgatanga (Mar.) *Morton* GC 8828! Kete Krachi *Morton* 9123! Ohawu (July) *Irvine* 4961! Kpong to Accra (Oct.) *Morton* A236! **Dah.:** Nioro to Kouandé *Chev.* 24004! **N.Nig.:** Nupe *Barter* 1469! Abinsi (fl. & fr. June–July) *Dalz.* 860! Yola (Apr.) *Dalz.* 237! **S.Nig.:** Oyo to Ilorin (Mar.) *Meikle* 1294! Widespread in drier savanna regions of tropical Africa.
3. **S. hostifolius** *Engl.* Bot. Jahrb. 36: 238 (1905).
    **Togo Rep.:** Sokodé-Basari, near Alédyo (Feb.) *Kersting* 316!
    [Further material is required for study to decide the status of this species: it is very close to *S. lancifolius* in the broad sense.]
4. **S. hypogaeus** *Lepr.* in Ann. Sci. Nat., Sér. 2, 2: 185, t. 5 (1834); F.T.A. 8: 192; Berhaut, Fl. Sén. 180. *S. similis* N. E. Br. in F.T.A. 8: 194 (1901). *S. barteri* N. E. Br. l.c. (1901); F.W.T.A., ed. 1, 2: 359; Chev. Bot. 684; Morton l.c. fig. 55. Rhizome rather thick; flowers appearing before or sometimes with the leaves; spathe purplish; in flood plains.
    **Sen.:** Dioubèl *Berhaut* 5306! Districts of Cayor, Oualo and Cape Verde *Leprieur*! **Iv.C.:** Zagoué to Sou-

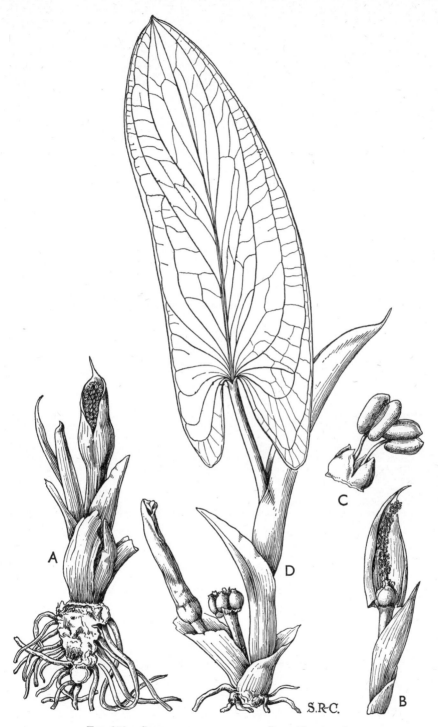

FIG. 358.—STYLOCHITON HYPOGAEUS *Lepr.* (ARACEAE).

A, habit with flower, × ⅔. B, spathe opened out exposing spadix with male flowers above and
female flowers below, × ⅔. C, male flower, × 10. D, habit, with infructescence, × ⅔.
From *Dalziel* 860.

couraba *Chev.* 21570! Mankono *Chev.* 21894! **Ghana:** Navrongo (Apr.) *Vigne* FH 3757! 4530! Tamale to Damongo (Mar.) *Adams* 3914! **Dah.:** Dassa-Zoumé *Chev.* 23660! Djougou *Chev.* 23864! **N.Nig.:** Nupe *Barter* 1472! **S.Nig.:** Oyo *Barter* 3424! Extending to C. African Republic.

A broad concept of the species has been adopted for this revision. Part of the difficulty is due to the presence of only leaves or flowers at the time of collection. Field workers should observe the range of variation within a population and obtain flowers and leaves from the same plant at different seasons.—F.N.H.

## 5. SAUROMATUM Schott, Melet. 1: 17 (1832); F.T.A. 8: 141; Hepper in Kew Bull. 21: 492 (1968). *Jaimenostia* Guinea & Gómez Moreno (1946).

Tuberous herb, tuber depressed globose with roots arising all round the top; leaf solitary, petiole stout smooth up to 1 m. long; lamina palmately and pedately lobed, segments broadly lanceolate, the largest lobe in the middle with the others decreasing in size outwards; inflorescence solitary; peduncle 5–15 cm. long, spathe elongate lanceolate up to 40 cm. long; spadix with basal female flowers about 5 cm. below male flowers, several sterile processes above the female portion, linear appendage up to 30 cm. long     ..     ..     ..     ..     ..     ..     ..     ..     ..     ..     *venosum*

**S. venosum** (*Ait.*) *Kunth* Enum. Pl. 3: 28 (1841). *Arum venosum* Ait. (1789). *Sauromatum guttatum* (Wall.) Schott (1832)—Engl., Pflanzenr. 4, 23 F: 123. *S. nubicum* Schott (1856)—F.T.A. 8: 141; Engl. l.c. 126. *Jaimenostia fernandopoana* Guinea & Gómez Moreno in Guinea, Ensayo Geobot. Guin. Continent. Españ. 248, figs. 39, 40 (1946), and in Ann. Jard. Bot. Madrid 6, 2: 466 (1946). A geophyte with a perennial tuber from which arises the short peduncle and purple-spotted spathe and later the single taller petiole with divided lamina 1–2 ft. across; fruits bright red; in shady places in upland.
**F.Po:** Moka *fide* Guinea *l.c.* Also in highland E. Cameroun and extending through much of the E. African mountains; also in India.

## 6. AMORPHOPHALLUS Blume ex Decne., Herb. Timor 38 (1835); F.T.A. 8: 144; Engl., Pflanzenr. 4, 23C: 16 (1911). *Nom. cons.*

Appendix of the spadix ovoid or ellipsoid, very short in comparison with its breadth, not more than 3 times as long as broad; leaves with narrow, more or less lanceolate segments:
Spathe open at the top, not much longer than the spadix, about 10 cm. long; appendix of spadix broadly ovoid, solid, 3–4(–5) cm. diam. at base; peduncle 11–20 cm. long     ..     ..     ..     ..     ..     ..     ..     ..     ..     1. *aphyllus*
Spathe hooded at the top, much longer than the spadix, about 30 cm. long, appendix of spadix ellipsoid 5–9 cm. long, about 3 cm. broad; peduncle 4–40 cm. long
2. *dracontioides*
Appendix of spadix cylindric or tapered to apex, long in proportion to the breadth, usually many times longer than broad, if only about 4 times as long as broad then about 6 mm. broad:
Spathe forming a tube at the base, tube broadly campanulate, 4–5 cm. long; whole spathe up to 15 cm. long; spadix long-stipitate, much shorter than the spathe; flowering part 1·5 cm. long, appendix 3 cm. long; female flowers very few; leaves not known     ..     ..     ..     ..     ..     ..     ..     ..     3. *elliotii*
Spathe not forming a tube at the base:
Spadix shorter than the spathe:
Leaf-segments shortly cuspidate-acuminate, obovate, decurrent on the rhachis, 6–8 cm. long, 3·5–4 cm. broad, with about 6 pairs of looped lateral nerves; peduncle up to 30 cm. long; spathe ovate, folded at the base, 15–20 cm. long, wrinkled inside towards the base but without hairs; spadix about 10 cm. long, appendix 4–5 cm. long     ..     ..     ..     ..     ..     4. *abyssinicus*
Leaf-segments gradually long-tailed-acuminate, oblanceolate, shortly decurrent on the rhachis, 15–17 cm. long, 4–5 cm. broad, with numerous looped lateral nerves; peduncle up to 50 cm. long, slender; spathe 12 cm. long, spadix a little shorter
5. *preussii*
Spadix longer than the spathe:
Spathe lacking stiff hairs within the basal folded portion, light green (3·5–)5–15 cm. long, broadly elliptic; spadix half as long again to twice as long as the spathe, tapering to the top; ovary finely verrucose; leaf-segments oblong-oblanceolate, cuspidate-acuminate, decurrent on the rhachis, 10–12 cm. long, 3–4 cm. broad, with numerous parallel nerves looped within the margin (see notes after this species)
6. *flavovirens*
Spathe with numerous stiff hairs within the basal portion:
Spadix not more than twice as long as the spathe:
Spathe broadly ovate, narrowed to the apex, about 17 cm. long and 15 cm. broad when opened out; leaf-segments oblanceolate, tailed-acuminate, about 12 cm. long and 3 cm. broad, thin     ..     ..     ..     ..     ..     7. *johnsonii*
Spathe ovate-orbicular, about 12 cm. diam. when spread out, not pointed; spadix slender, 18 cm. long; leaves not known     ..     ..     ..     8. *staudtii*
Spathe suborbicular, about 25 cm. diam. when spread out, deeply crenate; leaf-segments broadly elliptic, broadly acuminate, the terminal about 25 cm. long and 10 cm. broad     ..     ..     ..     ..     ..     ..     ..     9. *mannii*

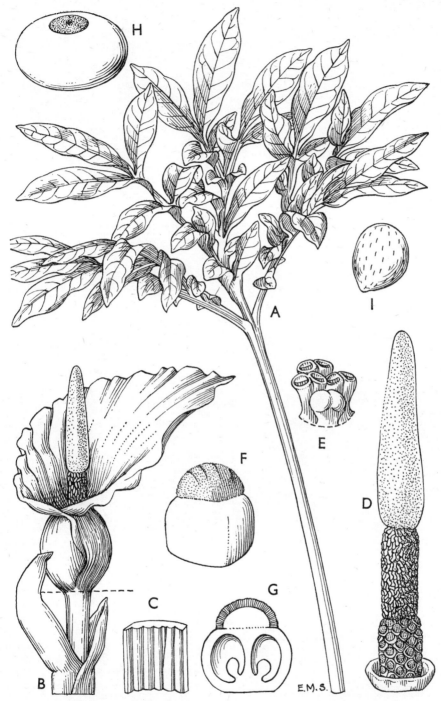

Fig. 359.—Amorphophallus abyssinicus (*A. Rich.*) *N.E. Br.* (Araceae).

A, leaf, × ¼. B, spathe and spadix, × ⅖. C, spathe opened at base to reveal lower portion of spadix, × 4. D, spadix, × 1. E, male flowers, × 8. F, female flower, × 8. G, L.S. female flower, × 8. H, fruit, × 4. I, seed, × 4. All from *FHI 22845*.

Spadix several times as long as the spathe:
  Appendix of spadix narrowed near its base and with its broadest expanded part
    well above the male flowers, very long; spathe broadly ovate when spread out,
    10–15 cm. long; male portion about 2 cm. long, 5–10 mm. diam. 10. *calabaricus*
  Appendix of spadix with its broadest part immediately above the male portion
    and without any further constriction:
    Spathe at the base within, clothed with filiform processes   ..          11. *zenkeri*
    Spathe at the base within, densely covered with stout hairs or soft bristles
                                                                       12. *accrensis*

1. **A. aphyllus** (*Hook.*) *Hutch.* F.W.T.A., ed. 1, 2: 362 (1936); Berhaut, Fl. Sén. 4, 11. *Arum aphyllum* Hook.
   in Gray, Travels in W. Afr. 386, t. A (1825). *Amorphophallus leonensis* Lem. (1846)—F.T.A. 8: 147,
   incl. vars.; Engl. l.c. 105, fig. 36; Bot. Mag. t. 7768; Chev. Bot. 680. *Corynophallus afzelii* Schott
   (1857). Inflorescence 6–12 in. high from a tuber 2 in. across; spathe dark red-purple; leaves produced
   after the inflorescence.
   **Sen.:** Kountaur (Mar.) *Berhaut* 5029! Kaolak (Mar.) *Berhaut* 274! Medina (May) *Trochain* 3446!
   **Mali:** Babela (Baléba) *Chev.* 394! Sinia to Diendénia (Feb.) *Chev.* 423! Banan *Chev.* 542! **Port Guin.:**
   Madina (Feb.) *Pereira* 2975! Campeana (Jan.) *Pereira* 2851! **Guin.:** Farana *Chev.* 13348! 13413!
   Timbo (Feb.) *Pobéguin* 1494! Bangadou, Kissi *Chev.* 20732! **S.L.:** *Afzelius*! *Masters*! *Sc. Elliot*! Kowama
   (May) *Deighton* 5779! Gbop, Nongoba Bullom (Jan.) *T. S. Jones* 390! **Togo Rep.:** Sokodé-Basari
   *Kersting* 311.
2. **A. dracontioides** (*Engl.*) *N. E. Br.* in F.T.A. 8: 148 (1901); Engl. l.c. 67, fig. 24, F–L; Chev. Bot. 680;
   Morton, W. Afr. Lilies & Orch. fig. 49. *Hydrosme dracontioides* Engl., Bot. Jahrb. 15: 461, t. 18 (1892).
   Tuber " the size of a child's head ", flattened above; inflorescence almost sessile or with a peduncle
   over 1 ft. high; spathe about 1 ft. long, hooded, pale outside, reddish-purple with yellowish streaks
   inside; in savanna.
   **Ghana:** Damongo (fl. & fr. Mar.) *Morton* GC 8673! Kpedsu (Dec.) *Howes* 1030! Kpong (Jan.) *Johnson*
   662! **Togo Rep.:** Bismarcksburg, Anjanga *Büttner* 419. **Dah.:** Massé to R. Ouémé *Chev.*! **N.Nig.:** Lokoja
   (fl. & fr. Oct.) *Dalz.* 564! Nupe *Barter* 1141! **S.Nig.:** Ilaro (Jan.) *Millen* 125! Ado-Awaye, Iseyin Dist.
   (Feb.) *Onochie* FHI 35278! Shepeteri, Oyo Dist. (Feb.) *Keay* FHI 22492!
3. **A. elliotii** *Hook f.* in Bot. Mag. t. 7349 (1894); F.T.A. 8: 147; Engl. l.c. 67, fig. 24, A–E. Peduncle about
   9 in. long; leaf about 1½ ft. diam.; spathe pinkish in the lower part, upper part green, mottled with
   brown, smelling strongly; in short grass.
   **S.L.:** Wallia (Jan.) *Sc. Elliot* 4640!
4. **A. abyssinicus** (*A. Rich.*) *N. E. Br.* in F.T.A. 8: 160 (1901); Milne-Redhead in Kew Bull. 5: 382, fig. 9
   map (1950); Verdoorn in Fl. Pl. Afr. t. 1251 (1957). *Arum abyssinicum* A. Rich., Tent. Fl. Abyss.
   2: 352 (1851). *Amorphophallus barteri* N. E. Br. in F.T.A. 8: 151 (1901); F.W.T.A., ed. 1, 2: 361;
   Morton, W. Afr. Lilies & Orch. fig. 50. *A. warneckei* (Engl.) Engl. & Gehrm. (1911). *A. chevalieri* (Engl.)
   Engl. & Gehrm. (1911). Tuberous rooted herb with very dark reddish purple spathe inside, paler out-
   side, markedly constricted below the middle, spadix nearly black; pungent smell; in rather damp
   places in savanna.
   **Ghana:** Pong Tamale (Apr.) *Bally* 142! Kete Krache (Apr.) *Morton* GC 9102! **Togo Rep.:** Lomé
   *Warnecke* 103! **N.Nig.:** Katagum Dist. *Dalz.* 237! Valley of Kaduna R., Bida Dist. (fl. & fr. May)
   *Keay* FHI 22845! Mamu, 2,500 ft. (May) *Lely* 159! Gudi to Keffi, Benue Dist. (Apr.) *Keay & Jones*
   FHI 37643! Abinsi (May) *Dalz.* 861! Nupe *Barter* 502! 1468! **S. Nig.:** Lagos *Foster* 211 **W.Cam.:**
   Bamessi to Bangola, 3,800 ft. (Apr.) *Brunt* 307! Widespread in savanna of tropical Africa to Ethiopia
   and Rhodesia.
5. **A. preussii** (*Engl.*) *N. E. Br.* in F.T.A. 8: 152 (1901); Engl., Pflanzenr. 23C: 95, fig. 34. *Hydrosme
   preussii* Engl., Bot. Jahrb. 15: 459 (1892). Spathe yellow-green, purple at the base; near upland stream
   in forest.
   **W.Cam.:** Buea, 3,000 ft. (fl. Dec.—Mar., leaves Sept.) *Preuss* 588 (leaves)! *Lehmbach* 127.
   [Further material needs to be examined before the value of this species can be assessed—F.N.H.]
6. **A. flavovirens** *N. E. Br.* in F.T.A. 8: 153 (1901); Engl., Pflanzenr. 4, 23C: 81. *A. baumannii* (Engl.)
   N. E. Br. in F.T.A. 8: 153 (1901); Engl. l.c. 97; Chev. Bot. 680. Inflorescence rather small on a slender
   peduncle up to 3 ft. high, spadix greenish, spathe grey, greenish yellow, sometimes tinged red, darker
   inside, with an objectionable smell; fruits red; in plantations and shady places.
   **Sen.:** (see note below). **Lib.:** Ganta (May) *Harley* 2036! Juarzon, Sinoe Co. (Mar.) *Baldwin* 11474!
   **U. Volta:** Diapaga to Fada (July) *Chev.* 24466! **Ghana:** Kuluwa (Mar.) *A. S. Thomas* D170! Agogo
   (May) *Irvine* 944! Kwahu (Mar.) *Johnson* 644! Aburi (Mar.–Apr.) *Johnson* 25! 873! Bame to Dzolok-
   puita, Ho Dist. (Mar.) *Hepper & Morton* A3031! **Dah.:** Agouagon (May) *Chev.* 23548, partly! Konkobiri
   (June) *Chev.* 24283! **S.Nig.:** Shasha (Omo) F.R., Ijebu Dist. (fl. & fr. Jan.) *Onochie* FHI 20689! Oke
   Iho to Iseyin (Apr.) *Keay & Polunin* FHI 37832! Extends to C. African Republic.
   [*A. consimilis* Blume (Rumphia 1: 149 (1835); F.T.A. 8: 154; Berhaut, Fl. Sén. 4, 11. *A. doryphorus*
   Ridl. (1886)—F.T.A. 8: 152.) was maintained as a species in Ed. 1 distinct from *A. flavovirens* in having
   a smaller spathe (7–9 cm. against 8–15 cm. long). It seems that it is a form of this species occurring in
   the NW. of our area (Sen.: *Leprieur*! *Chev.* 2601! Gam.: *Lester*! *Maxwell*!). *A. gracilior* Hutch. in
   Kew Bull. 1939: 245 (*A. gracilis* A. Chev., Bot. 653, name only, not of Engl.) was also maintained in
   Ed. 1 as distinct from *A. flavovirens* and *A. consimilis* in having a smaller spathe and short spadix—
   both inconstant characters (Dah.: *Chev.* 22957! an imperfect specimen). Certain specimens cited above
   (*Harley* 2036, *Baldwin* 11474) exhibit an expanded lip around the edge of the spathe.—F.N.H.]
7. **A. johnsonii** *N. E. Br.* in F.T.A. 8: 156 (1901); Engl. l.c. 80. *A. purpureus* (Engl.) Engl. & Gehrm. l.c.
   83 (1911); Chev. Bot. 681; Aké Assi, Contrib. 2: 21 [230, t. 20]. *Hydrosme purpurea* Engl. (1907).
   Tuber about the size of a flattened orange; spathe constricted, dark purple red, shortly stalked; leaves
   appearing after the inflorescence; petiole smooth brown-spotted; in forest undergrowth.
   **Mali:** Bamako, Doumbia (June) *Roberty* 2299! Sikasso (May) *Demange* 2096! **Guin:** Mamou to Kindia
   (May) *Chev.* 13585! **Lib.:** Javajai, Boporo Dist. *Baldwin* 12601! **Iv.C.:** Adiopodoumé (Jan.) *Aké Assi*
   7305! N'Douakro, Bouaké *Aké Assi* IA 2285. Tafiré *Mangenot & Aké Assi* IA 4285. **Ghana:** Tinte
   Bepo F.R. (Nov.) *Cox* 3215! Amosina, Cape Coast (Apr.) *Hall* 2582! Banka, Ashanti *Irvine* 482!
   Abetifi to Kumasi (Apr.) *Bally* 103! Kwahu (Mar.) *Johnson* 643! **S.Nig.:** Okomu F.R., Benin (Dec.)
   *Brenan* 8499! 8554! Sapoba (Apr.) *Richards* 3915!
8. **A. staudtii** (*Engl.*) *N. E. Br.* in F.T.A. 8: 154 (1901); Engl., Pflanzenr. 4, 23C: 97. *Hydrosme staudtii*
   Engl., Bot. Jahrb. 26: 420 (1899).
   **W.Cam.:** Johann-Albrechtshöhe (= Kumba) (Dec.) *Staudt* 767 (seen for Ed. 1).
9. **A. mannii** *N. E. Br.* in F.T.A. 8: 159 (1901); Engl. l.c. 100. *A. macrospadix* Font Quer in Cavanillesia 1:
   79, with fig. (1928). Spathe dark purple.
   **F. Po:** *Mann* 652!
10. **A. calabaricus** *N. E. Br.* in F.T.A. 8: 155 (1901); Engl. l.c. Peduncle 2–3 ft. high, mottled; spathe chocolate
    brown or pale yellow green, darkly spotted outside; spadix appendage hollow.
    **S.Nig.:** Ibadan South F.R. (Apr.) *Keay* FHI 22803! Calabar (Feb.) *Mann* 2336! Okuni, Cross R. (Jan.)
    *Holland* 256!

11. **A. zenkeri** (*Engl.*) *N. E. Br.* in F.T.A. 8: 159 (1901); Engl. l.c. *Hydrosme zenkeri* Engl., Bot. Jahrb. 26: 421 (1899). Spathe purple or purplish with white streaks inside and outside; pale and green spotted outside towards the base; spadix pale purple.
   **S.Nig.:** Oban *Talbot* 1300!
12. **A. accrensis** *N. E. Br.* in F.T.A. 8: 157 (1901); Engl. l.c. 86; Chev. Bot. 679. Spathe yellow-green on outside, fringed bright purple inside, spadix brown.
   **Iv.C.:** Dioandougou, Upper Sassandra *Chev.* 21515! **Ghana:** Accra *Sander & Co.*!
   [Apparently close to *A. johnsonii* and if no taxonomic significance is placed on the length of the appendix, it may be conspecific—F.N.H.]

## 7. COLOCASIA Schott in Schott and Endl., Melet. 1: 18 (1832); F.T.A. 8: 164; Engl., Pflanzenr. 4, 23E: 62 (1920).

A very stout rhizomatous herb; leaves peltate, more or less ovate, cordate at the base, basal lobes very obtuse, thin, glabrous; lateral nerves radiating from the base and pinnate, about 6–8 pairs; peduncle about 15 cm. long; spathe usually about 1·8 cm. long, narrow, acuminate, enfolding the spadix and articulating at the base of the male part in fruit; spadix much shorter than the spathe; tapered to the apex; berries enclosed in the persistent tube of the spathe    ..    ..    ..  *esculenta*

**C. esculenta** (*Linn.*) *Schott* l.c. (1832); A. F. Hill in Bot. Mus. Leafl. Harvard 7: 113 (1939); Morton, W. Afr. Lilies & Orch. fig. 60. *Arum esculentum* Linn. (1753). *Colocasia antiquorum* Schott (1832)—F.T.A. 8: 164; Bot. Mag. t. 7364; Engl. l.c. 65. Cultivated extensively in our area for the tuber (coco-yam) and frequently naturalized; the large leaves are seldom accompanied by flowers. Native of tropical Asia. (See Appendix, p. 481.)

## 8. XANTHOSOMA Schott in Schott & Endl., Melet. 1: 19 (1832); Engl., Pflanzenr. 4, 23 E: 41 (1920).

Stout rhizomatous herb; petiole up to 60 cm. long, usually less; lamina sagittate 30–45 cm. long and 20–35 cm. broad with large basal lobes not over-lapping; peduncle 30–40 cm. long; spathe about 20 cm. long; spadix slightly longer than the spathe
*mafaffa*

**X. mafaffa** *Schott* Arac. Betreff. 2: 5 (1855); Engl. l.c. 51, fig. 9J. *X. sagittifolium* of authors, not of (Linn.) Schott. Widely cultivated coco-yam, native of tropical S. America. Grown in the moister parts of our area and to about 5,000 ft in. W. Cameroun.
   [*X. mafaffa* appears to be the species in West Africa, as mentioned by Dalziel (Appendix, p. 483), although unfortunately he used the name *X. sagittifolium* for the main heading. It would be useful to know whether true *X. sagittifolium*, with overlapping basal leaf lobes and tapering male portion of the spadix, occurs in our area.]

## 9. ANUBIAS Schott in Oest. Bot. Wochenbl. 7: 398 (1857); F.T.A. 8: 182; Engl., Pflanzenr. 4, 23 DC: 2 (1915).

Leaves not hastately lobed, may be cordate (in No. 3):
  Spathe convolute, not opening at maturity, the spadix exserted; petioles not muricate at apex; lamina narrowly elliptic, acute at apex, 15–25 cm. long, 3–10 cm. broad, with numerous parallel lateral nerves spreading from the midrib at an angle of 45° with faint transverse tertiary nerves; spadix as long as or longer than the acuminate spathe ..    ..    ..    ..    ..    ..    ..    ..    ..    .. 1. *afzelii*
  Spathe opening at maturity, eventually partly reflexed:
  Spathe 3–5 cm. long; spadix as long as or longer than the spathe:
    Leaves broadest about the middle, lanceolate-elliptic, more or less narrowed to the base or slightly cordate, acute or slightly acuminate, 12–25 cm. long, 4–10 cm. broad, with numerous lateral nerves spreading from the midrib at an angle of about 80°; petioles more or less muricate, often densely so towards the base; spathe 3–5 cm. long, acuminate or cuspidate, narrow; spadix about as long as the spathe ..    ..    ..    ..    ..    ..    ..    .. 2. *lanceolata*
    Leaves broadest below the middle, cordate, truncate, rounded or exceptionally cuneate at the base, acute to acuminate at the apex, 15–25 cm. long, 5–10 cm. broad, glabrous except the puberulous midrib beneath, with numerous parallel lateral nerves spreading from the midrib at a wide angle; petiole slender, up to 30 cm. long, expanded and sheathing only towards the base, not tuberculate; peduncle very slender, about the same length as the leaves; spathe oblong-lanceolate, rather long-acuminate, 4–5 cm. long; spadix slender, longer than the spathe ..    ..    ..    ..    ..    ..    ..    ..    .. 3. *barteri*
  Spathe 1–1·5 cm. long, apiculate; spadix included, shorter than the spathe; petioles 3–17 cm. long; lamina narrowly elliptic 5–14 cm. long, about 1·5 cm. broad, cuneate at base, acute or obtuse at apex ..    ..    ..    ..    .. 4. *minima*
Leaves hastately lobed:
  Leaves shallowly hastately trilobed, triangular in outline, about 12 cm. long and 10 cm. broad, lobes ovate-oblong, rounded at the apex, glabrous, with numerous lateral nerves; petiole about 25 cm. long, deeply grooved on one side; peduncles slender, 15 cm. long; spathe 2·5 cm. long, broadly elliptic, mucronate, spadix as long as the spathe, the upper three-quarters occupied by the male flowers    5. *gracilis*

Leaves deeply and variably trilobed, sometimes nearly compound; middle lobe ovate-triangular to lanceolate, shortly acuminate, acute to obtuse at apex, 11–24 cm. long, up to 10 cm. broad, lateral lobes oblong-lanceolate, obtuse, up to 16 cm. long and 6 cm. broad, all glabrous with numerous arcuate nerves; petiole and peduncle up to 30 cm. long; spathe 3–4(–8) cm. long, mucronate; spadix longer than the spathe    ..    ..    ..    ..    ..    ..    ..    6. *gigantea*

1. **A. afzelii** *Schott* in Oest. Bot. Wochenbl. 7: 399 (1857); F.T.A. 8: 183; Engl. l.c. 3; Chev. Bot. 683. A gregarious herb of swamps or beside streams in forest; leaves with the lamina held erect, the petiole pulvinate at the apex; spathe green.
  **Sen.**: Kantora, Haute-Casamance *Etesse* 15! **Guin.**: Ditinn and Diaguissa *Chev.* 12653! 12681! 12905 *bis*! Kindia *Chev.* 12741! R. Kalendé (June) *Dybowski* 8! Timbo (Apr.) *Pobéguin* 1573! **S.L.**: *Afzelius*! *Barter*! Guma, Peninsula (Mar.) *Hepper & Pyne* 2504! Baiima, Gbo (May) *Deighton* 5767! Jigaya (fr. Sept.) *Thomas* 2829! Bafodeya (Apr.) *Sc. Elliot* 5560!

2. **A. lanceolata** *N. E. Br.* in F.T.A. 8: 183 (1901); Engl., Pflanzenr. 4, 23 DC: 4 (excl. syn A. *barteri* var. *glabra* N. E. Br.); Chev. Bot. 483. Terrestrial herb with creeping underground stem; spathe greenish white, spadix white; wet places in forest.
  **Iv.C.**: fide *Chev.* l.c. **S.Nig.**: Ikure (Jan.) *Holland* 167! Mfum, Cross R., Ikom to Mamfe (Dec.) *Keay* FHI 28315! British Ogbokum, Ikom Dist. (May) *Jones & Onochie* FHI 18880! Oban (Jan., Feb.) *Talbot* s.n.! 768! *Onochie* FHI 36302x! *Onochie & Okafor* FHI 36037x! **W.Cam.**: Rio del Rey *Johnston*! Also in E. Cameroun.

3. **A. barteri** *Schott* Prod. 159 (1860); F.T.A. 8: 185;. Engl. l.c. 4. A. *barteri* var. *glabra* N. E. Br. in F.T.A. 8: 185 (1901). Herb in dense masses in rocky streams; spathes green.
  **W.Cam.**: Ambas Bay *Mann* XV! Victoria (Jan.–Mar.) *Maitland* 1306! *Brenan* 9257! Kumba R., Barombi *Preuss* 422! 559! 1223! **F.Po**: (Dec.) *Mann* 104! *Barter* 2045! Musola (Jan.) *Guinea* 1441! Also in E. Cameroun and Gabon.
  [Engler placed var. *glabra* in A. *lanceolata*, but it seems to have more affinity with A. *barteri*. A. *nana* Engl. (1899)—type seen for Ed. 1—appears to be similar.—F.N.H.]

4. **A. minima** *A. Chev.* in Journ. de Bot. ser. 2, 2: 134 (1909); Bot. 684. A small herb with short rhizome; creeping on rocks in streams.
  **Guin.**: Kindia *Jac.-Fél.* 314! Nzo (Apr.) *Schnell* 1139! *Chev.* 21001! Nimba Mts. (June) *Schnell* 2815! Santa valley (Mar.) *Chev.* 12769! **Lib.**: Bili (Nov.) *Harley* 2079! Bilimu Mt. (Jan.) *Harley* 1337! Nyaake, Webo Dist. *Baldwin* 6118! Mnanulu *Baldwin* 6026! Nimba (Mar.) *Adam* 20296! 21071! **Iv.C.**: Nouba Mts. (Apr.) *Chev.* 2114! Grabo (Aug.) *Chev.* 19649! 19746! Grabo to Fété (Mar.) *Aké Assi* 7861!
  [There appears to be two forms; those plants with long petioles and leaves and with the lamina slightly acuminate, while others possess shorter petioles and leaves and with the lamina obtuse to acute.—F.N.H.]

5. **A. gracilis** *A. Chev. ex Hutch.* in Kew Bull. 1939: 246; F.W.T.A., ed. 1, 2: 366; Chev. Bot. 683. Slender herb; in savanna zone.
  **Guin.**: Macenta (Mar.) *Jac.-Fél.* 1561! Fassakoidou, Koniaukés country (Feb.) *Chev.* 20800! Souradou *Chev.* 20578! **S.L.**: Yifin to Bandakarafaia *Deighton* 5066! Tingi Mts. (Apr.) *Morton & Gledhill* SL 1929!

6. **A. gigantea** *A. Chev. ex Hutch.* in Kew Bull. 1939: 246; F.W.T.A., ed. 1, 2: 366; Chev. Bot. 683; Hepper in Kew Bull. 22: ined. (1968). A. *hastifolia* var. *robusta* Engl., Pflanzenr. 4, 23 DC: 9 (1915). Rhizomatous herb growing in shallow streams in high forest; spathes greenish brown, male flowers white, female flowers pale green.
  **Guin.**: Nionsomoridou *Chev.* 20857! 20858! **S.L.**: Mano (Apr.) *Deighton* 3373! 5027! Kondumbaia, N. Prov. (Mar.) *Morton & Gledhill* SL 1024! **Lib.**: Careysburg, Monrovia (fr. Feb.) *Dinklage*. Baila, St. John R. (Feb.) *Harley* 1478! **Togo Rep.**: *Baumann* 171!

## 10. AMAURIELLA Rendle in Cat. Talb. 115, t. 16 (1913).

Stemless herb; lateral nerves of leaves very numerous and closely parallel; blade hastate or 3-partite nearly to the base with the segments unequal, the two lateral narrowly lanceolate, the terminal elliptic to obovate-elliptic, 15–20 cm. long, up to 10 cm. broad; peduncle slender, about 8 cm. long; spathe 2·5–3·5 cm. long; spadix as long as the spathe; males in a cylindric mass about 2 cm. long, anthers clustered on top of the synandrium; ovary 2-celled, with a thick disk-like stigma .. *hastifolia*

**A. hastifolia** (*Engl.*) *Hepper* in Kew Bull. 22: 454 (1968). *Anubias hastifolia* Engl., Bot. Jahrb. 15: 462 (1892). and in Pflanzenr. 4, 23 DC: 9, fig. 9 (excl. K,L). *Amauriella obanensis* Rendle l.c. (caption A. *talbotii* Rendle in error) (1913). Stemless herb with petioles about 1 ft. high; inflorescences much shorter with a small spathe 1–2 in. long.
  **Iv.C.**: Gbépleu waterfall, Mt. Tonkoui (Feb.) *Aké Assi* 7366! **Ghana**: Vane, Trans-Volta Reg. (June) *Morton* A3439! Chilinga and Shiare, Buem-Krachi Dist. (Aug., June) *Morton* A3977! *Hall* 1355! **S.Nig.**: Oban *Talbot* 1297! 1532. Also in E. Cameroun.

## 11. ZANTEDESCHIA Spreng., Syst. Veg. 3: 765 (1826); Engl., Pflanzenr. 4, 23DC: 61 (1915). *Nom. cons.* *Richardia* Kunth (1815)—F.T.A. 8: 167, not of Linn.

Rhizomatous herb; leaves hastate, upper part elongate-oblong or elongate-deltoid, finely caudate-acuminate at apex, about 30 cm. long and 8 cm. broad, the basal lobes variable in shape and size with a marked sinus between them; petiole about 60 cm. long; peduncle as long as the petiole, smooth; spathe 8 cm. or more long, expanded slightly above, acutely and abruptly acuminate; spadix half as long as the spathe *angustiloba*

**Z. angustiloba** (*Schott*) *Engl.* Bot. Jahrb. 4: 64 (1883), and in Pflanzenr. 23DC: 65; Hepper in Kew Bull. 21: 493 (1968). *Richardia angustiloba* Schott (1865)—F.T.A. 8: 169. Marsh herb with leaves and inflorescence 2 ft. and more high; spathe conspicuous shining yellow.
  **N.Nig.**: Likitaba to Nguroje, Mambila Plateau, 5,500 ft. (July) *Chapman* 46! Also in Angola, Tanzania and Zambia.

**12. NEPHTHYTIS** Schott in Oest. Bot. Wochenbl. 7: 406 (1857); F.T.A. 8: 169; Engl., Pflanzenr. 4, 23C: 110 (1911).

Leaves with a rather deep and distinct sinus between the upper and lower lobes, lobes ovate-elliptic, acutely long-acuminate, 15–20 cm. long, 5–10 cm. broad, glabrous; petiole slender, 30–60 cm. high; peduncle about half as long as the petiole, slender; spathe decurrent on the peduncle, oblong-lanceolate, acutely acuminate, 3–5 cm. long; spadix subsessile, 1·5–2 cm. long   ..   ..   ..   ..    1. *constricta*
Leaves without or with only a very shallow sinus between the upper and basal lobes:
Spadix subsessile within the spathe, 4–5 cm. long; spathe oblong-elliptic, not acuminate, 5–7 cm. long, many-nerved; leaves broadly triangular in outline, lobes broadly acuminate, about 10–17 cm. long; petioles smooth, 40 cm. long   ..    2. *afzelii*
Spadix stipitate, stipe 1 cm. long; spathe abruptly acuminate, long-decurrent on the peduncle, finely spotted; leaves as in preceding but more narrowly acuminate, lobes 14–25 cm. long; petioles muricate, up to 120 cm. long   ..    3. *poissonii*

1. **N. constricta** *N. E. Br.* in Gard. Chron. 15: 790 (1881); F.T.A. 8: 170; Engl. l.c. 112, fig. 38 F–G. *N. talbotii* Rendle in Cat. Talb. 114 (1913). Herb with creeping rhizome, 1 in. thick, giving rise to several markedly sagittate leaves prominently veined; spathe pale green, reflexed in fruit; in forest. **S.Nig.:** Okomu F.R., Benin (fr. Feb.) *Brenan* 8527! **W.Cam.:** Oban *Talbot* 766! Rio del Rey *Johnston*! **F.Po:** (Dec.) *Mann* 106! Also in E. Cameroun.
2. **N. afzelii** *Schott* in Oest. Bot. Wochenbl. 7: 406 (1857); F.T.A. 8: 171; Engl. l.c. 110, fig. 38 A–E; Chev. Bot. 683. *N. liberica* N. E. Br. (1881). Creeping rhizome about 1 cm. thick, giving rise to 1 to 3 hastate leaves about 1 ft. high; fruits orange-yellow; in forest. **S.L.:** *Afzelius* (drawing)! York Pass (Dec.) *Deighton* 3325! **Lib.:** *Bull* (cult.)! Bushrod I. (fr. Aug.) *Baldwin* 13089! Duport (fr. Oct.) *Barker* 1440! Nyaake, Webo Dist. (fr. June) *Baldwin* 6146! **Iv.C.:** Baléko Forest (fr. Jan.) *Aké Assi* 7311! Abouabou Forest (fr. Jan.) *Leeuwenberg* 2364! *vide* Chev. *l.c.* **Ghana:** Begoro (fl. & fr. Nov.) *Morton* A238!
3. **N. poissonii** *(Engl.) N. E. Br.* in F.T.A. 8: 171 (1901); Engl. l.c. 112, fig. 39 G–N. *Oligogynium poissonii* Engl. (1883). *Nephthytis gravenreuthii* (Engl.) Engl. l.c. 112, fig. 39 A–F (1911). Herb with rather large hastate leaves about 4 ft. high from a rhizome 1 in. thick; spathe green, fruits fleshy, orange-red. **S. Nig.:** Okomu F.R. (fl. & fr. Jan.) *Brenan* 8862! Omo River F.R., Ijebu Dist. (fr. Jan.) *Onochie* FHI 20679! Afi River F.R., Obudu Dist. (fr. May) *Jones & Onochie* FHI 18762! Oban *Talbot* 767! **W.Cam.:** Banga, S. Bakundu F.R. (fl. & fr. Mar.) *Brenan* 9302! 9414! **F.Po:** Ureka, 1,000–2,000 ft. (fl. & fr. Aug.) *Thorold* TF82! Also in E. Cameroun and Gabon.

[Engler maintained that *N. gravenreuthii* has peduncles about half the length of the petioles whilst the peduncles of *N. poissonii* are much the same length as the petioles. The spathe is reputed to be smaller, more ovate and the stipe of the spadix shorter than in *N. poissonii*. Field observations would be useful. —F.N.H.]

*Imperfectly known species.*

**N. sp. A.** A fruiting specimen with a sagittate leaf and narrow lobes 21–23 cm. long and about 4 cm. broad. **Lib.:** Gbanga (fr. Sept.) *Linder* 573!

**13. ANCHOMANES** Schott in Oest. Bot. Wochenbl. 3: 313 (1853); F.T.A. 8: 161.

Male portion of spadix about equal to the female; spadix about ¼ as long as the spathe, lilac; spathe about 40 cm. long; petiole 1 m. or more high, prickly almost the entire length   ..   ..   ..   ..   ..   ..   ..   ..   ..    3. *nigritianus*
Male portion of the spadix much longer than the female:
Savanna woodland herb; spathe yellowish green; ovary smooth, green with prominent circular orange stigma; peduncle 60–120 cm. high; leaf developing after the inflorescence and about the same height, lamina up to 1 m. diam., much divided into numerous more or less rhomboidal lobes, each segment often with two apices    1. *welwitschii*
Rain forest herb; spathe dark purplish, ovary pink or purple with a rather small white stigma; peduncle 90–200 cm. high; leaf developing after the inflorescence and taller, petiole up to 3 m. high, lamina 1·5 m. diam.:
Ovary and style smooth or slightly warted   ..   ..    2a. *difformis* var. *difformis*
Ovary and style densely tuberculate   ..   ..   ..    2b. *difformis* var. *pallidus*

Besides the above Aké Assi (Contrib. 2: 20 [229]) lists *A. giganteus* Engl. (Bot. Jahrb. 26: 419 (1889)), described from the Congo, as native in Ivory Coast (*Aké Assi* IA 5722, 5724).

1. **A. welwitschii** *Rendle* in Cat. Welw. 2: 98 (1889); F.T.A. 8: 163; Morton, W. Afr. Lilies & Orch. 46. *A. difformis* of F.W.T.A., ed. 1, 2: 359, partly (syn. *A. dalzielii* N. E. Br.). *A. difformis* var. *welwitschii* (Rendle) Engl., Pflanzenr. 4, 23C: 55 (1911). *A. dalzielii* N. E. Br. (1913). Savanna herb similar to *A. difformis* but the leaf generally smaller, about 4 ft. high; spathes yellow-green, ovaries green and stigma orange. **Ghana:** Achimota (Apr.) *Bally* 9! Legon (Mar.) *Morton* GC 6488! Anyaboni to Jaketi (Feb.) *Morton* GC 8429! **N.Nig.:** Kontagora Prov. (fl. & fr. June) *Dalz.* 563! Olohengia (Mar.) *Meikle* 1255! Abinsi (Apr.) *Dalz.* 862! Mambila Plateau, 4,600 ft. (Jan.) *Hepper* 1839! **S.Nig.:** Ijaiye F.R., Ibadan (Mar.) *Keay* FHI 21195! Widespread in savanna zones of tropical Africa.
[Only those specimens where colour notes have enabled a positive identification to be made have been cited and the result may give an imperfect indication of the real distribution of this species.]
2. **A. difformis** *(Bl.) Engl.* in DC., Monogr. Phan. 2: 304 (1879), partly; Engl. l.c. fig. 21; F.W.T.A., ed. 1, 2: 359, partly (excl. syn. *A. dalzielii* N. E. Br.); Morton, W. Afr. Lilies & Orch. fig. 53. *Amorphophallus difformis* Bl. (1835). *Anchomanes dubius* Schott (1860)—F.T.A. 8: 163.
2a. **A. difformis** *(Bl.) Engl.* var. **difformis**—Hepper in Kew Bull. 21: 491 (1968). Moist forest herb with horizontal tuber up to 2½ ft. long and 8 in. across giving rise to one enormous divided leaf about 8 ft. high with a prickly petiole; spathe dark purplish, ovaries purplish or pink and stigma white. **S.L.:** Smeathman! Njala (May) *Deighton* 6066! **Lib.:** (cult., Sept.) *Bull* 1851a! **Iv.C.:** Sassandra (Dec.) *Leeuwenberg* 2230! **Ghana:** Assin-Yan-Kumasi *Cummins* 211! Bame Pass (Oct.) *Morton* GC 9340!

**N.Nig.:** Jemaa, Zaria Prov., 2,500 ft. (Feb.) *McClintock* 195! **S.Nig.:** Olokemeji F.R. (Mar.) *Hepper* 2297! Omo F.R. (Feb.) *Jones & Onochie* FHI 16987! Oban *Talbot*! **W.Cam.:** Bangola, Ndop Plain, 3,800 ft. (Mar.) *Brunt* 272!
    [See note above after *A. welwitschii*.]

2b. **A. difformis** var. **pallidus** (*Hook.*) *Hepper* in Kew Bull. 21: 491 (1968). *A. hookeri* var. *pallida* Hook., Bot. Mag. t. 5394 (1863). *A. hookeri* (Kunth) Schott (1853) —F.T.A. 8: 162. *A. difformis* var. *hookeri* (Kunth) Engl. l.c. 56 (1911). *A. petiolatus* (Hook.) Hutch., F.W.T.A., ed. 1, 2: 359 (1936).
    **F.Po:** *Boultbee*! (Dec.) *Mann* 107! El Pico, 4,800 ft., (Dec.) *Boughey* 163!
    [Typical var. *pallidus* appears to be endemic to Fernando Po. However, warting of the ovary is rather frequently found in specimens from the mainland (*A. obtusus* A. Chev. (1909). Ivory Coast *Chev.* 16745!) but apparently not to the same degree as in those from Fernando Po.]

3. **A. nigritianus** *Rendle* in Cat. Talb. 114 (1913). An inadequately known plant similar to the other species; leaf much divided with spiny peduncle 3–4 ft. high.
    **S.Nig.:** near Etara, Oban (Oct.) *Talbot* 1247!

## 14. REMUSATIA Schott, Melet. 1: 18 (1832); Gen. Aroid. t. 36 (1858).

A tuberous-rooted herb, the leaves accompanied by a peduncle bearing burr-like bulbils covered with hooked prickles; leaves ovate, peltate, cordate at the base, acutely acuminate, very variable in size, often tinged with purple below; spadix clothed with membranous leaf-sheaths, shortly pedunculate; spathe 8–10 cm. long, convolute around the female flowers, the male part of the short spadix exposed as the spathe reflexes; ovary 1-celled, ovules numerous  ..    ..    ..    ..    ..    *vivipara*

**R. vivipara** (*Roxb.*) *Schott* Melet. 1: 18 (1932); Hook. f., Fl. Brit. Ind. 6: 521; Aké Assi, Contrib. 2: 21 [229]. *Caladium viviparum* Roxb., Hort. Beng. 65 (1814). *Ditinnia rupicola* A. Chev., Bot. 279, name only. Epiphytic on trees by streams and associated with fern roots, or in moist cracks in rocks; flowers appearing rarely and during leafless stage, spathe greenish cream coloured, spadix whitish; bulbiliferous shoot leafless and brown.
    **Guin.:** Ditinn *Chev.* 12839; 18500; 18531. Iles de Los, Conakry *Mugnier-Serand* 11! **S.L.:** Musaia (Sept.) *Deighton* 4501! Roruks (Oct.) *Deighton* 4378! Kamasu (Dec.) *Deighton* 3824! Mafori (Oct.) *Adames* 138! Kameron to Kurubonla *Morton* SL 2503! Mano (Oct.) *Deighton* 4503! **Lib.:** (leaf May, bulbiliferous shoot Oct.) *Harley* 1638! **Iv.C.:** Mt. Tonkoui *Mangenot, Miège & Aké Assi* IA 2436. Tiapleu forest *Aké Assi* IA 3292; IA 5705. **W.Cam.:** Metschum Falls, Bamenda (Aug.) *Savory* UCI 307! Also in E. Cameroun, Tanzania and Zambia, and in the eastern tropics and sub-tropics.

## 15. CALADIUM Vent., Descr. Pl. Nouv. Jard. Cels. 30 (1800); F.T.A. 8: 165.

Stemless herb with a tuberous rootstock and radical leaves; petioles up to about 30 cm. long, lamina sagittate, peltate, with a short acute acumen at apex, basal lobes obtuse, midrib 10–15 cm. long, lobes 7–12 cm. long, glabrous, glaucous, usually with irregular spots of white and red; peduncle about ⅔ length of the petiole, glabrous; spathe glabrous; tube about 3 cm. long, ovoid, limb boat-shaped, about 7 cm. long and 2 cm. diam., caducous; spadix shorter than the leaves, lacking an apical appendage  *bicolor*

**C. bicolor** (*Ait.*) *Vent.* l.c. t. 30 (1800); F.T.A. 8: 166. *Arum bicolor* Ait., Hort. Kew. 3: 316 (1789). Thoroughly naturalized in several West African countries, occurring on roadsides and shady waste places in the forest region; leaves attractively mottled or coloured, appearing at the same time as the flowers; spathe greenish white. Introduced from tropical America.

## 16. RHEKTOPHYLLUM N. E. Br. in J. Bot. 20: 194 (1882); Engl., Pflanzenr. 4, 23C, Arac.: 119 (1911).

Climber; leaves large, unequally pinnatisect or perforated (the juvenile entire and hastate-triangular), lobes obliquely acuminate, glabrous; spathe about 10 cm. long, enfolding the slightly shorter spadix; stigma sessile, disk-like; fruiting spadix about 7 cm. long; fruits obovoid, about 1 cm. long, capped by the sessile discoid stigma  *mirabile*

**R. mirabile** *N. E. Br.* l.c. 195, t. 230 (1882); F.T.A. 8: 172. A stout climber in forest 20–30 ft. high with long tapering roots into the ground from the upper parts, and clasping roots round the host stem; spathe pale yellow on both sides; male portion of spadix creamy yellow, female pink.
    **Dah.:** Allada *Poisson* 03·6–5! **S.Nig.:** Epe *Stanfield* FHI 56575! Port Harcourt (Feb.) *Stubbings* 99! Afi River F.R., Ogoja Prov. (May) *Jones & Onochie* FHI 18747! Calabar to Mamfe (Mar.) *Onochie* FHI 34829! Calabar (May) *Monteiro*! **W.Cam.:** Mbaw Plain, Bamenda *Brunt* 468! Rio del Rey *Johnston* 2! **F.Po:** *Barter*! *Mann* 101! Also in E. Cameroun, Congo, Cabinda, Angola and Uganda.

## 17. CULCASIA P. Beauv., Fl. Oware 4 to ed. 3 (1803)*; F.T.A. 8: 173; Engl., Pflanzenr. 4, 23B, Arac.-Pothoid. 295 (1905).

Stem short and erect, not climbing (see also No. 9); with long stilt-like roots in No. 1: Upper end of the petiole sheath usually abruptly terminating within 5 mm. of the base of the lamina; petiole 3–5·5 cm. long; lamina oblanceolate (to linear) usually long-attenuate to the base, acute or gradually acuminate, 8–22 cm. long, (1·6–)4–8 cm. broad; main lateral nerves about 12 on each side with prominent venation between; peduncle solitary; spathe 3·5 cm. long; fruit obovoid-ellipsoid, up to 2·5 cm. diam. in the dry state; stilt-like roots conspicuous    ..    1. *striolata*

*Conservation of *Culcasia* has been proposed (Hepper in Regnum Veg. 40: 20 (1965)).

Upper end of the petiole sheath more or less merging with the petiole 1–3 cm. from
the base of the lamina; lamina broadly elliptic to ovate-elliptic, acute and not
acuminate, 14–20(–30) cm. long, 9–12(–18) cm. broad, with about 10 pairs of wavy
lateral nerves ascending within the margin, veins prominently reticulate beneath;
petiole 6–10 cm. long, spathe 4 cm. long; roots hardly stilt-like       .. 2. *mannii*
Stem more or less climbing:
  Leaves cordate at the base, gradually acuminate or acute at apex, broadly lanceolate
or oblong-lanceolate, 6–17 cm. long, 2·5–8 cm. broad, laxly reticulate beneath with
few scattered linear immersed glands, glabrous, with 3–4 pairs of lateral nerves;
petiole ½–⅓ as long as the lamina, sheathing nearly the whole length; peduncles
2–3 together, about 2 cm. long in flower, lengthening up to 5 cm. in fruit; spathe
apiculate, about 1·5 cm. long     ..     ..     ..     ..     ..     3. *parviflora*
  Leaves not cordate:
    Junction between the upper side of the petiole and the midrib interrupted by the
lamina or by a transverse swelling; fruits spherical:
      Lateral nerves looping near the margin; base of the lamina interrupting the junction
of the petiole and midrib:
        Stems slender, 2–3 mm. diam., with numerous short internodal roots; peduncles
solitary or paired, 3–6 cm. long, spathe 2·5–3 cm. long, spadix included; bract
(if apparent) about 1·5 cm. long; lamina broadly elliptic, unequal-sided, cuneate
or slightly rounded at the base, acute and hardly acuminate at the apex, 10–17
cm. long, 5–8 cm. broad, with about 6 principal lateral nerves on each side of
midrib     ..     ..     ..     ..     ..     ..     ..     4. *scandens*
        Stems rather stout, 5–10 mm. diam. or more; peduncles up to 8 together, 4–6 cm.
long; spathe 3–3·5 cm. long, spadix slightly longer; bracts several up to 5 cm.
long; lamina elliptic, narrowed to a slightly rounded base, acutely acuminate,
16–26 cm. long, 7–10 cm. broad, with numerous lateral nerves arising at nearly
right angles to the midrib     ..     ..     ..     ..     5. *liberica*
      Lateral nerves not looped near margin of leaves; junction of the petiole and midrib
interrupted by a small transverse swelling; stems usually stout:
        Leaves narrowed to the base, more than twice as long as broad, unequal-sided,
15–18 cm. long, 6–9 cm. broad, with numerous lateral nerves, about 8 pairs more
conspicuous than the others; spathe about 4·5 cm. long, articulating at the
base; stems about 5 mm. diam.     ..     ..     ..     ..     6. *barombensis*
        Leaves broadly rounded or truncate at the base, only about twice as long as broad,
up to at least 50 cm. long, with numerous lateral nerves, about 10 pairs more
conspicuous than the others; petiole about half as long as the lamina, broadly
sheathing in the lower three-quarters of its length; spathe about 7 cm. long,
articulating at the base; bracts up to 14 cm. long, sometimes longer than the
peduncle and spathe together     ..     ..     ..     ..     ..     7. *angolensis*
    Junction between petiole and the midrib uninterrupted, the lamina decurrent into
the petiole:
      Leaves with each main lateral nerve curving towards the apex but not joining one
another and not looping in a prominent intramarginal nerve, lamina lanceolate,
cuneate at base, gradually acute or slightly acuminate at apex ending in a spine-
like tip, 7–18(–23) cm. long, 1·5–8 cm. broad; petiole 2–9 cm. long; spathe about
2 cm. long     ..     ..     ..     ..     ..     ..     ..     ..     8. *lancifolia*
      Leaves with main lateral nerves looped near the margin, at least in the upper half:
        Aerial roots few or absent from the upper internodes; plant often semi-erect;
leaves broadly ovate to elliptic, shortly cuneate at the base, acuminate, 10–20 cm.
long, 3–10 cm. broad; petiole 3–7 cm. long, ¾ sheathed; peduncles 2–3, rather
longer than the petioles; spathe 3–4 cm. long, the spadix well exserted
                                          9. *saxatilis*
        Aerial roots conspicuous; plant always climbing, stems slender:
          Peduncles longer than the petioles; leaves abruptly long-acuminate; petioles
narrowly and shortly sheathed:
            Fruit and seed ellipsoid, stigmatic surface small, seed nearly 1 cm. long; peduncles
often numerous or solitary, up to 13 cm. long; spathe 4 cm. long; spadix
included, the female portion short; leaves oblong or oblong-elliptic, 15–28 cm.
long, 2·5–7 cm. broad, the looped marginal nerve very prominent for the
whole length of the lamina; peduncles up to 10 cm. long, slender, usually
¼ sheathed     ..     ..     ..     ..     ..     ..     ..     10. *seretii*
            Fruit spherical about 3 mm. diam. with a large stigmatic surface; peduncles
solitary, 6–7 cm. long; leaves elliptic, sometimes narrowly elliptic and
attenuate at each end, 9–14 cm. long, 2·5–5 cm. broad     ..     .. 11. *sapinii*
          Peduncles shorter than the petioles; petioles broadly sheathed:
            Fruit and seed ellipsoid, twice as long as broad, seed about 7 mm. long; spathe
3–3·5 cm. long, the spadix slightly exserted, male and female portions similar
in length; leaves oblong to oblong-elliptic, shortly cuneate at the base,
abruptly acuminate, somewhat unequal sided, 9–21 cm. long, 3·5–7 cm. broad,

intra-marginal nerve conspicuously looped, linear pellucid immersed glands inconspicuous few or rather numerous; petioles broadly sheathed almost to the base of the lamina and ending in expanded ligules    ..      12. *tenuifolia*
Fruit and seed hemispherical, seed about 5 mm. diam.; spathe 3–4 cm. long, the spadix inserted; leaves elliptic to obovate-elliptic, unequal sided, cuneate at base, acuminate, 11–14 cm. long, 4–6 cm. broad, conspicuously covered with numerous linear pellucid immersed glands; petioles sheathed nearly to the base of the lamina    ..    ..    ..    ..    ..    13. *glandulosa*

1. **C. striolata** *Engl.* Bot. Jahrb. 26: 417 (1899); F.T.A. 8: 179; Engl., Pflanzenr. 4, 23B: 297; Chev. Bot. 678. *C. gracilis* N. E. Br. in F.T.A. 8: 179 (1901); F.W.T.A., ed. 1, 2: 264. *C. englerana* A. Chev., Bot. 677, name only. Erect herb of the forest floor about 1 ft. high with stout, ribbed, pubescent stilt-like aerial roots; spathe and spadix pale yellow green, berries red or orange.
Guin.: Dantilia (fr. Mar.) *Sc. Elliot* 5293! Macenta (fr. Oct.) *Baldwin* 9811! S.L.: Gbinti (fr. July) *Deighton* 2500! Kasewe F.R. (Nov.) *Morton* SL 1479! Njala (fr. Nov.) *Deighton* 3093! Lib.: Bushrod I. (fr. Aug.) *Baldwin* 13088! Karmadhun, Kolahun Dist. (fr. Nov.) *Baldwin* 10167! Ganta, Sanokwele Dist. (Sept.) *Baldwin* 9226! Iv.C.: Adiopodoumé (Feb.) *Aké Assi* 7340! Béyo (fr. Jan.) *Leeuwenberg* 2446! Ghana: Janare (fr. Dec.) *Adams* 2115! Tano Suhein F.R. (Dec.) *Morton* A3634! Abomeng, S. Scarp F.R. (fr. Dec.) *Adams* 5149! S. Fomangsu F.R. (fr. Dec.) *Adams* 4454! Kibbi (fr. Dec.) *Morton* 8117! S.Nig.: Lagos *Batten-Poole* 13! Akure F.R., Ondo Dist. (fl. fr. Oct., Nov.) *Keay & Onochie* FHI 20250! *Keay* FHI 25564! Okomu F.R., Benin Dist. (fr. Dec.) *Brenan* 8526! W.Cam.: Metschum R., Wum Dist. (fr. July) *Ujor* FHI 29275! Kumba (fr. Sept.) *Rosevear* Cam. 104/37! Ambas Bay (fr. Feb.) *Mann* 781! *Kalbreyer* 86! Victoria to Bimbia *Preuss* 1161! Probably also in E. Cameroun.
[Linear-leaved plants from Liberia (*Harley* 1938, *Baldwin* 6728) with the lamina about 2 cm. broad appear to approach *C. lanceolata* Engl., except for much larger spathes (3·5 cm.)]

2. **C. mannii** (*Hook. f.*) *Engl.* in Gartenfl. 36: 84 (1887); F.T.A. 8: 178. *Aglaonema mannii* Hook. f. in Bot. Mag. t. 5760 (1869). Creeping underground stem ascending and becoming erect, 1–2 ft. high; spathe white or greenish, and pink stalk, male flowers of spadix whitish, female pink; fruits scarlet; in moist forest.
S.Nig.: Oban (Mar.) *Richards* 5201! Orem, Oban F.R. (Feb.) *Onochie* FHI 36294x! *Okafor* FHI 36271x! Afi F.R., Ogoja Prov. (fr. May) *Jones & Onochie* FHI 18646! W.Cam.: "Victoria Mts." *Mann*! Also in E. Cameroun.

3. **C. parviflora** *N. E. Br.* in F.T.A. 8: 176 (1901); Engl., Pflanzenr. 4, 23B: 299. Slender climber in forest with cordate leaves of variable size; spathe pale yellow-green; berries scarlet.
S.L.: *Deighton* 4135! Bumban (June) *Glanville* 439! Njala (fr. Dec.) *Pyne* 85! Njama, Kowa (Aug.) *Deighton* 6107! Lib.: Bushrod I. (Aug.) *Baldwin* 13075! Kle, Boporo Dist. (fr. Dec.) *Baldwin* 10566! Sanokwele (fr. Sept.) *Baldwin* 9541! Nyaake, Webo Dist. (June) *Baldwin* 6115! Iv.C.: Boutoubré *Chev.* 16334! Ghana: S. Scarp F.R. (fr. Jan.) *Moor* FH 2200! Ancobra R., Axim Dist. (Dec.) *Johnson* 872! S.Nig.: Shasha F.R., Ijebu Prov. (Mar., May) *Richards* 3466! *Ross* 82! W.Cam.: *Preuss* 1333! S. Bakundu F.R., Banga (fr. Mar.) *Brenan* 9287! Likoko to Bafia, Victoria Dist. (fr. Nov.) *Keay* FHI 37529! F.Po: (Dec.) *Mann* 105! Also in E. Cameroun.
[*C. parviflora* var. *obtusifolia* Engl. l.c. 300 (1905); F.W.T.A., ed. 1, 2: 364, is described as having the leaves obtusely rounded at the apex (W.Cam.: Victoria *Preuss*, seen for Ed. 1).]

4. **C. scandens** *P. Beauv.* Fl. Oware 4to ed.: 5, t. 3 (1803), fol. ed. 1: 4, t. 3 (1805); Hepper in Kew Bull. 21: 315, fig. 1, 1–4 (1967); Heine in Adansonia n.s., 7: 137 (1967). Epiphytic climbing herb with slender wiry stems producing numerous short adhering roots; spathe greenish.
Lib.: Nimba (Jan.) *Adam* 21633! Iv.C.: *Scaëtta* 3237! Ghana: Shama *P. Beauvois*! Assuantsi (fr. July) *Irvine* 5118! Koforidua (fl. & fr. Apr.) *Vigne* FH 4378! Akayao, Kwamikrom (Dec.) *Adams* 4504! Amedzofe (fr. Jan.) *de Wit & Morton* A2880! S.Nig.: Lagos (Apr.) *Dennett* 492! Sapoba *Kennedy* 1105! Idumuye (fr. Dec.) *Thomas* 21321 Ibadan to Ife (fl. & fr. Apr.) *Thorold* TN72! Also in E. Cameroun.

5. **C. liberica** *N. E. Br.* in J. Linn. Soc. 37: 115 (1905); Hepper in Kew Bull. 21: 324, fig. 4, 5–7 (1967). *C. scandens* of F.W.T.A. ed. 1, 2: 364, partly (as to syn.), not of P. Beauv. ?*C. barombiensis* of Aké Assi, Contrib. 2: 22 [230]. Climber up trees in forest; spathe greenish white.
S.L.: Gola Forest (Apr.) *Small* 581! Lib.: Koudessu, Boporo Dist. (fr. Dec.) *Baldwin* 10671! Karmadhun, Kolahun Dist. (fr. Nov.) *Baldwin* 10199! Firestone Plantations, Dukwia R. (fr. Oct.–Feb.) *Cooper* 19! 244! Ganta, Sanokwele Dist. (fr. Sept.) *Baldwin* 9283! Jabroke, Webo Dist. (fr. July) *Baldwin* 6693! Sinoe Basin *Whyte*! Jaurazon (fr. Mar.) *Baldwin* 10260! Iv.C.: Azaguié (fr. Sept.) *Chev.* B.22286! Byonouon to Soubiré (Mar.) *Chev.* 17696!

6. **C. barombensis** *N. E. Br.* in F.T.A. 8: 177 (1901). *C. angolensis* Welw. ex Schott var. *angustifolia* Engl. (1902)—Engl. l.c. 300. A climber.
W.Cam.: Barombi *Preuss* 388! Also in E. Cameroun.

7. **C. angolensis** *Welw. ex Schott* in Seem., J. Bot. 3: 35 (1865); F.T.A. 8: 178; Engl. l.c. 300; Chev. Bot. 677; Morton, W. Afr. Lilies & Orch. 57. Large climber up to 100 ft. high with masses of short inter-nodal roots adhering to tree trunks; spathe pale green.
Guin.: *vide* Chev. l.c. Kindia *Chillou* 1792! S.L.: Jau, near Gegbwema (fl. & fr. Jan.) *Deighton* 3890! Likuru, 3,000 ft. *Sc. Elliot* 4950! Lib.: Wohmen, Vonjama Dist. (Oct.) *Baldwin* 10062! Ganta, Sanok-wele Dist. (fr. Sept.) *Baldwin* 13260! Zeahtown, Tchien Dist. (July) *Baldwin* 6983! Iv.C.: Prolo to Bliéron (fr. Aug.) *Chev.* 19884! Béyo (fr. Jan.) *Leeuwenberg* 2436! Ghana: Agoja (fr. Apr.) *Adams* 2592! Pusu-pusu ravine (fr. Apr.) *Morton* A857! Aburi (Jan.) *Johnson*! F.Po: (fr. Dec.) *Mann* 102! Extending to Angola.

8. **C. lancifolia** *N. E. Br.* in F.T.A. 8: 175 (1901); Engl. l.c. 304; *C. insulana* N. E. Br. in F.T.A. 8: 175 (1901); Engl. l.c. 301; F.W.T.A., ed. 1, 2: 364; Aké Assi, Contrib. 2: 22 [231]. *C. tubulifera* Engl. l.c. 301 (1905). Slender climber on tree stems, up to 15 ft. long: spathe green.
N.Nig.: Vogel Peak, Sardauna Prov. (fl. & fr. Dec.) *Hepper* 1582! S.Nig.: Lagos (Oct.) *Moloney*! Sapoba *Kennedy* 1808! *Ajayi & Odukwe* FHI 26940! W.Cam.: Buea *Deistel* 148! Bambui, 4,000 ft. (Aug.) *Brunt* 1260! Bambui to Njinikom (fl. & fr. Aug.) *Nditapah* in *Brunt* 1209! Buea *Lehmbach* 159! F.Po: *Mann* 325! 651! Moka (Mar.) *Guinea* 2217! *Melville* 420! Also in E. Cameroun and Congo.

9. **C. saxatilis** *A. Chev.* in Journ. de Bot. 22: 133 (1909); Chev. Bot. 677; Hepper in Kew Bull. 21: 317, fig. 2, 1–3 (1967). More or less erect about 4 ft. high and often profusely branched, with few aerial roots; spathe pale green; spadix white or purplish; fruits red; in riverine forest mainly in savanna zone.
Sen.: Sédhiou, Casamance *Berhaut* 6074! Mali: Bamako (fl. & fr. Jan.) *Hb. IFAN* 2086! S.L.: Kambia (fr. Dec.) *Sc. Elliot* 4229! Robis (fl. & fr. Sept.) *Adames* 135! Mafinta, nr. Roruks (July) *Deighton* 4674! Njala (June) *Deighton* 716! Lib.: Péahtah (fr. Oct.) *Linder* 947! Firestone Plantation No. 3, Du R. (Aug.) *Linder* 247! Wohmen, Vonjama Dist. (fr. Oct.) *Baldwin* 10090! Tappita (Aug.) *Baldwin* 9069! Tchien (fr. Aug.) *Baldwin* 7047! Iv.C.: Adiopodoumé (fr. Nov.) *Leeuwenberg* 1888! Ghana: Gambaga (fr. Apr.) *Morton* GC 8983! Saboba (fr. Mar.) *Hepper & Morton* A3117! Nsemre F.R., Borku (Dec.) *Adams* GC 2828! Pokoase (fr. Apr.) *Irvine* 1590! Kete Krachi to Atebubu (Dec.) *Adams* GC 4617! Noratiem, nr. Axim (May) *Cudjoe* GC 137! Dah.: *Le Testu* 113! N.Nig.: Kontagora (Nov.) *Dalz.* 251! Nupe *Barter* 1471! Jebba (fr. Mar.) *Meikle* 1275! Shikaku, Kotokerifi Dist. (Oct.) *Daramola & Adebusuyi* FHI 38416! Lokoja (fr. Mar.) *Shaw* 28! Gangumi (fr. Dec.) *Latilo & Daramola* FHI 28821! S.Nig.:

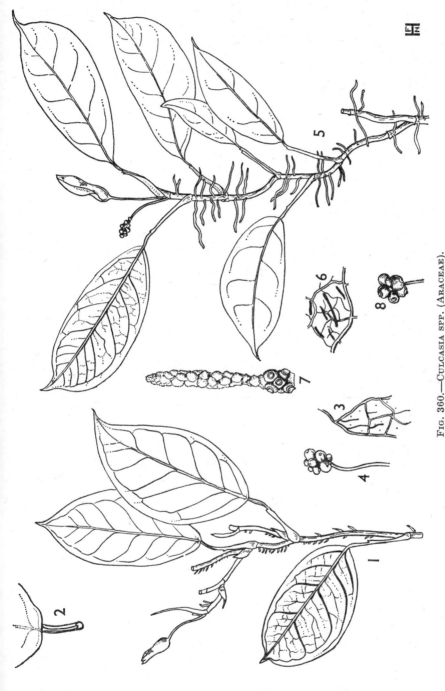

Fig. 360.—CULCASIA SPP. (ARACEAE).

*Culcasia scandens* P. Beauv.—1, habit, × ⅓; 2, detail of apex of petiole, × 1. 3, portion of lower leaf surface, × 4. 4, infructescence, × ⅓. *C. glandulosa* Hepper—5, habit, × ⅓. 6, portion of lower leaf surface, × 4. 7, spadix, × 1. 8, infructescence, × ⅓. 1, 3, 4 from *Thorold* TN 72. 2 from FHI 59875. 5, 7 from *Baldwin* 6791. 8 from *N. W. Thomas* 8684.

Agbelekaka, Benin Div. (Jan.) *Brenan* 8945! Stirling *T. Vogel* 204! Nun R. (Sept.) *Mann* 516! Ikom (fl. & fr. May) *Onochie* FHI 18868! Calabar to Mamfe (fr. Feb.) *Onyeachusim & Latilo* FHI 48166! Extending to the Congo.

10. **C. seretii** *De Wild.* Pl. Bequaert. 1: 172 (1922); Hepper in Kew Bull. 21: 320, fig. 2, 4–6 (1967). Climber or epiphyte with numerous adventitious roots; spathe green, fruits reddish.
    **Guin.:** Nimba Mts. (fr. Aug.) *Schnell* 3443! 3856! **S.L.:** Gola Forest (May) *Bakshi* 197! **Lib.:** Sanokwele (fr. Sept.) *Baldwin* 13224! Ganta (fl. & fr. Sept., Feb.) *Baldwin* 11012! Yoma, Maher R. (fr. Aug.) *Leeuwenberg* 4854! Gbanga (fr. Sept.) *Linder* 663! **Iv.C.:** mid-Sassandra to mid-Cavally (June) *Chev.* 19210! Guédéyo, E. of Soubré (fr. Dec.) *Leeuwenberg* 2182! **Ghana:** S. Fomangsu F.R. (Apr.) *Adams* 2723! Also in the Congo.

11. **C. sapinii** *De Wild.* Pl. Bequaert. 1: 170 (1922); Hepper l.c. 321, fig. 3 (1967). A slender climber with yellowish stems.
    **S.Nig.:** Okomu F.R. (fr. Jan.) *Brenan* 8902! 8902a! Also in the Congo.

12. **C. tenuifolia** *Engl.* Bot. Jahrb. 15: 447 (1893); F.T.A. 8: 176; Engl., Pflanzenr. 4, 23B: 301; Hepper l.c. fig. 4, 1–4 (1967). Climbing or epiphytic on trees; spathe pale green, spadix white, fruits red.
    **S.L.:** Walihun, Bumpe (Apr.) *Deighton* 1725! Fluima, Njala (May) *Deighton* 682! **Lib.:** Javajai, Boporo (fr. Nov.) *Baldwin* 10259! Du R. (fr. Aug.) *Linder* 314! Gbanga (fr. Sept.) *Linder* 453! Ganta (Aug.) *Harley* 944! Zeahtown (fr. Aug.) *Baldwin* 7000! **Iv.C.:** Grabo (July) *Chev.* 19657! **S.Nig.:** Oban *Talbot* 887! *Adebusuyi* FHI 43999! Afi River F.R. *Jones & Onochie* FHI 5805! Shasha F.R. *Richards* 3464! **W.Cam.:** Banga *Brenan* 9485! Barombi (Sept.) *Preuss* 542! Also in E. Cameroun and Cabinda.

13. **C. glandulosa** *Hepper* in Kew Bull. 21: 323, fig. 1, 5–8 (1967). *C. dinklagei* of Aké Assi, Contrib. 2: 22 [231], not of Engl. A slender climbing shrub.
    **S.L.:** *Thomas* 8684! **Lib.:** Gletown (fl. & fr. July) *Baldwin* 6791! Monroviatown (fr. Aug.) *Baldwin* 8001! Tappita (fr. Aug.) *Baldwin* 9032! Bony range (July) *Voorhoeve* 338! **Iv.C.:** Divo Forest (July) *Aké Assi* 5720!

## 18. CERCESTIS Schott in Oest. Bot. Wochenbl. 7: 414 (1857); F.T.A. 8: 180; Engl., Pflanzenr. 4, 23C: 114 (1911).

Leaves 3-lobed or 3-partite, midrib 12 cm. long, lateral nerves few and far apart, lobes oblong-elliptic, obtusely acuminate, held more or less vertical; peduncle about 4 cm. long; spathe shorter than the spadix, 5–6 cm. long, articulating and leaving a short cup at the base in fruit; fruit pentagonal ..  ..  ..  ..  1. *afzelii*
Leaves sagittate, cordate, or lamina obtuse or truncate at the base:
  Basal auricles oblong or broadly triangular:
    Auricles oblong, 4–5 cm. long, obtuse; lamina pandurate, acutely acuminate, 15–22 cm. long, with about 5 pairs of lateral nerves; peduncle 5–8 cm. long; bracts lanceolate, 5 cm. long, papery; spathe 5–6·5 cm. long ..  ..  2. *stigmaticus*
    Auricles broadly triangular but very obtuse-rounded, 4–5 cm. long; lamina not pandurate, narrowly triangular, about 20 cm. long and 9(–13) cm. broad with 4–5 pairs of nerves; peduncle in fruit 2 cm. long; infructescence 5 cm. long; fruit nearly 1 cm. diam.  ..  ..  ..  ..  ..  3. *sagittatus*
  Basal auricles shortly rounded or lamina truncate at the base:
    Leaves truncate at the base, oblong, slightly pandurate 20–25 cm. long, 6–8 cm. broad, papery, with 8–10 pairs of faint lateral nerves; petiole nearly as long as the lamina; peduncle 2–3 cm. long; spathe 6–8 cm. long (see also below) .. 4. *ivorensis*
  Leaves obtuse or cordate at the base:
    Leaves cordate at the base:
      Bracts 5–7 cm. long; leaves widely cordate or cordate and truncate on the same plant (see also above)  ..  ..  ..  ..  ..  ..  .. 4. *ivorensis*
      Bracts up to 5 cm. long; leaves shortly and closely cordate at the base; lamina oblong-lanceolate to obovate, acuminate, 15–26 cm. long, 6–10 cm. broad; peduncle 4 cm. long; spathe 5–6 cm. long  ..  ..  .. 5. *kamerunianus*
    Leaves obtuse at the base, acutely acuminate at apex, oblong-lanceolate, about 20 cm. long and 6–7 cm. broad, thin, with about 8 pairs of lateral nerves; petiole 6–10 cm. long; peduncle 3 cm. long; fruit subglobose, 7–8 mm. diam.
                                                      6. *lanceolatus*

1. **C. afzelii** *Schott* in Oest. Bot. Wochenbl. 7: 414 (1857); F.T.A. 8: 180; Engl. l.c. 118; Chev. Bot. 681; Berhaut, Fl. Sén. 180. *C. scaber* A. Chev., Bot. 682, name only. Creeping, climbing or epiphytic plant up to 50 ft. high in forest, with numerous adventitious roots; spathe green, spadix white; fruits red.
    **Sen.:** Niombato, Karang *Berhaut* 1782! **Guin.:** Farana (fr. Mar.) *Sc. Elliot* 5319! Nimba *Schnell* 875! Dentelé *Paroisse* 9! Bafing R. (June) *Pobéguin* 1633! **S.L.:** Wilberforce (fr. Mar.) *H. H. Johnston* 105! Bagroo R. (Apr.) *Mann* 906! Kambui Hills (fr. Mar.) *Small* 515! Gebwema (Oct.) *Pyne* 53! Gola Forest (fr. Apr.) *Small* 596! **Lib.:** Wohmen, Vonjama Dist. (fr. Oct.) *Baldwin* 10054! Kakatown *Whyte*! Gbanga (fr. Sept.) *Linder* 453! Ganta (fr. Dec.) *Harley* 381! Zeahtown, Tchien Dist. (fr. July) *Baldwin* 6994! **Iv.C.:** Montézo to Alépé (fr. Feb.) *Chev.* 17428! Grabo to Taté (fr. Aug.) *Chev.* 19769! Béyo (fl. & fr. Jan.) *Leeuwenberg* 2618! Abidjan Lagoon (fr. Aug.) *de Wit* 5604! Ganhoué to Bampleu (Apr.) *Chev.* 21180! Grabo *Chev.* 19727! **Ghana:** Assin-Yan-Kumasi *Cummins* 178! Kibbi Hills (Dec.) *Johnson* 258! Puso Puso to Asiakwa (fr. Jan.) *Adams* 3729! Southern Scarp F.R. (fr. May) *Adu* 48! **S.Nig.:** Boje-Ashoban to Katabong, Afi River F.R. (May) *Jones & Onochie* FHI 18753! Oban *Talbot* 1261!

2. **C. stigmaticus** *N. E. Br.* in F.T.A. 8: 181 (1901). *C. elliotii* Engl., Pflanzenr. 4, 23C: 116 (1911). *C. dinklagei* of Engl. l.c., partly, not of Engl., Bot. Jahrb. 26: 422 (1899). *C. alepensis* and *C. hastifolia* A. Chev., Bot. 681, names only. Slender forest climber with aerial roots; spathe purple at the base; spadix creamy coloured.
    **Guin.:** Moussadougou to Lola (fr. Mar.) *Chev.* 20961! **S.L.:** Morea (fr. Nov.) *Morton & Gledhill* SL 522! **Lib.:** Carder in Hort. *Bull* 1851a! Bilimu Mt. (Nov.) *Harley* 2078! Kitoma (Jan.) *Baldwin* 14065! **Iv.C.:** Alépé (Feb.) *Chev.* 17434! Kéeta (July) *Chev.* 19370! Lakota to Sassandra (Aug.) *de Wilde* 308! Yapo (fl. & fr. Oct.) *Leeuwenberg* 1818! **S.Nig.:** Oban F.R. (Mar.) *Onochie* FHI 36481x! 51 miles W. of Oban (Mar.) *Coombe* 169! *Brenan* 9492! **F.Po:** *Mann* 324!

3. **C. sagittatus** *Engl.* Pflanzenr. 4, 23C: 116 (1911). A climber with rather stout stems in forest; fruits red.
    **Lib.:** Careysburg, near Monrovia (fr. Feb.) *Dinklage* 2468!

4. **C. ivorensis** *A. Chev.* in Journ. de Bot., sér. 2, 2: 135 (1909).
   **Lib.:** Zolotown (fr. Dec.) *Adam* 16406! **Iv.C.:** Grabo, Cavally Basin (July) *Chev.* 19632! Ningué forest (Mar.) *Aké Assi* 3917! N'Zida forest (Dec.) *Aké Assi* 8401!
5. **C. kamerunianus** (*Engl.*) *N. E. Br.* in F.T.A. 8: 182 (1901); Engl., Pflanzenr. 4, 23C: 115, fig. 40. *Alocasiophyllum kamerunianum* Engl., Bot. Jahrb. 15: 449, t. 19 (1892). A slender climber up trees and becoming epiphytic later, spathe dull greenish purple.
   **S.Nig.:** Akure F.R., Ondo (fr. Aug.) *Jones* FHI 19526! Aponmu F.R., Ondo (fr. Aug.) *J. B. Gillett* 15315! Okomu F.R. (fl. & fr. Jan.) *Brenan* 8895! 9096! **W.Cam.:** Barombi (Apr.) *Preuss* 147. S. Bakundu F.R., Banga (Mar.) *Brenan* 9493! Also in Gabon.
   [Several Liberian specimens (*Linder* 1061, 1334, *Baldwin* 7085, 11445, 11484) are close to, if not conspecific with, *C. kamerunianus*. The leaves exhibit considerable variation in basal lobing from cordate to nearly sagittate and the leaves all possess thickened yellow margins. *Small* 585 (S.L., Gola Forest), with narrow cordate leaves lacking yellow margins, may also be this species.]
6. **C. lanceolatus** *Engl.* l.c. 115 (1911). An inadequately known species.
   **W.Cam.:** Buea *Reder* 1742!


# 187. LEMNACEAE

## By F. N. Hepper

Small to minute floating or submerged herbs without roots or roots simple and thread-like, capped. Flowers monoecious, naked or at first enclosed in a membranous sheath. Perianth absent. Male flowers: stamens 1–2, with slender filaments or the latter thickened in the middle or absent; anthers 1–2-locular. Female flowers: ovary sessile, 1-locular; style and stigma simple; ovules 1–7. Seeds with fleshy or no endosperm; embryo straight, axile.

Temperate and tropical regions; in fresh water. Duck-weeds are probably overlooked in many parts of W. Africa and they would repay careful attention in the field, especially for the presence of flowers which are seldom seen. (Hepper in Nig. Field 31: 18–21, with fig. (1966).)

Roots absent; flowers developing in upper surface of thallus; male flower solitary, anthers 1-locular:
  Thallus about 4 mm. long, thin, foliaceous, with internal air spaces; daughter thalli produced from linear transverse slit .. .. .. .. 1. **Wolffiella**
  Thallus about 1·5 mm. long, thick, more or less globular with fleshy parenchyma and without air spaces; daughter thalli produced from a round funnel-like pit
            2. **Wolffia**
Roots present; flowers developing in a lateral pouch; male flowers in pairs, anthers 2-locular:
  Roots solitary on each thallus; thallus obscurely 3-nerved; no basal leaf produced with daughter thalli .. .. .. .. .. .. .. .. 3. **Lemna**
  Roots several on each thallus, arising from a thickened part; thallus 5–11-nerved; young daughter thalli with a very thin leaf on each side at the base    4. **Spirodela**


## 1. WOLFFIELLA (Hegelm.) Hegelm. in Engl., Bot. Jahrb. 21: 303 (1895); Monod in Mém. Soc. Afr. Nord, Hors-Sér. 2: 229–242 (1949).

Thallus oblong, curved, concave, 4–6 mm. long, about 2 mm. broad, with undulate margin; daughter thallus budding from broader end; pair of flowers developed, one on each side of mid-line near broader end, 1 stamen with very short thick filament, 1 short style .. .. .. .. .. .. .. .. 1. *welwitschii*
Thallus in two parts; nearly opaque green floating portion, broadly elliptic, about 2 mm. long, 1·5 mm. broad, and hyaline oblong appendage 2–4 mm. long, 1 mm. broad, at right angles to the other and suspended in the water; budding pouch between the 2 parts arising from the thicker portion; flower apparently solitary near pouch, to one side of mid-line .. .. .. .. .. 2. *hyalina*

1. **W. welwitschii** (*Hegelm.*) *Monod* in Mém. Soc. Hist. Nat. Afr. Nord, Hors-Sér. 2: 229–242, figs. 1–41 (1949). *Wolffia welwitschii* Hegelm. in Seem., Journ. Bot. 3: 114 (1865); Hegelmaier, Lemnac. Monogr. 130, t. 4 figs. 1–10 (1868); F.T.A. 205; Koch in Ber. Schweiz. Bot. Ges. 41: 117 (1932); F. W. Andrews, Fl. Sudan 3: 284. Thallus usually curved in a semicircle, the broadest part slightly above the water surface and exposing stamen and style; daughter thallus budding from apex of broadest part and curving beneath older thallus and thereby more or less completing the circle.
   **Sen.:** near L. Ouörouai, Yombeul, Dakar *Pitot* (*fide* Monod l.c.). **Mali:** Soye, Bani Plain, Mopti (Dec.) *Monod*. **Ghana:** Akotokyir, Cape Coast (Sept.) *Hall* 3036! **S.Nig.:** Okomu F.R. *Brenan* 8788! Also in Sudan, Angola and reported from Venezuela, Cuba and S. Domingo.
2. **W. hyalina** (*Del.*) *Monod* l.c. 242 (1949). *Lemna hyalina* Del. (1813). *Wolffia delilei* Schleid. (1839)—F.T.A. 8: 204. *W. hyalina* (Del.) Hegelm., Lemnac. Monogr. 128, t. 4 figs. 11–19 (1868). *Lemna paucicostata* of Chev. Bot. 685, partly (?); of F.W.T.A., ed. 1, 2: 366 (*Chev.* 1143), not of Hegelm. Minute floating herb with colourless portion of the thallus suspended from the surface of the water.
   **Mali:** Djenné *Chev.* 1143! **N.Nig.:** Zaria *McFarlane*! Also in C. African Rep., Sudan and Kenya.

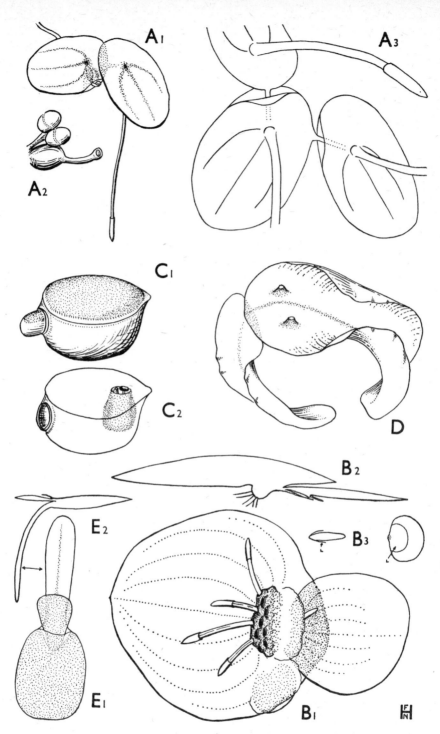

Fig. 361.—West african Lemnaceae.

A1, *Lemna paucicosta* Hegelm.—thallus, × 6.  A2, flowering, × 20.  A3, thalli from beneath.
B1, *Spirodela polyrhiza* (Linn.) Schleid.—thallus, × 8.  B2, section, × 8.  B3, resting bud,
× 6.  C1, *Wolffia arrhiza* (Linn.) Horkel ex Wimm.—thallus, × 20.  C2, flowering, × 20.
D, *Wolffiella welwitschii* (Hegelm.) Monod—thallus, × 12.  E1, *W. hyalina* (Del.) Monod—
thallus, × 12.  E2, in floating position.  A1, and A2, from *Hall* 3038.  A3, from *Brenan* 8768.
B1, and B2, from *Brenan* 4995.  C1, and C2, from *Hall* 3037.  D, from *Hall* 3036.  E1, from
*Chevalier* 1143.

**2. WOLFFIA** Horkel ex Schleid. in Linnaea 13: 389 (1839); F.T.A. 8: 203.

Free floating minute herb 1–1·5 mm. long, rather less in breadth and depth, upper surface flattened and dark green, paler elsewhere; flowers opening on to upper surface from a single median pit; style short, stigma concave; stamen solitary with filament 0·5 mm. long; fruit ellipsoid, 0·7 mm. long, erect  ..  ..  ..  ..  *arrhiza*

**W. arrhiza** (*Linn.*) *Horkel ex Wimm.* Fl. Schles. 140 (1857); Hegelmaier, Lemnac. Monogr. 124, t. 2 figs. 6–17, t. 3 figs. 1–12 (1868); Jumelle in Fl. Madag. 32: 2 (1937). *Lemna arrhiza* Linn. (1771). *Wolffia michelii* Schleid. (1844)—F.T.A. 8: 205. Locally abundant on permanent water, rarely flowering.
**Sen.:** Hann *Raynal* 5669! **Iv.C.:** Odienné *Aké Assi*! **Ghana:** Oduponkpehe, Accra to Winneba (fl. & fr. Sept.) *Hall* 3037! Mankessim *Hall* 2368! Weija *Hall* 3269! Akotokyir, near Cape Coast *Hall* 2379! **S.Nig.:** Yaba, Lagos *Killick* 288! Okomu F.R. *Brenan* 8596! Ibadan water works *Brenan* 8954! *Keay* FHI 22732! Almost cosmopolitan.

**3. LEMNA** Linn., Sp. Pl. 970 (1753); F.T.A. 8: 200.

Thallus flat, ovate-elliptic, 2–3·5 mm. long, 1·5–2 mm. broad, obscurely 3-nerved; solitary root arising towards one end; daughter thallus budding from near point of root insertion in the slit of the mother thallus; flowers in a lateral pouch; filament of stamen raises 2-locular anther above water surface; stigma concave

*paucicostata*

**L. paucicostata** *Hegelm. ex Engelm.* in Gray, Man. Bot., ed. 5, 681 (1867); Hegelm., Lemnac. Monogr. 139, t. 8 (1868); F.T.A. 8: 202; Chev. Bot. 685, partly (excl. *Chev.* 1143); Jumelle in Fl. Madag. 32: 2 (1937); Maheshawi in Am. J. Bot. 50: 677, 907 (1963). *L. aequinoctialis* of Chev. Bot. 685, not of Welw. Floating on the surface of still water, often in abundance; rarely flowering.
**Sen.:** Mare de Sill *Trochain* 3685! Yang Yang Naudi *Trochain* 4899! Sangalkam *Berhaut* 5267! **Mali:** Quignaka *Chev.* 259. Timbuktu *Chev.* 1288. Mopti *Chev.* 2604; 2605. **Guin.:** Koba *Jac.-Fél.* 7246! **S.L.:** Gbundapi *Adames* 72! **Iv.C.:** Mankono *Chev.* 21927! Rubino to Koubléké *Aké Assi* 7761! Soubré *Leeuwenberg* 4095! **U.Volta:** Azaguié (Sept.) *Aké Assi* 5953! **Ghana:** Akotokyir, near Cape Coast (Sept.). *Hall* 3038! Takoradi *Hall* 2421! Samreboi *Hall* 2489! Abura, near Cape Coast (Sept.) *Hall* 2349! **N.Nig.:** Jebba *Meikle* 1329! Babana to Kubi, Ilorin Prov. *Cook* 522! Zaria *McFarlane*! **S.Nig.:** Okomu F.R. *Brenan* 8768! Ibadan *Meikle* 955! Omo F.R. *Jones & Onochie* FHI 16688! Ikorodu to Lagos *Killick* 298! Widespread in the tropics and warm temperate zone.
[Hegelmaier segregated *L. paucicostata* from *L. perpusilla* Torrey (1843) on mainly ovule and seed characters which are usually not available. Following convention all our material is here named *L. paucicostata*. There may be some additional evidence for the maintenance of the species from recent chemotaxonomic studies—see McClure in Amer. J. Bot. 59: 849 (1966)—F.N.H.]

**4. SPIRODELA** Schleid. in Linnaea 13: 391 (1839). *Lemna* Linn. (partly)—F.T.A. 8: 200; F.W.T.A., ed. 1, 2: 266, partly.

Thallus orbicular-ovate, 3–8 mm. long, almost as broad, with 5–11 conspicuous nerves; roots 4 or more, simple; daughter thallus budding from near point of roots insertion in a slit in the mother thallus, with the thickened basal part of the thallus from which the roots also arise; flowers surrounded by a small open spathe in a lateral pouch with 1 pistillate flower and 2 or 3 staminate flowers each consisting of a solitary stamen; fruit wing-margined above, 1–2-ovuled  ..  ..  ..  ..  *polyrhiza*

**S. polyrhiza** (*Linn.*) *Schleid.* in Linnaea 13: 392 (1839); Lawalrée in Bull. Soc. Bot. Belg. 77: 27 (1945); Jumelle in Fl. Madag. 32: 3 (1937); Maheshwari in Nature 181: 1745 (1958). *Lemna polyrhiza* Linn. (1753)—F.T.A. 8: 201; Chev. Bot. 685; F.W.T.A., ed. 1, 2: 366; Clapham, Tutin & Warburg, Fl. Brit. Isles, ed. 2, 1053. Floating herb with roundish thallus grouped in colonies, often covering large areas of water; thallus green or pink tinged above, pink beneath.
**Sen.:** Sangalkam *Trochain* 3234! *Raynal* 5954! **Ghana:** Samreboi *Hall* 2486! **Dah.:** L. Azri, Zagnando *Chev.* 23044! **S.Nig.:** Okun-Owa, Ijebu-Ode *McFarlane*! Ikorodu to Lagos *Killick* 296! Onitsha *Barter* 583! Cosmopolitan.

## 188. TYPHACEAE

### By F. N. Hepper

Marsh or lake herbs with creeping rhizomes, often tall, with simple stems submerged at the base. Leaves mostly radical, elongated-linear, rather thick and spongy. Flowers unisexual, anemophilous, very numerous, densely crowded on a terminal spadix, the male and female similar, the male above, the female below, the two sexes contiguous or remote from each other. Perianth of very slender jointed threads or elongated spathulate scales mixed with imperfect ovaries or stamens. Male flowers: stamens 2–5; anthers linear, basifixed. Female flowers: ovary 1-locular, stipitate. Fruit dry, at length splitting. Seed with a striate testa and mealy endosperm.

The reed maces of marshes in temperate and tropical regions; only one genus.

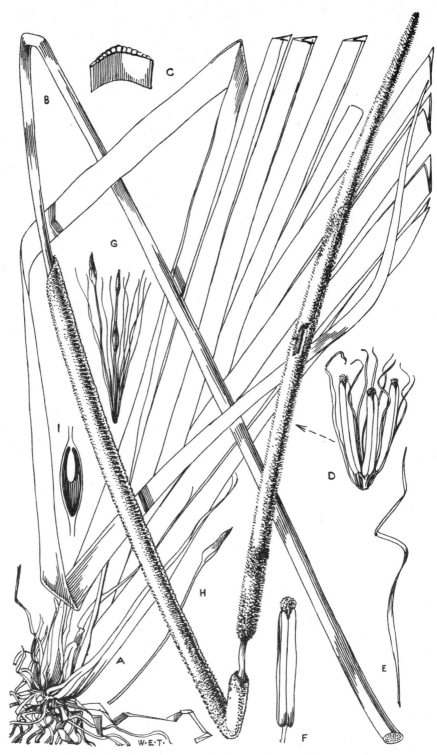

FIG. 362.—Typha australis *Schum. & Thonn.* (Typhaceae).

A, basal part of plant with most of leaves removed. B, spike. C, section of leaf. D, male flower.
E, perianth-segment. F, stamen. G, female flower. H, staminode. I, vertical section of
ovary.

130

**TYPHA** Linn., Sp. Pl. 971 (1753); F.T.A. 8: 135; Graebner in Engl., Pflanzenr. 4, 8 Typhac.: 8–18 (1900).

Bracteole absent from female flower; male and female portions of the inflorescence contiguous, female portion equalling or up to about twice as long as the male; anthers 2·5–3 mm. long; leaves smoothly rounded on the dorsal side, about 1 cm. broad
1. *latifolia*

Bracteole present on female flower; male and female portions of the inflorescence distinctly separated:

Dorsal side of leaf smoothly rounded, 1 cm. broad, elongate-linear, closely nerved, with thin acute margins; anthers 2 mm. long  ..    ..    ..    2. *australis*

Dorsal side of leaf angular especially towards the base where the leaf is more or less triangular in section, angle diminishing above, 1–1·5 cm. broad    3. *elephantina*

1. **T. latifolia** *Linn.* Sp. Pl. 971 (1753); Graebner in Engl., Pflanzenr. 4, 8 Typhac.: 8 (1900). Tall aquatic herb up to 9 ft. high; female inflorescence brown; in permanent water.
   **N.Nig.:** Katsina to Daura (Feb.) *Meikle* 1223! Widespread in the tropics and temperate regions.
2. **T. australis** *Schum. & Thonn.* Beskr. Guin. Pl. 401 (1827); F.T.A. 8: 135; Chev. Bot. 676; Berhaut, Fl. Sén. 186; Adam in Bull. I.F.A.N. 23: 399–405 (1961). *T. angustifolia* subsp. *australis* (Schum. & Thonn.) Graebn. l.c. 13 (1900). A tall aquatic herb up to 12 ft. high with very long narrow leaves; inflorescences brown on tall stiff unbranched peduncles; in permanent water.
   **Sen.:** *fide* Berhaut l.c. Dakar *Ndiaye* in *Hb. Chev.* 15855! *fide* Adam l.c. Thiès *de Wailly* 4597! **Mali:** Diré *Chev.* 43852! Diré to Bandiagara *Leclerq & Rogeon* 257! **Guin.:** Boké (Apr.) *Chillou*! **Ghana:** Quitta *Thonning*! Nkanfoa, Cape Coast (Dec.) *Hall* 2773! **Dah.:** Porto-Novo Lagoon (fr. Jan.) *Chev.* 22802! **Niger Rep.:** Kolo, nr. Niamey (Feb.) *Chev.* 42953! **N.Nig.:** Sokoto *Dalz.* 519! **S.Nig.:** Kradu Lagoon, Lagos *Barter* 3240! Lagos *Dalz.* 979! A tropical species, extending to E. and S. Africa. (See Appendix, p. 484.)
3. **T. elephantina** *Roxb.* Fl. Ind. 3: 566 (1832); Graebn. l.c. 11 (1900); Maire in Mém. Soc. Hist. Nat. Afr. Nord 3: 52 (1933); Adam l.c. (1961); A. Raynal in Ann. Fac. Sci. Dakar 9: 168 (1963). A herb growing on the edge of permanent water and invading bare or cultivated land.
   **Sen.:** M'Boro, near Dakar (Dec.) *Adam* 17806! Ouayembam *Trochain*. Other localities near Dakar *vide* Adam & Raynal l.c. A species with a subtropical distribution: also recorded from several localities in the Hoggar Mts. and Tibesti Mts., central Sahara, Egypt (?) and India.

## 189. AMARYLLIDACEAE

### By F. N. Hepper

Herbs with a tunicated bulbous rootstock or rarely a rhizome. Leaves few from the base of the stem or apex of the bulb, more or less linear, with parallel nerves and transverse secondary nerves. Flowers usually showy, bisexual actinomorphic, solitary to many and umbellate at the top of the scape, subtended by an involucre of two or more (rarely only one) usually membranous bracts. Perianth inserted below or usually above the ovary, petaloid, often withering and persisting, with or without a tube; segments or lobes 6, in 2 series, all equal and similar or the inner smaller or larger than the outer; corona often present. Stamens 6, opposite the segments, hypogynous or inserted on the tube or towards the base of the segments; filaments free or expanded at the base and connate and forming a " false " corona; anthers 2-locular, introrse, basifixed or versatile, opening by slits lengthwise. Ovary superior or inferior, 3-locular, with usually axile placentas; style slender, with a capitate or 3-lobed stigma. Ovules mostly numerous in each loculus. Fruit a capsule or a berry. Seeds with fleshy endosperm and small embryo, sometimes winged.

Temperate and warm-temperate regions, less frequent in the tropics.

Filaments free from each other to the base:
  Flowers numerous and rather small, scarlet, pedicellate; ovules 1–2 in each loculus
1. **Haemanthus**
  Flowers few and rather large, white or white with red stripes, sessile or subsessile; ovules usually several or numerous in each loculus  ..    ..    .. 2. **Crinum**
Filaments united with a basal cup to form a " false " corona; ovules numerous in each loculus of the ovary, in two series, spreading ..    ..    ..    3. **Pancratium**

Besides these indigenous genera the following are widely cultivated in our area as ornamental garden plants: *Eucharis grandiflora* Planch. & Lindl. (white, with a corona); *Hippeastrum equestre* (Ait.) Herb. the Harmattan Lily (red); *Hymenocallis littoralis* Salisb. the Spider Lily (white with a corona); *Zephyranthes citrina* Bak. (yellow), *Z. grandiflora* Lindl. (syn. *Z. carinata* Herb.), (pink), *Z. tubispatha* Herb. (white).

### 1. HAEMANTHUS Linn., Sp. Pl. 325 (1735); F.T.A. 7: 386.

Scape central, arising in the middle of the leaves but in No. 3 appearing lateral; rhizomatous:

Inflorescence remaining central among the leaves:
    Leaves almost acute, hardly acuminate at apex, cuneate at base, 18–33 cm. long,
      6–11 cm. broad, petioles not sheathed together; perianth-tube 4–8(–11) mm.
      long, perianth-segments (16–) 20–32 mm. long, (2–) 3–6 (–8) mm. broad; filaments
      (16–) 20–33 mm. long    ..    ..    ..    ..    ..    1. *cinnabarinus*
    Leaves cuspidate at apex, rounded at base, 18–24 cm. long, 5–8 cm. broad, petioles
      sheathed together into a false stem; perianth-tube 16–18 mm. long, perianth-
      segments 21–27 mm. long, 3–4 mm. broad; filaments 30 mm. long    ..    2. sp. *A*
Inflorescence-scape rupturing leaf sheath shortly above rhizome and thence appearing
    lateral; petioles closely sheathed together; lamina oblong, acute at apex; perianth-
    tube about 10 mm. long, perianth-segments 13 mm. long, 1·5 mm. broad; filaments
    20 mm. long    ..    ..    ..    ..    ..    ..    ..    ..    3. sp. *B*
Scape lateral to the leaves:
    Perianth-tube 18–28 mm. long, segments 27–30 mm. long, 2 mm. broad; filaments
    (23–) 26–33 mm. long, flowers numerous; bulb stout; leaves oblong-elliptic, acute
    to slightly mucronate, 13–23 cm. long, about 6 cm. broad, margin finely crinkled,
    sheathed together at the base    ..    ..    ..    ..    ..    4. *longitubus*
    Perianth-tube 5–16 mm. long; leaf margin not finely crinkled:
      Robust plant with bulb up to 4 cm. diam. and 6 cm. long; inflorescence with numerous
      flowers; perianth-tube 7–9 (–12) mm. long, segments 15–24 mm. long, 1·5 mm.
      broad; filaments 15–29 mm. long; leaves elliptic, 15–24 cm. long, 5–11 cm. broad,
      rounded and mucronate at apex    ..    ..    ..    ..    ..    5. *multiflorus*
      Slender plants with bulb about 1·5 cm. diam. and 2 cm. long; inflorescences few-
      flowered:
        Filaments as long as perianth-segments, 15–25 mm. long; perianth-tube 11–15 mm.
        long; leaves elliptic, about 11 cm. long and 6 mm. broad; forest species
                                                 6. *mannii*

        Filaments 16–23 mm. long, twice as long as perianth-segments, perianth-tube
        5–9 mm. long; leaves elliptic, 7–16 cm. long, 4–9 cm. broad, rounded and mucro-
        nate at apex, rounded at base; species of fringing forest in savanna    7. *rupestris*

1. **H. cinnabarinus** *Decne.* in Fl. des Serres t.1195 (1857); F.T.A. 7: 390; Morton, W. Afr. Lilies & Orch.
    fig. 34. *H. rotularis* Bak. (1877)—F.T.A. 7: 390. *H. longipes* Engl. (1886)—F.T.A. 7: 391. *H. brachy-*
    *andrus* Bak. in F.T.A. 7: 391 (1898); F.W.T.A., ed. 1, 2: 370. *H. germarianus* J. Braun & K. Schum.
    Mitth. Deutsch. Schutzgeb. 2: 146 (1889)—F.T.A. 7: 390. *H. kundianus* J. Braun & K. Schum. l.c.
    (1889)—F.T.A. 7: 389. Rhizomatous herb with stout central green scape up to 2 ft. high; inflorescence
    spherical with numerous pink or red flowers; fruits large, orange; in forest.
    **Iv.C.:** 61 km. N. of Sassandra, W. of Niapidou (Jan.) *Leeuwenberg* 2498! **Ghana:** *Burton & Cameron*!
    50 miles S.E. of Kumasi (Dec.) *Morton* A2548! Owabi (Dec.) *Lyon* 2881! Obuasi (July) *Chipp* 586!
    Asamankese (Dec., Jan.) *Plumptre* 61! 67! **S.Nig.:** Lagos *Schlechter* 13005! *Barter* 3416! Gambari, Oyo
    Prov. *W. D. MacGregor* 573! Akure Dist. (Feb.) *Ujor* FHI 32989! Orem, Calabar to Mamfe (Jan.)
    *Onochie & Okafor* FHI 36064x! **W.Cam.:** Rio del Rey *Johnston*! Ambas Bay (Feb.) *Mann* 779! Cam.
    Mt., 4,000 ft. (Feb.) *Mann* 1341! **F.Po:** Ureka (Feb.) *Guinea* 2367! Extending through much of tropical
    Africa. (See Appendix, p. 487.)
    [The type of *H. brachyandrus* Bak. is probably only a depauperate specimen of *H. cinnabarinus*—F.N.H.]
2. **H. sp. A.** *H. cinnabarinus* of F.W.T.A., ed. 1, 2: 370, partly (*Preuss* 874). Rhizomatous herb with
    slender central scape about 1 ft. high, inflorescence rather few-flowered, red; in upland forest.
    **S.Nig.:** Afi River F.R., Ogoja Prov. (May) *Jones & Onochie* FHI 17400! **W.Cam.:** *Preuss* 874! Buea
    (Apr.) *Hambler* 178! *Dundas* FHI 20387! *Gregory* 91!
3. **H. sp. B.** Inflorescence appearing lateral although arising centrally; in lowland forest.
    **S.Nig.:** Lagos *Phillips* 45! Ehor and Ibekewe, Benin Prov. *Fairbairn*! Shasha F.R., Oyo (Apr.) *Richards*
    3417! Iguobazowe F.R., Benin Prov. (May) *Onochie* FHI 34896! Usonigbe F.R., Benin Prov. (May)
    *Keay* FHI 37024!
    [Spp. A. and B. appear to be distinct species, but it is considered unwise to go further than this before
    the whole genus is revised.—F.N.H.]
4. **H. longit_bus** *C. H. Wright* in J. Linn. Soc. 37: 114 (1905). Stout bulbous herb with red flowers and
    spotted scape; in forest.
    **Lib.:** Bolahun, Montserrado Dist. (Jan.) *Bequaert* 43! Sinoe R. Basin *Whyte*! 18 miles N. of Tapeta
    (Feb.) *Voorhoeve* 168! Nimba (Jan.) *Adam* 20728! **Iv.C.:** Touba to Man (Feb.) *de Wit* 9186! Davo R.,
    E. of Beyo (Feb.) *Leeuwenberg* 2826! Gagnoa (Jan.) *de Wit* 9075! Angedédou Forest, nr. Abidjan (Feb.)
    *Leeuwenberg* 2647! Banco Forest (Dec.) *de Wit* 9856! **Ghana:** Dagou, nr. Axim (Apr.) *Chipp* 170!
    Eguafo (Apr.) *Hall* 2948! Nsawam to Aburi (Apr.) *Morton* A3652!
5. **H. multiflorus** *Martyn* Monogr., with fig. (1795); F.T.A. 7: 388; Bot. Mag. tt. 961, 1995; Chev. Bot.
    637, partly; Berhaut, Fl. Sén. 4, 182. Stout bulb giving rise to a lateral inflorescence 1–2 ft. high with
    red-blotched scape; flowers numerous, scarlet, in a spherical head; leaves sheathed together at the
    base into a false stem, fully expanding after flowering; in forest margin, secondary forest and in savanna
    woodland.
    **Sen.:** *Berhaut* 230. **Gam.:** Kuntaur *Ruxton* 9! Basse, W. Div. (June) *Frith* 82! **Guin.:** Los I. (June)
    *Kalbreyer* 230! **S.L.:** Freetown (May) *Morton* SL 1306! Mange, Port Loko Dist. (Apr.) *Hepper* 2608!
    Falaba (Mar.) *Sc. Elliot* 5462a! Roruks (Aug.) *Deighton* 5340! Njama, Kowa Dist. (May) *Deighton*
    5778! **Lib.:** Congotown (Mar.) *Barker* 1218! Sinoe Basin *Whyte*! **Iv.C.:** Mt. Mafa (Mar.) *Aké Assi*
    9533! **Ghana:** Sutawa (Mar.) *Thomas* D156! Elmina Plains (Feb.) *Morton* A1611! Legon Hill, Accra
    (Mar.) *Adams* 3809! **Togo Rep.:** *Warnecke* 92! **N.Nig.:** Kabba (July) *Clayton*! Zaria Prov. (May)
    *Elliott* 59! Kontagora (June) *Dalz.* 567! Musgu, Bornu Prov. (May) *E. Vogel* 100! Biu, Bornu Prov.
    (May) *Noble* 50! **S.Nig.:** Yoruba *Barter* 3423! *Millson* 65! Illaro (Jan.) *Millen* 113! Oyo to Ilorin
    (Mar.) *Meikle* 1293! Awka *Thomas* 130! **W.Cam.:** Bamali, Ndop Plain (Mar.) *Brunt* 156! Bamenda,
    5,000 ft. (Jan.) *Migeod* 319! **F.Po:** (Apr.) *Mann* 393! Widespread in tropical Africa. (See Appendix,
    p. 487.)
6. **H. mannii** *Bak.* in Bot. Mag. t. 6364 (1878); F.T.A. 7: 388. *H. cinnabarinus* of F.W.T.A., ed. 1, 2: 370,
    partly (syn., *Mann* 897). Small bulb giving rise to a slender lateral rather few-flowered scarlet
    inflorescence; in forest.
    **S.L.:** No. 2 River F.R., Peninsula (Apr.) *Hepper* 2523! Bagroo R. (Apr.) *Mann* 897!
7. **H. rupestris** *Bak.* in Gard. Chron. 7: 656 (1877); F.T.A. 7: 388; Morton, W. Afr. Lilies & Orch. fig. 33.

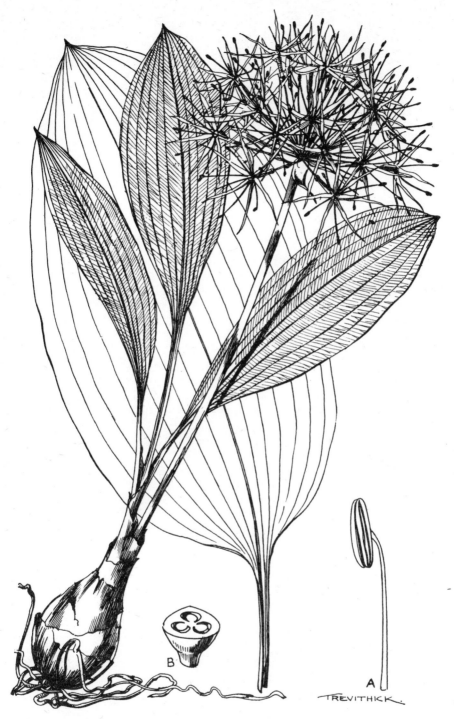

FIG. 363.—HAEMANTHUS RUPESTRIS *Bak.* (AMARYLLIDACEAE).
A, stamen. B, transverse section of ovary.

Small cylindrical bulb producing a lateral inflorescence 3–9 inches high, with rather few scarlet flowers; leaves small, expanding after the flowers; in fringing forest with seasonably flooded and sometimes rocky ground.

**Ghana:** Lawra (Sept.) *Hall* CC 773! Kpong to Akuse (Apr.) *Thomas* M17! Akuse (Apr.) *Morton* A4168! **N.Nig.:** Lokoja (Mar.) *Shaw* 6! Nupe *Barter* 1505! Anara F.R., Zaria Prov. (May) *Keay* FHI 19192! Jemaa to Jos, Assob R. (Apr.) *Keay & E. W. Jones* FHI 37621! Vango Malabu, Yola (May) *Dalz.* 252! (See Appendix, p. 487.)

## 2. CRINUM Linn., Sp. Pl. 291 (1753); F.T.A. 7: 393.

Plant submerged in running water; leaves linear, undulate, up to 100 cm. long and 5 cm. broad, usually much less; flowers 3–5 per umbel, held above water-level; spathe bracts narrowly triangular-lanceolate, 6–8 cm. long; perianth-tube 13–17 cm. long, lobes about 8 cm. long and 1 cm. broad .. .. .. .. 1. *natans*
Plant not normally submerged; leaves self-supporting:
  Perianth-segments 2–5 mm. broad, about 6 cm. long, less than half as long as the tube; anthers 15 mm. long; flowers 2–5 per umbel; spathe bracts linear-lanceolate, 7 cm. long, thin; leaves linear, 35–40 cm. long, 1·5–4 cm. broad    2. *purpurascens*
  Perianth-segments 1–4 cm. broad:
    Leaves 2·5–11 cm. broad; lanceolate to narrowly lanceolate-elliptic; flowers white or white and pink:
      Flowers white often tinged green; perianth-segments acute; leaf-margins smooth:
        Leaves glaucous, about 5 cm. broad, with numerous closely parallel nerves; perianth-segments 7–9 cm. long, 1·5–2·5 cm. broad, tube 6–9 (–13) cm. long
4. *glaucum*
        Leaves not glaucous, shining green when fresh, up to 11 cm. broad, with about 10 lateral nerves on each side of the midrib up to 5 mm. apart, tertiary nerves conspicuous; perianth-segments 7–9 cm. long, 2–4 cm. broad, tube 13–17 cm. long .. .. .. .. .. .. .. .. .. .. .. 5. *jagus*
      Flowers white with broad pink stripe along each perianth-segment, segments 8–12 cm. long, 2–4 cm. broad, cuspidate, tube 8–10 (–15) cm. long; anthers 18 mm. long; flowers 3–8 per umbel; leaves 3–4 cm. broad, with numerous close, parallel nerves, slightly scabrid-serrulate on the margins .. 3. *ornatum*
    Leaves 0·2–2 cm. broad, linear; flowers white with pink stripe along each perianth-segment:
      Perianth-tube 16–20 cm. long, segments 8–9 cm. long, 1·5–2·5 cm. broad; leaves not distichous, 9–15 mm. broad, up to 40 cm. long, margins slightly scabridulous; bulb about 3 cm. diam. .. .. .. .. .. .. 6. *scillifolium*
      Perianth-tube 6–11 cm. long:
        Leaves not distichous, very narrow, 2–7 mm. broad, about 25 cm. long; bulb about 3 cm. diam. with tubular neck; scape 10–17 cm. long; perianth-segments 6–8 cm. long, 1–1·5 cm. broad .. .. .. .. .. .. 7. *humile*
        Leaves distichous, 9–20 mm. broad, up to 70 cm. long but developing after flowering, margin serrulate-scabrid; bulb ovoid about 6 cm. diam.; scape up to 40 cm. long; perianth-segments 8–11 cm. long, 1–1·5 cm. broad .. 8. *distichum*

Besides the above native species, *C. lane-poolei* Hutch. (F.W.T.A. ed. 1, 2: 372 (1936), Engl. descr. only) is grown as an ornamental in Sierra Leone. It is a large plant with numerous white flowers, and it is probably an introduction from tropical America.

1. **C. natans** *Bak.* in F.T.A. 7: 396 (1898); Bot. Mag. t.7862; Chev. Bot. 636; Chev. in Rev. Bot. Appliq. 30: 624 (1950); Morton, W.Afr. Lilies & Orch. fig. 27. Aquatic, small bulb in mud at the bottom of perennial streams, with crinkled leaves submerged and floating; flowers large, white, fragrant, held well above the water-level.
  **Guin.:** Kolenté (June) *Chillou* 1581! **S.L.:** Port Loko (Apr.) *Hepper* 2669! Magburaka to Bendugu (Apr.) *Morton & Gledhill* SL 1161! Gondama, Maje (May) *Deighton* 5777! Franziga (Dec.) *Sc. Elliot* 4732! Mano (May) *Deighton* 3436! **Lib.:** *Linder* 63! Monrovia *Whyte*! Kakatown, Salala Dist. *Whyte*! (Feb.) *Harley* 1854! Sinoe Basin *Whyte*! **Iv.C.:** Adiopodoumé (Oct.) *J. de Wilde* 681! **Ghana:** *Burton & Cameron*! Axim to Takoradi (Mar.) *Hall* 2572! Antobam, nr. Prestea (July) *Vigne* 1273! Hintimbo, Dunkwa Dist. (Jan.) *C. J. Taylor* 5463! Nr. Dabouse, in a tributary of Prah R. (Mar.) *Fishlock* 86! **N.Nig.:** Maisamari, Mambila Dist., 5,000 ft. (Jan.) *Hepper & Chapman* 2805! **S.Nig.:** Ayisane, Lagos (Feb.) *Millen* 174! Sapoba Lake, Jamieson R. (Mar.) *Jones* FHI 737! Ethope R., nr. Sapele (Jan.) *Brenan* 9128! Calabar *Thomson* 84! Afi River F.R., Ogoja Prov. (May) *Jones & Onochie* FHI 18909! **W.Cam.:** Kumba (June) *Gregory* 332! Mbalange F.R., Kumba Dist. (Jan.) *Binuyo & Daramola* FHI 35464! Buea *Schlechter* 12851! **F.Po:** *Mann* 1416! Also in E. Cameroun. (See Appendix, p. 486.)
2. **C. purpurascens** *Herb.* Amaryllid. 250 (1837); F.T.A. 7: 396; Bot. Mag. t. 6525; Chev. l.c. 616 (1950). Small bulb; flowers with extremely narrow segments, white or purplish lined.
  **Gam.:** Dipa Kunda (July) *Fox* 147! **S.L.:** Kurusu, Limba (Apr.) *Sc. Elliot* 5545! **Ghana:** Ankobra R., Axim Dist. (Dec.) *Johnson* 867! **S.Nig.:** Bonny (Feb.) *Kalbreyer* 57! Ukpor (Jan.) *Kitson*! Abakaliki (Mar.) *Kitson*! Calabar *Robb*! Also in E. Cameroun and Angola.
3. **C. ornatum** *(Ait.) Bury* Hexandr. t. 18 (1834); Morton l.c. fig. 26. *Amaryllis ornata* Ait., Hort. Kew. 1: 418 (1789); Ker-Gawl. in Bot. Mag. t. 1171 (1809). *C. yucciflorum* Salisb., Parad. t. 52 (1806); F.T.A. 7: 399; F.W.T.A., ed. 1, 2: 372; Chev. Bot. 637; Chev. l.c. 622 (1950). *C. sanderanum* Bak. (1884)—F.T.A. 7: 400; F.W.T.A., ed. 1, 2: 372. Stout bulb giving rise to large more or less undulate leaves and an inflorescence 2–3 ft. high; flowers very ornamental, white with a broad pink stripe on each segment; beside streams in savanna.
  **Guin.:** Bafodeya to Farana (Mar.) *Sc. Elliot* 5155! **S.L.:** Hamilton (Apr.) *Hepper* 2530! Mahera, Kafu Bulom (cult. May) *Deighton* 5916! Kambia (May) *Morton & Gledhill* SL 1356! Njala (May) *Deighton* 1909! **Iv.C.:** Brafouédi Rock (Mar.) *Aké Assi* 8532! **Ghana:** Afram Plains (Feb.) *Cansdale* 11! Cape Coast (June) *Hall* 1467! Amedzofe (Jan.) *de Wit* 7802/612! Nungwa (May) *Irvine* 4819! Kpong (Apr.) *A. S. Thomas* M15! **Togo Rep.:** Lomé *Warnecke* 305! **Niger:** Niamey to Gao *Ryff*! **N.Nig.:** Nupe

FIG. 364.—CRINUM NATANS *Bak.* (AMARYLLIDACEAE).

*Barter* 1507! Anara F.R., Zaria Prov. (May) *Keay* FHI 22967! Panshanu Hills, Plateau Prov. (Apr.) *Okafor & Daramola* FHI 54651! Bichikki, Bornu Prov. (May) *Lely* 182! Abinsi *Dalz.* 858! **S.Nig.:** Badagry (cult. Jan.) *Millen* 90! Lagos (Sept.) *Dawodu* 43! **W.Cam.:** Bambalang, Ndop Plain (Apr.) *Brunt* 346! Extending throughout much of tropical Africa. (See Appendix, p. 486.)

4. **C. glaucum** *A. Chev.* in Mém. Soc. Bot. Fr. 2, 8: 212 (1910); Chev. Bot. 636. *C. yucciflorum* of F.W.T.A. ed. 1, 2: 372, partly (syn.). Bulbous plant with long glaucous leaves developing after flowering; inflorescence about 2 ft. high, flowers white; in moist places.
    **Port.G.:** Bambadinca (July) *Pereira* 3054! **Ghana:** Yeji (Apr.) *Hall* VBS 1275! **Dah.:** Savé to Ouémé R. Railway Bridge (May) *Chev.* 23581! **N.Nig.:** Jebba (Jan.) *Meikle* 1108! Nupe *Barter* 1503! Sherifuri, Bauchi Prov. (June) *Thornewill* 166!

5. **C. jagus** (*Thomps.*) *Dandy* in J. Bot. 77: 64 (1939); Chev. l.c. 623 (1950); Morton l.c. fig. 25. *Amaryllis jagus* Thomps. (1798). *C. giganteum* Andr., Bot. Rep. t. 169 (1801)—F.T.A. 7: 404; F.W.T.A. ed. 1, 2: 372; Chev. Bot. 635. *C. podophyllum* Bak. in Bot. Mag. t. 6483 (1880); F.T.A. 7: 403; Chev. Bot. 636. *C. suaveolens* A. Chev. in Mém. Soc. Bot. Fr., 2, 8: 212 (1912), and in Rev. Bot. Appliq. 30: 623, pl. 32 (1950). Stout bulb giving rise to broad bright green leaves and an inflorescence about 2 ft. high; flowers white tinged green; in swamps in forest.
    **S.L.:** Kamalu (May) *Thomas* 393! Njala (May) *Deighton* 684! Kangahun (Feb.) *Deighton* 6040! Hangha (Apr.) *Lane-Poole* 228! Gola Forest (Mar.) *Small* 544! **Lib.:** Kakata *Whyte*! Ganta *Harley*! **Iv.C.:** 15 km. S. of Taï (Mar.) *J. de Wilde* 3579! Bingerville to Potou Lagoon (Feb.) *Chev.* 20074! Lamé (Oct.) *Leeuwenberg* 1741! Yapo Forest (Feb.) *Aké Assi* 7350! **Ghana:** Ankaful, nr. Cape Coast (Mar.) *Hall* 2939! Assin-Yan-Kumasi *Cummins* 158! Swedru (Feb.) *Dalz.* 8301! Amedzofe (Jan.) *de Wit* 7808/608! Mampong (Mar.) *Deighton* 632! **N.Nig.:** Nupe *Barter* 1500! Likitaba to Nguroje, Mambila Plateau (July) *Chapman* 45! **S.Nig.:** Oyo to Iseyin (Apr.) *Keay & Polunin* FHI 37831! Ibadan (May) *A. H. Taylor* FHI 44564! Mawkawlawki, nr. Abeokuta (Dec.) *Burtt* 6! Port Harcourt to Owerri (Jan.) *Stubbings* 161! Calabar *Williams* 20!

6. **C. scillifolium** *A. Chev.* in Mém. Soc. Bot. Fr. 8: 210 (1912), and in Rev. Bot. Appliq. 30: 622 (1950). Narrow leaves arising from round bulb about 1 in. across with a characteristic tubular neck about as long as the bulb; scape about 9 in. high bearing 1 to 4 flowers with very long slender tubes, white with a pink median stripe; in seasonally flooded places in forest.
    **Lib.:** (Feb.) *Harley* 2209! **Iv.C.:** Taï (Mar.) *Leeuwenberg* 3041! Fort Binger, Mid. Cavally (July) *Chev.* 19499!

7. **C. humile** *A. Chev.* in Rev. Bot. Appliq. 30: 620, t. 29 (1950), as *humilis*. Grass-like leaves from a small round bulb with a tubular neck, flower solitary, white with pink median stripe; in savanna woodland in thin soil overlying rock.
    **Iv.C.:** Bouna Reserve (Apr.) *Aké Assi* 8653! Zaakro (Mar.) *Aké Assi* 8554! **U.Volta:** Fada-n-Gourma to Koupéla (July) *Chev.* 24530! **Ghana:** Bamboi to Bole (Apr.) *Morton* A3886!

8. **C. distichum** *Herb.* Amaryllid. 260 (1837); F.T.A. 7: 400. *Amaryllis ornata* of Ker-Gawl. in Bot. Mag. t. 1253, as to fig. only, not of Ait. *Crinum pauciflorum* Bak. (1878)—F.T.A. 7: 399; F.W.T.A., ed. 1, 2: 372. Narrow distichous leaves arising from a stout bulb; scape about 1 ft. high bearing 1–2 flowers, white with a pink median stripe; in seasonally flooded places with a thin soil layer over rock in transitional grassland.
    **Sen.:** Niokolo-Koba (June) *Adam* 14239! **Guin.:** Conakry (May) *Baldwin* 5792! **S.L.:** Freetown (May) *Morton* SL 1307! York (May) *Morton* SL 1319! *Deighton* 5542! Kambia (June) *Jordan* 881! 882! **Ghana:** Afram Plains (Mar.) *Johnson* 641! Kintampo Scarp (Mar.) *Hepper & Morton* A3186! Kwahu Tafo (Apr.) *Hall* 3007! Tamale (Mar.) *Adams* 3889! Bawku (Aug.) *Hall* CC 545! **N.Nig.:** Nupe *Barter* 1504! Mando, Birnin Gwari Dist. (June) *Keay* FHI 25864! Yola (Aug.) *Dalz.* 251! Extending to the Sudan. (See Appendix, p. 486.)


3. **PANCRATIUM** Linn., Sp. Pl. 290 (1753); F.T.A. 7: 406; Morton in Kew Bull. 19: 337–347 (1965).

Bulbs gregarious; leaves 5–15 per bulb, linear, 20–35 cm. long, erect, twisted, broadly U-shaped in section, 4–7 mm. broad, glabrous, leaf epidermal cells tapered to each end, about twice as wide in the middle with straight cell-walls and stomata wider than the tips of the epidermal cells; peduncle equalling or exceeding the perianth-tube (9·5–) 10–13 (–14) cm. long; perianth-lobes (5·5–) 6–7 (–7·5) cm. long, corona ⅔ length of the lobes with 2 triangular-acuminate lobes between each pair of stamens ..    ..    ..    ..    ..    ..    ..    ..    ..    1. *trianthum*
Bulbs solitary; leaves 1–5 per bulb, arching and not twisted, V-shaped in section, 5–10 mm. broad, often finely pubescent towards the base, leaf epidermal cells of equal width throughout their length with undulate walls and stomata equalling the epidermal cells in width; peduncle shorter than the perianth-tube; perianth-tube (9·5–) 12–13·5 (–15·5) cm. long; perianth-lobes (6·5–) 7–8·5 (–9·5) cm. long
                                                                          2. *hirtum*

1. **P. trianthum** *Herb.* in Ann. Nat. Hist. sér. 1, 4: 28 (1840); F.T.A. 7: 407; F.W.T.A., ed. 1, 2: 372, partly; Morton l.c. 340, fig. 2, pl. 13. A bulbous plant occurring in large numbers, with numerous erect dull and glaucous leaves; flowers white, opening at night, sweetly scented, pollen pale-yellow, perianth-tube glaucous green.
    **Sen.:** *Heudelot* 542! **Mali:** Timbuktu (July) *Hagerup* 237! **Iv.C.:** Zaakro (Mar.) *Aké Assi* 8553! **Ghana:** Damongo Scarp (Mar.) *Morton* GC 8690! Gambaga (Apr., May) *Morton* GC 7457! 9040! **N.Nig.:** Zungeru and Lokoja *Lugard*! Sherifuri (June) *Thornewill* 165! **S.Nig.:** Badagri (Jan.) *Millen* 128! Upper Ogun F.R., Oyo Dist. (Feb.) *Keay* FHI 23441! Widespread in Africa. (See Appendix, p. 487.)

2. **P. hirtum** *A. Chev.* in Mém. Soc. Bot. Fr. 2, 8: 88 (1908); Morton l.c. 337, fig. 1, pl. 12. *P. trianthum* of F.W.T.A., ed. 1, 2: 372, partly (*Warnecke* 93, *Sc. Elliot* 5213, 5287, *Dalz.* 253). Solitary bulbs 1–2 in. in diameter giving rise to a few arching glossy leaves; flowers white, opening at night, scented, pollen golden yellow.
    **Port. G.:** Bafata (May) *Esp. Santo* 2683! **S.L.:** Falaba (Mar.) *Sc. Elliot* 5213! 5287! **Iv.C.:** Bondoukou (Apr.) *Aké Assi* 8719! **Ghana:** Kintampo to New Longoro (Mar.) *Morton & Hepper* A3192! Ejura (Feb.) *Andoh* FH 5644! Dodowah to Senchi (Oct.) *Morton* A1687! Dodowah to Prampram *Irvine* 2797. **Togo Rep.:** Lomé *Warnecke* 93! **N.Nig.:** Nupe *Barter* 1501! Lokoja (Apr.) *Elliott* 225! Ancho, Plateau Prov. (Apr.) *Hepburn* 140! Magana, Jemaa Dist. (June) *Summerhayes* 55! 50 miles from Maifoni, Bornu Prov. *Parsons*! Yola (Apr.) *Dalz.* 253! **S.Nig.:** Abeokuta *Barter* 3328! Ibadan (Mar.) *Meikle* 1266! Iseyin (Apr.) *Hambler* 420! Also in C. African Republic.

Fig. 365.—Pancratium hirtum *A. Chev.* (Amaryllidaceae).
A, stamen.

137

## 190. IRIDACEAE

### By F. N. Hepper

Perennial herbs with underground rhizomes, corms or bulbs. Leaves often crowded at the base of the stem, mostly linear, flattened at the sides, sheathing at the base and equitant. Flowers hermaphrodite, actinomorphic, with a straight perianth-tube or the tube curved, or completely zygomorphic, usually ornamental and often mottled or spotted. Perianth petaloid, withering and persisting for some time; segments or lobes 6, 2-seriate, subequal and similar or different. Stamens 3, opposite the outer perianth-lobes; anthers 2-locular, opening by slits. Ovary inferior, 3-locular with axile, or 1-locular with parietal placentas; style 3-lobed, lobes sometimes petaloid. Ovules numerous. Capsule loculicidally dehiscent by valves. Seeds with copious endosperm.

Generally distributed throughout the world: most of the genera occuring in West Africa have many species in the southern part of the continent.

Style-branches petaloid; stamens opposite the style-branches; *Iris*-like flowers

　　　　　　　　　　　　　　　　　　　　　　　　　　　　　　　1. **Moraea**

Style-branches not petaloid; stamens alternate with the style-branches:
Spathes many-flowered; stems compressed or winged　　..　　..　　.. 2. **Aristea**
Spathes 1-flowered; stems not compressed:
　Flowers solitary, not spicate; leaves filiform-setaceous　..　　..　　3. **Romulea**
　Flowers spicate or cymose; leaves flat:
　　Perianth-tube elongated and slender, many times as long as the limb; segments
　　　subequal　..　　..　　..　　..　　..　　..　　..　　4. **Acidanthera**
　　Perianth-tube comparatively short:
　　　Style-branches bifid; flowers cymose　..　　..　　..　　5. **Lapeirousia**
　　　Style-branches or style entire:
　　　　Perianth-segments subequal; seeds globose, not winged　..　6. **Hesperantha**
　　　　Perianth-segments unequal; seeds usually winged:
　　　　　Perianth-limb oblique, not bilabiate　..　　..　　..　　..　　7. **Gladiolus**
　　　　　Perianth-limb bilabiate, the upper lobe hood-like, the 4 lower shorter and
　　　　　　recurved　..　　..　　..　　..　　..　　..　　..　　8. **Zygotritonia**

Besides the above indigenous genera, several others are cultivated and some are more or less naturalized, such as monbretia (*Crocosmia aurea* Planch.) on Cameroon Mt., *Neomarica caerulea* (Lodd.) Sprague, *N. gracilis* (Herb.) Sprague and *Trimezia martinicensis* (Jacq.) Herb. All are from the New World.

### 1. MORAEA Miller, Fig. Pl. 159, t. 238 (1758); F.T.A. 7: 338. *Nom. cons.*

Corm 2 cm. diam., giving rise to a peduncle about 30 cm. high and later to narrowly linear-lanceolate leaves up to 100 cm. long and 1 cm. broad, closely nerved, glabrous, rigid; stem-leaves bract-like, shortly overlapping each other and submembranous towards the apex; outer spathe-bract shorter than the inner, about 8 cm. long, closely nerved; ovary slightly exserted from the spathe, 2 cm. long; perianth 4–4·5 cm. long, mauve; capsule long-pedunculate, about 3·5 cm. long　　*schimperi*

M. schimperi (*Hochst.*) *Pichi-Serm.* in Webbia 7: 349 (1950). *Hymenostigma schimperi* Hochst. in Flora 27: 24 (1844). *Moraea diversifolia* Bak. (1877)—F.T.A. 7: 339. *M. zambesiaca* Bak. (1877)—F.T.A. 7: 339; F.W.T.A., ed. 1, 2: 374. Highland plant with 1–3 mauve Iris-like flowers on stiff stems about 1 ft. high; flowers appearing after fire in the dry season, followed by the leaves.
N.Nig.: Vom, Jos Plateau, 4,000 ft. (Feb.) *McClintock* 199! *Dent Young* 245! Vogel Peak, Sardauna Prov., 5,400 ft. (Nov.) *Hepper* 1504! Mambila Plateau, 4,900–6,000 ft. (Jan.) *Latilo & Daramola* FHI 34380! *Hepper* 1774! W.Cam.: Bamenda to Santa (Jan.) *Keay* FHI 28345! Widespread in the mountains of tropical Africa.

### 2. ARISTEA Ait., Hort. Kew. 1: 67 (1789); F.T.A. 7: 346; H. Weimarck in Lunds Univ. Årssk. N.F. Avd. 2, 36: 1–140 (1940).

Capsules ovoid or oblong-ellipsoid about 5–9 mm. long, beaked by the spirally twisted perianth:
Clusters of flowers 7–10 in an elongated spike, axis of inflorescence slender, only lower portion slightly winged, inflorescence sometimes branched; basal leaves linear, up to 30 cm. long, stem leaves 4–6 reducing in size upwards, bracts subtending flower cluster about 8 mm. long, as long as the cluster; capsule 5–7 mm. long, 3·5–4 mm. broad; (Sect. *Aristea*)　..　　..　　..　　..　　..　　1. *angolensis*
Clusters of flowers 1–2, stems compressed and conspicuously winged; leaves linear up to 20 cm. long, 5 mm. broad, closely nerved; outer subtending spathe-bract leafy, longer than the flower cluster; floral bracts membranous; capsule 6–9 mm. long, 4–6 mm. broad; (Sect. *Ancipites*)　..　　..　　.. 2a. *alata* var. *abyssinica*

Capsules cylindrical, elongate-oblong, 2 cm. long, valves with a deep groove in the middle; clusters of flowers 5–8 in an elongated spike; leaves about 30 cm. long, up to 1 cm. broad; (Sect. *Pseudaristea*) .. .. .. .. .. 3. *ecklonii*

1. **A. angolensis** *Bak.* in Trans. Linn. Soc., ser. 2, 1: 270 (1878); F.T.A. 7: 347; Weimarck l.c. 18, fig. 2 1-p (1940). *Anthericum djalonis* A. Chev., Bot. 648 (1920), name only, and Hutch., F.W.T.A., ed. 1, 2: 341 (1936), Engl. descr. only. *Aristea djalonis* Hutch., F.W.T.A., ed. 1, 2: 376 (1936), Engl. descr. only, and Kew Bull. 1939: 246. Slender herb 1–3 ft. high, with grass-like leaves, inflorescence sometimes branched; flowers bright or pale bluish; in rocky highland.
    **Guin.:** Dalaba-Diaguissa Plateau (fl. June, fr. Sept., Oct., Apr.) *Chev.* 12652. 18859. *Adames* 277! **S.L.:** Bintumane Peak, summit 6,000 ft. (May) *Deighton* 5099! **N.Nig.:** Jos Plateau (Aug.) *Lely* P631! Bukuru to Hepham, 4,300 ft. (July) *Lely* 342! Naraguta F.R. (Aug.) *Daramola* FHI 44139! Mambila Plateau, 5,500 ft. (June) *Chapman* 75! Also in Kenya, Rhodesia, Malawi, Angola, Swaziland, Transvaal and Natal.

2. **A. alata** *Bak.* in J. Linn. Soc. 21: 405 (1885); F.T.A. 7: 347; Weimarck l.c. 42, fig. 8a–e. In Tanzania and Kenya.

2a. **A. alata** subsp. **abyssinica** (*Pax*) *H. Weim.* l.c. 44 (1940). *A. abyssinica* Pax (1892). *A. johnstoniana* Rendle in Trans. Linn. Soc., ser. 2, 4: 48 (1894); F.T.A. 7: 346; F.W.T.A., ed. 1, 2: 374. Herb with a markedly winged inflorescence 6 in.–2 ft. high bearing solitary or paired clusters of bright blue or purple flowers; in rocky hills.
    **N.Nig.:** Jos Plateau (June) *Lely* P338! Vom *Dent Young* 240! Bukuru to Hepham (July) *Lely* 342a! Mambila Plateau (fl. & fr. June) *Chapman* 74! **W.Cam.:** Bapinyi Pass, Bamenda (May) *Ujor* FHI 30395! L. Bambulue (June) *Saxer* 48! Bum to Nchan (fl. & fr. June) *Maitland* 1600! Also in Ethiopia, the Congo and highland in E. Africa southwards to Rhodesia.

3. **A. ecklonii** *Bak.* in J. Linn. Soc. 16: 112 (1878); Fl. Capensis 6: 54; Weimarck l.c. 58, fig. 14 f–k. *A. maitlandii* Hutch., F.W.T.A., ed. 1, 2: 374 (1936), Engl. descr. only. In clumps 2 ft. high; in wet upland grassland.
    **W.Cam.:** above Buea, 4,500 ft. (fr. Mar., July) *Maitland* 27! *Morton* K459! Extending to Uganda and to many parts of S. Africa.

### 3. ROMULEA Maratti, Diss. Romul. et Saturn. Romae 13, t. 1 (1772); F.T.A. 7: 344. *Nom. cons.*

Corm subglobose, about 2 cm. diam., with smooth scales; leaves filiform-setaceous, up to 30 cm. long, ribbed, glabrous; spathe-bracts 2, equal, lanceolate, closely nerved, 1·5 cm. long, acutely pointed; perianth 2 cm. long; fruit about 1 cm. long; seeds broadly obovoid-globose, flattened on two sides, nearly black, 2 mm. diam.

*camerooniana*

**R. camerooniana** *Bak.* in J. Bot. 14: 236 (1876). Slender herb with fine leaves up to 1 ft. high; flowers white with blue margins.
    **W.Cam.:** Cam. Mt., 7,700–11,500 ft. (fl. Sept., fr. Nov., Dec.) *Mann* 2135! *Johnston* 19! *Maitland* 845! *Boughey*! Bafut-Ngemba F.R., Bamenda (July, Aug.) *Ujor* FHI 29996! *Brunt* 791! Also in E. Cameroun.

### 4. ACIDANTHERA Hochst. in Flora 27: 25 (1844); F.T.A. 7: 358.

Spathe-bract ¼–⅓ as long as the perianth-tube, linear, 4–6 cm. long, reddish; corm finely reticulate, ovoid; leaves very long and rather weak, 1–2 cm. broad; flowers few in a lax one-sided spike; nodes of inflorescence about 2·5–4 cm. apart; perianth white, the limb blotched with purple in the lower part; tube slender, about 12–14 cm. long, limb 4·5 cm. long; anthers exserted, 2 cm. long; capsule 2·5 cm. long
1. *aequinoctialis*
Spathe-bract about half as long as the perianth-tube, linear-lanceolate, 7–8 cm. long; leaves 1–1·5 cm. broad, with 3–5 prominent nerves; flowers about 5 in the spike; nodes 2–3 cm. apart; perianth white, with purple streaks; tube 12–13 cm. long, limb 4 cm. long; anthers exserted .. .. .. .. .. .. 2. *divina*

1. **A. aequinoctialis** (*Herb.*) *Bak.* in J. Linn. Soc. 16: 160 (1877); F.T.A. 7: 358; Bot. Mag. t. 7393 (1895). *Gladiolus aequinoctialis* Herb. in Bot. Reg., Misc. No. 97 (1842). Stem about 1 ft. high from a corm; flowers showy, 2 in. across, white with a mauve centre; in rock crevices.
    **Guin.:** Nimba Mts. (July) *Schnell* 1561! N'Zérékoré (Sept.) *Baldwin* 13285! **S.L.:** Sugar Loaf Mt., 1,500–3,000 ft. (Oct.) *Sc. Elliot* 3954! *T. S. Jones* 250! Gumah (Nov.) *Lane Poole* 81! Konta to Bumban (Aug.) *Deighton* 1244! Tingi Mts. (Feb.) *Morton & Gledhill* SL 1904! Bintumane Peak, 6,000 ft. (May) *Deighton* 5087! **Iv.C.:** Nimba Mts., 2,500–5,000 ft. (Aug.) *Boughey* GC 18051! Tonkoui Mt. (Oct.) *Aké Assi* 70151! **W.Cam.:** Bambili, Bamenda Dist. (Aug.) *Ujor* FHI 29970!

2. **A. divina** *Vaupel* in Notizbl. Bot. Gart. Berl. 7: 375 (1920).
    **F.Po:** above Basilé, north side of St. Isabel Peak, 7,000–8,500 ft. (Nov.) *Mildbr.* 7175! Also in E. Cameroun.
    [Although *A. divina* is distinguishable from *A. aequinoctialis* by the characters given in the key, they may in fact be taxonomically insufficient to maintain it as a distinct species. It should also be noted that the plant known as *A. bicolor* Hochst. (1844) is hardly distinct from *A. aequinoctialis*, which has the priority of name.—F.N.H.]

### 5. LAPEIROUSIA Pourr. in Mém. Acad. Toul. 3: 79, t. 6 (1788); F.T.A. 7: 350, as *Lapeyrousia.*

Stems flexuous, about 3-leaved, ribbed; leaves linear, very acute, up to 15 cm. long, 2 mm. broad, 3-nerved, glabrous; cymes few-flowered, dichotomously branched; spathe-bracts ovate-elliptic, 5 mm. long; perianth about 2·3 cm. long, lobed to the middle; capsule globose, shorter than the bracts .. .. .. .. *rhodesiana*

FIG. 366.—ACIDANTHERA AEQUINOCTIALIS (*Herb.*) *Bak.* (IRIDACEAE).
A and B, stamens.  C, style-arms.

140

**L. rhodesiana** *N. E. Br.* in Kew Bull. 1906: 169.  *L. montana* Hutch. in Kew Bull. 1921: 403; F.W.T.A., ed. 1, 2: 376, not of Klatt (1882).
**N.Nig.:** Liruwen-Kano Hills *Carpenter!* Jos Plateau (May) *Lely* P259! Vom, 3,000–4,500 ft. *Dent Young* 244! Naraguta (June) *Lely* 271! Zaranda Mt., 5,800 ft. (May) *Lely* 189! Rishi, Lame Dist., 3,200 ft. (June) *G. V. Summerhayes* 56! Also in Tanzania and Rhodesia.


## 6. HESPERANTHA Ker-Gawl. in Kon. & Sims, Ann. Bot. 1: 224 (1804); F.T.A. 7: 348.

Corm very small, subglobose; leaves few, cauline, narrowly linear, flat, up to 17 cm. long, 1-nerved, grass-like; inflorescence about 3-flowered; outer spathe-bract broadly lanceolate, 1·5–2 cm. long, several-nerved, shortly tubular at the base; perianth 2–2·5 cm. long, segments as long as the tube; capsule 1·5 cm. long, valves 3-nerved; seeds globose, 1·5 mm. diam.  ..  ..  ..  ..  .. *alpina*

**H. alpina** (*Hook. f.*) *Pax ex Engl.* in Abh. Preuss. Akad, Wiss. 1891: 174 (1892); F.T.A. 7: 348.  *Geissorhiza alpina* Hook. f. in J. Linn. Soc. 7: 223 (1864).  Simple erect herb nearly 1 ft. high; flowers pink.
**W.Cam.:** Cam. Mt., 8,000–10,000 ft. (fl. & fr. Nov.) *Mann* 2134! *Johnston* 18! *Preuss* 968.


## 7. GLADIOLUS Linn., Sp. Pl. 36 (1753); F.T.A. 7: 360.

Flowering stems with well-developed leaves:
  Leaves over 1 cm. broad; inflorescences about 1 m. high:
    Perianth (incl. tube) over 6 cm. long, yellow or yellow with orange spots or reddish areas, upper segment of inner whorl hooded; bracts about 6 cm. long; leaves 1·5–4 cm. broad; capsule 2–4·5 cm. long, transversely nerved..  1. *psittacinus*
    Perianth (incl. tube) up to 4 cm. long, purple, all segments spreading; bracts about 3 cm. long, closely nerved; leaves 1–1·5 cm. broad; capsule 2 cm. long, faintly transversely nerved ..  ..  ..  ..  ..  ..  ..  2. *klattianus*
  Leaves 0·5–1 cm. broad; inflorescences much less than 1 m. high; flowers not hooded:
    Perianth (incl. tube) about 4 cm. long, purple or white, segments obtuse; bracts 2–4 cm. long, closely nerved; capsule 1·5 cm. long, smooth; leaves nearly 1 cm. broad, rigid ..  ..  ..  ..  ..  ..  ..  3. *gregarius*
    Perianth (incl. tube) over 6 cm. long, pink, campanulate, segments acute; bracts 4–5 cm. long, with spaced nerves; capsule 2 cm. long, thin; leaves flaccid, up to 5 mm. broad ..  ..  ..  ..  ..  ..  ..  4. *oligophlebius*
Flowering stems with short scale-like leaves or very short leaves:
  Bracts about as long as the perianth-tube, 1·5 cm. long; flowers mauve-pink or white, usually numerous on the very slender spike; leaves very short and closely appressed to the stem; perianth about 3 cm. long; anthers 7 mm. long; capsule 1·5–2 cm. long, closely mottled with carmine, thin, not transversely nerved  5. *unguiculatus*
  Bracts much longer than the perianth-tube, about 3 cm. long; capsules strongly transversely nerved; leafy shoots arising alongside the inflorescences, often tomentose:
    Flowers not hooded, 3 (–6) cm. long (incl. tube), pale pink or vermillion, sometimes darker spotted  ..  ..  ..  ..  ..  ..  ..  6. *melleri*
    Flowers hooded, 6–8 cm. long (incl. tube), yellow with reddish markings and spots (see note after citations)  ..  ..  ..  ..  ..  ..  1. *psittacinus*

1. **G. psittacinus** *Hook.* Bot. Mag. t. 3032 (1830).  *G. quartinianus* A. Rich. (1851)—F.T.A. 7: 371; F.W.T.A., ed. 1, 2: 379.  *G. primulinus* Bak. (1890)—F.W.T.A., ed. 1, 2: 379; Morton, W. Afr. Lilies & Orch. fig. 45.  *G. buettneri* Pax (1892)—Chev. Bot. 633.  *G. occidentalis* A. Chev., Bot. 634, name only.  Erect rather robust herb 2–4 ft. high with sword-like leaves; flowers large and showy, pure yellow or variously spotted with red or heavily mottled (or rarely white); in rocky places.
   **Mali:** Bamako (July) *Waterlot* 1518!  **Guin.:** Dindea (June) *Pobéguin!*  Fabala, near Kérouané (July) *Collenette* 76!  Botola, near Kankan (July) *Collenette* 79!  **S.L.:** Freetown (Aug.) *Burbridge* 506!  Bintumane Peak, 5,000 ft. (May) *Deighton* 5086!  **Iv.C.:** Man to Touba (July) *Collenette* 53!  Gouékouma, Toura *Chev.* 21664!  Mt. Dourou (May) *Chev.* 21725!  **U.Volta:** Banfora (July) *Aké Assi* 5703!  Gourma (July) *Chev.* 24488!  **Ghana:** Yeji (Aug.) *Pomeroy* FH 1347!  Kumawu (Aug.) *Cansdale* FH 4429!  Logba Tota (June) *Howes* 927!  **Togo Rep.:** Bismarkburg (July) *Buettner* 8!  **Dah.:** Agouagon (July) *Annet* 20!  Savalou (June) *Annet* 125!  **N.Nig.:** Mando F.R., Birnin Gwari Dist. (Aug.) *Keay* FHI 28005!  Vom, Jos Plateau, 3,000–4,500 ft. *Dent Young* 241!  Zaranda, Jemaa Dist. (Sept.) *G. V. Summerhayes* 36!  Yola (Aug.) *Dalz.* 250!  Maisamari, Mambila Plateau (June) *Chapman* 51!  **S.Nig.:** Olokemeji (June) *Foster* 305!  Widespread in savanna in tropical and S. Africa.
   [Smaller plants with pure yellow flowers are conveniently known as *G. primulinus* Bak., which appears to be only a colour form of *G. psittacinus*.  Plants with leafless stems, sterile tomentose leafy shoots, commonly occur in the Bamenda area and also in highland E. Africa: they fall into the *G. psittacinus* complex.]
2. **G. klattianus** *Hutch.* F.W.T.A., ed. 1, 2: 379 (1936); Aké Assi, Contrib. 2: 24 [233].  *G. spicatus* Klatt (1867)—F.T.A. 7: 369, not of Linn. (1753).  Erect herb 2–4 ft. high with numerous reddish purple flowers in a long narrow spike; in seasonally moist savanna.
   **Sen.:** Bignona, Casamance (Nov.) *Berhaut* 6439!  **Mali:** Klela (Sept.) *Demange* 2754!  **Port.G.:** Piche (Sept.) *Pereira* 3246!  Piche to Canquelffá (Oct.) *Esp. Santo* 3108!  **Guin.:** Kindia (Oct.) *Jac.-Fél.* 1859!  Tinn 'ti Oule, near Béyla (July) *Collenette* 81!  **S.L.:** Suribolomia (Nov.) *T. S. Jones* 42!  Kambia, Magbema (Oct.) *Jordan* 945!  **Iv.C.:** Sémien to Sifié (Oct.) *Aké Assi* 6054!  **Ghana:** Larabanga (Oct.) *Rose Innes* GC 31006!  Banda (Nov.) *T. M. Harris!*  Ejura *Cox* 27!  Maliato, Togo Plateau (Nov.) *Morton!*  **Togo Rep.:** Büttner 143!  Banka *Barter!*  Zungeru (Sept.) *Dalz.* 252!  Anara F.R., Zaria (July) *Keay* FHI 25952!  Tangale to Waja (July) *G. V. Summerhayes* 71!  **S.Nig.:** Ago-Are F.R., Shaki Dist. (Oct.) *Sofoluwe* FHI 38175!  Olokemeji F.R. (Aug.) *Keay* FHI 25384!  Extends to Congo.

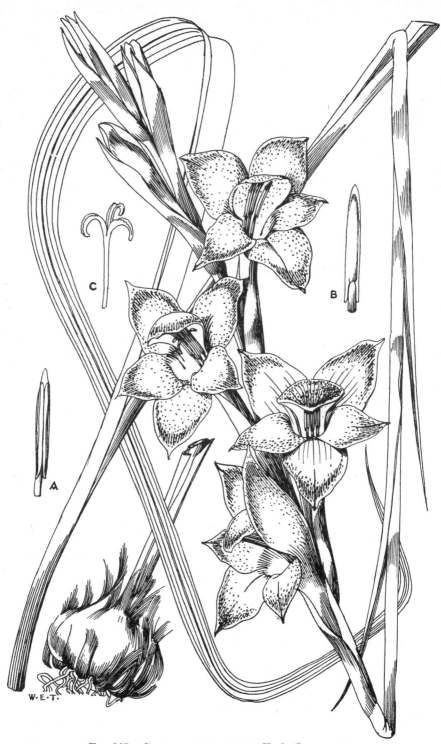

FIG. 367.—Gladiolus psittacinus *Hook*. (Iridaceae).
A and B, stamens.  C, style-arms.

Fig. 368.—Gladiolus klattianus *Hutch.* (Iridaceae).
A and B, stamens. C, style-arms. D, fruit.

143

3. **G. gregarius** *Welw. ex Bak.* in Trans. Linn. Soc. ser. 2, 1: 268 (1878); F.T.A. 7: 365; Hepper in Kew Bull. 21: 493 (1968). *G. pseudogregarius* Mildbr. ex Hutch., F.W.T.A., ed. 1, 2: 379 (1936), English descr. only. A slender herb with very narrow leaves; flowers purple or white with purple throat.
**Mali**: Sadam, near Sikasso (Sept.) *Jaeger* 5140! **U.Volta**: Takalédougou (Sept.) *Aké Assi* 6343! **Ghana**: *Lloyd Williams* 554! Tono, Navrongo (Aug.) *Irvine* 4663! **Togo Rep.**: Basari (July) *Kersting* 128! Baflo (Aug.) *Davidson* 22! **N.Nig.**: Kaciya to Zonkwa (July) *G. V. Summerhayes* 130! Amo, Jos Plateau, 4,000 ft. (Aug.) *King* in *Hb. Hepper* 2871! Vom, 3,000–4,500 ft. *Dent Young* 242! Hepham to Ropp, 4,600 ft. (July) *Lely* 357! Fobur to Korom, Jarawa Dist., 3,300 ft. (July) *G. V. Summerhayes* 15! **W.Cam.**: Bafut-Ngemba F.R. (Aug.) *Tamajong* FHI 26894! Binka, 6,500 ft. (Sept.) *Savory* UCI 368! Bamenda Nkwe to Bafut-Ngemba F.R. (Aug.) *Daramola* FHI 43210! Extends to Angola and possibly Rhodesia.
[The plants from W. Cameroun have white flowers with a purple throat; all the others have purple flowers.]
4. **G. oligophlebius** *Bak.* in Kew Bull. 1895: 73; F.T.A. 7: 367. *G. caudatus* Bak. (1895)—F.T.A. 7: 367. A herb with grass-like leaves 1–2 ft. long, the same length as the inflorescence; flowers 2 in. long, pink with yellow inside; in rock crevices and shallow soil.
**Mali**: Kita (July) *Duong* 519! **N.Nig.**: Mada Hills (Aug.) *Hepburn* 76! Amo, Jos Plateau, 4,000 ft. (Aug.) *King* in *Hb. Hepper* 2870! Jarawa Hills, Bauchi Prov., 5,000 ft. (Aug.) *Lawlor & Hall* 353! Extends to Rhodesia.
5. **G. unguiculatus** *Bak.* in J. Linn. Soc. 16: 178 (1877); F.T.A. 7: 372; Mildbr. in Engl., Bot. Jahrb. 58: 232, for full synonymy. *Antholyza labiata* Pax (1892)—F.T.A. 7: 374. *A. fleuryi, A. djalonensis* and *A. sudanica* A. Chev., Bot. 634, names only. Slender herb about 1 ft. high; flowers dull purple fading to white; in moist ground in savanna.
**Sen.**: Niokolo-Koba (June) *Adam* 14201! **Mali**: Simona to Toro (June) *Chev.* 977! **Guin.**: Near Pellel Bantam, Pita (July) *Adames* 299! Nzo Mt., 5,400 ft. (Mar.) *Chev.* 21024! **S.L.**: *Morson*! Fontani, Bombali Dist. (May) *Scotland* in *Hb. Deighton* 713! Yiffin to Banda Karafaia (Apr.) *Deighton* 5064! Bintumane Peak, 6,000 ft. (May) *Deighton* 5091! **Ghana**: Wa (Apr.) *Morton* A3316! Nangodi (June) *Vigne* FH 4512! Tamale (Apr.) *Vigne* FH 1699! Kete Krache (Apr.) *Cox* 111! Afram Plains *Johnson* 1071! Amedzofe Hill (May) *Morton* 7216! **N.Nig.**: Sokoto *Dalz.* 448! Anara F.R., Zaria (May) *Keay* FHI 22901! Vom, 3,000–4,500 ft. *Dent Young* 243! Tangale to Waja (May) *G. V. Summerhayes* 65! Abinsi & Katsina Ala (June, July) *Dalz.* 845! **S.Nig.**: Udi Highlands (May) *Kitson*! Abakaliki to Obubra (Apr.) *Kitson*! **W.Cam.**: Bum to Nehan, 4,000–5,000 ft. (June) *Maitland* 1777! Bambalong, Ndop Plain, 3,800 ft. (Apr.) *Brunt* 344! Bamessi, Ndop Plain, 3,800 ft. (Mar.) *Brunt* 262a! Bamenda, 5,000 ft. (Feb.) *Migeod* 460! Bamenda Ngure to Bafut-Ngemba (Apr.) *Ujor* FHI 30036! Widespread in tropical Africa.
6. **G. melleri** *Bak.* in J. Bot. 14: 334 (1876); Bot. Mag. t. 8626; F.T.A. 7: 362. A herb with almost leafless stems 2–3 ft. high and pale pink flowers; in rocky upland savanna.
**N.Nig.**: Jos Plateau (fl. & fr. May) *Lely* P284! Naranda Mt., summit 5,800 ft. (May) *Lely* 190!
[Plants occurring in Zaria Prov. (*Keay* FHI 25812, 25870; *G. V. Summerhayes* 192) have densely tomentose sterile shoots but they appear to be part of the *G. melleri* complex.]

## 8. ZYGOTRITONIA Mildbr. in Engl., Bot. Jahrb. 58: 230 (1923).

Leaves very narrow, at most 5 mm. broad, about 15 cm. long, those on the stem gradually smaller, acute; flowers in simple or branched spikes much longer than the basal leaves; perianth very zygomorphic, about 1·5 cm. long, the upper lobe very narrow and hood-like over the style and stamens, the remainder shorter and recurved; capsule 3-lobed, about 5 mm. long, slightly reticulate .. .. .. 1. *praecox*
Leaves about 2 cm. broad, up to 30 cm. long, mostly about 2 towards the base of the stem, acute; flowers in a simple or 3-branched panicle, very similar to the above, but a little larger; capsule warted .. .. .. .. .. 2. *crocea*

1. **Z. praecox** *Stapf* in Hook., Ic. Pl. sub. t. 3120 (1927). Corm about 1 in. diam., covered with reticulate-fibrous tunics; flowering while nearly leafless, flowers white; capsule slightly pustulate; in alluvial soil among grass.
**Mali**: Kita (Aug.) *Jaeger* 8! **Guin.**: Bissikrima *Pobéguin* 1123, partly! **N.Nig.**: Abinsi (June) *Dalz.* 847!
2. **Z. crocea** *Stapf* in Hook., Ic. Pl. t. 3120 (1927). Corm covered with a rather coarsely reticulate tunic; leaves and flowers appearing together; flowers yellow-green or tinged red; in shallow soil in savanna.
**Mali**: Sikasso (May) *Demange* 2214! **Guin.**: Bissikrima *Pobéguin* 1123, partly! **Ghana**: Nalerugu, Gambaga Dist. (June) *Akpabla* 689! **N.Nig.**: Zungeru (June) *Dalz.* 558! Mande F.R., Birnin Gwari Dist. (June) *Keay* FHI 25854! Ancho (May) *Hepburn* 142! Katsina (June) *Dalz.* 848! Magaji *de Leeuw* 812b! **S.Nig.**: Enugu *Floyer* 4! Nsukka (July) *Tuley* 801! Ngwo, Udi Dist. (Aug.) *Jones* FHI 6165!
[Very close to the Sudanese *Z. bongensis* (Pax) Mildbr. (1923) from which it differs in the curled lower perianth segments and, possibly, in having broader leaves—F.N.H.]

# 191. DIOSCOREACEAE

## By J. Miège

Climbers (at least the West African species), spiny or not, annual or perennial with tubers annually renewed or perennial. Tubers toxic or edible, often protected by thorny roots. Aerial tubers (bulbils) present or absent. Stems glabrous or pilose. Leaves alternate or opposite (sometimes both on the same plant), often cordate, entire or lobulate, more or less digitately nerved or palmately compound, acumen often large and glandulose. Petiole generally twisted and sometimes jointed at the base or with more or less leathery auricles. Basal leaves often reduced. Leaves moving, following the conditions of lighting. Plants dioecious; exceptionally on the same inflorescence are clustered male and female flowers*. Inflorescence spicate, racemose or paniculate. Flowers small, inconspicuous, actinomorphic. Perianth campanulate or spreading, 6-lobed, lobes

* For example *D. cayenensis* (*Thomas* 8; *Agric. Serv. Nigeria* 6).

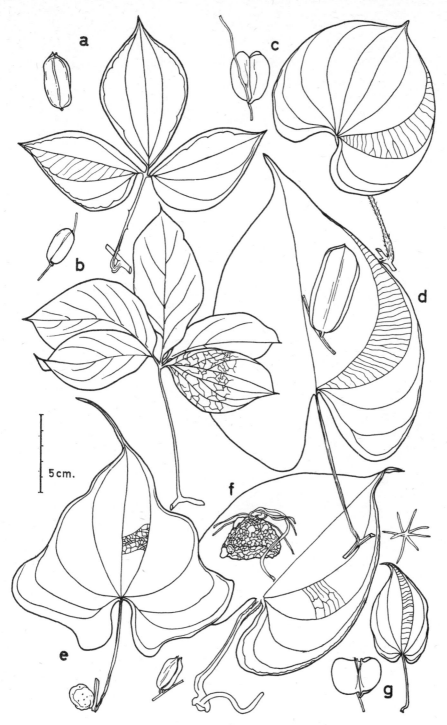

FIG. 369.—DIOSCOREA SPP. (DIOSCOREACEAE).

a, *D. dumetorum* (Kunth) Pax—leaf and fruit.  b, *D. quartiniana* A. Rich.—leaf and fruit.
c, *D. esculenta* (Lour.) Burkill—leaf and fruit.  d, *D. preussii* Pax—leaf and fruit.  e,
*D. sansibarensis* Pax—leaf and bulbil.  f, *D. bulbifera* Linn.—leaf, fruit and bulbil.  g, *D.*
*hirtiflora* Benth.—leaf, stellate hair and fruit.

145

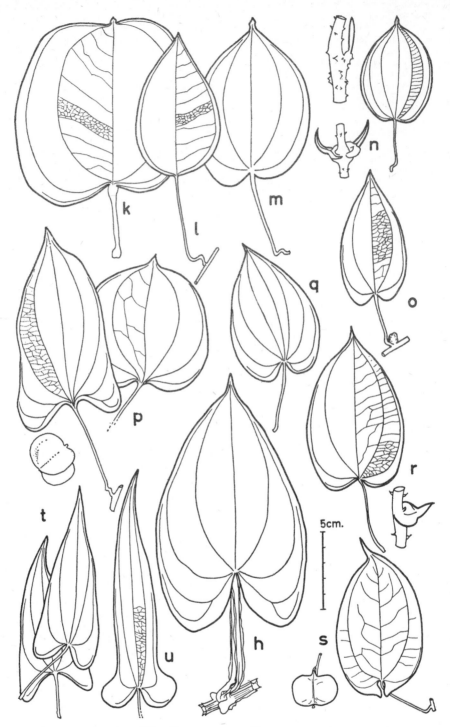

FIG. 370.—DIOSCOREA SPP. (DIOSCOREACEAE).

h, *D. alata* Linn.—leaf. k, *D. minutiflora* Engl.—leaf. l, *D. smilacifolia* De Wild.—leaf. m,
*D. burkilliana* J. Miège—leaf. n, *D. mangenotiana* J. Miège—normal and reduced basal
leaves. o, *D. togoensis* Knuth—leaf and bulbil. p, *D. cayenensis* Lam.—leaves and fruit.
q, *D. abyssinica* Hochst. ex Kunth—leaf. r, *D. praehensilis* Benth.—normal and reduced
leaves. s, *D. liebrechtsiana* De Wild.—leaf and fruit. t, *D. lecardii* De Wild.—leaves. u, *D.
sagittifolia* Pax—leaf.

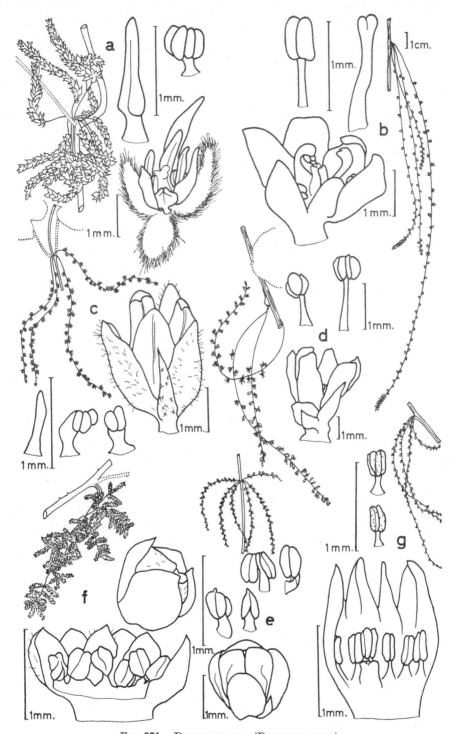

FIG. 371.—DIOSCOREA SPP. (DIOSCOREACEAE).

a, *D. quartiniana* A. Rich.— ♂ inflorescence, ♂ flower, stamen and staminode. b, *D. preussii* Pax— ♂ inflorescence, ♂ flower, stamen and staminode. c, *D. hirtiflora* Benth.— ♂ inflorescence, ♂ flower, stamen and staminode. d, *D. sansibarensis* Pax— ♂ inflorescence, ♂ flower, stamens. e, *D. liebrechtsiana* De Wild.— ♂ inflorescence, ♂ flower, stamens. f, *D. dumetorum* (Kunth) Pax.—♂ inflorescence, flower bud, flower. g, *D. bulbifera* Linn.— ♂ inflorescence, flower, stamens.

2-seriate, often connate at the base. *Male inflorescences:* spikes generally several in the leaf axils, sometimes clustered in racemes or compound panicles. Male flowers sessile or shortly pedicelled. Bracteole generally present in oblique position. Flowers placed singly along the axis or in short few-flowered lateral cymules. Stamens 6, or 3 with or without 3 staminodes. Filaments free or shortly connate; anthers 2-locular. Rudimentary ovary frequent. *Female inflorescences:* spikes looser, longer than the male ones, solitary or paired or sometimes more numerous, in the leaf axils. Staminodes 0, 3 or 6. Ovary inferior, 3-locular. Style 3, free or connate. Placentation axile: 2 anatropous ovules in each loculus. Fruits (in the tropical African species) 3-valved capsules. Seeds winged, with albumen.

Distributed in all the intertropical countries; also present in the subtropical and temperate regions. The tubers (yams) of some species of *Dioscorea* form an important food crop.

**DIOSCOREA** Linn., Sp. Pl. 1032 (1753); Kunth, Enum. Pl. 5: 325 (1850); Pax in E. & P., Pflanzenfam. 2, 5: 133 (1887); F.T.A. 7: 414 (1898); Knuth in Engl., Pflanzenr. Diosc. 4, 43 (1924); Prain & Burkill in Ann. Roy. Bot. Gard. Calcutta 14: 302 (1938); Burkill in Bull. Jard. Bot. Brux. 15: 345 (1939).

KEY TO PLANTS WITH FEMALE (♀) FLOWERS OR FRUITS OR ONLY IN VEGETATIVE CONDITION

Twining stems sinistrorse (twining to the left i.e. clockwise as seen from above); capsules longer than broad with negative geotropism:
  Leaves compound, palmate; seeds winged at one end only:
    Plants spiny and pubescent (some varieties glabrescent); leaves trifoliolate; lateral leaflets asymetrical, sometimes (principally the basal ones) lobulate or bilobed, median leaflet oblong, 8–18 cm. long, 5–10 cm. broad, 3-nerved from the base, setulose-pubescent above, softly tomentose beneath; petioles often spiny; ♀ flowers in slender axillary spikes, softly tomentose all over; fruits oblong, 4 cm. long, 2 cm. broad (Fig. 369a) ..    ..    ..    ..    1. *dumetorum*
    Plants not spiny, more or less pubescent; leaves (3–) 5 (–7) foliolate, central leaflet 4–10 cm. long, 2–5 cm. broad with one median nerve, minutely and sparsely pubescent above, shortly and sparsely pubescent beneath or glabrescent; petioles not prickly; ♀ flowers in short solitary or geminate-pedunculate spikes; fruits oblong-elliptic, 20–25 by 12–16 mm. (Fig. 369b) ..    ..    ..    2. *quartiniana*
  Leaves entire, generally deeply cordate; seeds winged almost all round:
    Plants with spiny stems and roots, pubescent (T-shaped hairs); leaves as broad as or broader than long, heart-shaped, 9–15-nerved, 6–10 cm. long, 8–10 cm. broad; petiole spinulescent; plants sterile in West Africa, otherwise ♀ flowers shortly pedicellate (1mm.) in long spikes (up to 40 cm.), perianth 3 mm. long, very small staminodes present; fruits reflexed 25–27 mm. long (Fig. 369c)    3. *esculenta*
    Plants not spiny:
      Stems often winged or angular; plants robust, tomentose (T-shaped hairs more or less caducous), some forms glabrescent; leaves broadly ovate, cordate at the base, obliquely acuminate, acumen elongate, 9 main nerves and numerous transverse nerves, 10–30 cm. long, 8–35 cm. broad, villous-tomentose beneath, more or less pubescent or glabrous above; ♀ inflorescences solitary, pendulous, up to 40 cm. long; fruits ascending, oblong, 50–60 by 30–40 mm. narrowly winged, pilose at first then glabrous; seeds with a prominent wing at each end (Fig. 369d)    4. *preussii*
      Stems terete sometimes slightly furrowed, unarmed; plants glabrous, with bulbils; ♀ inflorescence pendulous; ♀ flowers subsessile; fruits ascending:
        Leaves opposite in all mature plants except at the base of the stems; basal leaves often lobate or lobulate, the others entire, cordate at the apex bearing a long (up to 6 cm.) thick, glandular drip-tip; 9–11 main nerves, 10–20 cm. long, 12–45 cm. broad; petioles as long as the blades, with basal, thickened auricles; ♀ inflorescences 1–3 per node, up to 80 cm. long; ♀ flowers subsessile, with six staminodes; fruits large, 50–60 by 18–25 mm.; seeds winged at both ends (Fig. 371d) ..    ..    ..    ..    ..    ..    5. *sansibarensis*
        Leaves entire, always alternate, broadly ovate, widely cordate at the base and apiculate to acuminate at the apex, 6–20 cm. long, 6–20 cm. broad; petiole slightly winged at the margins and auriculate at the base; ♀ inflorescences 1–6 per node, up to 25 cm. long; fruits oblong-ellipsoid, 20–30 by 12–16 mm. (Fig. 371g) ..    ..    ..    ..    ..    ..    6. *bulbifera*

Twining stems dextrorse (twining to the right, i.e. anticlockwise as seen from above);
    fruits as broad as or broader than long; seeds attached near the middle of the placenta,
    annular more or less circular wing:
Plants pubescent, covered with stellate or branched hairs; lamina entire, cordiform
    with 7–9 main nerves; fruits reflexed (negative geotropism):
Leaves minutely puberulous above (sometimes glabrescent), lamina 6–12 cm. long and
    nearly as broad; ♀ spikes 1–5, 10–20 cm. long; perianth and ovary tomentose,
    fruits rounded, 18–20 by 20–25 mm.; seeds with rounded wings (Fig. 371c)
                                                                               7. *hirtiflora*
Leaves with stellate indumentum, lamina 6–15 cm. long and nearly as broad; ♀
    inflorescences 10–15 cm. long; fruits about 25 mm. long with trapezoidal wings;
    seeds winged all round    ..   ..   ..   ..   ..   ..   8. *schimperana*
Plants glabrous except for a few club-shaped hairs, sparse or very fugacious juvenile
    hairs; capsules not reflexed (positive geotropism):
Stems quadrangular, 4-winged; leaves opposite, sometimes alternate (at the base) or
    verticillate in 3's or 4's deeply cordate, elongate acuminate, broadly ovate, sinus
    deep, sometimes with overlapping lobes, 7-nerved, 6–22 cm. long, 5–18 cm. broad;
    ♀ spikes simple, solitary, axillary, glabrous; fruits emarginate, 20–25 mm. long,
    30–35 mm. broad; seeds suborbicular, winged all round (Fig. 370h) ..   9. *alata*
Stems terete, not winged, sometimes fluted or with narrow furrows:
Stems perennial; stems very spiny especially in the basal area, rooting at the
    nodes when these are in contact with the soil; leaves coriaceous, epidermis hard to
    remove; rootstock hard, woody at soil level but bearing tubers deep in the
    ground:
Leaves orbicular or suborbicular, 5–7-nerved, veins from each side of the midrib
    embracing a broad obovate area, apices often overlapping, principally on those of
    the base, acumen abrupt, short (2–5 mm.), cornet-shaped, glandulose at least on
    the basal leaves, sinus shallow or absent, margins rounded overlapping slightly;
    lamina 6–17 cm. long, 6–17 cm. broad; fruits pruinose, 25–30 by 40–45 mm.
    (Fig. 370k) ..   ..   ..   ..   ..   ..   ..   10. *minutiflora*
Leaves ovate or ovate-lanceolate or ovate-oblong, 3-ribbed, the two outer ones
    very close to the margin of the leaf, rounded or rounded-cuneate at the base,
    sometimes indentated; lamina 6–10 cm. long, 4–7 cm. broad; fruits pruinose,
    23–28 by 32–42 mm. (Fig. 370l) ..   ..   ..   ..   11. *smilacifolia*
Stems annual:
Tubers perennial; stems spiny:
Stems 15–25 mm. thick, very young branches with a light fugaceous pubescence;
    leaves firm to coriaceous, similar to those of No. 10 but less hard and with a flat
    acumen not rolled to form a cornet and more or less spathulate-lanceolate, thick,
    glandulose, reddish or yellowish; lamina 5–12 cm. long, 4–10 cm. broad, 5–7-
    ribbed, network less dense than that of No. 10 with which this species shows
    great resemblance (Fig. 370m) ..   ..   ..   ..   ..   12. *burkilliana*
Stems 25–50 mm. thick, towards the base; basal leaves very modified, reduced
    to a pair of leathery auricles terminating in a very long acumen with reflexed
    margins making a kind of glandulose and thickened spindle; leaves alternate at
    the base, then opposite, not coriaceous, shortly cordate, 5–7-ribbed, 5–12 cm.
    long, 4–10 cm. broad; ♀ flowers with 6 minute staminodes; fruits 20–27 mm.
    long, 30–36 mm. broad (Fig. 370n) ..   ..   ..   13. *mangenotiana*
Tubers annually replaced; leaves herbaceous or firm, not coriaceous:
Plants of small stature bearing bulbils, stems wiry, mottled with red; leaves
    glaucous, more or less cordate to rounded at the base, often mucronate at the
    apex; lamina firm, thickish, 6–10 cm. long, 3·5–7 cm. broad, 5–7-ribbed,
    venation hardly visible; fruits rather small, 17–21 by 22–26 mm. (Fig. 370o)
                                                                  14. *togoensis*

Plants without bulbils:
Cultivated plants with superficial tubers; leaves frequently modified, reduced,
    often alternate at the base, opposite above; lamina sinus rather deep but with
    a deltoid attachment to the petiole, 5–7 nerved, the veins on either side of the
    midrib forming an acute angle (less than 45–50°), 6–12 cm. long, 5–10 cm.
    broad; 1–2 short ♀ spikes (rarely exceeding 10 cm. in length), few-flowered,
    axillary (Fig. 370p) ..   ..   ..   ..   ..   15. *cayenensis*
Resting period short, two harvests in the season
Resting period long, one harvest in the season
                                  15a. *cayenensis* subsp. *cayenensis*
                                  15b. *cayenensis* subsp. *rotundata*
Wild plants, tubers deep-seated or else protected by spiny roots; leaves
    herbaceous:
Leaf length (measured along the midrib) less than 1·8 times the width, maximum
    width above the lowest quarter of the midrib (somewhere near the middle):

Plants non-spiny; leaves cordate-ovate, basal lobes rounded, venation on either side of the midrib fairly close; tubers deep-seated (Fig. 370q)
         16. *abyssinica*

Stems and roots spiny; basal leaves often modified and reduced to stipular formations (connivent when the leaves are opposite) and to an acumen of variable length:

Plants sturdy, stems often pruinose, brown, red or purplish, stem 15–20 mm. thick; leaves shortly cordate (Fig. 370r)    ..    17. *praehensilis*

Plants weaker, green; thickness of the stem generally less than 15 mm.; leaves deeply cordate more or less hastate, greatest width distinctly above the petiolar attachment (Fig. 370s)    ..    ..    18. *liebrechtsiana*

Leaves narrow, their length (measured along the midrib) equal to or more than 1·8 times the width, the maximum width in the lowest quarter of the lamina, or at the petiolar attachment, or even lower:

Stems unarmed; leaves lanceolate, generally opposite, 5–7-ribbed, 8–12 cm. long, 3·5–6 cm. broad (Fig. 370t) ..    ..    ..    19. *lecardii*

Stems spiny at the least at the base; leaves long, sagittate, generally alternate (but not always), 5–7-ribbed, basal lobes of the lamina divergent, 9–16 cm. long, 3–6 cm. broad (Fig. 370u)    ..    ..    ..    20. *sagittifolia*

KEY TO PLANTS WITH MALE (♂) FLOWERS

Male flowers with 3 fertile stamens and 3 staminodes; axis of the inflorescence pilose:

Inflorescences dense, formed of short axillary tomentose fasciculate spikes, axis not visible between the flowers; bracteoles absent; bract embracing the flower; staminodes linear or petaloid, with an entire or bifid apex; anthers subsessile, on very short filaments (Fig. 371a)    ..    ..    ..    ..    2. *quartiniana*

Inflorescences more or less lax, axis generally visible between the flowers; bract not embracing the flower:

Simple or branched racemes, 20–25 cm. long, clustered in the leaf axils; bract subulate; bracteole present; flowers 3–4 mm. distinctly pedicellate; perianth villose (T-shaped hairs); flowers opening wide at anthesis, possessing a small disk; stamens with short filaments curved over the disk; staminodes linear, at times bifid in the upper part (Fig. 371b) ..    ..    ..    ..    ..    4. *preussii*

Spikes 5–10 cm. long, fasciculate, 4–6 in the leaf axils; inflorescences forming long pendent more or less leafy racemes; axis and perianth covered, as on the stems, with characteristic stellate or dendroid hairs; flowers shortly or scarcely pedicelled, 3–5 mm. diam., solitary along the axis; at anthesis, sepals forming a roof, coherent at the tip, free in the middle part; stamens with a short filament and often a well-developed connective, staminodes long (Fig. 371c) ..    7. *hirtiflora*

Male flowers with six fertile stamens:

Inflorescence-axis pubescent:

Inflorescence an axillary or terminal much-branched panicle of dense spikes, spikes formed of the cymules of 2–6 flowers crowded together, sessile, obscuring the axis; flowers small with glabrous perianth, protected by very pubescent bracts; perianth opening little or very little; stamens 6, at least in the first flowers of the cymules, in the successive flowers 4 or 5 stamens or fewer, filaments short, shorter or equal to the anthers (Fig. 371f)    ..    ..    ..    ..    ..    1. *dumetorum*

Inflorescences lax, axis visible between the flowers; flowers sessile and isolated on the axis:

Perianth villous outside (hairs not stellate); stamens inserted on the tube, rudimentary ovary subpyramidal; flowers over 3 mm. long, sterile; bulbils absent (Fig. 369c) ..    ..    ..    ..    ..    ..    3. *esculenta*

Perianth hairs stellate; flowers 2·5 mm. long or less at anthesis; bracts equal to or more than half the flower length; bulbils present    ..    8. *schimperana*

Inflorescence axis glabrous:

Flowers elongate, more than 3 mm. long, white at anthesis; bulbils present:

Spikes 20–40 cm. long, single or geminate, between the purple bulbil and the branches or forming a terminal panicle, rhachis striate or angular often with nectariferous glands; flowers solitary or 2–4 in cymules, 5–6 mm., subsessile, parts free at the top but not opening wide at anthesis; filaments of the stamens long (Fig. 371d)
         5. *sansibarensis*

Spikes shorter, 3–12 cm. long, often united to form more complex inflorescences, aphyllous, about 20 cm. long or sometimes shorter; individual spikes pendulous or recurved toward the stem; flowers sessile, 3–6 mm. long; stamens of the outer whorl longer than those of the inner one (Fig. 371g) ..    6. *bulbifera*

Flowers spherical, more or less fleshy, sessile, glabrous, small, 2·5 mm. long or less, perianth not opening wide at anthesis; bulbils absent:

Stems winged, quadrangular; spikes in axillary panicles, aphyllous, 15 cm. long or more; spikes with 12–20 flowers alternately disposed on a zigzag axis; perianth

1·5–2 mm. long, disk thick; sepals ovate-elliptic; petals subspathulate; stamens
  short; anthers globose (Fig. 370h)     ..    ..    ..    ..    ..    9. *alata*
Stems terete:
  Male flowers on slender axes and rather less than their own diameter apart; leaves
    herbaceous or firm (plant not coarse); tertiary nervation relatively sparse,
    epidermis easily removed; spikes in clusters of 2–4 in the leaf axils; often
    forming large aphyllous inflorescences at least at their extremities:
    Stems spiny:
      Stem more than 2·5 cm. diam.; floriferous spikes with negative geotropism;
        tubers perennial; roots spiny; forest plant (Fig. 370n)  .. 13. *mangenotiana*
      Stems 2·5 cm. diam. or less; floriferous spikes with absent or positive geotropism;
        tubers replaced annually; roots spiny:
        Leaves cordiform or more or less triangular, less than 1·8 times as long as
          broad:
          Male flowers with a narrow base (Fig. 371e)     ..    .. 18. *liebrechtsiana*
          Male flowers with a broad base (Fig. 370r) ..    ..    .. 17. *praehensilis*
        Leaves sagittate, 1·8 times or more longer than broad (Fig. 370u)
                                                              20. *sagittifolia*
    Stems and roots not spiny; tubers annual:
      Leaves cordate or more or less triangular, less than 1·8 times as long as broad
        (Fig. 370q)  ..    ..    ..    ..    ..    ..    ..    16. *abyssinica*
      Leaves long, sagittate or hastate, 1·8 times or more longer than broad (Fig. 370t)
                                                              19. *lecardii*
  Male flowers generally on stout rigid axes of approximately the same diameter or
    rather more; geotropism nil:
    Leaves herbaceous or firm, tertiary network of veinlets relatively diffuse:
      Bulbils small, 4–16 mm., with 1 or rarely 2 buds between the leaves and spikes;
        stems mottled, green and red; spikes 3–6 cm. long, 2–4 spikes in leaf-axils;
        leaves generally opposite at the inflorescence level, not spiny (Fig. 370o)
                                                              14. *togoensis*

      Bulbils absent, plants unarmed or spiny, many cultivars and clones more or less
        hybrid; flowers in spikes, few in number (1–2 (–3) in the leaf-axils); spikes
        relatively short, rarely more than 5–6 cm. long, with a tendency to the
        reduction of sexual reproduction (Fig. 370p)     ..    .. 15. *cayenensis*
    Leaves coriaceous, nerve network dense, epidermis removed with difficulty or
      not at all; floriferous axes stout, divergent, usually 8 or more per axil; inflores-
      cences generally complex, aphyllous, sometimes branched; rootstock woody,
      hard, with fluted tubers at the base; stems spiny:
      Aerial system annual; stems not rooting at the nodes (Fig. 370m)
                                                              12. *burkilliana*
      Aerial system perennial; stems partially creeping on the ground, easily rooting at
        basal nodes:
        Primary veins of leaves 5–7, the first pair embracing a broadly lanceolate area;
          lamina cordate, orbicular to oblong (Fig. 370k)  ..    .. 10. *minutiflora*
        Primary veins of leaves usually 3, the first pair near the leaf margin, lamina
          elliptic or ovate, rounded at the base or with a small cleft (Fig. 370l)
                                                              11. *smilacifolia*

1. **D. dumetorum** (*Kunth*) *Pax* in E. & P., Pflanzenfam., 2, 5: 134 (1888); Dur. & Schinz, Consp. Fl. Afr. 5:
274 (1893); F.T.A. 7: 419; Knuth in Engl., Pflanzenr. Dioscoreac. 4, 43: 132 (1924); Burkill in Bull.
Jard. Bot. Brux. 15: 367 (1939); Miège, Contrib. Dioscor. Afr. occid. 75 (1952)*; Berhaut, Fl. Sén. 17.
*Helmia dumetorum* Kunth (1850). *Dioscorea buchholziana* Engl. (1886). *D. triphylla* var. *dumetorum*
(Kunth) Knuth, illegit. name. Spiny climber, 9 ft. high, with rather prickly bulbils; annual tubers,
several per plant and a common head, near the soil surface; the species includes more or less poisonous
wild varieties as well as cultivated edible varieties with either white or yellow-fleshed tubers.
**Sen.:** Ouassadou *Berhaut* 1264! 5480! **Mali:** Bamako *Waterlot* 1514! Sikasso *Miège*! **Guin.:** Kouroussa,
Timbo *Pobéguin* 748! 1076! Kouria *Caille* in *Hb. Chev.* 14941! Fouta Djalon *Chev.* 18267! 18772! **S.L.:**
Musaia (July) *Deighton* 4795! 4886! *Thomas* 2621! Kissy, Freetown, Baïma *Deighton* 2765! 3070! **Iv.C.:**
Mankono *Chev.* 21953! Odienné, Séguéla, Ferkessédougou, Dabakala & Oroumbo-boka, *Miège* 3038!
**U.Volta:** N. Ouahigouya *Chev.* 24807! **Ghana:** Accra *T. Vogel*! *Barter* 1537! Achimota *Irvine* 2581!
*Thomas* 36! Aburi Hills *Johnson* 482! Kibbi *Glover* 20! **Dah.:** Bassam Zoumé *Chev.* 23611! Konkobiri
*Chev.* 24362! **N.Nig.:** Zaria Prov. *Keay* FHI 22878! Jos Plateau *Batten-Poole* 283! *Lely* 309! Yola
*Dalz.* 226! **S.Nig.:** Oyo Prov. *Oladoyinbo* FHI 24253! Abeokuta *Irving* 108! **W.Cam.:** *Preuss* 1385!
(Dec.) *Latilo & Daramola* FHI 28873! Mungo *Buchholz.* Also in E. Cameroun, Chad, Congo and Angola.
(See Appendix, p. 491.)

2. **D. quartiniana** *A. Rich.* Tent. Fl. Abyss. 2: 316 (1851); Engl., Hochgebirgsfl. Trop. Afr. 172 (1892);
Knuth l.c. 151; Burkill l.c. 362; Aké Assi, Contrib. 2: 25 (1962). *D. anchietae* Harms ex De Wild.
(1914). *D. apiculata* De Wild. (1902). *D. beccariana* Martelli (1886). *D. cryptantha* Bak. (1887). *D.
dinteri* Schinz (1900). *D. excisa* Knuth (1924). *D. holstii* Harms (1895). *D. pentaphylla* of A. Rich. (1851),
not of Linn. *D. phaseoloides* Pax (1892). *D. schliebenii* Knuth (1932). *D. schweinfurthiana* Pax (1892).
*D. stuhlmannii* Harms (1895). *D. verdickii* De Wild. (1902). *Botryosicyos pentaphylla* Hochst. (1844).
A non-spiny climber 15–20 ft. high, without bulbils; annual tubers usually thin and joined together in
groups of 3–6.
**Gam.:** *Lecard* 175! **S.L.:** Musaia (July) *Deighton* 4812! **Iv.C.:** Madinani *Aké Assi* 4593! **Ghana:** Boro, nr.
Wa *Adams*! Pepease *Akpabla* 176! **N.Nig.:** Yola Prov. *Dalz.* 227! Vom, Jos Plateau *Dent Young* 247!
Naraguta (May) *Lely* 246! Extends to Ethiopia, Malawi and Angola; also in Madagascar.
[Numerous varieties have been distinguished by their pubescence and by the shape of the leaflets.]

\* Mimeographed, with limited distribution.

3. **D. esculenta** (*Lour.*) *Burkill* in Gard. Bull. Str. Settl. 1 : 396 (1917); Knuth l.c. 189; Prain & Burkill in Ann. Roy. Bot. Gard. Calcutta 14 : 80 (1936); Miège in Rev. Bot. Appliq. 28 : 509 (1948); Miège, Contrib. Dioscoreac. 35 (1952). *D. aculeata* Linn. (1724), not of (1753). *D. fasciculata* Roxb. (1832). *D. papuana* Warburg (1891). *D. spinosa* Roxb. ex Wall. (1830), name only. *D. tiliifolia* Kunth (1850). *Oncus esculentus* Lour. (1790). Spiny climber 30–40 ft. high, lacking bulbils and with numerous (up to 40) stalked edible tubers independent of each other. In Africa the introduced clones are sterile. Sparsely cultivated in coastal West Africa from Sierra Leone to E. Nigeria. Native of E. Asia. (See Appendix, p. 493.)

4. **D. preussii** *Pax* in Engl., Bot. Jahrb. 15 : 147 (1893); F.T.A. 7 : 417; Knuth l.c. 221; Burkill l.c. 351; Miège, l.c. 55 (1952). *D. andongensis* Rendle (1899). *D. chevalieri* De Wild. (1914). *D. longespicata* De Wild. (1914). *D. malchairii* De Wild. (1912). *D. pterocaulon* De Wild. and Th. Dur. (1899). *D. thonneri* De Wild. & Th. Dur. (1899). Robust non-spiny climber 20–30 ft. high, tubers generally deeply buried in the soil, often divided and lobed.
**Sen.:** Somone *Adam* 1654! 1658! Ouassadou, Toubacouta *Berhaut* 1263! 2195! 2196! Sokhone, Bignona & Sédhiou, *Miège & Doumbia* C506! C898! C959! **Guin.:** Macenta *Baldwin* 9624! Timbo *Maclaud* 226! Nzérékoré *Miège*! Kouroussa, Timbo *Pobéguin* 1075! 1077! 1859! **S.L.:** Binkolo *Thomas* 1678! 1699! 1718! 1760! Bumbuna *Thomas* 3202! Kasokora, Bwedu *Deighton* 1243! 3170! Masumbu *Glanville* 1424! Marampa *Tindall* 41! **Iv.C.:** Toura Prov. *Chev.* 21612! 21956! Man *Chev.* 34153! Taï, Boundiali, Daloa, Baoulé & Abengourou *Miège* 454! 598! 651! 714! 979! 1008! 1298! 1951! **Ghana:** *Boughey*! **Dah.:** Adja Ouéré *Le Testu* 160! 161! **N.Nig.:** Dogon Kurmi, Jemaa Div. *Killick* 3! **S.Nig.:** Onochie FHI 20201! *Kennedy* 2719! Gambari to Mamu (Aug.) *Jones* FHI 20256! Mamu F.R., Gambari (Aug.) *Keay* FHI 16231! Abeokuta *Irving* 103! Lagos *Millen* 17! Benin (Sept.) *Kennedy* 1614! Oban *Talbot* 780! Aguku *Thomas* 975! 1020! 1748! **W.Cam.:** Preuss 504! Victoria, Cam. Mt. *Maitland* 68! 314! 1311! Mamfe *Baldwin* 13862! Extends to the Congo, Gabon and C. African Republic. (See Appendix, p. 492.)

5. **D. sansibarensis** *Pax* in Engl., Bot. Jahrb. 15 : 146 (1892); F.T.A. 7 : 416; De Wildeman in Ann. Mus. Congo, Bot., sér. 5, 2 : 121 (1907); Knuth l.c. 87; Burkill in Blumea, suppl. 1 : 232 (1937); Burkill l.c. 348; Perr. de la Bâthier in Not. Syst. 12 : 200 (1946); Chevalier in Rev. Bot. Appliq. 32 : 14 (1952); Miège, Contrib. Discor. 42 (1952). *D. macabiha* Jumelle & Perr. de la Bâth. (1909). *D. macroura* Harms (1897)—F.T.A. 7 : 416; F.W.T.A., ed. 1, 2 : 382. *D. toxicaria* Bojer (1837), name only. *D. welwitschii* Rendle (1899). Glabrous non-spiny climber 15–20 ft. high, bearing purple or brownish toxic bulbils, the perennial tubers are also poisonous, female flowers white, veined purple.
**Iv.C.:** Mankona to Séguéla, Groumania *Miège* 244! 469! **Dah.:** Dassa Zoumé *Chev.* 23621! Savalou *Chev* 23692! **N.Nig.:** Jebba *Barter* 1533! **S.Nig.:** Olokemeji *Chev.* Extends to Angola, Congo, E. Africa, Mozambique and Madagascar.

6. **D. bulbifera** *Linn.* Sp. Pl. 1033 (1753); Knuth l.c. 88; Prain & Burkill l.c. 111; Burkill l.c. 357; Chevalier in Rev. Bot. Appliq. 32 : 16 (1952); Miège, Contrib. Dioscor. 63 (1952); Berhaut, Fl. Sén. 154. *D. anthropophagorum* A. Chev. (1913). *D. bulbifera* var. *anthropophagorum* (A. Chev.) Prain & Burkill (1931). *D. hoffa* Cordemoy (1895). *D. latifolia* Benth. (1849). *D. longipetiolata* Baudon (1913). *D. perrieri* Knuth (1924). *D. sativa* of F.T.A. 7 : 415, partly of Linn. *D. violacea* Baudon (1913). *Helmia bulbifera* (Linn.) Kunth (1850). Glabrous non-spiny climber 10–20 ft. high, with bulbils 1–8 cm. in size, toxic or edible according to the variety; tubers renewed annually, globose, absent in several varieties or according to the method of cultivation; female flowers, when produced, white turning pink and then purple when old. Some edible cultivated varieties with large leaves have lost their ability to produce flowers, the bulbils are then larger and composed of a larger number (4–5) of buds. Wild varieties with toxic, angular, greyish bulbils and medium-sized leaves occur and others with small purplish bulbils, small leaves with a red base to the petiole.
**Sen.:** Ouassadou, Ñiamboto, Dougar & Sangalkam *Berhaut* 717! 2202! 3613! 5395! Niokolo-Koba, Dougar *Miège*! *Adam* 14261! 15734! **Guin.:** Kouroussa, Kindia & Timbo *Pobéguin* 744! 1071! 1325! 1860! 1861! Kouria *Caille* in *Hb. Chev.* 14700! 14752! 15070! Fouta Djalon *Chev.* 18258! 18290! 18511! 18519! 18617! N'Zérékoré *Baldwin* 13303! *Miège* 713! **S.L.:** Jigaya, Kabala & Mabum *Thomas* 1538! 2220! 2679! Rokupr *Jordan* 143! 144! Musaia, Mapaki & Njala *Deighton* 1200! 1201! 1351! 4394! 4865! 4918! **Lib.:** Webo Dist. *Baldwin* 6680! **Iv.C.:** Bingerville *Chev.* 15204! Mankono *Chev.* 21941! Baleyo *Hédin* 25! Man *Nozerau*! Séguéla, Ferkessédougou, Mankono, Katiola & Sakasso *Miège* 654! 991! 1952! **Ghana:** Achimota *Irvine* 2578! Pra Anum F.R. (Aug.) *Darko* 953! Behwai *Vigne* FH 1316! Tamale *Sampson* 4! **U.Volta:** Banfora *Adam* 15431! *Miège*! **N.Nig.:** Jebba *Barter*! Jos Plateau *Batten-Poole* 291! Katsina *Dalz.* 669! **S.Nig.:** Lagos *Dawodu* 134! 145! *Dodd* 440! Ondo Prov. Onochie FHI 33417! Nun R. *T. Vogel* 30! Oban *Talbot* 131! 717! **W.Cam.:** Bamenda *Ujor* 29295! Widely distributed in the tropics. (See Appendix, p. 490.)

7. **D. hirtiflora** *Benth.* in Fl. Nigrit. 537 (1849); F.T.A. 7 : 416; Knuth l.c. 380; Burkill l.c. 374; Miège, Contrib. Dioscor. 84 (1952); Berhaut, Fl. Sén. 154. *D. anthropophagorum* var. *sylvestris* A. Chev. (1913), name only. *D. dusenii* Uline ex Knuth (1924). *D. polyantha* Rendle (1894). *D. rubiginosa* Benth. (1849). *D. sativa* var. *sylvestris* A. Chev., Bot. 642, partly, name only. *D. spinosa* Wall., name only. *D. stellato-pilosa* var. *cordata* De Wild. (1912). Pubescent climber 10–20 ft. high, with or without bulbils and when present the bulbils have white or pinkish flesh; tubers annual, more or less lobed, considered to be poisonous.
**Sen.:** Thiès, Toubakouta, M'bao *Berhaut* 1167! 2000! 2199! Niokolo-Koba, Kolda *Adam* 8! 18512! Ziguinchor *Berhaut* 6467! Bayotte Forest *Miège & Doumbia* C701! **Guin.:** Madina Tossékré (Oct.) *Adam* 12534! N'Zérékoré (Oct.) *Baldwin* 9695! Fouta Djalon (Sept.) *Chev.* 18269! 18649! Konkouré *Caille* in *Hb. Chev.* 14993! Kouria to Yombo *Caille* in *Hb. Chev.* 14979! Conakry *Maclaud* 69! 153! 154! Kindia, Lola & foot Mt. Nimba *Miège* 733! **S.L.:** Newton (Sept.) *Deighton* 5998! Bonganame (Oct.) *Deighton* 6009! Makump (Oct.) *Thomas* 3952! Bumbuna (Oct.) *Thomas* 3291! Yonibana (Oct., Nov.) *Thomas* 2164! 4680! 4704! 4756! Kameron *Morton* SL 2474! **Lib.:** Flumpa, Sanokwele Dist. (Sept.) *Baldwin* 9371! *Harley* 1243! Gbanga (Sept.) *Linder* 493! Peahtah *Linder* 1037! **Iv.C.:** Danané *Nozerau*! Taï, Odienné, Séguéla, Oroumbo-Boka, Abengourou & Man *Miège* 465! 471–2! 643–4! 701–2! 849! 990! 1006! **Ghana:** Shai *Morton* A4054! **N.Nig.:** Nupe *Barter* 561! "Quorra" (= Niger R.) *T. Vogel* 199! **S.Nig.:** Lagos *Dawodu* 211! *Millen* 29! Olokemeji F.R. (Sept.) *Keay* FHI 25424! Awba Hills F.R., Oyo (Oct.) *Jones* FHI 6302! Gambari F.R. (Nov.) Onochie FHI 34944! Oban *Talbot* 722! 724! 781! **W. Cam.:** Buea (Nov.) *Migeod* 95! Bamenda to Mamfe (Dec.) *Baldwin* 13860! Extends to E. Cameroun, Congo, Angola, Zambia and Tanzania. (See Appendix, p. 491.)

8. **D. schimperana** *Hochst. ex Kunth* Enum. Pl. 5 : 339 (1850); F.T.A. 7 : 419; Knuth l.c. 255; Burkill l.c. 371. *D. dawei* De Wild. (1914). *D. fulvida* Stapf (1906). *D. hockii* De Wild. (1911). *D. stellato-pilosa* De Wild. (excl. var. *cordata*) (1912). Pubescent climber 10–20 ft. high; in upland.
**N.Nig.:** Vom, Jos Plateau *Dent Young* 248! **W.Cam.:** Bambui *Brunt* 83! Extends to Central and E. Africa.

9. **D. alata** *Linn.* Sp. Pl. 1033 (1753); F.T.A. 7 : 417; Knuth l.c. 265; Prain & Burkill l.c. 308; Burkill l.c. 380; Miège in Rev. Bot. Appliq. 32 : 144 (1952); Miège, Contrib. Dioscor. 94 (1952). *D. colocasiifolia* Pax (1912), (descr. based on mixture : leaves of *D. alata*, ♀ infl. of *D. dumetorum*)—F.T.A. 7 : 417; F.W.T.A., ed. 1, 2 : 382. *D. sapinii* De Wild. (1912). Glabrous climber with winged stems 10–25 ft. high, rarely flowering in our area and usually reproducing vegetatively; bulbils developed in some forms; tubers annual, their shape, size and colour very variable according to the cultivars.
Asiatic species, native of E. India, formerly introduced into E. Africa by Arabs, later in W. Africa by the Portuguese. Widely distributed in W. Africa principally from Ivory Coast to W. Cameroun.

Numerous varieties producing many different types of tubers, digitate or not, straight, curved or coiled having white, pink or purple flesh. They play an important rôle in the agricultural economy of W. Africa. Much cultivated throughout the tropics. (See Appendix, p. 489.)

10. **D. minutiflora** *Engl.* Bot. Jahrb. 7: 332 (1886), and 15: 146 (1892); F.T.A. 7: 418; De Wildeman in Ann. Mus. Congo 3: 366 (1911–12); Knuth l.c. 300; Burkill l.c. 389; Burkill in Proc. Linn. Soc. 151: 57 (1939), and 159: 77 (1947); Miège in Rev. Bot. Appliq. 30: 428 (1950); Miège, Contrib. Dioscor. 138 (1952). *D. acarophyta* De Wild. (1904), partly. *D. armata* De Wild. (1914). *D. brevispicata* De Wild. (1912). *D. ealaensis* De Wild. (1912). *D. ekolo* De Wild. (1914). *D. engbo* De Wild. (1914). *D. grande-bulbosa* Knuth (1924). *D. hystrix* Knuth (1914). *D. lilela* De Wild. (1912). *D. litoie* De Wild. (1914). *D. multiflora* Pax (1892), in error for *D. minutiflora*. *D. praehensilis* var. *minutiflora* Bak. in F.T.A. 7: 418 (1898), partly. *D. pynaertioides* De Wild. (1914). Perennial climber with spiny stems 12–30 ft high and rounded several-nerved leaves; abundant principally on the forest edges.
**Sen.:** Kaémé Forest, Casamance *Berhaut* 6621! **Guin.:** Nimba Mt. *Miège* 816! *Schnell* 3351! 3799! **S.L.:** (Feb.) *Pyne* 107! Mabonto to Masumbiri Simiria (Nov.) *Glanville* 1525! Kinsuta, Samu (Dec.) *Sc. Elliot* 4235! Yonibana (Oct.) *Thomas* 3990! **Lib.:** Ganta (Jan.) *Baldwin* 14057! Bakratown (Oct.) *Linder* 857! Kakatown *Whyte*! **Iv.C.:** Bingerville, *Chev.* 16067! Guedéko *Chev.* 19012! Man *Chev.* 34174! Adiopodoumé *Leeuwenberg*! Yapo *Giovanetti* 324! Sassandra, Baléko, Agneby, Bingerville & Oroumbo-Boka *Miège* 915! 917! 922! 937! 942! Boulay I. *de Wilde* 815! **S.Nig.:** Lagos *Moloney*! Ibadan (Mar.) *Keay* FHI 25695! Ikom (Dec.) *Keay* FHI 28261! Oban *Talbot* 782! Nun R. *Barter* 2115! **W.Cam.:** Buea (Jan.) *Maitland* 289! Mungo (Sept.) *Buchholz*. Extends to the Congo, N. Angola, Uganda. (See Appendix, p. 491.)

11. **D. smilacifolia** *De Wild.* in Ann. Mus. Congo, Bot., sér. 2, 1: 58 (1899), and sér. 3, 1: 239 (1901); Knuth l.c. 303; Burkill l.c. 391; Miège, Contrib. Dioscor. 138 (1952). *D. demeusei* De Wild. & Th. Dur. (1901) partly. *D. echinulata* De Wild. (1911). *D. flamignii* De Wild. (1912). *D. praehensilis* of F.T.A. 7: 418, partly. Perennial climber with spiny stems 12 ft. high and trinerved ovate leaves.
**S.L.:** Kenema (Jan., Nov.) *Thomas* 7620! *Glanville* 1526! Regent (Dec.) *Sc. Elliot* 4026! Lester Peak (Dec.) *Sc. Elliot* 4157! 4171! Bo (Jan.) *Thomas* 7434! **Lib.:** Barclayville, Grand Bassa (Mar.) *Baldwin* 11136! Fortsville, Grand Bassa (Mar.) *Baldwin* 11151! Nyaake, Webo Dist. (June) *Baldwin* 6186! **Iv.C.:** Bouroukrou *Chev.* 16652! 16882! Basse Comoé *Chev.* 17591! Adzopé *Chev.* 22674! Abouabou & Taï *Leeuwenberg* 2382! 3581! Adiopodoumé, Baléko, Niégré & Singrobo *Miège* 935! 940! 941! 951! 1219! Grand Bassam *Pobéguin* 201! **Ghana:** Plumptre 87! *Irvine* 590! Assin-Yan-Kumassi (Feb.) *Irvine* 120! *Cummins*! Assuantsi road (Jan.) *Fishlock* 3! **Dah.:** Tohoué *Chev.* 22777! **S.Nig.:** Lagos (Dec.) *Millen* 48! Ibadan (Feb.) *Keay & Meikle* FHI 25663! Onochie FHI 35329! Gambari F.R. (Nov.) *Meikle* 1193! Calabar *Holland* 117! Mann 2283! **W.Cam.:** Bambe R., S. Bakundu (Jan.) *Binuyo & Daramola* FHI 35194! Tiko (Jan.) *Dunlap* 180! **F.Po:** Musola (Nov.) *Guinea* 1277! 1279! (See Appendix, p. 493.)
[A species closely related to *D. minutiflora*, though quite distinct. Distribution very similar.]

12. **D. burkilliana** *J. Miège* in Bull. I.F.A.N., sér. A, 20: 39 (1958). *D. minutiflora* Engl. (1886), partly. *D. acarophyta* De Wild. (1904), partly. Climber with spiny annual stems 18–25 ft. high from perennial fibrous tubers protected by a woody rootstock, stellate in growth; acarodomatia (swellings used by mites) are always present at the base of the lamina; forest species.
**S.L.:** Njala *Deighton* 2547! 2872! Sandugu, Makump & Pendembu *Thomas* 591! 722! 1001! **Iv.C.:** Bouroukrou *Chev.* 16710! Man *Chev.* 34174! Adiopodoumé, Ayamé, Akpouasso, Daoukro & Divo *Miège* 978! 989! 1987!
[This species is very difficult to distinguish from *D. minutiflora* on dried specimens: in the field the distinction is easier as it does not root at the nodes.]

13. **D. mangenotiana** *J. Miège* l.c. (1958). *D. odoratissima* Pax (1893), partly. *D. praehensilis* Benth. (1849), partly. The largest West African yam attaining a considerable size, 75–100 ft. in length annually from perennial tubers which become enormous reaching or exceeding 110 lbs; protective roots spiny; in forest.
**Lib.:** *Carder*! Morris farm, Monrovia (May) *Dinklage* 3045! Kakatown *Linder* (Aug.) 344! **Iv.C.:** Guédéko *Chev.* 16486! Divo, Yapo, Adiopodoumé & Assinie *Miège* 962! 969! 1318! 1880! 1881! **S.Nig.:** Lagos *Dawodu* 115! Oban *Talbot* 727!
[Near to *D. praehensilis* but male spikes with negative geotropism and stronger stems.]

14. **D. togoensis** *Knuth* in Engl., Pflanzenr. Dioscoreac. 4, 43: 299 (1924); Chevalier in Bull. Mus. Hist. Nat. Paris 8: 547 (1936). *D. abyssinica* Hochst. ex Kunth (1850), partly. *D. caillei* A.┋Chev. ex De Wild. (1914). *D. praehensilis* Benth. (1849), partly. Annual climber 2–6 ft. high, with rather weak development; bulbils small, grey-brown, reduced to a single bud, rarely two, 4–16 mm. falling off very easily and consequently seldom seen in herbarium specimens; tubers slender, deeply rooted in savanna or in rocky places in forest.
**Sen.:** M'Borouk *Miège*! **Gam.:** Dawe 69! **Guin.:** Macenta (Oct.) *Baldwin* 9818! Kouria *Caille* in *Hb. Chev.* 14973! Conakry *Maclaud* 130! Danané to N'Zerékoré *Miège* 706! **S.L.:** Rokupr (Sept.) *Jordan* 123! Yetaya (July) *Thomas* 2395! 2408! Njala (Oct., Nov.) *Deighton* 1340! 1345! 1780! Sumbuya (Oct.) *Thomas* 3167! Musaia (Oct.) *Thomas* 2645! **Iv.C.:** Issia rock, Kouroukouroun'ga, Niellé, Niangbo & Dabakala *Miège* 449! 587! 592! 600! 621! 972! **Ghana:** Djadje *Akpabla* 54! Achimota (Nov.) *Irvine* 2554! **Togo Rep.:** Lomé *Warnecke* 250! **Dah.:** Atacora Mts. *Chev.* 24148! **N.Nig.:** Nupe *Barter*!

15. **D. cayenensis** *Lam.* Encycl. Meth. 3: 233 (1789); Kunth, Enum. Pl. 5: 380 (1850); Knuth l.c. 298; Burkill l.c. 383; Miège, Contrib. Dioscor. 120 (1952); Miège in Rev. Bot. Appliq. 32: 144 (1952). *D. aculeata* Balbis ex Kunth (1850). *D. berteroana* Kunth (1850). *D. demeusei* De Wild. & Th. Dur. (1901), partly. *D. liebrechtsiana* De Wild. (1901), partly. *D. occidentalis* Kunth (1924). *D. praehensilis* of F.T.A. 7: 418, partly. *D. pruinosa* A. Chev. (1913). Cultivated plants with superficial tubers. In each subspecies numerous clones are characterized by the leaf-shape (more or less elongated, triangular or cordate); the shape, size and colour of the tubers and the persistence or not of rootlets on them; stem colour (lighter, dark or yellow green, pink or reddish, more or less pruinose); the base of the petiole is sometimes red; and the degree of spinescence of the stems and roots. Plants tending to a reduction of sexual reproduction. Some varieties with one annual harvesting, others with two.

15a. **D. cayenensis** *Lam.* subsp. subsp. **cayenensis.** Two annual harvests with a short resting period.
15b. **D. cayenensis** subsp. **rotundata** (*Poir.*) *J. Miège* stat. nov. *D. rotundata* Poir. (1813)—F.W.T.A., ed. 1, 2: 382. One annual harvest with a long resting period.
Species of hybrid origin, selected by the African farmers. Common in all West Africa but its cultivation is most intense from the Ivory Coast to S. Nigeria where it is the chief starchy root crop. (See Appendix, p. 490, 492.)

16. **D. abyssinica** *Hochst. ex Kunth* Enum. Pl. 5: 387 (1850); F.T.A. 7: 418; Knuth l.c. 294; Burkill l.c. 382. *D. lecardii* De Wild. (1903), partly. *D. mildbraedii* Knuth (1924), partly. *D. sagittifolia* Pax (1892). *D. togoensis* Knuth (1924), partly. Non-spiny climber 6–12 ft. high, with deep seated tubers.
**Mali:** Bamako *Waterlot* 1513! **Lib.:** Sanokwele Dist., Flumpo & Ganta *Baldwin* 9351! 12494! **Iv.C.:** Baoulé region, Sénoufo *Miège*! **Ghana:** Achimota *Irvine* 1421! 14401! N. Prov. *Patterson*! **N.Nig.:** Jos Plateau *Lely* 310! *Batten-Poole* 287! 290! Katsina *Dalz.* 225! 671!
[Probably one of the wild parents of *D. cayenensis*.]

17. **D. praehensilis** *Benth.* in Fl. Nigrit. 536 (1849); F.T.A. 7: 418, partly; Chevalier in Bull. Soc. Bot. Fr. 59: 223 (1912); Knuth l.c. 299; Miège, Contrib. Dioscor. 112 (1952). *D. odoratissima* Pax in Engl., Bot. Jahrb. 15: 146 (1893); Burkill l.c. 381. Sturdy climber 15–25 ft. high with spiny stems often pruinose, brown, red or purplish.
**S.L.:** *Sc. Elliot*! *T. Vogel* 21! Yonibana *Deighton* 5843! *Thomas* 4705! Bumba, Kaballa & Jagaya

*Thomas* 1904! 2243! 2605! 2796! **Lib.**: Peahtah *Linder* 899! Firestone Plantation, Sangwin R., Sinoe
Dist. *Linder* 182! **Iv.C.**: Mid. Sassandra *Chev.* 16481! 19057! Bouaké *Chev.* 22114! Taï (Aug.) *Schnell*
6039! Sangouiné, Baléko *Nozerau*! Divo, Sakasso, Béoumi, Toumodi, Bouaké, Abengourou & M'bahi-
akro *Miège* 840! 943! 947! 976! **Ghana**: Kibbi *Glover* 12! 15! **Togo Rep.**: Bismarckburg *Büttner* 104!
**N.Nig.**: Nupe *Barter* 1536! Naraguta *Lely* 492! **S.Nig.**: *Farquharson* 2! Asaba *Onochie* FHI 33297! Benin
(Sept.) *Kennedy* 1582! Oban *Talbot* 723! 728! (See Appendix, p. 492.)
    [Probably a parent of the several varieties of *D. cayenensis.*]
18. **D. liebrechtsiana** *De Wild.* in Bull. Herb. Boiss., sér. 2, 1: 53 (1900); in Ann. Mus. Congo, Bot., sér. 3:
      362 (1912), and sér. 5, 3: 58 (1909); Burkill l.c. 381; Miège, Contrib. Dioscor. 111 (1952). Spiny
      climber 15–25 ft. high, with tubers protected by spiny roots.
      **S.L.**: Njala & Musaia *Deighton* 1352! 4127! 4375! **Iv.C.**: Adiopodoumé, Ayame & Taï *Miège* 964! 975!
      **Ghana**: Dogoro Dosusu *Moor* 333! **S.Nig.**: *Farquharson* 3! 4! Isu Awin *Foster* 329!
      [Very near to *D. praehensilis* but plants less strong and with different leaf bases.]
19. **D. lecardii** *De Wild.* in Ann. Mus. Congo, Bot. sér. 5, 1: 19 (1903); Knuth l.c. 295; Burkill l.c. 387.
      *D. mildbraedii* Knuth (1924). *D. zara* Baufon (1913). Unarmed climber 15–25 ft. high with tubers deep
      and protected by spiny roots.
      **Sen.**: *Lecard* 214! 235! *Merlier* 439! **Mali**: *Bellamy* 186! **Guin.**: Kouroussa *Pobéguin* 1078! 1080!
      **S.L.**: Makump *Glanville* 1524! *Deighton* 4866! **Iv.C.**: N. Odienné, Boundiali *Miège*!
      [Perhaps played a rôle in the *D. cayenensis* pedigree.]
20. **D. sagittifolia** *Pax* in Engl., Bot. Jahrb. 15: 147 (1892); F.T.A. 7: 416; Knuth l.c. 295. *D. lecardii* De
      Wild. (1903). Climber with stem spiny at least towards the base, 10–15 ft. high, tubers deep or protected
      by spiny roots.
      **Sen.**: Tobor Forest, Bignona, Casamance & Tambacounda *Berhaut* 6429! 6749! **S.L.**: Makump
      *Glanville* 1524! **Iv.C.**: Odienné, Oddia, Niempurgué & Soriforo *Miège* 1927! 2018! **U.Volta**: Banfora
      Falls *Miège* 1914!

## 192. AGAVACEAE

### By F. N. Hepper

Rootstock a rhizome; stem short or well developed. Leaves usually crowded
on or at the base of the stem often thick and fleshy, entire or with prickly teeth
on the margin. Flowers bisexual, polygamous or dioecious, racemose to
paniculate or subcapitate, sometimes in a very large thyrse, bracteate. Perianth
segments free or united; corona never present. Stamens 6, inserted at the base
of the lobes or on the tube; filaments filiform or thickened, free; anthers
introrse, 2-locular, opening lengthwise. Ovary superior or inferior, often beaked,
3-locular, with axile placentas; style slender. Ovules numerous and superposed
in two series to solitary. Fruit a capsule or berry. Seeds with fleshy endosperm.

Those genera of this family which are native in our area possess superior ovaries.

Herbs, shrubs or trees; stems and leaves usually not fibrous; leaves flat, not greatly
   thickened; fruit a berry    ..    ..    ..    ..    ..    ..    1. **Dracaena**
Herbs with creeping rhizomes; stems and leaves usually fibrous; leaves erect, thick and
   fleshy; fruit with a thin pericarp falling away from the berry-like seeds
                            2. **Sansevieria**

The sisal plant *Agave sisalana* Perrine ex Engelm. and other *Agave* species, also the
Mauritius hemp, *Furcraea gigantea* Vent. and *F. selloa* Koch have been introduced into
our area and they occasionally become more or less naturalized. They are all American
plants and they possess inferior ovaries. *Cordyline terminalis* Kunth, from tropical Asia,
with superior ovaries, is also cultivated in gardens.

### 1. DRACAENA Vand., Diss. (1762); F.T.A. 7: 436.
*Pleomele* Salisb. (1796)—N. E. Brown in Kew Bull. 1914: 273.

Herbs or at most undershrubs; lamina ovate or ovate-lanceolate, broadest at or below
  the middle, rather abruptly narrowed into a usually long slender petiole:
Inflorescence spicate-racemose:
  Inflorescence about 10 cm. long, shorter than the leaves, simple; perianth 12 mm.
    long, erect; fruit 2–3-lobed, fleshy; stem hardly developed, nodes congested;
    leaves ovate-lanceolate, 20–31 cm. long, 5–7 cm. broad, with numerous close parallel
    nerves and distinct transverse veins; petiole nearly as long as the lamina, slender,
    gradually broadened into a striate sheath at the base    ..    ..    .. 1. *humilis*
  Inflorescence about 30 cm. long, usually longer than the leaves, simple or with a few
    short lateral branches; perianth about 25 mm. long; stem distinctly developed;
    leaves oblong-lanceolate to ovate-lanceolate, (12–) 18–39 cm. long, (3–) 5–8 cm.
    broad ..    ..    ..    ..    ..    ..    ..    ..    ..    2. *thalioides*
Inflorescence capitate; leaves ovate or ovate-elliptic:
  Petiole winged its full length, as long as or longer than the blade; blade ovate,
    acutely acuminate, 10–15 cm. long, 5–8 cm. broad, with numerous parallel nerves
    and distinct cross-veins; perianth 1·5 cm. long; fruiting head sessile    3. *elliotii*

Petiole winged only at the base and apex:
   Petiole nearly as long as or longer than the blade, slender; blade mottled, ovate-elliptic, acuminate, 15–25 cm. long, 6–10 cm. broad, faintly nerved; inflorescence capitate shortly pedunculate; bracts caudate-acuminate; fruit spherical, 1·5 cm. diam.   ..  ..  ..  ..  ..  ..  **4. phrynioides**
   Petiole much shorter than the blade, the latter transversely variegated, ovate-elliptic, very acutely acuminate, 12–25 cm. long, up to 14 cm. broad; inflorescence sessile, 5 cm. diam.; bracts long-acuminate; perianth 3 cm. long   **5. goldieana**
Shrubs or trees; lamina obovate to linear-oblanceolate or linear, broadest about the middle, if elliptic then subsessile or shortly petiolate:
Lamina more or less elongate-linear, linear-lanceolate or oblanceolate:
  Trees; inflorescences paniculate:
   Pedicels very short and inconspicuous, at most 2 mm. long; perianth 1 cm. long; lateral nerves of the dry leaves fine, numerous and contiguous or nearly so, oblique transverse veins also visible; leaves 100 cm. or more long, up to 12 cm. broad, sheathing at the base, acute and mucronate at the apex; inflorescence a large branched panicle; lowland tree  ..  ..  ..  ..  ..  **6. smithii**
   Pedicels conspicuous, leaving fruiting " pegs " 4–5 mm. long:
    Leaves 6–7 cm. broad, very long, with very numerous subcontiguous nerves; bracteoles persistent on the infructescence; perianth about 14 mm. long; fruits nearly 2 cm. diam.; tall tree  ..  ..  ..  ..  **7. arborea**
    Leaves 2–3 cm. broad; not sheathing at base:
     Perianth 11 mm. long; anthers 2 mm. long; S. Nigeria  ..  **8. mannii**
     Perianth 25–30 mm. long; anthers 3·5 mm. long; fruit at least 2 cm. diam.; widespread  ..  ..  ..  ..  ..  ..  **9. perrottetii**
  Shrubs; lateral nerves spaced 0·5–1 mm. apart usually with transverse veins between:
   Inflorescence very slender terminal panicle, about 25 cm. long, well branched, smooth; fruiting " pegs " 15–20 mm. long, slender; fruit about 1·5 cm. diam.
                                **10. scoparia**
   Inflorescence stouter, paniculate or simple, fruiting " pegs " up to 5 mm. long:
    Rhachis of inflorescence aculeolate; bracts 5 cm. or more long, broadly lanceolate, acute; dense clusters of flowers several cm. apart; flowers subtended by numerous lanceolate bracteoles about 1 cm. long; perianth 3 cm. long; leaves oblanceolate, up to 70 cm. long and 10 cm. broad, sheathing at base   **11. adamii**
    Rhachis of inflorescence smooth; perianth 1·3–3 cm. long:
     Inflorescence spicate, terminal or axillary:
      Inflorescence a dense continuous terminal spike furnished with long leafy bracts, the lowest bract about as long as the spike; perianth about 3 cm. long; leaves linear-oblanceolate, contracted towards the expanded base, up to 40 cm. long and 3 cm. broad, with numerous transverse nerves arcuately spreading from the midrib  ..  ..  ..  ..  ..  ..  **12. talbotii**
      Inflorescence terminal and/or axillary spike of few-flowered clusters subtended by rather short ovate inconspicuous bracts; perianth nearly 2 cm. long, slender; leaves broadly linear, very acute, sheathing and amplexicaul at the base, about 20 cm. long, 1·2–3·9 cm. broad, 8–10-nerved on each side of midrib
                              **13. viridiflora**
Inflorescence terminal, paniculate, with short stout branches or more or less simple and with large spherical clusters of flowers:
   Leaves acute and mucronate at the apex, lamina oblanceolate, 20–60 cm. long, 5–14 cm. broad, tufted at the end of the shoot, sheathing at the base, nerves about 1 mm. apart; inflorescence usually with many short branches and numerous clusters of flowers; bracteoles deciduous after flowering; perianth 1·5–2 cm. long; lowland shrub  ..  ..  ..  ..  **14. fragrans**
   Leaves attenuate and subulate-mucronate at the apex, lamina linear-lanceolate, 15–30 cm. long, 1·5–3 cm. broad, slightly narrowed towards the base and expanded into a clasping sheath 1 cm. long along the stem; inflorescence simple or with a few short branches and dense clusters of flowers; bracteoles persistent after flowering, conspicuous, 2 mm. long, scarious; perianth about 1·3 cm. long; montane shrub  ..  ..  ..  ..  **15. deisteliana**
Lamina elliptic to obovate or rarely broadly lanceolate:
  Flowers arranged in a lax panicle and not clustered on the axes; pedicels very slender, jointed at the top, about 1·5 cm. long; perianth about 1 cm. long in bud; leaves elliptic or oblong-elliptic, rather abruptly subulate; acuminate, cuneate at the base, 7–16 cm. long, 3–6 cm. broad, with spaced parallel nerves  16. **laxissima**
  Flowers arranged in dense spikes or small panicles of clusters:
   Leaves spirally arranged and crowded, with very short internodes; petioles about 8–17 cm. long, winged; lamina obovate to oblanceolate, about 20 cm. long, 5–9 cm. broad, shortly acuminate; inflorescences shortly cylindrical 5–10 cm. long, 5 cm. diam., very dense with numerous flowers and bracts  ..  17. **bicolor**

Leaves subverticillate, whorls distant from each other and with numerous scars or leaf sheath between:
Lamina broadly oblanceolate, gradually narrowed to the base, 12–19 cm. long, (3–) 5 cm. broad, with spaced parallel nerves and very oblique veins; upper stems 2 mm. diam.; inflorescence terminal or subterminal, reflexed, about 10 cm. long, slightly zigzag, with clusters of flowers sessile or shortly pedunculate on the slender axis; perianth about 2 cm. long; fruits up to 1·5 cm. diam.
　　　　　　　　　　　　　　　　　　　　　　　　　18. *camerooniana*
Lamina elliptic or oblong-elliptic to broadly elongate-oblanceolate, more or less abruptly narrowed at each end:
Perianth 3·5 cm. long; inflorescence sessile or subsessile, about 5 cm. long, with crowded flowers; fruits spherical, about 1·5 cm. diam.; upper stems 4 mm. diam.; leaves oblong-elliptic to broadly elongate-oblanceolate, 15–25 cm. long, 4–8 cm. broad, with spaced nerves   ..   ..   ..   ..   .. 19. *ovata*
Perianth 2 cm. long; fruits less than 1 cm. diam.; upper stems 2 mm. diam.; leaves elliptic, 8–16 cm. long, 3–7 cm. broad, very acutely acuminate, with spaced parallel nerves:
Inflorescences racemose with numerous flowers towards the apex and small bracteate clusters becoming well spaced in the middle of the axis
　　　　　　　　　　　　　　　　　　　20a. *surculosa* var. *surculosa*
Inflorescences capitate on slender peduncles about 10 cm. long
　　　　　　　　　　　　　　　　　　　20b. *surculosa* var. *capitata*

1. **D. humilis** *Bak.* in J. Bot. 12: 166 (1874); F.T.A. 7: 444; Chev. Bot. 646. *D. poggei* of Chev. Bot. 647, not of Engl. *Pleomele humilis* (Bak.) N. E. Br. (1914). A herb with imbricated leaf bases on a very short stem and an inflorescence about 1 ft. high, shorter than the leaves; flowers small and green; fruits greenish yellow or red; in forest.
 **Guin.:** Foot of Nimba Mt. *Schnell* 557! **S.L.:** Bagroo R. *Mann* 898! Roruks (July) *Deighton* 3264! Kenema F.R. (Mar.) *Deighton* 4121! **Lib.:** Dukwia R., Grand Cape Mount Dist. (fr. Feb.) *Cooper* 168! Nimba (fr. June) *Adam* 21484! Kitoma *Adam* 16621! **Iv.C.:** Béyo (fl. & fr. Jan.) *Leeuwenberg* 2464! Adiopodoumé (fr. Jan.) *Aké Assi* 7297! Guébo (Feb.) *Chev.* 17030! (See Appendix, p. 494.)
2. **D. thalioides** *Makog ex C. Morren* in Belg. Hortic. 10: 348, with fig. (1860); F.T.A. 7: 445. *D. humilis* of F.T.A. 7: 445, partly (*Soyaux* 96), not of Bak. A herb with the lower stem covered by imbricated leaf bases and an inflorescence 1–3 ft. high, usually rising above the leaves; flowers yellowish with reddish brown markings; in riverine forest.
 **W.Cam.:** Lus, Nkambe Div., 2,500 ft. (Feb.) *Hepper* 1875! Jua, Bamenda, 3,500 ft. (fr. Apr.) *Maitland* 1714! Also in E. Cameroun, C. African Rep., Gabon, Uganda and Angola.
3. **D. elliotii** *Bak.* in F.T.A. 7: 449 (1898); Aké Assi, Contrib. 2: 26 [234]. *D. ovata* of Chev. Bot. 464, not of Ker-Gawl. *Pleomele elliotii* (Bak.) N. E. Br. (1914). Stems erect, 6 in. to 1 ft. high sheathed by the expanded leaf bases; inflorescence sessile; fruits globose, orange; in forest.
 **Guin.:** Ziama (Feb.) *Adam* 3561! Nzo (fr. Mar.) *Chev.* 21017! Konkouré to Timbo (fr. Mar.) *Chev.* 12557! **S.L.:** York (Sept.) *Melville & Hooker* 623! Kukuna (Jan.) *Sc. Elliot* 4689! Sendugu (June) *Thomas* 568! Kambui F.R. (Feb.) *Lane-Poole* 346! Kenema (fr. Jan.) *Thomas* 7466! **Lib.:** Peahtah, Salala Dist. (Oct.) *Bequaert* in *Hb. Linder* 1063! Nimba Mts. (fr. July) *Leeuwenberg & Voorhoeve* 4791! Zeahtown, Tchien Dist. (fr. Aug.) *Baldwin* 6963! **Iv.C.:** Youkou *Schnell* 1686! **Ghana:** Kade (fr. June) *Irvine* 4897! Ankasa F.R. (fr. June) *Enti & Hall* GC 35599! (See Appendix, p. 493.)
4. **D. phrynioides** *Hook.* in Bot. Mag. t. 5352 (1862); F.T.A. 7: 447; Chev. Bot. 647. *Pleomele phrynioides* (Hook.) N. E. Br. (1914). *Dracaena humilis* of F.W.T.A., ed. 1, 2: 384, partly (*Irvine* 491, *Johnson* 763), not of Bak. A herb with a short tough stem rooting at the base, roots smelling of violets, leaves with pale mottling; fruits red; in forest.
 **Lib.:** Kitoma, Snokwele Dist. (Dec.) *Adam* 16705! **Iv.C.:** Guidéko to R. Zozro (fr. June) *Chev.* 19048! Mt. Bô, Danané (fr. Sept.) *Nozerau*! **Ghana:** E. Akim (fr. June) *Johnson* 763! Blanka, Ashanti *Irvine* 491! **S.Nig.:** Oshugoko to Oke Ibode (fr. Mar.) *Millson*! Oban *Richards* 5202! **F.Po:** *Mann* 417! Also in E. Cameroun.
5. **D. goldieana** *Bull* Cat. 1871: viii; Baker in J. Linn. Soc. 14: 535; F.T.A. 7: 449; Bot. Mag. t. 6630. *Pleomele goldieana* (Bull) N. E. Br. (1914). Herb about 1 ft. high, the leaves very ornamental with transverse bars of dark green and silver grey; flowers white in a sessile terminal head.
 **S.Nig.:** Uwet, Cross R. (Dec.) *Holland* 195! Calabar (cult. Kew) *Goldie*!
6. **D. smithii** *Bak. ex Hook. f.* in Bot. Mag. t. 6169 (1875); F.T.A. 7: 440; Morton, W. Afr. Lilies & Orch. fig. 21; Aké Assi, Contrib. 2: 27 [235]. *Pleomele smithii* (Bak. ex Hook. f.) N. E. Br. (1914). *D. fragrans* of Chev. Bot. 646, partly (Guin. specs.), not of (Linn.) Ker-Gawl. A shrub used for hedges or developing into a tree up to 30 ft. high, branched at the base and above, bearing near the top a large crown of leaves several feet long; inflorescence a panicle about 2 ft. long, arising amongst the leaves, flowers yellowish green or white tinged red, opening at night, very fragrant; in drier forest.
 **Guin.:** Labé *Chev.* 12391! Mamou to Kindia *Chev.* 13587. Longuery (Dec.) *Caille* in *Hb. Chev.* 14784! **S.L.:** Bathurst (Jan.) *Dalz.* 8254! York (cult., Jan.) *Deighton* 4586! Freetown (cult.) *Deighton* 2452! **Ghana:** Ankaful, Cape Coast (Feb.) *Hall* 1826! Tano-Ofin F.R. (Feb.) *Lyon* FH 2873! Adamsu (Dec.) *Vigne* FH 3500! **Togo Rep.:** *Kersting* A366!
 [Another species used as a hedge plant in Sierra Leone has shorter leaves than either *D. smithii* or *D. arborea* and it may be *D. fragrans* (q.v.). It has an unbranched weak trunk up to 15 ft. high and, according to notes with *Deighton* 4616, it has not been seen to flower.]
7. **D. arborea** (*Willd.*) *Link* Enum. Hort. Berol. 1: 341 (1821); F.T.A. 7: 439; Keay, Onochie & Stanfield, Nig. Trees 2: 440 (1964). *Aletris arborea* Willd. (1809). *Pleomele arborea* (Willd.) N. E. Br. (1914). Palm-like tree 30–45 ft. high, with several branches, leaves aggregated near branch ends; flowers creamy white in large shortly branched pendulous inflorescences 3–4 ft. long; fruits orange to red; often planted as a fence or ornamental tree, otherwise in upland forest often near streams.
 **S.L.:** Regent (Nov.) *Morton* SL 122! Konnoh country *Burbridge* 483! Panguma (fr. Feb.) *Deighton* 4001! Tingi Mts. (Dec.) *Morton & Gledhill* SL 3022! **Lib.:** Paynesville, 'common' (Aug.) *Voorhoeve* 395! **Ghana:** Kumasi *Cummins* 167! Togo Plateau F.R., Hohoe Dist. (Oct.) *St. Cl.-Thompson* FH 3611! **S.Nig.:** Idanre, Akure Dist. (fl. & fr. Jan.) *Bayo* in *Hb. Brenan* 8717! Nun R. *Mann* 454! Calabar (fr. Mar.) *Onyeachusim & Latilo* FHI 54266! Oban *Talbot* 740! **W.Cam.:** Sabga, 6,500 ft. (fl. & fr. Feb.) *Brunt* 966! Extending to Angola (?). (See Appendix, p. 493.)
8. **D. mannii** *Bak.* in J. Bot. 12: 164 (1874); F.T.A. 7: 438; Hepper in Kew Bull. 22: 451 (1968). *Pleomele mannii* (Bak.) N. E. Br. (1914). A tree about 30 ft. high; in forest.
 **S.Nig.:** Calabar *Thompson* 8! *Mann* 2329!

9. **D. perrottetii** *Bak.* in J. Bot. 12: 165 (1874), incl. var. *minor* Bak. in J. Linn. Soc. 14: 529 (1875); F.T.A. 7: 438. *Pleomele perrottetii* (Bak.) N. E. Br. (1914). *Dracaena mannii* of F.W.T.A., ed. 1, 2: 384, partly; of Sousa in Anais Junta Inv. Col. 6: 50 (1951); of Aubrév., Fl. For. C. Iv., ed. 2, 3: 320; of Keay, Onochie & Stanfield, Nig. Trees 2: 440 (1964), not of Bak. Tree 20–40 or up to 70 ft. high, often planted as a hedge; paniculate inflorescences with white, fragrant flowers.
   **Sen.:** Rio Nunez *Heudelot*! Casamance (Apr.) *Perrottet* 785! **Port.G.:** Cacheu *Esp. Santo* 1258. Safim, Bissau *Esp. Santo* 1911. Fonte, Bubaque *Esp. Santo* 2053. **Guin.:** Nzérékoré (Mar.) *Adam* 3903! **S.L.:** *Thomas* 9591! Yoni, Bonthe I. (Mar.) *Deighton* 2476! Bo (May) *Lane Poole* 189! **Iv.C.:** Moronou (Feb.) *Aké Assi* 8517! Man *Aubrév.* 1125! **Ghana:** Kumasi (Feb.) *Vigne* FH 1634! Akwapim (Apr.) *Murphy* 687! Aburi (Dec.) *Johnson* 870! **S.Nig.:** Lagos (Apr.) *Moloney* 2! Ibadan (Feb.) *Latilo* FH 22745! Calabar to Mamfe (Feb.) *Onyeachusim* FHI 54072! (See Appendix, p. 494, under *D. mannii*.)
   [Although most herbarium material named as *D. mannii* probably belongs to this species, only specimens bearing flowers have been cited above—see Hepper in Kew Bull. 22: 451, fig. 3 (1968).]

10. **D. scoparia** *A. Chev. ex Hutch.* in Kew Bull. 1939: 247; Chev. Bot. 647. A shrub 10–15 ft. high; slender panicle of white flowers, very fragrant; fruits red; in forest.
    **Iv.C.:** Erymacougnié (Jan.) *Chev.* 16967! Alépé (fr. Feb.) *Chev.* 17484! Yapo forest (Oct.) *de Wilde* 3141! Tienkula (fr. Feb.) *Bernardi* 8294! **Ghana:** *Irvine* 1097! Axim (fr. Feb.) *Irvine* 2181! **Tarkwa** (Feb.) *Kinloch* FH 3230!

11. **D. adamii** *Hepper* in Kew Bull. 22: 449, figs. 1 & 2 (1968). *D. densifolia* Bak., partly (*Mann* [8] from F.Po). Apparently a shrub; in forest.
    **Lib.:** Nimba reserve, 2,100 ft. (June) *Adam* 21103! 21220! 21367! 21512! **Ghana:** Ankasa F.R. (Jan.) *Enti & Hall* GC 36297! **F.Po:** *Mann* 8! (Also specimen cult. in Berg-Garten, Herrenhausen from same collection.)

12. **D. talbotii** *Rendle* in Cat. Talb. 112 (1913). *Pleomele talbotii* (Rendle) N. E. Br. (1914). Remarkable for its elongated sharply acuminate leaves and dense spikes with large bracts.
    **S.Nig.:** Oban *Talbot* 1532!

13. **D. viridiflora** *Engl. & K. Krause* in Engl., Bot. Jahrb. 45: 153 (1910); Hepper in Kew Bull. 21: 491 (1968). *D. vaginata* Hutch. in F.W.T.A., ed. 1, 2: (1936), English descr. only. *D. mannii* of F.W.T.A., ed. 1, 2: 384, partly (*Mildbr.* 10571; *Talbot* 729), not of Bak. A shrub with long slender branches and leaves sheathing at the base; lateral inflorescences about 4 in. long.
    **S.Nig.:** Oban *Talbot* 729! Obudu Plateau, 5,000 ft. (Nov.) *Tuley* 1042! **W.Cam.:** Likomba, Victoria (Oct.) *Mildbr.* 10571! Also in E. Cameroun.

14. **D. fragrans** (*Linn.*) *Ker-Gawl.* in Bot. Mag. 1081 (1808). *Aletris fragrans* Linn. (1762). *Pleomele fragrans* (Linn.) Salisb. (1796). Shrub about 5 ft. high; flowers white with pink lines, very fragrant; in drier lowland forest.
    **Ghana:** Aburi Hills (Mar.) *Johnson* 730! Tano-Ofin F.R. (Feb.) *Lyon* FH 2873! **S.Nig.:** Ubulubu (Feb.) *Thomas* 2305! Oban *Talbot* 1412! Aking, Mamfe to Calabar (Feb.) *Onyeachusim & Latilo* FHI 54003! Extending to E. Africa. (See Appendix, p. 493.)
    [The distinction between *D. fragrans* and *D. smithii* is not entirely clear and further study of both herbarium and living material is desirable.—F.N.H.]

15. **D. deisteliana** *Engl.* Bot. Jahrb. 32: 96 (1902). *D. fragrans* of F.W.T.A., ed. 1, 2: 384, partly (*Dalz.* 8348, *Lehmbach* 16), not of (Linn.) Ker-Gawl. Shrub of upland forest above 4,000 ft. and commonly planted as a hedge, with several stems 8 or 9 ft. high; flowers white with pink lines.
    **W.Cam.:** Buea, 4,000–5,000 ft. (Feb.) *Maitland* 357! *Lehmbach* 16. Cam. Mt., 5,000–6,000 ft. (Jan.–Mar.) *Dalz.* 8348! *Dundas* FHI 20374! *Breteler et al.* MC 178! *Maitland* 1328! Bambui, Bamenda (Feb.) *Brunt* 965! Also in E. Cameroun.
    [Plants from Jos Plateau, N. Nigeria (*W. D. MacGregor* 437, *McClintock* 202) have acute leaves and larger flowers like *D. fragrans*, but bracteoles as in *D. deisteliana*.]

16. **D. laxissima** *Engl.* Bot. Jahrb. 15: 478 (1892); F.T.A. 7: 446. *D. elegans* Hua (1898)—F.T.A. 7: 446. *Pleomele laxissima* (Engl.) N. E. Br. (1914). *P. elegans* (Hua) N. E. Br. (1914). Slender shrub with a slender lax pendulous panicle.
    **S.Nig.:** Ikpoba (June) *Farquhar* 13! **W.Cam.:** Su, Bamenda, 3,500 ft. (fr. May) *Maitland* 1686! Also in E. Cameroun, Rio Muni, S. Tomé, the Congos, Sudan, E. Africa, Zambia and Malawi. (See Appendix p. 494.)

17. **D. bicolor** *Hook.* in Bot. Mag. t. 5248 (1861); F.T.A. 7: 448; Hepper in Kew Bull. 22: 450 (1968). *D. cylindrica* Hook. f. in Bot. Mag. t. 5846 (1870); F.T.A. 7: 448. *D. preussii* Eng. (1892)—F.T.A. 7: 439. *Pleomele bicolor* (Hook.) N. E. Br. (1914). *P. cylindrica* (Hook. f.) N. E. Br. (1914). A shrub a few feet high or a small tree up to about 15 ft. high; inflorescence dense, terminal, with white flowers and purple-red bracts; in forest.
    **Ghana?:** cult. Kew 252–00! **S.Nig.:** Oban (Jan.) *Talbot* 1322! 2404! *Onochie* FHI 36261x! Calabar (Feb.) *Mann* 2328! *Thomson* 17! **W.Cam.:** Barombi *Preuss* 328 (seen for Ed. 1). **F.Po:** *Mann* 98! Also in E. Cameroun.

18. **D. camerooniana** *Bak.* in J. Bot. 12: 166 (1874); F.T.A. 7: 442; Chev. Bot. 646. *Pleomele camerooniana* (Bak.) N. E. Br. (1914). *Dracaena mayumbensis* of Aké Assi, Contrib. 2: 27 [235] (as to Iv.C. specs.) not of Hua. *D. lecomtei* Hua (1897)—Aké Assi l.c. (excl. W.Afr. specs.). A shrub of straggling habit up to 15 ft. high; flowers whitish green; in forest undergrowth.
    **Guin.:** Kouria *Chev.* 14967! Santa to Timbo *Chev.* 12439! Diaguissa (Apr.) *Chev.* 12658! **S.L.:** Yonibana (Oct.) *Thomas* 4169! Bayabaya, Scarcies (Feb.) *Sc. Elliot* 4551! Lumbaraya, Talla Hills *Sc. Elliot* 5009! **Lib.:** Gbanga (Sept.) *Linder* 765! **Iv.C.:** Abouabou forest (fr. Jan.) *Leeuwenberg* 2362! Taté, Cavally (Aug.) *Chev.* 19806! **Ghana:** *Burton*! Kumasi (Dec.) *Vigne* 4091! Asamankese (Jan.) *Plumptre* 76! **S.Nig.:** Sapoba (fl. & fr. May) *Kennedy* 1135! 1783! Oban *Talbot* 730! **W.Cam.:** Cam. Mt., 3,500 ft. (Jan.) *Mann* 1204! Barombi *Preuss* 146! Extending to Gabon, Congo, Zambia and Angola.

19. **D. ovata** *Ker-Gawl.* in Bot. Mag. t. 1179 (by error 1180) (1809); F.T.A. 7: 449. *D. afzelii* Bak. (1874)—F.T.A. 7: 448. *D. prolata* C. H. Wright (1905). *D. sessiliflora* C. H. Wright (1914). *Pleomele prolata* (C. H. Wright) N. E. Br. (1914). *Dracaena lecomtei* of Aké Assi, Contrib. 2: 27 [235], not of Hua. A shrub up to 5 ft. high with clusters of leaves; flowers creamy white, fragrant; in forest or planted as a hedge.
    **S.L.:** *Afzelius. Thomas* 8444! Heddle's Farm, Freetown *Lane-Poole* 155! (Jan.) *Dalz.* 987! Guma Dam, Peninsula (fr. Mar.) *Morton* SL 1182! Njala (cult., Feb.) *Deighton* 3364! Mano (fr. Nov.) *Deighton* 2412! **Lib.:** Dukwia R. (Feb.) *Cooper* 170! Monrovia *Whyte*! Zwedru (fr. Aug.) *Baldwin* 7012! Sinoe Basin *Whyte*! 18 miles N. of Tapeta (Feb.) *Voorhoeve* 169! **Iv.C.:** Yapo forest (fr. Oct.) *Leeuwenberg* 1837! Amitioro forest (fr. Dec.) *Aké Assi* 8381! **S.Nig.:** Eket Dist. *Talbot* 3191! Extending to the Congo. (See Appendix, p. 494.)
    [I have included several specimens (e.g. *Baldwin* 7012, *Aké Assi* 8381) with longer inflorescences which may be specifically distinct. Aké Assi (l.c.) regarded them as referable to *D. lecomtei* Hua, but this is a synonym of *D. camerooniana* Bak.—F.N.H.]

20. **D. surculosa** *Lindl.* in Bot. Reg. t. 1169 (1828); F.T.A. 7: 443. (See Appendix, p. 494.)

20a. **D. surculosa** *Lindl.* var. *surculosa*. *D. godseffiana* Sander ex Bak. (1894). *Pleomele surculosa* (Lindl.) N. E. Br. (1914). *P. godseffiana* (Sander ex Bak.) N. E. Br. (1914). A well branched shrub about 3 ft., or at the most 6 ft. high; flowers greenish white; in forest.
    **S.L.:** Freetown (Jan.) *Dalz.* 9447! S. Province (Mar.) *Dawe* 440! Kenema (Jan.) *Thomas* 7477! Joru (Jan.) *Deighton* 3859! **Lib.:** Duport, E. of Monrovia (Jan.) *Voorhoeve* 776! **Iv.C.:** Anguédédou forest, nr. Abidjan (Feb.) *de Wilde* 3433! Kassa forest (fr. Oct.) *de Wilde* 672! **Ghana:** New Jaubin (Mar.) *Johnson*

FIG. 372.—DRACAENA SURCULOSA *Lindl.* var. CAPITATA *Hepper* (AGAVACEAE).
A, flower laid open. B, perianth-segment and stamen. C, stamen. D, stigma. E and F, fruits.

628! Swedru (Feb.) *Dalz.* 8292! **S.Nig.**: Lagos (cult. Kew) *Millen*! Agoi F.R., Calabar (Jan.) *Binuyo* FHI 45464! Oban *Talbot* 1421! Also in E. Cameroun.

[Some plants heavily spotted with yellow on the leaf surface are cultivated as ornamentals in hot houses in temperate countries under the name *D. godseffiana.*]

20b. **D. surculosa** var. **capitata** *Hepper* in Kew Bull. 22: 453 (1968). *D. surculosa* of F.T.A. 7: 443, partly (*Barter* 2095, *Mann* 2327); of F.W.T.A., ed. 1, 2: 386, partly (fig. 316); of Morton, W. Afr. Lilies & Orch. t. 22, not of Lindl. A shrub with slender stems up to 25 ft. high, little branched, young shoots bearing white sheaths and attaining a considerable height before bearing leaves; flowers white in ball-like inflorescences, fragrant; fruits scarlet; in forest.
**S.L.**: Binkolo (Aug.) *Thomas* 1813! Jigaya (Sept.) *Thomas* 2827! Kanya (Oct.) *Thomas* 2990! Ninia (Feb.) *Sc. Elliot* 4815! Musaia (Aug.) *Haswell* 132! **Lib.**: Du R., Montserrado Dist. (July) *Linder* 74! Monrovia, Montserrado Dist. *Whyte*! Peahtah, Salala Dist. (Oct.) *Linder* 1025! Gbata Creek, Suakoko, Gbanga Dist. (July) *Leeuwenberg & Voorhoeve* 4587! Sinoe Basin, Sinoe Dist. *Whyte*! **Iv.C.**: Amitioro forest (Sept.) *de Wilde* 279! Sassandra (Aug.) *de Wilde* 317! **Ghana**: Simin road (fr. Dec.) *Hall* 2829! Tano-Ofin F.R. (Jan.) *Lyon* FH 2865! Amuni (July) *Vigne* FH 1318! Accra Plains (Oct.) *Brown* 370! Mampong Scarp (July) *Vigne* FH 2061! **S.Nig.**: Lagos *Millen* 144! Angiama, Niger Delta *Barter* 2095! Calabar (fl. & fr. Feb.) *Mann* 2327! Oban *Talbot* 146!

*Imperfectly known species*

**D. sp.** A herb with a very short stem bearing congested nodes and petiolate, lanceolate or linear-lanceolate leaves about 20 cm. long and 4–6 cm. broad; inflorescence apparently sessile and congested. Further fertile material is required.
**Lib.**: Mecca, Boporo Dist. *Linder* 10424! Bumbumi to Maola *Linder* 1337! Duo, Sinoe Country *Baldwin* 11356!

## 2. SANSEVIERIA Thunb., Prod. Pl. Cap. 65 (1794); F.T.A. 7: 332; N. E. Br. in Kew Bull. 1915: 185. *Nom. cons.*

Perianth about 8–10 cm. long; flowers in a dense continuous broad spike-like raceme; leaves flat, broadly oblanceolate, about 45–50 cm. long and 6 cm. broad, with hardened red-brown margins when dry; pedicels jointed above the middle; bracts 1–2·5 cm. long, thin and membranous .. .. .. 1a. *longiflora* var. *fernandopoensis*

Perianth at most 5 cm. long; flowers in interrupted racemes:

Perianth 5 cm. long; pedicels jointed above the middle; lower bracts broadly lanceolate, about 4 cm. long, persistent, upper bracts much smaller; leaves broadly oblanceolate, marked with transverse dark and light green bands, up to 10 cm. broad; margins red and white-scarious .. .. .. .. 2. *liberica*

Perianth at most 3 cm. long; bracts small and membranous; leaf margins green:

Leaves markedly banded with green and yellow, elongated-oblanceolate, up to 50 cm. long or more and 5 cm. broad; pedicels jointed a little above the middle
3. *trifasciata*

Leaves not banded or only slightly so, up to 40 cm. long and 5 cm. broad; pedicels jointed about the middle .. .. .. .. .. 4. *senegambica*

1. **S. longiflora** *Sims* in Bot. Mag. t. 2634 (1826); N. E. Br. in Kew Bull. 1915: 256. In the Congo and Angola.
1a. **S. longiflora** var. **fernandopoensis** *N. E. Br.* l.c. 257 (1915). Stemless with a creeping rootstock; lower bracts with sharp slender points; inflorescence about 1 ft. high.
**F.Po**: *Mann* 1169! *Barter* 2060!
2. **S. liberica** *Gér. & Labr.* in Bull. Mus. Hist. Nat. Paris 1903: 170, 173, fig. 4; N. E. Br. l.c. 247. A rather stout herb with several stiff red-margined leaves about 2 ft. high arising from the creeping rhizome, leaf-margins with a red line; inflorescence floriferous, flowers white; in the drier shady places by streams and rock outcrops.
**S.L.**: Coast (Dec.) *Dawe* 408! Peninsula (Nov.) *Morton & Gledhill* SL 144! Sasseni, Scarcies R. (Jan.) *Sc. Elliot* 4532! **Lib.**: cult. *Kew*! **Iv.C.**: Issia rock (May) *Leeuwenberg* 4134! Sifié to Touba (July) *Aké Assi* 9108! **Ghana**: Half Assinie (July) *Chipp* 296! Accra (Sept.) *Irvine* 864! Mankrong (Aug.) *Hall* CC 204! Akuse (June) *Dalz.* 162! Keta (June) *Thorold* 302! **Togo Rep.**: L. Togo, Lomé (July) *Davidson* 11! **N.Nig.**: *Imp. Inst.* 30470! Yola (May) *Dalz.* 231! Nupe *Barter* 1508! **S.Nig.**: Idanre (fr. Jan.) *Brenan* 8705! Extending to C. African Republic. (See Appendix, p. 494.)
3. **S. trifasciata** *Prain* Bengal Plants 2: 1054 (1903); N. E. Br. in Kew Bull. l.c. *S. guineensis* of Gér. & Labr. l.c., not of (L.) Willd. *Aloe guineensis* Jacq. (1762), partly (description; see Brenan in Kew Bull. 17: 174–177 (1963)). Erect herb with transversely barred leaves 2–2¼ ft. long.
**S.Nig.**: Enugu (Mar.) *Hepper* 2872! Oban *Talbot*! Also in the Congo (as var. *laurentii* N. E. Br.). (See Appendix, p. 494.)
4. **S. senegambica** *Bak.* in J. Linn. Soc. 14: 548 (1875); F.T.A. 7: 332; N. E. Br. in Kew Bull. 1915: 235; Sousa in Anais Junta Inv. Col. 6: 50 (1951); Schnell in Ic. Pl. Afr., I.F.A.N. 1: 22 (1953); Berhaut, Fl. Sén. 188. *Aloe guineensis* of F.T.A. 7: 333, partly; of Chev. Bot. 633, not of Willd. *Sansevieria liberiensis* M. Cornu ex A. Chev., Bot. 633, name only. *S. cornui* Gér. & Labr. (1903). A herb with leaves about 2 ft. long; flowers greenish white; in shady places.
**Sen.**: Cape Verde *Berhaut* 472. Youni, Casamance *Chev.* 2562. Itou, Casamance *Chev.* 2579. Perrottet 7821 Richard Tol *Richard* 72! **Gam.**: (Jan.) *Dalz.* 8253! Foni *Dawe* 75! **Port.G.**: Pussubé *Esp. Santo* 1108. Comura (Nov.) *Raimundo & Guerra* 169! **Guin.**: Kindia *Chev.* 12782. **S.L.**: Sugar Loaf Mt. (Dec.) *Sc. Elliot* 3980! **Iv.C.**: Siana to Nandala *Chev.* 21860. Mt. Lémélébou, Bomaké *Chev.* 22105. (See Appendix, p. 494.)

## 193. PALMAE

### By T. A. Russell

Stems stout or slender, sometimes climbing, sometimes very short or almost nothing, often covered by the persistent bases of the leaves; primary root soon disappearing and replaced by roots from the base of the stem. Leaves in a terminal cluster or in the climbing species scattered, sometimes very large, entire, pinnately or digitately divided, the segments or leaflets folded indupli- cately or reduplicately in bud, often sharp at the apex and prickly on the margins or midrib; rhachis often expanded at the base into a fibrous sheath. Flowers small, actinomorphic, bisexual, monoecious or dioecious, sometimes polygamous, arranged in an often paniculate inflorescence (spadix) either amongst or below the leaves. Spathes various, sometimes numerous and enclosing the peduncle and branches of the inflorescence, or few, leathery or membranous; bracteoles often connate below the flowers. Perianth double. Sepals 3, separate or connate, imbricate or open in bud. Petals 3, separate or connate, usually valvate in the male flowers and imbricate in the female. Stamens usually 6, in two series, rarely numerous; anthers 2-locular, loculi globose to linear, opening by slits lengthwise; pollen smooth or rarely echinulate. Ovary superior, rudimentary or absent in the male flowers, 1–3 locular, rarely 4–7 locular, or carpels 3 and distinct or connate only at the base; ovule solitary and erect or pendulous from the inner angle of each carpel or loculus of the ovary. Fruit a berry or drupe, 1–2-locular, or fruiting carpels distinct; exocarp often fibrous, sometimes covered by reflexed scales. Seeds free or adherent to the endocarp; endosperm present, sometimes ruminate; embryo small.

Tropics and warm temperate regions.

Leaves fan-shaped:
  Stem stout unbranched; flowering branches of female spadix very different from those
    of male ..      ..      ..      ..      ..      ..      ..      ..      10. **Borassus**
  Stem commonly branching dichotomously; flowering branches of male and female
    spadices not dissimilar      ..      ..      ..      ..      ..      11. **Hyphaene**
Leaves pinnate, pinnatisect or bifurcate:
  Stem not climbing, but erect, occasionally very short:
    Leaf-segments induplicate in vernation (V-shaped in section)      ..      12. **Phoenix**
    Leaf-segments reduplicate in vernation (Λ-shaped in section):
      Lateral segments rhomboid-oblanceolate or obliquely truncate:
        Lateral segments narrowly rhomboid-oblanceolate, doubly dentate in the upper
          part; terminal segment broadly cuneate-rhomboid   ..      ..   1. **Podococcus**
        Lateral segments obliquely truncate, curved as they meet the rhachis; stems very
          short, caespitose   ..      ..      ..      ..      ..      ..   2. **Sclerosperma**
      Lateral segments linear or ensiform:
        Fruit not cone-like and not covered with scales, but fibrous or spongy:
          Male and female flowers on separate spadices   ..      ..      ..      ..   3. **Elaeis**
          Male and female flowers on the same spadix   ..      ..      ..      ..   4. **Cocos**
        Fruit cone-like, covered with scales      ..      ..      ..      ..   5. **Raphia**
  Stem prostrate or climbing:
    Stem prostrate, branched; estuarine palm ..      ..      ..      ..      ..   13. **Nypa**
    Stem climbing:
      Mature leaf invariably terminating in leafy segments; inflorescence borne on slender
        prickly leafless branches   ..      ..      ..      ..      ..      ..   6. **Calamus**
      Mature leaf usually terminating in a hooked cirrus:
        Leaf-sheath with spines, or scars of caducous spines:
          Spadix terminal; ochrea elongate, pointed or torn   ..      7. **Ancistrophyllum**
          Spadix axillary; ochrea short, obliquely truncate   ..      ..   8. **Oncocalamus**
        Leaf-sheath striate-veined without spines or scars   ..      ..   9. **Eremospatha**

## 1. PODOCOCCUS Mann & Wendl. in Trans. Linn. Soc. 24: 426 (1864); F.T.A. 8: 99.

Leaves 6–9 in number, up to 2 m. long, pinnatisect; petiole 30–45 cm. long; leaf-segments 8–10, rhomboid-oblanceolate or fan-shaped, with many radiating nerves, the lower margins straight, the upper dentate; male spadix simple, pedunculate, the floriferous part about 30 cm. long, spike-like; male flowers many, commonly paired and sessile in pits of the spadix, stamens 6; female spadix similar with many flowers, usually solitary in pits of the spadix, the oblong ovary with 3 short recurved stigmas; fruit baccate, orange-red, oblong-cylindrical, 2·5 cm. long, 0·6 cm. wide      *barteri*

P. **barteri** *Mann & Wendl.* l.c. tt. 38, 40, 43 (1864); F.T.A. 8: 100. A slender elegant palm of the forest floor with reed-like annulate stem 6–9 ft. high, covered in many of its parts with a rusty scurf.
S.Nig.: Brass *Barter* 36! 325! 1837! Oloibiri, Brass *Tuley & Gottschalk* 600! Nun R. (Aug.) *Mann* 452! Stubbs Creek F.R., Eket *Onochie* FHI 32917! Ukpe-Sobo *Onochie* FHI 33423! Also in Gabon and Congo. (See Appendix, p. 510.)

## 2. SCLEROSPERMA Mann & Wendl. in Trans. Linn. Soc. 24: 427 (1864); F.T.A. 8: 100.

Leaves borne near ground-level, erect, unarmed, with fibrous basal-sheath up to 40 cm. long; lamina sometimes nearly entire, more commonly segmented, the segments sub-opposite, 9–12 on each side of the rhachis, unequal in width, about 50 cm. long, obliquely truncate, slightly narrowed and curved at junction with the rhachis, shining green above, whitish below with conspicuous parallel veining; spathes 2, fibrous, persistent; spadix erect, with flowers spirally arranged, female flowers interspersed with male flowers in the lower part, male flowers only above; fruit a large drupe with fibrous mesocarp, and thin hard endocarp; the seed 2 cm. across, with very hard albumen when ripe .. .. .. .. .. .. .. *mannii*

S. mannii *Wendl.* l.c. 427, tt. 38, 40 (1864); F.T.A. 8: 101. A slender palm forming clumps, with fronds up to about 16 ft. high, borne on a very short stem; in swampy places in areas of high rainfall.
Ghana: Ankasa F.R. *Hall & Enti* GC 36150! S.Nig.: Umon-Ndealachi F.R., Calabar Dist. *Iyizoba* FHI 20815! Ikot Okpora, Calabar Dist. *Tuley* s.n.! 654! Oban F.R. *Onochie* FHI 36356x! W.Cam.: Kembong F.R. *Richards* 5215! Also in Gabon and Angola.

## 3. ELAEIS Jacq., Select. Stirp. Am. Hist. 280 (1763); F.T.A. 8: 124.

Leaves many, in a terminal crown, up to 5 m. long, arching, pinnate; petiole short, thick, commonly spiny at margins; segments in 4 ranks, ensiform, acuminate, glabrous, the lowest spinescent; spadices interfoliaceous, the male pedunculate with many, crowded, short, lateral floriferous branches, 10–15 cm. long, ending in a sharp hard point; male flowers densely crowded, minute, with linear sepals, linear-oblong petals, and 6 stamens; female spadix with short peduncle and branches congested into a globose capitulum, the bracts rigid, acuminate; female flowers much larger than the male, with ovoid ovary and relatively large revolute stigmas; fruit ovoid or somewhat angular, often bright red and shining black when ripe; pericarp spongy and oily, fibrous inside; endocarp hard and often thick, enclosing the adnate seed *guineensis*

E. guineensis *Jacq.* l.c. t. 172 (1763); F.T.A. 8: 125. The African Oil Palm, easily recognizable by its arching, dark-green leaves and straight trunk clothed when young with petiole-bases, is cultivated and occurs spontaneously in much of the forest zone from Senegal to Cameroun, being particularly abundant near habitations, in land which has been tilled, and in river valleys. Several forms, differing in structure and colour of the fruit, are described in Kew Bull. 1909: 33 and 1914: 285, and in Holland 4: 734. Cultivated in other parts of the humid tropics.

A form occasionally seen, differing in vegetative characters, is the King Palm (*E. guineensis* var. *idolatrica* A. Chev.) with fronds held at an acute angle and leaf-segments remaining fused for the greater part. (See Appendix, p. 499.)

## 4. COCOS Linn., Sp. Pl. 1188 (1753); F.T.A. 8: 126.

Leaves pinnate, light-green, in a cluster at the top of the stem; petiole slightly sheathing, unarmed; segments almost equidistant, in 2 ranks, linear-lanceolate, mid-vein thick, yellowish; spadix 1 m. long or more, simply panicled, branches bearing usually few female flowers towards the base and many male flowers throughout their length; male flowers asymmetrical, small, with valvate sepals and petals; female flowers larger, 2·5 cm. long, globose, supported by broad bracteoles; fruit 20–30 cm. long, trigonously ovoid or globose, green or yellowish .. .. .. .. *nucifera*

C. nucifera *Linn.* l.c. (1753); F.T.A. 8: 126. The Coconut, distinguished by its smooth, rather slender trunk which is seldom straight, its crown of pinnate leaves, and its well-known fruit, is cultivated and grows chiefly in coastal areas from Senegal to Cameroun. Cultivated in similar situations throughout the tropics. (See Appendix, p. 497.)

## 5. RAPHIA P. Beauv., Fl. Oware 1: 75 (1806); F.T.A. 8: 104; Russell in Kew Bull. 19: 173–196 (1965).

Spadices curved, branching to form numerous hanging partial inflorescences:
Partial inflorescences short, solid, with conspicuous papery bracts from which protrude very short ultimate branchlets; fronds tall, erect, stiff, glaucous-green, segments thickly armed on margins and midvein with fine blackish spines; male flowers with calyx entire, lobed; stamens 10–12, filaments fused for half their length; fruit top-shaped, about 8 cm. long, 4·5 cm. wide, mahogany-red, shining, with stout blunt beak about 8 mm. long, and 9 or 10 vertical rows of rather flat or slightly convex scales with shallow furrow .. .. .. .. .. .. 4. *sudanica*
Partial inflorescence not thick and solid; the ultimate branchlets distinct, free, and spreading:
Ultimate branchlets in mature partial inflorescence crowded in close ranks, more or less in one plane:
Bract at base of partial inflorescence small, the exposed partial inflorescence roughly racquet-shaped, with stout, coarse, crowded branchlets; upper part of trunk clothed with a mass of black fibres; fronds tall, dark green shining above; corolla

segments distinctly thickened near the tip; male flower with (15–) 18–22 (–24) stamens; female flower with 12–15 narrowly sagittate staminodial points; fruit top-shaped or ellipsoid, variable in size from 6–12 cm. long, 4–5 cm. wide, with stout beak 1–1·5 cm. long, more or less obliquely tipped, and (11–) 12 (–14) vertical rows of scales, convex, slightly less broad than long, lightly and narrowly furrowed, chestnut-coloured, or cinnamon with darker point ..        .. 2. *hookeri*

Bract at base large, triangular, persistent, partly concealing the short partial inflorescence which has slender, light brown, polished branchlets; fronds tall, rising in a cluster from ground-level, or from a very short stem, dark green above, the segments armed on mid-vein and margins with many fine brown spines; male flowers with cup-shaped calyx, scarcely lobed; corolla with basal quarter fused with the filaments to form a solid column, the upper portion splitting into 3 narrow pointed segments scarcely thickened at the tip; stamens 6, the filaments thick, angular, joined for most of their length; female flowers numerous, much smaller than the male, with calyx slightly lobed enclosing the shorter sharply tridentate corolla; staminodes 6–9 with deltoid points; fruit top-shaped, 5·5–8·5 cm. long, 3·5 cm. wide, with rounded top ending abruptly in a fine beak 3–4 mm. long; scales in (9–) 10–11 (–12) vertical rows, convex with narrow median groove, chestnut or hazel, darker towards the point ..        5. *farinifera*

Ultimate branchlets in mature partial inflorescence lax, spreading on all sides:

Ultimate branchlets slender, tapering, flattened, with flowers superficially in 2 rows; fronds long, arching, with stout rhachis; male flower curved, calyx shortly tridentate, corolla segments little thickened at tip, with sharp, pricking point, stamens (6–) 9, filaments free for the greater part; female flower with usually 9 deltoid staminodes; fruit cylindrical-ellipsoid, 6·5–9 cm. long, 3·5–4 cm. wide, ending abruptly in a sharp beak 3–5 mm. long; scales in (8–) 9 vertical rows, rhomboidal, as broad as long, flat or slightly concave towards the point, with broad groove, hazel-brown or chestnut with darker margin, and membranous point lighter in colour     ..      ..      ..      ..      ..      ..     1. *vinifera*

Ultimate branchlets not flattened but rounded, rather knobby, with flowers obviously in 4 ranks; fronds light green, not shining; male flowers with calyx distinctly segmented, sepals separated for about one-third of their length, rather sharply dentate, stamens usually 9–12, free to the base; fruit elongate-ovoid, tapering towards the tip, 8–9·5 cm. long, 4 cm. wide, the beak narrow, sharp, about 5 mm. long, scales in 8 or 9 vertical rows, bigibbous, with deep furrows running the length of the fruit     ..      ..      ..      ..     3. *palma-pinus*

Spadix central, forming a large erect panicle, the partial inflorescences arching, strongly compressed, with distichous spikes bearing rather similar male and female flowers; calyx extending beyond the bracteole, deeply and acutely tridentate; corolla segments lanceolate, slightly falcate, sharply pointed; stamens of male flowers 6, filaments free to the base; female flower with conspicuous inner bracteole, extended on one side in a triangular point, staminodes 6, borne on the mid part of the corolla; fruit obovoid or fusiform, tapering towards both ends, up to 7·5 cm. long, 3·5 cm. at greatest breadth, the beak 5 mm. long, broadly based but narrowing to a fine point; scales in 9 vertical rows, flattish or slightly convex, with faint shallow furrow and chaffy fringe, dark mahogany, shining     ..      ..      ..      ..     6. *regalis*

1. **R. vinifera** *P. Beauv.* Fl. Oware 1: 77, tt. 44–46 (1806), and in Desvaux, Jour. Bot. 2: 79–87 (1809); Mann & Wendland in Trans. Linn. Soc. 24: 437 (1864); F.T.A. 8: 106; Beccari in Webbia 3: 88 (1910), and in Agric. Colon. 4: t. 6 (1910); A. Chev. in Rev. Bot. Appliq. 12: 208 (1932); Russell in Kew Bull. 19: 179, figs. 1A, 5D (1965). *R. gaertneri* Mann & Wendl. l.c. (1864) (excl. syn. *Sagus palma-pinus* Gaertn.); F.T.A. 8: 105, partly. *R. mannii* Becc. in Webbia 3: 70 (1910). *R. wendlandi* Becc. l.c. 81 (1910). A handsome palm of watery places, particularly abundant on the edge of creeks; the stout trunk of moderate size, attaining about 25 ft. in height, is crowned by large arching fronds up to 40 ft. long with sturdy midribs whose popular use as poles has prompted the common name of Bamboo-palm.
**Dah.:** Porto Novo *fide* Chev. l.c. **S.Nig.:** Okogbo, Benin *Redhead* FHI 48387! Nun R. *Tuley*! Okigwi *Tuley*! Oloibiri, Brass *Tuley*! Calabar *J. Irvine*! **F.Po:** *Mann*! Also in E. Cameroun and south to Congo. (See Appendix, p. 511.)

2. **R. hookeri** *Mann & Wendl.* in Trans. Linn. Soc. 24: 438 (1864); F.T.A. 8: 107; Beccari in Webbia 3: 109 (1910); Russell l.c. 181, figs. 2, 5E. *R. gigantea* A. Chev. in Rev. Bot. Appliq. 12: 198 (1932); F.W.T.A., ed. 1, 2: 388. *R. sassandrensis* A. Chev. l.c. 199 (1932). The common Wine Palm of the forest zone, with trunk up to 30 ft. high, and large dark green, shining fronds. Over its wide range it shows some variation in characters, particularly in size and shape of fruit; in E. Nigeria, where it is abundant and sometimes cultivated, several forms are distinguished.
**Guin.:** Boké *Jac.-Fél.* 7503! **S.L.:** Pendembu *Thomas* 858! Neikehun, Jama *Deighton*! Ngerihun, Jaiama *Jordan* 1015–1018! 1022–1028! Bonthe *McQuistan* 1! **Iv.C.:** Dabou *Chev.* 15496! Bouroukrou *Chev.* 16530! Guidéko, Sassandra *Chev.* 19095! **Ghana:** Cameron 86/1882! R. Volta *Rumsey* 8/82! Yenahin, W. Ashanti *Chipp* 134! Axim *Chipp* 422! Prapragum, Cape Coast *Hall* 2442! **Dah.:** Porto Novo *Chev.*! **S.Nig.:** Ibadan *Emwiogbon & Adebusuyi* FHI 30148! Umuahia *Tuley*! Abak *Tuley*! Calabar *Milne*! **W.Cam.:** Mamfe *Brunt* 1143! **F.Po:** *fide* Mann l.c.   Extends to Rio Muni and Gabon. (See Appendix, p. 510.)

3. **R. palma-pinus** *(Gaertn.)* Hutch. F.W.T.A., ed. 1, 2: 388 (1936); Russell l.c. 177, figs. 1B, 5A. *Sagus palma-pinus* Gaertn., Fruct. 1: 27, t. 10 (1788). *Raphia gaertneri* of F.T.A. 8: 105, partly, not of Mann & Wendl. *R. gracilis* Becc. in Agric. Colon. 4: t. 5 (1910), and in Webbia 3: 92 (1910); F.W.T.A., ed. 1, 2: 388. Palm forming clumps or thickets in fresh or slightly brackish water, abundant in swamps behind the mangrove area; the trunk is usually 6–10 ft. high, but sometimes considerably more, and several stems may arise from the one rootstock, bearing fronds distinctively light-green in colour.
**Sen.:** Casamance *Chev.*! **Gam.:** Kayi *Daniell*! **Port.G.:** Teixeira Pinto, Cacheu *Raimundo & Guerra*

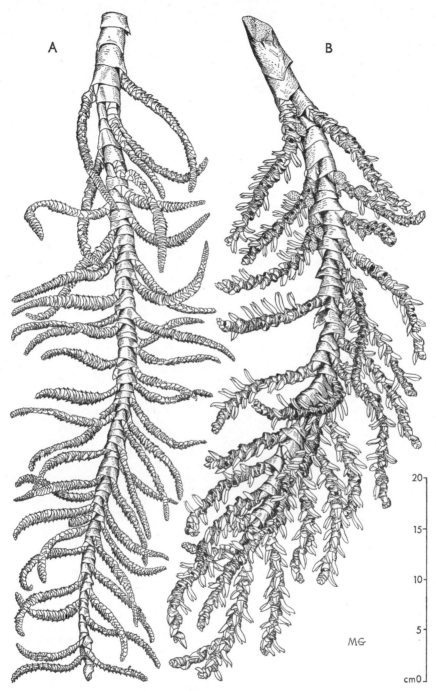

A

B

20

15

10

M.G

5

cm 0

FIG. 373.—RAPHIA SPP. (PALMAE).

A, *R. vinifera* P. Beauv.—partial inflorescence, showing lax arrangement of the tapering branchlets.  B, *R. palma-pinus* (Gaertn.) Hutch.—partial inflorescence with male flowers and young
fruits, the knobbly branchlets in lax arrangement.

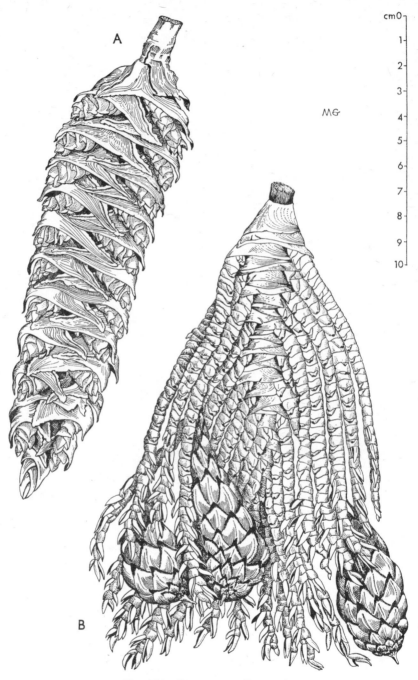

**Fig. 374.—Raphia spp. (Palmae).**

A, *R. sudanica* A. Chev.—partial inflorescence, the branchlets short and condensed. B, *R. farinifera* (Gaertn.) Hylander—partial inflorescence in fruiting state, old male flowers confined to tips of the branchlets.

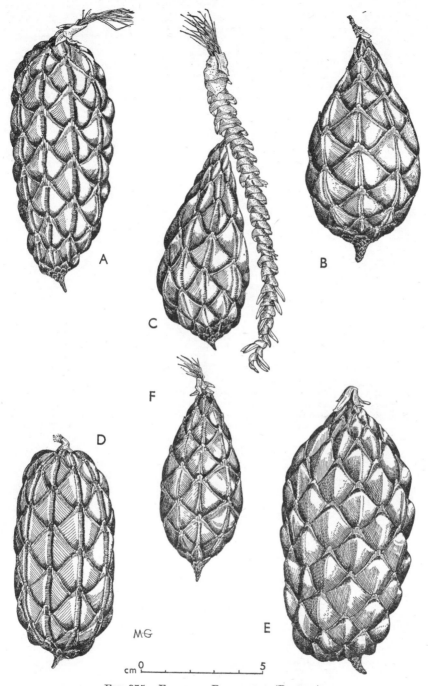

FIG. 375.—FRUITS OF RAPHIA SPP. (PALMAE).

A, *R. palma-pinus* (Gaertn.) Hutch.  B, *R. sudanica* A. Chev.  C, *R. farinifera* (Gaertn.) Hylander.
D, *R. vinifera* P. Beauv.  E, *R. hookeri* Mann & Wendl.  F, *R. regalis* Becc.

549! **Guin.:** Dubreka *Dybowski* 12! Kindia *Chev.* 13023! Friguiagbé *Chillou*! Koba *Jac.-Fél.* 7502! **S.L.:** Berria *Sc. Elliot* 5002! Dodo, Sembehun *Deighton* 2398! Panguma, Kenema *Deighton* 2979! Bandajuma, Small Bo *Jordan* 1010–1014! Waterloo *Lane-Poole* SLH 254! **Lib.:** *Dinklage* 1947. **Iv.C.:** Bingerville *Chev.* 1/490! Azaguié *Aké Assi* 8915! **Ghana:** Bonsaso, E. of Tarkwa *Chipp* 428! (See Appendix, p. 510.)

4. **R. sudanica** *A. Chev.* in Mém. Soc. Bot. Fr. 2, 8: 95 (1908), and in Rev. Bot. Appliq. 12: 206 (1932); Russell l.c. 182, figs. 3A, 5B. *R. heberostris* Becc. in Agric. Colon. 4: t. 5 (1910), and in Webbia 3: 96 (1910). A palm of swampy land in savanna country of the Sudan zone, often forming thickets, with a stout trunk, usually 6–10 ft. but up to 25 ft. high, and easily recognizable by its stiff, rather upright, leaves, its spadix with greatly condensed, solid branches, and its blunt-nosed, mahogany-red fruits.
**Sen.:** *fide* Beccari l.c. **Gam.:** Kayi *Daniell*! **Mali:** Kati *Chev.*! Koulikoro *Rogeon*! **Guin.:** Kollangui *Chev.* 12570! 12867! Timbo, Fouta Djalon *Chev.* 12531! **S.L.:** Musaia *Small* 500! *Jordan* 1049! Kabala *Glanville*! **Iv.C.:** Mankono *Chev.* 21946! Vavoua *Aké Assi* 9115! **Ghana:** Ejura *Vigne* FH 3462! Nsawkaw, Wenchi *Hall* 2040! Sampa *Tomlinson*! **Dah.:** *fide* Beccari l.c. **N.Nig.:** *Barter*! **S.Nig.:** Eruwa *Stanfield* FHI 54314! (See Appendix, p. 510.)

5. **R. farinifera** (*Gaertn.*) *Hylander* in Lustgarten 31: 91 (1952); Russell l.c. 187, figs. 3B, 5C; Hepper in Bull. I.F.A.N. 28, Sér. A: 126 (1966). *Sagus fariniferus* Gaertn. in Fruct. 2: 186, t. 120 (1791). *S. ruffia* Jacq., Fragm. 7, t. 4/2 (1801). *Raphia ruffia* (Jacq.) Mart., Hist. Nat. Palm. 3: 217 (1839). *R. pedunculata* P. Beauv., Fl. Oware 1: 78, t. 46 (1806). *R. kirkii* Engl., Pl. Ost. Afr. A: 10 (1895); Beccari in Webbia 3: 58 (1910). *R. humilis* of Portères in Rev. Bot. Appliq. 27: 204 (1947). This species, widely distributed in E. and Central Africa, is represented in highland W. Cameroun by a form without a clear trunk, but producing a great cluster of fronds at ground-level, or on a very short stem covered by petioles; common near villages at the edge of water courses, the midribs providing the chief local material for house building.
**N.Nig.:** S. end of Mambila Plateau *fide* Hepper l.c. **W.Cam.:** Bamenda *Keay* FHI 37938! Baforkum. Bambui *Brunt* 1139! 1140! Ndop Plain *Brunt* 1194–7!

6. **R. regalis** *Becc.* in Webbia 3: 125 (1910); Pellegrin, Fl. Mayombe 3: 56 (1938); Raponda-Walker & Sillans, Pl. Ut. Gabon 339; Russell l.c. 189, figs. 4, 5F. *R. monbuttorum* of Rendle, Cat. Talb. 149. Stem short and stout clothed with leaf-bases, the fronds tall with many segments (according to Beccari); the form of its spadix and, in particular, the zigzag arrangement of flowers on the distichous spike, render it distinctive from other species; in Gabon and Angola it is said to grow not in swamps but on forested slopes.
**S.Nig.:** Oban *Talbot* 948! Also in Gabon and Angola.
[This palm is represented from W. Africa by a single incomplete gathering in the British Museum.]

*Imperfectly known species.*

**R. humilis** *A. Chev.* in Rev. Bot. Appliq. 12: 204 (1932). The only herbarium specimen seen is a portion of leaf in the Paris Museum. Chevalier describes this as a palm forming no trunk or suckers, with short, very spiny leaves up to 3·5 m. long, bearing 60 pairs of segments. The short spadix is almost as broad as long, and the branches bear crowded spikes. Male flowers are large, up to 2·4 cm. in length, with 10–12 stamens, the filaments united for most of their length. The fruit, 4·5–7 cm. long and 2·5 cm. wide, has a large beak 1 cm. long, and scales in 10 (9–12) rows. It was found by watercourses in savanna country.
**Dah.:** Agouagon and Savé *fide* Chev. l.c.

## 6. CALAMUS Linn., Sp. Pl. 325 (1753); F.T.A. 8: 107; Beccari in Ann. Roy. Bot. Gard. Calcutta 11: 73 (1908).

Sheathed stem 1·8–2·5 cm. diam., seldom straight; leaf-sheath often thickly armed with straight spines and sometimes bearing a spiny lateral flagellum; ochrea similarly spiny, 4–5 cm. long, marcescent; leaves distant from each other, about 1 m. long; segments rather numerous, linear-lanceolate, up to 30 cm. long, 2·5 cm. broad, with fine subulate spines on margin and often on lower surface; rhachis, with sharp recurved hooks, not ending in a cirrus; spadix long, slender, tapering to a fine flagellum armed with recurved thorns, with partial inflorescences few, strict, bearing floriferous branchlets about 5 cm. long; flowers solitary in male spadix, paired in the fruiting spadix, a female flower with a male or neuter in the same bracteole; bracteoles obliquely truncate; fruit oblong-ellipsoid, pointed, about 1·5 cm. long, with 16–21 vertical rows of triangular-ovate, fimbriate scales ..     ..     ..     .. *deërratus*

C. **deërratus** *Mann & Wendl.* in Trans. Linn. Soc. 24: 429, t. 41 (1864); F.T.A. 8: 108; Chev. Bot. 673; Berhaut, Fl. Sén. 211. *C. barteri* Becc. ex Drude (1895)—F.T.A. 8: 109. *C. heudelotii* Becc. ex Drude (1895)—F.T.A. 8: 110. *C. leprieurii* Becc. (1902). *C. perrottetii* Becc. (1902). *C. akimensis* Becc. (1908). *C. falabensis* Becc. (1908). A slender climber with prickly stems, scrambling to tops of trees in wet forest; the male and fruiting spadices being borne on different plants.
**Sen.:** *Heudelot* 761! Cape Verde *Adam* 763! Sangalkam, Karang *Berhaut* 877. Near R. Casamance *Leprieur*! **Gam.:** *Ingram*! **Mali:** Banancoro *Chev.* 500! **Port.G.:** Catió *d'Orey* 216! **Guin.:** Farana *Chev.* 1342! Konyaya, Coyah *Chillou* 1905! **S.L.:** Bagroo R. (fr. Apr.) *Mann* 895! Musaia (Mar.) *Sc. Elliot* 5121! Kambia (fr. Jan.) *Sc. Elliot* 4738! Pepor, Taiama (fr. July) *Deighton* 1847! Jigaya *Thomas* 2753! **Lib.:** Peahtah *Linder* 1078! 1116! Banga *Linder* 1226! Kakatown *Whyte*! **Iv. C.:** Bouroukrou *Chev.* 16592; 16594; 16597. Soubré *Chev.* 19103. 18 Km. NW. of Sassandra *Leeuwenberg* 2882! Béréby *Oldeman* 589! **Ghana:** Bobiri F.R. (Dec.) *Tomlinson*! S. Fomangsu F.R. *Vigne* FH 3951! Kibbi, Akim (fr. Dec.) *Johnson* 242! Yenahin, Ashanti (Mar.) *Chipp* 127! Amentia, Ashanti (Mar.) *Vigne* FH 1868! **N.Nig.:** Koton Karifi (Nov.) *Keay* FHI 28091! *Allison* FHI 6994! **S.Nig.:** Ikorodu, Lagos *Tuley* 846! Okeluse, Ondo (fr. Mar.) *Unwin* 223! Ifon *Ayewoh* FHI 3854! Onitsha *Barter* 110! Ahoada *Imp. Inst.* 347! **W.Cam.:** Cam. R. *Mann* 2147! Extending eastwards to Uganda and south to Angola. (See Appendix, p. 497.)

## 7. ANCISTROPHYLLUM (Mann & Wendl.) Mann & Wendl. in Kerch., Palm. 230 (1878); F.T.A. 8: 113.

Leaf-segments numerous, almost equidistant on the rhachis, elongate, linear-lanceolate sometimes sigmoid, variable in length up to 60 cm., commonly about 30 cm. long, 2–3 cm. broad, with numerous fine spines on the margin, often with 1 prominent vein (but sometimes 2, or as many as 5) ciliate-spinose on the upper surface; rhachis without pinnae near the base where it bears many long straight spines, usually prolonged above into a stout unarmed cirrus, triangular in section, bearing several pairs of sharply triangular hooks thickened at the base ; spadix forming a large terminal panicle of many branches up to 50 cm. long, bearing numerous many-flowered branchlets about 25 cm. in length; bracts imbricate, pointed, glabrous; bracteoles cupular, truncate, 4 mm. long; flowers 1–1·2 cm. long, 3 mm. broad, slightly curved, narrowed to a rather obtuse tip; calyx campanulate 4 mm. long; fruit broadly ellipsoid, sharply mucronate, 1·5–1·8 cm. long, 1–1·2 cm. across, with 15–18 vertical rows of closely-packed, rhomboid-ovate, shining scales; seed smooth, flat on one side

1. *secundiflorum*

Leaf-segments mostly broad, sigmoid, often inequidistant or in irregular clusters on the rhachis :

Leaves tomentose at first; sheath tubular 30 cm. long, usually armed towards the top with crowded spines on laterally-flattened bases, extending into an ochrea 15–30 cm. long bearing similar spines; rhachis thickly armed with recurved spines throughout, usually produced into a fine cirrus bearing 4–6 pairs of hooks; segments lanceolate-sigmoid, marcescent, the longer 20–30 cm. long, 2½–10 cm. broad, opaque, slightly rough, firm when dry; with 1–4 (seldom 5) main nerves more or less prominent the, margin commonly with fine spinules; spadix terminal up to 60 cm. long with 5 or 6 primary branches bearing many-flowered branchlets 10–20 cm. long, slender, pendulous; bracts acuminate, bracteoles 2-horned, the points distinct and firm; flowers paired, divergent, fusiform; calyx campanulate with lobes 4 mm. long, ovate, acuminate; corolla twice as long, lobes elliptic-lanceolate; fruit red when fresh, globose 1·5 cm. diam., with scales in 12 vertical rows; seed subglobose warty, deeply grooved on one side .. .. .. .. .. .. 2. *opacum*

Like the preceding in many respects; the rhachis, which is usually produced into a cirrus and is armed throughout with recurved spines, has a rather persistent tomentum; leaf-segments sigmoid, gradually acuminate with a long fine point, smooth, shining on both surfaces, papery when dry, commonly, though not invariably, with 2 prominent main veins, margins unarmed; fruits sub-globose up to 1·5 cm. in length with 18 vertical rows of very small fimbriate scales; seed smooth, flattened on one side .. .. .. .. .. .. .. .. .. 3. *laeve*

1. **A. secundiflorum** (*P. Beauv.*) *Wendl.* in Kerch., Palm. 230 (1878); F.T.A. 8 : 115; Chev. Bot. 675. *Calamus secundiflorus* P. Beauv., Fl. Oware 1 : 15, t. 9–10. A large rattan, common in many areas of wet forest, with rather smooth stems, 100 ft. or more in length, scrambling over trees, and with handsome clusters of red fruits.
   **Port.G.:** Cacine *d'Orey* 262 ! **Guin.:** Ditinn *Chev.* 12630. Farana *Chev.* 13421. Irié *Roberty* ! **S.L.:** Kwaoma, Tunkia (fr. Mar.) *Deighton* 4119 ! Gorahun (fr. May) *Jordan* 2064 ! 2065 ! Tabe (Sept.) *Deighton* 3090 ! Njala *Deighton* 2593 ! Gola Forest, Bagbe *Small* 697 ! **Lib.:** Nimba *Adam* 20746 ! **Iv.C.:** Anguédédou *de Wilde* 3101 ! Bianouan *Leeuwenberg* 3954 ! Bouroukrou *Chev.* 16358. Fort Binger *Chev.* 19536. Makougnié *Chev.* 16596. **Ghana:** Mampong, Ashanti (Aug.) *Vigne* FH 2410 ! Bobiri *Tomlinson* ! Ankaful F.R. *Hall* 33711 **S.Nig.:** Ifon, Ondo (Mar.) *Ayewoh* FHI 3853 ! Okomu F.R., Benin (fr. Dec.) *Brenan* 8580 ! Nun R. (Aug.) *Mann* 453 ! Ikot Arna (Aug.) *Maggs* 150 ! Calabar (fr. June) *J. Smith* 53 ! Also in E. Cameroun, Congo, and Angola. (See Appendix, p. 495.)
   [Beccari, in Webbia 3 : 255 (1910), described from Cameroun a species, *A. acutiflorum* Becc. differing from *A. secundiflorum* in its more sharply apiculate flowers, less rigid leaf-segments, and other minor characters. Tuley, in correspondence, states there are two forms in Nigeria, one having drooping leaf-segments, the other more rigid segments held horizontally. Nevertheless all the flowering material at Kew, including collections of Tuley, seems referable to *A. secundiflorum*.]
2. **A. opacum** (*Mann & Wendl.*) *Drude* in Engl., Bot. Jahrb. 21 : 111 (1895); F.T.A. 8 : 115. *Calamus opacus* Mann & Wendl. in Trans. Linn. Soc. 24 : 431, tt. 41, 43 (1864). A scrambling palm, with stems up to 100 ft. long and bright red fruits; in forest.
   **Ghana:** Awoso, W.P. (Dec.) *Morton* A3618 ! Kakum F.R. *Hall* 2748 ! Awisa, Akim (Dec.) *Irvine* 2075 ! Banka, Ashanti *Vigne* FH 1365 ! Amentia (Mar.) *Vigne* FH 1875 ! **S.Nig.:** Calabar *Milne* ! **W.Cam.:** Victoria (fr. Oct.) *Maitland* 761 ! Mongu *Kwankam* Cam. 2/40 ! **F.Po:** *Barter* ! (fl. Apr., fr. Dec.) *Mann* 97 ! Also in Gabon. (See Appendix, p. 495.)
3. **A. laeve** (*Mann & Wendl.*) *Drude* in Engl., Bot. Jahrb. 21 : 111 (1895); F.T.A. 8 : 114; Chev. Bot. 674. *Calamus laevis* Mann & Wendl. in Trans. Linn. Soc. 24 : 430 (1864). A slender climbing palm of dense and swampy forests.
   **Lib.:** Banga *Linder* 676 ! 1228 ! **Iv.C.:** Attié (fr. Dec.) *Chev.* 22658 ! Bouroukrou *Chev.* 16796. Sanvi, Assinie *Chev.* 17859. Issia *Boughey* 14732 ! Anguédédou *de Wilde* 3432 ! Banco *Oldeman* 137 ! Also in E. Cameroun and Gabon.

## 8. ONCOCALAMUS (Mann & Wendl.) Hook f. in Benth. & Hook. f., Gen. Pl. 3 : 881, 936 (1883); F.T.A. 8 : 110.

Leaf-segments, usually about 16 on each side of the rhachis, linear-lanceolate, acute, or slightly falcate, up to 25 cm. long, 1·75–2 cm. across, rather stiff, with a prominent mid-rib; rhachis, leafy to the base without stout, hooked prickles, and extended above into a slender cirrus furnished with reflexed hooks; leaf-sheath with flattened spines, which are caducous, leaving a scar oblique to the axis; ochrea truncate 2–3 cm. long; flowers in groups, a female with male on each side; fruit unknown

1. *mannii*

Leaf-segments few, usually 5–7 on each side of the rhachis, broadly lanceolate or sigmoid, 10–15 cm. long, 2·5–5 cm. broad, thin, papery, the broader with 2 or more prominent veins; rhachis leafy almost to the base, where are a few stout hooked prickles on the margin, and extended upwards into a cirrus as in the previous species; leaf-sheath and ochrea as above; flowers unknown; fruits light-yellow .. 2. *wrightianus*

1. **O. mannii** (*Wendl.*) *Wendl.* in Kerch., Palm. 252. *Calamus (Oncocalamus) mannii* Wendl. in Trans. Linn. Soc. 24: 436, tt. 41, 43 (1864). A slender-stemmed rattan of swampy forest, climbing to 60 ft.
   **S.Nig.**: Calabar, Ikot Okpora *Tuley* 1078! Also in Gabon.
2. **O. wrightianus** *Hutch.* in Kew Bull. 17: 181 (1963). *O. mannii* of F.T.A. 8: 111 partly, not of (Wendl.) Wendl. A slender climbing palm of secondary forest, probably not uncommon in S. Nigeria but inadequately recognized and collected.
   **S.Nig.**: Lagos I. *Barter* 2220! Ebute Metta *Millen* 18! Sunmoge, Ijebu *Jones & Onochie* FHI 17419!
   (See Appendix, p. 508.)

## 9. EREMOSPATHA (Mann & Wendl.) Mann & Wendl. in Kerch., Palm. 244 (1878); F.T.A. 8: 111.

Leaf-segments less than 5 times longer than broad:
  Segments of leaves obovate, oblanceolate, or almost rhomboid but with side of the cuneate portion more or less rounded, with several sharp teeth on the margin; median nerve having on each side about 5 prominent veins nearly parallel; transverse veinlets irregular, not very close, crooked; rhachis produced into a slender cirrus armed with opposite hooks and usually with many black recurved prickles; spadix with several branches, slightly scabrid, branches with very short nodes; fruit ovoid-cylindrical with rounded apex, about 2·5 cm. long, 1·7 cm. wide, with 20–21 vertical rows of rhomboid scales, and with persistent perianth at the base, the inner segments short, 0·3 cm. long, minutely papillose on the outer side, dark-brown; seeds 1 or 2, peltate, ellipsoid, 2 cm. long, 1·2 cm. broad, 6 mm. thick
  1. *hookeri*
  Segments of leaves rhomboid or fan-shaped, the sides of the cuneate portion straight, the apex ovate-triangular, toothed, with several or many conspicuous straight veins radiating fan-wise from the base; transverse veinlets very crowded, nearly parallel and straight; rhachis prolonged into a cirrus with opposite hooks and armed with recurved prickles; fruit cylindrical, 2·3 cm. long, 1·3 cm. wide, terminating abruptly in a flat top bearing in the centre a conspicuous umbo, with scales small, rhomboid, in 19–20 vertical rows, and with persistent perianth at the base, the broad inner segments 6 mm. long, with pronounced thickening in the lower half; seeds up to 3, ovate, rather flat, 1·8 cm. long, 8 mm. broad, 3 mm. thick   2. *wendlandiana*
Leaf-segments more than 5 times longer than broad, usually narrow, 15–35 cm. long, 2–3 cm. broad, pointed at each end, with many prominent parallel nerves and numerous sharp subulate teeth on the margin; rhachis prolonged into a cirrus with opposite hooks but not armed with prickles; spadix with very few short lateral branches; flowers crowded, paired, sessile; calyx campanulate, sub-truncate, striate 4–5 mm. long; fruit oblong-ellipsoid, 3 cm. long, mucronate, with 18–22 vertical rows of thin closely-packed rhomboid scales   ..   ..   ..   ..   ..   3. *macrocarpa*

1. **E. hookeri** (*Mann & Wendl.*) *Wendl.* in Kerch., Palm. 244 (1878); F.T.A. 8: 112, partly. *Calamus hookeri* Mann & Wendl. in Trans. Linn. Soc. 24: 434, t. 41 (1864). A climbing palm up to 100 ft. high; in forest.
   **S.L.**: Mofari *Sc. Elliot* 4442! Njala *Deighton* 2591! Kambui Hills *Small* 832! **Iv.C.**: Yapo *Hallé* 272! **S.Nig.**: Nun R. (fr. Aug.) *Mann* 451! Ifon, Ondo *Ayewoh* FHI 3852! Kwa Falls, Calabar *Maggs* 160! Oban *Tuley* 651! **W.Cam.**: *Kalbreyer* 65!
2. **E. wendlandiana** *Dammer ex Becc.* in Webbia 3: 290 (1910). *E. hookeri* of F.T.A. 8: 112, partly, not of Wendland. A scrambling palm of high forest.
   **S.Nig.**: Ikot Okpora, Calabar *Tuley* 652! Oban (fr. Feb.) *Aninza* FHI 15402! **W.Cam.**: S. Bakundu F.R., Kumba *Dundas* FHI 8381! Badun *Rosevear* Cam. 30/38! Kembong *Richards* 5209! Barombi *Preuss* 460. Also in E. Cameroun.
3. **E. macrocarpa** (*Mann & Wendl.*) *Wendl.* in Kerch., Palm. 244 (1878); F.T.A. 8: 113. *Calamus macrocarpus* Mann & Wendl. in Trans. Linn. Soc. 24: 434, tt. 41, 43 (1864). *Eremospatha hookeri* of Chev. Bot. 674, not of Wendland. A climbing palm of swamp-forest, with smooth, polished stems of great length and 1·5 cm. diameter, and with conspicuous buff-yellow inflorescence.
   **S.L.**: Bagroo R. *Mann*! Giewahun *Deighton* 4118! **Lib.**: Bumbumi, Moala *Linder* 1341! Wanau forest (fr. Jan.) *Harley* 2174! **Iv.C.**: Niapidou *Leeuwenberg* 2515! Bingerville *Chev.* 15208; 15209; 15466. **Ghana**: Juaso (Feb.) *Vigne* FH 1829! Kumasi *Vigne* FH 4858! **N.Nig.**: Koton Karifi, Kabba (fr. Oct.) *Daramola & Adebusuyi* FHI 38415! **S.Nig.**: Osho (fr. Apr.) *Jones & Onochie* FHI 17237! Onipanu, Ondo (Dec.) *Onochie* FHI 5243! Uregin, Benin (fr. July) *Unwin* 107! Ahoada (Feb.) *Imp. Inst.* 346! Calabar (Feb.) *Mann* 2330! **W.Cam.**: Ndop *Brunt* 207! (See Appendix, p. 507.)

## 10. BORASSUS Linn., Sp. Pl. 1187 (1753); F.T.A. 8: 117.

Leaves flabellate; lamina divided less than halfway into broad and relatively short segments, bifid at the tip, transverse veinlets prominent giving the lower surface in dried material a corrugated appearance; petiole concave on upper surface, the margins flanged, thin, dark-brown, with irregular spine-like excrescences; male spadix branched from the base, branches poker-like, flowering part about 30 cm. long, 3 cm. across, male flowers several and densely packed within each bracteole; female spadix dissimilar, with few large flowers; fruit ovoid or globose with fibrous pericarp, enclosing usually 3 ovoid-compressed pyrenes; albumen horny, hollow   *aethiopum*

**B. aethiopum** *Mart.* Münch. Gel. Anzeig. 639 (1838), 46 (1839), and in Hist. Nat. Palm. 3: 220, tt. 108, 121, 162; Beccari in Webbia 4: 325 (1914), and in Palme Borass. 6 (1924). *B. flabellifer* var. *aethiopum* (Mart.) Warb. in Engl., Pfl. Ost. Afr. B: 20, C: 130 (1895); F.T.A. 8: 117. The tallest of African palms, often 70 ft. high and sometimes as much as 100 ft., with trunk swollen above the middle; in dry savanna. The palm is dioecious, the female conspicuous by its large orange fruits.

**Sen.:** *Brunner*! **Iv.C.:** Assinie *Chev.* 16312! Béréby *Oldeman* 625! **Ghana:** Ejura, Ashanti *Chipp* 773!
**N.Nig.:** Nupe *Barter* 792! Widely distributed in tropical Africa. (See Appendix, p. 496.)

## 11. HYPHAENE Gaertn., Fruct. 1: 28 (1788); F.T.A. 8: 118.

Leaves flabellate, segments divided for more than half their length, petiole armed on
the margins with black hooks upwardly curved; male spadix with several spiciform
flowering branches about 20 cm. long, 1 cm. diam.; male flowers in groups of three,
each group sunk in a pit enclosed by a ribbed bracteole, from which one flower at a
time projects at anthesis; branches of female spadix of similar length but stouter,
covered in the fruiting stage by dense tomentose cushions; fruit globose-quadrangular,
sometimes irregularly gibbous, about 6 cm. long, 5 cm. broad, with shiny brown coat,
fibrous pericarp, and woody endocarp to which the single seed is adnate; seed with
white albumen, hollow ..       ..       ..       ..       ..       ..       ..       ..       *thebaica*

**H. thebaica (***Linn.***)** *Mart.* Hist. Nat. Palm. 3: 226, tt. 131–3 (1839); F.T.A. 8: 120; Beccari in Agric. Col. 2:
152 (1908), and in Palme Borass. 23 (1924); Chev. Bot. 675; Chev. & Dubois in Rev. Bot. Appliq. 18:
93 (1938). *Corypha thebaica* Linn., Sp. Pl. 1187 (1753). A dioecious palm of hot dry localities, easily
recognized by its common, though not invariable, habit of branching. Although poorly represented in
herbarium collections, the Dum palm is widely distributed from Senegal and Gambia to N. Nigeria chiefly
between 12° and 18°N., and is abundant in many areas including the Upper Niger (*fide* Chevalier & Dubois
*l.c.*), the Northern Region of Ghana (Irvine), and Bornu (Drude). It extends through C. African Republic
to Egypt, Somali and Arabia. (See Appendix, p. 507.)
    [Further study may show other species to occur in W. Africa. Beccari (Palme Borass.) gives two
variants, *H. dahomeensis* Becc. and *H. togoensis* Dummer, and Thonning's *H. guineensis* may be in this
region. For lack of adequate herbarium material it has not been possible to ascertain how steadfastly
these differ from *H. thebaica* or what is their range. Until the situation is made plain it is considered prefer-
able to refer all to *H. thebaica* (Linn.) Mart.—T.A.R.]

## 12. PHOENIX Linn., Sp. Pl. 1188 (1753); F.T.A. 8: 102.

Fruit fleshy, ellipsoid, 3–5 cm. long, with thick sweet pericarp; fronds 3–6 m. long,
glaucous; segments lanceolate-linear, stiff; plants dioecious, male flowers rounded
at the tip     ..     ..     ..     ..     ..     ..     ..     ..     1. *dactylifera*
Fruit dry, oblong-ellipsoid, about 2 cm. long, with thin pericarp; fronds typically
arched, shining green; segments many, the lower spinescent, the remainder with
sharp tip, or sometimes bifid; flowers dioecious; male spadix with many flexuous
branches, flowers lanceolate, acute; female spadix with lower branches subverticillate
others spirally arranged, corolla cup-shaped persistent, calyx barely half as long, with
mucronate sepals; seed about 1·3 cm. long, ovoid, slightly curved, with deep furrow
                                                                                        2. *reclinata*

1. **P. dactylifera** *Linn.* l.c. (1753); F.T.A. 8: 102. The Date palm, with characteristic tall, straight trunk,
    clothed with rather persistent leaf-bases, and head of stiff fronds, is not uncommon in the hotter dry parts
    of the region, being planted for ornament as much as for food. (See Appendix, p. 509.)
2. **P. reclinata** *Jacq.* Fragm. 1: 27, t. 24 (1801); Beccari in Malesia 3: 349 (1890); F.T.A. 8: 103. *P. spinosa*
    Schum. & Thonn., Beskr. Guin. Pl. 437; Chevalier in Rev. Bot. Appliq. 32: 222 (1952). A tufted palm,
    often forming clumps, the stems occasionally as tall as 28 ft. but commonly much less; in sunny places
    where there is good moisture for its roots.
    **Sen.:** (Dec.) *Chev.* 2598! S. Senegambia *Brunner*! *Farmar*! **Gam.:** Bathurst (fr. Mar.) *Dawe* 70! *Brooke*
    1932! **Port.G.:** Uana-Porto d'*Orey* 208! Prabis *Pereira* 2121! **Guin.:** Candéca, Boké *Chillou*! **S.L.:**
    Bonthe I. (fr. Nov.) *Deighton* 2397! *Oldfield*! Tumbo I. *Pelly* 408! **Lib.:** Monrovia (fr. Nov.) *Linder*
    1439! **Iv.C.:** Foro-foro (Nov.) *Oldeman* 402! **Ghana:** Axim *Johnson*! Aburi Hills *Johnson*! Sampa,
    Ashanti (fr. Dec.) *Vigne* FH 3493! Kpedsu (fr. Dec.) *Howes* 1074! **N.Nig.:** Anara F.R., Zaria (Oct.)
    *Keay* FHI 21104! Zaria (fr. Oct.) *Cons. of For.*! Jos *Batten-Poole*! Mambila Plateau *Hepper* 1705!
    **S.Nig.:** Ibadan *Meikle* 973! Nun R. *Mann* 528! Aguku *Thomas* 342! Bonny (fr. Dec.) *Unwin*! Obudu
    Plateau *Tuley* 52! **W.Cam.:** Ninong, Kumba Dist. (fr. Mar.) *Unwin*! Widely distributed in Africa,
    eastwards to Uganda and Kenya, southwards to the Cape. (See Appendix, p. 509.)
    [Chevalier (l.c.) affirms that this palm in lowland habitats of W. Africa differs sufficiently from the
    S. African *P. reclinata* Jacq. to be regarded as a distinct species, for which he uses Schumacher and
    Thonning's name of *P. spinosa*. He further recognizes two other species, *P. djalonensis* and *P. baoulensis*,
    growing at higher elevations. In the material here examined, I have failed to find consistent characters
    separating the W. African from *P. reclinata*, and have preferred the earlier opinion of Beccari (l.c.) that
    the Phoenix occurring naturally in different parts of Africa are of one species, *P. reclinata* Jacq.—T.A.R.]

## 13. NYPA Steck, De Sagu 15 (1757).

Leaves pinnate, erect and recurved, 5–10 m. long, bright green above, glaucous beneath,
with many linear-lanceolate segments; spadix terminal, erect in flower, later drooping;
male flowers in catkin-like lateral branches, minute, surrounded by setaceous bracteoles; female flowers crowded in a terminal head, much larger; fruit a large globose
syncarp of many irregularly hexagonal 1-seeded carpels; pericarp fibrous, endocarp
spongy; seed ovoid, about 7 cm. long, 4 cm. wide, grooved on one side     *fruticans*

**N. fruticans** *Wurmb.* in Verh. Batav. Genootsch. 1: 349 (1779); Martius, Hist. Nat. Palm. 3: 305, t. 208
    (1839); Holland in Kew Bull., Add. Ser. 12, 4: 715 (1922). This prostrate-stemmed, gregarious palm
    growing in estuarine conditions similar to mangrove, was introduced from Singapore Botanic Gardens to
    Calabar in 1906 and Oron in 1912. It has since spread in the estuary of the Cross River and has more recently
    been transferred to the Niger Delta. Its natural range is from India to Queensland.

## 194. PANDANACEAE

### By F. N. Hepper

Trees or shrubs, trunk and branches often with aerial roots. Leaves in 4 rows or spirally arranged and crowded towards the top of the shoots, linear, sheathing at the base, keeled, mostly spinulose on the margins and keel. Flowers dioecious, paniculate or densely crowded into spadices, the latter axillary and terminal, fasciculate or paniculate, enclosed at first by spathaceous sometimes coloured or leafy bracts. Perianth rudimentary or absent. Male flower: stamens numerous; filaments free or connate; anthers erect, basifixed, 2-celled, the cells sometimes again once divided. Female flower: staminodes absent or small and hypogynous or adnate to the base of the ovary. Ovary superior, 1-celled, free or confluent with adjacent ovaries into bundles with separate or united stigmas; style very short or absent. Ovules solitary to many, basal or parietal. Syncarps oblong to globose; mature carpels woody, drupaceous or baccate, pulpy inside. Seeds minute, with fleshy endosperm and minute embryo.

Tropics and subtropics, especially in oceanic islands.

**PANDANUS** Linn. f., Suppl. 64 (1781); F.T.A. 8: 127.

Ovaries with solitary ovules, free or connate into clusters; placentas subbasal; fruit woody or drupaceous; staminodes absent from the female flowers.

A tree by water up to 10 m. high; leaves broadly linear, gradually narrowed to the apex, up to 1 m. long, 2·5–5 cm. broad, very closely nerved, with numerous upwardly directed sharp teeth on the margin; male inflorescences in the axils of large bracts, the lower bracts with leafy tops, the others thin and serrulate; flowering axes up to 15 cm. long, flowers very numerous; anthers about 2 mm. long; infructescence oblong-ellipsoid, about 16–17 cm. long and 10 cm. diam.; drupes angular at the top and bluntly pointed  ..    ..    ..    ..    ..    ..    ..    *candelabrum*

P. **candelabrum** *P. Beauv.* Fl. Oware 1: 37, tt. 21–22(1805); F.T.A. 8: 132; Berhaut, Fl. Sén. 187. *P. leonensis* Lodd. ex Wendl., Index Palm. 46 (1854), name only. *P. heudelotianus* (Gaud.) Balf. f. in J. Linn. Soc. 17: 49 (1878); Chev. Bot. 676. *P. barterianus* Rendle in J. Bot. 32: 324 (1894). *P. togoensis* Warb. and *P. kerstingii* Warb. in Notizbl. Bot. Gart. Berl., Append. 22, 2: 43 (1909), names only. *P. unwinii* Martelli in Webbia 2: 434 (1908). *P. umbellatus* Martelli in Webbia 4: 435 (1914). *Tuckeya candelabrum* (P. Beauv.) Gaud., Voy. Bonite, Bot. Atlas, t. 26, figs. 10–20 (1841). *Heterostigma heudelotianum* Gaud. l.c. t. 25, figs. 15–31 (1841). A tree 10–30 ft. high with conspicuous stilt-roots and thorny trunk; female inflorescences whitish. Screw-pine.
**Sen.:** Niokolo-Koba *Berhaut* 1516. Sedhiou *Chev.* **Mali:** Fincolo *Chev.* 775. Sikasso *Chev.* 802. **Guin.:** Timbo *Chev.* 12433; 12450. Kollangui *Chev.* 12868; 13539. **S.L.:** Guma R. *fide* Hepper. Yungeru *Thomas* 7318! Njala (♂ Dec.) *Deighton* 2579! Ninia *Sc. Elliot* 4918! Sasseni (♀) *Sc. Elliot* 4504! **Lib.:** Monrovia (♀ Jan.) *de Wit* 9128! Ganta (mature ♀ Feb.) *Harley* 834! **Iv.C.:** Niega Lagoon *Aubrév.* 1265. **N.Nig.:** Sanga River F.R., Jemaa Dist. *Keay* FHI 22255! Gashaka Dist. (♂ Feb.) *Latilo & Daramola* FHI 34478! Vogel Peak, 4,000 ft. (old ♂ Dec.) *Hepper* 1547! **S.Nig.:** Agolo *Thomas* 159! Various localities *vide* Hopkins in Nig. Field 27: 36, tt. 1–5 (1962). **W.Cam.:** Nkambe, 5,000 ft. (young ♀ Feb.) *Hepper* 1914! *Brunt* 849! Ambas Bay (♂ Feb.) *Mann* 780! (See Appendix, p. 513.)
   In spite of the additional collections since the first edition no progress has been made in the elucidation of the taxonomy of *Pandanus* in W. Africa. Further collections should show the *female inflorescence or fruit*, apical and basal portions of a leaf or the whole leaf may be pressed concertina-wise; male inflorescences are only useful when associated with females growing in the vicinity. Mature fruits may be collected and sectioned longitudinally.
   *P. utilis* Bory and *P. veitchii* Hort. are recorded by Berhaut (Fl. Sén. 187) as being cultivated in Senegal.

## 195. HYPOXIDACEAE

### By F. N. Hepper

Herbs with a tuberous rhizome or a corm. Leaves mostly all radical, usually prominently nerved and often clothed with long hairs. Flowers solitary, spicate, racemose or subumbellate, mostly white or yellow, actinomorphic. Perianth-tube nothing or very short or consolidated into a long beak on top of the ovary; segments 6, spreading, equal. Stamens 6 or rarely 3, opposite the perianth-segments and inserted at their base; anthers 2-celled, opening lengthwise. Ovary inferior, 3-celled, style short or 3 styles separate. Ovules numerous in 2 series on axile placentas, or rarely few. Fruit a capsule opening by a circular

FIG. 376.—Pandanus candelabrum *P. Beauv.* (Pandanaceae).

A, fertile shoot with male inflorescences. B, male flower. C, stamen. D, stigmas from female flower. (After P. de Beauvois).

slit or by vertical slits near the top, or indehiscent and fleshy.  Seeds small; embryo in abundant endosperm.

Mainly Southern Hemisphere and tropical Asia.

Ovary immediately below the perianth-segments, the latter free to the base; fruit
  dehiscent by a circular split around the middle or into valves   ..          1. **Hypoxis**
Ovary far below the perianth-segments which are separated from it by a long slender
  stipe-like tube; fruit indehiscent   ..   ..   ..   ..   ..          2. **Curculigo**

*Molineria capitata* (Lour.) Merrill from Assam, is commonly cultivated in Freetown.

1. **HYPOXIS** Linn., Syst. ed. 10, 986 (1759); F.T.A. 7: 377; Nel in Engl., Bot. Jahrb.
         51: 301 (1914).

Leaves about 1–2 cm. broad with about 10 principal nerves; rhizome massive, more than
  2 cm. diam., surmounted by old fibrous leaf-bases; anthers entire at apex; perianth-
  segments 1 cm. long, the inner whorl nearly glabrous:
Inflorescences ascending and leaves more or less markedly recurved; inflorescences
  densely pilose, with (2–)3–6 flowers; leaves developing just after the inflorescence and
  sometimes markedly recurved with the young ones appearing almost contorted, folded,
  finally attaining a length of 30–40 cm., ciliate or pilose at the margins and on the
  midrib outside ..   ..   ..   ..   ..   ..   ..   ..   .. 1. *recurva*
Inflorescences and leaves nearly straight and erect; inflorescences densely long-
  pilose, with 4–8(–10) flowers; leaves finally attaining a length of up to 80 cm.,
  long-pilose at the margins and on the midrib outside, and sometimes also on the
  other surface  ..   ..   ..   ..   ..   ..   ..   .. 2. *urceolata*
Leaves up to 5 mm. broad, from a few cm. to 30 cm. long, with about 4 principal nerves,
  margins and midrib outside villous; rhizome fusiform, about 1 cm. diam. usually
  without old fibrous leaf-bases; inflorescences 1–3-flowered, densely appressed-
  pilose; perianth segments 4–5 mm. long, the inner whorl glabrous or pilose in the
  middle; anthers slightly divided at apex  ..   ..   ..   .. 3. *angustifolia*

1. **H. recurva** *Nel* in Engl., Bot. Jahrb. 51: 325 (1914). *H. ledermannii* Nel l.c. 314. *H. thorbeckei* Nel l.c. 328.
  *H. lanceolata* Nel l.c. 325. *H. suffruticosa* Nel l.c. 335. *H. villosa* " var. foliis recurvis "—Hook. f. in J. Linn.
  Soc. 7: 223 (1864). *H. villosa* of F.T.A. 7: 379, partly, not of Linn. f. Subterranean rhizome giving rise
  to several massive erect rhizomes bearing succulent white roots and masses of old leaf fibres; inflorescences
  appearing during the dry season before the leaves, which develop before flowering is completed; flowers
  yellow, frequented by bees; in montane grassland.
      **S.Nig.:** Obudu Plateau, 5,000 ft. (Feb.) *Tuley* 583! **W.Cam.:** Cam. Mt., 7,000–8,000 ft. (Nov.–Feb., Apr.)
  *Mann* 1224! 2133! *Maitland* 1023! *Brenan* 9567! Bafut-Ngemba F.R., Bamenda, 6,000 ft. (Mar.)
  *Richards* 5311! L. Bambuluwe, Bamenda, 7,000 ft. (Feb.) *Thorbecke* 231! 275! *Hepper* 2111! Kufum
  (Dec.) *Ledermann* 2007! *Brunt* 840! Also in E. Cameroun. (See Appendix, p. 514.)
      [Several species described by Nel can hardly be maintained as distinct in this confusing genus. The
  appearance of the plants changes as the leaves develop. However, field workers should observe the variation within populations to determine the taxonomic worth of the various characteristics. Nel seems to
  have been in error in attributing divided anthers to *H. ledermannii*.—F.N.H.]
2. **H. urceolata** *Nel* l.c. 336 (1914); Morton, W. Afr. Lilies & Orch. fig. 35. Large rhizome bearing fibrous leaf
  bases; inflorescences up to 10 ins. high appearing before and with the leaves; flowers yellow; in upland
  savanna.
      **N.Nig.:** Jos Plateau (Apr.) *Lely* P202! Naraguta (July) *Lely* 14! *Lawlor & Hall* 121! Zelau (Apr.) *Lely*
  114! Werran, Plateau Prov. (Apr.) *Gregory* 25! Extending to E. Africa.
      [Typically with 6 perianth lobes but 4 may be found. Some plants approach *H. angustifolia* Lam. in
  having narrow leaves with 2–4 principal nerves, but larger flowers and rhizome.]
3. **H. angustifolia** *Lam.* Encycl. 3: 182 (1789); F.T.A. 7: 378; Nel l.c. 303; Morton l.c. fig. 36. *H. camerooniana* Bak. in F.T.A. 7: 577 (1898); Nel l.c. 302; F.W.T.A., ed. 1, 2: 394. *H. djalonensis* Hutch. in
  F.W.T.A., ed. 1, 2: 394 (1936), English descr. only. *Curculigo djalonensis* A. Chev., Bot. 635 (1920), name
  only. A small herb with narrow grass-like leaves; flowers yellow; in upland savanna.
      **Guin.:** Passo, nr. Pita, Fouta Djalon (July) *Adames* 293! **S.L.:** Musaia, Koinadugu Dist. (Aug.) *Haswell*
  37! Bintumane Peak, Loma Mt., 6,000 ft. (May, Aug.) *Deighton* 5090! *Jaeger* 1048! **N.Nig.:** Anara F.R.,
  Zaria Prov. (May) *Keay* FHI 25779! Jos Plateau (July) *Lely* P700! Zaranda Mt., Jos Plateau, 5,800 ft.
  (May) *Lely* 193! Mambila Plateau, Sardauna Prov., 5,500 ft. (July) *Chapman* 66! **S.Nig.:** Obudu Plateau,
  Ogoja Prov., 5,500 ft. (Aug.) *Stone* 47! **W.Cam.:** Buea *Preuss* 848! Jango, Cam. Mt., 6,900–7,000 ft.
  (Apr.) *Brenan* 9581! Mann's Spring to Buea, 8,000 ft. (Apr.) *Morton* GC 6795! Bafut-Ngemba F.R.,
  6,000 ft. (Mar.) *Richards* 5308! Bum, Bamenda (May) *Maitland* 1526! Bamenda to Banso (Mar.) *Onochie*
  FHI 34879! Extending to Ethiopia, E. Africa and Madagascar.
      [Note 1: The dwarf plants occurring on Cameroon Mt. may be separated under the name of *H.
  camerooniana* Bak. and similar ones in the mountains of Guinée and Sierra Leone as *H. djalonensis* Hutch.,
  but they are hardly distinct from *H. angustifolia*. Note 2: Plants from Kwahu Tafo, Ghana, (*Hall* CC 136)
  have very long peduncles and nearly glabrous flowers and they also fall into the *H. angustifolia* aggregate—
  F.N.H.]

2. **CURCULIGO** Gaertn., Fruct. 1: 63, t. 16 (1789); F.T.A. 7: 377.

Perianth-segments about 7 mm. long, 2–3 mm. broad, nearly glabrous; flowers solitary;
  anthers 2 mm. long on filaments half as long; leaves folded, half width 2–4 mm. broad,
  up to 30 cm. long usually much less, nearly glabrous or with some long weak hairs
                                        1. *minor*
Perianth-segments about 20 mm. long, 5–6 mm. broad, pilose outside; flowers solitary
  or paired, the peduncle short and hidden in the leaf-sheaths; anthers 5–6 mm. long

Fig. 377.—Curculigo pilosa (*Schum. & Thonn.*) *Engl.* (Hypoxidaceae).
A, stamen.  B, stigma.

173

on filaments half as long; fruit oblong, sessile, crowned by the persistent tube of
the perianth; leaves folded, half width 4–15 mm. broad, up to 50 cm. long very acute
at apex, thinly pilose with weak hairs   ..    ..    ..    ..    ..   2. *pilosa*

1. **C. minor** *E. Guinea* Ensayo Geobot. Guin. Continent Espan. 258 (1946), and in An. Jard. Bot. Madrid 6, 2:
   471 (1946). A small herb with narrow leaves and small yellow flowers; in moist sandy places in savanna.
   **S.L.:** Lumley (May) *Deighton* 2682! Hoya (Apr.) *Adames* 40! **Lib.:** Paynesville (Feb., Apr., May)
   *Harley* 15621  1858!  1900! *Leeuwenberg* 4893! Mt. Barclay (June) *Bunting* 36! **S.Nig.:** Eket Dist.
   *Talbot* 3268! Calabar *Robb*! Also in Rio Muni, Uganda and the Congos.
2. **C. pilosa** (*Schum. & Thonn.*) *Engl.* in Engl. & Drude, Veg. der Erde 9, 2: 353 (1908); Morton, Lilies & Orch.
   fig. 37. *Gethyllis pilosa* Schum. & Thonn., Beskr. Guin. Pl. 172 (1827). *Curculigo gallabatensis* Schweinf.
   ex Bak. (1878)—F.T.A. 7: 383; Chev. Bot. 635. *Hypoxis villosa* of Chev. Bot. 635, partly (*Chev.* 21988),
   not of Linn. f. Common herb with stout erect rhizome of seasonally marshy savanna; flowersg olden
   yellow, towards the end of the dry season, together with the developing leaves which eventually reach a
   length of 2 ft.
   **Gam.:** *Hayes* 540a! **Mali:** Diagara to Sienso *Chev.* 1019! **Guin.:** Kouroussa *Brossart* in *Hb. Chev.* 15638.
   Farana, 3,300 ft. (Mar.) *Sc. Elliot* 5374! **S.L.:** York (May) *Deighton* 5535! Rokupr (July) *Jordan* 47!
   Gbap (Mar.) *Adames* 21! Kabala (Mar.–Apr.) *Hargreaves* in *Hb. Deighton* 1910! **Iv. C.:** Buandougon to
   Marabadiassa *Chev.* 21988! Touba (Mar.) *Bouquet* in *Hb. Leeuwenberg*! **U.Volta:** Banfora to Sindou
   (June) *Leeuwenberg* 4305! **Ghana:** Navrongo (May) *Andoh* FH 5181! Kintampo (Mar.) *Dalz.* 76! *Hepper &
   Morton* A3185! Salaga *Krause*! Afram Plains (Mar.) *Johnson* 701! Kpeshi Lagoon, nr. Labadi (Mar.)
   *Adams* 3832! **Togo Rep.:** Bismarckburg (Jan.) *Büttner* 381! **N.Nig.:** Nupe *Barter* 1506! Zungeru (May)
   *Dalz.* 272! Falingo Gabo, Kano Prov. (July) *Onwudinjoh* FH 22398! Biu (May) *Noble* 47! **S Nig.:**
   Ogboro (May) *Denton* 25! Upper Ogun F.R. (Apr.) *Oriafo* FHI 38203! Olokemeji to Iseyin (Apr.) *Hambler*
   423! Olla hills F.R. (Mar.) *Binuyo* FHI 36917! **W.Cam.:** Mamfe (Mar.) *Richards* 5226! Widespread in
   tropical Africa and in Madagascar. (See Appendix, p. 513.)

# 196. VELLOZIACEAE

## By F. N. Hepper

Stems woody and fibrous, dichotomously branched, covered with the persistent
bases of the fallen leaves; habit arborescent or shrubby. Leaves crowded in a tuft
at the ends of the branches, narrow, often pungent-pointed. Flowers solitary on
each peduncle, white, yellow, or blue, sometimes very handsome, actinomorphic,
bisexual. Perianth-tube very short or absent; segments equal, spreading.
Stamens 6, or numerous and in 6 bundles of 2–6; anthers linear, basifixed,
opening by longitudinal slits. Ovary inferior, 3-locular; style slender, with a
capitate stigma or 3 short arms. Ovules very numerous on axile, stalked placentas.
Fruit a dry or hard capsule, often flat or concave on the top, crowned with the
scar of the perianth, or 6-toothed, sometimes spiny, loculicidally dehiscent.
Seeds numerous, embryo small in copious, rather hard endosperm.

A small family occurring in S. America, Southern and tropical Africa, Madagascar and
Arabia.

**VELLOZIA** Vand., Fl. Lusit. Bras. Sp. 32, t. 2, fig. 11 (1788).

Perennial herb, stems numerous up to 50 cm. long, 1·5–2 cm. diam.; leaves linear,
coriaceous, acute at apex, up to 32 cm. long and 6 mm. broad, glabrous but the
margin and dorsal nerves minutely and inconspicuously aculeate towards the apex;
peduncles solitary in leaf axils, shorter than the leaves, setose in the upper part;
perianth-lobes linear-lanceolate, acute at apex, 3 outer ones about 3·5 cm. long and
8 mm. broad, 3 inner ones about 3 cm. long and 10 mm. broad; anthers 17–20 mm.
long; style 16 mm. long, ovary 6 mm. long, obscurely 3-angled; ovules numerous;
fruit subglobose 1·7 cm. long, 1·5 cm. diam.; seeds subcylindrical, 3 mm. long, nearly
1 mm. diam.    ..    ..    ..    ..    ..    ..   *schnitzleinia* var. *occidentalis*

**V. schnitzleinia** (*Hochst.*) *Bak.* in F.T.A. 7: 409 (1898). *Hypoxis schnitzleinia* Hochst (1844). Var. *schnitzleinia*
   is widespread in eastern Africa and var. *somaliensis* Terrac. occurs in Somalia.
**V. schnitzleinia** var. **occidentalis** *Milne-Redh.* in Kew Bull. 5: 381 (1951). A stout perennial herb with stems
   about 1½ ft. long, covered with old leaves; flowers white; forming thick mats on bare rock in savanna
   woodland.
   **N.Nig.:** Anara F.R., Zaria (May) *Keay* FHI 22903! 25758! *Olorunfemi* FHI 24356! Gawu Hills, Niger
   Prov. (June, Aug.) *Onochie* FHI 35936! *E. W. Jones* FHI 42224! NE. of Kaduna (July) *Jackson* FHI
   55409!

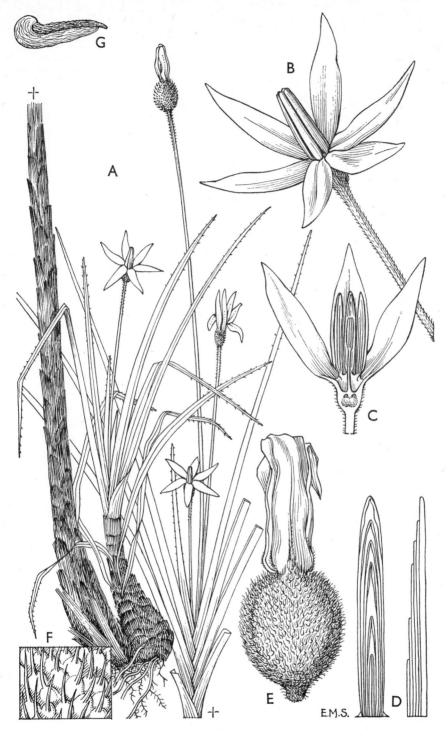

FIG. 378.—VELLOZIA SCHNITZLEINIA (*Hochst.*) *Bak.* var. OCCIDENTALIS
*Milne-Redh.* (VELLOZIACEAE).

A, habit, × ⅓. B, flower, × 2. C, section of flower, × 2. D, stamens, × 3. E, ovary, × 2.
F, portion of surface of ovary, × 4. G, seed, × 10. All from *Keay* FHI 22903.

175

# 197. TACCACEAE

## By F. N. Hepper

Perennial herbs with a tuberous or creeping rhizome. Leaves all radical, large, entire or much lobed. Flowers actinomorphic, hermaphrodite, umbellate; bracts forming an involucre, the inner often thread-like. Perianth with a short tube and 6 lobes, lobes 2-seriate, mostly somewhat corolline. Stamens 6, inserted on the perianth; filaments short; anthers 2-celled, opening lengthwise. Ovary inferior, 1-celled, with 3 parietal placentas; style short, the 3 stigmas often petaloid and reflexed over the style; ovules numerous. Fruit a berry or rarely opening by valves. Seeds numerous, with copious endosperm and minute embryo.

Tropical regions and China, with two genera.

**TACCA** J. R. & G. Forst., Char. Gen. Pl. 69 (1776); F.T.A. 7: 413. *Nom. cons.* Characters of the family; fruit a berry.

Leaves erect, shining, 3-partite, each segment 2-forked, pinnatipartite, the lower lobes separated, the upper ones connected, sometimes one segment not 2-forked, glabrous; peduncle long and rather slender; involucral bracts about 6, obovate-spathulate, strongly nerved outside, 3–4 cm. long, acuminate and sometimes lobed at the apex; innermost bracts up to 10 cm. long and thread-like; fruit ovoid-ellipsoid, about 3·5 cm. long; seeds ribbed ..    ..    ..    ..    ..    *leontopetaloides*

T. **leontopetaloides** (*Linn.*) *O. Ktze.* Rev. Gen. Pl. 2: 704 (1891); Carter in F.T.E.A. Taccac. 1, fig. 1 (1962). *Leontice leontopetaloides* Linn. (1753). *Tacca involucrata* Schum. & Thonn., Beskr. Guin. Pl. 177 (1827); Limpricht in Engl., Pflanzenr. Taccac. 29 (1902); Chev. Bot. 638; F.W.T.A., ed. 1, 2: 396; Berhaut, Fl. Sén. 11; Morton, W. Afr. Lilies & Orch. fig. 48. *T. pinnatifida* J. R. & G. Forst. (1776)—F.T.A. 7: 413; Chev. Bot. 638. Ovoid tuber several inches across usually giving rise to one or two erect leaves 2–4 ft. high and an inflorescence 3–6 ft. high; involucre greenish, flowers yellowish with inner bracts thread-like, purplish; in thickets and amongst grass. **Sen.:** *Berhaut* 250. Kaolak 84 *Kaichinger*. **Gam.:** Mungo Park! Bathurst (fr. Sept.) *Frith* 114! **Mali:** Banankalidoro to Bama *Chev.* 933. San (Sept.) *Chev.* 2397. **Port.G.:** Buruntuma, Nova Lamego (fr. July) *Pereira* 3081! **Guin.:** Kaba *Chev.* 13243. **S.L.:** Coast near Freetown (May, Aug.) *Deighton* 2684! *Melville & Hooker* 251! Musaia (Oct.) *Thomas* 2654! **Iv.C.:** Gouékouma *Chev.* 21689. Bouaké (Apr.) *Leeuwenberg* 3289! Kouroukourounga (Apr.) *Aké Assi* 8746! **Ghana:** Cape Coast *Don*! Accra (June) *Johnson*! Bosomoa F.R., Kintampo (June) *Enti* FH 6724! Mampong (Apr.) *Vigne* FH 1924! Nkoranza, N. Ashanti (Apr.) *Irvine* 916! **Dah.:** Agouagon *Chev.* 23538. Kouandé to Konkobiri *Chev.* 24270. **N.Nig.:** Nupe *Barter* 1541! Katagum *Dalz.* 238! Anara F.R., Zaria (May) *Keay* FHI 22869! Nabardo, 2,300 ft. (May) *Lely* 209! Vom, 4,000 ft. *Dent Young* 246! **S.Nig.:** Lagos *W. MacGregor* 62! Abeokuta (Apr.) *Baldwin* 12006! Obu Dist. *Thomas* 452! Obokoffia to Oborotta *Talbot* 3802! **W.Cam.:** Mbaw Plain (May) *Brunt* 401! Throughout tropical Africa, Madagascar and the Old World Tropics, China, N. Australia and the Pacific Islands. (See Appendix, p. 514.)

# 198. BURMANNIACEAE[1]

## By F. N. Hepper

Stamens 3; flowers actinomorphic:
  Perianth with 3 broad lobes and 3 much smaller lobes or which may be absent, perianth persistent:
    Ovary 1-celled, with 3 parietal placentas; saprophytic ..    ..    1. **Gymnosiphon**
    Ovary 3-celled, with axile placentas; saprophytic or perianth-tube winged in some
      species    ..    ..    ..    ..    ..    ..    ..    2. **Burmannia**
  Perianth with 6 similar subulate lobes, deciduous; saprophytic; ovary 1-celled with 3
    parietal placentas    ..    ..    ..    ..    ..    3. **Oxygyne**
Stamens 6; flowers zygomorphic with 6 similar narrow perianth-lobes; saprophytic;
  ovary 1-celled with 3 parietal placentas; perianth deciduous    ..    4. **Afrothismia**

1. **GYMNOSIPHON** Blume, Enum. Pl. Jav. 1: 29 (1827); F.T.A. 7: 11; Jonker, Monogr. Burmanniac. 168 (1938).

Saprophytic herb 7–18 cm. high; stems with a few minute scale leaves; cymes laxly bifurcate with the arms up to 7 cm. long and flowers arranged along them; pedicels slender, 2–4 mm. long; perianth-tube urceolate, 4 mm. long, shortly 3-lobed; fruit subglobose, capped by the persistent perianth    ..    ..    ..    *longistylus*

---

[1] Including Thismiaceae which was separated as a distinct family in Ed. 1 and is still maintained as such by Hutchinson (Fam. Fl. Pl. ed. 2 (1959)). Jonker, however, shows (Monogr. 9–10 (1938)) that there are intermediate genera which prevent its separation from Burmanniaceae.

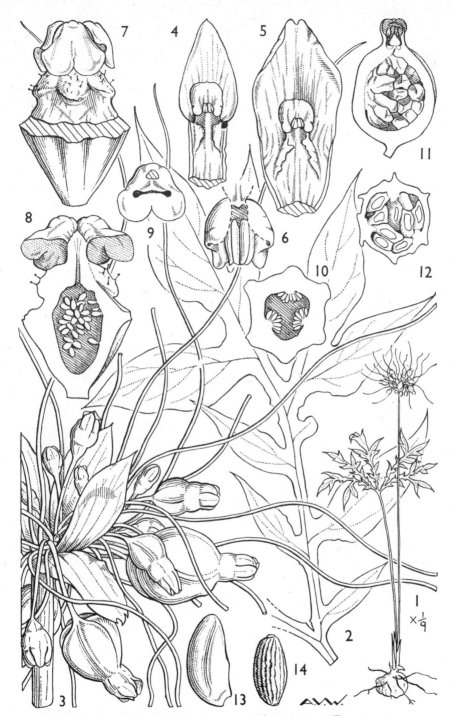

FIG. 379.—TACCA LEONTOPETALOIDES (*Linn.*) *O. Ktze.* (TACCACEAE).

1, plant, × ⅑. 2, leaf, × ⅖. 3, inflorescence, × 1. 4, outer per.-seg., × 4. 5, inner per.-seg., × 4. 6, stamen from inside hood, × 8. 7, ovary and style, × 6. 8, longitudinal section of same, × 6. 9, stigma-lobe from below, × 6. 10, transverse section of ovary, × 6. 11, longitudinal section of mature fruit, × 1. 12, transverse section of mature fruit, × 1. 13, seed surrounded by the aril, × 3. 14, seed, × 3. 1 from *Milne-Redhead and Taylor* 8263, 8263*a* and a drawing by *H. Faulkner.* 2 from *Milne-Redhead and Taylor* 8263. 3–10 from *Milne-Redhead & Taylor* 8263*a*. 11–14 from *Milne-Redhead & Taylor* 8263*b*. (Reproduced from F.T.E.A.)

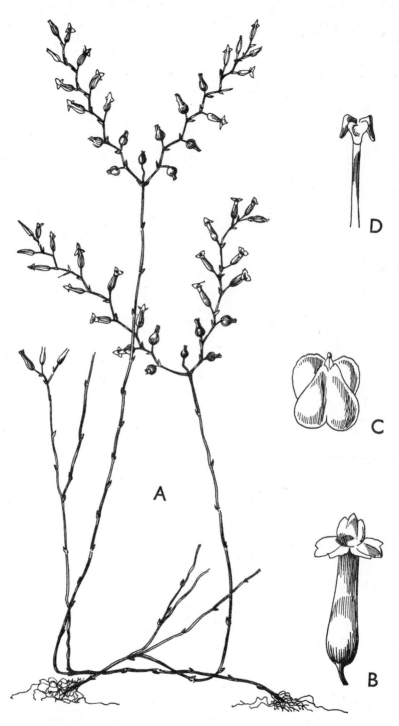

Fig. 380.—Gymnosiphon longistylus (*Benth.*) *Hutch.* (Burmanniaceae).
A, habit, × 1. B, flower, × 7. C, stamens. D, style and stigmas.

**G. longistylus** (*Benth.*) *Hutch.* F.W.T.A., ed. 1, 2: 399 (1936); Jonker l.c. 201; Aké Assi, Contrib. 2: 28. *Dictyostegia longistyla* Benth. (1849)—F.T.A. 7: 12. *Gymnosiphon squamatus* Wright in F.T.A. 7: 12 (1897). Colourless saprophyte on the forest floor amongst decaying leaves.
**S.L.:** John Obey, Peninsula (Sept.) *Melville & Hooker* 393! **Lib.:** Banga (Oct.) *Linder* 1217! Zuie, Boporo Dist. (Nov.) *Baldwin* 10231! Bobei Mt., Sanokwele Dist. (Sept.) *Baldwin* 9582! Bilimu Mt. (Sept.) *Harley* 1531! 3 miles N. of Getti Rwaji, Gola (June) *Bunting*! **Iv.C.:** Tiapleu Forest *Mangenot & Miège* IA 2694. Yapo Forest *de Wilde* 3145! *Aké Assi* 6894! **Ghana:** Ankasa F.R. (Dec.) *Vigne* FH 3216! **S.Nig.:** Apapa, Lagos (Jan.) *Dalz.* 1277! Nun R. *Mann* 515, partly. Annye (Sept.) *Unwin* 144! Oban *Talbot* 700! 715! Also in E. Cameroun and Gabon.

## 2. BURMANNIA Linn., Sp. Pl. 287 (1753); F.T.A. 7: 11; Jonker, Monogr. Burmanniac. 57 (1938).

Perianth-tube without distinct wings; small plants about 3(–13) cm. high, without basal rosette:
Flowers with 6 ridges on the 3·5 mm. long perianth-tube; usually only 1–2-flowered; stem-leaves 2–3 mm. long and 3–5 mm. apart  ..    ..    ..    1. *hexaptera*
Flowers without ridges on the 3·5–6 mm. long perianth-tube; usually with a terminal cluster of flowers:
Flowers 2–7; stem leaves scale-like about 2 mm. long and 5 mm. apart; stem 3–10 cm. high; margin of outer perianth-lobes not crenate, inner lobes obovate    2. *congesta*
Flowers numerous, up to 27; stem leaves 2–6 mm. long, up to 15 mm. apart; stem up to 32 cm. high, with 1–2-branched inflorescence; margin of outer perianth lobes broadly spathulate  ..    ..    ..    ..    ..    ..    ..    3. *densiflora*
Perianth-tube distinctly winged; plants (7–)10–32 cm. high, basal rosette usually present:
Wings 3–5 mm. broad, 7–13 mm. long, half-elliptic to half-obovate, usually purple; anther connective with 2 broad crests and acute basal spur  ..    .. 4. *latialata*
Wings 1·5–3 mm. broad, 5–7 mm. long, bluish or yellow:
Anther connective with 2 broad crests; wings half-obovate, decurrent along the basal part of the ovary, about 3 mm. broad; inner perianth-lobes nearly as long as the outer  ..    ..    ..    ..    ..    ..    ..    ..    5. *liberica*
Anther connective with 2 acute processes; wings half-elliptic to half-obovate, about 1·5 mm. broad; inner perianth-lobes much shorter than the outer  6. *welwitschii*

1. **B. hexaptera** *Schltr.* in Engl., Bot. Jahrb. 38: 143 (1906); Jonker, Monogr. Burmanniac. 93 (1938). Small erect saprophyte; flowers white, the lower part orange yellow.
   **W.Cam.:** Man O' War Bay (Oct.) *Schlechter* 15785! 15786! Also in E. Cameroun.
2. **B. congesta** (*Wright*) *Jonker* l.c. 94 (1938). *Gymnosiphon congestus* Wright in F.T.A. 7: 12 (1898); F.W.T.A., ed. 1, 2: 399. *Burmannia aptera* Schltr. (1906). Small erect yellow or white saprophyte with a terminal cluster of white flowers; in forest in damp silty soil.
   **Lib.:** Gbanga (Oct.) *Linder* 1216. Wanau (fr. Nov.) *Harley* 2060! **Ghana:** Ankasa F.R. (Jan.) *Hall & Enti* GC 36371! **S.Nig.:** Nun R. (Sept.) *Mann* 515, partly! Orem, Oban F.R. (Jan.) *Onochie & Okafor* FHI 36132z! Oban *Talbot* 716! **W.Cam.:** Moliwe (Sept.) *Schlechter* 15787! Also in E. Cameroun, Congo and Angola.
3. **B. densiflora** *Schltr.* in Engl., Bot. Jahrb. 38: 141 (1906); Jonker l.c. 95. Erect saprophyte up to 1 ft. high; in forest.
   **W.Cam.:** Moliwe *Stammler*! Also in E. Cameroun.
4. **B. latialata** *Hua ex Pobéguin* Ess. Fl. Guin. Fr. 166 (1906); Jonker l.c. 104. *B. le-testui* Schltr. (1925). *B. bicolor* of F.W.T.A., ed. 1, 2: 399, partly, not of Mart. Slender erect simple annual herb 3½–11 in. high; flowers whitish with bluish wings; in moist places on rock outcrops.
   **Guin.:** Farmer 200! Mt. Gangan *Schnell* 748! Bouria *Chev.* 14704! Fon Massif *Schnell* 6628! Dalaba (Oct.) *Adames* 406! Kindia (Oct.) *Pobéguin* 1392! **S.L.:** Kanya (Oct.) *Thomas* 2979! **Lib.:** Genna Loffa, Kolahun Dist. (Nov.) *Baldwin* 10083! **Ghana:** Sampa (Apr.) *Morton* A3260! Kwahu Tafo (Aug.) *Hall* CC 133! **N.Nig.:** Mada Hills *Hepburn* 85! Kontagora (Dec.) *Dalz.* 263! **S.Nig.:** Mt. Orosun, Idanre Hills (Oct.) *Keay* FHI 22599! Also in E. Cameroun, the Congos, Uganda, Zambia, Rhodesia and Angola.
   [See note after sp. No. 5.]
5. **B. liberica** *Engl.* Bot. Jahrb. 48: 505 (1913); Jonker l.c. 103. *B. bicolor* of F.W.T.A., ed. 1, 2: 399, partly, not of Mart. *B. bicolor* var. *micrantha* Engl. ex Gilg (1903). Erect very slender often wavy stems, 7–10 in. high; flowers whitish with corolla-lobes sometimes reddish, wings bluish or white; in moist places on sandy soil or rock outcrops.
   **Guin.:** Macenta (Oct.) *Baldwin* 9801! **S.L.:** Juring to Blama (Dec.) *Deighton* 298! Bumban (Aug.) *Deighton* 1305! Mapotolon, Samu (Sept.) *Jordan* 919! Mabonto (Aug.) *Jordan* 509! **Lib.:** Duport (Nov.) *Linder* 1491! Monrovia savanna (Aug., Oct.) *Dinklage* 2831! *Harley* 1682! *Baldwin* 5856! 9189! 13038! Mt. Barclay (June, Nov.) *Bunting* 21! 32! **Iv.C.:** Mossou, Grand Bassam *Schnell* 6552! Also in E. Cameroun, the Congos and Angola.
   [Note: I have maintained the species concept of Jonker in his Monograph, but *B. latialata* and *B. liberica* are not satisfactorily separable on the wing characters. Other tropical African species can also be included in the complex.—F.N.H.]
6. **B. welwitschii** *Schltr.* in Fedde, Rep. 21: 84 (1925). A small erect annual 2½–5 in. high, sometimes branched; flowers in a terminal cluster or solitary; in wet places.
   **Guin.:** near Pita *Pobéguin*! Widespread in tropical Africa.

## 3. OXYGYNE Schltr. in Engl., Bot. Jahrb. 38: 140 (1906); Jonker, Monogr. Burmanniac. 260 (1938).

A tiny saprophyte about 4 cm. high; stem 1-flowered, covered with several scales; bract similar to the upper scales, rounded; flower erect, about 2·2 cm. long; perianth narrowly campanulate, tube about 1 cm. long, lobes broadly triangular-acuminate, narrowly caudate-acuminate  ..    ..    ..    ..    ..    ..    ..    *triandra*

**O. triandra** *Schltr.* l.c. 140, fig. 1 G-M (1906), and in Notizbl. Bot. Gart. Berl. 8: 45 (1921); Jonker l.c.
   **W.Cam.:** Moliwe (Sept.) *Schlechter* 15790!

**4. AFROTHISMIA** (Engl.) Schltr. in Engl., Bot. Jahrb. 38: 138 (1906); Jonker, Monogr. Burmanniac. 222 (1938).

Perianth with a right-angle bend in the middle, perianth-lobes 6, narrowly triangular or filiform and spreading from the open throat, 4·5–15 mm. long; plants 2·5–12 cm. high, 1–3-flowered, stem usually simple with the underground part possessing spherical clusters of tubers or bulbils bearing fine rootlets .. .. 1. *winkleri*
Perianth obovoid sometimes bent at the base, perianth-lobes 6, shortly subulate and clustered over the small throat, about 2 mm. long; habit similar to the last
2. *pachyantha*

1. **A. winkleri** (*Engl.*) *Schltr.* l.c. 139 (1906), and in Notizbl. Bot. Gart. Berl. 8: 44 (1921); Jonker l.c. 223. A small white saprophyte in the forest floor litter; " ovary and fruit cream, base of perianth-tube cream with six white (or purple?) V-shaped marks, centre of perianth white with crimson patch above, mouth and lobes of perianth yellow, stamens cream " (Keay).
   **S.Nig.:** Aponmu, Akwe F.R. (Nov.) *Keay* FHI 25540! **W.Cam.:** New Tegel, near Buea (July) *Winkler* 225! Moliwe (Sept.) *Schlechter* 15788! Also in E. Cameroun and Uganda.
2. **A. pachyantha** *Schltr.* in Engl., Bot. Jahrb. 38: 139, fig. 1, A-F (1906), and in Notizbl. Bot. Gart. Berl. 8: 44; Jonker l.c. 224. A tiny saprophyte up to 1¼ in. high; perianth about ½ in. long; in forest.
   **W.Cam.:** Moliwe (Sept.) *Schlechter* 15789!

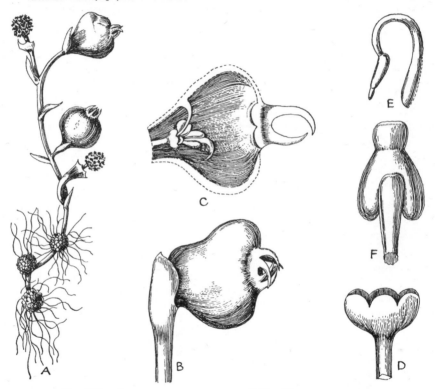

Fig. 381.—Afrothismia pachyantha *Schltr.* (Burmanniaceae).
A, habit.  B, flower.  C, vertical section through same.  D, style.  E and F, stamens.  (After Schlechter.)

## 199. ORCHIDACEAE

### By V. S. Summerhayes

Perennial, terrestrial, epiphytic or saprophytic herbs with rhizomes or tuberous roots or rootstock; stem leafy or scapose, frequently thickened at the base into pseudobulbs and bearing aerial assimilating roots. Leaves undivided, alternate and often distichous, rarely opposite, sometimes all reduced to scales, often fleshy, sheathing at the base. Flowers bracteate, hermaphrodite or very rarely polygamous or monoecious, zygomorphic; inflorescence spicate, racemose or paniculate, or flowers solitary. Perianth epigynous, composed of 6 petaloid segments (*tepals*) in 2 whorls, or the outer whorl calyx-like and the inner corolla-like, or the outer rarely corolla-like and the inner minute, free or variously connate in each whorl; outer segments (*sepals*) imbricate or subvalvate, the middle segments of each whorl generally different in size and colour from the lateral ones, especially the middle petal which is often extremely complicated in structure and is termed the lip or *labellum*; the basal part of the labellum, the *hypochile*, is often articulated to the base of the column or is much constricted, when it is termed the claw; the middle part, the *mesochile* and the apical part, the *epichile*, may be variously lobed and often bear outgrowths. On account of the twisting of the ovary through 180°, the labellum is often placed in an abaxial position; frequently the labellum or more rarely the odd sepal is prolonged into a sac or spur, sometimes very long. Stamens 2 or 1; stamens and style united to form a special structure (*column*), the apex of which may be produced vertically into *stelidia* or laterally into wings, and the base of which may be produced downwards to form a foot; anther or anthers 2-locular, introrse, opening by a slit lengthwise; often operculate, i.e. can be lifted like a little cap; pollen granular or generally agglutinated into mealy, waxy or bony masses (*pollinia*); at one end the pollinium may be extended into a sterile portion (*caudicle*); the pollinia may be free in the anther-loculi or more or less loosely united. Ovary inferior, 1-locular with 3 parietal placentas or very rarely 3-locular with axile placentas, usually produced at the apex to form the column; stigmas 3 fertile, or more frequently the lateral 2 fertile, the other sterile and transformed into a small outgrowth (*rostellum*) which lies between the anther and the stigmas; a portion of the rostellum is sometimes modified into a viscid disk or disks (*viscidia*) to which the pollinia are attached, often by a stalk or stipes. Ovules very numerous and minute. Fruit usually a capsule, mostly opening laterally by 3 or 6 longitudinal slits. Seeds very numerous, minute, often drawn out at each end, or rarely winged, without endosperm; embryo not differentiated.

Widely distributed, most numerous and of very diverse form in the tropics; main centres of distribution Indo-Malaya and tropical America.

### Key to the Genera[1]

Anther attached to the column by its base, loculae adnate to the column and persistent; pollinia granular, with caudicles and 2 (or rarely 1) viscidia; mostly terrestrial with erect annual leafy stem and terminal inflorescence, a few epiphytic or saprophytic:
Lip with 1 or 2 distinct but sometimes rather short spurs:
  Spurs 2; lip at top of flowers (flowers not resupinate); sepals and petals similar, more or less united to one another and to the lip; column slender, somewhat curved; stigma superior, cushion-like, fleshy; anther pendulous **9. Satyrium**
  Spur 1:
    Lip more or less united to the column; leaves radical, orbicular **1. Holothrix**
    Lip quite free from the column:
      Stigmas sessile; stems leafy all the way up; bracts leaf-like .. **3. Brachycorythis**
      Stigmas borne on or forming club-shaped processes, projecting from the front of the column, free or partly united to the lateral lobes of the rostellum:
        Stigmatic processes partly united to the lateral lobes of the rostellum; rhachis and ovary frequently glandular pubescent:
          Flowers bluish, side lobes of lip entire .. .. .. .. **4. Cynorkis**
          Flowers green, side lobes of lip fimbriate .. .. .. **5. Habenaria**
        Stigmatic processes free from the rostellum:
          Middle lobe of rostellum concave, frequently placed some way in front of the anther; lip entire or with a small tooth-like lobe on each side at the base; petals entire or with a small lobe at the base, more or less falcate; dorsal sepal large, convex, laterals sharply deflexed; inflorescence few-flowered **6. Platycoryne**
          Middle lobe of rostellum usually flat or subulate, more or less adpressed to the anther; other characters not associated as in last .. .. **5. Habenaria**
Lip sometimes concave at the base but with no distinct spur:
  Dorsal sepal drawn out into an erect or pendulous spur; lip entire or with short tooth-like lateral lobes; rostellum small, tooth-like:

[1] It should be pointed out that this key is drawn up mainly from the West African species, and some of the minor characters used may therefore not apply to species from other areas.

Petals free from dorsal sepal; anther erect or horizontal     ..     ..   **7. Disa**
Petals united to dorsal sepal; anther horizontal ..     ..     ..   **8. Brownleea**
Dorsal sepal without a spur:
Rostellum large, more or less orbicular, forming a shield-like process covering the anther; lip united to the column at the base; dorsal sepal and petals united to form a variously shaped hood; lateral sepals each with a small spur-like sac near the inner margin   ..     ..     ..     ..     ..   **10. Disperis**
Rostellum rather small, cucullate, shorter than the anther; lip free from the column; dorsal sepal and petals free:
Plants with numerous leaves; inflorescence usually many-flowered
                             **3. Brachycorythis**
Dwarf leafless plants; inflorescence up to 4-flowered   ..   ..**2. Schwartzkopffia**
Anther attached to the column by its apex (usually at the back of the column), either operculate (i.e. the anther can be lifted like a little cap, the pollinia often being carried with it) or erect and persistent, the caudicle and viscidium being then at the top:
Pollinia granular or much divided into small masses attached to a common axis:
Tall climbers with fleshy green stems and usually with leaves; flowers in short axillary racemes, rather large (perianth at least 2 cm. long); lip more or less united to the long curved column, disk bearing either scales or lines of hairs
                             **11. Vanilla**
Terrestrial, erect or rarely somewhat decumbent herbs with usually terminal but rarely lateral inflorescences:
No leaves at time of flowering or saprophytic and non-chlorophyllose; flowering stems arising from underground tuber; rostellum short or minute:
Lip with a distinct spur; saprophytic   ..     ..     ..     **13. Epipogium**
Lip without a spur:
Slender saprophytic plants with short raceme of small flowers (less than 1 cm. long); lip simple with a slender claw; petals and sepals united to one another in the lower part; capsule borne on elongated pedicels   ..   **14. Auxopus**
Flowers appearing on leafless scapes before the leaves; leaf solitary, reniform or ovate with a long or short petiole; flowers over 1 cm. long, usually few; lip more or less trilobed, bearing either numerous hairs or 2 distinct keels; petals and sepals free   ..     ..     ..     ..     **12. Nervilia**
Leafy plants usually with slender creeping rhizome or fibrous rootstock; rostellum elongate:
Stems up to 6 ft. high, rigid, leafy; leaves plicate, lanceolate or narrowly elliptical, lamina 15–35 cm. long; inflorescences axillary or terminal, paniculate; perianth-segments over 5 cm. long, linear; lip very like other segments but broadly ovate at apex   ..     ..     ..     ..     ..     ..   **20. Corymborkis**
Stems less than 3 ft. high including inflorescence, not very rigid, often decumbent at the base; leaves soft and thin, frequently with a long petiole, lamina less than 20 cm. long, usually much shorter; flowers less than 1 cm. long:
Lateral sepals united to one another and to the petals for at least half their length; column with 2 terminal appendages:
Stem elongated, decumbent below, erect and leafy at apex; lip free from the sepals, with a spreading limb, narrow claw and concave base; appendages of column entire   ..     ..     ..     ..     ..   **17. Cheirostylis**
Stem very short with a bunch of radical leaves; lip united to the sepals, at the base of the free part provided with 2 small reflexed marginal appendages; appendages of column serrate ..   ..     ..     ..   **15. Manniella**
Lateral sepals free or only united at the very base:
Column much longer than broad; arms of rostellum long, subulate, acute
                             **16. Platylepis**
Column short, nearly as broad as long; arms of rostellum short and acute, or if larger linear-spathulate, obtuse:
Column with 2 usually parallel longitudinally placed keels on the front; flowers usually not resupinate   ..     ..     ..     ..   **19. Hetaeria**
Column without keels on the front; flowers resupinate   ..   **18. Zeuxine**
Pollinia waxy, entire, 2, 4 or 8, either free, or adhering at one end by viscid outgrowths (the caudicles), or attached to 1 or 2 sticky gland-like structures (the viscidia); anther operculate:
Growth sympodial, consisting either of annual growths from a tuberous underground root or stem, or of a series of erect or creeping lateral shoots, the stem in each growth being often partly thickened to form a fleshy pseudobulb; inflorescence terminal on the year's growth, or axillary:
Pollinia 8:
Lip adnate to the column, with a long slender spur; inflorescence tall, many-flowered, apparently terminal; column short; leaves plicate     **26. Calanthe**

Lip free from column, spur absent or broadly sac-like ; inflorescence short, 1–3-flowered :
Inflorescence lateral from base of leafy pseudobulb ; leaves plicate ; column long and slender ; pollinia equal, elongate, attached to a sticky appendage
27. **Ancistrochilus**
Inflorescence terminal between leaves at apex of pseudobulb ; leaves conduplicate ; column short ; pollinia 4 large and 4 small, pyriform, free ; low creeping plants
25. **Stolzia**
Pollinia 2 or 4, in the latter case sometimes united in pairs :
Anther 2-locular ; pollinia free or joined together by a sticky appendage, neither attached to the rostellum nor possessing a distinct viscidium :
Column not produced downwards into a foot ; lip continuous with the base of the column ; inflorescence terminal on the annual leafy growths ; pseudobulbs not well defined and often absent :
Anther attached at the back below the apex of the column ; rostellum terminal ; column short ; sympodial growths usually forming a slender creeping rhizome, the erect stems being distant from one another ; pseudobulbs absent ; leaves some way from the base of the stem ..    ..    ..    ..       21. **Malaxis**
Anther attached at the apex of the column ; rostellum subterminal ; column rather long, often curved ; sympodial growths usually closely placed, often pseudobulbous ; leaves usually near the base of the stem ..      .. 22. **Liparis**
Column usually short, produced downwards into a foot often nearly as long as itself ; lip articulated on the column-foot and freely moving ; inflorescences arising from the base of the pseudobulbs ; pseudobulbs well defined, bearing 1–2 (rarely 3) leaves at the apex    ..    ..    ..    .. 28. **Bulbophyllum**
Anther incompletely 2-locular ; pollinia attached by a stalk (stipes) to the viscidium (both stipes and viscidium are part of the rostellum), which comes away with them :
Flowers not resupinate (lip uppermost) ; mostly epiphytic or on rocks :
Terrestrial with underground tubers but without pseudobulbs ; column without a foot ; leaves plicate ; viscidium large, transversely rhomboid
32. **Pteroglossaspis**
Mostly epiphytic or on rocks, no tubers but often with pseudobulbs ; column with a distinct foot ; leaves not plicate ; viscidium small, often poorly developed :
Inflorescence terminal on the leafy shoots ; petals easily visible
24. **Polystachya**
Inflorescence axillary from the base of the distantly placed pseudobulbs ; petals minute ..    ..    ..    ..    ..    ..    .. 29. **Genyorchis**
Flowers resupinate (dorsal sepal uppermost) ; mostly terrestrial, a few epiphytic :
Lip without a spur ; epiphytic ; pseudobulbs tall, fusiform, many-leaved ; inflorescence paniculate ; sepals and petals similar ..    ..    23. **Ansellia**
Lip with a distinct spur or very concave sac at the base :
Epiphytic ; rostellum long, beak-like ; inflorescence paniculate
33. **Graphorkis**
Terrestrial, or very rarely epiphytic ; rostellum very short and broad :
Lip 4-lobed or 3-lobed with clearly retuse middle-lobe ; pseudobulbs well-developed, 1–2-leaved ; petioles articulated irregularly above junction with pseudobulb ; leaves fleshy, conduplicate, often variegated
30. **Eulophidium**
Lip entire to 3-lobed, middle lobe scarcely retuse ; pseudobulbs often lacking, but plant with tuberous underground stem ; leaves plicate, often not developed at time of flowering    ..    ..    ..    .. 31. **Eulophia**
Growth monopodial, continued by a relatively slow-growing apical bud ; no pseudobulbs ; leaves when present more or less distichous, fleshy, conduplicate, very frequently unequally bilobed at the apex ; inflorescences always axillary ; pollinia with stipes and viscidium ; nearly all epiphytic :
Leaves absent ; roots very abundant, assimilating :
Sepals and petals united to one another in the lower two-thirds ; lip with a cucullate apex and retrorse point on upper surface at apex ; spur almost spherical
37. **Taeniophyllum**
Sepals and petals free to the base :
Rostellum long and slender, sword-like, projecting forwards from the column ; stipes long and narrow ; viscidium linear-ligulate    ..    .. 40. **Encheiridion**
Rostellum not sword-like, usually quite short :
Lip lamina reduced to a small point in front of the spur ; mouth of spur with tall erect 3-angled tooth-like callus ; apex of stipes much enlarged, partially enveloping the pollinia ; viscidium short    ..    ..    .. 41. **Chauliodon**
Lip lamina variously developed, usually quite large ; no vertical tooth in mouth of spur ..    ..    ..    ..    ..    ..    ..    .. 39. **Microcoelia**

Leaves present:
Rostellum not elongated, so deeply cleft that the apex of the column appears to be
  bifid in front ; lip usually very concave, entire, the base more or less enveloping
  the column, frequently with a central longitudinal keel ; mouth of spur wide, the
  disk of the lip gradually passing into the spur..    ..      ..    36. **Angraecum**
Rostellum elongated, easily distinguishable, often deeply bifid, sometimes with an
  additional central lobe:
Lip with a small tooth-like or rim-like callus in the centre at the mouth of the spur;
  mouth of spur narrow ; pollinia with separate stipes and common viscidium or
  each with its own stipes and viscidium        ..      ..      .. 42. **Diaphananthe**
Lip without any tooth-like callus:
Column long and slender, terete below the stigma, 3·5–4 cm. long ; perianth-
  segments all very similar, the lip broader than the others ; spur over 15 cm.
  long    ..      ..      ..    ..      ..      ..      ..      ..    48. **Barombia**
Column usually rather thick, 1 cm. or less in length, rarely terete ; spur very
  rarely over 15 cm. long, usually much shorter:
Pollinia on a single stipes and viscidium, the stipes very rarely bilobed in the
  upper part and bearing one pollinium on each lobe:
Stipes of pollinia elliptical, somewhat constricted below the middle, the apex
  running down into a deep depression at the back of the androclinium, the
  pollinia attached near the centre of the ovate upper part ; viscidium rectangular, attached to the anticous margin of the stipes ; lip trilobed ; stem
  long ; inflorescences several-flowered    ..      ..      ..    52. **Dinklageëlla**
Stipes of pollinia linear or somewhat widened in the upper part, the pollinia
  attached near or at the apex:
Rostellum produced downwards and then sharply reflexed parallel to itself
  so that the apex points upwards, bipartite almost from the base ; stipes
  somewhat shortly bifid at the apex ; viscidium long and narrow, attached
  to nearly the whole length of the rostellum ; inflorescences short and dense-
  flowered, arising at the base of the plant in the axils of the oldest leaves or
  leaf-bases    ..      ..      ..      ..      ..      ..    54. **Ancistrorhynchus**
Rostellum not sharply recurved upwards in the apical part:
Stems elongated, bearing leaves at more or less regular intervals usually for
  the greater part of their length:
Spur with a narrow mouth, easily distinguishable from the limb of the lip,
  almost straight, not or only slightly thickened in the apical part:
Viscidium large, nearly as long as the stipes, elliptical or oblong ; lip
  entire, without thickened basal auricles    ..      ..    56. **Eggelingia**
Viscidium small, much shorter than the long narrow stipes ; lip frequently
  3-lobed, the side-lobes sometimes tooth-like, in most species with thickened auricles on each side of the mouth of the spur      57. **Tridactyle**
Spur with a wide mouth, gradually merging into the limb of the lip, narrowest in the middle and swollen at the apex:
Spur abruptly recurved near the apex, about as long as the rest of the lip ;
  lip usually distinctly 3-lobed ; perianth-segments acuminate
                                            35. **Calyptrochilum**
Spur straight or very gently curved, much longer than the rest of the lip;
  lip indistinctly 3-lobed or entire ; perianth-segments obtuse
                                            51. **Solenangis**
Stems short, bearing all the leaves closely together:
Spur with a narrow mouth, easily distinguishable from the limb of the lip,
  usually much longer than the rest of the lip and not or scarcely thickened
  in the apical part ; lip acute, very similar in shape to the other perianth-
  members ..      ..      ..      ..      ..      ..      ..    47. **Aërangis**
Spur with a wide mouth, gradually merging into the limb of the lip, about
  the same length as the rest of the lip, distinctly thickened and recurved in
  the apical part ; lip obtuse, obscurely trilobed, much broader than the
  other perianth-segments..    ..      ..      ..      ..    53. **Eurychone**
Pollinia with separate stipites and either a common viscidium or 2 separate
  viscidia:
Leaves terete, or Iris-like, radiating like a fan from a common point, the 2
  surfaces facing sideways, quite entire at the apex, closely imbricate at the
  base, fleshy ; pollinia with a common viscidium:
Leaves terete ; stems elongated ; inflorescences short, few-flowered ; lip
  bilobed with elliptical rounded lobes ; stipites slender, shorter than the
  pollinia    ..      ..      ..      ..      ..      ..      ..    58. **Nephrangis**
Leaves Iris-like ; lip entire or more or less trilobed:
Inflorescences much shorter than the leaves, rather dense-flowered, flowers
  with long pedicels; spur much longer than the limb of the lip, swollen at the
  apex, nearly straight ; lip entire..    ..      ..      ..    34. **Podangis**

Inflorescences about as long as, or longer than, the leaves, sometimes a little shorter ; spur not swollen towards the apex :

Flowers with long pedicels ; sepals over 6 mm. long ; lip entire, ovate or lanceolate ; spur longer than the limb of the lip, incurved at the apex only        ..        ..        ..        ..        ..        ..        ..        49. **Rangaëris**

Flowers with very short pedicels ; sepals less than 4 mm. long ; lip with rounded side-lobes and acute middle-lobe ; spur equal to or shorter than the limb of the lip, sharply curved forward under the lip

44. **Bolusiella**

Leaves not terete nor Iris-like, the surfaces usually placed horizontally and frequently much reflexed, or if vertically placed the apices more or less unequally bilobed :

Flowers in very dense almost spherical subsessile inflorescences at the base of the leaves ; bracts large, as long as the flowers ; rostellum produced downwards and then recurved sharply parallel to itself, deeply bifid ; viscidium elongated, attached to nearly the whole length of the rostellum

54. **Ancistrorhynchus**

Flowers in long and usually relatively lax inflorescences ; bracts much shorter than the flowers ; rostellum never as above :

Rostellum long and beak-like, longer than the short column, bifid for two-thirds of its length ; viscidium elongated, either linear or with a broad rolled-up portion to which the stipites are attached and a narrow hyaline portion adnate to the apical part of the rostellum; lip similar to perianth-segments, all narrow and acuminate ; spur somewhat widened towards the mouth, at least twice as long as the lip      ..        ..        50. **Cyrtorchis**

Rostellum shorter than the column ; viscidium as broad as long or rarely shortly rectangular :

Lip deeply 3-lobed, the lateral lobes narrow and spreading ; petals more or less deltoid, the anticous margin rounded and projecting a long way forward        ..        ..        ..        ..        ..        ..        55. **Angraecopsis**

Lip entire, very shortly 3-lobed at the apex only, or rarely with almost orbicular side lobes at the base :

Mouth of spur some distance from base of lip, base of column and attachment of lateral sepals ; inflorescence erect, very dense ; flowers small ; leaves with parallel margins, closely imbricate at the base ; viscidium common ..        ..        ..        ..        ..        ..        38. **Listrostachys**

Mouth of spur immediately below base of column or attachment of lateral sepals :

Viscidium 1, common to 2 pollinia and stipites :

Spur swollen at the apex ; flowers small (sepals 3–6 mm. long), sometimes opposite or whorled ..        ..        ..        ..        45. **Chamaeangis**

Spur tapering towards the apex, often much longer than the lip ; flowers rather small (sepals 7 mm. long) to large :

Lip indistinctly 3-lobed at the apex, lateral lobes rounded, somewhat toothed, middle lobe much longer, acuminate, at the base of the lip a toothed projection on each side of the mouth of the spur ; rostellum long, subulate ; viscidium small, rounded        46. **Plectrelminthus**

Lip quite entire, with no toothed projection on the sides at the base, rarely 3-lobed with almost orbicular side lobes ; rostellum shorter, usually bifid or trifid        ..        ..        ..        ..        49. **Rangaëris**

Viscidia 2, each with pollinium and stipes :

Lip indistinctly 2–3-lobed, obtuse ; sepals and petals obtuse :

Lip much longer than broad, distinctly constricted in the middle ; petals oblong        ..        ..        ..        ..        43. **Sarcorhynchus**

Lip usually broader than long, sometimes length and breadth about equal ; petals nearly orbicular or broadly ovate      42. **Diaphananthe**

Lip quite entire, acute ; sepals and petals acute or acuminate

49. **Rangaëris**

**Additional genus:** since this account was written **Diceratostele gabonensis** Summerh. has been discovered in the Ivory Coast (Tiapleu Forest (Sept.)*Aké Assi* 3297!)

Many species of orchids from tropical America and tropical Asia are cultivated as ornamentals throughout Africa. Occasionally a species may become more or less naturalized and a well-known example of this is the Asiatic *Spathoglottis plicata* Bl.

## 1. HOLOTHRIX Rich. ex Lindl., Gen. & Sp. Orch. Pl. 257, 283 (1835) ; F.T.A. 7 : 190.

*Nom. cons. Deroemera* Rchb. f.(1852)—F.T.A. 7 : 195 ; F.W.T.A., ed. 1, 2 : 405.

Scape densely and retrorsely pubescent, without sheaths, 6–16 cm. long ; leaves 2, 1–1·5 cm. long, upper smaller ; lip 7-toothed ; flowers in a short spike, white, tinged with mauve ; sepals and petals 3-toothed at apex ..        ..        ..        1. *tridentata*

Scape glabrous, with numerous lanceolate acuminate sheaths in the lower part, up to
18 cm. high; leaf solitary, withering just before or at flowering time; lip elliptical,
3-lobed, side lobes rounded, middle lobe narrow, acute, 4–5 mm. long; inflorescence
10–30-flowered; petals oblong, triapiculate, the middle lobe longest    .. 2. *aphylla*

1. **H. tridentata** (*Hook. f.*) *Rchb. f.* Otia Bot. Hamburg. 119 (1881); F.T.A. 7: 193.  *Peristylus tridentatus*
   Hook. f. in J. Linn. Soc. 7: 221 (1864).  *Holothrix platydactyla* Kraenzl. (1893)—F.T.A. l.c.  A dwarf herb
   up to 5 in. high with 2 basal flattened leaves and mauve-white flowers.
     **W.Cam.**: Cam. Mt., 7,000–11,000 ft. (Oct.–Nov.) *Mann* 2128! *Preuss* 1036! *Johnston*! Also in Ethiopia.
2. **H. aphylla** (*Forsk.*) *Rchb. f.* Otia Bot. Hamburg. 119 (1881).  *Orchis aphylla* Forsk., Fl. Aegypt. Arab.
   156 (1775).  *Holothrix ledermannii* Kraenzl. in Engl., Bot. Jahrb. 48: 385 (1912).  *H. calva* Kraenzl. l.c
   386.  *Deroemera ledermannii* (Kraenzl.) Schltr. in Engl., Bot. Jahrb. 53: 486 (1915), in obs.; F.W.T.A.,
   ed. 1, 2: 405.  *D. calva* (Kraenzl.) Schltr. l.c. (1915), in obs.  A dwarf herb up to 6 in. high with 1 basal
   flattened leaf and white or pale bluish flowers.
     **N.Nig.**: Ropp, Plateau Prov., 4,500 ft. (Nov.) *King* 102! Mambila Plateau (Jan.) *Hepper* 1785!
     **W.Cam.**: Kufum, Banso Mts., 6,700 ft. (Dec.)  *Ledermann* 2005! Mbai, 6,000 ft. (Feb.) *Brunt* 969!
   Bamenda, 8,000 ft. (Jan.–Feb.) *Daramola* FHI 40506! Bafut-Ngemba F.R., Bamenda (Feb.) *Hepper*
   2159! Kumbo, 4,000 ft. (Dec.) *Ledermann* 1988! Eastwards to Ethiopia, Kenya and Arabia.

## 2. SCHWARTZKOPFFIA Kraenzl. in Engl., Bot. Jahrb. 28: 177 (1900).

Terrestrial, probably saprophytic; leaves absent; scape up to 9 cm. high, glabrous,
covered with large overlapping acute sheaths; flowers 2–3, subcapitate, lilac; lip
nearly equally 3-lobed, with a sac-like base surrounded by 2 wing-like calli running
down from the column and uniting in front, disk naked ..      ..      ..    *pumilio*

**S. pumilio** (*Lindl.*) *Schltr.* Die Orchid. 63 (1914).  *Penthea pumilio* Lindl. (1862).  *Brachycorythis pumilio* (Lindl.)
   Rchb. f. (1882)—F.T.A. 7: 203.  *B. rosea* A. Chev., Bot. 621 (1920).  *Schwartzkopffia buettneriana* Kraenzl.
   in Engl., Bot. Jahrb. 28: 177 (1900).  A dwarf leafless herb up to 4 in. high with lilac flowers.
     **S.L.**: Bagroo R. (Apr.) *Mann* 904! Jepihun (Jan.) *Smythe* 220! S. Kambui Hills (Mar.) *Dawe* 452! **Lib.**:
   Jui, Gola Forest (Apr.) *Bunting*! **Iv.C.**: Middle Cavally R., Fort Binger to Mt. Niènokoué (July) *Chev.*
   19502! **Togo Rep.**: Bismarckburg *Büttner* 40! **N.Nig.**: Kwarra, Wamba *King* 110! Nr. Jemaa, Zaria
   Prov. *King* 110a!

## 3. BRACHYCORYTHIS Lindl., Gen. & Sp. Orch. Pl. 363 (1838); F.T.A. 7: 200; Summerhayes in Kew Bull. 9: 226 (1955). *Diplacorchis* Schltr. (1921)—F.W.T.A., ed. 1, 2: 405. *Phyllomphax* Schltr. (1919)—F.W.T.A., ed. 1, 2: 407.

Leaves softly velvety; lip without obvious spur; lamina of lip flat, trilobed towards
apex, lobes about equal or middle shorter than laterals ..      ..      10. *pubescens*
Leaves quite glabrous; lip spurred or unspurred:
  Spur of lip conical with an acute apex or elongated and obtuse, often over 4·5 mm. long:
    Leaves broadly lanceolate or elliptical-lanceolate, up to 13 cm. long and 5 cm. broad,
      more or less spreading, not very stiff; all but lowest flowers considerably longer
      than the bracts; lamina of lip broadly obcordate, 1·5–2 cm. long and broad; spur
      with a narrow acute apex from a broader swollen base, 7·5–10 mm. long; petals
      quite free from column  ..    ..    ..    ..    ..    ..      1. *macrantha*
    Leaves narrowly lanceolate, up to 9 cm. long and 1·5 cm. broad, suberect, rather stiff;
      flowers equalling or shorter than the bracts; petals united at their base to the
      column:
        Lamina of lip broadest at the base, sometimes with a lobe on each side, apical part
          narrow, tongue-shaped, 5–8·5 mm. long; spur equally broad to apex or somewhat
          thickened in apical part, apex obtuse, 4·5–8·5 mm. long    ..    .. 2. *tenuior*
        Lamina of lip broadest above the middle, obovate, truncate in front, tapering
          towards the base; spur distinctly conical with a broad mouth and narrow acute
          apex, 2·5–6 mm. long   ..    ..    ..    ..    ..    ..    .. 3. *conica*
  Spur of lip very broad and rounded or absent, the base then boat-shaped:
    Largest leaves in lower part of stem, upper part with few usually much smaller
      leaves; middle lobe of lip usually much longer than laterals, latter quite rounded
      and not at all sickle-shaped; lower bracts shorter than the flowers; spur of labellum
      short and rounded  ..    ..    ..    ..    ..    ..    .. 4. *paucifolia*
    Largest leaves in middle or upper part of stem, lower part of stem usually covered
      with sheaths; middle lobe of lip equalling or shorter than laterals and always
      much smaller, laterals more or less sickle-shaped or incurved and acute; lower
      bracts usually longer than the flowers:
        Epichile (apical part) of lip with a conical projecting callus just in front of the sac-like
          hypochile, reniform, the middle lobe shorter and smaller than the lateral lobes;
          dorsal sepal 4–5 mm. long; spike dense, 1–2 cm. diam.; sides of basal part of lip
          broadly angled   ..    ..    ..    ..    ..    ..    .. 6. *buchananii*
        Epichile of lip without a conical basal callus but usually with a longitudinal ridge
          terminating in the middle lobe; dorsal sepal 5–14 mm. long; spike over 2 cm.
          diam.:
            Lip scarcely longer than the sepals, about 1 cm. long, the middle lobe about the
              same length as the laterals; plant about 70–100 cm. high with many-flowered
              inflorescence 10–18 cm. long and 3–4 cm. diam.; spur rounded, up to 4·5 mm.
              long ..    ..    ..    ..    ..    ..    ..    ..    .. 5. *sceptrum*

Lip considerably longer than the sepals, the middle lobe much shorter than the lateral lobes; spur absent, base of lip boat-shaped:
Hypochile of lip short, 1·5–2 mm. long from back to front, sides triangular; leaves terminating in a very long fine point, lanceolate or narrowly lanceolate; petals lying forward from column, semi-orbicular or broadly obovate; stem and leaves becoming black on drying; leaves and flowers numerous .. 7. *pleistophylla*
Hypochile of lip 3·5–6 mm. long from back to front, sides sloping down from column, not or scarcely angled; leaves acuminate but usually not very finely pointed, lanceolate or broadly lanceolate; petals ovate or elliptical-ovate; stem and leaves not blackening on drying:
Plant usually 40–100 cm. high, terrestrial; leaves numerous, closely overlapping, stiff and rigid; flowers numerous in a long spike, rather fleshy; petals erect on either side of column; lip 1–2 cm. long 8a. *ovata* subsp. *schweinfurthii*
Plant usually 20–40 cm. high, frequently epiphytic; leaves less than 15, rather thin in texture, not closely overlapping; flowers in a loose spike, less than 22 in number, thin in texture; petals set forward from column or spreading; lip 2–3 cm. long.. .. .. .. .. .. .. 9. *kalbreyeri*

1. **B. macrantha** (*Lindl.*) *Summerh.* in Kew Bull. 9: 236 (1955). *Gymnadenia macrantha* Lindl. (1835). *Platanthera helleborina* (Hook. f.) Rolfe (1898)—F.T.A. 7: 204. *Phyllomphax helleborina* (Hook. f.) Schltr. (1919)—F.W.T.A., ed. 1, 2: 407. A stout herb up to 15 in. high, the upper half bearing up to 20 green and mauve flowers.
   **Guin.:** Fouta Djalon, Ditinn (Sept.) *Chev.* 18533! Manfara to Tendou, Kissidougou (July) *Martine* 345! Kouria to Languery (Aug.) *Caille* in *Hb. Chev.* 14652! Timbo (July) *Chev.* 18517! Nimba Mts. (Sept.) *Schnell* 1857! **S.L.:** Sugarloaf Mt. (Dec.) *Sc. Elliot* 4061! Jigaya (Sept.) *Thomas* 2738! Bagwema, Bafi R. (July–Aug.) *Dawe* 547! Mahinto, Tonko Limba (Sept.) *Adames* 195! junction Port Loko and Kambia roads (Sept.) *Adames* 244! **Lib.:** Nimba Mts., 2,500 ft. (July) *Leeuwenberg & Voorhoeve* 4811! Sanokwele, Central Prov. (Sept.) *Baldwin* 7536! **N.Nig.:** Anara F.R., Zaria (July) *Keay* FHI 25960! **S.Nig.:** Idanre Hills, Ondo *Hoskyns-Abrahall* FHI 36068! (Aug.) *Gillett* 15312! Iseyin, Oyo Prov. *Latilo* FHI 53918! **W.Cam.:** Fonfaka, Bamenda, 3,000 ft. (June) *Maitland* 1639! Bambui, 4,800 ft. (May) *Brunt* 1136! Also in E. Cameroun, Gabon and C. African Republic.
2. **B. tenuior** *Rchb. f.* in Flora 48: 183 (1865). *Platanthera tenuior* (Rchb. f.) Schltr. (1895)—F.T.A. 7: 205. *P. engleriana* (Kraenzl.) Rolfe in F.T.A. 7: 204 (1898). *Diplacorchis engleriana* (Kraenzl.) Schltr. (1921)—F.W.T.A., ed. 1, 2: 407. A slender herb up to 3 ft. high; raceme 4 in. long or less, bracts not much projecting; flowers reddish, sepals with greenish tips.
   **Guin.:** Nimba Mts., 5,300 ft. (June, July, Aug.) *Schnell* 1558b! 2997! 3379! Fon Massif, 5,300 ft. (July, Aug.) *Schnell* 3098b! 3317! **S.L.:** Bintumane Mt., 5,000–6,000 ft. (July) *Bakshi* 227! 230! **Iv.C.:** Nimba Mts., 2,500–5,000 ft. (Aug.) *Boughey* GC 18075! 18077a! 18093! **N.Nig.:** Katsina Ala (Aug.) *Dalz.* 835! Generally distributed in tropical Africa, southwards to Transvaal and Natal.
3. **B. conica** (*Summerh.*) *Summerh.* in Kew Bull. 9: 244 (1955). *Diplacorchis conica* Summerh. in Kew Bull. 1938: 141. A slender herb up to 18 in. high; with a short raceme of purple or violet flowers.
   **N.Nig.:** Mambila Plateau, Sardauna Prov. (June) *Chapman* 2! Also in Gabon, Congo and Zambia.
4. **B. paucifolia** *Summerh.* in Kew Bull. 1: 123 (1948). A herb 8–20 in. high with a rather short and dense flower spike.
   **Guin.:** Man, Nimba Mts. (Apr.) *Roberty* 3107! Nzérékoré, Mt. Nimba, 6,000 ft. (Oct.) *Jac.-Fél.* 1936! Nimba Mts. (Aug.–Sept.) *Schnell* 1853! 3372! 3405! **S.L.:** Bintumane Pk., Loma Mts., 5,000–5,300 ft. (Aug.) *Jaeger* 375! 1027! 1132! Da-Oulen Mt., 5,200 ft. (Aug.) *Jaeger* 1278! 1327!
5. **B. sceptrum** *Schltr.* in Beih. Bot. Centralbl. 38, II: 114 (1921). *Diplacorchis ashantensis* Summerh. in Kew Bull. 1931: 378 (1931); F.W.T.A., ed. 1, 2: 407. A slender leafy herb 2½–3 ft. high with white and purple flowers.
   **Sen.:** Kaème Forest, Oussouye (Sept.) *Doumlia* 849! **Guin.:** Kindia (Aug.) *Pobéguin* 1375! **Ghana:** Amoma (July) *Chipp* 531! **N.Nig.:** Kagarko, Zaria (July) *G. V. Summerhayes* 122! Jemaa, Zaria (Aug.) *King* 145! Kabba (Apr.) *Westwood*! **S.Nig.:** Oyo (July) *Sanford* 1785/65! Also in E. Cameroun.
6. **B. buchananii** (*Schltr.*) *Rolfe* in F.T.A. 7: 570 (1898). *B. parviflora* Rolfe in F.T.A. 7: 202 (1898). A herb 1–2½ ft. high with a dense narrow spike of small pink, mauve or purple flowers.
   **N.Nig.:** Mambila Plateau, 5,500 ft. (June) *Chapman* 4! **S.Nig.:** Obudu Plateau, Ogoja Prov., 5,300 ft. (June) *Horwood* in *Hb. King* 158! Also in E. Africa from Uganda and Kenya southwards to Rhodesia.
7. **B. pleistophylla** *Rchb. f.* in Otia Bot. Hamburg. 104 (1881); F.T.A. 7: 202. A herb 1–3½ ft. high, with numerous very finely pointed leaves and a spike of purple, mauve or violet flowers.
   **N.Nig.:** Ropp, Plateau Prov., 4,300 ft. (May) *King* 42! Sara Hills, Bauchi Prov., 5,400 ft. *King* s.n! Generally in tropical Africa, southwards to Angola.
8. **B. ovata** *Lindl.* Gen. & Sp. Orch. 363 (1838). Widespread in tropical Africa.
8a. **B. ovata** subsp. **schweinfurthii** (*Rchb. f.*) *Summerh.* in Kew Bull. 9: 257 (1955). *B. schweinfurthii* Rchb. f. (1878)—F.T.A. 7: 201; F.W.T.A., ed. 1, 2: 405. A slender leafy herb with purple-spotted flowers. The only subspecies so far recorded from the area of this flora.
   **Sen.:** Oussouye (Nov.) *Berhaut* 6549! **Iv.C.:** Kong Dist. *Bouet* 2560! **N.Nig.:** Lapai, Niger Prov. Onochie FHI 35386! Zungeru (June) *Dalz.* 561! Abinsi (June–July) *Dalz.* 836! Wana, 1,700 ft. (June) *Hepburn* 119! Anara F.R., Zaria (July) *Keay* FHI 25968! **W.Cam.:** Lakom, Bamenda, 6,000 ft. (Apr.) *Maitland* 1787! Bali-Ngemba F.R. (May) *Ujor* FHI 30324a! Mbaw Plain, Sabonuri, 2,600 ft. (May) *Brunt* 405! Eastwards to Sudan and Kenya.
9. **B. kalbreyeri** *Rchb. f.* in Flora 61: 77 (1878); F.T.A. 7: 201. A slender herb with large lilac or violet flowers in a short raceme; epiphytic or on rocks.
   **Guin.:** Bandakene, SW. of Pita (July) *Adames* 290! **S.L.:** Heddles Farm (June) *Lane-Poole* 376! Makump (July) *Thomas* 894! Ndijajula, near Njala (May, June) *Deighton* 700! 4315! **Lib.:** St. Paul R., Dobli Isl. (Apr.) *Bequaert* 173! Ganta (July) *Harley* 2138! Nimba *Adam* 21547! **W.Cam.:** Cam. Mt. 5,500–6,000 ft. (Mar.) *Kalbreyer* 145! Also in Congo and Kenya.
10. **B. pubescens** *Harv.* Thes. Cap. 1: 35, t. 54 (1859); F.T.A. 7: 201. *B. sudanica* Schltr. in Beih. Bot. Centralbl. 38(2): 111 (1921); F.W.T.A., ed. 1, 2: 405. A slender herb 18–30 in. high with pink, bluish or reddish-purple, or rarely white, flowers.
    **Mali:** Folo (May) *Chev.* 827! **Guin.:** Sabodougou, near Touba (July) *Collenette* 61! **Iv.C.:** Séguéla (Oct.) *Aké Assi* 5633! **Togo Rep.:** Bassari *Thienemann*. Fasugu (May) *Büttner* 647! **N.Nig.:** Lokoja *Lugard*! Zungeru (July) *Dalz.* 571! Vom *Dent Young* 232! Naraguta, 4,000 ft. (May) *Lely* 237! Tangale-Woja, Bauchi Prov. (May) *G. V. Summerhayes* 66! Mambila Plateau, 5,500 ft. (June) *Chapman* 8! **W.Cam.:** Lakom, Bamenda, 6,000 ft. (Apr.) *Maitland* 780! Jua, Bamenda, 3,000–3,500 ft. (Apr.) *Maitland* 1625! Jakiri (June) *Gregory* 144! Also generally in tropical Africa, southwards to Transvaal and Natal.

FIG. 382.—BRACHYCORYTHIS PUBESCENS *Harv.* (ORCHIDACEAE).

A, upper part of stem and inflorescence. B, base of stem and roots. C, flower, without tepals.
D, lip. E, lateral sepal. F, column and petals. G, pollinium.

**4. CYNORKIS** Thou. in Nouv. Bull. Sci. Soc. Philom. Paris 1: 317 (1809); F.T.A. 7: 259.

Lip 5-lobed, lowest 2 lobes reduced to teeth, other 3 lobes oblong, subacute, middle longer than laterals; leaves 1–2, lanceolate, oblanceolate or oblong, up to 10·5 cm. long and 3 cm. broad; spike dense, up to 8 cm. long; intermediate sepal 2–2·5 mm. long, spur 2–3 mm. long ... .. .. .. .. .. .. 1. *debilis*
Lip entire, ligulate or narrowly oblong, obtuse; leaves 2–5, oblanceolate or narrowly oblanceolate, acuminate, up to 14 cm. long and 1·5 cm. broad; spike dense, usually 2–4 cm. long, but exceptionally up to 9 cm.; intermediate sepal 2–3 mm. long, spur about the same length or a little longer .. .. .. .. 2. *anacamptoides*

1. **C. debilis** (*Hook. f.*) *Summerh.* in Kew Bull. 1933: 246. *Habenaria debilis* Hook. f. (1864)—F.T.A. 7: 213. A terrestrial herb up to 9 in. high with small white flowers with wine-coloured spotted lip.
**W.Cam.:** Cam. Mt., 5,000–7,000 ft. (Nov.) *Mann* 2127! **F.Po:** Moka, 4,550 ft. (Sept.) *Wrigley & Melville* 616! Also in S. Tomé and Congo.
2. **C. anacamptoides** *Kraenzl.* in Engl., Pflanzenr. Ost-Afrika C: 151 (1895), and in Engl., Bot. Jahrb. 22: 18 (1896); F.T.A. 7: 260. A slender terrestrial herb 6–20 in. high with a dense spike of small pink or purple flowers.
**W.Cam.:** Jakiri, Bamenda Div. (Feb.) *Hepper* 1958! **F.Po:** Las Corteras (Jan.) *Guinea* 2125! Carretera Moka (Jan.) *Guinea* 1895! Mioka, 5,000 ft. (Dec.) *Boughey* 156! Also in Congo, Ethiopia, Uganda, Kenya, Tanzania, Angola and Rhodesia.

## 5. HABENARIA Willd., Sp. Pl. 4: 44 (1805); F.T.A. 7: 206.

Lip superior, trilobed at apex only; sepals and ovary hairy; spur erect, somewhat recurved, 10–15 mm. long; petals shortly bidentate at apex; scape 15–30 cm. tall; leaf radical, ovate, more or less adpressed to the ground, variegated with reticulate markings .. .. .. .. .. .. .. .. .. .. 47. *occidentalis*
Lip inferior:
Petals undivided: (to p. 191)
  Lip entire:
    Dorsal sepal 6·5–11 mm. long; anther canals elongated, over 2 mm. long; stigmas free:
      Leaves narrowly lanceolate, up to 18 cm. long and 5 cm. broad; petals ovate or elliptical, obtuse or rounded, posterior margin slightly infolded; lip 9–12·5 mm. long; anther canals straight, 2–3 mm. long .. .. .. 8. *leonensis*
      Leaves broadly or oblong-lanceolate, up to 12 cm. long and 4 cm. broad; petals broadly lanceolate, acute, posterior margin not infolded; lip 11–13 mm. long; anther canals much incurved, 3·5–4 mm. long .. .. .. 9. *stenochila*
    Dorsal sepal 4–6 mm. long; anther canals very short; posterior half of petal infolded; stigmas usually connate .. .. .. .. .. .. 10. *zambesina*
  Lip trilobed:
    Side lobes of lip pectinate or fimbriate:
      Flowers small; dorsal sepal 5–7 mm. long; lip 1–1·5 cm. long, middle lobe flabellate, emarginate with a central apiculus, front margin pectinate, spur 2–3 cm. long; plant 7–15 cm. high, 1–2-flowered; leaves linear-lanceolate, 1–4 cm. long, 2–4 mm. broad .. .. .. .. .. .. .. 21. *jacobii*
      Flowers large; dorsal sepal over 12 mm. long; lip at least 2 cm. long, middle lobe linear or linear-oblong, entire, subacute; plant 15–70 cm. high, usually over 4-flowered; leaves 5–25 cm. long, 0·75–7 cm. broad:
        Side lobes of lip semi-ovate or fan-shaped, shortly pectinate; spur over 5 cm. long; anther-loculi approximate; leaves broadly oblong-lanceolate, 3–7 cm. broad:
          Lip about 2·5–4 cm. across, lobes divergent; side lobes fan-shaped, deeply cordate on each side of claw; spur 12–15 cm. long .. .. .. 19. *englerana*
          Lip about 2·5 cm. across, lobes nearly parallel; side lobes semi-ovate, bases at right angles with claw; spur about 5 cm. long .. 20. *prionocraspedon*
        Side lobes of lip linear or slightly broadened upwards, divided into many narrow segments on the outer edge; anther loculi distant from one another at the ends of the narrow ribbon-shaped connective; leaves linear-lanceolate, 0·75–3 cm. broad:
          Spur 1·2–2·2 cm. long; side lobes of lip only slightly widened upwards, thread-like segments rising pinnately; receptive portions of stigmas stalked; anther canals 3–5 mm. long .. .. .. .. .. .. .. 23. *mannii*
          Spur 3·5–7·5 cm. long; side lobes of lip considerably widened upwards, thread-like segments arising almost palmately; receptive portions of stigmas sessile; anther canals only 2 mm. long .. .. .. .. 24. *jaegeri*
    Side lobes of lip entire:
      Leaf suborbicular, radical, closely appressed to the ground, 2–6·5 cm. broad; perianth-segments ovate-lanceolate, acuminate, 7–10 mm. long; lip-segments narrowly lanceolate, of equal length, the middle one broader; spur 3–6 cm. long .. .. .. .. .. .. .. .. .. .. 48. *lelyi*

Leaves cauline, or if radical neither suborbicular nor closely appressed to the ground:

Column elongated, over 13 mm. high, the anther at the apex; anther-canals free from the side lobes of the rostellum; stigmas cushion-like, partially confluent; leaves in a tuft at the base, lanceolate, up to 25 cm. long; lobes of lip filiform; dorsal sepal 1·8–3 cm. long ..    ..    ..    ..    7. *macrandra*

Column short, less than 6 mm. high; anther-canals united to the side lobes of the rostellum; stigmas stalked or nearly sessile, but not cushion-like nor confluent:

Spur 4 cm. or more long; dorsal sepal 8–12 mm. long:

Epiphytic plant (rarely in humus); leaves lanceolate or narrowly ovate, up to 30 cm. long, 2–7 cm. broad; spur 6–10 cm. long; inflorescence rather dense, usually 10–30-flowered    ..    ..    ..    ..    ..    ..    11. *procera*

Terrestrial plants; leaves lanceolate or narrowly lanceolate, 0·5–4 cm. broad; inflorescence rather lax, almost always less than 12-flowered:

Leaves linear- or narrowly lanceolate, up to 1 cm. broad; spur 4–5 cm. long; growing on rocks by streams    ..    ..    ..    ..    12. *weilerana*

Leaves narrowly lanceolate or lanceolate, 1–4 cm. broad; spur 5–12 cm. long; growing in humus or grassy places among rocks ..    ..    13. *gabonensis*

Spur less than 3 cm. long; dorsal sepal up to 6·5 mm. long:

Leaves in a tuft at the base of the stem with often a smaller leaf on the scape above:

Plant 15–55 cm. high; leaves ascending, elliptical-lanceolate, shortly stalked, 4–14 cm. long, 1·5–3·5 cm. broad; inflorescence 5–20-flowered; dorsal sepal 5–7 mm. long; lobes of lip all narrowly linear; spur decurved, inflated in distal half    ..    ..    ..    ..    ..    ..    ..    ..6. *buntingii*

Plant up to 15 cm. high; leaves spreading or more or less recurved, broadly linear or linear-lanceolate, 1–2·5 cm. long, 2–4 mm. broad; inflorescence 1–2-flowered; dorsal sepal 4–5·5 mm. long; side lobes of lip narrow, middle lobe suddenly fan-shaped from a narrow base; spur pendulous, almost straight, scarcely inflated    ..    ..    ..    ..    ..    ..    22. *parva*

Leaves at intervals along stem, but often closer in lower part:

Spur 1–1·5 mm. long, shorter than the lip; anther canals very short:

Lip broad, 2–2·5 mm. long, shortly trilobed, three lobes about equal in length, middle lobe triangular with an elevated keel; stigmas nearly sessile; leaves elliptic-lanceolate, acuminate, up to 10 cm. long and 4 cm. broad; rostellum small, tooth-like; spike slender, up to 1 cm. diam.; dorsal sepal 2–2·5 mm. long    ..    ..    ..    ..    ..    ..    5. *microceras*

Lip longer than broad, 7·5–9·5 mm. long, the middle lobe much longer than the spreading laterals, ligulate, base of lip deeply cordate with 2 rounded auricles; stigmas oblong or club-shaped, 1·5–2 mm. long; leaves narrowly lanceolate, acute, up to 25 cm. long and 2 cm. broad; rostellum large, folded, shield-like when flattened; spike rather stout, 1·5–3 cm. diam.; dorsal sepal 4–6·5 mm. long    ..    ..    ..    ..    ..    18. *peristyloides*

Spur 10 mm. or more long, equalling or longer than the lip:

Spur almost filiform, not much thickened near the apex, always much longer than the lip; dorsal sepal up to 5 mm. long; flowers numerous in a long spike:

Petals ovate or narrowly ovate; segments of lip ligulate or narrowly oblong, up to 5·5 mm. long; middle lobe of rostellum short, hooded:

Leaves 1 or 2 near base of stem, with several much smaller bract-like ones above; pedicels (and ovaries) more or less appressed to rhachis so that spike is very narrow, 1 cm. diam.; dorsal edge of petals folded inwards; side lobes of rostellum very short, middle lobe obtuse, clearly hooded; side lobes of lip up to 2·5 mm. long    ..    ..    ..    1. *attenuata*

Leaves 4–8 all along stem below inflorescence; pedicels (and ovaries) arcuate so that spike is 1·5–2·5 cm. diam.; dorsal edge of petal not folded inwards; side lobes of rostellum relatively long and narrow, middle lobe very short, slightly hooded; side lobes of lip 2·5–5·5 mm. long    ..    ..    2. *bracteosa*

Petals linear or narrowly lanceolate; segments of lip linear, up to 9 mm. long; middle lobe of rostellum narrow, elongated, more or less acute:

Leaves oblong or narrowly lanceolate, about 5–6 times as long as broad, up to 10 cm. long; stigmas papillose underneath; rostellum middle lobe without tall keel ..    ..    ..    ..    ..    ..    3. *filicornis*

Leaves linear, 10–20 times as long as broad, up to 17 cm. long; stigmas smooth underneath; rostellum middle lobe with tall keel    4. *chlorotica*

Spur stouter, more or less markedly club-shaped near apex:

Dorsal sepal 2·5–4 mm. long; spur shortly bifurcate at apex, about 7 mm. long; leaves linear to lanceolate-linear, at least 8 times as long as broad,

7–13 cm. long, 5–10 mm. broad; inflorescence 4–17-flowered; anther-loculi divergent    ..    ..    ..    ..    ..    17. *nigrescens*
Dorsal sepal 5–7 mm. long; spur rounded or entire at apex, not bifurcate; leaves lanceolate or oblong-lanceolate, at most 7 times as long as broad:
Leaves longest immediately below the inflorescence, lower half of stem almost covered by 4–5 rather loose sheaths; leaves lanceolate, 2·5–8 cm. long, 7–12 mm. broad; lower bracts longer than the flowers; petals curved lanceolate; side lobes of lip shorter than middle lobe; spur about equalling or longer than lip    ..    ..    ..    ..    16. *nigerica*
Leaves all along stem or in the lower part only, basal sheaths very inconspicuous; lower bracts shorter than the flowers; petals obovate, rounded; side lobes of lip equalling or longer than middle lobe; spur considerably longer than lip:
Stigmas club-shaped; leaves mostly near base of the stem, up to 9 cm. long and 1·7 cm. broad; spur much dilated at apex; anther loculi some distance apart on the broadened connective..    ..    ..14. *obovata*
Stigmas awl-shaped; leaves all along the stem, 5–15 cm. long, 2–4 cm. broad; spur moderately dilated at apex; anther loculi contiguous
15. *dinklagei*
Petals bilobed or bipartite, the posterior lobe often adnate to the dorsal sepal: (from p. 189)
*Leaves borne along the stem, neither orbicular nor adpressed to the ground:
Spur 5 cm. or more in length:
Lobes or lip and anterior petal-lobe lanceolate; anther canals much shorter than stigmas; lateral sepals not rolling up lengthwise; spur 13–17 cm. long; dorsal sepal 11–15 mm. long    ..    ..    ..    ..    ..    ..    ..46. *walleri*
Lobes of lip and anterior petal-lobe linear or linear-lanceolate, latter horn-like; anther canals slightly longer than stigmas; lateral sepals rolling up lengthwise; dorsal sepal over 15 mm. long:
Anterior petal-lobe lanceolate-linear, 2–3·5 cm. long; spur up to 7 cm. long, swollen at apex    ..    ..    ..    ..    ..    ..    ..    ..42. *holubii*
Anterior petal-lobe filiform or narrow-linear, 4·5–7·5 cm. long; spur 5–17 cm. long:
Spur 12–17 cm. long, moderately swollen near end, tapering again towards apex itself    ..    ..    ..    ..    ..    ..    ..    ..43. *cirrhata*
Spur 5–8 cm. long, suddenly swollen at apex to form a rounded obtuse sac
44. *laurentii*
Spur less than 5 cm. in length:
Anterior petal lobe slender, fleshy, subulate, rather like a long curved horn, 2–4 times as long as the erect dorsal sepal; dorsal sepal over 5 mm. long; spur much swollen at apex; stigmas suddenly broadened at apex; leaves mostly in lower part of stem, 1–3 cm. broad:
Stigmas 8–12 mm. long; dorsal sepal 11–17 mm. long; spur 3–5 cm. long; side lobes of lip very slender, entire, longer than the anther canals    ..41. *clavata*
Stigmas 4·5–6 mm. long; dorsal sepal 6–9 mm. long; spur 1·5–2·5 cm. long; side lobes of lip narrowly lanceolate, often with several comb-like teeth or threads on the outside edge ..    ..    ..    ..    ..    ..    ..    ..45. *cornuta*
Anterior petal-lobe not as above:
Dorsal sepal 1 cm. long or more:
Petal-segments linear or almost filiform; spur less than 2 cm. long; leaves 1·5–7 cm. broad, up to 12 cm. long:
Petals divided nearly to base, segments glabrous; side lobes of lip longer than the middle lobe, side lobes spreading, middle lobe hanging vertically downwards, narrow; leaves in a group just above the middle of the stem; stigmas not swollen at apex; inflorescence up to 10-flowered    ..    ..26. *barrina*
Petals divided two-thirds down, segments ciliate; side lobes of lip shorter than the middle lobe; leaves in lower half of stem; stigmas club-shaped; inflorescence 15–70-flowered    ..    ..    ..    ..    ..    25. *longirostris*
Petals divided two-thirds down, posterior segment lanceolate, anterior curved, oblong or linear, much shorter or equal in length; spur 3–4 cm. long; leaves narrower, 1–2 cm. broad, up to 6 cm. long ..    ..    ..40. *phylacocheira*
Dorsal sepal less than 9 mm. long:
Leaves lanceolate, oblong-lanceolate or oblanceolate, 1·5–5 cm. broad, mostly in central part of stem, lower part of stem more or less covered by sheaths; stigmas obliquely club-shaped or scarcely thickened upwards:
Lamina of leaves lanceolate or oblong-lanceolate, the basal part broadest and usually more or less enfolding the stem, up to 14 cm. long, 1·5–3·5 cm. broad; lower leafless part of stem 5–10 cm. long, sheaths loose, funnel-shaped; dorsal sepal 4·5–6·5 mm. long; inflorescence dense-flowered; spur 10–15 mm. long    ..    ..    ..    ..    ..    ..    ..    ..    27. *papyracea*
Lamina of leaves oblanceolate, narrowing gradually to the base, 5–16 cm. long,

1·5–5 cm. broad; lower leafless part of the stem 5–20 cm. long, sheaths tight; inflorescence rather loose-flowered; dorsal sepal 3·5–5·5 mm. long:

Pedicel and ovary bending out from rhachis so that flowers are nearly horizontal, inflorescence 3–5 cm. diam.; spur 1·2–2·5 cm. long, slightly thickened in distal half; anterior petal-lobe equalling, shorter than or slightly longer than the posterior .. .. .. .. .. 28. *buettnerana*

Pedicel and ovary diverging from rhachis at about 45°, flowers bending forward slightly or vertical, inflorescence 2–3 cm. diam.; spur 9–16 mm. long, thickest in centre and tapering towards both ends, much incurved; anterior petal-lobe longer and narrower than the posterior .. 29. *malacophylla*

Leaves lanceolate, linear-lanceolate or linear, never more than 2 cm. broad and usually much narrower, when lanceolate the largest near the base of the stem:

Anterior petal-lobe shorter and narrower than posterior; stigmas club-shaped, not obliquely truncate-capitate at apex; dorsal sepal erect, about 4 mm. long; inflorescence dense, many-flowered, about 2 cm. diam.; leaves lanceolate or narrowly lanceolate, in lower part of stem; spur descending, slightly inflated in distal half, 8–10 mm. long .. .. .. 30. *bongensium*

Anterior petal-lobe longer than or rarely equalling the posterior, usually broader; stigmas slender in lower part, obliquely truncate-capitate at apex; dorsal sepal more or less reflexed; leaves usually linear or linear-lanceolate:

Spur spirally twisted in middle and sharply incurved, much swollen at apex, 9–13 mm. long; lobes of lip narrowly linear, middle lobe 9–11 mm. long, longer than laterals .. .. .. .. .. .. 31. *genuflexa*

Spur straight or slightly curved, not spirally twisted or if so not incurved:

Petals divided for two-thirds of the length only, anterior segment narrower than posterior; spur very slender, slightly incurved, not thickened at end, 5–7 mm. long; lobes of lip narrowly linear; stigmas 6–7 mm. long
32. *dalzielii*

Petals divided almost to the base; lateral sepals obliquely obovate, laterally apiculate; lobes of lip narrowly linear, not very acute, middle distinctly longer than laterals:

Bracts equalling or longer than pedicel + ovary; dorsal sepal 5–8 mm. long; spur 15–20 mm. long, thickened in lower half, rather acute
36. *huillensis*

Bracts considerably shorter than pedicel + ovary:

Anterior segment of petals about the same length as, or shorter than, the posterior segment; spur 6–8 mm. long:

Posterior petal-segment much broader than anterior, about the same length; dorsal sepal 5–6 mm. long; inflorescence often rather lax, 1·5–2·5 cm. diam.; leaves up to 13 cm. long and 8 mm. broad; anther canals 4–5 mm. long .. .. .. .. .. 33. *angustissima*

Posterior petal segment narrower and longer than anterior; dorsal sepal 3–4 mm. long; inflorescence dense, 1·5–2 cm. diam.; leaves up to 23 cm. long and 1 cm. broad; anther canals about 2·5 mm. long
35. *pauper*

Anterior segment of petals much longer and usually broader than the posterior segment; spur 1–3 cm. long:

Staminodes tongue-shaped, entire; leaves rather short, 4–10 cm. long; stigmas obliquely truncate at apex; spur 18–22 mm. long, distinctly swollen in the lower half and narrowly club-shaped 37. *linguiformis*

Staminodes bilobed; leaves longer, up to 30 cm. long:

Spur equal in thickness nearly to apex where there is a very slight swelling, 12–21 mm. long; staminodes large, deeply bilobed with a distinct stalk; racemes rather densely many-flowered 38. *chirensis*

Spur distinctly club-shaped and swollen in the apical half or third; staminodes about 1 mm. long, sessile:

Leaves linear, up to 20 cm. long and 8 mm. broad; ovary 5–6 mm. long, much shorter than the long slender pedicel; spur 1–3 cm. long, swollen in the apical third; inflorescence rather lax
34. *ichneumonea*

Leaves narrowly lanceolate, up to 10 cm. long, 10–14 mm. broad; ovary 8–9 mm. long, often nearly as long as the pedicel; spur 10–11 mm. long, swollen in the apical half .. .. 39. *maitlandii*

*Leaves 1–2, radical, suborbicular or heart-shaped and adpressed to the ground:

Rhachis and flowers hairy; leaf on flowering stem 1, rarely with a much smaller additional one near the base of the scape or young flowering plants with 2 leaves; dorsal sepal 5–8 mm. long; scape up to 45 cm. high:

Spur about 6 cm. long; leaf about 3–4 cm. long and broad, margins densely hairy, otherwise almost glabrous; dorsal sepal about 6 mm. long; segments of lip and anterior petal-segments filiform .. .. .. .. .. 49. *lecardii*

Spur 10–14 mm. long; leaf (or leaves) (1·5–)4–12 cm. long, including basal lobes,
(1–)4–11 cm. broad, margins hairy, often with scattered long hairs on upper
surface; dorsal sepal 5–8 mm. long; segments of lip and anterior petal-segment
linear or very narrowly strap-shaped  ..    ..    ..    ..    .. 50. *keayi*
Rhachis and flowers glabrous; leaves on adult plants 2; dorsal sepal 8 mm. long or
more; spur 5–20 cm. long; scape 30–70 cm. high:
  Segments of lip and anterior petal-segments oblanceolate, up to 2 cm. long, 2·5–
  9 mm. broad; leaves 3–8 cm. long; scape almost covered with large sheaths;
  stigmatic arms thick all the way up, truncate at apex, 3–5 mm. long
                                                            51. *macrura*
  Segments of lip and anterior petal-segments filiform, 2·5–4·5 cm. long; leaves
  4–16 cm. long, 9–20 cm. broad; sheaths on scape small, distant; stigmatic arms
  clavate-capitate from a very slender base, 6–10 mm. long  .. 52. *armatissima*

1. **H. attenuata** *Hook. f.* in J. Linn. Soc. 7: 221 (1864); F.T.A. 7: 216; F.W.T.A., ed. 1, 2: 410 (excl. syn.
*H. clarencensis* Rolfe, partly). A slender terrestrial herb up to 18 in. high with small green flowers in a
narrow spike.
    **W.Cam.**: Cam. Mt., 7,000 ft. (Nov.) *Mann* 2188! 8,000–10,000 ft. *Johnston* 30! 33! **F.Po:** Clarence
Peak, on summit, 10,000 ft. (Dec.) *Mann* 645 (partly)! Also in Congo, Ethiopia and Uganda (Mt. Elgon).
2. **H. bracteosa** *Hochst. ex A. Rich.* Tent. Fl. Abyss. 2: 292 (1851); F.T.A. 7: 217. *H. clarencensis* Rolfe (1898)
—F.T.A. 7: 216, partly. An erect leafy terrestrial plant up to 3½ ft. high with a long dense spike of
green or greenish-yellow flowers.
    **W.Cam.**: *Preuss* 970! Lakom, Bamenda, 6,000 ft. *Maitland*! **F.Po:** Clarence Peak, on summit, 10,000 ft.
(Dec.) *Mann* 645 (partly)! Also in Ethiopia, Uganda, Kenya and Tanzania.
3. **H. filicornis** *Lindl.* Gen. & Sp. Orch. Pl. 318 (1835); F.T.A. 7: 216; Chev. Bot. 620; F.W.T.A., ed. 1, 2:
410 (excl. *Dalz.* 232). A slender herb up to 18 in. high; flowers small, green, in an open spike; in damp
grassy plains, etc.
    **Iv.C.**: Kodiokoffi (Aug.) *Chev.* 22345. **Ghana:** *Thonning* 58! Nandom (Sept.) *Hall* CC 779! Lawra to
Wa (Sept.) *Hall* CC 632! **N.Nig.**: Jos Plateau (June) *Lely* P330! Hepham to Ropp, 4,600 ft. (July)
*Lely* 374! Vom, 3,000–4,500 ft. *Dent Young*! Also in Ethiopia, Sudan, Uganda, Kenya, Tanzania,
Angola, Congo, Malawi and Rhodesia.
4. **H. chlorotica** *Rchb. f.* in Flora 48: 178 (1865); F.T.A. 7: 217. A slender herb up to 2 ft. high, with long
narrow leaves; flowers small, green, in an open spike; in marshy or damp grassy places.
    **Guin.**: Timbi, Fouta Djalon (June) *Adam* 14580! Labré (July) *Adames* 304! **S.L.**: N. of Loma Mansa,
Loma Mts., (Nov.) *Jaeger* 578! **N.Nig.**: Zungeru (Sept.) *Dalz.* 232! Dorofi, Mambila Plateau, 5,500 ft.
(June) *Chapman* 19! Eastwards to Uganda and Kenya, southwards in E. Africa to Natal, also Angola.
5. **H. microceras** *Hook. f.* in J. Linn. Soc. 7: 221 (1864); F.T.A. 7: 213. *Peristylus preussii* (Kraenzl.) Rolfe
in F.T.A. 7: 199 (1898). A leafy terrestrial or epiphytic herb up to 2½ ft. high; flowers small, green,
in a narrow spike 2–10 in. long.
    **W.Cam.**: Cam. Mt., 7,000–10,000 ft. (Sept.–Nov.) *Mann* 2116! *Johnston* 31! *Preuss* 967! **F.Po:** Moka,
4,600 ft. (Sept.) *Melville* 417!
6. **H. buntingii** *Rendle* Cat. Talb. 109 (1913). Terrestrial up to 18 in. high; leaves in a basal tuft; flowers
white or pale pinkish green; in forests or shady places.
    **S.L.**: Sennihun to Gene (Nov.) *Deighton* 305! Makene (Jan.) *Deighton* 3610! **Lib.**: Tawata, Boporo
(Nov.) *Baldwin* 10306! Nekabozu, Vonjama (Oct.) *Baldwin* 10038! Dzipla, Webo (July) *Baldwin* 6307!
Toué Tié-Tié (Dec.) *Adam* 16318! Ganta (Dec.) *Harley* 1846! **Iv.C.**: Péhiri to Kopréagui, Sassandra R.
(Nov.) *de Wilde* 3309! **Ghana:** Aburi Hills (Oct.) *Johnson*! Santrokofi, Hohoe (Oct.) *Thompson* 1489!
Also in Gabon and Congo.
7. **H. macrandra** *Lindl.* in J. Linn. Soc. 6: 139 (1862); Cat. Talb. 109. *Podandria macrandra* (Lindl.) Rolfe
in F.T.A. 7: 206 (1898). A terrestrial herb to 2 ft. high, with thick woolly roots, a tuft of radical leaves
and a lax raceme of star-like white and pale green flowers; in forests.
    **Guin.**: Nimba Mts., in summit forest, 4,000 ft. (Sept.) *Schnell* 3514! **S.L.**: Konina (Nov.) *Glanville* 122!
Sennihun to Gene (Nov.) *Deighton* 306! Bedu Falls, Kailahun (Nov.) *Deighton* 4671! Gola Forest,
Lalahun (Dec.) *King* 90b! Bage to R. Pampana (Nov.) *Marmo* 41! **Lib.**: Wohmen, Vonjama *Baldwin*
10053! Nyalai (Oct.) *Linder* 1007! Bobei Mt., Sanokwele (fr. Sept.) *Baldwin* 9600! Bobei (Oct.) *Harley*
1832! Ganta (Nov.) *Harley* 1758! **Iv.C.**: Mt. Tonkoui (Sept.) *Schnell* 1745! **Ghana:** Popokyere, Tarquan
(Apr.) *Chipp* 198! Aburi Hills (Oct.) *Johnson* 454! Assuantsi *Fishlock* 30! Konango, Ashanti (Nov.)
*Cansdale* 131! Tano Suhein F.R., W. Prov. (Dec.) *Morton* A3632! Afegame Falls (Sept.) *Westwood* 44!
**S.Nig.**: Bonny R. (Oct.) *Mann* 518! Owam F.R., Benin (Oct.) *Keay* FHI 28059! Sapoba (Oct., Nov.)
*Ujor* FHI 23923! *Keay* FHI 28062! Oban *Talbot* 775! **W.Cam.**: Cam. Mt., 3,000 ft. (Nov.) *Mann* 2117!
Also in E. Cameroun, Congo, Uganda, Tanzania and Angola.
8. **H. leonensis** *Dur. & Schinz* Consp. Fl. Afr. 5: 80 (1895); F.T.A. 7: 212. Usually terrestrial, often on
rocks, but sometimes epiphytic; leaves broad and thin; flowers white.
    **Guin.**: Dalaba–Diaguissa Plateau, 3,000–4,000 ft. (Oct.) *Chev.* 18787b! Mt. Gangan, Kindia *Brun* 622!
Mt. Nimba, 6,000 ft. (Sept., Oct.) *Jac.-Fél.* 1934! **S.L.**: Sugar Loaf Mt. (Oct.) *T. S. Jones*
261! Yakala (Sept.) *Thomas* 2364! Bunbuna (Aug.) *Deighton* 1212! Sakasakala (Sept.) *Deighton* 5142!
S. of Loma Mansa, Bintumane Mt. (Aug.) *Jaeger* 1279! Da-Oulen Mt., summit (Aug.) *Jaeger* 1353!
**Lib.**: Sanokwele (Sept.) *Baldwin* 9522! **Iv.C.**: Tonkoui Mt. (Aug.–Sept.) *Schnell* 1728! 1747! *Roberty*
3097!
9. **H. stenochila** *Lindl.* in J. Linn. Soc. 6: 139 (1862); F.T.A. 7: 212. Terrestrial or epiphytic herb 18–24 in.
high with leafy stem; flowers white.
    **W.Cam.**: Metchum Falls, Bamenda, 2,000 ft. (Aug.) *Savory* 361! Also in Principe and C. African
Republic.
10. **H. zambesina** *Rchb. f.* Otia Bot. Hamburg. 96 (1881); F.T.A. 7: 211. *H. baoulensis* A. Chev., Bot. 620,
name only. A ground orchid up to 4 ft. high with leafy stem; flowers white in dense spike; in meadows
or marshes.
    **Sen.**: *Adam* 18152! **Port.G.**: Bafata to Capé (Aug.) *Esp. Santo* 3318! Bafata *Esp. Santo* 3366! Bissora
to Cuale (Aug.) *Esp. Santo* 3233! **Guin.**: Férédougouoa R., Beyla, 1,750 ft. (July) *Collenette* 67!
Kissinkoro, Kissidougou (July) *Martine* 350! Fon Massif, summit, 5,300 ft. (July, Aug.) *Schnell* 3098!
3309! Nimba Mts., crest (Aug.) *Schnell* 3377! **S.L.**: Gloucester Saddle (Aug.) *Lane-Poole* 383! Jigaya
(Sept.) *Thomas* 2711! Binkolo (Aug.) *Thomas* 1797! Mayaki, Mambolo (Sept.) *Adames* 243! Nasadou
to Kamaro (Aug.) *Jaeger* 894! **Iv.C.**: Marabadiassa to Gottoro (July) *Chev.* 22034! Toumodi, Dimbokro
(Aug.) *Roberty* 3085! Tonkoui, 3,300–4,000 ft. (Aug.) *Boughey* GC 18331! **Ghana:** Kintampo, Ashanti
(June, July) *Vigne* FH 3060! 3934! Kumasi to Tamale (July) *Cox* 21! Pepease (July) *Akpabla* 178!
Ejura (Aug.) *Andoh* FH 5061! Kpandu to Soloknati (July) *Westwood* 18! Amedzofe, 2,500 ft. (Aug.)
*Westwood* 18a! **Togo Rep.**: Sokodé (Aug.) *Davidson* 21! **N.Nig.**: Gawa to Abuja, Niger Prov. *Onochie*
FHI 35749! Katsina Ala (June) *Dalz.* 837! Naraguta (July) *Lely* 461! Mongu, 4,300 ft. (July) *Lely*
405! Agameti (Aug.) *Hepburn* 147! Lekitaba to Gembu, Mambila Plateau, 5,500 ft. (June) *Chapman*
16! **S.Nig.**: S. of Oyo (Aug., Sept.) *Keay* FHI 28038! *J. B. Gillett* 15410! Forest Hill, Ibadan (Oct.)

*Jones & Keay* FHI 13721! Ogoja *Rosevear* 69/29! **W.Cam.:** Nchan, Bamenda, 5,600 ft. (June) *Maitland* 1387! Jakiri (June) *Gregory* 145! Nkom-Wum F.R. (July) *Ujor* FHI 29287! Generally in tropical Africa from E. Cameroun, Uganda and Kenya southwards to Angola and Rhodesia.

11. **H. procera** (*Sw.*) *Lindl.* Gen. & Sp. Orch. Pl. 318 (1835); F.T.A. 7: 220; Chev. Bot. 621; Cat. Talb. 147. Usually an epiphyte (often on oil palms) up to 2 ft. high with leafy stem and white flowers.
    **S.L.:** Mapaki (Aug.) *Deighton* 1217! Gbap (Sept.) *Adames* 80! Madina (Sept.) *Adames* 77! Zimi, Makpele (Sept.) *Deighton* 5170! Kinto to Sinikoro, Lomo Mts. (Sept.) *Jaeger* 1816! **Iv.C.:** Mt. Niénoué, 1,700 ft. (July) *Chev.* 19467! **Ghana:** Efiduasi, Ashanti (Aug.) *Cox* 3! Afoso (Aug.) *Fishlock* 82! Mpraeso, 1,500 ft. (Sept.) *Vigne* FH 4253! **S.Nig.:** Mt. Orosun, Idanre (Oct.) *Keay* FHI 22589! Oluwa F.R., Ondo (July) *Onochie* FHI 33419! **W.Cam.:** Victoria (Sept.) *Ngongi* FHI 15084! Buea *Deistel*! Masama Camp, Buea, 5,000 ft. (May) *Maitland* 736! Mamfe (July) *Swarbrick* 2050! Also in Gabon and Uganda.

12. **H. weilerana** *Schltr.* in Engl., Bot. Jahrb. 38: 149 (1906); Cat. Talb. 147. Herb up to 1 ft. high, with leafy stem and white flowers; among rocks by streams.
    **S.Nig.:** Ndebbige, Oban (Oct.) *Talbot* 844! **W.Cam.:** Bibundi (Nov.) *Weiler*! *Mildbr.* 10660! Tsoni, Cam. Mt. (Nov.) *Jungner* 179! Also in Gabon.

13. **H. gabonensis** *Rchb. f.* in Bot. Zeit. 10: 934 (1852); F.T.A. 7: 220. A terrestrial herb up to 2½ ft. high, leafy in lower part of stem; flowers white; in humus among rocks.
    **S.L.:** Kafoko (Sept.) *Thomas* 2115! **Lib.:** Wanau (July) *Harley* 2039! **Iv.C.:** Mt. Niénoué (Aug.) *Schnell* 1650! Issia Rock (Aug.) *de Wilde* 429! **N.Nig.:** Gembu, Mambila Plateau, 5,500 ft. (July) *Chapman* 22! **W.Cam.:** Mamfe (Sept.) *Gregory* 322! **F.Po:** Ureka (Feb.) *Guinea* 2497! Also in E. Cameroun, Principe, Gabon and Congo.

14. **H. obovata** *Summerh.* in Kew Bull. 1932: 191. *Roeperocharis occidentalis* Kraenzl. in Engl., Bot. Jahrb. 17: 67 (1893). A terrestrial herb up to 18 in. high; flowers small, green, in a dense spike.
    **W.Cam.:** Cam. Mt., 7,000–10,000 ft. (Sept.–Dec.) *Maitland* 804! *Kingsley*! *Johnston* 29! *Preuss* 980!

15. **H. dinklagei** *Kraenzl.* in Engl., Bot. Jahrb. 51: 374 (1914). A terrestrial herb 12–18 in. high with leafy stem; leaves broadly lanceolate, acute; flowers white in short spike.
    **Lib.:** Sinoe (Nov.) *Dinklage* 2330! **S.Nig.:** Idanre, Ondo (Aug.) *Jones* FHI 20721!

16. **H. nigerica** *Summerh.* in Kew Bull. 7: 575 (1953). A terrestrial herb 6–12 in. high; flowers yellowish green in a leafy bracteate spike.
    **N.Nig.:** Mando F.R., Zaria (July) *Keay* FHI 25993! 25986! Zongon Katab, Zaria (June, July) *G. V. Summerhayes* 109! 128! Mongu F.R., Plateau Prov. (Aug.) *King* 29!

17. **H. nigrescens** *Summerh.* in Kew Bull. 20: 165 (1966). A terrestrial herb 9 in.–2 ft. high; flowers pale or yellow-green in a short spike.
    **W.Cam.:** Bamenda-Nkwe to Bafut-Ngemba (May, July) *Daramola* FHI 41177! 41568! Bafut-Ngemba F.R. (July) *Brunt* 774! *Lowe* 72!

18. **H. peristyloides** *A. Rich.* in Ann. Sci. Nat. sér. 2, 14: 270, t. 17, fig. III (1840); F.T.A. 7: 214. *H. cardio-chila* Kraenzl. (1892)—F.T.A. 7: 214. *H. rendlei* Rolfe in F.T.A. 7: 213 (1898). A terrestrial rather leafy herb up to 4 ft. high with numerous green or yellowish green flowers in long dense spikes.
    **N.Nig.:** Kan Iyaka, Mambila Plateau, 5,500 ft. (July) *Chapman* 3! **S.Nig.:** Obudu Plateau, 5,500 ft. (Aug.) *Stone* 54! Also in eastern Congo, Ethiopia, Sudan, Uganda, Kenya and Tanzania.

19. **H. englerana** *Kraenzl.* in Engl., Bot. Jahrb. 17: 68, t. VA (1893); F.T.A. 7: 222. *H. hunteri* Rolfe in Kew Bull. 1918: 238. A herb up to 2 ft. high with lanceolate acute leaves up to the inflorescence; flowers large, white.
    **Guin.:** Guéchédou (July) *Jac.-Fél.* 1022! **S.L.:** Kimadougou, Loma Mts. (Aug.) *Jaeger* 933! **Ghana:** *Hunter*! Southern Scarp F.R., Obomen (July) *Darko* FH 5091! **Togo Rep.:** Bismarckberg (Apr.) *Büttner* 692! **N.Nig.:** Kagoro R., Jemaa, Zaria Prov. (June) *King* 186! **S.Nig.** Ibadan (July) *Westwood* 268! Mt. Orosun, Idanre Hills *E. W. Jones* FHI 25342!

20. **H. prionocraspedon** *Summerh.* in Kew Bull. 1932: 342. A terrestrial herb with lanceolate acute leaves all up the stem, and a dense raceme of rather large white flowers.
    **S.Nig.:** Boshi, Ogoja, 3,500 ft. *Rosevear* 61/29!

21. **H. jacobii** *Summerh.* in Kew Bull. 1935: 196. A terrestrial herb about 6 in. high with narrow leaves and 1 or 2 white and green flowers.
    **Guin.:** Kindia (Aug.) *Jac.-Fél.* 140! 7040! Passo, Pita (July) *Adames* 292! **S.L.:** Kponkponto to Waie, Sule Mts. (July) *Marmo* 158!

22. **H. parva** (*Summerh.*) *Summerh.* in Kew Bull. 1938: 148. *Cynorchis parva* Summerh. in Kew Bull. 1932: 338; F.W.T.A., ed. 1, 2: 407. A small ground orchid 2–6 in. high with 1 or 2 green and white flowers; grassy plains or open woodland.
    **Guin.:** Timbo (July) *Pobéguin* 1592! **N.Nig.:** Jos Plateau (June) *Lely* P 337! Hepham to Ropp, 4,600 ft. (July) *Lely* 361! Randa (July) *Hepburn* 121! Mando F.R., Zaria (July) *Keay* FHI 25989! Zongon Katab, Zaria (June) *G. V. Summerhayes* 111! **W.Cam.:** Bum, Bamenda, 4,000 ft. (May, June) *Maitland* 1398! 1669! Also in C. African Republic.

23. **H. mannii** *Hook. f.* in J. Linn. Soc. 7: 222 (1864); F.T.A. 7: 225. An erect herb up to 2 ft. high with a leafy stem and a dense raceme of a few to 25 large green or whitish green flowers.
    **N.Nig.:** Zonkwa, Zaria (July) *G. V. Summerhayes* 132! Vom, 3,000–4,500 ft. *Dent Young* 234! Naraguta and Jos (Sept.) *Lely* 560! Jos (July, Aug.) *Keay* FHI 20184! *Lawlor & Hall* FHI 45581! **S.Nig.:** Obudu Cattle Ranch, Ogoja Prov. (Aug.) *Stone* 82! **W.Cam.:** Cam. Mt., 5,000–9,000 ft. (Nov., Feb.) *Mann* 2119! Basenako-Lakom, Bamenda, 6,000 ft. (June) *Maitland* 1617! Nchan, Bamenda, 5,000 ft. (June) *Maitland* 1643b! Bafut-Ngemba F.R. (Aug.) *Ujor* FHI 29956! Manenguba Mts., 6,000 ft. (June) *Gregory* 300! **F.Po:** Moka (Sept.) *Tessmann* 2829! *Wrigley & Melville* 623! Also in E. Cameroun.

24. **H. jaegeri** *Summerh.* in Kew Bull. 1: 124 (1948). An erect terrestrial herb 1–2 ft. high with a leafy stem and a broad raceme of up to 12 large white flowers.
    **Guin.:** Fon Massif, on crest, 5,300 ft. (June–Aug.) *Schnell* 3100! 3316! **S.L.:** Bintumane Peak, Loma Mts. (Aug.) *Jaeger* 976! Koya Mt., Bafodia, Koinadugu, 2,400 ft. (Aug.) *Haswell* 73!

25. **H. longirostris** *Summerh.* in Kew Bull. 1932: 192. An erect leafy herb 1–3 ft. high; flowers fragrant, green and white, in a many-flowered raceme; grassy plains and open woodland.
    **Port.G.:** Gabu, near Oco *Esp. Santo* 2! **N.Nig.:** Naraguta (July, Aug.) *Lely* 462! *Keay* FHI 20170! Vom, 3,000–4,500 ft. *Dent Young*! Wana (Aug.) *Hepburn* 120! Jarawa & Gumau, Bauchi Prov. (July, Aug.) *G. V. Summerhayes* 16! 22! Tangale Waja, Bauchi Prov. (Sept.) *G. V. Summerhayes* 74! Also in Uganda.

26. **H. barrina** *Ridl.* in Bol. Soc. Brot. 5: 202, t. D, fig. B (1887); F.T.A. 7: 229, in syn. under *H. thomana* Rchb. f.; Cat. Talb. 147. A leafy terrestrial herb up to 18 in. high with a few-flowered loose racemes, in forests.
    **Iv.C.:** Keéta, Cavally Basin (July) *Chev.* 19364! **S. Nig.:** Ilaro (Sept.) *Punch*! Ibadan (Sept.) *Ahmed & Chizea* FHI 19779! Mamu F.R., Onitsha Dist. (Aug.) *Ajayi* FHI 26974! Also in S. Tomé and Congo.

27. **H. papyracea** *Schltr.* in Engl., Bot. Jahrb. 53: 498 (1915). A terrestrial herb 1–2½ ft. high with leafy stem and numerous small green flowers in a rather dense spike.
    **N.Nig.:** Mongu F.R., Plateau Prov. (Nov.) *King* 100! Naraguta F.R., Plateau Prov. (Sept.) *King* 100a! Jos (Sept.) *King* 100b! *Batten-Poole* 345! 354! Also in Congo, Tanzania, Zambia and Rhodesia.

28. **H. buettnerana** *Kraenzl.* in Engl., Bot. Jahrb. 16: 68 (1892); F.T.A. 7: 237. *H. graminea* A. Chev., Bot. 620, name only, partly. A leafy herb up to 3½ ft. high with a long rather lax raceme of small green flowers; in forests.
    **Port.G.:** Tulacunda (Oct.) *Esp. Santo* 2218! Cambasse to Geba (Aug.) *Esp. Santo* 3297! Bafata to Tantam Cossé (Sept.) *Esp. Santo* 3346! **Guin.:** Mamou (Aug.) *Pobéguin* 1676! Kouria to Irébeléya (Sept.)

FIG. 383.—Habenaria longirostris *Summerh.* (Orchidaceae).

1, flowering stem, lower leafless part and upper 10 cm. of raceme omitted, × 1. 2, dorsal sepal and petals, × 3. 3, column, lateral view, × 6 :—A, anther; CV, caudicle and viscidium; Rm, median lobe of rostellum; Rl, lateral lobes of rostellum; S, staminode; St, stigmatic processes. 4, rostellum spread out, × 6. 5, one pollinium, × 8. (Reprod. from Hook., Ic. Pl. t. 3211, by permission of Bentham-Moxon Trustees).

195

*Chev.* 18232! Kindia *Brun* 1295! **Lib.:** Sanokwele (Sept.) *Baldwin* 9552! **Ghana:** Santrokofi (Oct.) *St. Clair-Thompson* 1482! **Togo Rep.:** Bismarckburg (June) *Büttner* 213! 288! **N.Nig.:** Nindam F.R., Zaria (Sept.) *King* 96! **S.Nig.:** Olofin, Akure (Sept.) *Savory* 31! Ibadan, Awba Hills F.R. (Oct.) *Jones* FHI 6981! Also in Uganda.

29. **H. malacophylla** *Rchb. f.* Otia Bot. Hamburg. 97 (1881); F.T.A. 7: 230. A slender terrestrial herb, 1–3½ ft. high, leafy in the middle of the stem, with numerous small green flowers in a long loose raceme; in forest.
**S.L.:** Mt. Fuen-Koli, Loma Mts. (Nov.) *Jaeger* 1424! **N.Nig.:** Dogon Kurmi, Sanga F.R., Jemaa Dist. (Aug.) *Killick* 21! **S.Nig.:** Mamu F.R., Ibadan (Sept.) *Latilo* FHI 7927! Eket *Talbot* 3355! **W.Cam.:** Cam. Mt., above Buea, 4,000–5,000 ft. (Sept.) *Dundas* FHI 15309! Bafut-Ngemba F.R., Bamenda (July) *Daramola* FHI 41572! Also in Congo and E. Africa from Ethiopia and Uganda, southwards to Natal and the eastern Cape Province.

30. **H. bongensium** *Rchb. f.* Otia Bot. Hamburg. 58 (1878); F.T.A. 7: 233. A slender terrestrial herb up to 18 in. high with a dense spike of small white flowers.
**N.Nig.:** Randa (July) *Hepburn* 122! Mongu, 4,300 ft. (July) *Lely* 407! *Keay & King* FHI 37089! Mando, Zaria (July) *Keay* FHI 25981! Abinsi *Dalz.* 854! Also in E. Cameroun and Sudan.

31. **H. genuflexa** *Rendle* in J. Bot. 33: 279 (1895); F.T.A. 7: 242. *H. anaphysema* of F.W.T.A., ed. 1, 2: 412, not of Rchb. f. An erect terrestrial herb up to 2 ft. high; flowers greenish white in lax racemes; in grassland and among rocks.
**Sen.:** Casamance (Feb.) *Berhaut* 6857! Sedhiou, *Portères*! **Port.G.:** Pussube, Bissau (Nov.) *Esp. Santo* 1391! Peluba, Bissau (Jan.) *Esp. Santo* 1706! **Guin.:** Kouria to Ymbo (Oct.) *Caille* in *Hb. Chev.* 14966! Timbo (Oct.) *Jac.-Fél.* 1888! Mamou, Bilima, 3,300 ft. (Sept.) *Jac.-Fél.* 1820! Nzérékoré (Oct.) *Jac.-Fél.* 1951! Nimba Mts., on crests (Sept.) *Schnell* 1854! 3727! **S.L.:** Kanya (Oct.) *Thomas* 3058! Gbap, Nongoba Bullom (Oct.) *Jordan* 607! Sérélen-Konko Massif, Loma Mts. (Sept.) *Jaeger* 1552! Kailahun (Sept.) *Deighton* 3766! Rokupr, Magbema (Oct.) *Adames* 252! **Lib.:** Devilbush-Duport, 10 miles E. of Monrovia (Sept.) *Voorhoeve* 507! Duport (Nov.) *Linder* 1506! Poro-bush, Duport (Sept.) *van Dillewijn* 100! Gbau, Sanokwele (Sept.) *Baldwin* 9434! **Iv.C.:** Man (Sept, Oct.) *Chev.* 2555! Sassandra & Cavally Rivers *Chev.* 34172! Banfora (Sept.) *Adam* 15230! **Ghana:** Takoradi (Aug.) *Vigne* FH 4785! Jato, near Ejura (Sept.) *Westwood* 151! **N.Nig.:** Vom, 3,000–4,500 ft. *Dent Young*! Naraguta (Aug.) *Lely* 508! Dogon Kurmi, Jemaa Dist. (Aug.) *Killick* 42! Jos (Aug.) *Keay* FHI 20175! Mada Hills (Sept.) *Hepburn* 152! Jawando, Bauchi Prov. (Aug.) *G. V. Summerhayes* 21! Mambila Plateau (July) *Chapman* 21! **S.Nig.:** Ibadan to Oyo (July) *Savory* 191! Igbosere (Aug.) *Savory* 267! Badagry (Aug.) *Onochie* FHI 33476! Apkpo, Ogoja Prov. (Aug.) *Stone* 33! **W.Cam.:** Bum, Bamenda, 4,000 ft. (June) *Maitland* 1640! 1641! Eastwards to Uganda and Tanzania, also in Angola.
[Many of the specimens, especially from Sierra Leone, bear only abnormal flowers in which the lip is more or less reduced, e.g. *Deighton* 3766, *Adames* 252.]

32. **H. dalzielii** *Summerh.* in Kew Bull. 1932: 339. An erect terrestrial herb up to 3 ft. high with a loose raceme of green flowers.
**S.L.:** Hill Station, Freetown (Oct.) *Deighton* 2173! **N.Nig.:** Kurum, Plateau Prov. (June) *King* 92! Kilba country, Yola (Aug.) *Dalz.* 222!
[These may possibly be abnormal specimens of some other species.]

33. **H. angustissima** *Summerh.* in Kew Bull. 1933: 249. A slender terrestrial herb 6–15 in. high with rather small white flowers in a loose raceme 1½–5 in. long.
**Guin.:** Mamou (Aug.) *Pobéguin* 1680! Kinsom, near Kindia (July) *Jac.-Fél.* 1802! Timbi, Fouta Djalon (June) *Adam* 14579! Bondakuri, SW. of Pita (July) *Adames* 291!

34. **H. ichneumonea** (*Sw.*) *Lindl.* Gen. & Sp. Orch. Pl. 313 (1835); F.T.A. 7: 240. *Orchis ichneumonea* Sw. in Schrad., Neues Journ. 1: 21 (1805). A very slender herb up to 2½ ft. high with an open raceme of small green or reddish green flowers; in damp grasslands or open rocky hills.
**Sen.:** Badiouré to Touloumé (July) *Doumlia* 410! Ziguinchor, Casamance (Sept.) *Broadbent* 93! **Mali:** Folo (May) *Chev.*! **Guin.:** Kouria to Irébéléya (Sept.) *Chev.*! Macenta, 2,000–2,500 ft. (Oct.) *Baldwin* 9802! Timbi, Fouta Djalon *Adam* 14745! below Nimba Mts. (June) *Schnell* 3027! **S.L.:** *Afzelius*! Waterloo (Aug.) *Deighton* 2064! *Melville & Hooker* 280! Wellington (Aug.) *Harvey* 56! **N.Nig.:** Vom, Plateau Prov., 3,000–4,500 ft. *Dent Young*! Delimi, Plateau Prov. (Aug.) *King* 175! Eastwards to Ethiopia and Uganda, south to Angola and Rhodesia.

35. **H. pauper** *Summerh.* in Kew Bull. 1932: 341. A slender terrestrial herb 12–18 in. high with very small white flowers in a rather dense raceme; damp or stony grasslands.
**N.Nig.:** Mambila Plateau, 5,500 ft. (July) *Chapman* 1! **W.Cam.:** Bamenda, 5,000 ft. (May) *Maitland* 1562! Also in E. Cameroun, Congo, Uganda and Tanzania.

36. **H. huillensis** *Rchb. f.* in Flora 48: 179 (1865); F.T.A. 7: 240. An erect herb 1–3 ft. high; flowers green or whitish in rather dense, many-flowered racemes up to 1 ft. long.
**Sen.:** Near Kaème Forest, Oussouye (Sept.) *Doumlia* 867! **Port.G.:** Nova Lamego to Sonaco (July) *Esp. Santo*! **Ghana:** Amedzofe, 3,000 ft. (July, Aug.) *Westwood* 27! 27a! Abamansu (Sept.) *Rose Innes* GC 31331! **N.Nig.:** Zaria (Sept.) *Bryant*! Jos Plateau (Sept.) *Lely* P732! Vom, 3,000–4,500 ft. (Aug.) *Dent Young*! Rahama, Zaria Prov. (Aug.) *G. V. Summerhayes* 29! Wanda (Sept.) *Hepburn* 124! **S.Nig.:** Ibadan (Sept.) *Keay* FHI 22442! *Collier* FHI 22440! Also in Sudan, Uganda, Kenya and Angola.

37. **H. linguiformis** *Summerh.* in Kew Bull. 1932: 340. An erect herb 12–18 in. high; flowers green and white.
**N.Nig.:** Zongon Katab, Zaria Prov. (June) *G. V. Summerhayes* 110! Bukuru to Hepham, 4,300 ft. (July) *Lely* 343a! Kuru, Plateau Prov. (July) *G. V. Summerhayes* 160! N'gell, Plateau Prov. (June) *King* 27!

38. **H. chirensis** *Rchb. f.* Otia Bot. Hamburg. 99 (1881); F.T.A. 7: 238. An erect herb up to 3 ft. high; flowers green and white in rather dense racemes; in grasslands.
**N.Nig.:** Jos Plateau (June) *King* 28! *King* 28! Bukuru to Hepham, 4,300 ft. (July) *Lely* 343! Kuru Station, Plateau Prov. (July) *G. V. Summerhayes* 148! Also in E. Cameroun, Ethiopia, Uganda, Kenya and Tanzania.

39. **H. maitlandii** *Summerh.* in Kew Bull. 1932: 341. A terrestrial herb up to 1 ft. high; flowers white in short racemes.
**W.Cam.:** Nchan, Bamenda, 5,600 ft. (June) *Maitland* 1386!

40. **H. phylacocheira** *Summerh.* in Kew Bull. 1932: 190. An erect herb up to 18 in. high; flowers green and white in a short few-flowered raceme.
**N.Nig.:** Ropp, 4,600 ft. (July) *Lely* 451! Vom, 3,000–4,500 ft. *Dent Young*! Heipan, 4,000 ft. (July) *G. V. Summerhayes* 150! *King* 16!
[It is possible that these specimens are abnormal examples of *H. clavata* (Lindl.) Rchb. f. or some similar species.]

41. **H. clavata** (*Lindl.*) *Rchb. f.* in Flora 48: 180 (1865). *Bonatea clavata* Lindl. in Hook. Comp. Bot. Mag. 2: 208 (1837). An erect herb 1–2½ ft. high with leafy stem and rather loose raceme of green or greenish-white flowers; in grassland and open woodland.
**N.Nig.:** Zonkwa, Zaria (July) *G. V. Summerhayes* 134! Also in E. Cameroun, Congo, Tanzania, Rhodesia and southwards to Cape Province.

42. **H. holubii** *Rolfe* in F.T.A. 7: 249 (1898). *H. valida* Schltr. in Engl., Bot. Jahrb. 38: 148 (1906). An erect herb 1–2½ ft. high with leafy stem and rather loose raceme of large greenish-white flowers; in grassland and open woodland.
**Port.G.:** Gabu, Nova Lamego to Canjadude (Oct.) *Esp. Santo* 3538! **Togo Rep.:** Quamikrum (Mar.) *Thienemann*! **N.Nig.:** Zungeru & Lokoja *Lugard*! Zungeru (Aug.) *Dalz.* 229! Yelwa, Zaria (Aug.) *Keay* FHI 28015! Mando F.R., Zaria (Aug.) *Keay* FHI 28002! Abinsi (Oct., Nov.) *Dalz.* 839! Also in C. African Rep., Uganda, Kenya, Tanzania, Congo, Angola and the Zambesi valley.

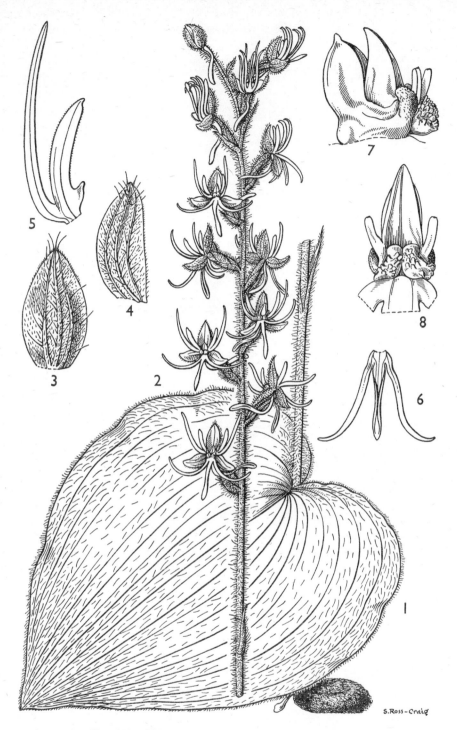

Fig. 384.—Habenaria keayi *Summerh.* (Orchidaceae).

1, base of plant, × 1. 2, apex of scape, × 1. 3, intermediate sepal, × 4. 4, lateral sepal, × 4. 5, petal, × 4. 6, lip, excluding spur, × 2. 7 & 8, column in side and front views respectively, × 8. All from *Keay* FHI 25395. (Reprod. from Hook., Ic. Pl. t. 3562, by permission of Bentham-Moxon Trustees.)

43. **H. cirrhata** (*Lindl.*) *Rchb. f.* in Flora 48: 180 (1865); F.T.A. 7: 248. An erect herb 1½–4½ ft. high with leafy stem and loose raceme of large greenish-white flowers; in grasslands.
**Guin.:** Bonhouri (July) *Pobéguin* 375! Kindia (July) *Jac.-Fél.* 1! **S.L.:** N. of Mongo, Musaia (Aug.) *Haswell* 65! **Ghana:** Vane (July) *Harris*! **N.Nig.:** Jos Plateau (Aug.) *Lely* P615! Kagarko, Zaria Prov. (July) *G. V. Summerhayes* 125! Anara F.R., Zaria Prov. (July) *Keay* FHI 25958! Nindam F.R., Zaria Prov. (July) *King* 46a! Kilba country, Yola (Aug.) *Dalz.* 221! Also in E. Cameroun, Congo, Sudan, Ethiopia, Uganda, Tanzania, Malawi, Zambia and Madagascar.

44. **H. laurentii** *De Wild.* Not. Pl. Util. Congo 2: 325 (1904). A leafy erect herb 1–3 ft. high, with a loose raceme of large pale green and white flowers.
**Mali:** Kita Massif (July) *Jaeger*! **Guin.:** Kinsam Plateau, Kindia (July) *Jac.-Fél.* 1801! **S.L.:** Bintumane Peak, Loma Mts. (July) *Bakshi* 232! **N.Nig.:** Naraguta (June) *Lely* 319! Jos Plateau (July) *Lely* 397! Kontagora (June) *Dalz.* 565! Ancho (Aug.) *Hepburn* 145! Mando, Zaria (July) *Keay* FHI 25980! Abinsi & Katsina Ala (Sept.) *Dalz.* 840! **S.Nig.:** Ago-Are F.R., Oyo Dist. (Aug.) *Onochie* FHI 40879! Also in E. Cameroun, C. African Rep., Congo, Kenya and Rhodesia.

45. **H. cornuta** *Lindl.* in Hook., Comp. Bot. Mag. 2: 208 (1837). *H. ceratopetala* A. Rich. (1840)—F.T.A. 7: 231. *H. ruwenzoriensis* Rendle (1895)—F.T.A. 7: 233; F.W.T.A., ed. 1, 2: 412. An erect herb up to 2½ ft. high; flowers green in a rather dense raceme; in damp grasslands and marshes.
**N.Nig.:** Zungeru (Sept.) *Dalz.* 228! Naraguta, Jos Plateau (June) *King* 176! Eastwards to Ethiopia and Kenya, southwards in E. Africa to the eastern Cape Province and Natal.
[*H. ruwenzoriensis* Rendle consists entirely of plants with abnormal flowers in which the anterior petal-lobe is very much reduced.]

46. **H. walleri** *Rchb. f.* Otia Bot. Hamburg. 98 (1881); F.T.A. 7: 247. A slender terrestrial herb 1–4 ft. high; flowers large, white, up to 12, in narrow racemes.
**Guin.:** Yaoulalay, Pita (July) *Adames* 300! **N.Nig.:** Minna (June) *Dudgeon*! Jos (June) *Lely* 315! Naraguta F.R. *Kennedy* FHI 8062! Shitikan, Kwaya (Aug.) *Daggash* FHI 24864! Mando, Zaria (July) *Keay* FHI 25979! Kilba country, Yola (Aug.) *Dalz.* 220! Eastwards to Kenya, southwards in E. Africa to Malawi.

47. **H. occidentalis** (*Lindl.*) *Summerh.* in Kew Bull. 1933: 246. *Amphorchis occidentalis* Lindl. (1862)—F.T.A. 7: 262. *A. atacorensis* A. Chev., Bot. 621, name only. A ground orchid 6–12 in. high; leaf with white markings; flowers orange or yellow.
**Iv. C.:** Bauna Reserve (Apr.) *Aké Assi* 8651! **Dah.:** Atacora Mts. (June) *Chev.* 24194! **N.Nig.:** Jebba *Barter* 148! Anara F.R., Zaria (May) *Keay* FHI 22906! Mando F.R., Zaria (June) *Keay* FHI 25834! Bonu, Niger Prov. (June) *E. W. Jones* 224! Also in C. African Republic.

48. **H. lelyi** *Summerh.* in Kew Bull. 1932: 188. An erect herb 1–2 ft. high with single basal leaf; raceme 7–13-flowered; flowers white.
**Guin.:** Gali to Hinde, Fouta Djalon (July) *Adames* 318! **N.Nig.:** Ropp, 4,600 ft. (July) *Lely* 457! Anara F.R., Zaria (July) *Keay* FHI 25967! Kujuru, Zaria (July) *G. V. Summerhayes* 131! Kagarko, Zaria (July) *G. V. Summerhayes* 121! Heipan, Plateau Prov., 4,000 ft. (July) *G. V. Summerhayes* 149!
[In many plants the flowers are abnormal, the spur being more or less completely suppressed and/or the lateral lobes of the lip are very short and sometimes tooth-like.]

49. **H. lecardii** *Kraenzl.* in Engl., Bot. Jahrb. 16: 150 (1892); F.T.A. 7: 228. Slender herb up to 15 in. high with single basal leaf; flowers white, hairy, in short lax racemes.
**Sen.:** Mahina *Lécard* 190! **U.Volta:** Firon to Konkobiri, Gourma (July) *Chev.* 24350! **N.Nig.:** Tegina, Zungeru (June) *Dalz.* 566!

50. **H. keayi** *Summerh.* in Bot. Mus. Leafl. Harv. Univ. 14: 217 (1951). A terrestrial herb 3–18 in. high with a basal heart-shaped leaf (young plants with 2 leaves); flowers hairy, green, in a rather long loose raceme.
**N.Nig.:** Kaduna (July) *Cole*! Dogon Kurmi, Jemaa (Sept.) *King* 30a! Naraguta F.R., Plateau Prov. (July) *King* 30! Sara Mts., Bauchi Prov. (July) *King* 30b! **S.Nig.:** Ibadan (Sept.) *Keay* FHI 25395!

51. **H. macrura** *Kraenzl.* in Engl., Bot. Jahrb. 16: 152 (1892); F.T.A. 7: 229. A slender terrestrial herb 1–2½ ft. high with 2 basal leaves; flowers greenish white in few-flowered racemes.
**N.Nig.:** Vom, Jos Plateau, 3,000–4,500 ft. *Dent Young*! Eastwards to Ethiopia and Kenya and south to Angola and Rhodesia.

52. **H. armatissima** *Rchb. f.* Otia Bot. Hamburg. 98 (1881); F.T.A. 7: 227. *H. eburnea* Ridl. in J. Bot. 34: 293 (1886). *H. yatengensis* A. Chev., Bot. 621, name only. A slender terrestrial herb about 1 ft. high, with 2 large round basal leaves; flowers white in few-flowered racemes.
**Mali:** Yatenga (Aug.) *Chev.* 24809! **N.Nig.:** Mandara *E. Vogel*! East to Ethiopia and Kenya, south to SW. Africa and Mozambique.

## 6. PLATYCORYNE Rchb. f. in Bonplandia 3: 212 (1855); F.T.A. 7: 255.

Lip quite entire; leaves evenly scattered along the stem, up to 5 cm. long and 1 cm. broad:
Inflorescence usually 1–3-, rarely 4–5-flowered; middle lobe of rostellum narrowly lanceolate, either short and tooth-like or up to nearly as long as the anther and closely appressed to it, side lobes slender, not much wider than the anther-canals, the latter 2·8–4 mm. long; dorsal sepal 6·5–9·5 mm. long; petals entire; lip usually 5–7 (sometimes to 9) mm. long, spur 8–11·5 mm. long, sometimes shorter by partial abortion; stigmas club-shaped, 2–2·8 mm. long .. .. .. 1. *paludosa*
Inflorescence 3–9-flowered; middle lobe of rostellum lanceolate, much broadened at the base, overtopping the anther, and placed some distance in front of it, side lobes stout from a broad base, much wider than the anther-canals, the latter 2–2·5 mm. long; dorsal sepal 6–7 mm. long; petals entire; lip 6–7 mm. long, spur much thickened in lower part, 12–16 mm. long; stigmas with broad flattened ovate apex, 2·2–5 mm. long .. .. .. .. .. .. .. .. .. 2. *megalorrhyncha*
Lip with a short tooth-like lobe on each side near the base; leaves mostly in a bunch at base of the stem, but a few scattered all along, up to 2·5 cm. long and 4·5 mm. broad; inflorescence 2–6-flowered; dorsal sepal 6·5—7·5 mm. long; petals entire; lip 5–7 mm. long, lateral lobes 0·3–0·5 mm. long, middle lobe 4–5 mm. long, much broader, spur much thickened in the lower part, 9–11·5 mm. long; anther shortly apiculate, canals 1·5–2 mm. long; rostellum middle lobe narrowly lanceolate, acute, shorter than the anther, side lobes rather stout; stigmas club-shaped, 1·5–2 mm. long .. .. .. .. .. .. .. .. 3a. *crocea* subsp. *elegantula*

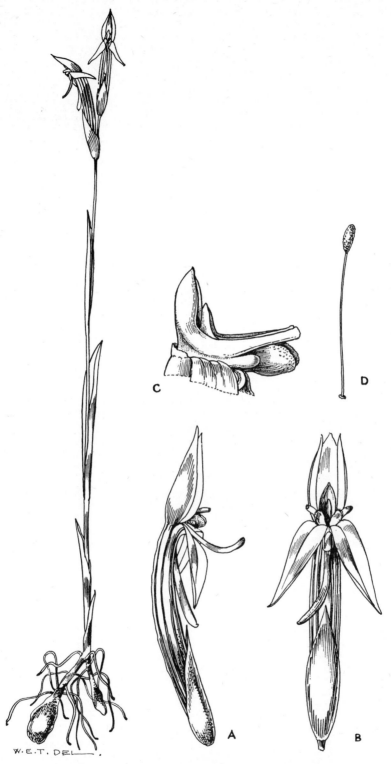

Fig. 385.—Platycoryne paludosa (*Lindl.*) *Rolfe* (Orchidaceae).
A, flower in side view. B, the same, front view. C, column. D, pollinium.

199

1. **P. paludosa** (*Lindl.*) *Rolfe* in F.T.A. 7: 256 (1898); Chev. Bot. 622. *P. aurea* (Kraenzl.) Rolfe in F.T.A. l.c. (1898). *P. wilfordii* (Ridl.) Rolfe in F.T.A. l.c. (1898). A slender terrestrial herb up to 18 in. high with a very short raceme of usually 1–3 deep yellow or orange flowers; in marshes or damp grassy places. **Sen.**: Upper Senegal *Lécard* 204! Djebelor, Ziguinchor (Sept.) *Doumlia* 752! *Broadbent* 1021 Casamance (Aug.) *Berhaut* 6347! **Mali**: Mt. Kita, 5,300 ft. (July) *Jaeger*! **Port.G.**: Farim to Begene (Aug.) *Esp. Santo* 2400! Pitche to Buruntuma, Gabu (July) *Esp. Santo* 2715! Buruntuma, Nova Lamego (July) *Pereira* 3084! **Guin.**: Timbo (July) *Pobéguin* 1653! Sabodougou, Touba (July) *Collenette* 60! Timbi, Fouta Djalon (June) *Adam* 14558! Dalaba (June) *Adames* 275! **S.L.**: Waterloo *Burbridge* 149! Brookfields, Freetown (July) *Deighton* 3990! Wellington (June) *Deighton* 5547! Sendugu (June) *Thomas* 692! **Iv.C.**: Boundiali to Korhogo (July) *Aké Assi* 5613! **Ghana**: Bawku (Aug.) *Cansdale* 161! **Dah.**: Konkobiri to Diapaga (July) *Chev.* 24403! **N.Nig.**: Nupe *Barter* 1479! Zungeru (July) *Dalz.* 233! Neill's Valley, Naraguta (June) *Lely* 269! Mada Hills, 2,000 ft. (June) *Hepburn* 67! Anara F.R., Zaria (July) *Keay* FHI 25955! Birni, Bornu (Aug.) *Daggash* FHI 24858! Also in E. Cameroun.
2. **P. megalorrhyncha** *Summerh.* in Kew Bull. 1933: 250. A terrestrial somewhat leafy herb 6–12 in. high with yellow flowers; in marshes, swamps or on wet rocks. **N.Nig.**: Likitaba to Gembu, Mambila Plateau (June) *Chapman* 11! **W.Cam.**: Wum (June) *Daramola* FHI 41084! Fougom, Bamenda, 3,500 ft. (Apr.) *Maitland* 1509! Bafut, Bamenda (June) *Ujor* FHI 30450!
3. **P. crocea** (*Schweinf. ex Rchb. f.*) *Rolfe* in F.T.A. 7: 257 (1898). Widespread in tropical Africa.
3a. **P. crocea** subsp. **elegantula** (*Kraenzl.*) *Summerh.* in Kew Bull. 17: 532 (1964). *P. elegantula* Kraenzl. in Engl., Bot. Jahrb. 51: 376 (1914). A slender terrestrial herb 5–10 in. high with white or yellow flowers; in grasslands or marshes. **W.Cam.**: Bum, Bamenda, 4,000 ft. (May, June) *Maitland* 1443! 1538! Also in E. Cameroun and Zambia.

## 7. DISA Berg., Pl. Cap. 348, t. 4, fig. 7 (1767); F.T.A. 7: 275.

Dorsal sepal with a long narrow claw, 2–2·5 cm. long, spur usually curved upwards, 5–6 mm. long; petals with an ear-shaped lower part and a narrow twisted upper portion, nearly as long as the dorsal sepal; lip simple, thread-like, 1·5 cm. long; anther erect  ..    ..    ..    ..    ..    ..    ..    ..    **4. *erubescens***
Dorsal sepal narrowed at the base but with no distinct claw, 5–10 mm. long:
  Spur upright or curved downwards in the upper part only; petals entire; anther horizontal:
    Dorsal sepal, including spur, upright, slightly swollen at apex, 6–13 mm. long; slender plant 20–70 cm. high; inflorescence rather lax:
      Bracts narrowly lanceolate, scarcely encircling the pedicel and ovary, 4–10 mm. long; leaves on stem narrowly lanceolate, up to 8 mm. broad    ..    .. **2. *nigerica***
      Bracts oblong- or elliptical-lanceolate, more or less encircling the pedicel and ovary; leaves on stem lanceolate, up to 15 mm. broad ..    ..    .. **1. *equestris***
    Spur curved downwards in the upper part, tapering from a broad base, 7–8 mm. long; rather stout plant, 35–80 cm. high; inflorescence dense, bracts large, 1–2 cm. long; leaves linear-lanceolate, 8–15 cm. long, 1·5–2 cm. broad    .. **3. *hircicornis***
  Spur pendulous from its base; petals bilobed, anterior lobe shorter, ovate or semi-orbicular, posterior lobe upright, narrow, acute or obtuse; anther erect:
    Flower spike very dense, 13–30 cm. long, 1·5 cm. diam.; flowers yellow; sepals 5–8 mm. long, spur 7–8·5 mm. long; petals 4–6·5 mm. long, anterior lobe much shorter than posterior, more or less truncate, narrowed towards base
                                                            **5. *ochrostachya***
    Flower spike rather dense, 10–18 cm. long, 2–3 cm. diam.; flowers pink or purple; sepals 6–12 mm. long, spur 5–8 mm. long; petals 6–8 mm. long, anterior lobe nearly as long as or equalling posterior, semi-orbicular, posterior oblong-lanceolate, acute or obtuse    ..    ..    ..    ..    ..    ..    ..    **6. *welwitschii***

1. **D. equestris** *Rchb. f.* in Flora 48: 181 (1865); F.T.A. 7: 284. A slender terrestrial herb 1–2 ft. high with a rather lax inflorescence of small purple flowers. **N.Nig.**: Jos (June) *King* 90! Daffo, 4,200 ft. (July) *Horsman* in *Hb. King* 90a! **W.Cam.**: Ndop Plain, 3,800 ft. (Mar.) *Brunt* 260a! Also in E. Cameroun, C. African Rep., Congo, Tanzania, southwards to Angola and Mozambique.
2. **D. nigerica** *Rolfe* in Kew Bull. 1914: 214. A slender terrestrial herb 8–12 in. high with a rather lax inflorescence about 4 in. long of small purple flowers. **N.Nig.**: Ropp, 4,600 ft. (July) *Lely* 446! Sara Mt., Bauchi Prov. (May) *King* 52! 52a! **W.Cam.**: Bafut-Ngemba F.R., Bamenda, 6,000 ft. (Mar.) *Richards* 5313!
3. **D. hircicornis** *Rchb. f.* Otia Bot. Hamburg. 105 (1881); F.T.A. 7: 283. A slender to rather stout terrestrial herb, 1–2½ ft. high, with a dense cylindrical flower spike 2–7 in. long; flowers purple with darker markings; in damp grassy places and marshes. **N.Nig.**: Vom, Plateau Prov., 3,000–4,500 ft. (July) *Dent Young*! *King* 161! Kan Iyaka, Mambila Plateau, 5,500 ft. (July) *Chapman* 23! **W.Cam.**: Lakom, Bamenda, 6,000 ft. (June) *Maitland* 1781! East to Kenya, south to Angola and Rhodesia.
4. **D. erubescens** *Rendle* in J. Bot. 33: 297 (1895); F.T.A. 7: 277. A terrestrial herb 1–3 ft. **high**, with the lower sheaths spotted; leaves on separate barren shoots, linear or narrowly lanceolate, 6–18 in. long; spike about 2–6 in. long, rather lax, 3–10-flowered; flowers orange, flame-coloured or deep red, spotted in parts; in grassy places and marshes. **N.Nig.**: Lekitaba to Gembu, Mambila Plateau, 5,500 ft. (June) *Chapman* 12! **W.Cam.**: Bafut-Ngemba F.R., 6,200 ft. (July) *Brunt* 838! Basenako, Bamenda, 6,000 ft (June) *Maitland* 1503! Manenguba Mts., 6,000 ft. (June) *Gregory* 301! Also eastern Congo, E. Africa, from Sudan southwards to Angola and Zambia.
5. **D. ochrostachya** *Rchb. f.* in Flora 48: 181 (1865); F.T.A. 7: 279. An erect terrestrial herb, 1–3 ft. high, with sheathing lanceolate leaves on flowering stems; foliage leaves on separate shoot, narrow, up to 15 in. long; flower spike long and slender; flowers yellow with orange markings; in grassland. **W.Cam.**: Nchan, Bamenda, 5,000 ft. (Apr.) *Maitland* 1779! Also in eastern Congo, E. Africa from Uganda and Kenya south to Angola and Rhodesia.
6. **D. welwitschii** *Rchb. f.* in Flora 48: 181 (1865); F.T.A. 7: 280. *D. subaequalis* Summerh. in Kew Bull. 1936: 221; F.W.T.A., ed. 1, 2: 414. *D. scutellifera* of F.W.T.A., ed. 1, 2: 414, not of A. Rich. An

erect terrestrial herb, 1–3 ft. high, with linear leaves up to 18 in. long on sterile shoots and sheaths on the scape; flowers pink or purple; in grasslands and swamps.
**Guin.**: Nimba Mts., 5,300 ft. (June–Sept.) *Schnell* 1491! 1558! 1830! 3394! **S.L.**: summit Bintumane Peak, Loma Mts. (Aug.) *Jaeger* 1042! **Iv.C.**: Nimba Mts., 2,500–5,000 ft. (Aug.) *Boughey* GC 18047! **N.Nig.**: Vom, 3,000–4,500 ft. *Dent Young*! Heipan, Plateau Prov., 4,000 ft. (July) *G. V. Summerhayes* 152! *King* 18! Kan Iyaka, Mambila Plateau, 5,500 ft. (July) *Chapman* 17! **W.Cam.**: Basenako, Bamenda, 5,000–6,000 ft. (June) *Maitland* 1545! Basenako–Lakom, 5,000–6,000 ft. (June) *Maitland* 1616b! Bum to Nchan, Bamenda, 4,000–5,000 ft. (May) *Maitland* 1788! Bali-Ngemba F.R., Bamenda (May) *Ujor* FHI 30324! Widespread in tropical Africa.

## 8. BROWNLEEA Harv. ex Lindl. in Hook., Lond. J. Bot. 1: 16 (1842); F.T.A. 7: 287.

Slender terrestrial herb 20–60 cm. high; leaves 2–3, erect, the lower one 7–22 cm. long, narrowly lanceolate, acute, the upper much smaller but otherwise similar; spike dense, 3–7 cm. long, 1–1·5 cm. diam.; flowers small, purple; dorsal sepal 3 mm. long, hooded, with a short curved spur; petals united to dorsal sepal, obliquely elliptical; lip very small, entire; anther horizontal; rostellum erect, notched or cleft at apex
*parviflora*

**B. parviflora** *Harv. ex Lindl.* l.c. (1842). *B. alpina* (Hook. f.) N. E. Br. in F.T.A. 7: 287 (1898); F.W.T.A., ed. 1, 2: 417.
**W.Cam.**: Cam. Mt., 6,000–8,000 ft. (Sept.–Nov.) *Mann* 2120! *Johnston* 86! *Preuss* 973! Also in E. Africa from Kenya southwards to eastern Cape Province.

## 9. SATYRIUM Sw. in Vet.-akad. Handl. Stockholm, ser. 2, 21: 214 (1800); F.T.A. 7: 262. *Nom. cons.*

Leaves 2, radical, broadly orbicular, more or less adpressed to the ground, up to 6 cm. long and 9·5 cm. broad; inflorescence short, up to 18-flowered; sepals and petals a little longer than the lip; lip 1–1·7 cm. long, spurs rapidly tapering from a swollen base, shorter than or equalling the lip; column bow-shaped; stigma oblong; rostellum middle lobe shovel-shaped, side lobes rounded .. .. .. .. 1. *carsonii*
Leaves cauline, or if radical, not orbicular nor adpressed to the ground:
Spurs slender, only slightly tapering from the base, much longer than the lip; bracts sharply reflexed at time of flowering:
Lip rather fleshy with narrow opening; leaves on flowering-stem sheath-like, true foliage leaves on a separate short stem:
Flowers white or purple; inflorescence rather dense, up to 15 cm. long, lip more or less globose, 5 mm. long, spurs 9–13 mm. long; lateral sepals slightly longer than the lip; dorsal sepal and petals glabrous; middle lobe of rostellum triangular, acute .. .. .. .. .. .. .. .. 2. *coriophoroides*
Flowers green or brownish-green; inflorescence rather lax, many-flowered, up to 40 cm. long; lip flattened, ellipsoid, 4·5–6 mm. long, spurs 1–2 cm. long; lateral sepals about as long as the lip; petals densely pubescent; middle lobe of rostellum broadly triangular-oblong with a narrow claw .. .. .. 3. *volkensii*
Lip rather membraneous with a broad opening; leaves on flowering stem narrowly or oblong-lanceolate, acute, up to 30 cm. long and 5 cm. broad; inflorescence rather dense, many-flowered, up to 40 cm. long; flowers white, rose-pink or purplish; lip about 5 mm. long, lateral sepals and petals about the same length; spurs 8–12 mm. long; rostellum middle lobe ovate or orbicular with a narrow claw
4. *crassicaule*
Spurs rather stout, tapering towards the apex, shorter than or at most equalling the lip, 4–6 mm. long; inflorescence dense, up to 17 cm. long, with large often spreading bracts; lip 6–8 mm. long with a rather broad opening; sepals and petals about as long as the lip; rostellum with 3 almost equally long lobes, the middle one retuse or more or less bifid .. .. .. .. .. .. .. .. 5. *atherstonei*

1. **S. carsonii** *Rolfe* in F.T.A. 7: 265 (1898). *S. leucanthum* Schltr. in Engl., Bot. Jahrb. 53: 525 (1915); F.W.T.A., ed. 1, 2: 417, fig. 327. *S. nigericum* Hutch. in Kew Bull. 1921: 402. A terrestrial herb up to 2 ft. high with white or pink-tinged rather waxy flowers; in grasslands.
**N.Nig.**: Jemaa (June) *G. V. Summerhayes* 109! Zonkwa to Kaciya, Zaria (July) *G. V. Summerhayes* 126! Vom, 3,000–4,500 ft. (July, Aug.) *Dent Young* 237! *Lely* P398! Bukuru to Hepham, 4,300 ft. (July) *Lely* 344! Also in E. Cameroun, Congo, Kenya, Tanzania, Malawi and Zambia.
2. **S. coriophoroides** *A. Rich.* in Ann. Sci. Nat. sér. 2, 14: 274, t. 18, fig. 11, 1–5 (1840); F.T.A. 7: 269. A terrestrial herb up to 3 ft. high; in grasslands.
**W.Cam.**: Bubaaki, Bamenda, 4,500 ft. (Apr.) *Maitland* 1786! Banso, 5,000 ft. (June) *Gregory* 140! Also in E. Cameroun, Kenya and Ethiopia.
3. **S. volkensii** *Schltr.* in Engl., Bot. Jahrb. 24: 425 (1897). *S. dizygoceras* Summerh. in Kew Bull. 1932: 508; F.W.T.A., ed. 1, 2: 417. A slender terrestrial herb up to 3 ft. high; in grasslands.
**N.Nig.**: Jos Plateau (May) *Lely* P318! Naraguta, 4,000 ft. (May) *Lely* 243! Kassa, Plateau Prov., 4,200 ft. *King*! **W.Cam.**: Lakom, Bamenda, 6,000 ft. (May) *Maitland* 1366! Bali-Ngemba F.R. (May) *Ujor* FHI 303246! Babungo, Ndop Plain, 3,800 ft. (May) *Brunt* 451! Also in Kenya, Congo, Tanzania, Malawi and Rhodesia.
4. **S. crassicaule** *Rendle* in J. Bot. 33: 295 (1895); F.T.A. 7: 271. A terrestrial moderately stout to stout herb up to 3½ ft. high; in wet grassy places, marshes or even in water.
**N.Nig.**: Nguroje, Mambila Plateau, 5,500 ft. (July) *Chapman* 20! **W.Cam.**: Nchan, Bamenda, 5,000 ft. (June) *Maitland* 1385! Also in E. Cameroun, northern Congo, Ethiopia, Kenya and Tanzania.
5. **S. atherstonei** *Rchb. f.* in Flora 64: 328 (1881). *S. occultum* Rolfe in F.T.A. 7: 273 (1898); F.W.T.A., ed. 1, 2: 417. *S. djalonis* A. Chev., Bot. 621. A terrestrial herb up to 3½ ft. high with 2–3 lanceolate, acute or subacute leaves on the stem and white flowers; bracts often whitish; in marshes and grasslands.

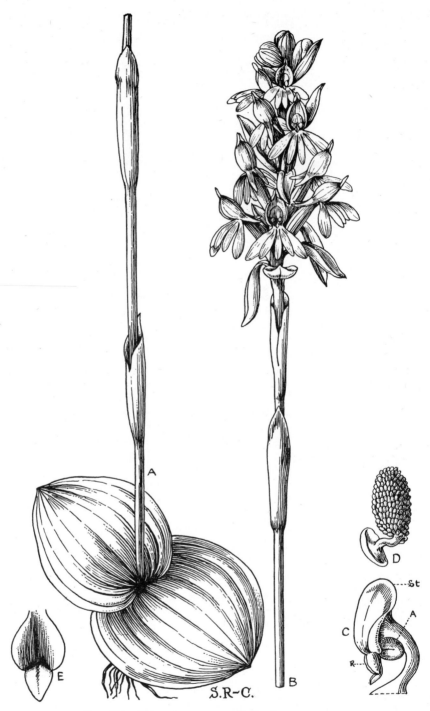

Fig. 386.—Satyrium carsonii *Rolfe* (Orchidaceae).

A, base of scape and leaves. B, inflorescence. C, column (a, anther; r, rostellum; st, stigma).
D, pollinium. E, rostellum.

**Guin.:** Dalaba-Diaguissa Plateau, 3,000–4,000 ft. (Oct.) *Chev.* 18814! Dalaba (Aug.) *Jac.-Fél.* 7059! Boula, Baizia (Apr.) *Adam* 12067! Fon Massif, Simandou, 5,300 ft. (July, Aug.) *Schnell* 3099! 3322! **S.L.:** Bintumane Peak, Loma Mts. (May, July, Aug.) *Jaeger* 1049! *Bakshi* 233! *Deighton* 5092! **N.Nig.:** Maisamari, Mambila Plateau, 5,500 ft. (June) *Chapman* 9! **S.Nig.:** Obudu Plateau, Ogoja Dist., 5,300 ft. (June) *Howard* in *Hb. King* 157! **W.Cam.:** Bum, Bamenda, 4,000 ft. (June) *Maitland* 1575! Generally in tropical Africa; southwards to Transvaal, Natal and eastern Cape Province.

## 10. DISPERIS Sw. in Vet.-akad. Handl. Stockholm, ser. 2, 21: 218 (1800); F.T.A. 7: 288.

Leaves 1–3, alternate, sessile:

Hood formed by the dorsal sepal and petals helmet-shaped, narrow, 4–6 mm. long; appendage of lip entire, linear, slightly broadened at the apex, placed vertically within the hood; lamina of lip sharply reflexed, narrowly oblong, about 0·5 mm. long; arms of rostellum short, apices sharply incurved; leaves ovate, up to 7 mm. long and 4 mm. wide   ..    ..    ..    ..    ..    ..    1. *parvifolia*

Hood formed by dorsal sepal and petals broad and shallow, not helmet-shaped; appendage of lip bifid with apparently two divergent lobes; lamina of lip much broadened at apex, longer than the appendages:

Petals cordate at the base; hood 6–8 mm. long, bent forward, narrowed towards the apex; lateral sepals 8–10 mm. long, united in lower half or below, obliquely ovate, apices turned outwards, sub-acute; lip 5–6 mm. long, claw narrow, lamina transversely elliptical with tall narrow keel in lower centre, appendages elliptical-oblong; leaves ovate or lanceolate, cordate at base, up to 3 cm. long and 2 cm. broad   ..    ..    ..    ..    ..    ..    ..    2. *togoensis*

Petals narrowed at the base, not cordate; hood 8–10 mm. long, much bent forward, more or less rounded at apex; lateral sepals 8–10 mm. long, united in lower third, semi-elliptical, apices rather obtuse; lip about 6 mm. long, claw narrow, lamina ovate with high keel in centre of lower part, appendages linear-oblong; leaves orbicular-ovate, elliptical-ovate or lanceolate, cordate at base, up to 2·5 cm. long and 1·5 cm. broad; bracts much smaller, lanceolate or ovate   ..    3. *johnstonii*

Leaves 2, opposite, broadly ovate, acute, cordate at base; hood formed by dorsal sepal and petals cylindrical or helmet-shaped; petals variously lobed:

Hood of flowers rather short and rounded, 4·5–5·5 mm. long; appendages of lip 2, parallel, cuneate, bilobed in upper part, 2·3–3·7 mm. long:

Spurs on lateral sepals much projecting, 1·7–3 mm. long; lateral sepals 5–6 mm. long; basal lobe of petals usually short and narrow; free part of claw of lip much longer than appendages, the latter 2·7–4 mm. long; rostellum arms 1·5–2 mm. long   ..    ..    ..    ..    ..    ..    4. *thomensis*

Spurs on lateral sepals scarcely projecting, 0·7–1 mm. long; lateral sepals about 9·5 mm. long; basal lobe of petals large and rounded; free part of claw of lip only equalling or slightly longer than the appendages, the latter about 2·5 mm. long; rostellum arms 0·5–0·7 mm. long   ..    ..    ..    ..    ..    5. *nitida*

Hood of flowers narrow, cylindrical, over 6·5 mm. long:

Anterior or upper petal-lobes slender, acuminate, projecting from front of hood like two horns about 4 mm. long; claw of lip 12 mm. long, dilated in the middle to form two small rounded projections; appendix at apex pendulous, 2·8–3·5 mm. long, divided at apex only or almost to base, apex of claw reflexed beneath appendages, about 1 mm. long; hood 12–16 mm. long   ..    ..    ..    6. *mildbraedii*

Anterior or upper petal-lobes triangular or semi-elliptical, acute or rounded, 2 mm. long; claw of lip not dilated in middle:

Apex of lip widened to form a triangular lamina about 3 mm. long bearing at its base on the upper surface two hairy retrorse clavate outgrowths; upper petal-lobe triangular, acute; claw of lip 6·5–7 mm. long; hood 6·5–10 mm. long

                                          7. *kamerunensis*

Apex of lip shortly bent over, oblong, entire, appendage also pendulous, a little longer than the apex itself, divided irregularly to form a sort of fringe about 1·5 mm. long; upper petal-lobe semi-elliptical, rounded; claw of lip 8–12 mm. long; hood 10–17 mm. long   ..    ..    ..    ..    8. *anthoceros*

1. **D. parvifolia** *Schltr.* in Engl., Bot. Jahrb. 53: 547 (1915). Flowers pink or yellowish. **W.Cam.:** Basenako to Nchan, Bamenda, 5,000 ft. *Maitland!* Also in Tanzania.

2. **D. togoensis** *Schltr.* in Engl., Bot. Jahrb. 38: 2 (1905). *D. cardiopetala* Summerh. in Hook. Ic. Pl. 33: t. 3270 (1935); F.W.T.A., ed. 1, 2: 418. *D. atacorensis* A. Chev., Bot. 622, name only. A dwarf terrestrial herb 2–6 in. high; leaves green with whitish veins above, purplish beneath; flowers 2–10 in a short raceme, rose, pink or pale mauve; among rocks or at edge of forests. **Ghana:** Akpafu, 2,000 ft. (Apr.) *Scholes* 20! **Togo Rep.:** Mt. Agome, near Ashanti-Kpoeta (Mar.) *Schlechter* 12990! **Dah.:** Forfa, Atacora Mts. (June) *Chev.* 24060! **W.Cam.:** Bum, Bamenda, 4,000 ft. (Apr.) *Maitland* 1519! Also in E. Cameroun, Uganda and Tanzania.

3. **D. johnstonii** *Rchb. f. ex Rolfe* in F.T.A. 7: 291 (1898). A slender terrestrial herb 2–8 in. high; leaves veined on upper surface, purple beneath; flowers 2–6 in a short raceme, mauve and white or pink; among rocks. **N.Nig.:** Anara F.R., Zaria Prov. *Cole!* Naraguta (Aug.) *Lely* 485! *Keay* FHI 20169! Jos *Batten-Poole* 385! *Daramola* FHI 44137! Also in E. Cameroun and Tanzania.

4. **D. thomensis** *Summerh.* in Kew Bull. 1937: 458. A terrestrial or epiphytic herb 3–8 in. high; leaves green or bronzy above, purplish beneath; flowers 1–6, in a short raceme, white with faint purplish markings; in forests.

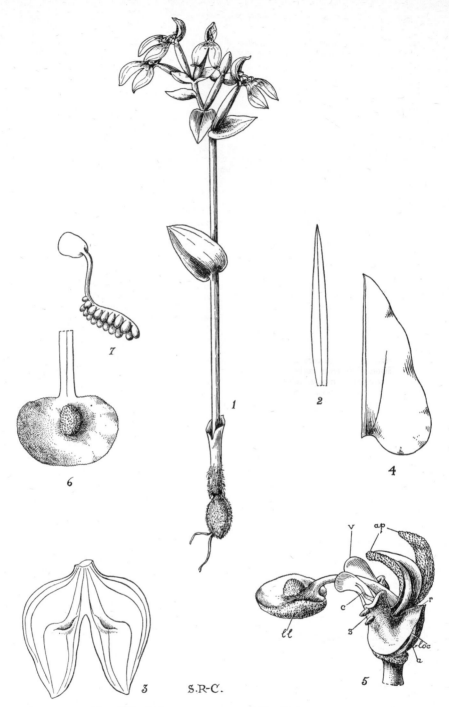

Fig. 387.—Disperis togoensis *Schltr.* (Orchidaceae).

1, habit, × 1. 2, dorsal sepal, × 6. 3, lateral sepals, × 4. 4, petal, × 6. 5, lip and column, × 12.
6, apex of lip, × 12. 7, pollinium, × 12. a, anther; ap, appendages of lip; c, caudicle;
ll, lamina of lip; loc, anther loculus; r, middle lobe of rostellum; s, stigma with pollen
masses; v, viscidium. All from *Chev.* 24060. (Reprod. from Hook., Ic. Pl. t. 3270 by permission of Bentham-Moxon Trustees).

**Guin.:** Mamou, Mt. Bilima, 3,300 ft. (Sept.) *Jac.-Fél.* 1833! Dalaba, Fouta Djalon (Aug.) *Adames* 328! Nimba Mts., crests, 4,000 ft. (Sept.) *Schnell* 3534! **S.L.:** Sugar Loaf Mt., 2,000 ft. (fl. Sept., fr. Nov.) *T.S. Jones* 273! *Milne-Redhead* 5191! Da-Oulen, Loma Mts. (Aug.) *Jaeger* 1289! **Lib.:** Bilimu (Sept.) *Harley* 1537! **Ghana:** Mt. Ejuanima, Mpraeso (Aug.) *Hall* CC 258! **F.Po:** Moka, 4,600 ft. (Sept.) *Melville* 419! Also in S. Tomé and Angola.

5. **D. nitida** *Summerh.* in Kew Bull. 10: 222 (1956). An epiphytic or terrestrial herb 5–12 in. high; leaves dark green, satiny; flowers 1–4 in a short raceme, white; in montane forest in deep shade.
   **W.Cam.:** L. Bambulue, Bamenda, 6,000–7,000 ft. (Sept.) *Savory* UCI 475! above L. Oku, Bamenda, 6,000–7,000 ft. (Sept.) *Savory* UCI 451!

6. **D. mildbraedii** *Schltr. ex Summerh.* in Kew Bull. 1933: 253; Schltr. in Wiss. Ergebn. 1910–11, 2: 180 (1922), name only. A terrestrial herb 4–11 in. high; flowers 1–4, in short raceme, white; in forests.
   **W.Cam.:** L. Bambulue, Bamenda (Aug.) *Ujor* FHI 29965! **F.Po:** S. Isabel Mt., above Basilé, 3,700–4,700 ft. (Aug.) *Mildbr.* 6312!

7. **D. kamerunensis** *Schltr.* in Engl., Bot. Jahrb. 24: 431 (1897); F.T.A. 7: 575. *D. preussii* Rolfe l.c. 291. A terrestrial herb 4–8 in. high; flowers 1–3, in a short raceme.
   **W.Cam.:** Cam. Mt., above Buea, 3,600–3,900 ft. (Sept.) *Preuss* 609! Musake Camp, 6,000 ft. *Maitland* 106!

8. **D. anthoceros** *Rchb. f.* Otia Bot. Hamburg. 103 (1881); F.T.A. 7: 292. A terrestrial herb 2–11 in. high; flowers 1–4, in a short raceme, white or pale pink with lilac or pale purple markings.
   **N.Nig.:** R. Bauna F.R., Plateau Prov., 4,200 ft. (Aug.) *King* 187! Eastern Congo and eastern Africa from Ethiopia southwards to Natal.

## 11. VANILLA Mill., Gard. Dict. Abr. ed. 4 (1754); F.T.A. 7: 176.

Sepals and petals 2–3 cm. long; outgrowths on lip composed of several thin flat overlapping cuneate jagged scales:
  Leaves oblong-lanceolate or oblong-elliptical, acuminate in the distal half, rounded at the base, up to 12 cm. long and 2·5 cm. broad; sepals and petals 2–2·5 cm. long; lip united to column for three-quarters of length of latter forming a wide tube; side lobes broadly rounded, projecting forward from the point of union with, and front margin approximately at right angles to, the column; total length of lip 2 cm., crest at base of middle lobe composed of 4–5 scales, below this a rough narrow keel to the base of the lip; column 1·6 cm. long, slightly incurved    ..    .. 1. *africana*
  Leaves elliptical or oblong-elliptical, shortly acuminate or broadly apiculate in the distal third or quarter, rounded at the base, up to 20 cm. long and 8 cm. broad; sepals and petals 2–3 cm. long; lip united to column for two-thirds of length of latter, forming a wide tube; front margin of side lobes forming a backwardly directed acute angle with the column, edges somewhat reflexed, both side and front lobes somewhat crenulate; crest composed of 4–5 scales, 2–4 mm. long, a raised smooth keel running from this towards base of the lip:
    Inflorescences many-flowered, often branched at the base; bracts very small, triangular, never leaf-like; lip 1·4–1·6 cm. long, front lobe 6–7 mm. broad; column 1·3–1·8 cm. long   ..    ..    ..    ..    ..    .. 2. *ramosa*
    Inflorescences many-flowered, always simple; upper bracts small, triangular, lower ones often leaf-like, orbicular or elliptical, apiculate or shortly acuminate, up to 6 cm. long and 4·5 cm. broad, or the inflorescence-branch bearing small leaves below the flowers; lip 1·5–2 cm. long, front lobe 7–11 mm. broad; column 1·8–2·3 cm. long ..    ..    ..    ..    ..    ..    .. 3. *crenulata*
Sepals and petals 3·5 cm. or more long; outgrowths on lip consisting of hair-like structures:
  Bracts of inflorescence small, 4 mm. long; sepals and petals 3·5–4 cm. long, white; tube formed by lip and column broad; leaves narrowly oblong-lanceolate, bluntly acuminate, 15–25 cm. long, 3·5–4·5 cm. broad; inflorescence up to 6 cm. long, 4–5-flowered; lip united to column for 1·2 cm., 3 cm. long, margins crenulate, coarsely hairy near the mouth ..    ..    ..    ..    ..    ..    .. 4. *nigerica*
  Bracts of inflorescence ovate, imbricating, 1–3 cm. long; sepals and petals 6–8 cm. long, yellow or cream; tube formed by lip and column narrow and cylindrical at base, much widened above, blotched with purple or rose; leaves elliptical or oblong-obovate, shortly but broadly apiculate, 12–25 cm. long, 5–12 cm. broad; inflorescence up to 15 cm. long, densely many-flowered; lip united to column nearly to anther, 5–6 cm. long, obscurely trilobed, side lobes broadly rounded enveloping the apex of the column, middle lobe acute, front margin of lip very crenulate and folded; callus in centre of lip consisting of a dense tuft of fine hairs, tube of flower hairy inside middle lobe of lip with thicker long papillae especially at extreme apex
                                                            5. *imperialis*

  Besides the above indigenous species, *V. planifolia* Andr., is cultivated in our area as it is throughout the tropics for the sake of the vanilla flavouring obtained from the capsules.

1. **V. africana** *Lindl.* in J. Linn. Soc. 6: 137 (1862); F.T.A. 7: 176. A high-climbing herb with short racemes of fragrant white or yellow flowers with purplish markings on the lip; in rain forest.
   **Guin.:** Nzérékoré (May) *Schnell* 2761! Peahtah (Oct.) *Linder* 1108! Sinoe Basin *Whyte*! Nimba (Dec.) *Adam* 20267! **Iv.C.:** Béréby (Nov.) *Oldeman* 575! Guessabo to Duékoué (Nov.) *de Wilde* 843! **Ghana:** Akropong Hills, 1,500 ft. (Oct.) *Johnson* 859! Aburi (Sept.) *Johnson* 798! **S.Nig.:** Brass R. *Barter* 47! Agol-Ibami F.R., Abakaliki Prov. (Mar.) *Sanford* 531/66! Also in Rio Muni.

2. **V. ramosa** *Rolfe* in J. Linn. Soc. 32: 457 (1896); F.T.A. 7: 177. *V. ovalifolia* Rolfe (1896)—F.T.A. 7: 178. A fleshy climbing herb; flowers white or creamy white, lip marked with purple or mauve; in rain forest.
   **Ghana:** Fumisua, Ashanti (Apr.) *Andoh* FH 5257! Axim (Feb.) *Irvine* 2197! **S.Nig.:** Lagos *Rowland*! Sapoba *Kennedy* 220! 2081! Angiama *Barter* 20134! Calabar *Thomson* 132! Oban *Talbot*! **W.Cam.:**

Mbalange F.R., Kumba (Mar.) *Binuyo & Daramola* FHI 35648! Also in E. Cameroun, Rio Muni, Gabon and the Congos.

**V. crenulata** *Rolfe* in J. Linn. Soc. 32: 477 (1896); F.T.A. 7: 178. *V. crenata* A. Chev., Bot. 624, name only. A climbing herb with long hanging racemes of white or yellow flowers with purple markings on the lip; in rain forest.

**Guin.**: base of Nimba Mts. (Mar.) *Schnell* 799! 3086! **S.L.**: Bumbuna (Apr.) *Thomas* 3432! Robis to Koinadugu (Mar.) *Deighton* 2963! Bo, Kakua (Oct.) *Deighton* 6151! Bumban to Lokko (Apr.) *Sc. Elliot* 5733! S. Province (Mar.) *Dawe* 439! **Lib.**: Barclayville, Grand Bassa (Mar.) *Baldwin* 11116! Brewersville, Montserrado (Dec.) *Baldwin* 10966! Firestone Plantation 3, Du R. (July) *Linder* 76! Road to White Plains (Apr.) *Barker* 1265! Mt. Barclay (May) *Bunting*! **Iv.C.**: Yapo F.R. (Sept.) *de Wilde* 576! Agboville (Nov.) *Chev.* 22383! Tai to Guiglo (Mar.) *Leeuwenburg* 3054! Alépé, Attié (Mar.) *Chev.* 17504! Man (Sept., Oct.) *Chev.* 34171! **Ghana**: Bunsu, Akim (Mar.) *Irvine* 1808! Begoro, Akim (Apr.) *Irvine* 1186! Kumasi, 1,000 ft. (Aug.) *Vigne* FH 3062! Mensahkrom, W. Prov. (Oct.) *Darko* 1018! R. Fia, Techiman to Nkoranza (Feb.) *Westwood* 112! **S.Nig.**: Aguna (Dec.) *Miles*! Also in Principe. (See Appendix, p. 515.)

4. **V. nigerica** *Rendle* Cat. Talb. 108 (1913). A fleshy climbing herb; flowers white with rosy markings on the lip.

**S.Nig.**: Oban *Talbot* 776! Also in E. Cameroun.

5. **V. imperialis** *Kraenzl.* in Notizbl. Bot. Gart. Berl. 1: 155, t. 1 (1896); F.T.A. 7: 178. A fleshy climbing herb with thick stems and dense racemes of large cream or yellow flowers with purple or rosy markings on the lip.

**Iv. C.**: Bafing Forest (July) *Aké Assi* 9041! **Ghana**: Ashanti, Mfrim, 1,500 ft. (Aug.) *Vigne* FH 2412! R. Fia, Techiman to Nkoranza (May) *Westwood* 113! Also in E. Cameroun, Congo, Uganda and Tanzania.

**12. NERVILIA** Commerson ex Gaud. in Frey., Voyage, 421 (1829); F.T.A. 7: 186 (as *Pogonia* Juss.). *Nom. cons.*

Inflorescence 1-flowered, much elongating in fruit:
Lip more or less wedge-shaped, trilobed at or near the apex, side lobes small, rounded, middle lobe much broader, margins erose, fimbriate or broadly toothed and much undulate; sepals and petals 8–19 mm. long; column 6–8 mm. long:
Middle lobe of lip erose or fimbriate, not much longer than the side lobes, with long hair-like outgrowth often projecting beyond the margins; leaf purple beneath; lip 8–18 mm. long .. .. .. .. .. .. .. .. 1. *petraea*
Middle lobe of lip shortly toothed or crenate, much undulate, distinctly longer than the side lobes, with short outgrowths along the centre and fine hairs scattered all over lip; leaf green on both sides; lip 14–17 mm. long.. .. 2. *reniformis*
Lip trilobed at or below the middle, side lobes enveloping the column and about the same length, middle lobe much larger and longer, margin entire, with fleshy median pubescent callus; sepals and petals 12–20 mm. long:
Middle lobe of lip much longer than basal entire part, broadly elliptical, 4–6 mm. broad; side lobes distinct, triangular, subacute or obtuse, about 1 mm. long; lip 13–17 mm. long; leaves irregularly marked with paler and darker green on upper surface; column 5–6 mm. long .. .. .. 3. *fuerstenbergiana*
Middle lobe of lip about equalling or shorter than basal entire part, narrowly elliptical, 2–4 mm. broad; side lobes rounded, obscure; lip 9–17 mm. long; column 7–10·5 mm. long; leaves uniformly green.. .. .. .. .. .. 4. *adolphii*
Inflorescence 2–12-flowered; lip trilobed in upper half, side lobes short, acutely obtuse, middle lobe larger, triangular, recurved, disk with 2 longitudinal keels extending to the base of the middle lobe and a thickened vein between:
Inflorescence 10–30 cm. high, rather slender; tubers spherical, 1–2 cm. diam.; leaf broadly or reniform-ovate, apiculate, cordate at base, with short petioles, lamina 6–17 cm. long, 9–13 cm. broad, mature petiole 1–6 cm. long; sepals ligulate, acute, 11–19 mm. long; petals similar, but relatively broader, 9–17 mm. long; lip 10–19 mm. long, side lobes 1–2 mm. broad, middle lobe 4–6 mm. long .. 5. *kotschyi*
Inflorescence 20–40 cm. high, rather stout; tubers 1·5–2·5 cm. diam.; leaf ovate-orbicular, long-apiculate, deeply cordate at base, with long petiole, lamina when full-grown 10–15 cm. long, 12–17 cm. broad, mature petiole 17–22 cm. long; sepals ligulate, obtuse, 2·3–4 cm. long; petals similar but a little shorter; lip 2·5–4 cm. long, side lobes 3–6 mm. broad, middle lobe 7·5–10 mm. long, 9–15 mm. broad 6. *umbrosa*

1. **N. petraea** (*Afzel. ex Pers.*) *Summerh.* in Bot. Mus. Leafl. Harv. Univ. 11: 249 (1945). *Arethusa petraea* Afzel. ex Pers., Syn. Pl. 2: 512 (1807). *Nervilia afzelii* Schltr. (1911)—F.W.T.A., ed. 1, 2: 420. *Pogonia thouarsii* Rolfe in F.T.A. 7: 187 (1898), not of Bl. *P. fineti* A. Chev., Bot. 620, name only. A small terrestrial orchid about 2–5 in. high; flower with yellowish-green tepals and whitish lip; leaf small, reniform, purple underneath.
**Guin.**: Farana (May) *Chev.* 13406! **S.L.**: *Afzelius*! York (May) *Deighton* 5537! Kassewe to Moyamba (May) *Jordan* 879! **N.Nig.**: Bukuru (May) *Baldwin* 12003! Kwoi, Zaria Prov. (May) *King* 81! Abinsi (June) *Dalz.* 844! **W.Cam.**: Bum, Bamenda, 4,000 ft. (May) *Maitland* 1528! Also in C. African Rep., Congo, Uganda and Tanzania.
2. **N. reniformis** *Schltr.* in Engl., Bot. Jahrb. 53: 551 (1915). A small terrestrial orchid 2–10 in. high; tepals white to pale yellow, lip with reddish veins and centre yellow streak; leaves pale green, reniform.
**N.Nig.**: Sanga River F.R., Jemaa, Zaria Prov. (May) *E. W. Jones* 69! *King* 111! Richa, Plateau Prov. (May) *King* 111a! 111b! **S.Nig.**: Olla Hills F.R., Ibadan (Apr.) *Binuyo* FHI 36921! Also in Tanzania.
3. **N. fuerstenbergiana** *Schltr.* in Fedde Rep. 9: 331 (1911). A small terrestrial orchid 2–5 in. high; tepals white or greenish, lip with reddish markings; leaf reniform, mottled above.
**S.L.**: S. Kambui Hills (Mar.) *Dawe* 451! **N.Nig.**: Nindam F.R., Zaria Prov. *King* 95! Also in Cameroun.
4. **N. adolphii** *Schltr.* in Engl., Bot. Jahrb. 53: 552 (1915), and in Fedde Rep., Beih. 68: t. 46, No. 182 (1932). A small terrestrial orchid 2–8 in. high; flowers greenish-yellow or yellowish, tinged and marked mauve; leaves reniform, green above, tinged purple beneath.

**N.Nig.:** Dogon Kurmi, Jemaa, Zaria Prov. *King.* 112*b*! Monguna, Plateau Prov. (Apr.) *King* 112*a*! Also in Uganda, Tanzania and Zambia.

5. **N. kotschyi** (*Rchb. f.*) *Schltr.* in Engl., Bot. Jahrb. 45: 404 (1911). *Pogonia kotschyi* Rchb. f. (1864)—F.T.A. 7: 187. *Nervilia purpurata* (Rchb. f. & Sond.) Schltr. (1911)—F.W.T.A., ed. 1, 2: 420. A slender terrestrial herb 4–12 in. high, with a short raceme of 2–5 pale olive-green flowers with purple veins on the lip; leaf broadly ovate, short-petioled, lamina 3–7 in. long.
**Mali:** Kita Massif *Jaeger* 2221! **N.Nig.:** Naraguta F.R., Plateau Prov. (Apr.) *King* 107! Zelau, 3,200 ft. (Apr.) *Lely* 117! Amban, Plateau Prov. (Apr.) *King* 107*a*! Anara F.R., Zaria (May) *Keay* FHI 22949! Kaciya to Zonkwa, Zaria Prov. (Mar.) *G.V. Summerhayes* 141! Lame Dist., Bauchi Prov. (May) *G.V. Summerhayes* 50! Eastwards to Sudan and Uganda, south to Angola, Mozambique and Transvaal.

6. **N. umbrosa** (*Rchb. f.*) *Schltr.* Westafr. Kautsch.-Exped. 274 (1900). *Pogonia umbrosa* Rchb. f. (1867)—F.T.A. 7: 186. *Nervilia shirensis* of F.W.T.A., ed. 1, 2: 420, not of (Rolfe) Schltr. A terrestrial herb 8–16 in. high with 2–12 greenish or yellowish flowers with purplish-red or brown veins on the lip; leaf often purple beneath.
**Sen.:** Sedhiou, Casamance (July) *Berhaut* 6134! **Port.G.:** Bafata (June) *Esp. Santo* 2685! Bananto to Canjambarim (June) *Esp. Santo* 2398! **Guin.:** Dalaba, Fouta Djalon *Caille* in *Hb. Chev.* 18146! Diaguissa, 4,000 ft. (Apr.) *Chev.* 12651! **S.L.:** Falaba (Mar.) *Sc. Elliot* 5119! **Ghana:** cult. specimen. **Togo Rep.:** Kewe (Mar.) *Schlechter* 12947! **N.Nig.:** Nupe (leaves only) *Barter* 1540! Zelau, 3,200 ft. (Apr.) *Lely* 117*a*! Randa (Mar.) *Hepburn* 131! Naraguta F.R., Plateau Prov. (Apr.) *King* 38! Abinsi (leaves only, June) *Dalz.* 843! **S.Nig.:** between Igboho and Ogun F.R., Oyo (Feb.) *Keay* FHI 23430! Ibadan (Mar.) *Keay* FHI 37569! Also in C. African Rep., Congo, Angola, Tanzania, Malawi and Zambia.

## 13. EPIPOGIUM R. Br., Prodr. Fl. Nov. Holl. 330, 331 (1810); F.T.A. 7: 188.

Saprophytic leafless terrestrial herb up to 45 cm. high; tuber ovoid, up to 4 cm. long and 2·5 cm. diam.; scape erect, rather fleshy, with numerous blunt loosely sheathing scales on the lower part; raceme up to 25-flowered, 3–15 cm. long; flowers usually pendulous, whitish or cream with small purplish or pink spots; sepals and petals narrow, acute, nearly 1 cm. long; lip entire, narrowly ovate, with the cordate base enveloping the column, acute, about 1 cm. long, with 2 lines of short hairs running from base to apex, the latter somewhat thickened; spur up to 5 mm. long; column with swollen stigmatic lobes at the base, and much thickened at the apex   *roseum*

**E. roseum** (*Don*) *Lindl.* in J. Linn. Soc. 1: 177 (1857). *Limodorum roseum* Don (1825). *Epipogium nutans* (Bl.) Rchb. f. (1857)—F.T.A. 7: 188. *E. africanum* Schltr. in Engl., Bot. Jahrb. 45: 399 (1911). **Ghana:** Puso Puso, Asiakwa (Feb.) *Morton* A3866! **W.Cam.:** Ambas Bay (Feb.) *Mann* 784! Moliwe, Buea (Feb.) *Dalz.* 8205! Mimbia, 4,000 ft., Cam. Mt. *Richards* 9346! Ebie to Boviongo, Kumba (Jan.) *Keay* FHI 37388! **F.Po:** 4,000–5,000 ft. (Jan.) *Exell* 795! Also in Annobon, Congo, Angola, Uganda, Indo-Malaysia, Australia and New Hebrides.

## 14. AUXOPUS Schltr., Westafr. Kautsch. Exped. 275 (1901), and in Engl., Bot. Jahrb. 38: 3 (1905).

Perianth segments less than 4 mm. long; labellum cuneate, emarginate at apex; tuber globose or cylindrical, up to 5·5 cm. long and 6–7 mm. diam.; stems slender, rather weak and flexuous, up to 27 cm. high, with a few very small sheathing scales; capsules 1–1·5 cm. long  ..  ..  ..  ..  ..  ..  ..  1. *kamerunensis*
Perianth segments 5·5–7 mm. long; labellum oblanceolate, obtuse at apex; tubers narrowly cylindrical, up to 5 cm. long and 4–8 mm. diam.; stems slender, flexuous at the base only, 15–30 cm. high, with a few sheathing scales; capsules about 1·5 cm. long  ..  ..  ..  ..  ..  ..  ..  ..  2. *macranthus*

1. **A. kamerunensis** *Schltr.* in Engl., Bot. Jahrb. 38: 4, fig. 2 (1905); Cat. Talb. 147. Whole plant pale brown or yellowish; in deep shade in forest.
**Lib.:** Kle, Boporo (Dec.) *Baldwin* 10629! **Iv.C.:** Adzopé to Boudepé (Dec.) *Chev.* 22681! **Ghana:** Tano Anwia F.R. (Dec.) *Adams* 2259! **N.Nig.:** Nupe *Barter*! **S.Nig.:** Sapele to Benin (Dec.) *Sanford* WS/49! BC/66! Okomu F.R., Benin (Dec.) *Brenan* 8483! 8535! Oban *Talbot* 1450! Itu, Calabar (Jan.) *Jones* 2545 (FHI 6873)! **W.Cam.:** S. Bakundu F.R., Kumba (Jan.) *Keay* FHI 28570! Also in E. Cameroun.

2. **A. macranthus** *Summerh.* in Kew Bull. 5: 467 (1952). *A. kamerunensis* var. *grandiflora* Summerh. in F.W.T.A., ed. 1, 2: 420 (1936); Kew Bull. 1936: 222. Plant brown, flowers orange or olive; in forests.
**Iv.C.:** Makougnié (Jan.) *Chev.* 17025! **Ghana:** Kumasi (Jan.) *Baldwin* 14033! Wenchi to Chiraa (Dec.) *Adams* 3187! Koti Pare, N. of Accra (Jan.) *West-Skinn* 320! Nsawam to Aburi (Jan.) *Morton* GC 25318! **S.Nig.:** Okomu F.R., Benin (Jan.) *Brenan* 8535*a*! 8777! Also in Congo and Uganda.

## 15. MANNIELLA Rchb. f., Otia Bot. Hamburg. 109 (1881); F.T.A. 7: 185.

Terrestrial herb 50–90 cm. high; leaves radical, long stalked, lamina obliquely ovate, shortly acuminate, 4·5–16 cm. long, 2·5–7 cm. broad, green usually with white spots, petiole somewhat sheathing at the base, 5–14 cm. long; scape slender, erect, with a few large membranous sheaths; spike slender, many-flowered, rather lax, 15–40 cm. long; flowers small, suberect, brownish or pink; sepals 6 mm. long  ..  *gustavii*

**M. gustavii** *Rchb. f.* Otia Bot. Hamburg. 109 (1881); F.T.A. 7: 185; Cat. Talb. 147.
**S.L.:** Sankan Biriwa Massif (Jan.) *Cole* 159! **Lib.:** Bobei Mt., Sanokwele (Jan.) *Harley*! Bili (Nov.) *Harley* 2080! Nimba reserve (Dec.) *Adames* 841! Nimba *Adam* 20334! 20780! **Ghana:** Tumfa Hills, Akim (Dec.) *Johnson* 274! **S.Nig.:** Owhy, Cross R. *Holland* 194! Oban *Talbot* 1339! **W.Cam.:** Cam. Mt., 4,000 ft. (Jan.) *Mann* 1336! **F.Po:** Ruiché to Caldera San Carlos, 4,000 ft. (Jan.) *Sanford* 4326! Also in S. Tomé, E. Cameroun, the Congos, Uganda and Tanzania.

**16. PLATYLEPIS** A. Rich. in Mém. Soc. Hist. Nat. Paris 4 : 34 (1828) ; F.T.A. 7 : 184. *Nom. cons.*

Terrestrial herb, 15–50 cm. high ; stem creeping at base, bearing many tomentose roots ; leaves rather close together on the lower part of the stem, petiolate, lamina obliquely elliptical, shortly acuminate, 4–11 cm. long, 2–5 cm. broad, petiole sheath-like below, 2–5 cm. long ; scape with several sheaths ; raceme densely many-flowered, 5–12 cm. long, bracts broad, glandular-pilose ; flowers white ; sepals and petals 7–9 mm. long ; lip united to the column for half its length, 7 mm. long, with a short broad reflexed lobe at the apex, 2 narrow calli in the upper part, and forming at the base 2 short outwardly directed spurs ; rostellum about equalling the lip ; anther acuminate
*glandulosa*

P. **glandulosa** (*Lindl.*) *Rchb. f.* in Linnaea 41 : 62 (1877) ; F.T.A. 7 : 184. *Notiophys glandulosa* Lindl. (1862)· *Platylepis talbotii* Rendle, Cat. Talb. 147 (1913).
    **Port.G.:** Ihla Formosa, Mato de Amadi (Apr.) *Esp. Santo* 1972! **Guin.:** Ditinn to Dalaba (Sept.) *Chev.* 18524! 18542! **Lib.:** Wanau (July) *Harley* 2042! Suacoco, Gbanga, Central Prov. (Aug.) *Daniel* 241! **Iv.C.:** Soubré to Peturi (June) *Chev.* 19185! **Ghana:** *Johnson* 1072! Nsuaem, Agona to Tarkwa (Aug.) *Hall* 3348! **S.Nig.:** Lagos (July) *Killick* 246! Oban *Talbot* 1463! **F.Po:** 2,000 ft. *Mann* 1481! Also in S. Tomé, Annobon, Principe, Congo, C. African Rep., Sudan, Angola, Zambia, Mozambique and Natal.

## 17. CHEIROSTYLIS Blume, Bijdr. 413, t. 16 (1825) ; F.T.A. 7 : 182.

Leaves ovate, shortly acuminate, rounded or slightly cordate at base, 1–4 cm. long, 0·8–2·5 cm. broad, petiole shorter than lamina ; labellum equalling the perianth or slightly longer, 3·5–5·5 mm. long ; herb 10–30 cm. high, stem creeping at base ; raceme dense, short, up to 20-flowered ; perianth 3·5–4·8 mm. long ; labellum lobes broad, divergent, entire or obscurely toothed    ..    ..    ..    1. *lepida*
Leaves broadly ovate or orbicular-ovate, very shortly acuminate, distinctly cordate at base, 7–18 mm. long, 6–16 mm. broad, petiole about equalling lamina ; labellum distinctly longer than the perianth, 4·8–5·8 mm. long ; herb 7–16 cm. high, creeping at base ; raceme short, 3–10-flowered ; perianth 3·5–4·2 mm. long ; labellum lobes broad, divergent, distinctly toothed ..    ..    ..    ..    ..    2. *divina*

1. **C. lepida** (*Rchb. f.*) *Rolfe* in F.T.A. 7 : 182 (1897) ; F.W.T.A., ed. 1, 2 : 421 (excl. S.L. specimen) ; Cat. Talb. 147. *Monochilus lepidus* Rchb. f. (1881). A small creeping plant rooting in lower part with a short raceme of white flowers ; in shade in forest.
    **S.Nig.:** Ikwette Plateau, Sonkwala, 5,300 ft. (Dec.) *Savory & Keay* FHI 25168! Oban *Talbot* 870! **W.Cam.:** Cam. Mt., 3,000–6,000 ft. (Nov.–Jan.) *Dunlap* 95! *Mann* 2130! *Maitland* 899! *Schlechter* 12845! **F.Po:** Ruiché to Caldera San Carlos 4,000 ft. (Jan.) *Sanford* 4324! Also in S. Tomé, Congo, Uganda and Tanzania.
2. **C. divina** (*Guinea*) *Summerh.* in Kew Bull. 7 : 131 (1953). *Mariarisqueta divina* Guinea, Ensayo Geobot. Guinea Contin. Espan. 268, fig. 46 (1946). A small creeping plant, rooting in the lower part, with a short raceme of white flowers with buff spots in the throat ; on mossy or damp rocks.
    **S.L.:** Sugar Loaf Mt., 2,000–2,500 ft. (Nov., Dec.) *Sc. Elliot* 4027! *T. S. Jones* 281! Picket Hill, 2,000–2,800 ft. (Nov.) *T. S. Jones* 186! 209! **Lib.:** Nimba, 1,500 ft. (Nov.) *Adames* 788! **Ghana:** Nyinahin Range, Chairaiso, 1,500 ft. (Nov.) *Morton* A2812! Also in Rio Muni.

## 18. ZEUXINE Lindl., Collect. Bot. App. (1825) ; F.T.A. 7 : 180. *Nom. cons.*

Plant 15–45 cm. high ; lower part of stem decumbent, rooting, upper part erect ; leaves in a bunch at base of the erect part, shortly stalked from a sheathing base, lamina lanceolate-ovate or lanceolate, acute, rounded at base, up to 7 cm. long and 3 cm. broad, petiole and sheath 1–2·5 cm. long ; raceme slender, rather laxly many-flowered, 4–13 cm. long ; flowers small, green and white ; sepals and petals 2–3 mm. long ; lip the same length, with a concave claw below bearing 2 hooked calli at the base, and a transversely elliptical or semi-orbicular often more or less bilobed lamina at the apex
1. *elongata*
Plant 8–18 cm. high, very shortly decumbent at base, otherwise erect ; leaves all up the stem, linear, acute, somewhat sheathing at base, 1–3·5 cm. long, 2–4 mm. broad ; spike short, 1–2·5 cm. long, densely up to 20-flowered ; sepals and petals about 2·5 mm. long ; lip the same length, more or less oblong, somewhat cordate at base, narrowed towards apex, obtuse, the margins incurved and more or less erose, the whole surface strongly papillose    ..    ..    ..    ..    ..    .. 2. *africana*

1. **Z. elongata** *Rolfe* in Bol. Soc. Brot. 9 : 142 (1892) ; F.T.A. 7 : 181 ; Cat. Talb. 147. Flowers greenish-white, sometimes tinged pink, lip with orange or reddish claw ; in plantations or forest.
    **S.L.:** Lomaburu, Talla (Feb.) *Sc. Elliot* 5020! Taiama (Jan.) *Deighton* 5738! Kangahun, Gardima (Feb.) *Deighton* 6039! **Iv.C.:** Tos Forest, Bouaflé to Sinfra (Dec.) *Aké Assi* 7222! Yaokro (Jan.) *Aké Assi* 7307! **Ghana:** Cadbury Hall, Kumasi (Mar.) *Cox* 115! Atewa Range F.R., Boma (Jan.) *de Wit & Morton* A2938! Saunders Falls, Kintampo (Jan.) *Westwood* 116! **S.Nig.:** Mawkawlawki, near Abeokuta (Dec.) *Burtt* B8! Ikene, Ijebu-Ode (Jan.) *Burtt* B25! Oban *Talbot* 1360! **W.Cam.:** Basuma to Bafia, Victoria (Feb.) *Keay* FHI 37512! Lus, Nkambe (Feb.) *Hepper* 1883! Also in S. Tomé, Principe, Annobon, E. Cameroun, Congo, C. African Rep., Uganda and Tanzania.
2. **Z. africana** *Rchb. f.* in Flora 50 : 103 (1867) ; F.T.A. 7 : 181. A short herb up to 6 in. high ; in marshes.
    **N.Nig.:** Sokoto (Jan.) *Dalz.* 447! Also in Angola and Natal.

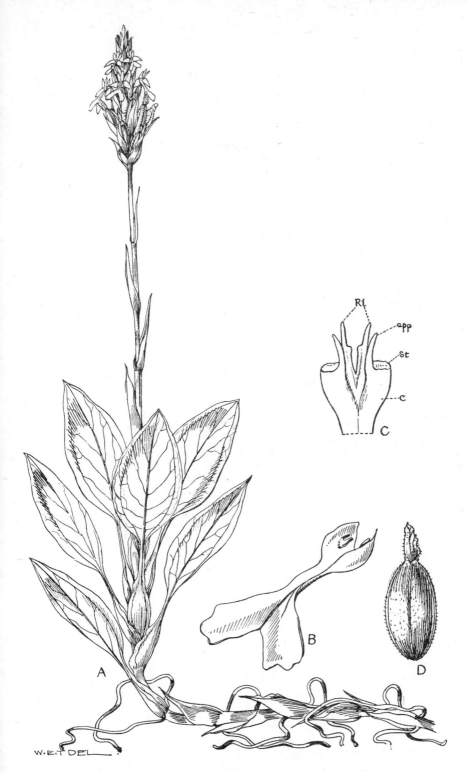

Fig. 388.—Cheirostylis lepida *Rolfe* (Orchidaceae).

A, flowering plant.  B, lip.  C, column (st, stigma; app, appendage; rl, rostellum-lobes).  D, capsule.

**19. HETAERIA** Blume, Bijdr. 409 (1825), as *Etaeria*, corr. Blume, Praef. Fl. Java vii (1828); F.T.A. 7: 183.

Lip inferior; raceme 1–6 cm. long, rather densely 3–20-flowered; plant up to 20 cm. high; leaves aggregated in middle of stem; lamina lanceolate, acute, rounded or subcordate at base, 1·5–4·5 cm. long, 6–18 mm. broad; petiole and sheath 7–12 mm. long; dorsal sepal and petals 3–5 mm. long, lateral sepals a little shorter; lip 4–5 mm. long with a concave claw with 2 hooked calli at the base and 2 obovate divergent lobes at the apex  ..      ..      ..      ..      ..      ..      ..      1. *heterosepala*
Lip superior; raceme 3–20 cm. long, many-flowered:
Lamina of lip entire, broadly deltoid or somewhat diamond-shaped with the long axis transverse, claw furnished just below lamina with a thickened incurved wing on each side and with a series of hooked calli on each side at the base; outgrowths on face of column united inwards so as to form a pouch below the rostellum; plant 15–35 cm. high, leaves 3–7 in the middle of the stem, lamina obliquely oblong-ovate, shortly acuminate, 3–8 cm. long, 1–3·5 cm. broad, petiole and sheath 1·5–2·5 cm. long; raceme 3–9 cm. long, dense; sepals and petals 3–4·5 mm. long; lip 2·5–4 mm. long  ..      ..      ..      ..      ..      ..      ..      ..      2. *stammleri*
Lamina of lip divaricately bilobed, lobes narrowest at base, claw unthickened below lamina; outgrowths on face of column free from one another; leaves all along upper part of stem, lamina over 3 cm. long, petiole and sheath over 1·5 cm. long; raceme rather lax, 5–18 cm. long:
Rostellum longer than the column, 2·5–3 mm. long; leaves obliquely elliptical-ovate or ovate, 3–10 cm. long, 1·5–4 cm. broad; inflorescence 5–15 cm. long; sepals 5–7·5 mm. long; lip 6·5 mm. long (to apex of middle lobe or central apiculus), lateral lobes 4–6 mm. long, 1·6–3 mm. broad  ..      ..      ..      3. *tetraptera*
Rostellum equalling or shorter than the column, 0·9–1·2 mm. long; lobes of lip 1·5–4 mm. long:
Lobes of lip reflexed, oblong or narrowly oblanceolate but not much widened above, 1·5–2 mm. long; claw 3–4 mm. long, base forming two rounded sacs not projecting more than 0·5 mm. below the insertion of the lip; sepals 3–4·5 mm. long; arms of rostellum 0·9 mm. long; plant 30–45 cm. high; leaves obliquely lanceolate, lamina 9–14 cm. long, 2–5 cm. broad  ..      ..      ..      ..      ..      4. *mannii*
Lobes of lip spreading, broadly cuneate from a narrow base, rounded at apex, 2·5–4 mm. long, 2–3 mm. broad; claw 4–5·5 mm. long, base forming two distinct rounded spurs 1–2 mm. long; sepals 4–6·5 mm. long; arms of rostellum 0·9–1·2 mm. long; plant 30–60 cm. high; leaves obliquely lanceolate or elliptical-lanceolate, lamina 6–16 cm. long, 2–5 cm. broad  ..      ..      5. *occidentalis*

1. **H. heterosepala** (*Rchb. f.*) *Summerh.* in Kew Bull. 1934: 207. *Cheirostylis heterosepala* Rchb. f. (1881)—F.T.A. 7: 183. *Zeuxine commelinoides* A. Chev., Bot. 619, name only. A herb with the lower part of the stems creeping and rooting, and small green and white flowers; in forests.
   **Lib.:** Krahn Bassa Forest (Jan.) *van Harten* 309! Nimba *Adam* 20803! **Iv.C.:** Cavally Basin, Grabo (July) *Chev.* 19614! **W.Cam.:** Cam. Mt., 3,000 ft. (Nov.) *Mann* 2130a! Also in S. Tomé, E. Cameroun, Congo and Tanzania.
2. **H. stammleri** (*Schltr.*) *Summerh.* in Kew Bull. 1934: 208. *Zeuxine stammleri* Schltr. in Engl., Bot. Jahrb. 38: 151 (1906); Cat. Talb. 147. Stem creeping and rooted at the base, erect above; sepals pinkish brown or green, petals white; in shade in forest.
   **Iv.C.:** Assikasso, Moyen Comoe (Dec.) *Chev.* 25601! Yapo Forest, N. of Abidjan (Dec., fr. Feb.) *de Wilde* 1017! *Aké Assi* 7348! **S.Nig.:** Shasha F.R., Ijebu (Feb.) *Richards* 3072! Okomu F.R., Benin (Dec.–Jan.) *Brenan* 8603! 8781! 8781a–d! Oban *Talbot* 922! 1364! **W.Cam.:** Moliwe *Stammler*. Tiko (Jan.) *Dunlap* 245! S. Bakundu F.R., Kumba (Jan.) *Keay* FHI 28569! **F.Po:** Basilé, 2,000 ft. (Dec.) *Sanford* 4052! 4054! Also in Principe and C. African Republic.
3. **H. tetraptera** (*Rchb. f.*) *Summerh.* in Kew Bull. 1934: 207. *Monochilus tetrapterus* Rchb. f. (1881). *Zeuxine tetraptera* (Rchb. f.) Dur. & Schinz (1895)—F.T.A. 7: 181. *Z. batesii* Rolfe in F.T.A. 7: 182 (1897). Herb with stem creeping and rooting below, erect above, 8–12 in. high; flowers white, greenish outside; in damp ground in forest.
   **S.Nig.:** Baba Eko, Ijebu (May) *Ross* 284! **W.Cam.:** S. Bakundu F.R., Kumba (Mar., Apr.) *Brenan* 9447! *Daramola* FHI 41006! Also in E. Cameroun, Gabon and Congo.
4. **H. mannii** (*Rchb. f.*) *Dur. & Schinz* Consp. Fl. Afr. 5: 57 (1895); F.T.A. 7: 184. Herb with stem creeping and rooting at base, erect above.
   **W.Cam.:** Bafut Ngemba F.R., Bamenda Dist. (June) *Daramola* FHI 41067! Also in E. Cameroun (*Mann* 2131) and Gabon.
5. **H. occidentalis** *Summerh.* in Kew Bull. 1934: 206. *Zeuxine elongata* of Chev. Bot. 619, not of Rolfe. *Z. batesii* of Cat. Talb. 147, not of Rolfe. Herb with stem creeping and rooting in lower part, erect above; flowers white tinged pink, lip with yellow markings; in forests or damp ravines.
   **Port.G.:** Dandum (Mar.) *Esp. Santo* 470! **Guin.:** Mangata, Farana (Jan.) *Chev.* 20424! **S.L.:** Mabandu (Dec.) *Marmo* 68! **Lib.:** Bilimu (Dec.) *Harley* 2085! **Iv.C.:** Macoupué (Jan.) *Chev.* 16999! Banco Forest, Abidjan (Jan.) *de Wit* 9019! *Leeuwenberg* 2330! *Aké Assi* 7270! Maféré (fr. Feb.) *Aké Assi* 7334! **Ghana:** Kibbi, Akim (Dec.) *Johnson* 592! Atewa Range F.R., Boma (Jan.) *de Wit & Morton* A2938! Afao Hills F.R., W. Prov. (Dec.) *Adams* 1969! Wenchi to Chiraa (Dec.) *Adams* 3186! **N.Nig.:** Kontagora (Jan.) *Dalz.* 231! **S.Nig.:** Oban *Talbot* 920! 921! Shagamu, Ijebu-Ode (Jan.) *Burtt* B19! Afi River F.R., Ogoja (Dec.) *Keay* FHI 28242! Iloro (Sept.) *Punch*! Okomu F.R., Benin (Dec., Jan.) *Brenan* 8603a! 8603b! *Richards* 3854! Also in E. Cameroun, Congo and C. African Republic.

**20. CORYMBORKIS** Thou. in Nouv. Bull. Sci. Soc. Philom. Paris 1: 318 (1809); F.T.A. 7: 179.

Terrestrial erect herb; leaves with a narrow sheathing base and a short petiole; lamina 15–35 cm. long, 3·5–8·5 cm. broad, many-veined; inflorescence up to 7 cm. long, up

to 20-flowered, bracts small; flowers erect; perianth-segments 5–7 cm. long,
broadened near apex; column long and slender, partially enveloped by the lip
*corymbosa*

C. corymbosa *Thou.* Orch. Iles Austr. Afr. Prem. Tabl. Synopt. tt. 37, 38 (1822); F.T.A. 7: 180. *C. welwitschii*
Rchb. f. (1865)—F.T.A. 7: 180; Chev. Bot. 624; F.W.T.A., ed. 1, 2: 423. Flowers white or greenish;
in forests.
**Guin.:** Macenta (May) *Jac.-Fél.* 912! Nimba Mts. *Schnell* 2868! **S.L.:** Freetown *Burbridge* 521! **Iv.C.:**
Guidéko (Mar.–May) *Chev.* 19014! Yapo Forest *Giovannetti* 380! **Ghana:** *Farmar* 580! Kibbi Hills, Akim
(Dec.) *Johnson* 272! Kwabeng, Atewa Range (Dec.) *Adams* 4340! Kade (May) *Morton* A4182! **S.Nig.:**
Gambari F.R., Ibadan (July) *Onochie* FHI 19106! Ifi F.R., Oyo *Olorunfemi* FHI 41504! Akure, Ondo
(fr. Oct.) Agoi-Ibami, Abakaliki Prov. *Sanford* WS/603/66! *Ujor* FHI 26199! Ikom (May) *Jones* FHI
4516! **W.Cam.:** S. Bakundu F.R., Kumba *Akuo* FHI 15162! **F.Po:** *Barter* 1478! (June) *Mann* 430!
Generally distributed in tropical Africa, southwards to Natal and eastern Cape Province, Réunion and
Madagascar.

## 21. MALAXIS Soland. ex Sw., Prodr. Veg. Ind. Occ. 119 (1788); F.T.A. 7: 17, as *Microstylis* (Nutt.) Eaton.

Flowers close together at apex of inflorescence, forming a false umbel; petals almost as
wide as sepals; margin of lip denticulate or fimbriate:
Lip bearing 2 parallel ridges at the base, often with a needle-like projecting point just
in front on the median vein; sepals and petals 4–6 mm. long, the latter ciliolate; lip
flabellate or transversely elliptical, emarginate at apex, 3·5–5·5 mm. broad; stems
3–6 cm. high, 3–4-leaved; leaves ovate or lanceolate-ovate, 4–7 cm. long, 2·5–5 cm.
broad .. .. .. .. .. .. .. .. 1. *maclaudii*
Lip bearing a single oblong entire callus at the base; sepals and petals 2·5–3 mm.
long, all entire; lip quadrate-cuneate, emarginate at apex, basal margins much
thickened, 2·5–3 mm. long, 2·5–3 mm. broad; stems 2·5–5 cm. high, 3-leaved;
leaves narrowly ovate, 4–7 cm. long, 1·5–3 cm. broad .. .. 2. *chevalieri*
Flowers in an elongated raceme or spike; petals much narrower than sepals; margin
of lip entire:
Lip with a single central pubescent cushion at the base; sepals and petals 1·5–2·5 mm.
long; column very short (0·3–0·7 mm. long), anther horizontal:
Leaves at about the middle of the total height (10–16 cm.) of the plant, the vegetative
stem more or less equalling the inflorescence; leaves 2–5, elliptical-lanceolate,
1–6 cm. long, 1–3 cm. broad; rhizome slender, creeping .. 3. *prorepens*
Leaves near the base of the plant (total height 7–30 cm.), often almost on the soil, the
vegetative stem (up to 4 cm. long) being much shorter than the inflorescence;
leaves 2–3, elliptical to ovate, 2·5–8 cm. long, 1–6 cm. broad; no creeping rhizome,
root a tuber 1–2 cm. diam. .. .. .. .. .. .. 4. *katangensis*
Lip with 2 lateral pubescent cushions:
Lip oblong, longer than broad, with small rounded purple auricles, otherwise pale
yellowish, 3·5 mm. long (auricles included); sepals and petals 3 mm. long; column
0·8 mm. long; stems 3–8 cm. high, 4–5-leaved; leaves lanceolate or broadly
lanceolate, 2–5 cm. long, 1–2 cm. broad .. .. .. .. 5. *melanotoessa*
Lip more or less quadrate or broadly orbicular, broader than long, narrowed at the
front, auricles large, almost forming distinct side lobes, lip 2·5–3 mm. long (auricles
included); sepals and petals 3–4 mm. long; column 1·6 mm. long; stems 3–12 cm.
high, 3-leaved; leaves ovate, 3–5 cm. long, 1–3 cm. broad .. 6. *weberbauerana*

1. **M. maclaudii** (*Finet*) *Summerh.* in Kew Bull. 1934: 208. *Microstylis maclaudii* Finet (1907). Flowers flesh
to deep rose-coloured.
**Guin.:** Macenta, Mbalasso (June) *Jac.-Fél.* 891! Songoya *Maclaud* 81! foot of Nimba Mts. (June) *Schnell*
3093! **S.L.:** Loma Mts. (July) *Jaeger* 6840! **Lib.:** S. Nimba (June) *Adam* 21468! **Ghana:** R. Fia,
Techiman, Ashanti (June) *Westwood* 139! **N.Nig.:** Sanga River F.R., Zaria Prov. (June, July) *King* 103!
*Keay* FHI 37122! **S.Nig.:** Idanre Hills, Ondo Prov. (fl. cult. Ibadan May, June) *Sanford* WS/1503/65!
**F.Po:** Pico S. Isabel, 4,000 ft. (fr. Jan.) *Sanford* 4116! Also in E. Cameroun, C. African Republic and
Sudan.
2. **M. chevalieri** *Summerh.* in Kew Bull. 1934: 208. *Liparis sassandrae* A. Chev., Bot. 613, name only. Flowers
greenish crimson.
**Sen.:** Diantène, Casamance (July) *Doumbia* 520! Oussouye (Aug.) *Berhaut* 6239! **Iv.C.:** Toura, Sassandra
R. valley (May) *Chev.* 21786! **S.Nig.:** Olokemeje F.R., Abeokuta (July, fr. Dec.) *Keay & Jones* FHI
14188! *Keay* FHI 16229! Also in C. African Republic.
3. **M. prorepens** (*Kraenzl.*) *Summerh.* in Kew Bull. 1934: 208. *Microstylis prorepens* Kraenzl. (1893)—F.T.A.
7: 18. Flowers green, purple, or brownish.
**Guin.:** Kindia *Jac.-Fél.* 135! Mt. Gangan (Aug.) *Jac.-Fél.* 7037! Tangama F.R., Dalaba (Aug.) *Adames*
321! **S.L.:** Sugar Loaf Mt. (June) *Preuss*! **N.Nig.:** Jemaa, Zaria Prov. (June) *King* 117! Bonu Kurmi,
Niger Prov. (June) *E. W. Jones* 168! **S.Nig.:** Idanre Hills, Ondo Prov. (May) *Sanford* 1601/65! Also in
C. African Republic.
4. **M. katangensis** *Summerh.* in Bot. Mus. Leafl. Harv. Univ. 14: 221 (1951). Flowers pale green, turning brown
or orange.
**S.L.:** Musaia (Aug.) *Haswell* 33! **N.Nig.:** Jemaa, Zaria Prov. (Aug.) *King* 98! Nimbia F.R., Zaria Prov.
(July) *King* 98a! Also in Congo and Tanzania.
5. **M. melanotoessa** *Summerh.* in Kew Bull. 1934: 209. Flowers pale greenish-yellow, lip dark purple at base.
**Lib.:** Gola Forest, S. of Ba (May) *Bunting*!
6. **M. weberbauerana** (*Kraenzl.*) *Summerh.* in Kew Bull. 1934: 208. *Liparis weberbauerana* Kraenzl. (1908).
Flowers brown or pink.
**W. Cam.:** Cam. Mt. *Weberbauer* 42! Bali-Ngemba, Bamenda (June) *Ujor* FHI 30424! Bafut-Ngemba F.R.,
Bamenda (May) *Keay* FHI 37935! Nkom Wum F.R. ,Wum (June) *Daramola* FHI 41083! **F.Po:** (cult.
Ibadan, June) *Sanford* 4331!

**22. LIPARIS** L. C. Rich. in Mém. Mus. Paris 4: 43, 52 (1818); F.T.A. 7: 19. *Nom. cons.*

Stems swollen at base to form an almost globose pseudobulb; stem 1·5–2 cm. long;
    leaves ovate, almost sessile, in a tuft at the base, 2–4 cm. long, 1–3 cm. broad; scape
    erect, 7–11 cm. long, flowering nearly to the base; fruit ellipsoid, 6–8 mm. long,
    about 4 mm. diam.; flowers unknown    ..    ..    ..    ..    1. *kamerunensis*
Characters not combined as above:
  Lip distinctly 3-lobed, about 4 mm. long, side lobes shortly triangular, middle lobe
    long and narrow, the margins inrolled so as to make it appear subulate, sharply
    inflexed at the middle, 2·5–3 mm. long, lip bearing a short bilobed callus at the base;
    leaves in a tuft at the base, narrowly oblong-oblanceolate, acute, up to 7 cm. long
    and 1·8 cm. broad, margins sometimes very crinkled; inflorescence overtopping the
    leaves, many-flowered, bracts rather large, almost cordate at the base..    2. *tridens*
  Lip bilobed, or if trilobed the middle lobe reduced to an apiculus, the side lobes much
    larger and rounded:
    Stem long and slender, lower part creeping and rooting, upper part erect, leafy;
      leaves with short sheathing base and rather long petiole, lamina ovate or lanceolate-
      ovate, cordate or rounded at the base, 1–4 cm. long, 1–2 cm. broad; inflorescence
      about 4 cm. long, several-flowered; sepals and petals about 5 mm. long, the lateral
      sepals united shortly at the base; lip transversely oblong, about 4 mm. long and
      7 mm. broad, side lobes almost orbicular, front lobe triangular, much smaller, lip
      with a large tooth at the base    ..    ..    ..    ..    3. *goodyeroides*
    Primary stem short, consisting of a sympodial series of upright flowering shoots
      close together, the shoots sometimes thickened to form a pseudobulb:
      Petals amost twice as long as the dorsal sepal; leaves ovate or lanceolate, up to
        6 cm. long, and 12 mm. broad; pseudobulbs 12–20 mm. long, 2–4-leaved; in-
        florescence flaccid, pendulous, up to 14–18 cm. long, many flowered, rhachis
        sharply 2–4 winged, 5 mm. broad; lower flowers aborted with up to 4 linear
        segments; fertile flowers, dorsal sepal 5–10 mm. long, petals narrowly linear,
        8–15 mm. long; lip obovate or oblong from a cuneate base, 3·5–5 mm. long,
        bilobed, lobes rounded    ..    ..    ..    ..    ..    ..    4. *caillei*
      Petals equal to or only slightly longer than the dorsal sepal; leaves lanceolate or
        lanceolate-ovate, mostly over 3 cm. long and usualiy at least 3 times as long as
        broad:
        Flowers large, lip over 1 cm. wide, transversely elliptical, with a broad retuse
          tooth-like callus in front of the column and a yellowish rounded swollen area on
          each side at the base; pseudobulbs ovoid, up to 2 cm. long and 2 cm. broad;
          leaves lanceolate to elliptical-lanceolate, up to 17 cm. long and 1·8 cm. broad;
          sepals 8–13 mm. long, the laterals united nearly to the apex; petals very narrow,
          11–14 mm. long    ..    ..    ..    ..    ..    ..    5. *platyglossa*
        Flowers rather small, lip under 7 mm. broad, suborbicular, elliptical, obcordate
          or obcordate-flabellate, with 2 small teeth or calli at the base; inflorescence
          usually much overtopping the leaves, many-flowered:
          Terrestrial plants, very rarely epiphytic; inflorescences 15–60 cm. high; lamina
            of leaves 5–25 cm. long, 0·9–8·5 cm. broad; lip obcordately bilobed, somewhat
            recurved, 2·5–4 mm. long, 2·5–4·5 mm. broad, lobes rounded; sepals 4–6
            mm. long; petals narrower, 4·5–5·5 mm. long:
            Leaves lanceolate, narrowly lanceolate or lanceolate-linear, with short petiole,
              7–25 cm. long, 0·9–5 cm. broad, usually less than half the height of the in-
              florescence; lip rather fleshy; stem distinctly swollen at base forming a corm-
              like structure    ..    ..    ..    ..    ..    ..    ..    6. *rufina*
            Leaves elliptical or lanceolate-elliptical with a long petiole, 5–20 cm. long,
              2–8·5 cm. broad, often more than half the height of the inflorescence; lip thin
              in texture; stem hardly swollen at the base    ..    ..    7. *guineensis*
          Epiphytic plants, rarely terrestrial on rocks; inflorescences 5–13 cm. high; lamina
            of leaves 2–10 cm. long, 0·4–1·8 cm. broad; lip suborbicular, elliptical, or
            obcordately flabellate:
            Leaves narrowly oblong or oblong-lanceolate, 4–12 mm. broad, tapering gradually
              to the petiole, obtuse or rounded and emarginate at apex, 2–11 cm. long;
              scape usually longer than the leaves; flowers green; dorsal sepal 5–6 mm.
              long; petals narrower, 5–6 mm. long; lip elliptical-obcordate, side lobes
              rounded, middle lobe reduced to an apiculus, lip 3·5–4·5 mm. long, 3–4 mm.
              broad, calli small, tooth-like ..    ..    ..    ..    ..    8. *epiphytica*
            Leaves lanceolate or elliptical-lanceolate, usually over 15 mm. broad, acute,
              3–10 cm. long:
              Lip flabellate, deeply emarginate, lobes rounded, whole lip 6 mm. long and
                broad; dorsal sepal 8–9 mm. long; petals very narrow, 9 mm. long; scape
                much exceeding the leaves    ..    ..    ..    ..    ..    9. *deistelii*
             Lip suborbicular, narrowed at the base, trilobed or truncate at the apex,
              side lobes large, rounded, middle lobe an apiculus, or scarcely visible,

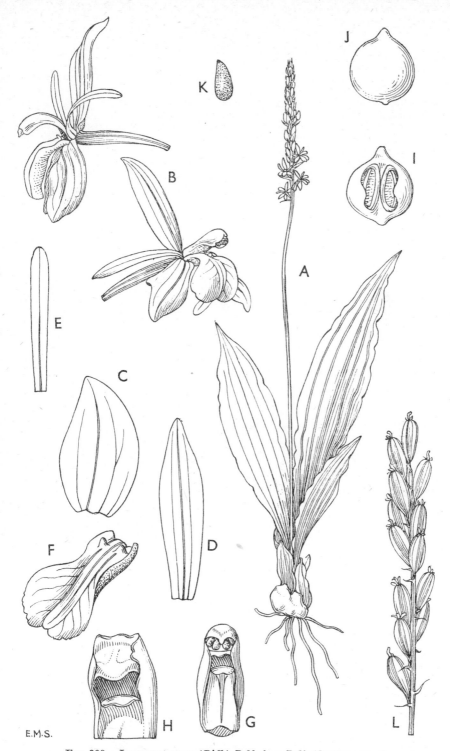

Fig. 389.—Liparis rufina (*Ridl.*) *Rchb. f. ex Rolfe* (Orchidaceae).

A, habit, × ½.  B, flower, × 6.  C, lateral sepal, × 9.  D, dorsal sepal, × 9.  E, petal, × 9.
F, lip, × 9.  G, column, complete, × 9.  H, column, anther cap and pollinia removed,
× 20.  I, anther cap, ventral view, × 20.  J, anther cap, dorsal view, × 20.  K, pollinium,
× 20.  L, fruiting spike, × 1.  A, from *Linder* 345.  B—J, from *FHI* 22404.  K and L
from *Hepburn*.

whole lip 4–5 mm. long, 4–5 mm. broad; dorsal sepal about 7 mm. long;
petals narrower, equal in length; scape longer than the leaves

10. *suborbicularis*

1. **L. kamerunensis** *Schltr.* in Engl., Bot. Jahrb. 53: 560 (1915). *L. capensis* of F.T.A. 7: 22, not of Lindl.
**W.Cam.:** Cam. Mt., 6,000–7,000 ft. (Nov.) *Mann* 2129!
2. **L. tridens** *Kraenzl.* in Engl., Bot. Jahrb. 28: 162 (1900); Cat. Talb. 145. A small epiphytic herb 2–5 in.
high with small yellow or yellow-green flowers.
**S.Nig.:** Oban *Talbot* 872! 873! Sapoba *Kennedy* 2635! Akure F.R., Ondo Prov. (July) *Onochie* FHI
23417! Calabar *Wright* 102! **F.Po:** Moka, 4,200 ft. (fr. Sept.) *Wrigley* 652! A fruiting specimen from
**Iv.C.** (Tiapleu (fr. Oct.) *Aké Assi* 6027!) is probably referable to this species. Also in E. Cameroun and
Uganda.
3. **L. goodyeroides** *Schltr.* in Engl., Bot. Jahrb. 38: 152 (1906).
**W.Cam.:** Moliwe *Stammler*!
4. **L. caillei** *Finet* in Bull. Soc. Bot. Fr. 56: 97, t. I, fig. 1–13 (1909). An epiphytic herb with a few small
leaves and a long flattened rhachis with rather widely spread yellowish or brownish-maroon flowers.
**Guin.:** *Caille.* Nzérékoré (fr. Sept.) *Jac.-Fél.* 1126! **S.L.:** Jama (fr. Sept.) *Deighton* 3062! Musaia (July)
*Deighton* 4832! **Lib.:** Du R. (July) *Linder* 78! St. John R., Gbanga (fr. Aug.) *Baldwin* 9138! Fo R.,
Kpalala (July) *Harley* 1507! **N.Nig.:** Jemaa, Zaria Prov. (June, July) *King* 86! **W.Cam.:** N'dian (May)
*Wright* 126! Also in Uganda.
5. **L. platyglossa** *Schltr.* in Engl., Bot. Jahrb. 38: 152 (1906). *L. winkleri* Schltr. l.c. 153 (1906). Flowers
reddish.
**S.Nig.:** cult. *spec.*! **W.Cam.:** Victoria (May) *Stossel* 5b! 21a! Victoria to Buea (July) *Winkler* 240!
M'bonge *Wright* 60/58!
6. **L. rufina** (*Ridl.*) *Rchb. f. ex Rolfe* in F.T.A. 7: 19 (1897); Chev. Bot. 613. *L. elata* Lindl. var. *rufina* Ridl.
(1886). Flowers green or yellow to reddish- or purplish-brown; damp places in grassland or among
rocks, sometimes in scrub.
**Sen.:** Borofaye, Zinguinchor (Sept.) *Doumlia* 624! **Port.G.:** Bissau, Antula (Sept.) *Esp. Santo* 1342!
**Guin.:** Benty (June) *Jac.-Fél.* 1680! Hollandé Tossékré (fr. Oct.) *Adam* 12729! Yanaya, Kissidougou
(July) *Martine* 351! Timbi, Fouta Djalon (June) *Adam* 14484! 14559! **S.L.:** Mabonto to Bumban (July)
*Deighton* 2218! Roboli, Rokupr (July) *Jordan* 290! Brookfields, Freetown (Aug.) *Jordan* 73! Nyandehun,
Valunia (July) *Deighton* 5799! Kakama, Gbo (Aug.) *Deighton* 6106! **Lib.:** Memmeh's Town (Aug.)
*Linder* 345! Monrovia (fr. Aug.) *Baldwin* 9203! SW. of Kaka Town (fr. Aug.) *Linder* 349! Paynesville
(July) *Harley* 2156! Ganta (July) *Harley* 1817! **Iv.C.:** Kodiokoffi to Tiégoualcro *Chev.* 22346! Oroumba
Boka, S. of Toumodi (fr. Oct.) *de Wilde* 461! **Ghana:** Ofinso (Aug.) *Cox* 100! **N.Nig.:** Heipan, 3,900 ft.,
Plateau Prov. (July) *G. V. Summerhayes* 158! Omo, Plateau Prov. (June) *King* 24! **S.Nig.:** Lagos
*Barter* 20202! Mt. Orosun, Idanre (fl. May, fr. Aug.) *Jones* FHI 20722! *Keay & Brenan* FHI 22404!
Benin *Unwin*! Nun R. *Barter* 20122! Generally in tropical Africa southwards to Angola and Rhodesia.
7. **L guineensis** *Lindl.* in Bot. Reg. 20: t. 1671 (1835); F.T.A. 7: 20; Cat. Talb. 145. *L. atacorensis* A. Chev.,
Bot. 613, name only. Flowers green, yellow, brown or purple; among rocks in forest or open woodlands,
rarely epiphytic.
**Sen.:** *Schnell* F183! Fogni, Casamance (Dec.) *Berhaut* 6767! **Guin.:** Macenta (June) *Jac.-Fél.* 997!
Ziama Massif, 4,000–4,300 ft. (fr. May) *Schnell* 2631! **S.L.:** Sugar Loaf Mt., 2,400 ft. (fr. Oct.) *T. S. Jones*
256! Rowala (July) *Thomas* 1074! Bumban, N. Prov. (July) *Deighton* 2219! Sumbuya (fr. Oct.) *Deighton*
4900! Da Oulen, 5,230 ft., Loma Mts. (Aug.) *Jaeger* 1344! **Lib.:** So (fr. Oct.) *Linder* 1127! E. of Zwedru,
Tchien Dist. (Aug.) *Baldwin* 7049! Tawata, Boporo Dist. (fr. Nov.) *Baldwin* 10289! Bobei Mt.,
Sanokwele Dist. (Sept.) *Baldwin* 9571! Ganta (Oct.) *Harley* 1839! **Iv.C.:** Tonkoui, 3,600–4,000 ft. (Aug.)
*Boughey* GC 18215! Issia Rock (Aug.) *de Wilde* 421! Adiopodoumé (July) *Leeuwenberg* 4581! **Ghana:**
Efiduase (June) *Cox* 22! **Dah.:** Atacora Mts., 1,300–2,000 ft. (June) *Chev.* 24193! **S.Nig.:** Mt. Orosun,
Idanre (June) *Charter* FHI 38713! Utanga, Ogoja Prov. (July) *Ejiofor* FHI 25341! Uzuakoli, Ogoja
(July) *Stone* 28! Oban *Talbot* 713! 777! **W.Cam.:** Mankom, Bamenda (June) *Daramola* FHI 41553!
R. Ife, Bamenda (July) *Ujor* FHI 30461! **F.Po:** lava flow (Sept.) *Wrigley* 544! Also in E. Cameroun,
Principe, (?) Annobon, Congo, C. African Republic, Tanzania, Zambia and Malawi.
8. **L. epiphytica** *Schltr.* in Engl., Bot. Jahrb. 38: 6, fig. 3 (1905). *L. lloydii* Rolfe in Kew Bull. 1906: 31.
**Iv.C.:** Tai Forest (Oct.) *Aké Assi* 6029! **S.Nig.:** Calabar (Aug.) *Lloyd* 3! Eket *Talbot*! **W.Cam.:** M'Bonge
*Wright* 58/58! Also in E. Cameroun, Gabon, Congo and Uganda.
9. **L. deistelii** *Schltr.* l.c. 151 (1906). Flowers yellow.
**W.Cam.:** Cam. Mt., 6,000 ft. (June) *Deistel*. **F.Po:** Moka (Sept.) *Wrigley* 652a! Also in (?) Uganda and
Kenya.
10. **L. suborbicularis** *Summerh.* in Kew Bull. 1934: 210. *L. epiphytica* of Cat. Talb. 145, not of Schltr. Flowers
greenish yellow; in rock crevices.
**S.Nig.:** Oban *Talbot* 871! **W.Cam.:** Nkambe to Binka, 6,500 ft. (Sept.) *Savory* UCI 380! Also in E.
Cameroun.

## 23. ANSELLIA Lindl., Bot. Reg. 1844, sub. t. 12 (1844); F.T.A. 7: 100.

Petals and sepals elliptical, petals often broader than sepals; lip broad, usually broader
than long, front lobe more or less orbicular and somewhat retuse, keels 2, the central
one either absent or very slightly developed; spots on flowers usually large and dark

1. *africana*

Petals and sepals elliptical or more frequently narrowly elliptical, petals never broader
than sepals; lip narrow, usually longer than broad, front lobe elliptical or almost
oblong, obtuse, keels 3, the central one smaller but usually quite well developed;
spots on flowers usually small and pale     ..     .. 2a. *gigantea* var. *nilotica*

1. **A. africana** *Lindl.* Bot. Reg. 1844, sub. t. 12; F.T.A. 7: 101. *A. confusa* N. E. Br. (1886)—F.T.A. 7: 102.
A robust epiphyte, up to 3–5 ft. high; flowers yellow with red or purple spots; in forest or bush.
**Lib.:** Lamco road, Nimba Mts. *Harley* 2248! **Iv.C.:** Kagbe (Nov.) *Aké Assi* 9332! **Ghana:** Abene, Kwahu
(Jan.) *Chipp* 626! Adum Bansu, W. Prov. (Jan.) *Fishlock* 84! Abetifi, Kwahu, 2,000 ft. (Dec.) *Irvine* 1820!
Mampong, Ashanti (Dec.) *Vigne* FH 4116! Degoro *Moore* 167! **S.Nig.:** Oban *Talbot* 1387! Atijere,
Ondo Prov. *Dawodu*! **F.Po:** Clarence (Oct.) *T. Vogel* 25! *Ansell*! Also in E. Cameroun, Gabon,
Congo, Angola, Uganda and Kenya.
2. **A. gigantea** *Rchb. f.* in Linnaea 20: 673 (1874).
2a. **A. gigantea** var. **nilotica** (*Bak.*) *Summerh.* in Kew Bull. 1937: 462. *A. africana* Lindl. var. *nilotica* Bak.
(1875). *A. nilotica* (Bak.) N. E. Br. (1886)—F.T.A. 7: 101; F.W.T.A., ed. 1, 2: 425. An epiphyte about
2–3 ft. high; flowers yellow with brown or orange spots; in open forest and savanna.
**N.Nig.:** Sokoto (Dec., Jan.) *Kennedy* 2994! near Zungeru and Lokoja *Lugard*! Katagum *Dalz.* 432!
Vom, 4,500 ft. *Dent Young* 231! Kurmi-Sauchi, Zaria Prov. (Feb.) *Daggash* FHI 31418! Eastward to
Kenya and south to Rhodesia and Mozambique.

**24. POLYSTACHYA** Hook., Exot. Fl. 2: t. 103 (1824); F.T.A. 7: 103. *Nom. cons.*

Pseudobulbs or stems narrowly cylindrical or rarely swollen at apex, superposed, each one arising some distance above the base of the preceding one and continuing the growth in the same direction :

Flowers rather large, sepals over 6 mm. long :

Flowers solitary, almost sessile ; leaves knife-shaped, fleshy, often with many overlapping sheaths at base ; stems widely creeping or hanging ; sepals 8–9 mm. long, hairy ; lip more or less oblong, side lobes small or very small, front lobe oblong, slightly retuse at apex, 10–11 mm. long, 5–6 mm. broad :

Leaves 8–15 cm. long, 6–8 mm. broad ; each pseudobulb or stem arising from apex of preceding one ; all stems 1-leaved ; side lobes of lip distinct      46. *crassifolia*

Leaves 2–4·5 cm. long, 5–9 mm. broad ; each pseudobulb or stem arising from upper part or apex of preceding one ; some stems with numerous distichous leaves, others 1-leaved ; side lobes of lip obscure    ..    ..    ..    47. *mystacioides*

Flowers in simple or branched inflorescences ; leaves lanceolate-linear or lanceolate, gradually tapering towards the acute apex, up to 22 cm. long, 4–14 mm. broad ; each pseudobulb arising some distance below apex of the preceding one, up to 30 cm. long, with 3 or more leaves ; front lobe of lip triangular, acute :

Dorsal sepal 10–16 mm. long, narrowly lanceolate, acuminate, sepals sparsely hairy ; inflorescence simple, laxly 3–11-flowered ; stems erect ; lip in general shape lanceolate, side lobes shortly acutely triangular, front lobe long-triangular, sharply acute, lip 8–13 mm. long, 5–8 mm. broad ; pseudobulbs arising from upper part of preceding one, 3–9-leaved, 5–20 cm. long    ..    .. 43. *microbambusa*

Dorsal sepal 6–9 mm. long, lanceolate to broadly lanceolate, sepals sparsely hairy ; inflorescence branched with up to 4 branches, many-flowered, branches up to 2 cm. long ; stems pendulous ; lip when spread out more or less fanshaped with a short claw, distinctly 3-lobed in the upper part, side lobes broad, rounded, middle lobe broadly triangular, acute, lip 6·5–8 mm. long and broad ; each pseudobulb arising from lower part of preceding one, 3–7-leaved, 12–30 cm. long

2b. *albescens* subsp. *angustifolia*

Flowers small, sepals 5 mm. long or less :

Raceme simple ; pseudobulbs arising from one-third way up the preceding one, 3–7·5 cm. long, 2–4-leaved ; leaves linear-ligulate, apex shortly bilobed, 3–12 cm. long, 3–8 mm. broad ; raceme 3–13 cm. long, many-flowered ; sepals 2·5–3·5 mm. long ; lip 3-lobed, more or less hastate in outline, about 3·5 mm. long, side lobes acute, recurved, middle lobe lanceolate, acuminate    ..    ..    23. *calluniflora*

Inflorescence a panicle, often many-flowered ; pseudobulbs arising from upper part of preceding one :

Dorsal sepal 2–3 mm. long, obtuse ; lip distinctly trilobed, hastate or cruciform ; upper internode of stem not much swollen :

Pseudobulbs 3·5–8 cm. long, 2–4-leaved ; leaves linear-ligulate, 4–9 cm. long, 4–8 mm. broad ; inflorescence 3–5 cm. long, branches simple ; sepals 2–2·5 mm. long ; lip 2 mm. long, broadly hastate, side lobes recurved, middle lobe broadly ovate, lip with small callus at base    ..    ..    ..    ..    44. *superposita*

Pseudobulbs up to 22 cm. long, 4–7-leaved ; leaves oblong-lanceolate, unequally bilobed at apex, lamina 5–16 cm. long, 6–16 mm. broad ; inflorescence 4–8 cm. long, branches compound ; sepals 2–3 mm. long ; lip 2–2·5 mm. long, cruciform, side lobes slightly incurved, middle lobe oblong-ovate, no callus at base of lip

45. *fusiformis*

Dorsal sepal 4·5–5 mm. long, acuminate ; lip obscurely trilobed, rhomboid-oblanceolate, 7 mm. long ; upper internode of stem much swollen to form an ovoid or ellipsoid pseudobulb ; stem with 4–5 leaves along its length and a single one at apex of the terminal pseudobulb ; leaves 1·5–3·5 cm. long and 1–1·5 cm. broad ; inflorescence very short, branched almost from the base    .. 42. *camaridioides*

Pseudobulbs or stems tufted or closely placed, their bases forming a short creeping rhizome :

*Pseudobulbs much flattened, more or less prostrate on the substratum, 1 or more leaved at apex :

Inflorescence developing after the leaves at the apex of the mature pseudobulb ; pseudobulbs orbicular or broadly elliptical, up to 5 cm. long and broad ; stems erect, slender, 1–3-leaved, up to 12 cm. long ; leaves oblanceolate, lamina 6–20 cm. by 2–5 cm., often purplish beneath ; inflorescence with a few branches, usually longer than the leaves ; bracts large, frequently longer than flowers, ovate-lanceolate, acuminate ; flowers hairy ; sepals 6–7·5 mm. long ; lip obscurely 3-lobed, 5·5–8 mm. long, side lobes rounded, front lobe obtuse, lip with a pubescent keel in the lower part    ..    ..    ..    ..    ..    ..    ..    ..    40. *affinis*

Inflorescence developing with the leaves on the young pseudobulb ; pseudobulbs elliptical, 2·5–5 cm. long, 1–2 cm. broad, with a single leaf at the apex and a leaf on each side at the base ; leaves narrowly oblong, with a rounded slightly bilobed apex, 5–15 cm. long, 0·8–2·5 cm. broad; inflorescence branched, up to 17 cm. high; sepals 6·5–15 mm. long; lip 3-lobed, 7·5–11 mm. long, side lobes obtuse, middle lobe shortly acuminate, recurved, lip with short keel at base thickened into a callus in centre of middle lobe    ..    ..    ..    ..    ..    **41. *obanensis***

*Pseudobulbs or stems vertically placed, usually much longer than broad, and never much flattened in the plane of the substratum:

†Mature pseudobulbs or stems 1-leaved at apex, terete or almost so, sometimes gradually tapering upwards:

Lateral sepals each produced at the base into a hollow spur 2 mm. long; stems very slender, terete, 2–7 cm. long; leaf linear, tapering very gradually upwards, apex shortly bilobed, 4–16 cm. long, 2–4 mm. broad; inflorescence paniculate, 3–6·5 cm. long, peduncle and rhachis very slender, branches 1–4, very short, almost umbellate; dorsal sepal 3·5–4 mm. long; lip 3-lobed, about 6 mm. long, side lobes small, oblong, front lobe cuneate; lip with a small callus at base

**53. *bicalcarata***

Lateral sepals not produced into spurs at base:

Dorsal sepal 8–17 mm. long; laterals united with foot of column to form a mentum 13–22 mm. long; lip 3-lobed, 13–25 mm. long, side lobes small, ovate, middle lobe ovate to orbicular, all lip pubescent inside; stems 5–15 cm. high, rather stout; leaf elliptical-oblong, apex rounded or subacute, 10–25 cm. long, 1–3 cm. broad; inflorescence 4–18 cm. long, simple or with a few very short branches

**49. *galeata***

Dorsal sepal 7 mm. long or less:

Leaves oblong-oblanceolate, oblong or elliptical-oblong, suddenly narrowed at base and thus appearing cordate, 1–5 cm. broad, up to 23 cm. long; pseudobulbs tapering from the base, narrowly cylindrical to conical, 0·7–17 cm. long; inflorescence up to 22 cm. long, usually paniculate in the upper part; bracts with a broad thin base, suddenly narrowed into a slender point, up to 5 mm. long; dorsal sepal 4–6·5 mm. long; lip obscurely trilobed, 4–6 mm. long, side lobes rounded, middle lobe ovate, apiculate or acuminate, disk with callus running from base to centre    ..    ..    ..    ..    ..    ..    **50. *cultriformis***

Leaves 12 mm. or less broad, narrowed gradually at base and not at all cordate; stems very slender, not tapering, up to 18 cm. long; bracts small, triangular, of even texture, not exceeding 2 mm. in length:

Dorsal sepal 4–7 mm. long; mentum 9–12 mm. long; side lobes of lip rounded, obscure; leaves narrowly elliptical-oblong, gradually tapering towards base, 5–16 cm. long, 6–12 mm. broad; inflorescence much shorter than the leaves, 2–6-flowered; lip 9–11 mm. long, with a long claw, middle lobe more or less orbicular ..    ..    ..    ..    ..    ..    **48. *supfiana***

Dorsal sepal 1·5–3 mm. long; mentum 2·5–4·5 mm. long; lip ecallose, side lobes nearly as long as front lobe, incurved, acute or obtuse; inflorescence up to 20 cm. long, with 4–8 short branches:

Leaves more or less terete with a groove along the upper surface, 1–3 mm. diam., 12–22 cm. long; lip 2·5–4 mm. long, 2·5 mm. broad, front lobe broadly ovate or orbicular    ..    ..    ..    ..    ..    ..    **51. *tenuissima***

Leaves narrowly oblong or linear, unequally bilobed at apex, 5–14 cm. long, 3–6 mm. broad; claw of lip longer than the limb, front lobe orbicular

**52. *inconspicua***

†Pseudobulbs or stems 2 or more leaved, leaves sometimes fallen off at time of flowering:

Pseudobulbs ellipsoid to longly cylindrical, 5–19 cm. long, more or less flattened, thicker in the middle, 1–1·5 cm. thick, 3–5-leaved in the upper half; leaves oblong or oblong-elliptical, rounded and shortly bilobed at apex, 8–30 cm. long, 1·5–4 cm. broad; inflorescence paniculate, 9–25 cm. long, with several large sheathing bracts at the base, branches long, only rarely branched again; flowers yellowish to reddish-orange; sepals 3–4 mm. long; lip entire, elliptical-ovate, subacute, 3 mm. long    ..    ..    ..    ..    ..    ..    ..    **1. *paniculata***

Pseudobulbs or stems, if elongated, not more than 1 cm. diam., if thickened, short and often somewhat conical, the leaves arising in a tuft:

Bracts longly setaceous-subulate from a very short broad base, usually spreading or reflexed; flowers small, dorsal sepal 2–4 mm. long:

Inflorescence simple with no arrested branches in the axils of the lower bracts; leaves strap-shaped:

Lateral sepals long-acuminate; callus on lip absent or consisting of a fleshy central cushion; petals linear, lanceolate or oblanceolate, acute:

Side lobes of lip oblong-elliptical or elliptical-orbicular, broader than front lobe, latter sharply recurved, lanceolate, acuminate, disk setose or glabrous; pseudobulbs 1–5 cm., rarely to 9 cm., long; leaves 4–20 cm. long, 4–14 mm. broad; inflorescence 3–14 cm. long; dorsal sepal 2·5–4 mm. long

**26. *adansoniae***

Side lobes of lip spreading, narrowly triangular-oblong, acute, much smaller

than the middle lobe, latter broadly triangular, oblong or almost orbicular,
  acute or apiculate; pseudobulbs 3–18 cm. long; leaves 7–32 cm. long,
  1–2·2 cm. broad; inflorescence 7–22 cm. long; dorsal sepal 2–2·5 mm. long
                                                          27. *polychaete*
Lateral sepals acute only; calli on lip basal, paired, bow-shaped, converging at
  base and in front; petals ovate or elliptical, obtuse; pseudobulbs 3–7·5 mm.
  long, 2–4-leaved; leaves 3–12 cm. long, 3–8 mm. broad; raceme 3–13 cm.
  long; sepals 2·5–3·5 mm. long; lip hastate in outline, side lobes small, acute,
  recurved, middle lobe lanceolate        ..        ..        ..        ..        23. *calluniflora*
Inflorescence branched, the branches sometimes represented by buds in the axils of
  the lower bracts; lip without calli:
Stem more or less elongated, the leaves arising some distance from the base, lower
  part covered with sheaths; leaves strap-shaped or lanceolate-oblong, widest
  about or below the middle, up to 15 cm. long, 5–16 mm. broad; dorsal sepal
  2–2·5 mm. long; lip obscurely 5-lobed, the lowermost pair of lobes infolded,
  densely pubescent outside, the front lobe broadly deltoid or rounded triangular;
  plant up to 30 cm. tall ..        ..        ..        ..        ..        ..        .. 19. *elegans*
Stem short, the leaves arising close together in a basal tuft, their bases covered
  by overlapping chaffy sheaths; leaves oblanceolate, widest in upper part,
  3–8 cm. long, 4–20 mm. broad; dorsal sepal 3–4 mm. long; lip 3-lobed in
  upper half, pubescent inside, side lobes rounded, front lobe triangular-ovate,
  acute or shortly acuminate; plant up to 12 cm. tall        ..        20. *coriscensis*
Bracts short, lanceolate or very broad, obtuse or acute or sometimes with a short
  slender point, if with a long point then the dorsal sepal over 4 mm. long; flowers
  large or small:
Stems slender, elongated, not thickened at the base, with no foliage leaves in the
  lower part which is covered by sheaths and the leaves scattered more or less
  regularly along the whole length above:
Plant small, 4–11 cm. high, including inflorescence; leaves narrowly oblong,
  1–2 cm. long, 1–2 mm. broad, erect; dorsal sepal 1 mm. long; stems slender
  1·5–4 cm. high, 2–6-leaved; inflorescence overtopping the leaves, peduncle
  thread-like, simple or much branched, racemes very short, up to 10-flowered;
  lip entire, rhombic, acute, 1·75 mm. long        ..        ..        ..        22. *seticaulis*
Plant much taller, 12–50 cm. high, including inflorescence, or with hanging stems
  up to 75 cm. long; leaves over 4 cm. long and usually much longer, over 5 mm.
  broad; dorsal sepal over 2·5 mm. long:
Rhachis of branches of inflorescence and of terminal spike markedly zigzag with
  the very broad bracts distichously arranged at the angles; inflorescence
  usually with few branches, entirely glabrous; leaves more or less oblanceolate
  or elliptical-oblanceolate, broader above the middle; lip (spread out) as
  broad as long or broader than long, with broad central callus; capsules
  3–3·5 cm. long:
Leaves 1·5–5 cm. broad, 6–17 cm. long, thin in texture and somewhat undulate,
  much tapering towards base; dorsal sepal 8–11 mm. long; lip about as broad
  as long, side lobes large, rounded; stem 15–35 cm. tall, 3–6-leaved
                                                          3. *caloglossa*
Leaves 0·8–2 cm. broad, 6–12 cm. long, thinly coriaceous, not undulate; dorsal
  sepal 6 mm. long; lip broader than long, side lobes triangular-ovate; stem
  5–18 cm. tall, 3–5-leaved        ..        ..        ..        ..        ..        5. *fractiflexa*
Rhachis of branches of inflorescence and of terminal spike not zigzag, bracts not
  distinctly in two rows:
Flowers pubescent; bracts of inflorescence small, spreading, acuminate;
  vegetative stems 7–25 cm. long, 2–6-leaved; lip with large rounded side-
  lobes and triangular or ovate acute middle lobe:
Inflorescence usually a spreading panicle with rather long slender branches,
  4–30 cm. long; branches rarely reduced to short tufts; leaves oblanceolate
  or oblanceolate-elliptic, 1·5–5·5 cm. broad, 8–26 cm. long; dorsal sepal
  5–9 mm. long; lip as broad as long        ..        ..        ..        6. *laxiflora*
Inflorescence a narrow panicle with short dense-flowered branches, 7–30 cm.
  long; leaves narrowly lanceolate, oblong-lanceolate or ligulate, 1–3 cm.
  broad, narrowing gradually towards apex; dorsal sepal 6·5–10 mm. long;
  lip longer than broad        ..        ..        ..        ..        ..        ..        7. *stricta*
Flowers glabrous:
Lip middle lobe triangular, acute or obtuse, not retuse, callus in centre of lip,
  side lobes rounded; rhachis glabrous; inflorescence simple or with up to 4
  branches:
Leaves very narrowly to elliptical-lanceolate, up to 22 mm. broad; lip up
  to 8 mm. long, mid lobe not long pointed:
Leaves lanceolate or elliptical-lanceolate, acute at apex, 5·5–14 cm. long,

6–22 mm. broad; stems erect, 11–30 cm. tall; lip distinctly longer than broad, about 6 mm. long .. ..      2a. *albescens* subsp. *albescens*
Leaves very narrowly lanceolate, very acutely 2–3-lobed at apex, 8–17 cm. long, 4–11 mm. broad; stems pendulous, up to 30 cm. long; lip about as broad as long, 6·5–8 mm. long     2b. *albescens* subsp. *angustifolia*
Leaves elliptic or lanceolate-elliptic 20–30 mm. broad; lip 16 mm. long
                                                     4. *cooperi*
Lip middle lobe rounded or retuse at apex, callus basal; lip distinctly longer than broad, side lobes much smaller than middle lobe, triangular or ligulate :
   Mentum or chin of flower longer than the dorsal sepal, conical, obtuse, 5·5–8 mm. long; stems more or less stiffly erect, 10–50 cm. high, slender, 4–12-leaved; leaves linear-lanceolate, 4–16 cm. long, 4–15 mm. broad, apex broadly if sometimes acutely bilobed; inflorescence unbranched, 4–12 cm. long; dorsal sepal 2·5–4·5 mm. long; lip oblong, 3-lobed above the middle, 6–9 mm. long, side lobes small, acute, front lobe oblong, lip with large more or less orbicular hairy callus at base and small tooth in centre just in front    ..    ..    ..    ..    ..    8. *bifida*
   Mentum shorter than the dorsal sepal, rounded conical, 3–5 mm. long; stems rather flaccid and flexuous, 12–40 cm. long, 4–11-leaved; leaves narrowly lanceolate, 5–21 cm. long, 5–23 mm. broad, apex acute, very narrowly bilobed; inflorescence simple or more usually somewhat branched, 4–12 cm. long; dorsal sepal 5–7 mm. long; lip 3-lobed at base, 5–7·5 mm. long, lateral lobes spreading, narrow, middle lobe much larger, broadly cuneate    ..    ..    ..    ..    ..    9. *rhodoptera*
Stems usually short, often thickened at base to form a more or less conical pseudobulb, internodes short so that the leaves are borne in a tuft near the base; if stem elongated the lowest foliage leaf-base arises from base of stem (this leaf may sometimes be somewhat reduced):
Inflorescence branched, the branches sometimes arrested and represented by short shoots in the axils of the lower bracts (rarely quite simple inflorescences may be found on individual shoots):
Stems leafless or almost so at time of flowering, swollen at the base to form a conical pseudobulb; leaves linear or ligulate, up to 2 cm. broad; scape enveloped in tight chaffy sheaths:
Dorsal sepal 6–7·5 mm. long:
   Rhachis, ovary and sepals almost or quite glabrous; lip oblong in shape with a very obscure rounded lobe on each side at the middle, bearing numerous long clavate hairs; flowers thin in texture, purplish red; scape very slender, 15–50 cm. high, branches of inflorescence slender, sometimes branched again at the base, up to 7 cm. long; leaves linear, up to 4 mm. broad
                                             54. *bequaertii*
   Rhachis and sepals pubescent, ovary densely so; lip trilobed with large rounded side-lobes, with short clavate hairs on the disk; flowers rather fleshy, greenish white, middle lobe of lip yellow; scape rather stout, branches stout, up to 7 cm. long; leaves ligulate, 1·5–1·8 cm. broad
                                             25. *kingii*
Dorsal sepal 3·5–4 mm. long; lip distinctly trilobed, more or less cruciform, middle lobe much recurved, side lobes shorter, erect, centre with whitish mealy hairs; flowers rather fleshy, yellow or greenish-yellow, sometimes with red markings; scape rather stout, 4–25 cm. long, branches of inflorescences up to 7, rather stout, less than 2 cm. long; leaves ligulate, 3–10 mm. broad; rhachis pubescent, flowers almost glabrous    ..    ..    24. *steudneri*
Stems leafy at time of flowering; leaves oblong, lanceolate or oblanceolate, usually over 1 cm. broad; flowers rather fleshy; lip usually distinctly 3-lobed, never bearing long clavate hairs:
Branches of inflorescence secund (all turning to the same side of the rhachis) or almost so :
   Lip without a distinct longitudinal keel in the lower part but with a cushion of short mealy hairs; leaves usually 0·5–1·5 cm. broad, rarely up to 3 cm. :
     Leaves fleshy, narrowly oblong with almost parallel sides, V-shaped in section, apex broad, obtusely bilobed, 4–18 cm. long, 5–30 mm. broad; inflorescence 6–35 cm. high with up to 8 laterals; laterals up to 5 cm. long, many-flowered; upper part of sheaths on rhachis tight; dorsal sepal 1·5–3 mm. long; lip trilobed, 2·5–3 mm. long, side lobes ovate, rather incurved, middle lobe elliptical, retuse    ..    ..    ..    34. *golungensis*
     Leaves thin in texture, narrowly oblanceolate or oblong-oblanceolate, not markedly V-shaped in section, apex narrow, rounded or acute, up to 20 cm. long, 5–22 mm. broad; inflorescence up to 15 cm. high with up to 6 laterals; laterals up to 2 cm. long; upper part of sheaths on rhachis rather loose;

dorsal sepal 2·5–3·5 mm. long; lip trilobed, 3–4·5 mm. long, side lobes rounded, middle lobe elliptical-quadrate, bullate    ..    ..33. *modesta*

Lip with a distinct keel running from the base to the middle or above; leaves 1–6 cm. broad, usually over 1·5 cm.; inflorescence up to 80 cm. high, usually over 15 cm.:

    Leaves narrowly elliptical-lanceolate, acutely bilobed at apex, 1–3·5 cm. broad, 7–30 cm. long; dorsal sepal 3·5–5 mm. long; lip 3-lobed at apex only, 4–7 mm. long; lobes almost the same in length, middle lobe very broadly deltoid-semiorbicular, apiculate, margins very undulate, with a small keel in the centre separate from the basal one; stem somewhat elongated; branches of inflorescence up to 7, rarely branched again

                                                  32. *mukandaensis*

    Leaves oblanceolate or oblong-elliptical, obtusely or subacutely bilobed at apex, 1·5–6 cm. broad, 6–30 cm. long; dorsal sepal 2·5–4·5 mm. long; lip 3-lobed at the middle, 4–5·5 mm. long, middle lobe oblong or almost orbicular, much longer than the laterals, with the margin undulate towards the base, no keel at the apex; branches of inflorescence up to 14, frequently branched again near the base    ..    ..    ..    ..    31. *tessellata*

Branches of inflorescence not secund, sometimes represented by a single short branch:

    Lip entire, narrowly oblanceolate, apex acuminate, 5–6 mm. long, scarcely 2 mm. broad, with a bilobed basal callus; sepals densely pubescent, narrowly lanceolate, about 7 mm. long; leaves lanceolate or narrowly lanceolate to 6 cm. long, 3–14 mm. broad, very acutely bilobed at apex

                                           36. *oblanceolata*

    Lip trilobed, lobes sometimes obscure but lip then nearly as broad as long, callus entire or absent:

    Lip without a callus; fruits 4–9 mm. long:

        Dorsal sepal 2·5–3·5 mm. long; inflorescence with 1–6 short side branches not much diverging from the main axis and up to 2 cm. long; lip 3–4·5mm. long, side lobes rounded, middle lobe elliptical-quadrate, bullate; leaves narrowly oblanceolate or oblong-oblanceolate, 5–22 mm. broad

                                         33. *modesta*

        Dorsal sepal 1·8–2 mm. long; leaves oblong-elliptical or oblanceolate, acute, 4–18 cm. long, 1–3 cm. broad; inflorescence a spreading compound panicle with slender many-flowered branches, 6–25 cm. high; lip 3-lobed in the upper half, 2·5–4 mm. long, margins incurved below, ecallose, side lobes spreading, rounded-triangular, middle lobe elliptical, smaller than the side lobes but projecting in front of them, 0·7–0·8 mm. long; fruits 4–8 mm. long    ..    ..    ..    ..    ..    28. *ramulosa*

    Lip with a callus at the base or in the centre; fruits 9–18 mm. long:

        Panicle large with spreading branches, 13–40 cm. high; side lobes of lip incurved, acute, middle lobe oblong, retuse at apex; leaves oblanceolate or rarely almost linear, acute, 5–30 cm. long, 0·5–4·5 cm. broad:

            Dorsal sepal 3–4 mm. long; lip 4–4·5 mm. long, front lobe nearly quadrate or longer than broad, callus short, central, not continuous to the base; fruits 9–12 mm. long    ..    ..    ..    ..    ..    29. *puberula*

            Dorsal sepal 4–5·5 mm. long; lip 6–7·5 mm. long, front lobe transversely oblong, broader than long, callus a linear keel extending from the base to the centre; fruits 12–18 mm. long ..    ..    ..    ..30. *odorata*

        Panicle small with erect or suberect branches, 4–24 cm. high; side lobes of lip rounded, sometimes rather obscure, not incurved, middle lobe triangular, ovate or elliptical, acute or apiculate at apex:

        Rhachis and flowers densely pubescent; leaves of young growth not properly developed at time of flowering; leaves narrowly oblong, apex obtusely bilobed, 8–30 cm. long, 1·2–2·2 cm. broad; dorsal sepal 3·5–6 mm. long; lip 5–7·5 mm. long, 3–4·6 mm. broad; fruits 10–15 mm. long..    ..    ..    ..    ..    ..    35. *dolichophylla*

        Rhachis more or less pubescent, flowers almost or quite glabrous; leaves of young growth fully expanded at time of flowering; leaves ligulate or oblanceolate, apex acute, acutely bilobed, 6–21 cm. long, 0·6–3 cm. broad; dorsal sepal 3·5–5·5 mm. long; lip 4–5·5 mm. long, 2·7–4·5 mm. broad; fruits 9–12 mm. long ..    ..    ..    ..    37. *subulata*

Inflorescence simple, no arrested branches in axils of lower bracts:

    Sepals densely pubescent, narrowly lanceolate, about 7 mm. long; mentum very short; lip oblanceolate, acuminate, with bilobed basal callus and central hair-cushion, 5–6 mm. long; plant up to 7 cm. high; leaves lanceolate-linear or lanceolate, up to 6 cm. long and 3–14 mm. broad, very acutely 2-lobed at apex    ..    ..    ..    ..    ..    ..    ..    ..    36. *oblanceolata*

Sepals almost or entirely glabrous; mentum quite prominent, cylindrical, conical or rounded, often longer than the dorsal sepal:

Flowering scapes 30 cm. or more in height, the peduncle almost covered with chaffy sheaths, the rhachis many-flowered; flowers pale mauve with one or more yellow cushions of hairs on the lip, thin in texture; lip bent back like a knee in the middle; dorsal sepal 2·5–4 mm. long:

Lip with a short keel-like callus just above the base and two parallel almost touching hair cushions in the centre; stems leafless at time of flowering; sepals each with a horn-like projection on the back just below the tip; apical part of lip (epichile) bent upwards, broadly orbicular or transversely elliptical with a wavy margin; leaves narrowly strap-shaped, up to 10 cm. long and 1 cm. broad       ..       ..       ..       ..       ..       11. *pseudo-disa*

Lip without any callus in basal part but with a single hair-cushion stretching right across the centre; leaves present at time of flowering; sepals without horn-like projections at the back; epichile of lip not bent upwards, more or less bifid or bilobed; leaves 2–4 to each stem, very narrowly lanceolate or strap-shaped, 10–16 cm. long, 4–8 mm. broad ..       ..       12. *geniculata*

Flowering scape 20 cm. or less in height, usually much shorter:

Flowers borne on leafless pseudobulbs:

Sepals rather fleshy, hairy outside; ovary densely pubescent; flowers greenish white; middle lobe of lip yellow, disk and lobes with numerous hairs but these not forming dense hair cushions; sepals 6–7 mm. long; lip distinctly 3-lobed, lobes nearly equal; leaves ligulate, 3–12·5 cm. long, 1·5–1·8 cm. broad ..       ..       ..       ..       ..       ..       25. *kingii*

Sepals thin in texture, quite glabrous, frequently with a horn-like projection just below the apex; ovary glabrous or rarely with very few scattered hairs; flowers white, rose or lilac; disk of lip with 1 or 2 dense cushions of yellow hairs; roots often flattened, frequently green and assimilating:

Lip not sharply reflexed, 5–5·5 mm. long, lower part (hypochile) more or less concave, orbicular from a narrow base, 4–4·5 mm. long, with 2 hairy cushions in the upper part and a tooth-like callus at the base, apical part (epichile) more or less inflexed, transversely oblong or semiorbicular, margins much undulate; dorsal sepal 3·5–4 mm. long; mentum 6·5 mm. long       ..       ..       ..       ..       ..       ..       ..       10. *dalzielii*

Lip sharply reflexed below the middle:

Lip without a callus or keel at the base and with a single cushion of hairs in the upper part:

Lip not hollowed out at apex, 7 mm. long, 3-lobed, side lobes spreading, linear, 1·5 mm. long, middle lobe much larger, oblong-elliptical, rounded at apex, with a single cushion of hairs in the centre; dorsal sepal 3 mm. long; mentum 4–4·5 mm. long; inflorescence 5–8 cm. long, many-flowered ..       ..       ..       ..       ..       13. *monolenis*

Lip hollowed out at apex into a sac or hood, 4 mm. long, entire, pandurate-oblong with a narrow claw, sac furnished with a large transversely elliptical cushion of hairs; dorsal sepal 2·5 mm. long; mentum 3 mm. long; inflorescence 4–12 cm. long ..       ..       ..       .. 14. *saccata*

Lip with a callus or keel at the base, reflexed just at base of cushions of hairs, hair-cushions 2, elongated, contiguous or separated:

Mentum of flower curved, 10–11 mm. long; lip 15–17 mm. long (stretched straight), claw narrow, 4·5–5 mm. long, middle part (mesochile) with obtusely triangular side lobes, hair-cushions contiguous, epichile orbicular, more or less emarginate at apex, inflexed, 4–5 mm. diam.; dorsal sepal 6–7 mm. long ..       ..       ..       ..       15. *elastica*

Mentum scarcely curved, 6–9 mm. long; lip 8·5–13 mm. long:

Claw widened or lobed below the hair-cushions:

Callus at base of lip forming a high quadrate keel; hair-cushions separate, reaching nearly to the edges of the mesochile; epichile flabellate, not truncate at the base       ..       ..       16. *pobeguinii*

Callus at base of lip tooth-like, not much elevated; claw of lip frequently with narrow lateral lobes; hair-cushions contiguous, with a broad glabrous border to the mesochile; epichile transversely oblong or semi-orbicular, basal angles truncate       .. 17. *reflexa*

Claw not widened below the hair-cushions, which are at broadest part of lip and do not reach the margins; epichile more or less orbicular, margin dentate       ..       ..       ..       ..       ..       18. *victoriae*

Flowers borne on leafy pseudobulbs, usually rather fleshy; sepals without a horn-like projection below the apex:

Mentum narrow, more or less cylindrical, slightly longer than or equalling the dorsal sepal; leaves narrowly ligulate or linear-oblong, rounded and shortly bilobed at apex, up to 1·2 cm. broad; inflorescence short, up to 10-flowered :

Claw (narrow basal part) of lip hairy outside, upper part ovate or elliptical with a large 2–4–6-lobed flat callus wider than the lamina; dorsal sepal 5·5–6·5 mm. long; stems erect, 2–7 cm. high, swollen (0·5–1 cm. diam.) at base; leaves 2·5–9 cm. long, 3–12 mm. broad; inflorescence up to 9 cm. long ..    ..    ..    ..    ..    ..    .. 21. *alpina*

Claw of lip glabrous, upper part distinctly 3-lobed, middle lobe broadly ovate or almost orbicular, side lobes rounded triangular, decurrent on to the longer middle lobe, no callus; dorsal sepal 4–5·5 mm. long; stems 1–2 cm. high, forming ovoid pseudobulbs in the lower part; leaves 2–7 cm. long, 1–7 mm. broad; inflorescence up to 5 cm. long 39. *parva*

Mentum broad, rounded-conical, considerably longer than the dorsal sepal; leaves lanceolate or oblanceolate, acute at apex, 1–3 cm. broad, 8–20 cm. long; inflorescence 9–30 cm. long, many-flowered, bracts much shorter than the ovary and pedicel; stem up to 5 cm. long, 3–5-leaved; lip more or less obovate, acutely 3-lobed at apex, 4·5–6·5 mm. long 38. *leonensis*

1. **P. paniculata** *(Sw.) Rolfe* in F.T.A. 7: 113 (1897). Flowers orange with red markings.
    **Guin.:** Nzo (fr. Mar.) *Chev.* 21051! **S.L.:** *Afzelius*! Kékédou to Kissidougou (Oct.) *Jaeger* 2162! **Lib.:** Ganta, Busi road (Nov.) *Harley* 1836! **Ghana:** Aburi Hills (Nov.) *Johnson* 483! Ofinso (Oct.) *Cox* 104! Effiduase *Cox* 57! Kadjakpe, Togo Plateau F.R. (Oct.) *St. Cl.-Thompson* 1509! Amedzofe (Sept.) *Westwood* 45! **S.Nig.:** Akure F.R., Ondo (Aug.) *Keay* FHI 25371! Usonigbe F.R., Sapoba (fr. Nov.) *Ejiofor* FHI 24653! Degema *Talbot* 3610! Kundeve, Sonkwala, Ogoja (Sept.) *Keay* FHI 25381! Ikom to Mamfe, Ogoja Prov. (Mar.) *Sanford* 310/65! **W.Cam.:** Victoria (Sept.) *Ngongi* FHI 15341! Tombel, Kumba (fr. Mar.) *Thorold* CM17! Also in E. Cameroun, Gabon, the Congos and Uganda.
2. **P. albescens** *Ridl.* in Bol. Soc. Brot. 5: 199 (1888); F.T.A. 7: 111. *P. imbricata* Rolfe in Kew Bull. 1893: 172; F.T.A. 7: 112.
2a. **P. albescens** subsp. **albescens**—Summerh. in Kew Bull. 12: 74 (1958). Flowers greenish or whitish, sometimes with reddish veins, lip tinged or veined red.
    **S.Nig.:** Obudu Plateau, Ogoja, 3,500–5,000 ft. (May) *Head* 96! *Cooper* 84! **W.Cam.:** Bambui, Ndop Plain, 5,000 ft. (May) *Brunt* 450! **F.Po:** Basilé to Estrada, 3000 ft. (Dec.) *Sanford* 4008! Also in S. Tomé, Principe and Annobon.
2b. **P. albescens** subsp. **angustifolia** *(Summerh.) Summerh.* in Kew Bull. 12: 75 (1958). *P. imbricata* Rolfe subsp. *angustifolia* Summerh. in Kew Bull. 10: 227 (1956). Flowers white or cream, the lip with brownish markings within.
    **W.Cam.:** Buea, 3,000 ft. (Apr., July) *Gregory* 165! 284! *Keay* FHI 22411! *Ejiofor* FHI 25340!
3. **P. caloglossa** *Rchb. f.* Otia Bot. Hamburg. 111 (1881); F.T.A. 7: 128. *P. excelsa* Kraenzl. in Engl., Bot. Jahrb. 36: 117 (1905). *P. rolfeana* Kraenzl. in Engl., Bot. Jahrb. 33: 61 (1902), not Kraenzl. (1900). Flowers cream, yellow or orange.
    **W.Cam.:** Cam. Mt., 5,000 ft. (Nov.) *Mann* 2110! Buea, 3,200–4,000 ft. (May, Nov.) *Maitland* 728! *Migeod* 25! *Deistel*! **F.Po:** SW. of island *Mildbr.* 7110. Moka, 4,600 ft. (Aug.) *Melville* 412! Also in E. Cameroun, Gabon, Congo and Uganda.
4. **P. cooperi** *Summerh.* in Kew Bull. 17: 553 (1964). Flowers greenish-white or cream.
    **S.Nig.:** Obudu Plateau, Ogoja (Nov.) *Cooper* 2a!
5. **P. fractiflexa** *Summerh.* in Kew Bull. 10: 224 (1956).
    **W.Cam.:** Tombel (fr. Nov.) *Thorold* TN37! Also in Congo.
6. **P. laxiflora** *Lindl.* in J. Linn. Soc. 6: 129 (1862); F.T.A. 7: 111; Cat. Talb. 146. *P. dixantha* Rchb. f. (1882)—F.T.A. 7: 126. Flowers white, yellow or orange-yellow, sometimes with red markings on the side lobes of the lip.
    **Guin.:** Macenta, 2,000 ft. (May) *Collenette* 7! *Jac.-Fél.* 865! Dalaba (Mar.) *Caille* in *Hb. Chev.* 18107! Kouria (Apr.) *Pobéguin* 1559! Nimba Mts. (fr. Apr.) *Schnell* 1204! **S.L.:** Mt. Horton (Jan.) *Deighton* 6035! Musaia (July) *Deighton* 4837! Masingbe to Mamansu (Mar.) *Austin* in *Hb. Deighton* 5748! Ndilajula, Kori (Oct.) *Deighton* 6021! Loma Mts., 5,000 ft. (Feb.) *Jaeger* 4236! **Lib.:** Bumbuna (Oct.) *Linder* 1324! Suen (Nov.) *Linder* 1324a! Dobli I., St. Paul R. (Apr.) *Bequaert* 169! Brewersville (Sept., Oct.) *Harley* 1831! Ya R., Sanokwele (Sept.) *Baldwin* 9515! Nimba (Dec.) *Adames* 839! **Iv.C.:** Man to Danané (Nov.) *de Wilde* 849! Mt. Tonkoui, SW. of Man (Mar.) *Leeuwenberg* 2956! Yapo Forest (Aug.) *Oldeman* 249! **Ghana:** Bunsu (Mar.) *Westwood* 131! **S.Nig.:** Mt. Orosun, Idanre, Ondo (Apr.) *Symington* FHI 3372! Nun R. *Barter* 2126! Brass *Barter* 1055! Newi, Onitsha (July) *Jones* FHI 6129! Oban *Talbot* 918! Afi River F.R., Ogoja (June) *Jones & Onochie* FHI 17344! **W.Cam.:** Moliwe (Aug.) *Schlechter* 15778! Victoria (June) *Preuss* 1216! *Rosevear* 51/37! S. Bakundu F.R., Kumba (Mar., Apr.) *Onochie* FHI 31198! *Daramola* FHI 29809! **F.Po:** (June) *Mann* 437! Also in E. Cameroun and Gabon.
7. **P. stricta** *Rolfe* in Kew Bull. 1909: 63. Flowers pale yellow, often more or less tinged with pink, rarely mauve-pink.
    **N.Nig.:** Mambila Plateau, 3,000–4,600 ft. (Jan., Feb.) *Wimbush* 61! (= *King* 35!) *Hepper* 2882! Also in E. Cameroun, Uganda, Kenya and Tanzania.
8. **P. bifida** *Lindl.* in J. Linn. Soc. 6: 129 (1862); F.T.A. 7: 108. *P. farinosa* Kraenzl. (1893)—F.T.A. l.c. Flowers white tinged mauve, lip mostly mauve.
    **S.Nig.:** Obudu Plateau, Ogoja, 3,500 ft. (Nov.) *Cooper* 84! **W.Cam.:** Cam. Mt., 4,000–5,000 ft. (Nov.–Jan.) *Mann* 1339! 2115! *Johnston* 100! *Migeod* 32! Buea, 3,300 ft. (Oct., Jan.) *Preuss* 1064! *Gregory* 613! **F.Po:** 4,000 ft. (Dec.) *Mann* 649! Also in S. Tomé, Gabon and Congo.
9. **P. rhodoptera** *Rchb. f.* in Hamburg. Gartenz. 14: 214 (1858); F.T.A. 7: 109. *P. ensifolia* Lindl. (1862)—F.T.A. 7: 108. *P. pyramidalis* Lindl. (1862)—F.T.A. 7: 109; Cat. Talb. 146. *P. subcorymbosa* Kraenzl. in Kew Bull. 1926: 288. Flowers white or yellow, often tinged rose.
    **S.L.:** *Sc. Elliot*! Niawa (Oct., Nov.) *Deighton* 4919! **Lib.:** Moylakwelli–Totokwelli (Oct.) *Linder* 1290! Duport (Nov.) *Linder* 1496! Wohmen, Vonjama, Western Prov. (Oct.) *Baldwin* 10063! Tawata, Boporo (Nov.) *Baldwin* 10337! Ganta (Oct.) *Harley* 2050! Nimba (Dec.) *Adames* 842! **Iv.C.:** *Guillaumet* 282! **S.Nig.:** Sapoba *Kennedy* 1917! Nun R. (Sept.) *Mann* 522! Oban *Talbot* 903! Also in Principe, E. Cameroun, Gabon and Congo.
10. **P. dalzielii** *Summerh.* in Kew Bull. 1927: 418. *P. sp.* F.T.A. 7: 123, under *P. angularis* Rchb. f. Flowers white to lilac or pale purple.
    **Guin.** Dalaba (Mar.) *Dalz.* 8433! Gangan, Kindia, 3,700 ft. (Jan.) *Jac.-Fél.* 424! **S.L.:** Bintumane Peak, Loma Mts., 5,400 ft. (Feb., Mar.) *T. S. Jones* 169! Loma Mts., 5,300 ft. (Feb.) *Jaeger* 4249! 4256! **Lib.:** Grand Bassa (fr. Apr.) *T. Vogel* 104! Mt. Wolagwisa, near Pandamai, 4,500 ft. (Mar.) *Bequaert* 100! Bili (Feb.) *Harley* 1773! Bilimu (Mar.) *Harley* 1999! Nimba *Adam* 21128! **Iv.C.:** Mt. Gbon, 3,300 ft. (fr. May) *Chev.* 21412! Mt. Tonkoui, SW. of Man, 3,900 ft. (Mar.) *Leeuwenberg* 2969!
11. **P. pseudo-disa** *Kraenzl.* in Kew Bull. 1926: 293. Flowers mauve or purplish with yellow hairs on lip; on rocks or on *Calagyna pilosa*.

FIG. 390.—POLYSTACHYA COOPERI *Summerh.* (ORCHIDACEAE).

1, flowering plant, × 1. 2, flower, front view, × 2. 3, flower, with one lateral sepal and lip removed, lateral view, × 2. 4, intermediate sepal, × 1½. 5, petal, × 1½. 6 and 7, lip, lateral view and from inside respectively, × 1½. 8, lip, cut through longitudinally, × 1½. 9 and 10, anther with pollinarium, front and back views respectively, × 4. 11, pollinarium, back view, × 4. All from *Cooper 2a.*

**S.L.:** Sefadu (Mar.) *Deighton* 3517! 3569! 3626! *T. S. Jones* 170! **Lib.:** Ballowallah, Kailahun to Bolahun (Feb.) *Bequaert* 69! **Iv.C.:** Mt. Tonkoui, 3,400 ft. (Apr.) *Leeuwenberg* 3858!
[The original locality of Uganda given for this species was probably an error—V.S.S.]

12. **P. geniculata** *Summerh.* in Kew Bull. 10: 229 (1956). Flowers mauve or yellowish.
**W.Cam.:** Mamfe (May, Sept.) *Gregory* 124! 323! Banyang, Mamfe (Aug.) *Eyeku* FHI 22304! Manenguba Mts., 6,000 ft. (June) *Gregory* 302!

13. **P. monolenis** *Summerh.* in Kew Bull. 1935: 198. *P. expansa* Ridl., partly—F.T.A. 7: 122, partly.
**S.L.:** *Wilford*!

14. **P. saccata** (*Finet*) *Rolfe* in Orch. Rev. 26: 107 (1918). *Epiphora saccata* Finet in Notulae Syst. 2: 30 (1911). Flowers white and rose, hair-cushion on lip yellow, fragrant.
**Port.G.:** Chitole, Cosselinta (June) *Esp. Santo* 3192! **Guin.:** Pita (Apr.) *Pobéguin* 2294! 2302b! *Adam* 11641! Ditinn (Apr.) *Chev.* 12985! Gangan, Kindia (Apr.) *Adam* 11916! **N.Nig.:** Kagarko, Zaria (Apr.) *G. V. Summerhayes* 90! Gimi River F.R., Zaria (Apr.) *Keay & Jones* FHI 37620! Sanga River F.R., Zaria (Apr.) *Keay & Jones* FHI 37639!

15. **P. elastica** *Lindl.* in J. Linn. Soc. 6: 131 (1862); F.T.A. 7: 129. Flowers lavender or rose with yellow hairs on lip.
**S.L.:** *Afzelius.* Bagroo R. (Apr.) *Mann* 902! **Lib.:** Ganta (May) *Harley* 1924!

16. **P. pobeguinii** (*Finet*) *Rolfe* in Orch. Rev. 26: 107 (1918). *Epiphora pobeguinii* Finet in Notulae Syst. 2: 29 (1911). Flowers rose or lilac with yellow hairs on lip.
**Guin.:** Ninkan Plateau (Apr.) *Pobéguin* 2087! Labé *Pobéguin* 2089. Dalaba (Apr.) *Caille* in *Hb. Chev.* 18141! Gangan (May) *Roberty* 17763! Nimba Mts., 3,000–5,300 ft. (Feb., June) *Schnell* 463! 3023! **S.L.:** Bandakarafaia to Yalumbu (May) *Deighton* 5080! Loma Mts., 5,500 ft. (Jan.) *Jaeger* 4185! **Lib.:** Nimba (Mar.) *Adam* 21156! **Iv.C.:** Brafouédi (cult. Abidjan, Dec.) *Aké Assi* 9353!

17. **P. reflexa** *Lindl.* Bot. Reg. 27: Misc. 18 (1841); F.T.A. 7: 127, partly. *P. smytheana* Rolfe in Kew Bull. 1908: 71. *P. liberica* Rolfe l.c. 72 (1908). *P. elastica* of Chev. Bot. 616, partly, not of Lindl. Flowers pale rose-purple with yellow hair-cushions on lip.
**S.L.:** *Whitfield*! Hamilton (Feb.) *Smythe* 56! Mano Bonjema, near L. Mape (Jan.) *Adames* 6! 114! Kambui Hills (Apr.) *Small* 905! Bureh Town, 30 miles from Freetown (Jan.) *Deighton* 5961! **Lib.:** *Johnston*! **Iv.C.:** Mt. Dou, Dyola country, 4,500 ft. (May) *Fleury* in *Hb. Chev.* 21470! Adiopodoumé (Nov.) *Aké Assi* 4468! **Ghana:** Amedzofe (Apr ) *Rose Innes* GC 30556!

18. **P. victoriae** *Kraenzl.* in Engl., Bot. Jahrb. 28: 165 (1900). Flowers white or pale lilac with yellow hairs.
**W.Cam.:** Victoria (Feb.) *Simon* 14! *Deistel* 190! Nkolanjeng (Feb.) *Schorkopf* 30f!

19. **P. elegans** *Rchb. f.* Otia Bot. Hamburg. 113 (1881); F.T.A. 7: 119. *P. mannii* Rolfe in F.T.A. 7: 117 (1897). Flowers greenish white or yellowish white, sometimes tinged mauve.
**S.Nig.:** Sonkwala, Ogoja Prov. (Aug.) *Keay* FHI 25376! **W.Cam.:** Cam. Mt. 4,000–6 000 ft. (Nov.–Jan.) *Mann* 1338! 2113! *Johnston* 99! Banso to Jikiri, Bamenda, 5,600 ft. (Sept.) *Savory* UCI 385! **F.Po:** nr. S. Isabel (fr. Dec.) *Sanford* 4013!

20. **P. coriscensis** *Rchb. f.* l.c. 112 (1881); F.T.A. 7: 120. *P. kiessleri* Schltr. in Engl., Bot. Jahrb. 38: 153 (1906). Flowers pale green or yellow, with brown or rose-coloured stripes.
**S.Nig.:** Sapoba, Benin Prov. (fl. cult. Ibadan, May) *Sanford* WS/185/66! Eket *Talbot*! **W.Cam.:** Moliwe (Aug.) *Schlechter* 15780! M'bonge *Wright* 58/11! Also in E. Cameroun, Gabon and Congo.

21. **P. alpina** *Lindl.* in J. Linn. Soc. 6: 131 (1862); F.T.A. 7: 128. *P. preussii* Kraenzl. (1893)—F.T.A. 7: 128; Cat. Talb. 146. *P. winkleri* Schltr. in Engl., Bot. Jahrb. 38: 154 (1906). *P. talbotii* Rolfe in Kew Bull. 1910: 282; Cat. Talb. 146. Flowers white or rose.
**S.Nig.:** Oban, Niagi (Oct.) *Talbot* 835! 919! **W.Cam.:** Cam. Mt., 5,000–6,750 ft. (Apr., May) *Preuss* 934! 1021! *Maitland* 726! *Winkler* 1267! Nkom-Wum F.R. (June) *Daramola* FHI 41078! Bafut-Ngemba F.R. (July) *Daramola* FHI 41566! **F.Po:** 6,000 ft. (Dec.) *Mann* 647!

22. **P. seticaulis** *Rendle* in Cat. Talb. 104 (1913). Flowers white or pale yellow, tinged rose or purple.
**S.Nig.:** Oban *Talbot* 926! **W.Cam.:** Lohe, M'bonge (Sept.) *Wright* 164! Also in Gabon and Congo.

23. **P. calluniflora** *Kraenzl.* in Engl., Bot. Jahrb. 28: 166 (1900). Flowers white and rose.
**W.Cam.:** Cam. Mt., Buea, 3,000 ft. (Apr., May) *Deistel* 75! *Maitland* 734! *Gregory* 288! *Preuss* 1009. Also in E. Cameroun and Uganda.

24. **P. steudneri** *Rchb. f.* Otia Bot. Hamburg. 113 (1881); F.T.A. 7: 117. Flowers yellow or greenish yellow with red stripes or markings.
**N.Nig.:** Rukuba, Jos Plateau (Jan., Feb.) *King*! Pankshin to Mongu, Plateau Prov. *Cole* 18! Gembu, Mambila Plateau (Jan.) *Hepper* 1825! Also in Ethiopia, Sudan, Uganda and Kenya.

25. **P. kingii** *Summerh.* in Kew Bull. 17: 555, fig. 18 (1964). Flowers greenish white with yellow lip.
**W.Cam.:** Mambila Plateau (Apr.) *Howard* in *Hb. King* 141! *Nash* in *Hb. King* 181!

26. **P. adansoniae** *Rchb. f.* in Flora 48: 185 (1865); F.T.A. 7: 121; Cat. Talb. 146. *P. albo-violacea* Kraenzl. (1893)—F.T.A. 7: 120; F.W.T.A., ed. 1, 2: 431. *P. nigerica* Rendle, Cat. Talb. 103. *P. caillei* Guillaum. in Bull. Mus. Nat. Hist. Paris, 26: 672 (1920).
**Guin.:** Kouria (Aug.) *Pobéguin* 1675! Seoua, Pita (Sept.) *Adames* 3711 Nzérékoré (May) *Collenette* 33! *Schnell* 2785! Nimba Mts. (June) *Schnell* 2832b! **S.L.:** Yagoi, Jong R. (Aug.) *Adames* 66! Seli R., Kamadugu (June) *Deighton* 5938! Bafodia, Koinadugu (fr. Aug.) *Haswell* 93! **Lib.:** Sanokwele (fr. Sept.) *Baldwin* 9521! Yasona (Sept.) *Harley* 1928! Ganta (May) *Harley* 2132! Nimba (Sept.) *Adames* 517! (Feb.) *Adam* 20928! **Iv.C.:** Yaou (Mar.) *Chev.* 17760! Goureni (May) *Chev*! W. of Niapidou, N. of Sassandra (fr. Feb.) *Leeuwenberg* 2779! **Ghana:** Aburi (May, June) *Johnson* 223! *Irvine* 2831! Ofinso (May) *Vigne* FH 4867! Kamawu *Cox* 34! Juaso, Ashanti (May) *Cansdale* 94! Amedzofe (fr. Sept.) *Westwood* 46! **N.Nig.:** Bonu, Gwari, Niger Prov. (Jan.) *Onochie* FHI 40169! *E. W. Jones* 223! **S.Nig.:** Boyogbe, Benin Prov. (fl. cult. Ibadan, June) *Sanford* WS/216a, 221/66! Okomu F.R., Benin Prov. (Jan.) *Keay* FHI 22438! Oban *Talbot* 867! 929! Kundeve, Ogoja Prov. (May) *Ejiofor* FHI 25317! **W.Cam.:** Buea (Apr., May) *Maitland* 727! *Gregory* 293! Bibundi (Apr.) *Schlechter* 12409! Barombi (June) *Preuss* 298! Kake II, Kumba (fr. July) *Thorold* TN31! Also in Gabon, Congo, Uganda, Kenya, Tanzania and Angola.

27. **P. polychaete** *Kraenzl.* in Engl., Bot. Jahrb. 17: 50 (1893); F.T.A. 7: 120; Cat. Talb. 146. Flowers greenish white or greenish yellow, sometimes tinged reddish.
**S.L.:** Vaama, Barri (Dec.) *Deighton* 3848! Jaluahun, Barri (Nov.) *Deighton* 4391! **Lib.:** Peterstown (Oct.) *Bunting* 104! Tawata, Boporo Dist. (Nov.) *Baldwin* 10322! Ganta (Nov.) *Harley* 2067! Nimba (Oct.) *Adames* 726! **Iv.C.:** Yapo Forest, N. of Abidjan (Oct.) *de Wilde* 806! **Ghana:** Aburi (Oct.) *Johnson* 224! Agogo-Akim, Ashanti (Nov.) *D. Gillett* 8! **S.Nig.:** Oban *Talbot* 927! 928! **W.Cam.:** Buea, 3,000ft. (Mar.–May) *Maitland* 731! *Deistel* 89! Cam. Mt., 4,300–5,000 ft. *Preuss* 881! *Gregory* 89! Bamuko F.R., Kumba (Nov.) *Keay* FHI 37698! **F.Po:** Las Cortesas (Jan.) *Guinea* 2130! Bokoko, above Basilé, 3,300–4,200 ft. *Mildbr.* 6445! 6954. Moka (fr. Sept.) *Wrigley* 650! Also in E. Cameroun, Gabon, Congo, Uganda and Tanzania.

28. **P. ramulosa** *Lindl.* in Bot. Reg. 24: Misc. 76 (1838); F.T.A. 7: 118; Cat. Talb. 146. Flowers white, yellow or orange, tinged with pink.
**S.L.:** Sugar Loaf Mt. *cult. by Loddiges*! Picket Hill, 2,900 ft. (Nov.) *Jones* 172! Njala (July) *Deighton* 1742! Sanbaia (Nov.) *Deighton* 4335! **Lib.:** Loffa R., Boporo Dist. (fr. Dec.) *Baldwin* 10665! Bomi Hills (Sept.) *Voorhoeve* 438! Zuie, Boporo (fr. Nov.) *Baldwin* 12108! Gondolahun, Kolahun Dist. (fr. Nov.) *Baldwin* 10107! Nimba (June) *Adam* 21548! **Ghana:** Kanyankov, W. Prov. (Nov., Dec.) *Miles* 22! Assin Yan Kumasi (fr. Nov.) *Cummins* 73! **S.Nig.:** Sapoba, Benin Prov. (fl. cult. Ibadan, Apr.) *Sanford* WS/184/66! Degema *Talbot* 3647! Oban *Talbot* 864! Also in E. Cameroun, the Congos, Gabon and Uganda.

29. **P. puberula** *Lindl.* in Bot. Reg. 10: t. 851 (1824); F.T.A. 7: 114. Flowers yellow or yellow-green.
**Port.G.:** Catio, Quebil (May) *Esp. Santo* 2057! **Guin.:** Konikouré (June) *Pobéguin* 1624! Nzérékoré (Sept.) *Jac.-Fél.* 1248! **S.L.:** Njala (June) *Deighton* 4828! Ndilajula, Njala (May) *Deighton* 712! Kamakwie (June) *Deighton* 4829! Above Véré near Kinto, Loma Mts. (fr. Sept.) *Jaeger* 1778! Dilleh-Juleh rocks, R. Taia (June) *Bunting* 58! **Lib.:** Mt. Barclay (June) *Bunting* 5! Kailahun, Kolahun Dist. (fr. Nov.) *Baldwin* 10141! Ganta (May, Nov.) *Harley* 1944! 2133! **Iv.C.:** Adiopodoumé *Boyco* K223!
30. **P. odorata** *Lindl.* in J. Linn. Soc. 6: 130 (1862); F.T.A. 7: 113.
30a. **P. odorata** *Lindl.* var. **odorata**—Flowers white or yellow, tinged rose.
**Iv.C.:** Bouroukrou *Chev.* 16604! Sanrou to Ouodé (May) *Chev.* 21607! **U. Volta:** nr. Koualé, 3,000 ft. (May) *Chev.* 21740! **Ghana:** Offin Valley (fr. Oct., Dec.) *Miles* 13! Ksi-Mampong, Kumasi (Apr.) *Westwood* 106! Efiduasé *Cox* 58! **N.Nig.:** Mambila Plateau (Apr.) *Howard* in *Hb. King.* 154! **S.Nig.:** Onitsha *Barter* 1483! Sapoba *Kennedy* 1686! Shasha F.R., Ijebu (May) *Richards* 3409! Orukim to Unyene, Eket (May) *Onochie* FHI 33176! Lower Enyong F.R., Itu (May) *Onochie* FHI 33209! **W.Cam.:** Victoria (May, fr. Sept.) *Maitland* 729! *Ngongi* FHI 15085a! Bibundi (Apr.) *Schlechter* 12431! Buea, 3,000 ft. (May) *Preuss* 1218! Bamenda (Apr.) *King* 5! **F.Po:** (June) *Mann* 436! Also in E. Cameroun, Gabon, C. African Republic, Congo and Uganda.
30b. **P. odorata** var. **trilepidis** *Summerh.* in Kew Bull. 3: 433 (1949). Leaves very narrowly lanceolate; inflorescence narrow with short branches; side lobes of lip more obtuse than in type, middle lobe recurved; flowers yellow with paler lip; at edges of patches of *Trilepis pilosa.*
**S.Nig.:** Mt. Orosun, Idanre, Ondo Prov. (Apr., Aug., Oct.) *Jones* FHI 20729! *Keay & Onochie* FHI 21559! *Symington* FHI 3374!
31. **P. tessallata** *Lindl.* l.c. (1862); F.T.A. 7: 114; Cat. Talb. 146. *P. lehmbachiana* Kraenzl. in Engl., Bot. Jahrb. 28: 166 (1900). *P. praealta* Kraenzl. l.c. 36: 118 (1905); Stapf in Johnston, Lib. 654. Flowers greenish to pale yellow with orange to purple markings.
**Guin.:** Songuèta (June) *Pobéguin* Kindia *Jac.-Fél.* 10! Mt. Benna, 3,300 ft. (June) *Jac.-Fél.* 1773! Macenta, 2,000–2,500 ft. (fr. Oct.) *Baldwin* 9822! **S.L.:** Gloucester (Aug.) *Deighton* 150! Dilleh-Juleh rocks, R. Taia (Aug.) *Deighton* 4333! Njala (Aug.) *Deighton* 4332! Bonthe (Aug.) *Adames* 60! **Lib.:** Grand Bassa (Sept.) *Dinklage* 2069! Woeme, Vonjama Dist. (fr. Oct.) *Baldwin* 10109! Flumpa, Sanokwele Dist. (fr. Sept.) *Baldwin* 9370! Ganta (July) *Harley* 1646! White Plains Road (June) *Barker* 1343! **Iv.C.:** Brafouédi, NW. of Abidjan (fr. Dec.) *Leeuwenberg* 2312! Danané (June) *Collenette* 42! 50! 51! **Ghana:** Aburi (May, June) *Irvine* 3031! Sanrou to Ouodé (May) *Chev.* 21626! Efiduase (May) *Cox* 23! Tarkwa (July) *Vigne* FH 4467! Juaso, Ashanti (May) *Cansdale* 95! Amedzofe (May) *Scholes* 71! **N.Nig.:** Kushanka, Zaria Prov. (Apr.) *King* 54! Omewo (July) *Elliott* 80! **S.Nig.:** Nun R. (Aug.) *Mann*! Akure F.R., Ondo (Aug.) *Keay* FHI 25363! Sonkwala, Ogoja (July) *Ejiofor* FHI 25344! Stubbs Creek F.R., Eket (May) *Onochie* FHI 33190! Calabar (Sept.) *Keay* FHI 13348! **W.Cam.:** Buea, 3,000 ft. (Apr., May) *Maitland* 724! *Dundas* FHI 15339! *Gregory* 101! Victoria (Sept.) *Ngongi* FHI 15085! Bapinyi, Bamenda (fr. June) *Ujor* FHI 30423! **F.Po:** Bokoko *Mildbr.* 6917. Throughout tropical Africa, (? Kenya), southwards to Angola, Zambia and Malawi.
32. **P. mukandaensis** *De Wild.* Not. Pl. Util. Congo 1: 139 (1903). *P. plehniana* Schltr. in Engl., Bot. Jahrb. 38: 8 (1905). *P. dorotheae* Rendle, Cat. Talb. 103 (1913). Flowers yellow-green or whitish green with brown-purple markings.
**Ghana:** Kumasi (Aug., fr. Dec.) *Vigne* FH 1522a! *Cox* 101! Bechem (Aug., Sept.) *Westwood* 111! Mampong, Ashanti (Oct.) *Cansdale* 129! Alavanyo F.R., Hohoe (Sept.) *St. Cl.-Thompson* 1473! Odomi, Jasikan (Aug.) *Westwood* 26! **S.Nig.:** Onika Olono (fr. Oct.) *Thomas* 1902! Olokemeji F.R., Abeokuta (June) *Keay* FHI 22410! Onitsha (fr. Feb.) *Jones* FHI 6128! Oban *Talbot* 861! Obudu scarp, Ogoja *Cooper* 80! Also in E. Cameroun, Congo, Angola and Uganda.
33. **P. modesta** *Rchb. f.* in Flora 50: 114 (1867); F.T.A. 7: 116. Flowers whitish or greenish yellow, sepals often pink or purplish.
**Ghana:** *Cox* 52! **N.Nig.:** Zaria *Cole* 13! Mongu F.R., Plateau Prov. (July) *King* 31! Sanga River F.R., Zaria Prov. (Sept.) *Keay* FHI 22443! **S.Nig.:** Eleyele, Ibadan (Aug.) *Keay* FHI 22429! East to Uganda, south to Angola, Zambia and Malawi.
34. **P. golungensis** *Rchb. f.* in Flora 48: 185 (1865); F.T.A. 7: 118. *P. johnsonii* Kraenzl. in Kew Bull. 1926: 291. Flowers yellow or greenish yellow, sometimes tinged pink or brown, colour intensifying with age.
**Mali:** Diaragonéla (Feb.) *Chev.* 473! **Lib.:** Beidin (Dec.) *Harley* 1844! **Iv.C.:** Bouroukrou (Dec., Jan.) *Chev.* 16906! **Ghana:** Aquapim Hills (Dec.) *Johnson* 588! Juaso, Ashanti (Nov.) *Cansdale* 131! Awaso *Blayney* 16! Osien (Nov.) *Westwood* 61! **N.Nig.:** Kogin Delli, Zaria Prov. (Jan.) *King* 22! Kaciya, Zaria Prov. (Jan.) *G. V. Summerhayes* 146! Kabba Prov. (Mar.) *Westwood*! **S.Nig.:** Olokemeji F.R., Abeokuta (Oct.) *Keay* FHI 22712! Aponmu, Akure F.R., Ondo (Dec.) *Keay* 28294! Obudu (Jan.) *Savory & Keay* FHI 25281! Tropical Africa southwards to Angola, Zambia and Tanzania.
35. **P. dolichophylla** *Schltr.* in Engl., Bot. Jahrb. 38: 8 (1905). *P. hamiltonii* W.W. Sm. in Notes Bot. Gard. Edin. 8: 347 (1915). *P. guezorum* A. Chev., Bot., 617 (1920), name only. *P. simoniana* Kraenzl. in Fedde Rep., Beih. 39: 61 (1926). *P. oxychila* Schltr. ex Kraenzl. l.c. 98 (1926). Flowers yellow or cream-coloured; sometimes terrestrial in *Trilepis* mats.
**Guin.:** Moribadou to Nionmoradougou (Mar.) *Chev.* 20950! **S.L.:** Musaia (May) *Deighton* 4309! 5054! Makene (May) *Deighton* 4637! Njala (Mar.–May) *Deighton* 4599! 4672! **Ghana:** Kumawu, Ashanti (Mar.) *Cansdale* 80! FH 4369! Jasikan (Apr.) *Westwood* 66! **N.Nig.:** S. of Benue R. *Hamilton*! Nindam F.R., Zaria (Apr.) *King* 39! **S.Nig.:** Osun F.R., Ijebu Prov. (Apr.) *Latilo* FHI 26118! Akure F.R., Ondo (fr. Aug.) *Jones* FHI 19543! Mt. Orosun, Idanre Hills, Ondo (Mar., Apr.) *Symington* FHI 3373! *Keay* FHI 25304! *Coombe* 154! **W.Cam.:** Man o' War Bay, S. of Victoria *Baum*! Victoria (Feb.) *Simon* 14! Buea, 3,000 ft. (Jan.) *Schlechter* 12837! *Dunlap* 145! Also in E. Cameroun and Gabon.
36. **P. oblanceolata** *Summerh.* in Kew Bull. 10: 231 (1956). Flowers bright pink; on boulders.
**Guin.:** Gangan, Kindia (May) *Roberty* 17764! *Jac.-Fél.*! **S.L.:** Loma Mts., 3,000 ft. (Mar.) *Morton & Gledhill* SL 1048!
37. **P. subulata** *Finet* in Notulae Syst. 2: 26 (1911). *P. inaperta* Guillaum. in Bull. Mus. Hist. Nat. 29: 543 (1923). Flowers cream, white or greenish white, mentum or callus sometimes violet.
**Guin.:** Tabelli *Mann*! Boulou Kountou (June) *Pobéguin* 1625! **S.L.:** Sefadu (July) *Deighton* 4830! 5053! **Iv.C.:** Sanrou to Ouodé, Toura country (May) *Chev.* 21601! Ouodé to Gouréni (May) *Chev.* 21637! **N.Nig.:** Ribako F.R., Zaria Prov. (May) *Keay* FHI 22898! Mongu F.R., Plateau Prov. (May, June) *King*! Kaciya to Zonkwa, Zaria (Apr.) *G. V. Summerhayes* 91!
38. **P. leonensis** *Rchb. f.* Otia Bot. Hamburg. 112 (1881); F.T.A. 7: 107. Flowers brownish or yellowish green, lip white.
**Guin.:** Nimba Mts., 5,000 ft. (Feb., fr. Apr.) *Schnell* 309! 309b! 1024! **S.L.:** (May) *Barter*! Sugar Loaf Mt. (May) *Sc. Elliot* 5822! *Luke* 2010! *Tindall* 64! Picket Hill, 1,300 ft. (June) *Deighton* 5917! **Lib.:** Mt. Wolagwisa, Pandamai, 4,500 ft. (Mar.) *Bequaert* 99! **Iv.C.:** Abidjan *Aké Assi* L439! Mt. Tonkoui, SW. of Man, 3,300–3,800 ft. (Feb., Mar.) *Leeuwenberg* 2937! *de Wit* 9149! **W.Cam.:** Ambas Bay *Lord Scarborough*! Also in E. Cameroun.
39. **P. parva** *Summerh.* in Bot. Mus. Leafl. Harv. Univ. 10: 285 (1942). Flowers white, tinged greenish, lip sometimes orange or purple, anther purple.
**Ghana:** Amedzofe (Aug.–Oct.) *Westwood* 39! 39b! **N. Nig.:** Mambila Plateau (June) *Nash* in *Hb. King* 198! Also in Zambia.
40. **P. affinis** *Lindl.* Gen. & Sp. Orch. Pl. 73 (1830); F.T.A. 7: 126. Flowers yellow with red or brownish markings, fragrant.

**Guin.:** Sambadougou, Faranah (Jan.) *Chev.* 20550! Faranah (Jan.) *Chev.* 20478! **S.L.:** Makuta *Thomas* 483! Kenema Town (May) *Lane-Poole* 226! Karina (Jan.) *Glanville* 150! Wallia (fr. Jan.) *Sc. Elliot* 4627b! Makali (Jan.) *Deighton* 4567! **Lib.:** *Adam* 20761! Bolahun (Feb.) *Bequaert* 61! Ganta (Feb.) *Harley* 1772! **Iv.C.:** Tiapleu Forest (cult. Abidjan, Feb.) *Aké Assi* 9418! **Ghana:** Brafu Edru (Jan.) *Cummins* 15! Efiduase (May) *Cox* 31! Amedzofe (Jan.) *Vohringer*! *Westwood* 32! **S.Nig.:** Abeokuta to Ijebu Ode (Jan.) *Burtt* B27! Usonigbe F.R., Sapoba (Nov.) *Keay* FHI 24666! Owo, Ondo Prov. (Sept.) *Keay* FHI 22714! Onitsha *Barter* 1863! Olibo, Degema *Talbot* 3776! Also in E. Cameroun, Gabon, C. African Republic, Congo and Angola.

41. **P. obanensis** *Rendle* Cat. Talb. 102, t. 13, fig. 1–2 (1913). Flowers cream or yellow.
**Lib.:** Ganta (June) *Harley* 1645! **Iv.C.:** Mt. Tonkoui (Oct.) *Aké Assi* 6026! **S.Nig.:** Oban *Talbot* 930! Also in E. Cameroun.

42. **P. camaridioides** *Summerh.* in Kew Bull. 11: 111 (1957). About 2–3 in. tall; flowers very pale yellow or greenish yellow.
**S.Nig.:** Calabar (July) *Wright* 101!

43. **P. microbambusa** *Kraenzl.* in Kew Bull. 1926: 245. *P. cyperacearum* A. Chev., Bot. 616 (1920), name only. *Nienokuea lutea* A. Chev. l.c. 622, name only. *N. bambusoides* A. Chev. (1945). *N. microbambusa* (Kraenzl.) A. Chev. (1945). A yellow-flowered bamboo-like herb growing on granite outcrops often on or among the roots of *Trilepis pilosa.*
**Guin.:** Nzo Mt., 5,450 ft. (Mar.) *Chev.* 21037! Macenta, 2,000–2,500 ft. (Oct.) *Baldwin* 9810! Nimba Mts. (Feb.) *Schnell* 323! **S.L.:** Selu, Valunia (July) *Deighton* 5817! Makeni Hill (Aug.) *Jordan* 500! Sefadu (May–Sept.) *Small* 66! Mandu, Bo to Mongeri (July) *Deighton* 1965! Bintumane Peak, Loma Mts., 6,000 ft. (May–Aug.) *Jaeger* 1046! Kukunia Mt., 4,300 ft. (Sept.) *Jaeger* 1676! **Lib.:** Palilah, Gbanga (Aug.) *Baldwin* 9150! Sanokwele (Sept.) *Baldwin* 9520! 20 miles E. of Zwedru, Tchien (Aug.) *Baldwin* 7076! Nimba (Feb.) *Adam* 20881! **Iv.C.:** Mt. Nienokué, 1,700 ft. (Sept.) *Chev.* 19466! Bouaké, Mt. Lémélébon (July) *Fleury* in *Hb. Chev.* 22093! Mt. Dou, 4,000 ft. (May) *Chev.* 21471! Toulepleu (May) *Schnell* 1254! Nimba Mts., 2,500–5,000 ft. (Aug.) *Boughey* GC 18088!

44. **P. superposita** *Rchb. f.* Otia Bot. Hamburg. 111 (1881); F.T.A. 7: 129. Flowers reddish.
**W. Cam.:** Cam. Mt., 5,000 ft. (Nov.) *Mann* 2125! Buea (Oct.) *Deistel.* **F.Po:** Baca Peals (Jan.) *Sanford* 4281!

45. **P. fusiformis** (*Thou.*) *Lindl.* in Bot. Reg. 10: sub. t. 851 (1824); F.T.A. 7: 129. *Dendrobium fusiforme* Thou. (1822). *Polystachya composita* Kraenzl. in Fedde, Rep. Beih. 39: 103 (1926); F.W.T.A., ed. 1, 2: 429. Flowers greenish suffused with purple, to deep purple or plum-coloured.
**W.Cam.:** Buea, 3,000 ft. (June) *Preuss* 1072! *Gregory*! **F.Po:** Moka, SE. of island, 4,000–6,000 ft. (Nov.) *Mildbr.* 7068! Las Corteras (fr. Jan.) *Guinea* 2138! Also in Congo, Uganda, Kenya, Tanzania, Zambia, Madagascar and Réunion.

46. **P. crassifolia** *Schltr.* in Engl., Bot. Jahrb. 38: 7 (1905). A creeping epiphyte with pale brownish flowers.
**W.Cam.:** Moliwe (Jan.) *Schlechter* 12841!

47. **P. mystacioides** *De Wild.* Not. Pl. Util. Congo 1: 133 (1903); Kraenzl. in Fedde Rep., Beih. 39: 10 (1926). A creeping or hanging epiphyte; flowers white with red or purplish markings.
**Iv.C.:** Tai Forest (Oct.) *Aké Assi* 6030! Also in E. Cameroun and Congo.

48. **P. supfiana** *Schltr.* in Engl., Bot. Jahrb. 38: 10, fig. 4 (1905). Flowers yellow with brown markings.
**S.Nig.:** Oban *Talbot*! Kwa R. (Mar.) *Talbot* 440! Eket *Talbot* 3306! **W.Cam.:** Bibundi (Apr.) *Schlechter* 12415! Also in Gabon.

49. **P. galeata** (*Sw.*) *Rchb. f.* in Walp. Ann. 6: 637 (1863). *Dendrobium galeatum* Sw. (1805). *Polystachya grandiflora* Lindl. (1839)—F.T.A. 7: 127; Chev. Bot. 617. Flowers large, helmet-shaped, white, green, yellow or yellow-green with red or purplish markings.
**Guin.:** Benna, Kindia, 3,000 ft. (June) *Jac.-Fél.* 1776! Milo, Macenta (Apr.) *Adam* 11876! Diani R. (fr. Jan.) *Schnell* 182! **S.L.:** Ndialajula, Njala (May) *Deighton* 711! Benduma, Mongeri (July) *Deighton* 2220! Musaia (Apr.) *Deighton* 4636! Mesima (Apr.) *Deighton* 3697! Baiima, Gbo (May) *Deighton* 5765! **Lib.:** Ganta (Mar.) *Harley* 1633! Begwai, St. Johns R. (Sept.) *Bunting* 9! Peter's Town (Oct.) *Bunting* 93! Yah R., Sanokwele Dist. (Sept.) *Adames* 537! **Iv.C.:** Man, Dyolas country (May) *Ripert* in *Hb. Chev.* 21533! Mt. Nienokoué, 1,500 ft. (July) *Chev.* 19495! Banco F.R. (fl. Nov., fr. Sept.) *Paulian* 47! *de Wilde* 3210! **S.Nig.:** Okomu F.R., Benin (Feb.) *Brenan* 9116! *Keay* FHI 22738a! Ajagbodudu, Sapele (May, June) *Wright* 128! Degema *Talbot* 3726! Ogoja Plain (July) *Cooper* 82! Also in Gabon and Congo.

50. **P. cultriformis** (*Thou.*) *Spreng.* Syst. Veg. 3: 742 (1826). *Dendrobium cultriforme* Thou. (1822). *Polystachya cultrata* Lindl. (1824)—F.T.A. 7: 109. *P. appendiculata* Kraenzl. in Notizbl. Bot. Gart. Berl. 3: 238 (1903); F.W.T.A., ed. 1, 2: 431. Flowers white, yellowish, pink or purplish.
**W.Cam.:** Buea, Cam. Mt., 3,000–4,000 ft. (Nov.) *Preuss* 1009! *Migeod* 75! *Gregory* 574! *Lehmbach.* Bafut-Ngemba F.R. (Dec.) *Dioh* FHI 41109! Mbiami, 6,000 ft. (June) *Brunt* 762! **F.Po:** Moka, 4,600 ft. (fr. Sept.) *Melville* 431! Lava flow (Sept.) *Melville* 449! Also in Gabon, Congo, Ethiopia and E. Africa, southwards to Rhodesia and Mozambique, Madagascar and Mascarene Islands.

51. **P. tenuissima** *Kraenzl.* in Engl., Bot. Jahrb. 19: 250 (1894); F.T.A. 7: 110. Flowers pale yellow.
**Iv.C.:** Mt. Bono, Zoaulé (fr. May) *Chev.* 21467! **Ghana:** Amedzofe, 2,800 ft. (Aug.) *Westwood* 39d! Also in E. Cameroun.

52. **P. inconspicua** *Rendle* in J. Linn. Soc. 37: 218 (1905). *P. ashantensis* Kraenzl. in Kew Bull. 1926: 294; F.W.T.A., ed. 1, 2: 431. Flowers yellow or greenish-yellow marked with purple or brown, or almost entirely purplish.
**Iv.C.:** Tiapleu Forest, (Oct.) *Aké Assi* 9227! **Ghana:** Agogo (Sept.) *Chipp* 578! Nyinahin range (June) *Morton* A3401! Mampong, Ashanti (Oct.) *Cansdale* 127! Kamawu *Cox* 43! Begoro, Eastern Prov. (Sept.) *Westwood* 39a! Amedzofe (Nov.)*Westwood* 53! *Morton* A3415! Also in Congo, Uganda, Kenya and Tanzania.

53. **P. bicalcarata** *Kraenzl.* in Engl., Bot. Jahrb. 36: 118 (1905). A small densely-tufted epiphyte with white and purple or rose flowers.
**W.Cam.:** Cam. Mt., Buea and above, 3,000–6,000 ft. (Aug., fr. May) *Deistel* 62c! 79! *Maitland* 730! *Dundas* FHI 15303! **F.Po:** Biao Crater, 6,300 ft. (Sept.) *Melville* 495!

54. **P. bequaertii** *Summerh.* in Kew Bull. 1: 131 (1948). Flowers purplish red.
**S.L.:** Mt. Da-Oulen, Loma Mts., 5,000 ft. (Nov.) *Jaeger* 660! Sankan Biriwa summit, 6,000 ft. (Jan.) *Cole* 138! **Lib.:** Ballawallah, Kailahun to Bolahun (Feb.) *Bequaert* 70! Mt. Mpaka Fossa, Kolahun, 1,960 ft. (Jan.) *Bequaert* 40!

## 25. STOLZIA Schltr. in Engl., Bot. Jahrb. 53: 564 (1915).

Peduncle very slender, setaceous, 2–3 cm. long; sepals and petals very long acuminate, 5·5–7 mm. long; leaves oblanceolate, 1–2 cm. long, distinctly narrowed towards base; flowers greenish-white; pseudobulbs more or less globose with depressed centre so as often to appear bilobed .. .. .. .. .. .. 1. *elaidum*

Peduncle rather stout, very short, scarcely exceeding 5 mm.; sepals and petals obtuse or acute, 3–7 mm. long; leaves elliptical or nearly orbicular, almost sessile, 5–12 mm. long, not at all narrowed at base; flowers orange, ochre-coloured or reddish, often with darker red stripes; pseudobulbs scarcely thickened, club-shaped 2. *repens*

1. **S. elaidum** (*Lindl.*) *Summerh.* in Kew Bull. 17: 557 (1964). *Bulbophyllum elaidum* Lindl. (1862)—F.T.A. 7: 29; F.W.T.A., ed. 1, 2: 439. A low creeping epiphyte with spaced-out pseudobulbs bearing pairs of leaves and single greenish white flowers.
   **Lib.:** Duport (Nov.) *Linder* 1495! Begwai to Peter's Town (Oct.) *Bunting* 100! Jaurazon, Sinoe (Mar.) *Baldwin* 11473! **S.Nig.:** Brass R. *Barter* 73! 1841! Also in Principe.

2. **S. repens** (*Rolfe*) *Summerh.* in Kew Bull. 7: 141 (1953). *Polystachya repens* Rolfe in Kew Bull. 1912: 132. A low creeping plant with pairs of leaves at intervals and single orange or red flowers.
   **S.Nig.:** Sapoba, Jamieson R. (Sept., Dec.) *Onochie* FHI 34269! *Sanford* 145/66! **W.Cam.:** Bamenda, 6,000 ft. *Gregory* 257! Also in Congo, Uganda, Kenya, Tanzania and Zambia.

## 26. CALANTHE R. Br. in Edw., Bot. Reg. t. 573 (1821); F.T.A. 7: 45. *Nom. cons.*

Rhizome stout; leaves 3–5, long-petiolate, lanceolate or elliptical-lanceolate, acuminate, lamina 15–35 cm. long, 4–13 cm. broad, petiole 10–17 cm. long, dilated at the base; inflorescence erect, 30–75 cm. high; flowers white, mauve or pink; sepals oblong-lanceolate, acute, 12–18 mm. long; lip trilobed at the base, 8–14 mm. long, side lobes short, rounded, front lobe divaricately bilobed with a short claw, side lobules oblong, more or less toothed on the outside, disk with a warty crest in front of the column; spur 12–28 mm. long, slender .. .. .. .. .. .. .. *corymbosa*

**C. corymbosa** *Lindl.* in J. Linn. Soc. 6: 129 (1862); F.T.A. 7: 46. *C. delphinioides* Kraenzl. (1893)—F.T.A. l.c. Terrestrial, in forests.
   **Guin.:** Dalaba to Diaguissa, 3,000–4,000 ft. (Sept., Oct.) *Chev.* 18852! **W.Cam.:** Cam. Mt., 5,000 ft. *Johnston* 107! Mimbia, 3,600 ft. *Preuss* 1061! Fonfuka, Bamenda, 3,000 ft. (June) *Maitland* 1785! Bafut-Ngemba to Bamenda Nkwe (May) *Daramola* FHI 41063! **F.Po:** 5,000–6,000 ft. (Dec.) *Mann* 392! *Boughey* 133! Moka, 4,600 ft. (Sept.) *Wrigley* 567! Also in S. Tomé, Annobon, Gabon, Congo, Uganda and Angola.

## 27. ANCISTROCHILUS Rolfe in F.T.A. 7: 44 (1897).

Pseudobulbs orbicular, flattened, up to 2·5 cm. diam.; leaves narrowly lanceolate or elliptical-lanceolate, up to 21 cm. long and 3 cm. broad; sepals and petals lanceolate, acuminate, 3·5–4·5 cm. long, white, greenish at base; lip middle lobe 16–25 mm. long, very slender, often more or less recurved, bright purple, side lobes green with brown markings .. .. .. .. .. .. .. 1. *thomsonianus*
Pseudobulbs conical or pyriform, up to 5 cm. diam.; leaves lanceolate, elliptical-lanceolate or oblanceolate, up to 40 cm. long and 7·5 cm. broad; sepals broadly lanceolate to oblanceolate, acute or obtuse, 2–3 cm. long; petals oblanceolate, distinctly narrower than the sepals but only slightly shorter; all tepals rose-coloured to mauve, rarely almost white; lip middle lobe 8–12 mm. long, purple with a yellow apex, side lobes green with brown markings .. .. .. 2. *rothschildianus*

1. **A. thomsonianus** (*Rchb. f.*) *Rolfe* in F.T.A. 7: 44 (1897); Cat. Talb. 146, partly. *Pachystoma thomsonianum* Rchb. f. (1879). In forests.
   **S.Nig.:** Calabar *Kalbreyer*! Oban *Talbot* 88! **W.Cam.:** Victoria (Sept.) *Schlechter* 15762! Also in E. Cameroun.

2. **A. rothschildianus** *O'Brien* in Gard. Chron. ser. 3, 41: 51, fig. 24 (1907). In forests.
   **Guin.:** (Oct., Nov.) *Harley* 1605! 1840! **S.L.:** Denkali R., Loma Mts., 2,700 ft. (Oct.) *Jaeger* 271! (?) Kono *Bowden*! **Lib.:** Wohmen, Vonjama Dist. (Oct.) *Baldwin* 10077! near Bolahun *Earthy* 26! **Iv.C.:** Touba to Man (Nov.) *Aké Assi* 2126! Gagnoa to Sassandra (Dec.) *de Wilde* 3341! **S.Nig.:** Jamieson R., Sapoba (Nov.) *Ross* 227! Agba to Urbehue, Usonigbe, Benin (Oct.) *Ujor* FHI 15290! Usonigbe F.R., Sapoba (Nov.) *Ejiofor* FHI 24654! Sapoba (Dec.) *Ujor* FHI 23941! Owam F.R., Benin (Nov.) *Keay* FHI 22688! **W.Cam.:** M'bonge *Wright* 58/4! Also in Uganda.

## 28. BULBOPHYLLUM Thou., Orch. Iles. Austr. Afr. tt. 93–97 (1822); F.T.A. 7: 22, including *Megaclinium* Lindl. *Nom. cons.*

Pseudobulbs normally 1-leaved: (to p. 230)
   Leaves terete, over 15 cm. long, 1·5–2 mm. diam.; pseudobulbs narrowly cylindrical, 2·5–3 cm. long; scapes much shorter than leaves, flowering nearly from base, rhachis rather swollen but not much flattened, bracts 2 mm. long .. 14. *teretifolium*
   Leaves flat:
      Lip with pink, red or brown hairs at least as long as half the width of the lip and usually much longer, linear or tongue-shaped; sepals nearly similar to each other, narrowly lanceolate, 6–18 mm. long; petals linear-subulate, quite glabrous; column with long subulate stelidia:
         Longer hairs on lip with club-shaped swollen ends; pseudobulbs much flattened, elliptical or circular in outline, 1·5–3 cm. long, 1·5–2·7 cm. broad; inflorescence 9–20 cm. long, laxly many-flowered, bracts spreading; leaves oblong or elliptical, 2·5–11 cm. long, 1–2·8 cm. broad .. .. .. .. .. 28. *barbigerum*
         Longer hairs on lip not club-shaped or swollen at ends; bracts not spreading:
            Inflorescence shorter than the leaves, 5–9 cm. long, densely flowered in the upper half or two-thirds; flowers all open at once; sepals 6–7 mm. long; pseudobulbs ovoid, or conical-ovoid, flattened, obscurely 3–4-angled; leaves elliptical or oblong, 4·5–12 cm. long, 1·5–2·5 cm. broad; bracts 5–7 mm. long; lip with a shallow U-bend in the middle .. .. .. .. .. 23. *saltatorium*
            Inflorescence longer, usually much longer, than the leaves, 15–60 cm. long, flowering in the upper half or quarter only; flowers opening singly or a few out together; sepals over 7 mm. long; lip without a shallow U-bend in the middle:

Fig. 391.—Ancistrochilus rothschildianus *O'Brien* (Orchidaceae).

A, habit, × ⅔. B, flowering spike, × ⅔. C, flower, × ⅔. D, column and lip, side view, × 1. E, column and lip, front view, tepals removed, × 2. F, column, anther cap and pollinia removed, × 6. G, anther cap, dorsal view, × 6. H, anther cap with pollinia, ventral view, × 12. I, pollinia, × 2. F, column, anther cap and pollinia removed, × 6. G, anther cap, dorsal view, × 6. H, anther cap with pollinia, ventral view, × 12. I, pollinia, × 6. A from *Keay* 22688. B from *Eggeling* 2307 and 5462. C—I from *Keay* 22688.

227

Bracts 4–15 mm. long, 2–4 mm. broad, greenish; flowers opening singly; leaves 7–19 cm. long, 1·5–4 cm. broad; hairs on lip deep red or purplish; pseudobulbs conical or ovoid, those first produced lenticular, later ones 3–4-angled, 1–4 cm. long, 1–2·5 cm. diam.; base of lip fleshy:

    Longer hairs on lip arising from underside of margins only, margins densely pubescent, upper surface glabrous; lip tongue-shaped, fleshy 25. *calamarium*

    Longer hairs on lip arising from margins both above and beneath and also from the surfaces; lip linear, not very fleshy, often with 2 reflexed papillose more or less hyaline auricles on the fleshy base just above the point of attachment
                                  24. *distans*

Bracts 17–30 mm. long, 6–10 mm. broad, yellow, pink or orange; several flowers open at once, or rarely flowers opening singly; leaves 15–30 cm. long, 2·5–5·5 cm. broad; hairs on lip red-brown, purplish-brown or chocolate; pseudobulbs ovoid, 3–4-angled, 2–4 cm. long, 1·5–3 cm. diam.:

    Petals 6·5–7·5 mm. long ..    ..    ..    ..    ..    ..    26. *phaeopogon*
    Petals 4·5–5·2 mm. long ..    ..    ..    ..    ..    ..    27. *schinzianum*

Lip glabrous, papillose, or if ciliate or hairy the hairs colourless:

    Sepals with long hairs, 8–11 mm. long; petals lanceolate or spathulate-lanceolate, densely papillose, 3–4 mm. long; lip very fleshy, densely papillose; stelidia of column obtuse; rhachis more or less swollen and fleshy, pendulous:

        Sepals hairy over the whole surface, only slightly keeled at back; scape 6–13 cm. long; leaves 10–24 cm. long, 1·3–2·7 cm. broad    ..    ..    16. *comatum*

        Sepals with prominent hairy keels on outside, otherwise only pubescent; scape 3·5–7·5 cm. long; leaves 8–11 cm. long, 1·6–4 cm. broad    ..    17. *inflatum*

    Sepals glabrous or very shortly papillose, or pubescent:

        Rhachis fleshy and more or less flattened; bracts reflexed after flowering; scape 15–45 cm. long, robust:

            Rhachis flattened, not very fleshy, 5–12 mm. wide; bracts very narrow, subulate from a broader base, 3–6 mm. long; pseudobulbs flattened, ovoid or ellipsoid, 3–4-angled, 1–6 cm. long; leaves oblong-elliptical, 4–21 cm. long, 8–34 mm. broad; sepals acuminate, dorsal 1 cm. long; lip very fleshy with often serrulate keels; scape 20–45 cm. long; column with broad rounded wings
                                  57. *colubrinum*

        Rhachis fleshy, not much flattened, 4–7 mm. broad; bracts as broad as the rhachis, deltoid-ovate, obtuse; lip much curved, with rounded basal lobes, and a thin longitudinal keel underneath; lateral sepals falcate-ovate, recurved:

            Lip entire at the base; petals 1·8 mm. broad, 3-nerved, not papillose; dorsal sepal lanceolate, acute, 8·5 mm. long; pseudobulbs ovoid, 3-angled, 1–2 cm. long; leaves 7–9 cm. long, 1·5 cm. broad    ..    62. *magnibracteatum*

            Lip pectinate at the base; petals 0·6 mm. broad, 1-nerved, papillose; dorsal sepal lanceolate, 7·5 mm. long; pseudobulbs elongate-ovoid, 3–4-angled, 2–4 cm. long; leaves 10–20 cm. long, 1·3–2·2 cm. broad    ..    .. 65. *linderi*

    Rhachis terete, not or only slightly thicker than the peduncle; bracts sometimes spreading but not reflexed:

        Pseudobulbs very small, not exceeding 1 cm. in length, usually spaced out on a slender creeping rhizome; leaves linear or narrowly ligulate, 3–7 mm. broad:

            Scape very slender, up to 18 cm. long, flowers up to 12, sometimes only 2 or 3, usually widely spaced; bracts usually very small, up to 3 mm. long, shorter than the flowers; leaves 1–10 cm. long; sepals 2·5–6 mm. long; petals oblong or broadly oblanceolate; lip entire or more or less trilobed at the base, the side lobes and margins often ciliate    ..    ..    ..    15. *intertextum*

            Scape stout, 2–5 cm. long, peduncle almost covered by overlapping sheaths; flowers in a dense spike; bracts 3–6 mm. long, nearly as long as the flowers; leaves 1·5–5 cm. long; sepals 4·5–6·5 mm. long; petals linear or linear-lanceolate; lip entire, papillose but not ciliate    ..    ..    .. 8. *pipio*

    Pseudobulbs exceeding 1 cm. in length, adjacent to or at short distances from one another (occasionally small plants will have pseudobulbs less than 1 cm. long); leaves lanceolate, ligulate, oblong or elliptical, usually more than 1 cm. wide:

        Lip triloded at base, side lobes separated from larger middle lobe by narrow acute sinuses; column furnished with a sharp tooth-like projection at the base of each stelidium, and with a rounded knob-like projection in the centre of the foot; pseudobulbs 1·5–4 cm. long, conical-ovoid; leaves narrowly strap-shaped, 8–20 cm. long, 5–14 mm. broad; peduncle very short, inflorescence flowering nearly to base; sepals 3·5–6 mm. long; petals narrowly lanceolate, acute, 1–2·2 mm. long; lip lobes ciliolate or denticulate
                                  7. *nigritianum*

Lip entire or with rounded very obscure basal lobes:

    Margins of lip entire or shortly papillose, not ciliolate:

        Petals lanceolate or broadly lanceolate with a short stalk and markedly papillose margins; lip rounded and cordate at the base, very acute at the

apex, with 2 calli near the base; peduncle very short, inflorescence flowering nearly to base, 8–24 cm. long; leaves elliptical, rounded at apex, 2–7 cm. long, 8–20 mm. broad; sepals 4–6·5 mm. long, white to pale yellow
                                                                9. *buntingii*
Petals not stalked; lip not cordate at base nor sharply acute at apex:
  Sepals 2·5–5 mm. long, broadly lanceolate; petals oblong or elliptical, 1·5–2·5 mm. long; inflorescence flowering in upper half only; leaves elliptical-oblanceolate, obtuse or acute at apex, 3–12 cm. long, 1–2 cm. broad; lip elliptical or ovate, rounded at base, papillose at apex ..        10. *recurvum*
Sepals exceeding 5 mm. in length, narrowly lanceolate:
  Stelidia shorter than the column (the latter measured from its base to the apex of the filament); petals oblong-lanceolate, acute:
    Leaves with a marked petiole 1·5–3 cm. long, 11·5–20 cm. long, 1–3 cm. broad; pseudobulbs narrowly ovoid or conical-ovoid, 2–3 cm. long; sepals orange in apical part, lip pale orange; sepals 8–12 mm. long; inflorescence 15–35 cm. long; petals 3·5–4·5 mm. long        4. *josephii*
    Leaves with a short petiole 5–10 mm. long; pseudobulbs ovoid, up to 2·5 cm. long; sepals magenta, red-purple or pale yellow-green, lip deep magenta or red; inflorescence 11–20 cm. long:
      Sepals 6–7 mm. long, 1·5–2 mm. broad; leaves 7–12 cm. long, 1 cm. broad; petals 2·5–3 mm. long; sepals magenta or red-purple with white base, lip deeper magenta; pseudobulbs 1–1·5 cm. diam...        5. *modicum*
      Sepals 10 mm. long, 2·5–4 mm. broad; leaves 12–19 cm. long, 3–4·5 cm. broad; petals 5·5 mm. long; sepals pale yellow-green, lip red with yellow base; pseudobulbs 2–3 cm. diam. ..    ..    .. 6. *calvum*
  Stelidia longer than the column; leaves oblanceolate or oblong, often narrowed for some distance at both ends; petals oblong, obtuse or subacute:
    Flowers white, lip sometimes yellowish; sepals 6·5–10 mm. long, laterals suddenly widened at base; lip about the same width all along, more or less V-shaped in section; pseudobulbs ovoid, 1–2 cm. long; leaves oblanceolate or oblong, 4·5–12 cm. long, 1·3–2·3 cm. broad, obtuse or rounded at apex; scape 9–19 cm. long    ..    .. 2. *schimperanum*
    Flowers greenish-yellow, sometimes tinged reddish or purplish or entirely red-purple; sepals 5–9 mm. long, widened gradually at base; lip narrowing towards the front and papillose; pseudobulbs conical or narrowly conical, 1–2·5 cm. long; leaves oblanceolate or narrowly oblong, 5–14 cm. long, 1–3 cm. broad, acute at apex; scape 7–18 cm. long    1. *flavidum*
Margins of lip ciliolate or ciliate, keels sometimes also ciliate or pubescent:
  Sepals 2·5–5 mm. long:
    Sheaths on peduncle with loose swollen apex; flowers very densely placed; pseudobulbs ovoid, reddish with wrinkled surface; leaves lanceolate to oblanceolate, acute, 4–16 cm. long, 1–2·5 cm. broad; inflorescence shorter than the leaves; sepals triangular-lanceolate, laterals with much dilated front margin at base, 2·5–4·5 mm. long, pale at base, red or purple at apex; petals elliptical-oblong, apiculate, 1·5–2·5 mm. long; stelidia slightly longer than the column ..    ..    ..    ..    ..        12. *winkleri*
    Sheaths on peduncle tight, not swollen at apex:
      Peduncle much shorter than the rhachis; inflorescence slightly longer, or shorter than the leaves; leaves oblong to oblong-lanceolate, 4–10·5 cm. long, 1–1·5 cm. broad; sepals narrowly lanceolate, 4–5 mm. long, cream-coloured; petals narrowly oblong, 1·8–2·5 mm. long; stelidia shorter than the column ..    ..    ..    ..    ..    .. 13. *porphyroglossum*
      Peduncle distinctly longer than the rhachis, often much longer; inflorescence 3–20 cm. long, as long as or usually longer than the leaves; stelidia longer than the column; petals oblong, 1·5–2·5 mm. long:
        Keels of lip glabrous; sepals greenish-yellow, tinged reddish, 2·3–5 mm. long; pseudobulbs ovoid or conical-ovoid, obscurely 4-angled, 1–2 cm. long; leaves elliptical-oblanceolate, obtuse or acute, 3–12 cm. long 1–2 cm. broad; lip very shortly ciliolate in the lower part, papillose near the apex    ..    ..    ..    ..    ..        10. *recurvum*
        Keels of lip pubescent or long-papillose; sepals dark red, obtuse, 3·2–4·7 mm. long; pseudobulbs ovoid, somewhat flattened, obscurely 2–3-angled, 1·3–2·5 cm. long; leaves oblong-elliptical, 4·5–15 cm. long, 1·3 cm. broad; lip densely ciliate all round ..        11. *pavimentatum*
  Sepals 5–17 mm. long; inflorescence slender, 7–45 cm. in length, many-flowered:
    Inflorescence 14–45 cm. long; sepals long acuminate, 8·5–14 mm. long, white or cream-coloured, often flushed pink or purple; pseudobulbs ovoid, biconvex or 3–4-angled, 1·5–4 cm. long; leaves lanceolate or oblong-

lanceolate, acute, 8–27 cm. long, 1–3 cm. broad; petals oblanceolate, denticulate, 2·5–3·5 mm. long  ..    ..    ..    ..    **3.** *cocoinum*

Inflorescence 7–18 cm. long; sepals acute or shortly acuminate, 5–9 mm. long, greenish yellow, sometimes tinged reddish or wine-coloured, or even almost entirely wine-coloured; pseudobulbs conical or narrowly conical, 1–2·5 mm. long; leaves oblanceolate or narrowly oblong, 5–14 cm. long, 1–3 cm. broad; petals oblong, obtuse or sub-acute, 1·7–2·8 mm. long  **1.** *flavidum*

Pseudobulbs normally 2-leaved, rarely 3-leaved (weak plants may have some of the pseudobulbs 1-leaved): (from p. 226)

Lip hairy, hairs often very long; rhachis slender, not thicker or wider than the peduncle:

Bracts short, 3–4 mm. long, not imbricate; hairs on lip in length less than half width of lip; sepals 4–6 mm. long; petals lanceolate or ligulate, acute or subacute, 1·5–2·5 mm. long; lip tongue-shaped from a broader fleshy base, 1·5–2·5 mm. long, ciliate; pseudobulbs narrowly conical or cylindrical, 2–4 cm. long, 3–13 mm. diam.; leaves oblong or narrowly oblong, 5–14 cm. long, 5–13 mm. broad; scape 10–16 cm. long  ..    ..    ..    ..    ..    ..    ..    ..    ..    **18.** *tenuicaule*

Bracts large, broad and boat-shaped, 6–11 mm. long, spreading, and more or less imbricate; hairs on lip longer than half the width of the lip; lip ligulate from a narrower base:

Scape 10–24 cm. long; leaves 4–20 cm. long:

Pseudobulbs elongate-ovoid, 4-angled, 2–3·5 cm. long; leaves 6–12 cm. long, 5–8 mm. broad; sepals 6–7 mm. long; petals oblong-ligulate, 2·3 mm. long; lip 5·5 mm. long, 1·5 mm. broad, base very thick and fleshy  **19.** *gravidum*

Pseudobulbs narrowly cylindrical, sometimes slightly flattened, 2·5–10 cm. long, 3–15 mm. diam.; leaves 4–16 cm. long, 4–15 mm. broad; sepals 4–7·5 mm. long; petals subfalcate-lanceolate, fleshy, 1–2·5 mm. long; lip 2·7–5 mm. long, less than 1 mm. broad  ..    ..    ..    ..    ..    **20.** *cochleatum*

Scape 35–60 cm. long; leaves 11–23 cm. long, 11–18 mm. broad; pseudobulbs conical-cylindrical, 8–12 cm. long, 7–12 mm. diam. at base; sepals 7–7·5 mm. long; petals falcate-ligulate, 2–2·7 mm. long; lip 4–4·5 mm. long  ..    **21.** *mannii*

Lip glabrous, rarely papillose or with a few short cilia along the margins:

Bracts large, spreading, strictly distichous, often imbricate, more or less rigid, 6 mm. or more long, equalling or much longer than the flowers; rhachis often more or less angular or more or less flattened, but only a little wider than the peduncle:

Lip very broad and short, almost orbicular, with fleshy centre and thin toothed margins:

Bracts closely overlapping at time of flowering, boat-shaped, covering the flowers; lateral sepals connivent, not spreading; flowers white or greenish yellow with red or purple markings, or deep purple:

Rhachis and bracts with numerous short blackish or purplish scaly hairs; bracts fawn-coloured or pale brown, the margins straight or slightly convex; central part of lip not very fleshy, wings rather broad; leaves 10–23 cm. long, 1·8–4·5 cm. broad; scape 15–40 cm. long  ..    ..    ..    ..    **38.** *lupulinum*

Rhachis and bracts glabrous; bracts usually purple but rarely green the margins slightly concave so that apices appear incurved; central part of lip very fleshy, wings very narrow, apex tuberculate; leaves 6–16 cm. long, 1·3–2·3 cm. broad; inflorescence (including peduncle) 8–16 cm. long  ..    **39.** *porphyrostachys*

Bracts not closely overlapping at time of flowering, only slightly convex, the flowers easily visible; lateral sepals deflexed; flowers yellow with purple markings; pseudobulbs 4–6-angled, 4·5–6 cm. long; leaves elliptical-ligulate, 13–16 cm. long, 2–2·5 cm. broad  ..    ..    ..    ..    ..    **37.** *wrightii*

Lip distinctly longer than broad, margins entire, sometimes with a few cilia:

Pseudobulbs very narrowly conical or cylindrical, 4–12 cm. long, 3–8 mm. diam.; inflorescence distinctly secund with all flowers facing towards one side; bracts 6–7 mm. long; flowers nearly as long as the bracts, easily visible, reddish-brown or deep red; labellum tongue-shaped from a short fleshy base, sometimes very shortly ciliate in the lower part; sheaths of peduncle tight, acute; leaves linear-ligulate, 7–21 cm. long, 6–22 mm. broad; sepals blunt, 4–5·5 mm. long  **22.** *bequaertii*

Pseudobulbs ellipsoid or ovoid, with several sharp angles, 1–3 cm. diam.; flowers distichous; lip fleshy, recurved in middle or upper part:

Petals orbicular-oblong, the apex rounded, entire; bracts rather rigid, straw-coloured, distinctly longer than the flowers; pseudobulbs 1·5–3·5 cm. long; leaves linear or linear-oblong, 3–12 cm. long, 6–10 mm. broad; flowers whitish or rose-coloured  ..    ..    ..    ..    ..    ..    ..    **35.** *bifarium*

Petals broadly elliptical or oblong, 3-lobed at apex, the central lobe or tooth longest; bracts not very rigid, buff-coloured, about as long as the flowers; pseudobulbs up to 2·5 cm. long; leaves oblong-elliptical, up to 5·5 cm. long and 12 mm. broad; flowers green and purple  ..    ..    ..    **36.** *nigericum*

Bracts less than 9 mm. long, spreading or reflexed; rhachis unthickened, or fleshy and thickened, or much flattened and broad:

Pseudobulbs narrowly cylindrical, slightly tapering upwards, 5–10 cm. long, 6 mm. diam.; leaves broadly oblanceolate, 9–20 cm. long, 2–4 cm. broad; scape 30–40 cm. long, rhachis flattened into broad wings with undulate margins, about 1·5 cm. broad, flowers in a row along centre of each side of rhachis; bracts small, triangular, reflexed; floral parts unknown    ..    ..    ..    47. *longibulbus*

Pseudobulbs ellipsoid, ovoid or conical-ovoid, frequently flattened or angular, up to 7 cm. long, but then not cylindrical, and much thicker:

Lip pectinate in the lower half, entire above:

Rhachis much flattened with broad leaf-like wings on one or both sides of the mid-rib, 5–15 mm. broad; bracts narrow, triangular, acuminate; pseudobulbs sharply 3–4-angled, 2–7 cm. long; scape 18–40 cm. long; leaves 5–18 cm. long, 1·3–4 cm. broad; sepals acuminate, laterals recurved; petals glabrous, linear, curved:

Margins of rhachis markedly undulate; rhachis 1–1·5 cm. broad, the apex abruptly narrowed into a short point, the flowers arranged along a central line; dorsal sepal caudate-acuminate; labellum not much narrowed in front 59. *maximum*

Margins of rhachis not or scarcely undulate; rhachis 0·5–1·5 cm. broad, the apex gradually narrowed into a long blunt point, the flowers along the centre or frequently much nearer one edge; dorsal sepal acuminate; labellum distinctly narrowed in front, the apex much recurved    ..    58. *oxypterum*

Rhachis fleshy, only slightly flattened, 4–10 mm. broad; bracts almost as broad as the rhachis, ovate-triangular, obtuse; petals more or less papillose, linear, curved; dorsal sepal lanceolate, acute:

Rhachis whitish-green, flowers orange and yellow; petals nearly as long as dorsal sepal, latter 6–9 mm. long; pseudobulbs ovoid, 3–4-angled, 3–7 cm. long; leaves 11–18 cm. long, 1–3 cm. broad; scape 15–35 cm. long 64. *leucorrhachis*

Rhachis green or purplish or blackish, flowers dull purple- or black-spotted on a green ground; petals distinctly shorter than dorsal sepal, latter 7–11 mm. long; pseudobulbs ovoid, rather flattened, 3-angled, 3–7 cm. long; leaves 8–21 cm. long, 1–3·5 cm. broad; scape 18–65 cm. long    ..    63. *imbricatum*

Lip not pectinate in the lower half:

Bracts almost white, scarious, spreading, almost as long as the flowers, 4–8 mm. long; sheaths on peduncle similar, adpressed; lip fleshy, bent like a knee just below the middle; rhachis not widened, angular; dorsal sepal triangular-ovate, apiculate, 4–4·5 mm. long; petals obliquely orbicular-elliptical, rounded at apex, 4–4·5 mm. long, 3–3·5 mm. broad; leaves 1·5–8 cm. long, 5–12 mm. broad    ..    ..    ..    ..    ..    ..    ..    34. *scariosum*

Bracts and sheaths not white and scarious; lip either fleshy and usually not curved (if so, then petals denticulate), or thin and often much recurved; petals linear, oblanceolate, oblong or ovate, obtuse to acuminate but not rounded at apex:

Lip very fleshy, not much curved; lateral sepals not recurved in the upper part; flowers distichous; rhachis slender or if slightly flattened the flowers borne on the edges; scapes up to 15 cm. long but frequently much shorter:

Petals ovate, acute, denticulate, 4 mm. long; lip dorsally flattened, papillose at the base underneath, much folded down the centre; scape slender, rhachis quite unthickened, nodding; pseudobulbs ovoid, 4–5-angled, 3 cm. long; leaves oblong, 7–9 cm. long, 1–1·5 cm. broad; sepals 5–6 mm. long, papillose

29. *denticulatum*

Petals linear, obtuse, entire, 2–3·5 mm. long; scape erect or decurved, somewhat flattened and often winged; pseudobulbs oblong-ellipsoid, 4-angled, 1·5–4·5 cm. long; leaves elliptical or oblong-elliptical, 1·5–11 cm. long, 0·5–2 cm. broad; lip with 2 prominent smooth keels in the lower part:

Rhachis, and to a less extent the sepals, more or less densely pubescent-papillose or velvety; lip slightly dorsally compressed; sepals 3–5 mm. long; stelidia entire, acute, curving forwards    ..    ..    ..    32. *fuscoides*

Rhachis quite glabrous, the margins on which the flowers are borne more or less deeply canaliculate:

Lip dorsally compressed, distinctly 3-lobed at the base, the side lobes triangular-ovate and slightly incurved; sepals 2·5–3·5 mm. long; flowers in inflorescence all open at same time    ..    ..    ..    ..    30. *fuscum*

Lip laterally compressed, side lobes reduced to very low ridges on the outside of the lip near the base; sepals 3–7 mm. long; only some flowers open at any time:

Rhachis angled or narrowly winged, the wings quite entire; flowers somewhat incurved    ..    ..    ..    ..    ..    ..    ..    31. *oreonastes*

Rhachis broadly winged, the wings toothed and continuous with the toothed base of the bract above; flowers spreading or somewhat recurved; pedicel

and base of flowers usually deeply sunk in hollows between the wings
                         33. *zenkeranum*

Lip relatively thin and flat, often much curved, sometimes with a longitudinal keel running along the middle of the lower surface; lateral sepals usually, and frequently abruptly, reflexed in the upper half, more or less falcate; rhachis slender or variously flattened, in the latter case with the flowers borne along the flat surfaces; dorsal sepal frequently spathulate:

Petals filiform with a thickened club-shaped apex, resembling the antennae of a butterfly; dorsal sepal lanceolate-linear, acute, 10–12 mm. long, 1–2·5 mm. broad; pseudobulbs elongate-conical-ovoid, biconvex or 3–4-angled, 3–5 cm. long; leaves linear-oblong, 7–16 cm. long, 7–20 mm. broad; scape 12–25 cm. long with short loose sheaths on peduncle; rhachis swollen and more or less flattened, up to 14 mm. broad; bracts broadly ovate, often nearly as wide as the rhachis; lip with erect rounded side lobes ..    ..     46. *tentaculigerum*

Petals linear, oblanceolate or oblong, obtuse or acuminate, not like antennae:

Rhachis fleshy and swollen, somewhat flattened, not leaf-like; bracts broad, triangular-ovate, nearly as broad as the rhachis; leaves over 1·5 cm. broad; dorsal sepal linear-lanceolate, acute, never spathulate or oblanceolate; petals linear, nearly as long as the dorsal sepal:

Rhachis nearly terete, 2–3 mm. broad, 3–4 cm. long; petals glabrous, sub-obtuse; lip ovate, obtuse, 2 mm. long; pseudobulbs ovoid, flattened, distinctly 4-angled, 6–7 cm. long, 2–2·5 cm. diam.; leaves oblong, obtuse, 16–22 cm. long, 2–3 cm. broad; scape 18–19 cm. long; dorsal sepal 8 mm. long ..    ..    ..    ..    ..    ..    ..    60. *bibundiense*

Rhachis flattened, 6–8 mm. broad, about 10 cm. long; petals papillose, acute; lip oblong, acuminate, 3 mm. long; pseudobulbs ovoid, 3-angled, 3·5–4 cm. long, 1·5–2 cm. diam.; leaves elliptical-oblong, 7·5–9 cm. long, 1·5–2 cm. broad; scape 27–47 cm. long; dorsal sepal 5 mm. long   61. *kamerunense*

Rhachis either slender unthickened or somewhat flattened, or widened on each side to form a flat thin leaf-like wing; bracts in the latter two cases distinctly narrower than the rhachis:

Dorsal sepal more or less spathulate or widened above the middle, the sides of the widened part thick and fleshy, apiculate or rounded at the apex, rarely acute:

Petals obtuse or rounded and somewhat thickened at the apex which is often yellow, falcate, linear or oblong, 1·7–2·7 mm. long, a little less than half as long as the dorsal sepal; pseudobulbs ovoid or elongate-ovoid, 3–4-angled, 2–6 cm. long; leaves oblong-lanceolate or oblanceolate, 4–16 cm. long, 6–23 mm. broad; scape up to 35 cm. long; rhachis thin, a little sinuate at the margin, up to 14 mm. broad; dorsal sepal 4·5–7 mm. long; wings of column terminating in a rounded lobe below the very short stelidia ..    ..    ..    ..    ..    ..    51. *falcatum*

Petals not thickened at the apex, either very acute or almost subulate and usually more than half as long as the dorsal sepal, or obtuse or subacute and from a third to a fifth of the length of the dorsal sepal:

Dorsal sepal 2½–1½ times as long as the very acute or acuminate petals, not exceeding 6·5 mm. in length; wings of column with a distinct lobe below the stelidia; lip ovate or elliptical, not much narrowed in the apical part; leaves ligulate, oblong or elliptical, not more than 10 times as long as broad; ovary and sepals with many short reddish hairs:

Flowers 1–2·5 mm. apart (both sides of rhachis included); dorsal sepal 2·5–3·5 mm. long; rhachis 1–8 mm. broad; leaves oblong-oblanceolate, obtuse, 1–9 cm. long, 2–13 mm. broad; pseudobulbs narrowly ovoid, 1–3·5 cm. long; scape 2–9 cm. long ..    ..    50. *melanorrhachis*

Flowers 2·5–8 mm. apart; dorsal sepal 3·5–6·5 mm. long; scape 5–21 cm. long; pseudobulbs biconvex or obtusely 3-angled:

Rhachis flattened, 4–15 mm. broad; flowers 2·5–4 mm. apart; pseudobulbs ellipsoid or conical-ovoid, 1·5–4·5 cm. long; leaves ligulate or narrowly oblanceolate, 4–15 cm. long, 5–25 mm. broad; scape 5–13 cm. long ..    ..    ..    ..    ..    49. *velutinum*

Rhachis not flattened, very slender, rather flexuous; flowers 3–8 mm. apart; pseudobulbs ovoid or ellipsoid, 1–2·5 cm. long; leaves elliptical-oblong or linear-oblong, 2–8 cm. long, 5–11 mm. broad; scape 9–21 cm. long ..    ..    ..    ..    ..48. *simonii*

Dorsal sepal 3–5 times as long as the short, obtuse or subacute petals:

Leaves 1–3·7 cm. broad, 6·5–20 cm. long, oblanceolate; wings of column terminating in a distinct rounded lobe below the stelidia; pseudobulbs ovoid or conical-ovoid, biconvex, 3–7 cm. long; scape 16–30 cm. long, erect; rhachis variously flattened, sometimes no wider than the peduncle,

up to 11 mm. broad; dorsal sepal 6–9 mm. long, usually broadly
spathulate; petals lanceolate-oblong, subacute, 1·5–2·2 mm. long
53. bufo
Leaves less than 1·5 cm. broad, usually less than 1 cm., linear or narrowly
    ligulate; wings of column not terminating in a distinct lobe below the
    stelidia, but gradually merging into the column:
Rhachis horizontal or recurved below the pseudobulb away from the
    light, much flattened, 1·5–2 cm. broad, crinkled at the edges; pseudo-
    bulbs conical-ovoid, obtusely 4-angled, 2–4 cm. long; leaves narrowly
    ligulate or linear-ligulate, 20–24 cm. long, 6–9 mm. broad; scape
    8–12·5 cm. long; dorsal sepal broadly spathulate, about 6·5 mm. long;
    petals obliquely narrowly ovate, about 1·5 mm. long      56. <i>lucifugum</i>
Rhachis erect, not at all recurved, variable in width:
Leaves ligulate or linear-ligulate, 5–19 cm. long, 4–13 mm. broad, from
    10–33 times as long as broad; scape 16–65 cm. long; rhachis slender
    or broadly flattened, up to 15 mm. broad; flowers reflexed after
    opening; pseudobulbs ovoid, conical or cylindrical-conical, more or
    less 3-angled, 1–6 cm. long; dorsal sepal broadly spathulate, 5·5–
    8·3 mm. long, often somewhat recurved; petals oblong, 1–2 mm. long
55. calyptratum
Leaves linear, grass-like, 9–32 cm. long, 3–10 mm. broad, from 26–80
    times as long as broad; scape 10–28 cm. long; rhachis slender or only
    very slightly flattened; flowers not reflexed after opening, usually
    somewhat incurved; pseudobulbs conical-cylindrical to conical-
    ovoid, 3–4-angled, 1–5 cm. long; dorsal sepal 6–7·5 mm. long, more or
    less incurved; petals oblong or lanceolate-oblong, 1·7–2·6 mm.
    long   ..    ..    ..    ..    ..    ..    .. 54. <i>graminifolium</i>
Dorsal sepal acute, usually narrowed gradually from the base, not thickened
    at the sides in the apical part; leaves usually over 1 cm. broad:
Scape over 8 cm. long, usually much longer; rhachis slender or flattened:
Flowers densely hairy; rhachis up to 4·5 cm. broad, often twisted spirally,
    heavily blotched purple; leaves oblanceolate, 13–25 cm. long, 3–5 cm.
    broad, purple speckled beneath; peduncle with several short loose sheaths;
    pseudobulbs oblong, flattened, 5–8 cm. long; petals narrowly lanceolate,
    acuminate ..    ..    ..    ..    ..    ..    ..      52. <i>purpureorhachis</i>
Flowers glabrous or sparsely puberulous:
Petals falcate-oblanceolate, wider in the upper part, 0·9–1·5 mm. broad;
    leaves oblanceolate, 10–24 cm. long, 3–5·5 cm. broad; pseudobulbs
    elongate ellipsoid, 3–5-angled, 4–10 cm. long; scape 16–25 cm. long,
    rhachis flattened, 7–20 mm. broad, sinuate at the edge; dorsal sepal
    lanceolate, acute, sides inrolled, 6–8 mm. long   45. <i>fuerstenbergianum</i>
Petals falcate linear or very narrowly lanceolate, not broader in the upper
    part, less than 0·8 mm. broad; leaves ligulate, or oblong-elliptical, 8–35
    mm. broad:
Dorsal sepal triangular-lanceolate, 3–3·5 mm. long; scape 35–55 cm.
    long, rhachis very slender; petals curved oblong, 2 mm. long; pseudo-
    bulbs elongate ovoid, angular, 2·5–4 cm. long; leaves oblong-elliptical,
    6–7 cm. long, 10–16 mm. broad; wings of column terminating below
    stelidia in a broad short rounded lobe; lateral sepals apiculate
43. filiforme
Dorsal sepal narrowly lanceolate, acute, 5·5–8 mm. long; scape 8–66 cm.
    long, rhachis more or less flattened, up to 10 mm. broad; petals linear,
    much curved, 3–5 mm. long; pseudobulbs elongate ovoid or conical,
    3–5-angled, 2–7·5 cm. long; leaves ligulate or oblong, 6·5–16 cm. long,
    8–35 mm. broad; wings of column terminating below stelidia in a
    narrow subacute or obtuse lobe; lateral sepals acuminate
44. congolanum
Scape less than 8 cm. long, often much shorter; rhachis always slender, with
    the flowers closely placed; dorsal sepal less than 5 mm. long, lanceolate;
    petals narrow and acute; leaves up to 1 cm. broad:
Plant very small; pseudobulbs less than 1 cm. long; leaves 6–9 mm. long,
    3–4 mm. broad; scape less than 2 cm. high, few-flowered; dorsal sepal
    about 3–5 mm. long; petals 2·3 mm. long, falcate-linear, very acute;
    lip curved, ovate, 1·5 mm. long; wings of column terminating in narrow
    rounded-triangular lobes      ..    ..    ..    ..    .. 42. <i>parvum</i>
Plant larger, pseudobulbs 1–3·5 cm. long, biconvex; leaves 1–6 cm. long,
    3–10 mm. broad; scape 3–8 cm. long; petals falcate-ligulate, acute;
    wings of column terminating in broad rounded lobes:
Flowers 1·5–2 mm. apart, bronze-yellow to purple, the bracts equalling or
    longer than the internodes; lip ovate, not narrowed much in front;

pseudobulbs up to 3·5 cm. long; leaves 1·5–6 cm. long; scape 4–8 cm.
long, many-flowered; dorsal sepal 2·5–5 mm. long; petals 1·5–3 mm.
long    ..    ..    ..    ..    ..    ..    ..    41. *rhizophorae*
Flowers 2–3·5 mm. apart, yellow, the bracts distinctly shorter than the
internodes; lip distinctly narrowed in front from an ovate base; pseudobulbs 1–2·5 cm. long; leaves 1–4·5 cm. long; scape 3–6 cm. long, few-
flowered; dorsal sepal 4–5 mm. long; petals 2·5–3·5 mm. long
40. *falcipetalum*

1. **B. flavidum** *Lindl.* in Bot. Reg. 26: Misc. 83 (1840); F.T.A. 7: 30. (?) *B. herminiostachys* (Rchb. f.)
Rchb. f. (1861)—F.T.A. 7: 35. *B. molivense* Schltr. in Engl., Bot. Jahrb. 38: 157 (1906). Flowers yellow
or yellowish-green, marked or tinged with red or pink, or rarely deep purple.
  **Guin.:** Macenta (July) *Jac.-Fél.* 1044! **S.L.:** Musaia (July, Aug.) *Deighton* 4327a! Koidu, Kisi Kama
(Aug.) *Deighton* 5913! Bumpe Peri (July) *Deighton* 4836! Fanima, Soro (July) *Deighton* 4640! 4653!
Kasasa, Tonko Limba (Sept.) *Adames* 194! **Lib.:** Yoma, Montserrado Dist. (Aug.) *Leeuwenberg* 4830!
**Iv.C.:** *Lecoufle* K509! **Ghana:** Amedzofe (June, July) *Scholes* 86! *Morton* A3413! **S.Nig.:** Idanre Hills,
Ondo (May) *Keay & Brenan* FHI 22405! Usonigbe F.R., Benin (July) *Keay* FHI 26720! Abure F.R.
(July) *King* 184! **W.Cam.:** Moliwe (Aug.) *Schlechter* 15757! **F.Po:** lava flow (Sept.) *Wrigley* 543! Also
in Gabon and S. Tomé.
2. **B. schimperanum** *Kraenzl.* in Engl., Bot. Jahrb. 33: 71 (1902). *B. xanthoglossum* Schltr. in Engl., Bot.
Jahrb. 38: 158 (1906). Flowers white.
  **S.Nig.:** Ajagbodudu, Warri Prov. (Sept.) *Wright* 135! 161! **W.Cam.:** Victoria (Sept.) *Schimper* 341!
Moliwe (Aug.) *Schlechter* 15755! Buea (Mar.) *Deistel* 54. Also in Congo and Uganda.
3. **B. cocoinum** *Batem. ex Lindl.* in Bot. Reg. 23: t. 1964 (1837); F.T.A. 7: 31. *B. vitiense* Rolfe in Kew Bull.
1893: 5. Flowers white, cream-coloured or greenish, tinged pink or with pink apices.
  **S.L.:** *cult. spec.*! **Lib.:** (July) *Harley* 2046! Mt. Barclay (June) *Bunting* 10! **Ghana:** Ofinsu (May)
*Vigne* FH 4864! Bunsu (May) *D. Gillett*! Efiduase (May) *Cox* 25! Boanim to Mampong (Apr.) *Westwood*
134! Mampong to Nsula *Westwood* 103! Also in Annobon and (?) Uganda.
4. **B. josephii** (*Kuntze*) *Summerh.* in Bot. Mus. Leafl. Harv. Univ. 11: 250 (1945). *Phyllorchis josephii* Kuntze,
Rev. Gen. Pl. 676 (1891). *Bulbophyllum aurantiacum* Hook. f. (1864)—F.T.A. 7: 30, not of F. Muell.
*B. gustavii* Schltr. in Fedde Rep. 9: 165 (1911), in obs.; F.W.T.A., ed. 1, 2: 439. Flowers orange or
orange-red and whitish green.
  **Guin.:** Nimba Mts., 5,300 ft. (Oct.) *Schnell* 3845! **S.L.:** Loma Mts., *Jaeger* 1191! **Iv.C.:** Mt. Tonkoui
(Oct.) *Aké Assi* 5667! **S.Nig.:** Obudu Plateau *Horwood* 165! Mambila Plateau *Wimbush* in *Hb. King*
149! **W.Cam.:** Cam. Mt., 5,000–6,000 ft. (Nov.) *Mann* 2124! Buea, 3,000 ft. *Gregory* 597! 890!
5. **B. modicum** *Summerh.* in Kew Bull. 11: 114 (1957). Flowers magenta and white, column green.
  **W.Cam.:** Buea, 3,000 ft. (Oct.) *Gregory* 193! 558! *Keay* FHI 25439!
6. **B. calvum** *Summerh.* in Kew Bull. 20: 185, fig. 8 (1966). Sepals pale yellow-green, lip red with yellow base
and white flecks above.
  **N.Nig.:** Maisamari, Mambila Plateau, 6,000 ft. (Oct.) *Nash & Wimbush* in *Hb. King* 178!
7. **B. nigritianum** *Rendle* Cat. Talb. 99 (1913). Flowers white or pale yellow, sometimes tinged green.
  **S.L.:** Gegbwema (Dec.) *Deighton* 2823! Kebawana, Gaura (Nov.) *Deighton* 4649! 4946! **Lib.:** Suen
(Nov.) *Linder* 1393! Tawara, Boporo Dist. (Nov.) *Baldwin* 10307! **Iv.C.:** Tiapleu (Oct.) *Aké Assi*
9124! **Ghana:** Awaso *Blayney* 2! **S.Nig.:** Oban *Talbot* 933! Also in Congo.
8. **B. pipio** *Rchb. f.* in Linnaea 41: 92 (1877); F.T.A. 7: 32. *B. milesii* Summerh. in Kew Bull. 1935: 200;
F.W.T.A., ed. 1, 2: 439. Flowers cream or yellow, purplish in centre.
  **S.L.:** York I. (Nov.) *Adames* 106! Victoria Creek (Oct.) *Adames* 99! Yagoi, Jong R. (Oct., Nov.)
*Adames* 100! **Ghana:** Sekondi (Apr.) *Cox* 129! Western Prov. (Nov., Dec.) *Miles* 19! Also in E·
Cameroun.
9. **B. buntingii** *Rendle* Cat. Talb. 99 (1913). Pseudobulbs often reddish; flowers white, cream-coloured or
very pale pink.
  **Guin.:** Sérédou *Pujol* L363! **Lib.:** Begwai (Oct.) *Bunting* 29! Tawata, Boporo Dist. (Nov.) *Baldwin*
10300! Ganta (Oct.) *Harley* 1759! Kitoma (Oct.) *Harley* 1833! Yah R., Sanokwele Dist. (Oct.) *Adames*
664! **Iv.C.:** Taï Forest (Oct.) *Aké Assi* 6033! **N.Nig.:** Kwarra, Plateau Prov. (Aug.) *King* 200! **S. Nig.:**
Oban *Talbot* 935! Kundeve to Bagga, Obudu (Oct.) *Keay* FHI 25430! **W.Cam.:** M'bonge *Wright* 58/56!
Also in the Congos and Uganda.
10. **B. recurvum** *Lindl.* Gen. & Sp. Orch. Pl. 53 (1832); F.T.A. 7: 31; F.W.T.A., ed. 1, 2: 439 (excl. Lagos
specimen). Sepals yellowish green, sometimes tinged purple, lip dark red or purple.
  **S.L.:** Sugar Loaf Mt. (Apr.) *Sc. Elliot* 5778! Njala (Aug.) *Deighton* 1148! Messima, Messi Krim (June)
*Adames* 46! Roruks (May) *Deighton* 4652! Kasewe F.R. (May, June) *Deighton* 4651! **Lib.:** Ganta
(Nov.) *Harley* 1762! **W.Cam.:** M'bonge *Wright* 58/57! **F.Po:** S. Isabel to Basilé, 3,000 ft. (fr. Dec.)
*Sanford* 4035!
11. **B. pavimentatum** *Lindl.* in J. Linn. Soc. 6: 128 (1862); F.T.A. 7: 32. *B. dorotheae* Rendle, Cat. Talb. 100
(1913). *B. recurvum* of F.W.T.A., ed. 1, 2: 439, partly (Lagos specimen), not of Lindl. Flowers deep
crimson, violet or purple.
  **S.Nig.:** Lagos *Moloney*! Nun R. (Sept.) *Mann* 519! Okomu F.R., Benin (Sept.) *Keay* FHI 22445!
FHI 25420! Ajagbodudu, Sapele (Sept.) *Wright* 130! Oban *Talbot* 934! Also in E. Cameroun, Gabon
and Congo.
12. **B. winkleri** *Schltr.* in Engl., Bot. Jahrb. 38: 158 (1906). *B. imogeniae* K. Hamilt. in Trans. Bot. Soc.
Edinb. 27: 228 (1917). Leaves often purplish beneath; sepals pale, white or greenish at base, upper
part red or purple, apices sometimes yellow or brownish.
  **Lib.:** Ganta (Mar.) *Harley* 1634! 2025! **N.Nig.:** N. of Katsina R. *Hamilton*! Jemaa, Zaria Prov. (June)
*King* 10! *Cole* 25! Sha, Plateau Prov. *King* 10a! Hoss, Plateau Prov. *King* 10b! **S.Nig.:** Kundeve,
Obudu (Apr.) *Keay & Savory* FHI 25309! **W.Cam.:** Buea (July) *Winkler* 157! Also in E. Cameroun
and Congo.
13. **B. porphyroglossum** *Kraenzl.* in Engl., Bot. Jahrb. 22: 24 (1895); F.T.A. 7: 34. *B. calabaricum* Rolfe in
Kew Bull. 1906: 114. Sepals cream to yellowish-green; lip dull red or purple.
  **S.Nig.:** Calabar *Holland*! Eket *Talbot* 3288! Stubbs Creek F.R., Eket (May) *Latilo* FHI 32916!
**W.Cam.:** Bibundi (Apr.) *Schlechter* 12361! Victoria (May) *Preuss* 1299! *Rosevear* Cam. 53/37! Also in
E. Cameroun.
14. **B. teretifolium** *Schltr.* in Engl., Bot. Jahrb. 38: 18 (1905).
  **W.Cam.:** Bibundi *Schlechter* 12362!
15. **B. intertextum** *Lindl.* in J. Linn. Soc. 6: 727 (1862); F.T.A. 7: 29. *B. viride* Rolfe in Kew Bull. 1893:
170; Cat. Talb. 146. *B. triaristellum* Kraenzl. & Schltr. in Orchis 2: 98 (1908). *B. amauryae* Rendle,
Cat. Talb. 101, t. 12, fig. 11, 12 (1913). Flowers pale green with or without red markings, or entirely
red or purple.
  **S.L.:** Sugar Loaf Mt. (May, Oct.) *Barter*! *Deighton* 6022! Roruks (July–Sept.) *Deighton* 4329! **Lib.:**
Du R. (Aug.) *Linder* 334a! Tawata, Boporo Dist. *Baldwin* 10301! Truo, Sinoe *Baldwin* 11380! **Ghana:**
Kumasi road, Accra *Westwood* 170! Nkawkaw (Oct.) *Westwood* 206! **N.Nig.:** Assob R., Jemaa, Zaria
(Sept., Oct.) *King* 58! **S.Nig.:** Nun R. (Sept.) *Mann* 527! Oban *Talbot* 952! 955! Ikwette, Obudu

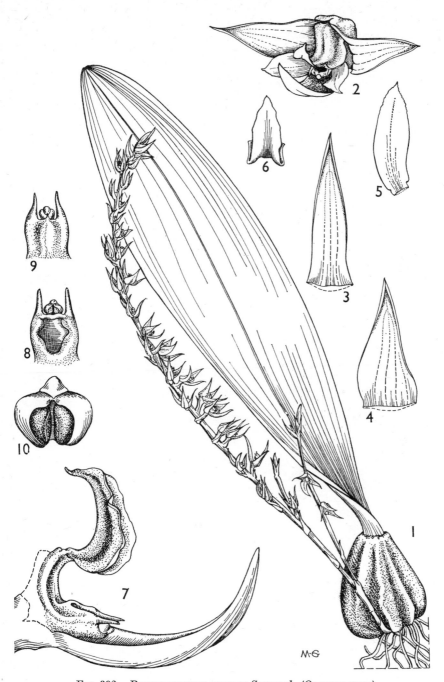

FIG. 392.—BULBOPHYLLUM CALVUM *Summerh.* (ORCHIDACEAE).

1, habit, showing branched inflorescence, × 1. 2, flower, × 4. 3, dorsal sepal, × 4. 4, lateral sepal, × 4. 5, petal, × 4. 6, lip, × 4. 7, flower with sepals and petals removed showing column with lip raised, × 6. 8, column, face view, × 6. 9, column, back view, × 6. 10, anther cap with pollinia, × 24. All from *Nash & Wimbush* in *Hb. King* 178.

(Sept.) *Keay & Savory* FHI 25398! **W.Cam.:** Victoria (June) *Rosevear* Cam. 57/37! Bibundi (May) *Keay* FHI 42336! Mimbia, 3,600 ft. (Mar.) *Brenan & Richards* 9350! Moliwe (Aug.) *Schlechter* 15756! N'dian (June–Sept.) *Wright* 138! **F.Po:** Moka, 4,600 ft. (Sept.) *Melville* 437! Also in Gabon, Angola, Kenya, Tanzania, Malawi and Seychelles.

16. **B. comatum** *Lindl.* in J. Linn. Soc. 6: 128 (1862); F.T.A. 7: 32. Flowers purple with long white hairs on perianth.
     **S.L.:** Maloloka (July) *Thomas* 1299! Ikwette, Ogoja Prov., 5,200 ft. (Dec.) *Savory & Keay* FHI 25245! **Iv.C.:** Mt. Momi (Oct.) *Aké Assi* 9198! 9200! **F.Po:** 2,000 ft. (Dec.) *Mann* 642! Pico S. Isabel, 3,000 ft. (Dec.) *Sanford* 4055!

17. **B. inflatum** *Rolfe* in Gard. Chron. ser. 3, 9: 334 (1891); F.T.A. 7: 33. Flowers yellowish-green with purple flecks.
     **S.L.:** *cult. specimens!* **Lib.:** Bili (Dec.) *Harley* 2083!

18. **B. tenuicaule** *Lindl.* in J. Linn. Soc. 6: 126 (1862); F.T.A. 7: 27. Flowers red-purple, whitish at base.
     **N.Nig.:** Mambila Plateau *Wimbush* in *Hb. King* 150! **W.Cam.:** Buea, 3,000 ft. (Sept.) *Gregory* 612! Cam. Mt., 6,000 ft. *Cole* 24! **F.Po:** 5,000 ft. (Dec.) *Mann* 648! (?) Also in S. Tomé.

19. **B. gravidum** *Lindl.* in J. Linn. Soc. 6: 126 (1862); F.T.A. 7: 27.
     **W.Cam.:** Cam. Mt., 5,000 ft. (Nov.) *Mann* 2126! **F.Po:** 3,000 ft. (Dec.) *Mann* 650! Pico S. Isabel, 4,000 ft. *Sanford* 4112! 4163!

20. **B. cochleatum** *Lindl.* in J. Linn. Soc. 6: 125 (1862); F.T.A. 7: 28. *B. talbotii* Rendle, Cat. Talb. 100 (1913). *B. jungwirthianum* Schltr. (1922). Pseudobulbs and leaves often purplish; sepals yellow-green with purple lip or whole flower purple or red.
     **Guin.:** Diaguissa to Daraba, Fouta Djalon, 3,000–4,500 ft. (Sept.–Dec.) *Chev.* 18817! 20280! *Des Abbayes* 1325! Nimba Mts., 3,300 ft.(Apr.) *Schnell* 1224! **S.L.:** Bintumane Mt., 5,000–6,000 ft. (Oct.–Feb.) *Jaeger* 294! 4232! Deighton 5719! *T. S. Jones* 167! 168! **Lib.:** Jaurazon, Sinoe (Mar.) *Baldwin* 11478! Truo, Sinoe *Baldwin* 11382! Bili (Mar.) *Harley* 1786! **Iv.C.:** Mt. Tonkoui (Feb.) *Aké Assi* 6943! **N.Nig.:** Vogel Peak, 4,900 ft. (Nov.) Hepper 1507! Kakara, Mambila Plateau, 4,800 ft. (Jan.) *Hepper* 1798! *Wimbush* 128! 134! **S.Nig.:** Oban *Talbot* 931! 1268! Koloishe, Obudu, 6,000 ft. (Dec.) *Savory & Keay* FHI 25098! Kwa Falls, Calabar *Maggs*! **W. Cam.:** Lake Mbuluwe, Bamenda, 7,200 ft. (fl. & fr Jan.) *Migeod* 376! *Keay & Lightbody* FHI 28374! **F.Po:** 4,000 ft. (Dec.) *Mann* 643! Also in E. Cameroun, Sudan, Uganda, Kenya, Tanzania, Zambia.

21. **B. mannii** *Hook. f.* in J. Linn. Soc. 7: 218 (1864); F.T.A. 7: 28.
     **S.L.:** Loma Mts., 5,500 ft. (Jan.) *Jaeger* 4161! **W.Cam.:** Cam. Mt., 4,000–5,000 ft. (Dec.) *Mann* 1337! 2111! Buea, 2,700 ft. (Jan.) *Schlechter* 12836! **F.Po:** Moka, 4,600 ft. (Sept.) *Melville* 461!

22. **B. bequaertii** *De Wild.* in Rev. Zool. Afr. 9, suppl. bot. 28 (1921). Flowers red or reddish brown.
     **W.Cam.:** Bum, Wum Div., 3,500 ft. (Feb.) *Hepper* 1918! Bamenda (Apr.) *Sanford* WS/695/65! **F.Po:** Pico S. Isabel, 3,000 ft. (fr. Dec.) *Sanford* 4109! Also in eastern Congo, Uganda, Kenya and Tanzania.

23. **B. saltatorium** *Lindl.* Bot. Reg. 23: t. 1970 (1837); F.T.A. 7: 34. Flowers reddish or pink.
     **S.L.:** Ndilajula, Njala (Apr.–June) *Deighton* 4317! 4638! Kambui F.R., Kenema (May) *Jordan* 2085! **Ghana:** Mansu (Dec.) *Miles* 31!

24. **B. distans** *Lindl.* in J. Linn. Soc. 6: 125 (1862); F.T.A. 7: 34. *B. kindtianum* De Wild. (1904)—F.W.T.A., ed. 1, 2: 437. *B. nudiscapum* Rolfe in Kew Bull. 1909: 365; Cat. Talb. 146 (1913). Sepals yellowish green, or variously tinged maroon or purplish, lip maroon or purplish.
     **Lib.:** Peahtah (Oct.) *Linder* 1020! Bomi Hills (Sept.) *Voorhoeve* 436! Mt. Barclay (June) *Bunting* 11! Ganta (Nov.) *Harley* 1841! Nimba (Oct.) *Adames* 686! **Iv.C.:** Tiapleu (cult. Abidjan, Nov.) *Aké Assi* 9224! **Ghana:** Nsudem (Nov., Dec.) *Miles* 23! **S.Nig.:** Sapoba (Mar.–Apr.) *Onyeagocha* FHI 7131! *Jones* FHI 1085! Nun R. (Sept.) *Mann* 525! Oban *Talbot* 779! Eket *Talbot* 3309! **W.Cam.:** Man of War Bay *Cole* 35! **F.Po:** Las Corteras (Jan.) *Guinea* 2135! Also in E. Cameroun, Gabon, Congo and Uganda.

25. **B. calamarium** *Lindl.* in Bot. Reg. 29: Misc. 70 (1843); F.T.A. 7: 33, partly; F.W.T.A., ed. 1, 2: 437, partly (both as to S.L. specimen only). Sepals green or yellow-green, marked reddish, lip greenish or red with red or purplish hairs.
     **S.L.:** *Fielding*! Roruks (Oct.) *Deighton* 4371! **Lib.:** Ganta (Apr.) *Harley* 2130! Bilimu (July) *Harley* 2038! **S.Nig.:** Agagbodudu, Sapele (May–Sept.) *Wright* 137! Sonkwala Mts., Obudu Dist. (Dec.) *Keay* FHI 28687!

26. **B. phaeopogon** *Schltr.* in Engl., Bot. Jahrb. 38: 157 (1906); Cat. Talb. 146. Flowers yellow with brown or purple spots.
     **Iv.C.:** Taï Forest (cult. Adiopodoumé, June) *Aké Assi* 9654! **Ghana:** Efiduase *Cox* 35! Tafo (Mar.) *Westwood* 73! **S.Nig.:** Degema *Talbot* 3724! Eket *Talbot* 3308! Oban *Talbot* 778! **W. Cam.:** Bimbia road, Victoria (May) *Preuss* 1225! *Maitland* 733! Victoria (May) *Winkler* 14a! Nyoke (July) *Schlechter* 15791! Also in E. Cameroun and Gabon.

27. **B. schinzianum** *Kraenzl.* in Bull. Soc. Bot. Belg. 38: 57 (1900). *B. calamarium* of F.T.A. 7: 33, partly and F.W.T.A., ed. 1, 2: 437, partly (S. Nig. specimen). Sepals pale yellow, marked red; lip with chocolate or purple hairs.
     **Lib.:** Wohmen, Vonjama Dist. (Oct.) *Baldwin* 10100! **N.Nig.:** Nsube, Niger Dist. *Barter* 1482! Also in Gabon and Congo.

28. **B. barbigerum** *Lindl.* in Bot. Reg. 23: t. 1942 (1837); F.T.A. 7: 34. Flowers deep red or wine-coloured, sepals tinged greenish.
     **S.L.:** confluence of Gbagbi and Bafi rivers (Aug.) *Smythe* 117! Koindu, Kissi Kama (Nov.) *Deighton* 4657! Panguma (Oct.) *Deighton* 5609! **Lib.:** Genne Loffa, Kolahun Dist. (Nov.) *Baldwin* 10079! Cocopa, Yasono (Sept.) *Harley* 1926! **Iv.C.:** *Merle*! Sanguiné Forest (Nov.) *Aké Assi* 9223! **Ghana:** Juaben, Ashanti (July) *Cox* 127! **S.Nig.:** Lagos *Moloney*! Niger Delta *Dalton*! Insofan (Jan.) *Holland* 297! Afi River F.R., Ogoja Prov. (May) *Jones* FHI 1878! **W.Cam.:** Victoria (May) *Keay* FHI 43914! Also in Congo.

29. **B. denticulatum** *Rolfe* in Kew Bull. 1891: 197; F.T.A. 7: 25. Sepals brownish green; lip orange.
     **S.L.:** *cult. specimen!*

30. **B. fuscum** *Lindl.* in Bot. Reg. 25: Misc. 3 (1839); F.T.A. 7: 24, partly. Flowers mustard yellow, brownish red or brown.
     **Guin.:** Pita, Fouta Djalon (Dec.) *Jac.-Fél.* 664! **S.L.:** Njala (Nov.) *Deighton* 5244! Kasewe F.R. (Nov.) *Deighton* 4923! Kamaranka Bridge, Magbema *Adames* 198! **Lib.:** Suen, Montserrado (Nov.) *Baldwin* 10454! **Iv.C.:** Kagbé Forest (cult. Abidjan, Jan.) *Aké Assi* 9411! **N.Nig.:** Tof, Plateau Prov., 3,800 ft. *King* 191! **S.Nig.:** Agbadi, Sapoba (Nov.) *Meikle* 599! Ajagbodudu, Sapoba *Wright* 141! Calabar, *Cooper* 42/2! **W. Cam.:** Little Cam. Mt., 3,000 ft. *Cole* 36! Also in the Congos and Tanzania.

31. **B. oreonastes** *Rchb. f.* Otia Bot. Hamburg. 118 (1881); F.T.A. 7: 24. *B. fuscum* of F.T.A. 7: 24, partly; Cat. Talb. 146 (1913). *B. obanense* Rendle, Cat. Talb. 101, t. 12, fig. 9–10 (1913). Flowers yellow, orange, brown or reddish brown.
     **Guin.:** Pita, Fouta Djalon (Dec.) *Jac-Fél.* 663! Célimélé (June) *Jac-Fél.* 1785! **S.L.:** Sugar Loaf Mt. (May–Nov.) *Sc. Elliot* 4023! *Barter*! *Deighton* 5648! Kasewe F.R. (Nov., Dec.) *Deighton* 4922! Kambui F.R., Kenema (May) *Jordan* 2086! **Lib.:** Banga (Oct.) *Linder* 1249! Tawata, Boporo Dist. (Nov.) *Baldwin* 10298! 10313! 10343! Boporo (Nov.) *Baldwin* 10389! **Iv.C.:** Grabo, Cavally Basin (July) *Chev.* 19696! **Ghana:** *cult. specimen!* **N.Nig.:** Vogel Peak, 3,500 ft. (Nov.) *Hepper* 1430! Mambila Plateau *Howard* in *Hb. King* 152! **S.Nig.:** Shasha F.R., Ijebu (May) *Richards* 3457! Afi River F.R. (May) *Jones* FHI 18780! Niaji Peak, Oban (Oct.) *Talbot* 925! Ikwette Plateau & Mt. Koloishe, Sonkwala, 5,400–6,200 ft. (Apr.) *Olorunfemi* FHI 25313! FHI 28293! *Tuley* 616! **W.Cam.:** Cam. Mt., 5,000 ft. (Nov.) *Mann*

Fig. 393.—Bulbophyllum barbigerum *Lindl.* (Orchidaceae).
A, flower. B, lip. C, column and petals. D, apex of hair from lip.

237

FIG. 394.—BULBOPHYLLUM PORPHYROSTACHYS *Summerh.* (ORCHIDACEAE).

1, flowering plant, × 1.   2, flower with one lateral sepal cut off, × 8 :—a, anther;   ds, dorsal
sepal;  ls, lateral sepal;  p, petal;  l, lip.  3, lip, side view and front view, × 12.  4, dorsal
sepal, × 8.  5, petal, × 8.  6, column with anther still in position, lip removed, × 12.
7, anther and the two pollinia, × 16.  8, the two pollinia, × 32.  All from *Meikle* 600.
(Reprod. from Hook., Ic. Pl. t. 3547 by permission of Bentham-Moxon Trustees).

238

2122! near Buea, 5,700 ft. (June) *Preuss* 943! **F.Po**: Las Corteras (Jan.) *Guinea* 2145! Throughout tropical Africa eastwards to Uganda and Tanzania, south to Angola and Mozambique.

32. **B. fuscoides** *J. B. Petersen* in Bot. Tidsskr. 49: 163, fig. 3–5 (1952). Flowers dark red, lip bright red or yellow.
 **W.Cam.**: M'bonge *Wright* 58/1! **F.Po**: Pico S. Isabel 2,500 ft. (Dec.) *Sanford* 4002! Also in E. Cameroun.
33. **B. zenkeranum** *Kraenzl.* in Engl., Bot. Jahrb. 48: 391 (1912). Flowers orange-yellow to orange-brown.
 **S.L.**: Lalehun, Gaura (May) *Bakshi* 192! **Lib.**: Kitoma (Apr.) *Harley* 1795! **Ghana**: Puso-puso (Jan.) *Hall* 2449! **N.Nig.**: Mambila Plateau *Wimbush* in *Hb. King* 135! **W.Cam.**: Mimbia, 3,900 ft. *Brenan* 9568! Buea, 3,000 ft. (Apr.) *Schlechter* 12377! Cam. Mt., path to Mann's Spring (Mar.) *Hambler* 153! Also in E. Cameroun, C. African Republic and Zambia.
34. **B. scariosum** *Summerh.* in Kew Bull. 7: 145 (1953). Flowers creamy white or flesh coloured.
 **Guin.**: Dalaba–Diaguissa Plateau, 1,300 ft. (Nov.) *des Abbayes* 1326! Nimba Mts., 5,300 ft. (Apr.) *Schnell* 1094! **S.L.**: Bintumane Peak, Loma Mts., 5,300–5,400 ft. (Jan.) *T. S. Jones* 64! *Jaeger* 568! 4145! **Lib.**: Nimba Mts. (Jan.) *Harley* 2204!
35. **B. bifarium** *Hook. f.* in J. Linn. Soc. 7: 219 (1864); F.T.A. 7: 27. *B. pallescens* Kraenzl. in Engl., Bot. Jahrb. 51: 385 (1914). Flowers whitish or bluish rose; bracts straw coloured.
 **Iv.C.**: Mt. Tonkoui (May) *Aké Assi* 5606! **W.Cam.**: Cam. Mt., 5,000 ft. (Nov.) *Mann* 2121! Buea, 3,300 ft. (Nov.) *Maitland* 781! Bimbia road, Victoria (fr. Apr.) *Maitland* 732! Bibundi (Apr.) *Schlechter* 12418. Nyassossa, 2,700 ft. *Schlechter* 12896! Mfongu, Bagangu, 5,700–6,300 ft. (Oct.) *Ledermann* 5860!
36. **B. nigericum** *Summerh.* in Kew Bull. 16: 307, fig. 21 (1962). Flowers green and purple, always on rocks.
 **Iv.C.**: Mt. Momi (Oct.) *Aké Assi* 9218! **N.Nig.**: Garawuri Hills, Plateau Prov. (Oct.) *King* 124! Also recorded from several other localities in Plateau Province.
37. **B. wrightii** *Summerh.* in Kew Bull. 16: 309, fig. 22 (1962). Flowers yellow with purple markings.
 **W.Cam.**: M'bonge, Kumba Div. *Wright* 58/18!
38. **B. lupulinum** *Lindl.* in J. Linn. Soc. 6: 126 (1862); F.T.A. 7: 28. (?) *B. tetragonum* Lindl. (1830)—F.T.A. 7: 32. *B. urbanianum* Kraenzl. in Engl., Bot. Jahrb. 28: 163 (1900). Flowers yellow with red spots, red or dark purple; bracts pale brown with purplish tinge.
 **Guin.**: Télimélé (Jan.) *Roberty* 16450! **S.L.**: Niénankolla (Jan.) *Chev.* 20569! Ninia (Feb.) *Sc. Ellio* 4915! Makali to Mamansu, Kunike (cult., Dec.) *Capstick* in *Deighton* 5718! Fromager to Kamaro (Jan.) *Jaeger* 3902! Alikalia (Feb.) *Bunting* 31! **Ghana**: Amedzofe (fl. & fr. Nov.) *Westwood* 3! **N.Nig.**: Karamti, Gashaka Dist. (Dec.) *Latilo & Daramola* FHI 28943! **S.Nig.**: Mt. Orosun, Idanre (Jan.) *Brenan et al.* 8643! Koloishe Mt., Sonkwala, 5,000 ft. (Dec.) *Savory & Keay* FHI 25272! **W.Cam.**: Ambas Bay (Feb.) *Mann* 783! Victoria (Jan.) *Deistel* 79! Buea, 3,700 ft. (Jan.) *Schlechter* 12844! Bum, Wum Div. (Feb.) *Hepper* 1923! Also in E. Cameroun and Congo.
39. **B. porphyrostachys** *Summerh.* in Bot. Mus. Leafl. Harv. Univ. 14: 230 (1951). Flowers green or yellow with red or purple spots or markings; bracts crimson, chocolate, red or purple, rarely green.
 **S.Nig.**: Okomu F.R. (Oct.) *Keay* FHI 22711! Sapoba F.R. (Oct.) *Keay & Onochie* FHI 21590! Agbadi, Sapoba (Nov.) *Meikle* 600! Urhehue, Usonigbe F.R. (Nov.) *Meikle* 601! Akamkpa Rubber Estate, Calabar Dist. (Mar.) *Latilo* FHI 41339! **W.Cam.**: Cam. Mt. (Nov.) *Keay & Brenan* FHI 28133! Victoria *Rosevear* Cam. 60/37!
40. **B. falcipetalum** *Lindl.* in J. Linn. Soc. 6: 128 (1862); F.T.A. 7: 26. *Megaclinium lutescens* Rolfe in Kew Bull. 1910: 158. Flowers yellow or yellow-green; usually on mangroves.
 **S.L.**: Yagoi, Jong R. (Oct.) *Adames* 111! **Iv.C.**: Yapo Forest, N. of Abidjan (Oct.) *de Wilde* 688! **Ghana**: *Band*! **S.Nig.**: Nun R. (Sept.) *Mann* 526! **W.Cam.**: Man of War Bay *Cole* 33! Also in E. Cameroun.
41. **B. rhizophorae** *Lindl.* in J. Linn. Soc. 6: 125 (1862); F.T.A. 7: 26. Flowers bronze-yellow to reddish violet or deep purple.
 **S.L.**: Yagoi, Jong R. (Nov.) *Adames* 110! **Iv.C.**: Yapo Forest, N. of Abidjan (Dec.) *de Wilde* 980! **Ghana**: Tarquah *Miles*! **S.Nig.**: Nun R. *Barter* 20118! Oban *Talbot*! **W.Cam.**: M'bonge, Kumba Div. *Wright* 58/5! Also in Congo.
42. **B. parvum** *Summerh.* in Kew Bull. 11: 122 (1957). Flowers purple.
 **S.L.**: Picket Hill, 2,300 ft. (Nov.) *T. S. Jones* 235!
43. **B. filiforme** *Kraenzl.* in Engl., Bot. Jahrb. 22: 25 (1895); F.T.A. 7: 27. *B. longispicatum* Kraenzl. & Schltr. in Orchis, 2: 98 (1908).
 **W.Cam.**: Victoria to Bimbia (Apr.–Sept.) *Preuss* 1242! Man of War Bay *Schlechter* 15759! Also in E. Cameroun.
44. **B. congolanum** *Schltr.* in Engl., Bot. Jahrb. 38: 14 (1905). *Megaclinium clarkei* Rolfe (1891)—F.T.A. 7: 41. *M. pobeguinii* Finet in Notulae Syst. 1: 167 (1910). *Bulbophyllum pobeguinii* (Finet) De Wild., Pl. Bequaert. 1: 95 (1921); F.W.T.A., ed. 1, 2: 440. *B. chevalieri* De Wild. l.c. 80 (1921). **Rhachis** green variously mottled red or crimson, or entirely crimson; flowers yellow, greenish or crimson.
 **Guin.**: Komassa *Pobéguin* 636! Banko *Pobéguin* 925! Ditinn (Apr.) *Chev.* 12838! Farana (Apr.) *Sc. Elliot* 5341! **S.L.**: Musaia (Feb.–Apr.) *Deighton* 4274! Kpindu, Kissi Kama (Nov.) *Deighton* 4656! Yengema, Kono (Nov.) *Deighton* 4655! 5131! **Lib.**: Ganta (Nov.) *Harley* 1837! 1943! **Iv.C.**: Bouroukrou *Chev.* 16839! Moossou (Nov.) *de Wilde* 801! **Ghana**: Ninting, Ashanti (Oct.) *Vigne* FH 4869! Awaso, Blayney 17! Mampong (Oct.) *Cansdale* 126! 136! Juaso, Ashanti (Oct.) *Cansdale* 04! Dayi R., Hohoe (Oct.) *Westwood* 48! **N.Nig.**: Assob R., Zaria Prov. (July) *King* 21! Tof, Plateau Prov. (Feb.) *King* 180a! Tibba Plat. *Horsman*! Delimi R. Gorge, Naraguta (Oct.) *Westwood*! Vogel Peak, 3,500 ft. (Nov.) *Hepper* 1431! **S.Nig.**: Mamu F.R., Ibadan (Oct.) *Keay* FHI 25426! Akure F.R., Ondo (Oct.) *Keay* FHI 22668! 25473! Mt. Orosun, Idanre (Nov.) *Keay & Brenan* FHI 28130! Obudu, Ogoja Prov. (Dec.) *Keay* FHI 28296! Also in E. Cameroun, Gabon, C. African Republic, Congo and Uganda.
45. **B. fuerstenbergianum** (*De Wild.*) *De Wild.* Pl. Bequaert. 1: 86 (1921). *Megaclinium fuerstenbergianum* De Wild. in Cogn. Dict. Ic. Orch. Megacl. t. 1 (1905). Rhachis green, mottled brown; flowers green, mottled brown, petals crimson, purple or chocolate.
 **N.Nig.**: Mambila Plateau *Wimbush* in *Hb. King* 129! **S.Nig.**: Ikwette, Sonkwala, 5,200 ft. (Dec.) *Savory & Keay* FHI 25183! *Cooper* 38! **F.Po**: Pico S. Isabel 4,000 ft. (Dec.) *Stanford* 4134! Also in Congo.
46. **B. tentaculigerum** *Rchb. f.* in Flora 61: 77 (1878). *Megaclinium tentaculigerum* (Rchb. f.) Dur. & Schinz (1895)—F.T.A. 7: 42. *Bulbophyllum stenopetalum* Kraenzl. in Engl., Bot. Jahrb. 22: 25 (1895). Flowers yellow, yellow with purple markings or purple.
 **Lib.**: Sinoe Basin *Whyte*! **Iv.C.**: 35 km. S. of Guéyo (Mar.) *Leeuwenberg* 3776! **Ghana**: Ashanti *Miles*! **S.Nig.**: Akamkpa Estate, Calabar R. (Mar.) *Latilo* FHI 41331! **W.Cam.**: Mopanya, 5,000–6,000 ft. (Mar.) *Kalbreyer*. Bibundi (Apr.) *Schlechter* 12358! Victoria *Preuss* 1217! Cam. Mt., 5,700–6,100 ft. (Feb.–Nov.) *Brenan* 9569! *Keay* FHI 37501! Tombel (Mar.) *Thorold* CM16! Also in E. Cameroun and Uganda.
47. **B. longibulbus** *Schltr.* in Engl., Bot. Jahrb. 38: 17 (1905).
 **W.Cam.**: Nyassosso, 2,700 ft. *Schlechter* 12893!
48. **B. simonii** *Summerh.* in Kew Bull. 1935: 204. *Megaclinium lasianthum* Kraenzl. in Engl., Bot. Jahrb. 48: 393 (1912). Flowers yellowish with brown or purple markings or entirely brown or purple.
 **S.Nig.**: Akamkpa Estate, Calabar (Mar.) *Latilo* FHI 41342! **W.Cam.**: Victoria (Feb.–Apr.) *Schlechter* 12373! *Simon* 11! *Rosevear* Cam. 56/37! 58/37! Bakangili, Victoria (Mar.) *Thorold* CM15! Also in E. Cameroun.
49. **B. velutinum** (*Lindl.*) *Rchb. f.* in Walp., Ann. 6: 258 (1861). *Megaclinium velutinum* Lindl. (1847)—F.T.A. 7: 39. *M. melanorrhachis* of F.T.A. 7: 41, partly (excl. spec. cult. Saunders). Rhachis green variously tinged purple, or entirely purple; flowers red or purple.

FIG. 395.—BULBOPHYLLUM VELUTINUM (*Lindl.*) *Rchb.f.* (ORCHIDACEAE).

A, flowering plant. B, flower. C, column.

**S.L.:** Jaluahun, Barri (Dec.) *Deighton* 4375! Moria, Mapaki (Dec.) *MacDonald* in *Hb. Deighton* 5714! Jagbwema, Luawa (Nov.) *Oliphant* in *Hb. Deighton* 6011! **Lib.:** Banga (Oct.) *Linder* 1255! Totokwelli—Moylakwelli (Oct.) *Linder* 1289! Ganta (Nov.) *Harley* 2066! **Iv.C.:** Abidjan *Aké Assi* L77! Adiopodoumé (Dec.) *Aké Assi* 7149! Gribo rapids, Sassandra R. (Nov.) *de Wilde* 3292! **Ghana:** Kanyankor, W. Prov. (Nov.–Dec.) *Miles* 20! Ancobra R. (Dec.) *Johnson* 855! Assin Yan Kumasi (Jan.) *Cummins* 78! Awaso *Blayney* 24! Efiduase *Cox* 17! **W.Cam.:** M'bonge *Wright* 58/39! Also in S. Tomé.

50. **B. melanorrhachis** (*Rchb. f.*) *Rchb. f. ex De Wild.* Pl. Bequaert. 1: 93 (1921). *Megaclinium minutum* Rolfe (1893)—F.T.A. 7: 40. *M. millenii* Rolfe in F.T.A. 7: 40 (1897); Cat. Talb. 146. *M. angustum* Rolfe in Kew Bull. 1922: 26. Rhachis purple or brown; flowers greenish or yellow variously marked purple, or entirely purple or reddish.
    **Guin.:** Forécaria *Jac.-Fél.*! **S.L.:** Sugar Loaf Mt., 3,000 ft. *Sc. Elliot*! Maswari (Apr.) *Adames* 209! **Lib.:** Bumbuma to Moala (Nov.) *Linder* 1347! Tawata, Boporo Dist. (Nov.) *Baldwin* 10339! **Ghana:** Jasikan *Westwood* 63a! **S.Nig.:** Lagos *Millen*! Jamieson R., Sapoba (Nov.) *Keay* FHI 28074! Calabar *Holland*! Oban *Talbot* 1576! Eket *Talbot* 3288! Also in E. Cameroun, Annobon, Gabon and Congo.

51. **B. falcatum** (*Lindl.*) *Rchb. f.* in Walp., Ann. 6: 258 (1861). *Megaclinium falcatum* Lindl. (1826)—F.T.A. 7: 41. *M. endotrachys* Kraenzl. in Engl., Bot. Jahrb. 36: 115 (1905). *Bulbophyllum leptorrhachis* Schltr. in Engl., Bot. Jahrb. 38: 17 (1905). Rhachis green to purple; flowers green or variously marked purple, or entirely red or purple, apex of petals often yellow.
    **Guin.:** Sérédou *Pujol* L364! **S.L.:** Sugar Loaf Mt., 1,800 ft. *Deighton* 5653! Mafari, Dixing (Nov.) *Adames* 200! Pendembu (Dec.) *Deighton* 4376! Sefadu (Nov.) *Deighton* 4658! Bintumane Peak, Loma Mts., 3,000 ft. (Nov.) *Deighton* 5910! **Lib.:** Begwai (Oct.) *Bunting* 25! Grand Bassa *Dinklage* 1852! Vahun, Kolahun Dist. (Nov.) *Baldwin* 10174! Buchanan (Nov.) *Adam* 16033! Kitoma (Oct.) *Harley* 1838! **Iv.C.:** Taï Forest (Oct.) *Aké Assi* 6035! **Ghana:** Kanjankor, W. Prov. (Nov., Dec.) *Miles* 21! Ninting, Ashanti (Oct.) *Vigne* FH 4870! Efiduase *Cox* 11! 14! Hohoe, Ntumada, 1,000–2,250 ft. (Nov.) *St.-Cl. Thompson* 1607! Jasikan (Nov.) *Lange* in *Hb. Westwood* 63! **S.Nig.:** Okomu F.R., Benin (Oct.) *Brenan* FHI 22713! Agbadi, Sapoba (Nov.) *Meikle* 603! Akure F.R., Ondo (Nov.) *Keay* FHI 28127! Koloishe Mt., Sonkwala, 6,000 ft. (Nov.) *Savory & Keay* FHI 28131! Kundere, Sonkwala, 2,800 ft. (Nov.) *Savory & Keay* FHI 28132! **W.Cam.:** Cam. Mt. *Brenan* FHI 28295! Onochie FHI 32436! Moliwe (Jan.) *Schlechter* 12992! Batoki *Cole* 42! **F.Po:** S. Isabel to San Carlos (fr. Jan.) *Sanford* 4220! Also in Principe, Congo and Uganda.

52. **B. purpureorhachis** (*De Wild.*) *Schltr.* Die Orchid. 328 (1914). *Megaclinium purpureorhachis* De Wild., Not. Pl. Util. Congo 126 (1903); Gard. Chron. ser. 3, 45: 293, fig. 126 (1909); Rolfe in Bot. Mag. 135: t. 8273 (1909). Rhachis closely and heavily blotched purple with numerous blackish hairs; flowers purplish or dark brown.
    **Iv.C.:** Taï Region (July) *Aké Assi* 5639! Hiré Region (Dec.) *Aké Assi* 9338! Also in E. Cameroun, Gabon and Congo.

53. **B. bufo** (*Lindl.*) *Rchb. f.* in Walp., Ann. 6: 258 (1861). *Megaclinium bufo* Lindl. (1844)—F.T.A. 7: 43. *M. deistelianum* Kraenzl. in Engl., Bot. Jahrb. 33: 72 (1902). *Bulbophyllum bakossorum* Schltr. in Engl., Bot. Jahrb. 38: 13, fig. 6A, 1–11 (1905). Rhachis green, blotched crimson or entirely crimson or purple; flowers cream or pale yellow, spotted crimson, or entirely crimson or purple.
    **Guin.:** Plateau de Benna *Jac.-Fél.*! Sérédou *Pujol* L365! **S.L.:** cult. *specimen*! **Lib.:** Gbanga (Oct.) *Linder* 1179! 1210! Peter's Town (Oct.) *Bunting* 102! Ganta (Oct.) *Harley* 1761! Zogowi (Nov.) *Harley* 1835! **Iv.C.:** Mt. Momi (Oct.) *Aké Assi* 9195! **Ghana:** cult. *specimen*! **S.Nig.:** Usonigbe F. R., Benin (Nov.) *Keay* FHI 28134! Akure F.R., Ondo (Oct.) *Keay* FHI 25474! Utanga, Obudu (Oct.) *Keay & Savory* FHI 25425! **W.Cam.:** Buca *Deistel*! Nyassosso, 2,700 ft. (Jan.) *Schlechter* 12898! **F.Po:** S. Isabel to San Carlos (fr. Jan.) *Sanford* 4219! Also in Congo.

54. **B. graminifolium** *Summerh.* in Kew Bull. 11: 117 (1957). *B. calyptratum* of F.W.T.A., ed. 1, 2: 440, partly, not of Kraenzl. Flowers greenish white or greenish yellow, sometimes with small red spots near base.
    **Guin.:** Bembaya, Farana (Feb.) *Chev.* 20670! **S.L.:** Musaia (Feb.) *Deighton* 4273! Seli R., Kamadugu (Feb.) *Glanville* in *Hb. Deighton* 5744! Karina (Feb.) *T. S. Jones* in *Hb. Deighton* 5745! **Lib.:** Vahun, Kolahun Dist. *Baldwin* 10173! Soplima, Vonjama Dist. *Baldwin* 10036! Kongba (Dec.) *Harley* 1950! Ganta (Nov.) *Harley* 2064! **Ghana:** Aburi (Nov.) *Johnson* 217! Agogo (Nov.) *D. Gillett* 7! Efiduase *Cox* 20! Awaso *Blayney* 20! Osiem, near Tafo (Nov.) *Westwood* 62!

55. **B. calyptratum** *Kraenzl.* in Engl., Bot. Jahrb. 22: 24 (1895); F.T.A. 7: 26; F.W.T.A., ed. 1, 2: 440, partly (Nigerian specimens). *B. lindleyi* (Rolfe) Schltr., Die Orchid. 327 (1914); F.W.T.A., ed. 1, 2: 440. *Megaclinium lindleyi* Rolfe (1897)—F.T.A. 7: 43. *M. lepturum* Kraenzl. in Ann. Natur. Holmus. Wien 30: 61 (1916). Rhachis green, sometimes spotted or tinged with red or purple; flowers greenish or greenish-yellow with small red or purple spots.
    **S.L.:** Yagoi, Jong R. (Oct.) *Adames* 84! 98! Kamasu, Tunkai (Jan.) *Deighton* 4576! Kebawana, Gaura (Mar.–Apr.) *Deighton* 4598! 5026! **Iv.C.:** *Merle* K519! **Ghana:** Techiman to Nkoranza (Jan.) *Westwood* 114! **S.Nig.:** Okomu F.R., Benin (Feb.) *Brenan* FHI 22735! Eket *Talbot* 3302! Obaniko-Unya, Obudu (Mar.) *Keay & Savory* FHI 25307! Oban F.R. (Nov.) *Latilo* FHI 45812! **W.Cam.:** Victoria (Apr.) *Schlechter* 12369! Preuss 1215! Rosevear Cam. 59/37! Tombel (Mar.) *Thorold* CM18! Bibunde (Mar.) *Rosevear* 114/36 (FHI 11136)! Also in Gabon and Congo.

56. **B. lucifugum** *Summerh.* in Kew Bull. 11: 119 (1957). Rhachis green spotted purple; flowers green and yellow with purple spots.
    **S.L.:** Makali, Kunike Barina (Nov.) *Deighton* 4377! Gegbwema, Tunkia (Dec.) *Deighton* 6154!

57 **B. colubrinum** (*Rchb. f.*) *Rchb. f.* in Walp., Ann. 6: 257 (1861). *Megaclinium colubrinum* Rchb. f. (1855)—F.T.A. 7: 40. *M. imschootianum* Rolfe (1895)—F.T.A. 7: 39. *Bulbophyllum imschootianum* (Rolfe) De Wild., Pl. Bequaert. 1: 86 (1921). Rhachis green with brown or purplish spots, or entirely brown; flowers yellow, speckled purple or entirely purple.
    **S.L.:** Great Scarcies R. (Sept.) *Adames* 128! **Iv.C.:** Bouroukrou *Chev.* 16524! **S.Nig.:** Usonigbe F.R., Sapoba (Nov.) *Ejiofor* FHI 24652! Akure F.R., Ondo Prov. (Nov.) *Keay* FHI 28129! *Sanford* 1567/65! Also in Gabon and Congo.

58. **B. oxypterum** (*Lindl.*) *Rchb. f.* in Walp., Ann. 6: 258 (1861). *Megaclinium oxypterum* Lindl. (1839)—F.T.A. 7: 39, partly (excl. Principe specimen). *Bulbophyllum ciliatum* Schltr. in Engl., Bot. Jahrb. 38: 156 (1906). *B. maximum* of F.W.T.A., ed. 1, 2: 439, partly, not of (Lindl.) Rchb. f. Rhachis green, variously spotted purple or almost entirely purple; flowers yellow with purple spots or entirely purple.
    **S.L.:** cult *specimen*! **Ghana:** Efiduase *Cox* 13! **N.Nig.:** Kagoma, Zaria Prov. (Mar.) *King* 126! Jos to Wamba (July) *Sanford* 1368/65! **S.Nig.:** Ajagbodudu, Sapele (May, June) *Wright* 134! Akure F.R., Ondo (Oct.) *Keay* FHI 21552! **W.Cam.:** near Bimbia (Sept.) *Schlechter* 15758! Also in Annobon, Congo, Kenya, Tanzania, Zambia, Rhodesia and Mozambique.

59. **B. maximum** (*Lindl.*) *Rchb. f.* in Walp., Ann. 6: 259 (1861), excl. descr.; F.W.T.A., ed. 1, 2: 439, partly. *Megaclinium maximum* Lindl. (1830)—F.T.A. 7: 38. Rhachis and flowers green, spotted purple, or entirely purple or red.
    **S.L.:** Smeathman! *Afzelius*! Mano Bonjema (Nov.) *Adames* 109! Bonthe (Nov., Dec.) *Adames* 112! **Lib.:** Wumbi (Sept.) *Linder* 827! Ganta (Oct.) *Harley* 1932! **Iv.C.:** Brafouédi, Dabou to N'Douci (fr. Dec.) *de Wilde* 1012! Abouabou Forest, Abidjan to Grand Bassam (fr. Dec.) *de Wilde* 1013! **Ghana:** Mansu (Dec.) *Miles* 32! Begoro, E. Prov. (Nov.) *Westwood* 156! **S.Nig.:** Brass R. *Barter* 1854! **Bagga-Balinge**, Sonkwala (Dec.) *Savory & Keay* FHI 25163! Also in Congo.

60. **B. bibundiense** *Schltr.* in Engl., Bot. Jahrb. 38: 155 (1906).
    **W.Cam.:** Moliwe (Oct.) *Schlechter* 15784!

**61. B. kamerunense** *Schltr.* in Engl., Bot. Jahrb. 38: 15, 16, fig. 6B and C (1905).
  **S.Nig.:** Sapoba F.R., Benin Prov. (fl. cult. Ibadan Feb.) *Sanford* 136/66! Also in E. Cameroun.
**62. B. magnibracteatum** *Summerh.* in Kew Bull. 1935: 203.
  **Ghana:** Imbraim (Oct.–Dec.) *Miles* 11! **S.Nig.:** Ijebu to Benin (Aug.) *Sanford* 1726/65!
**63. B. imbricatum** *Lindl.* Bot. Reg. 1841: Misc. 37. *B. stenorhachis* Kraenzl. in Engl., Bot. Jahrb. 22: 25
  (1895). *Megaclinium imbricatum* (Lindl.) Rolfe in F.T.A. 7: 37 (1897). *M. triste* Rolfe in F.T.A. 7: 38
  (1891). Flowers blackish purple, rarely with green markings.
  **S.L.:** Kabala to Musaia (Dec.) *Deighton* 4392! Giema, Dama (Nov.) *Deighton* 4954! **Lib.:** Peahtah
  (Oct.) *Linder* 938! Gbanga (Sept.) *Linder* 669! Ganta (Oct.) *Harley* 1827! **Iv.C.:** Abouabou Forest,
  Abidjan to Grand Bassam (Dec.) *de Wilde* 1025! **Ghana:** *Blayney*! **S.Nig.:** Ajagbodudu (Apr.) *Wright*
  149! **W.Cam.:** Victoria (Apr.) *Preuss* 1241! M'bonge *Wright* 58/44! Also in Gabon and Congo.
**64. B. leucorrhachis** (*Rolfe*) *Schltr.* in Engl., Bot. Jahrb. 38: 17 (1905). *Megaclinium leucorrhachis* Rolfe
  (1891)—F.T.A. 7: 37. Rhachis pale green; flowers yellow and orange.
  **S.L.:** Smythe 114! **S.Nig.:** Lagos *Moloney*! Millen 189! Okomu F.R., Benin (Dec.) *Keay* FHI 22715!
  Also in Gabon and Congo.
**65. B. linderi** *Summerh.* in Kew Bull. 1935: 202. Rhachis and flowers cream coloured or whitish.
  **S.L.:** Mesima, Pukumu Krim (Nov.) *Deighton* 4921! **Lib.:** Bumbuma (Oct., Nov.) *Linder* 1325! *Harley*
  1947! Nimba (Jan.–Oct.) *Adam* 20685! *Adames* 685! **Iv.C.:** Tiapleu (Oct.) *Aké Assi* 9129!

### 29. GENYORCHIS Schltr. in Engl., Bot. Jahrb. 38: 11 (1905).

Pseudobulb 2-leaved, scarcely flattened:
  Lateral lobes of the lip as long as, or longer, than the remainder of the lip below the
    recurved middle lobe; mentum of flower very short and rounded, the margins of the
    lateral sepals equal; pseudobulbs up to 12 mm. long; leaves 1–3 cm. long, 2–3 mm.
    broad; scape 2–4 cm. long, very slender; rhachis zigzag, up to 10-flowered
                                                                      1. *micropetala*
  Lateral lobes of the lip much shorter than the remainder of the lip below the recurved
    middle-lobe; mentum of flower helmet-shaped or conical, the upper margin of the
    lateral sepals much longer than the lower:
    Middle lobe of lip small, tooth-like; stelidia lanceolate, acute, much overtopping the
      androclinium, scarcely angled on each side of the stigmatic cavity:
      Dorsal sepal over 4 mm. long, lateral sepals somewhat spreading so that flower
        appears open, upper margins over 5·5 mm. long; apex of lip middle lobe incurved,
        quite free from the lower surface; wings of column somewhat curved backwards;
        pseudobulbs 10–12 mm. long; leaves broadly or elliptical-oblong, 1–2 cm. long,
        5–7 mm. broad; scape very slender, 8 cm. long, 5-flowered in the upper 1 cm.
                                                                      2. *apertiflora*
      Dorsal sepal 2·5–3 mm. long; lateral sepals more or less parallel in position, upper
        margins 3·5–4·5 mm. long; apex of lip middle lobe not incurved, whole of lobe
        adnate to the lower surface; wings of column straight; pseudobulbs 6–15 mm. long;
        leaves ligulate or oblong, 7–20 mm. long, 2–5 mm. broad; scape up to 17 cm. long,
        up to 12-flowered in the upper third or quarter  ..    ..    .. 3. *pumila*
    Middle lobe of lip large, almost orbicular, entirely adnate to the lower surface;
      stelidia scarcely overtopping the androclinium, forming triangular projections on
      each side of the stigmatic cavity; mentum 3 mm. long, upper margins of lateral
      sepals over 6·5 mm. long; lip curved, 4·5–5 mm. long, side lobes very short; pseudo-
      bulbs 1–2 cm. long; leaves 1–3 cm. long, 3–7 mm. broad; scape to 6·5 cm. long, up
      to 7-flowered ..    ..    ..    ..    ..    ..    ..    .. 4. *macrantha*
Pseudobulbs 1-leaved, somewhat flattened, 1–1·7 cm. long, 5–10 mm. broad; leaves
    oblong-elliptical, 1·5–4·5 cm. long, 4–9 mm. broad; scape slender, 10–14 cm. long,
    5–17-flowered in the upper half; mentum helmet-shaped, 2·5–3 mm. long; lip about
    3 mm. long, front lobe recurved, tooth-like  ..    ..    .. 5. *platybulbon*

**1. G. micropetala** (*Lindl.*) *Schltr.* in Engl., Bot. Jahrb. 38: 11 (1905). *Bulbophyllum micropetalum* Lindl. in
  J. Linn. Soc. 6: 127 (1861). *Polystachya micropetala* (Lindl.) Rolfe in F.T.A. 7: 131 (1897).
  **F.Po:** 4,000 ft. (Dec.) *Mann* 644! Pico Boca, 5,400 ft. *Sanford* 4267! Also in (?) Congo.
**2. G. apertiflora** *Summerh.* in Kew Bull. 11: 123 (1957). Sepals white, lip pale yellow.
  **S.Nig.:** Mt. Orosun, Idanre Hills, Ondo (Nov.) *Keay* FHI 25545!
**3. G. pumila** (*Sw.*) *Schltr.* in Engl., Bot. Jahrb. 38: 12 (1905). *Dendrobium pumilum* Sw. (1805). *Polystachya
  bulbophylloides* Rolfe (1891)—F.T.A. 7: 131. Flowers white or greenish white.
  **S.L.:** *Afzelius*! Bagroo R. *Mann*! Benkia, Port Loko Creek (Mar.) *Glanville* in Hb. *Deighton* 1816!
  Maswari, Peninsula (Mar.) *Adames* 203! Yagoi, Jong R. (Nov.) *Adames* 101! **Iv.C.:** Abouabou (Dec.)
  *Aké Assi*! Yapo Forest (Dec.) *de Wilde* 978! **Ghana:** Agma, W. Prov. (Dec.) *Cox* 107! **S.Nig.:** Brass R.
  *Barter* 72! Okomu F.R., Benin (Jan.) *Brenan* 8754! 8914! Sapoba F.R., Benin (Jan.) *Keay* FHI 37736!
  Akilla, Ijebu Ode (Jan.) *Onochie & Ladipo* FHI 20669! Mt. Orosun (Dec.) *Gregory*! **F.Po:** Basilé to
  Estrada (fr. Dec.) *Sanford* 4009! Also in E. Cameroun, Principe, the Congos and Uganda.
**4. G. macrantha** *Summerh.* in Kew Bull. 11: 124 (1957). Sepals white or very pale pink; lip pink with yellow
  apex.
  **W.Cam.:** Mann's Spring, Cam. Mt., 7,500 ft. (Apr.) *Brenan & Richards* 9570!
**5. G. platybulbon** *Schltr.* in Engl., Bot. Jahrb. 38: 155 (1906).
  **W.Cam.:** Moliwe *Stammler*! Victoria *Unknown Coll.* (491) FHI 11155!

### 30. EULOPHIDIUM Pfitz., Entw. Nat. Anord. Orch. 87 (1887).

Labellum without any calli at the base, 7 mm. long, 4–6 mm. broad, broadly 3-lobed, the
  middle lobe clearly emarginate; leaves 2–3 to the pseudobulb, lamina 5–13 cm. long,
  1·5–4·5 cm. broad, rounded at base; pseudobulbs long and slender, up to 13 cm. long,
  usually less than 5 mm. diam.; inflorescence slender, shorter than the leaves, nearly

always branched; sepals and petals ligulate, a little broader above, 7–9 mm. long, the petals a little shorter than the sepals; spur slightly swollen, 3–4 mm. long

                                                     1. *latifolium*

Labellum with 2 calli close together at the base, as broad as long or a little broader than long, almost equally 4-lobed, the two parts of the middle lobe separated by a deep cleft; leaves 1–2 (rarely 3) to the pseudobulb, lamina almost always over 9 cm. long and often much longer; inflorescence usually longer than the leaves:

Leaves always 1 to a pseudobulb, lamina narrowly lanceolate, thick and leathery, tapering gradually into a broad folded petiole, pale green with irregular transverse bands of darker spots, 8–32 cm. long, 2–5·5 cm. broad, petiole much shorter than the lamina; pseudobulbs narrowly conical or cylindrical, up to 4 cm. long and 1·5 cm. diam.; tepals about 1 cm. long, the petals a little broader than the sepals; spur 4–4·5 mm. long .. .. .. .. .. .. .. 2. *maculatum*

Leaves 1–3 to a pseudobulb, lamina broadly lanceolate, leathery but not very thick, dark green in colour, 10–22 cm. long, 4–8 cm. broad, narrowing rather abruptly into an almost terete channelled petiole which is nearly as long as the lamina; pseudobulbs narrowly conical to narrowly cylindrical, up to 20 cm. long and 1·5 cm. diam.; tepals about 1 cm. long, the petals distinctly shorter and broader than the sepals; spur 5 mm. long .. .. .. .. .. .. .. 3. *saundersianum*

1. **E. latifolium** (*Rolfe*) *Summerh.* in Bull. Jard. Bot. Brux. 27: 396 (1957). *Eulophia latifolia* Rolfe (1891, pub. 1892)—F.T.A. 7: 50. *E. ugandae* Rolfe in Kew Bull. 1913: 339. Flowers white, with faint reddish or purplish markings.
    **Ghana**: Breniesi, E. Krachi, Ashanti (fr. Dec.) *Westwood* 147! Also in S. Tomé, Congo and Uganda.
2. **E. maculatum** (*Lindl.*) *Pfitz.* Entw. Natur. Anordn. Orch. 88 (1887). *Angraecum maculatum* Lindl., Collect. Bot. t. 15 (1821). *Eulophia ledienii* Stein ex N. E. Br. (1889–90)—F.T.A. 7: 50. *Eulophidium ledienii* (Stein ex N.E. Br.) De Wild. in Ann. Mus. Congo, sér. 5, 1: 115 (1904); F.W.T.A., ed. 1, 2: 440. *E. warneckeanum* Kraenzl. in Engl., Bot. Jahrb. 33: 70 (1902). Flowers greenish white with pink blotches on the lip.
    **Sen.**: Boukitengo Forest, Oussouye (Nov.) *Berhaut* 6527! **Port.G.**: Bissau, Cumura (fr. Nov.) *Esp. Santo* 2226! **S.L.**: Njala (Aug.) *Deighton* 2208! R. Seli, Kombile (Oct.) *Small* 818! Banana Is. (Aug.) *Melville & Hooker* 375! York (Sept.) *Melville & Hooker* 384! Bunce Is. (Sept.) *Milne-Redhead* 5143! **Lib.**: Ganta (Aug.) *Harley* 1825! Gbanga, Central Prov. (July) *Daniel & Barker* 215! **Iv.C.**: Issia rock (Aug.) *de Wilde* 417! **Ghana**: Aburi (Apr.) *Cox* 96! Ashanti *cult. plant!* Bana Hill, Krobo *Irvine* 2883! Kpeve (Aug.–Oct.) *Westwood* 15! 15a! **Togo Rep.**: Lomé (July) *Warnecke* 196! **N.Nig.**: Bonu, Niger Prov. (Aug.) *Onochie* FHI 18259! **S.Nig.**: Ukungu (fr. Jan.) *Thomas* 219! Ibadan (Oct.) *Jones* FHI 13733a! Milliken Hill, Enugu, *Cole* 51! Ofumbougha, Obubra, Abakaliki Prov. *Sanford* 667/66! Also in S. Tomé, the Congos, Sudan, Uganda, Tanzania and Brazil.
3. **E. saundersianum** (*Rchb. f.*) *Summerh.* in Bull. Jard. Bot. Brux. 27: 401 (1957). *Eulophia saundersiana* Rchb. f. (1866)—F.T.A. 7: 50; F.W.T.A., ed. 1, 2: 444. Flowers green, yellow-green or yellow, with purplish veins and markings; in forest.
    **S.L.**: Kambui F.R., Kenema (Apr.) *Jordan* 2035! Loma Mts. (Feb.) *Jaeger* 4311! **Lib.**: Sanokwele (Sept.) *Baldwin* 7513! Zuole, Central Prov. (Apr.) *de Wilde* 3755! Ganta (Apr.) *Harley* 2026! **Iv.C.**: Bouroukrou (Dec.–Jan.) *Chev.* 16905! Banco F.R. *de Wit* 8168! Yaokro, Bouaflé to Sinfra (Dec.) *Aké Assi* 7223! **Ghana**: Assin-Yan-Kumasi *Cummins* 222! Aburi *Johnson* 219! Mirouam, Ashanti (Jan.) *Chipp* 668! Efiduase (Dec.) *Cox* 51 (FH 4133)! **S.Nig.**: Ijebu-Ode (Jan.) *Tamajong* FHI 21052! Ikene, Ijebu-Ode (Jan.) *Burtt* B28! Ijaiye F.R., Oyo (Jan.) *Keay* FHI 21158! Ibadan North F.R. (Dec.) *Jones* FHI 7177! Oban *Talbot* 924! **W.Cam.**: Barombi (Sept.) *Preuss* 96! 546. **F.Po**: Musola (Jan.) *Guinea* 1140! 1231! Also in E. Cameroun, Gabon, Congo, Uganda, Tanzania and Angola.

---

31. **EULOPHIA** R. Br. ex Lindl., Bot. Reg. t. 686 (1823); F.T.A. 7: 47, 70. *Nom. cons.*
            *Lissochilus* R. Br. (1821)—F.T.A. 7: 70.

Spur 1·5–3 cm. long, very slender; sepals and petals similar, green or brownish, lanceolate acute, spreading and somewhat reflexed, 2–3 cm. long; lip 3-lobed, 2–4 cm. long, without calli, pink, side lobes small, middle lobe much larger, ovate to suborbicular, acute to retuse; scape 40–90 cm. high; leaves up to 45 cm. long, 4–13 cm. broad:

Leaves fully developed at time of flowering; lip ovate, elliptical or obovate, more or less acute at apex; usually in forest or woodlands.. .. .. 1. *guineensis*

Leaves developing after the inflorescence; lip broadly obovate or almost orbicular, broadly rounded or slightly emarginate at apex; usually in savanna 2. *quartiniana*

Spur up to 1·5 cm. long, usually less than 1 cm., if over 1 cm. then very wide in lower part:

Lip truncate, shorter than the spur, the front margin longly ciliate, callus tooth-like near margin in centre; leaves opening with flowers, lanceolate, 1–5 cm. broad; scape 40–95 cm. high, laxly many-flowered; sepals lanceolate, acute, 10–12 mm. long; petals similar but smaller; spur clavate at the apex, widened at the base, 8 mm. long; rarely epiphytic .. .. .. .. .. .. .. .. 3. *gracilis*

Lip not truncate, usually considerably longer than the spur, margin sometimes slightly toothed but never ciliate:

Bracts tapering from the base or just above, usually linear-lanceolate and long-acuminate, or if broader less than 1 cm. long: (to p. 246)

*Sepals more or less connivent or somewhat spreading, not reflexed, in most species very similar to the petals in shape and colour (see species 12, 15, 16, 17 & 18 where petals are broader): (to p. 245)

    A leafless pale yellow saprophyte; underground stem rather thin, fleshy, with numerous obtuse scale-leaves and semiglobose whitish tubercles (? arrested roots);

scape 10–25 cm. high, up to 5-flowered; sepals and petals oblanceolate, 1·5–
2·5 cm. long; lip 3-lobed, 1·5–2 cm. long, side lobes very small, rounded, middle
lobe much larger, emarginate, lip covered in the front part with small purplish
papillae    ..        ..        ..        ..        ..        ..        ..        ..        19. *galeoloides*
Plants with leaves, which may expand after the flowers; stems fleshy tubers or
pseudobulbs:
Sepals 4–12 mm. long; leaves in most species not expanding until after flowering:
Inflorescence branched; scape 40–70 cm. high; no leaves at time of flowering;
mature leaves 5–6, linear, spreading; bracts very short; sepals narrow, acute,
about 7 mm. long; petals similar, a little shorter; lip 3-lobed, 6–7 mm. long,
side lobes obtuse, middle lobe orbicular, disk with 5 slightly rough keels;
spur 2–3 mm. long        ..        ..        ..        ..        ..        ..        14. *ramifera*
Inflorescence not branched, but sometimes 2 inflorescences arise from the same
stem (rarely a single branch in No. 13):
Leaves 4–6 on the stem, narrowly linear, up to 20 cm. long but less than 4 mm.
broad, usually well-developed at time of flowering; stems swollen at base to
form narrowly conical aerial pseudobulbs; inflorescences often paired, up
to 25 cm. long, rather lax; flowers spreading, becoming pendulous in fruit;
sepals ligulate, 8–11 mm. long, petals a little shorter and broader; lip 3-
lobed in upper part, middle lobe elliptical, veins with several rows of quadrate
calli or bearing interrupted keels    ..        ..        ..        ..        ..        9. *lindiana*
Leaves 1 on the flowering shoot, or 2–4 on a separate sterile shoot, or absent at
time of flowering; stems in form of underground tubers:
Calli of lip consisting of verrucose keels which may be more or less confluent,
not of hairs:
Petals distinctly shorter than the sepals but much broader, 4·5–6 mm. long;
leaves partly developed at time of flowering, 2–4 on distinct sterile shoots,
broadly lanceolate or elliptical, when mature up to 30 cm. long and 7 cm.
broad; scape 30–50 cm. tall, many-flowered in upper part; sepals 6–8 mm.
long; lip 3-lobed, side lobes short, adnate to column, middle lobe much
longer, keels of callus confluent into a single mass, spur 4–5 mm. long
                                                17. *leonensis*
Petals equalling or longer than the sepals; no leaves at time of flowering;
scape 30–40 cm. long, many-flowered:
Spur narrow and elongated from a broad base, total length, from apex of
ovary, 6–8·5 mm.; sepals 6·5–8 mm. long, ligulate; lip distinctly 3-lobed,
side lobes erect, semi-elliptical, middle lobe elliptical, 4–6·5 mm. long with
5–7 verrucose keels; petals much broader than sepals and usually a little
longer, broadly elliptical or almost orbicular    ..        ..        18. *stenoplectra*
Spur very short, less than 2 mm. long; sepals less than 6·5 mm. long:
Lip side lobes very small, triangular, not much diverging from the much
larger orbicular front lobe; keels of lip rather indistinct, the thickenings
transversely arranged; sepals 4·5–6·5 mm. long, oblong-lanceolate,
apiculate; petals broader than sepals and about the same length,
elliptical; leaves on separate shoots, linear acute    ..        15. *sordida*
Lip side lobes basal, spreading, broadly triangular; keels of lip prominent
with a group of irregular taller calli just in front of the spur opening;
sepals 3·5–6·5 mm. long, oblong, acute; petals broader than sepals, the
same length or longer, elliptical to nearly orbicular; leaves on separate
shoots, linear-lanceolate, fleshy        ..        ..        ..        ..16. *parvula*
Calli of lip consisting mainly of hair-like outgrowths:
Petals about half to two-thirds as long as the sepals, narrowly elliptical;
sepals narrowly oblanceolate, subacute, 7·5–14 mm. long; lip 3-lobed, side
lobes very small, more or less incurved, middle lobe elliptical or almost
orbicular, much larger; spur swollen at the end, about 3–4 mm. long, incurved; scape 30–60 cm. tall; raceme lax, 6–15-flowered, bracts very
small ..        ..        ..        ..        ..        ..        ..        ..        13. *brevipetala*
Petals about the same length as the sepals:
Raceme short (3–7 cm. long), dense; peduncle completely encased in overlapping chaffy brown sheaths; bracts similar, longer than the pedicels;
scape 40–80 cm. tall; sepals and petals very similar, ovate or lanceolate,
9–12 mm. long; lip 3-lobed, side lobes triangular, middle lobe oblong,
disk with 2 thickened keels below the hairy veins; leaves 1 or 2 on a
separate sterile shoot, narrowly oblanceolate, 30–90 cm. long, 7–19 mm.
broad    ..        ..        ..        ..        ..        ..        ..        8. *shupangae*
Raceme relatively long, lax; sheaths of peduncle not overlapping; sepals
narrow or oblong-lanceolate:
Sepals linear-ligulate, acute, 10–12 mm. long; petals ligulate, a little
shorter and broader; lip side lobes obtusely triangular, middle lobe
obovate, lip about 9 mm. long; raceme 15 cm. long..        ..        12. *monile*

Sepals oblong or oblong-lanceolate, 4–7 mm. long; petals similar but a little shorter; lip 3-lobed, middle lobe rectangular or slightly wider in front, emarginate, margins irregular, veins bearing numerous hair-like outgrowths; scape 15–40 cm. tall, raceme to 10 cm.; leaf single or rarely 2, almost thread-like or very narrowly linear, up to 33 cm. long:

Side lobes of lip not diverging from middle lobe; disk with thickened callus or calli; spur 2–4·5 mm. long, rather slender     11. *warneckeana*

Side lobes of lip divergent; disk without thickened calli but 3 central veins only slightly thickened; spur about 1·5–2 mm. long, slightly widened at the apex     ..     ..     ..     ..     ..     ..     10. *milnei*

Sepals 13–35 mm. long; bracts narrowly lanceolate or linear, acuminate:

Spur represented by a broad rounded sac; lip bearing 2 quadrate calli in the centre and hairs on the veins of the broadly ovate or semiorbicular middle lobe; petals a little shorter and broader than the sepals; leaves lanceolate, 3–10 cm. broad, appearing with the flowers; scape 40–120 cm. high, many-flowered, bracts very narrow ..     ..     ..     ..     ..     ..     20. *alta*

Spur acute or, if obtuse, much narrowed:

Leaves more or less well-developed at time of flowering, 1·5–6 cm. broad; lip 12–25 mm. long; sepals rather longer:

Leafy stems thickened at base to form tall cylindrical somewhat tapering green pseudobulbs; lip elliptical in general outline, side lobes small, triangular, middle lobe ovate, crenulate, central nerves thickened to form 2–3 low keels; spur 6–8 mm. long, wider at apex; scapes 40–200 cm. high; raceme 20–40 cm. long; sepals and petals similar in shape and colour     4. *euglossa*

Leafy stems thickened only at the very base, not forming pseudobulbs; scapes 30–70 cm. high:

Sepals and petals differently coloured, 1·5–2 cm. long, the petals broader than the sepals; lip nearly as broad as long, disk and front lobe with 3–5 verrucose keels sometimes bearing hair-like outgrowths; raceme rather lax, 10–20 cm. long; spur 3–4 mm. long, tapering towards apex

5. *stachyodes*

Sepals and petals the same colour, 2–3·5 cm. long, the petals narrower than the sepals; lip longer than broad, disk with 2 low keels, front lobe bearing numerous slender hairs; raceme very dense, usually less than 10 cm. long; spur 2–4 mm. long, more or less incurved, slender not tapering

6. *zeyheri*

Leaves scarcely developed at time of flowering, apparently much less than 1 cm. broad; plants with irregular tubers:

Petals about half to two-thirds as long as the sepals; sepals narrowly oblanceolate, subacute, 7·5–14 mm. long; spur swollen at end, 3–4 mm. long; lip 3-lobed, side lobes much smaller than the elliptical or almost orbicular middle lobe; scape 30–60 cm. tall; raceme lax, 6–15-flowered ..     13. *brevipetala*

Petals nearly as long as the sepals; sepals lanceolate, acute; spur tapering towards the apex:

Flowers green with brownish, purple or crimson markings on lip; petals oblong, narrower or not broader than the sepals, 13–20 mm. long; lip 10–20 mm. long, disk with 2 low keels, veins of front lobe bearing short hairs or much-interrupted keels; scape 28–80 cm.     ..     ..     7. *adenoglossa*

Flowers pale yellow or cream-coloured with sometimes the veins of the lip purplish; petals oblong-elliptical, broader than the sepals; sepals 17–29 mm. long; lip 20–27 mm. long, the orbicular crenulate middle lobe with 5–9 low verrucose keels; scape 30–100 cm.     ..     .. 22. *flavopurpurea*

*Sepals sharply reflexed, in most species markedly different from the petals in shape and colour: (from p. 243)

Sepals more or less spathulate, widest near the rounded apex, 1·3–2·5 cm. long; petals elliptical or oblong-elliptical, about as long as the sepals and at least 2–3 times as wide; scapes 0·5–1·5 m. high; lip indistinctly 3-lobed, side lobes broad, rounded:

Petals white to red; leaves very narrow, 5–15 mm. broad; central veins of lip bearing numerous short hair-like outgrowths and 3 low keels towards the base

24. *caricifolia*

Petals yellow; leaves broader, 12–45 mm. broad; lip with 3 semicircular or lamellae-like keels in the centre and 3 lower ones at the base     25. *angolensis*

Sepals not wider at the apex than below, obtuse or acuminate:

Flowers yellow, petals with reddish veins or entirely red beneath; sepals 6–11 mm. long, rarely to 14 mm., obovate-elliptical, shortly acuminate; petals ovate-orbicular, 11–20 mm. long and almost as broad; lip 3-lobed, middle lobe ovate more or less convex, side-lobes erect, adnate to the foot and also the base of the column:

Leaves more or less developed at time of flowering, long and round, almost terete in section, rush-like, less than 5 mm. diam.; spur upcurved, total length to apex of ovary 3–7·5 mm.; sides and apex of lip middle-lobe more or less incurved and somewhat undulate, keels 5, rather low; sepals 6·5–9 mm. long; petals 11–14·5 mm. long    ..    ..    ..    ..    33. *juncifolia*

Leaves developed after flowering, strap-shaped or lanceolate, 1–3 cm. broad; spur almost straight, often with long cylindrical apex, total length to apex of ovary 11–19 mm.; lip with 3 tall keels and 1 or 2 additional lower ones on each side; sepals 7·5–14 mm. long; petals 12·5–20 mm. long ..    34. *orthoplectra*

Flowers various shades of pink or mauve; sepals 12–33 mm. long:

Sepals and petals similar, more or less oblong, the latter a little broader; lip relatively narrow, 1·3–2·2 cm. long, trilobed, the middle lobe elliptical or elliptical-ovate, disk with 2 semicircular entire keels towards the base and 5–7 much-broken-up keels in front; spur narrowed into an acute somewhat up-turned apex; raceme 20–45 cm. long, 13–40-flowered; leaves narrowly lanceolate, 2–6 cm. broad, appearing after the flowers    ..    23. *cristata*

Sepals and petals very dissimilar, the latter much wider, broadly elliptical or almost orbicular; lip broad, trilobed, the middle lobe more or less bilobed or retuse:

Spur relatively narrow, conical; lip less than 2 cm. long, middle lobe with several low rugulose keels but no erect calli    ..    ..    ..    21. *calantha*

Spur very broad and rounded, scarcely distinct from rest of lip; lip 2–3·5 cm. long, ecallose or with 2 erect calli at the base of the middle lobe and a few lower ones in front:

Lip with no calli, front lobe with a distinct narrow claw; sepals more or less oblong, obtuse; petals retuse at apex; leaves appearing after the flowers    28. *buettneri*

Lip with 2 erect narrow quadrate calli in the centre and the veins just in front variously keeled and/or with small horn-like outgrowths, front lobe not much narrowed at base; petals rounded or minutely apiculate at apex:

Leaves appearing after the flowers, linear, up to 1·5 cm. broad 26. *cucullata*

Leaves appearing with the flowers, lanceolate, 2·5–6 cm. broad    27. *dilecta*

Bracts broad from a much narrower base, apiculate to acuminate in the upper part, over 1 cm. long and 7–30 mm. broad; scapes 1–2 m. high; leaves at least 3 cm. broad: (from p. 243)

Petals much longer and broader than the narrow sepals, 2·5–3 cm. long, suborbicular; sepals narrowly spathulate-oblong, 1·5–2 cm. long, 5–7 mm. broad; bracts obovate, apiculate; spur more or less inflated underneath in the lower part, suddenly narrowed near the apex; lip about 3 cm. long, 3-lobed, side lobes rounded, middle lobe ovate, obtuse, disk with 3 rather short keels in the centre 31. *oedoplectron*

Petals not much longer than or equalling the sepals in length, but broader; bracts broadly lanceolate or ovate, acuminate or apiculate:

Lip indistinctly 3-lobed, 2·5–4 cm. long and broad, front lobe much broader than long, disk with 3–5 keels; sepals obovate, obtuse, 2·5–3 cm. long, 1–2 cm. broad; spur conical with a narrow upcurved apex; bracts 2–5 cm. long, 1·5–2·5 cm. broad; petals not much spreading, orbicular or broader than long 32. *latilabris*

Lip distinctly 3-lobed, front lobe longer than broad or length and breadth about equal, disk with 3 keels, semicircular at the base and running out into crenulate lamellae in front; petals broadly elliptical or almost orbicular, usually longer than broad:

Sepals oblanceolate or obovate, apiculate or shortly acuminate, much narrowed in the lower part, 1·7–2·7 cm. long; lip 2–4 cm. long, 1·6–4 cm. broad across the side lobes    ..    ..    ..    ..    ..    29. *horsfallii*

Sepals oblong-oblanceolate, obtuse, not much narrowed in the lower part, 1·3–1·7 cm. long; lip 1·6–2·5 cm. long, 1·5–2·5 cm. broad across the side lobes 30. *barteri*

1. **E. guineensis** *Lindl.* in Bot. Reg. 8: t. 686 (1823); F.T.A. 7: 69, partly; F.W.T.A., ed. 1, 2: 444, partly (excl. syn. *E. quartiniana* A. Rich.). A terrestrial herb 1–3 ft. high with several leaves arising from the base and spike of pink and greenish flowers; in forest.
**Gam.:** Kudang (June) *Brooks* 58! Genieri, Dyjia Kunda (July) *Fox* 145! **Mali:** Massif de Kita (fr. Sept.) *Jaeger* 2638! **Guin.:** Timbi Toumi, Fouta Djalon (Sept.) *Adames* 370! **S.L.:** Yonibana (Nov.) *Thomas* 4676! Njala (Nov.) *Deighton* 2405! foot of Mt. Péran-Konko, Loma Mts. (Sept.) *Jaeger* 1749! Bafodia to Kamba (Sept.) *Deighton* 5186! Musaia, Koinadugu (Aug.) *Haswell* 128! **Iv.C.:** Oroumba Boca, Baoulé (Aug.) *de Wit* 1017! **Ghana:** Tanosu, W. Ashanti (Jan.) *Chipp* 70! Aburi (Nov.) *Irvine* 800! NE. of Obenyemi (Nov.) *Morton* GC 6068! Kumasi (Sept.) *Andoh* 4464! Mampong Scarp (Dec.) *Adams* 4530! **N.Nig.:** Kabba (Oct.) *Parsons* L106! **S.Nig.:** Akilla F.R., Ijebu Dist. (Oct.) *Emwiogbon* FHI 37953! Abo to Iddah (Sept.) *T. Vogel* 1! Sapoba *Kennedy* 2651! Akpaka F.R., Onitsha (Sept.) *Onochie* FHI 33449! Eket *Talbot* 3650! Also in Congo, Uganda, Tanzania and Angola.

2. **E. quartiniana** *A. Rich.* Tent. Fl. Abyss. 2: 284, t. 81 (1851). *E. guineensis* of F.W.T.A., ed. 1, 2: 444, partly, not of Lindl. (1823). A terrestrial herb 1–3 ft. high with a bunch of leaves at the base and a spike of pink flowers; in savanna and among rocks.
**Sen.:** Hann F.R., Cape Verde Peninsula (July) *Naegelé* 4490! **Mali:** Dogoma, Ségou N. (July) *Roberty* 2510! **Ghana:** Gambaga (June) *Harris*! **N.Nig.:** Nupe *Barter* 1485! Fatika, Zaria Prov. (May) *Daggash*

FHI 31439! Zungeru (June) *Dalz.* 234! Tilde Filani, 3,300 ft. (May) *Lely* 232! Anara F.R., Zaria Prov. (May) *Keay* FHI 22904! **S.Nig.**: Old Oyo F.R. (May) *Keay* FHI 16210! Also in E. Cameroun, C. African Republic, Congo, Sudan, Ethiopia, Uganda, Kenya and Tanzania.

3. **E. gracilis** *Lindl.* in Bot. Reg. 9: t. 742 (1823); F.T.A. 7: 51. *E. preussii* Kraenzl. (1893). *E. virens* A. Chev., Bot. 614 (1920), name only. Leaves basal, long and narrow, flowers green in a long rather lax inflorescence.
**Sen.**: *Frey* M1682! Londia Forest, Oussouye (July) *Berhaut* 6210! **Port.G.**: Quitafine, Cacine (Jan.) *Raimundo & Guerra* 739! **S.L.**: Sugar Loaf Mt. (Sept.) *Preuss* ! Near Jama (Apr.) *Deighton* 3724! Njala (May) *Deighton* 680! Bagroo R. (Apr.) *Mann* 903! Bumban to Loko (Apr.) *Sc. Elliot* 5738! Gola Forest (Apr.) *Small* 610! **Lib.**: Gola National Forest (Apr.) *de Wilde* 3802! Ganta (Apr.) *Harley* 1812! Paynesville, Monrovia (Apr.) *Bequaert* 185! Nyaake, Webo Dist. (June) *Baldwin* 6170! Duo, Sinoe Co. (Mar.) *Baldwin* 11345! **Iv.C.**: Bingerville to Potou Lagoon (Feb.) *Chev.* 20077! Banco F.R., Abidjan (May) *Oldeman* 59! *de Wit* 8169! **Ghana**: *Thonning* 261! Dixcove (Mar.) *Morton* A255! Hohoe (Apr.) *Morton* GC 9176! **N.Nig.**: Kabba to Ilorin (Mar.) *Westwood* 276! **S.Nig.**: Apapa, Lagos (Jan.) *Dalz.* 1284! Ijaiye F.R., Oyo Prov. (Apr.) *Onochie* FHI 21982! R. Ogun, Oyo Prov. (Mar.) *Keay* FHI 22535! Sapoba F.R. (Mar.) *Emwiogbon* FHI 45340! Oban *Talbot* 313! Also in E. Cameroun, Congo and Angola.

4. **E. euglossa** (*Rchb. f.*) *Rchb. f.* in Bot. Mag. 92: t. 5561 (1866); F.T.A. 7: 57; Cat. Talb. 146; Chev. Bot. 613. *E. dusenii* Kraenzl. (1894)—F.T.A. 7: 52. Terrestrial with tall cylindrical leafy pseudobulbs and tall inflorescence; flowers greenish, lip white with pink or purple veins.
**S.L.**: Sugar Loaf Mt. (May) *Barter*! **Iv.C.**: Mt. Goula, Danané (Apr.) *Chev.* 21228! Dahoukro (July) *Hallé*! Brafouédi (Apr.) *Leeuwenberg* 3343! **Ghana**: Aburi (July) *Johnson* 1065! Bensu, W. Prov (Aug.–Dec.) *Miles* 3! Efiduase (June) *Cox* 24! Agu to Anum (Aug.) *Westwood* 81! **N.Nig.**: Ake to Opin, Osi Dist. (Dec.) *Ajayi* FHI 19300! **S.Nig.**: Idanre Hills, Ondo Prov. (May) *Keay & Brenan* FHI 19200! Oban *Talbot* 162! **W.Cam.**: Bande, Bamenda (Aug.) *Ujor* FHI 29951! Bamenda (June) *Daramola* FHI 41556! Also in Gabon, Congo and Uganda.

5. **E. stachyodes** *Rchb. f.* Otia Bot. Hamburg. 66 (1878); F.T.A. 7: 58. *E. lambii* Rolfe in Kew Bull. 1914: 212. Herb to 2½ ft. high with broad leaves and a rather close spike of brown and white flowers.
**N.Nig.**: Vom (June) *Dent Young*! *Lamb*! Kassa, Plateau Prov. (June) *King* 119! **W.Cam.**: Nchan, Bamenda, 5,000 ft. (June) *Maitland* 1783! Basenako, Bamenda, 5,000 ft. (June) *Maitland* 1507! Also in E. Cameroun, Congo, Sudan, Ethiopia, Uganda and Kenya.

6. **E. zeyheri** *Hook f.* in Bot. Mag. 119: t. 7330 (1893). *E. milanjiana* Rendle (1895)—F.T.A. 7: 63. A herb up to 2½ ft. high with broad leaves and a short dense spike of yellow flowers which often have a deep orange, brown or plum-coloured centre.
**N.Nig.**: Vom, Plateau Prov., 4,200 ft. (May) *King* 155! Also in Congo and eastern Africa from Sudan southwards to the Eastern Cape Province.

7. **E. adenoglossa** (*Lindl.*) *Rchb. f.* Otia Bot. Hamburg. 66 (1878); F.T.A. 7: 59. A herb up to 2½ ft. high with very narrow leaves developing after the flowers, the flowers green and brown.
**N.Nig.**: Nupe *Barter*! Kontagora (June) *Dalz.* 562! Keana, Nassarawa Prov. (May) *Hepburn* 23! Randa (June) *Hepburn* 61! Source of Jarawa R., Bauchi Prov. *King* 109a! Mando F.R., Zaria Prov. (June) *Keay* FHI 25815! Also in Kenya.

8. **E. shupangae** (*Rchb. f.*) *Kraenzl.* in Engl., Pflanzenw. Ost-Afr. C: 157 (1895); F.T.A. 7: 66. *E. propinqua* Hutch. in Kew Bull. 1921: 410; F.W.T.A., ed. 1, 2: 446. *E. baoulensis* A. Chev., Bot. 613, name only. Flowers in a dense head, white, yellow with brown or purple markings, or entirely brownish or purplish.
**Guin.**: Massif de Fon (Aug.) *Schnell* 3320! 3321! Nimba Mts. (Aug.) *Schnell* 3382! **S.L.**: Kortright (Sept.) *Gledhill* 50! Sugar Loaf Mt. (Aug.) *Melville & Hooker* 57! Kasokora (Aug.) *Deighton* 1234! Bintumane Peak, Loma Mts. (Aug.) *Jaeger* 971! Koya Mt., Bafodia (Aug.) *Haswell* 68! **Iv.C.**: Baoulé (July) *Chev.* 22175! Nimba Mts. (Aug.) *Boughey* GC 18155! **Ghana**: Gemmi, Amedzofe (Aug.) *Westwood* 28! 28a! **N.Nig.**: Hepham to Ropp (July) *Lely* 353! Vom *Dent Young* 233! Jos (Aug.) *Keay* FHI 20087! Nguroje to Lekitaba, Mambila Plateau (June) *Chapman* 14! **S.Nig.**: Ogoja *Rosevear* 67/29! **W.Cam.**: Bum to Nchan, Bamenda (June) *Maitland* 1638! Bafut-Ngemba F.R. (May) *Keay* FHI 37927! Manenguba Mts. (June) *Gregory* 304! Ndop to Sabgo (June) *Brunt* 769! Generally distributed in tropical Africa.

9. **E. lindiana** *Kraenzl.* in Engl., Bot. Jahrb. 51: 388 (1914). A herb up to 1 ft. high with narrow grass-like leaves; flowers greenish, lip white or cream with purple veins.
**Ghana**: Dutukpene (Mar.) *Hepper & Morton* A3061! **N.Nig.**: Anara F.R., Zaria Prov. (May) *Keay* FHI 22863! Mando F.R., Zaria Prov. (June) *Keay* FHI 25827! Kulfana, Bauchi Prov. (May) *G. V. Summerhayes* 51! Also in Tanzania, Malawi and Rhodesia.

10. **E. milnei** *Rchb. f.* Otia Bot. Hamburg. 116 (1881); F.T.A. 7: 52. A slender herb up to 15 in. high with very narrow leaves and small yellow flowers in a rather short spike.
**S.Nig.**: Calabar *Robb*! **F.Po**: *Milne*! Also in E. Cameroun, Rio Muni, Gabon, C. African Republic, Congo, Zambia and Rhodesia.

11. **E. warneckeana** *Kraenzl.* in Engl., Bot. Jahrb. 33: 67 (1902). *E. lutea* Lindl. (1862)—F.T.A. 7: 52; F.W.T.A., ed. 1, 2: 446; Chev. Bot. 614; not of Blume. *E. pusilla* Rolfe in Kew Bull. 1914: 212. *E. microdactyla* Kraenzl. in Engl., Bot. Jahrb. 51: 389 (1914). A slender herb up to 18 in. high with very small leaves and narrow, pale yellow flowers.
**Sen.**: Santiaba, Oussouye (July) *Berhaut* 6176! **Port.G.**: Gabu, Canguelifa (July) *Esp. Santo* 2938! **Guin.**: Nzo, Nimba Mts. (Mar.) *Schnell* 872! Beyla (Mar.) *Chev.* 20876! Siredougou, NW. of Beyla (Apr.) *Collenette* 1! Héremakon (= Herimankuna) (Mar.) *Sc. Elliot* 5248! **Lib.**: Paynesville (Feb.–Apr.) *Bequaert* 164! *Harley* 1857! Monrovia (Feb.–June) *Baldwin* 5916! 11083! **Ghana**: Sesiamang (Feb.) *A. S. Thomas* D137! Afram Plains (Feb.–May) *Burbridge* 245! *Johnson* 853! Jati-zonzo, Ashanti (Mar.) *Westwood* 127! Winneba Plains (Apr.) *Morton* A2018! **Togo Rep.**: Lomé (May) *Warnecke* 328! Badja (June) *Schlechter* 12971! **N.Nig.**: Kontagora *Dalz.* 446! Nupe *Barter* 1480! Maska, Katsina Prov. (June) *Keay* FHI 25889! **S.Nig.**: Upper Ogun Estate, Oyo Prov. (May) *Sanford* WS/899/65! WS/765/66! Also in E. Cameroun, Congo, Sudan, Uganda, Kenya and Zambia.

12. **E. monile** *Rchb. f.* in Flora 50: 105 (1867); F.T.A. 7: 53. Flowers greenish, in a rather lax spike.
**N.Nig.**: Keana *Hepburn* 55! Also in Angola and C. African Republic.

13. **E. brevipetala** *Rolfe* in F.T.A. 7: 53 (1897). Flowers greenish or brownish with reddish markings, in a long rather lax spike.
**Guin.**: valley of R. Kaba (May) *Chev.* 13179! **S.L.**: above Falaba (Mar.) *Sc. Elliot* 5224! **N.Nig.**: Bonu, Abuja, Niger Prov. (June) *E. W. Jones* 139! Anara F.R., Zaria Prov. (May) *Keay* FHI 25776! Mando F.R., Zaria Prov. (June) *Keay* FHI 25851! Mayere, Jemaa Dist. *King* 139! Also in C. African Republic.

14. **E. ramifera** *Summerh.* in Kew Bull. 12: 80 (1958). *E. elliotii* Rolfe in F.T.A. 7: 54 (1897); F.W.T.A., ed. 1, 2: 446; not of Rolfe (1891).
**Sen.**: Niombato (Mar.) *Adam* 10425! **S.L.**: Falaba (Mar.) *Sc. Elliot* 5116! **N.Nig.**: Andaha (Mar.) *Hepburn* 113! Jos (Apr.) *Cole* 36! *King* 108!

15. **E. sordida** *Kraenzl.* in Engl., Bot. Jahrb. 23: 69 (1902). Flowers small, yellow and brownish in a long raceme.
**Ghana**: Achimota (Feb., Mar.) *Irvine* 1368! 1999! 2853! Legon Hill (Feb.) *Adams* FH 3751! Afram Plains (Feb.) *Cansdale* 7! **Togo Rep.**: Lomé (Mar.) *Warnecke* 95! **N.Nig.**: N'gell, Plateau Prov. (Mar.) *King* 61! Kumbul, Plateau Prov. (Feb.) *King*! **S.Nig.**: Ilorin to Jebba (Feb.) *Richards* 5026!

16. **E. parvula** (*Rendle*) *Summerh.* in Kew Bull. 11: 125 (1957). *Lissochilus parvulus* Rendle (1895)—F.T.A. 7: 74. Flowers small, yellow and brown or maroon, in a long raceme.
**N.Nig.**: Kaciya to Zonkwa, Zaria Prov. (Mar.) *G. V. Summerhayes* 142! 143! Eastwards to Kenya and southwards in E. Africa to Rhodesia.

FIG. 396.—EULOPHIA WARNECKEANA *Kraenzl.* (ORCHIDACEAE).

A, scape.  B, leaf.  C, dorsal sepal.  D, lateral sepal.  E, petal.  F, lip and column, side view.
G, lip with spur removed, front view.  H. column.

17. **E. leonensis** *Rolfe* in F.T.A. 7: 51 (1897). Leaves at base of plant, developing with the flowers, broad; flowers in long raceme, greenish brown, tinged purple.
**Gam.:** Genieri (July) *Fox* 119! **Port.G.:** Madina de Mamadi Alfa (June) *Esp. Santo!* **Guin.:** Boulivet to Toukan (Apr.) *Chev.* 12942! **S.L.:** Bafodeya (Apr.) *Sc. Elliot* 5536! Musaia (Apr.) *Deighton* 4735! **N.Nig.:** Anara F.R., Zaria Prov. (May) *Keay* FHI 25777! Kabba (Apr.) *Westwood!* Vom *Dent Young!* Naraguta (May) *G. V. Summerhayes* 156! 157! Rukuba, Plateau Prov. (May) *King* 73! Also in C. African Republic and Uganda.

18. **E. stenoplectra** *Summerh.* in Kew Bull. 12: 81 (1958). Flowers in long raceme, red or brownish, lip yellow.
**Ghana:** Afram Plains (Feb.) *Cansdale* 5! 12! Jaketi to Anyaboni (Feb.) *Morton* A3021! Also in Sudan and Uganda.

19. **E. galeoloides** *Kraenzl.* in Engl., Bot. Jahrb. 24: 508 (1898). Saprophyte with brown scales on paler stem and whitish or cream flowers.
**Ghana:** Tano-Ofin F.R. (Feb.) *Lyon* FH 2874! **S.Nig.:** Etemi, Ijebu Ode (Dec.) *Tamajong* FHI 20295! Also in Kenya and Tanzania.

20. **E. alta** (*Linn.*) *Fawcett & Rendle* Fl. Jam. 1: 112 (1910). *Limodorum altum* Linn., Syst. Nat. ed. 12, 2: 594 (1767). *Eulophia woodfordii* (Sims) Rolfe in F.T.A. 7: 68 (1897). *E. longifolia* (H.B.K.) Schltr. (1914)—F.W.T.A., ed. 1, 2: 446. Leaves lanceolate, up to 4 ft. long and 4 in. broad; scape up to 4 ft. high, flowers numerous, green and purple.
**Sen.:** M'bidjem (Sept.) *Berhaut* 6348! **Guin.:** Madina Tossekré (Oct.) *Adam* 12539! **S.L.:** Kimadougou, Loma Mts. (Oct.) *Jaeger* 177! **Lib.:** Gbanga (Aug.) *Traub* 251! Ganta (May) *Harley* 2134! Nimba (Dec.) *Adam* 20097! **Iv.C.:** Sassandra (Aug.) *de Wilde* 358! **Ghana:** *Burton & Cameron!* Nkawkaw *Morton* A3406! **S.Nig.:** Lagos *Millen!* **W.Cam.:** Mamfe (July) *Nditapah* FHI 51337! Also in Gabon, Congo, C. African Republic, Sudan, Uganda, Zambia, Rhodesia and tropical America and the W. Indies.

21. **E. calantha** *Schltr.* in Warb., Kun.-Samb. Exped. 215 (1903). Leaves narrow, linear; scape 1–3½ ft. high; flowers up to 12, white and pink or purple.
**Guin.:** Mirire, Fouta Djalon (July) *Adames* 301! Also Congo, Uganda, Kenya, Tanzania, Zambia and Angola.

22. **E. flavopurpurea** (*Rchb. f.*) *Rolfe* in F.T.A. 7: 65 (1897). *Lissochilus millsoni* Rolfe in F.T.A. 7: 79 (1897). *L. lacteus* Kraenzl. in Engl., Bot. Jahrb. 43: 398 (1909). *L. andersoni* Rolfe in Kew Bull. 1910: 159. *L. johnsoni* Rolfe l.c. 160 (1910). *Eulophia millsoni* (Rolfe) Summerh. in F.W.T.A., ed. 1, 2: 446 (1936). No leaves at time of flowering; flowers greenish or yellowish, lip sometimes with purple veins or streaks.
**Iv.C.:** Adioukrou, Dabou (Jan., Feb.) *Chev.* 17119! *Portères* 618! *de Wilde* 1062! **Ghana:** Accra Plains (Apr.) *Johnson* 854! Afram Plains (Apr.) *Johnson* 861! Achimota (May) *Irvine* 1614! **Togo Rep.:** Misahöhe (Mar.) *Baumann* 24! **N.Nig.:** Ilorin *Millson* 86! *Rowland!* Zungeru, Niger Prov. (May) *Keay* FHI 22847! Anara F.R., Zaria Prov. (May) *Keay* FHI 22937! Kontagora (Apr.) *Dalz.* 445! Hoss, Plateau Prov. (Apr.) *Gregory* 269a! Abinsi (May) *Dalz.* 841! **S.Nig.:** Ikiri *Foster* 209! Ugbogin, Benin (Apr.) *Dundas* FHI 21460! Enugu (Mar.) *Abrahall* FHI 27567! Udi to Enugu Ngwo (Feb.) *Jones* FHI 7368! **W.Cam.:** Boviongo, Kumba (Jan.) *Keay* FHI 37392! Jua, Bamenda (Apr.) *Maitland* 1626! Ndop village, Ndop Plain (Apr.) *Brunt* 363! Eastwards to Sudan and Uganda, Malawi, Zambia, Rhodesia and Mozambique.

23. **E. cristata** (*Sw.*) *Steud.* Nomen. Bot. ed. 2, 1: 605 (1840). *Limodorum cristatum* Sw. in Schrad., Neues Journ. 1: 86 (1805). *Lissochilus purpuratus* Lindl. (1862)—F.T.A. 7: 79. *L. heudelotii* Rchb. f. (1878). *L. uliginosus* Rolfe in Kew Bull. 1913: 340. Terrestrial herb up to 4 ft. high with long raceme of pink or purplish flowers with darker lip.
**Sen.:** *Heudelot!* **Gam.:** (July) *Brooks* 30! Kuntaur (June) *Brooks* 57! N. bank Gambia R. (July) *Ozanne* 21! **Port.G.:** Abu, Formosa I. (Apr.) *Esp. Santo* 1951! Boé (June) *Esp. Santo* 3205! **S.L.:** Mange, Port Loko Dist. (Apr.) *Hepper* 2606! Bonjema (July) *Deighton* 5953! Njala (Mar.) *Deighton* 1117! Kondembaia (Apr.) *Deighton* 5074! Falaba to Sinkunia (Apr.) *Deighton* 5413! **Lib.:** Pandamai (Mar.) *Bequaert* 98! **Iv.C.:** Yamoussokro to Bouaflé (Jan.) *Aké Assi* 7263! Bouna to Bondoukou (Mar.) *de Wilde* 3503! **Ghana:** Achimota (Apr.) *Bally* 11! Afram Plains (Feb.) *Cansdale* 11 Legon Hill (Mar.) *Morton* GC 6487! Chama (Apr.) *Chipp* 192! Gambaga (May) *Vigne* 4698! **Togo Rep.:** Lomé *Warnecke* 96! Kpeve (Feb.) *Westwood* 84! Yendi (Apr.) *Williams* 172! **Dah.:** Massé to Ketou, Zagnanado Dist. (Feb.) *Chev.* 23014! **N.Nig.:** Zungeru to Zaria (May) *Dalz.* 287! Zongon Katab, Zaria (Mar.) *G. V. Summerhayes* 81! Vom, Jos Plateau *Dent Young!* Kulfana Lame, Bauchi Dist. (May) *G. V. Summerhayes* 52! Biu (Apr.) *Noble* 45! **S.Nig.:** Lagos to Abeokuta (Feb.) *Burtt* B33! Abeokuta *Barter* 333! Olokemeji F.R., Egba Div. (Mar.) *Hepper* 2302! Oyo to Iseyin (Mar.) *Onochie* FHI 31547! Odongele to Okuri (Feb.) *Rosevear* 25/31! **W.Cam.:** Bu to Mudele, Bamenda (Feb.) *Gregory* 251! Also in E. Cameroun, C. African Republic, Congo, Sudan, Ethiopia and Uganda. (See Appendix, p. 515.)

24. **E. caricifolia** (*Rchb. f.*) *Summerh.* in F.W.T.A., ed. 1, 2: 444 (1936). *Lissochilus caricifolius* Rchb. f. (1877) —F.T.A. 7: 77. *L. longifolius* Benth. (1849)—F.T.A. 7: 77; Stapf in Johnston, Lib. 654. Leaves very long and narrow; sepals red or purplish, petals white to red, crests on lip yellow.
**Guin.:** Benty (June) *Jac.-Fél.* 1694! Nzérékoré (June) *Jac.-Fél.* 959! **S.L.:** Blama to Sendumi (Mar.) *Dawe* 460! Falaba, Southern Prov. (Apr.) *Deighton* 1626! Mano Salija (Nov.) *Deighton* 432! Perewahun, Nongoba Bullom (Mar.) *Adames* 14! Rhombe (Mar.) *Adames* 153! **Lib.:** Sangwin, Sinoe (Mar.) *Baldwin* 11325! Monrovia (June) *Baldwin* 5863! Paynesville (Apr.) *Barker* 1237! Grand Bassa (July) *T. Vogel* 8! **Iv.C.:** Abouabou (Mar.) *Raynal* 13648! NE. of Grand Bassam (Sept.) *de Wit* 1212! Moossou, Grand Bassam (Aug.) *de Wilde* 212! **Ghana:** Asientum (July) *Chipp* 284! Atwabo *Fishlock* 55! **S.Nig.:** Ikoyi Plains, Lagos (May) *Dalz.* 964! Epe (June) *Harper* FHI 36089! Sapele, Sapoba *Kennedy* 2389! Uke to Nobi, Onitsha Dist. (May) *Onochie* FHI 35801! Eket *Talbot* 3160! Also in Gabon, Congo, Uganda and Tanzania.

25. **E. angolensis** (*Rchb. f.*) *Summerh.* in Kew Bull. 12: 76 (1958). *Lissochilus angolensis* (Rchb. f.) Rchb. f. (1878)—F.T.A. 7: 76. *L. lindleyanus* Rchb. f. (1878)—F.T.A. 7: 77. *Eulophia lindleyana* (Rchb. f.) Schltr. in Westafr. Kauts.-Exped. 279 (1900); F.W.T.A., ed. 1, 2: 444. Leaves narrowly lanceolate, 1–2¼ in. wide; flowers with reddish or brownish sepals and yellow petals and lip.
**Sen.:** Ziguinchor (Sept.) *Broadbent* 99! *Doumbia* 625! **Gam.:** Kuntaur *Ruxton* 98! **Port.G.:** S. Domingos (Aug.) *Esp. Santo* 3074! 3076! Farim to Bigerre (Aug.) *Esp. Santo* 3081! **S.L.:** Fogbo, Peninsula (July) *T. S. Jones* 413! Rokupr (Aug.) *Harvey* 8! Rokon (July) *Adames* 239! Gbinti (July) *Deighton* 2511! Njala (June) *Deighton* 2525! **Iv.C.:** Férédougouoa R. to Touba (July) *Collenette* 64! Mankono (July) *Chev.* 22005! **Ghana:** Gura (June) *Vigne* FH 3911! Ejura (fl. & fr. Sept.) *Westwood* 189! Kpandu to Golokuati (Aug.) *D. Gillett* 67! **N.Nig.:** Nupe *Barter* 1486! Ibaji Ojoko F.R., Igala Dist. (June) *Howard* FHI 47764! Samaru, Zongon Katab Dist. (June) *G. V. Summerhayes* 108! Maska, Katsina Dist. (June) *Keay* FHI 25898! Kilba, Yola (Aug.) *Dalz.* 219! **S.Nig.:** Ogoya, Lagos (Feb.) *Richards* 5087! Oyo (July) *J. B. Gillett* 15192! Orle F.R., Kukuruku Dist. (Aug.) *Onochie* FHI 33293! Igbosere, Kuramo (Aug.) *Savory* UCI 265! Mamu R., Awka Dist. (Aug.) *Jones* FHI 6692! **W.Cam.:** Wum, Bamenda (June) *Daramola* FHI 41076! *Ujor* FHI 29252! Mbaw Plain (May) *Brunt* 419! Eastwards to Kenya and in E. Africa south to Rhodesia.

26. **E. cucullata** (*Sw.*) *Steud.* Nomen Bot. ed. 2, 1: 605 (1840). *Limodorum cucullatum* Sw. in Schrad. Neues Journ. 1: 86 (1805). *Lissochilus arenarius* Lindl. (1862)—F.T.A. 7: 82. Leaves narrow, appearing after the flowers; flowers few, large, pink or purplish with yellow or white inside the broad spur; in savanna.
**Sen.:** Zankoré, Casamance (Aug.) *Berhaut* 6343! Ziguinchor (Sept.) *Broadbent* 100! **Gam.:** Aboku (June) *Brooks* 59! **Mali:** Kroukoto, Kita (July) *Jaeger!* Fana, Bamako East (June, July) *Roberty* 2426! 2475! **Port.G.:** Cacheu, Chorubrigue (Aug.) *Esp. Santo* 2495! Ingoré to Barro (Aug.) *Esp. Santo* 3075! Bafalá to Capé (Aug.) *Esp. Santo* 3319! **Guin.:** Bonhouri (July) *Pobéguin* 375! Kindia (June) *Jac.-Fél.*

1772! **S.L.**: Port Loko (May) *Deighton* 4780! Yifin to Bandakarafaia (Apr.) *Deighton* 5065! Samaia (May) *Deighton* 4792! Bambaia, Samu (July) *Adames* 212! **Iv.C.**: Baoulé-Nord (Aug.) *Chev.* 22234! Séguéla (Apr.) *Leeuwenberg* 3271! **Ghana**: Weila, N. Ashanti (Apr.) *Kinloch* FH 3275! Attabubu to Krachi (Mar.) *Westwood* 133! Damongo (Mar.) *Adams* 3978! **Togo Rep.**: Lomé (June) *Davidson* 2! **N.Nig.**: Nupe (Apr.–June) *Barter* 1488! Gawu to Abuja, Gwari Dist. *Onochie* FHI 18698! Anara F.R., Zaria *Keay & Mutch* FHI 22932! Zungeru *Dalz.* 230! **S.Nig.**: Borgu Bariba, Lagos, (May) *Punch*! Orle River, F.R., Benin Prov. (June) *Umana* FHI 29132! Ikom *Rosevear* 42/29! **W.Cam.**: Bamessi (Mar.) *Brunt* 271! Bamali to Bambalang (Apr.) *Brunt* 354! Bambuluwe L., Bamenda (May) *Daramola* FHI 41062! Throughout the savanna areas of tropical Africa, also in Natal and Comoro Is.

27. **E. dilecta** (*Rchb. f.*) *Schltr.* in Westafr. Kauts.-Exped. 279 (1900). *Lissochilus dilectus* Rchb. f. (1878)— F.T.A. 7: 83. Similar to *E. cucullata* but leaves broader and developed at time of flowering; flowers a little larger.
　　**Ghana**: Nsawkaw, Wenchi (Aug.) *Hall* 2043! **Togo Rep.**: between Golokuati and Kpandu (July, Aug.) *D. Gillett* 64! *Westwood* 20! **N.Nig.**: Okwoga to Idah (July) *Kitson*! Kabba (May) *Westwood*! Bokkos, Plateau Prov. *King* 94! **S.Nig.**: Ogoja *Talbot*! Also in Congo, Cabinda and Angola.

28. **E. buettneri** (*Kraenzl.*) *Summerh.* in F.W.T.A., ed. 1, 2: 446 (1936). *Lissochilus buettneri* Kraenzl. (1893)— F.T.A. 7: 84. Similar to *E. cucullata* but flowers smaller, fragrant.
　　**Guin.**: Nimba Mts. (Mar.) *Tournier* 1! 2! **S.L.**: Falaba (Mar.) *Sc. Elliot* 5144! **Lib.**: Mt. Mpaka Fossa, Kolahun (Jan.) *Bequaert* 39! **Ghana**: Afram Plains, Ashanti (Mar.) *Johnson* 852! *Cansdale* 2! Dodi to Dain, Trans-Volta Prov. (Mar.) *Westwood* 82! Kpandu (Feb.) *Robertson* 125! **Togo Rep.**: Anganje, Bismarckburg (Feb.) *Büttner* 415! **N.Nig.**: Amban, Plateau Prov. *King* 195! **S.Nig.**: Old Oyo F.R., Oyo Prov. (Feb.) *Keay* FHI 23436! **W.Cam.**: Bambuko F.R., Kumba (Jan.) *Keay* FHI 37460! Also in E. Cameroun.

29. **E. horsfallii** (*Batem.*) *Summerh.* in F.W.T.A., ed. 1, 2: 446. *Lissochilus horsfallii* Batem.—F.T.A. 7: 84. *L. roseus* Lindl. (1843)—F.T.A. 7: 85, partly; Stapf in Johnston, Lib. 654 (all excl. syn. *Dendrobium roseum* Sw.). Leaves in a tuft, up to 6 ft. high and 5 in. broad; scapes to 8½ ft. tall, with a long raceme of white and pink or purple flowers.
　　**Port.G.**: Bafata, Capé (Aug.) *Esp. Santo* 2943! **S.L.**: Port Loko (Apr.) *Sc. Elliot* 5744! Tawia (Jan.) *Sc. Elliot* 4472! Yataya (Sept.) *Thomas* 2425! Kasawa to Moyamba (Dec.) *King* 63b! Waterloo (Aug.) *Melville & Hooker* 174! **Lib.**: Grant's Farm, Sinoe R. *Johnston*! Dukwia R., Monrovia (Oct.–Nov.) *Cooper* 35! Zwedru, Tchien Dist. (Aug.) *Baldwin* 7042! Monrovia (June) *Baldwin* 5866! Blazie (Dec.) *Adam* 16292! **Iv.C.**: Béreby (Nov.) *Oldeman* 662! Danané (June) *Collenette* 49! Banco Forest, N.W. of Abidjan (Aug.) *de Wilde* 207! Adiopodoumé (Oct.) *de Wilde* 684! **Ghana**: Oda (Aug.) *Howes* 969! Aburi (Apr.) *Johnson*! Niru (Nov.) *Vigne* FH 2574! Assin *West-Skinn* 36! Busua Bay (Dec.) *Morton* A2537! **Dah.**: Porto-Novo (Jan.) *Chev.* 22724! **N.Nig.**: Kabba (Oct.) *Parsons* L107! Bida to Zungeru (Jan.) *Meikle* 1023! Vom *Dent Young* 236! Abinsi (June) *Dalz.* 834! **S.Nig.**: Lagos *Phillips* 46! Mawkawlawki, Abeokuta (Dec.) *Burtt* B11! Ibadan (Jan.) *Latilo* FHI 8187! Sapoba, Benin Prov. (Jan.) *Richards* 3916! Onyeagocha FHI 7620! **W.Cam.**: Buea (Apr.) *R. H. Brown* 105a! Bamenda (May) *Ujor* FHI 30400! Kumbo to Oku, 6,000 ft. (Feb.) *Hepper* 1999! **F.Po**: Moka (Aug., Dec., Jan.) *Exell* 852! *Boughey* 54! *Melville* 403! Eastwards to Kenya and Tanzania, Zambia and Malawi.

30. **E. barteri** *Summerh.* in F.W.T.A., ed. 1, 2: 442, 444 (1936), and in Kew Bull. 1936: 224. *Lissochilus roseus* of Rolfe F.T.A. 7: 85, partly, not of (Sw.) Lindl. Similar to *E. horsfallii* but flowers smaller, lilac and reddish or brownish purple.
　　**Guin.**: Kindia (Sept.) *Jac.-Fél.* 458! **S.L.**: Mapaki, N. Prov. (Aug.) *Deighton* 1213! Giema (Nov.) *Deighton* 364! Sawula, Kagbaro R. (Feb.) *Deighton* 3560! Newton (June) *Deighton* 4786! **Lib.**: Paynesville, Monrovia (Apr.) *Bequaert* 184! Gbanga, Central Prov. (July) *Daniel & Barker* 206! **Iv.C.**: Touba, Férédougou R. (July) *Collenette* 66! **Ghana**: Kintampo (July) *Vigne* FH 3935! **N.Nig.**: Nupe *Barter* 1481! Sare *Barter* 3429!

31. **E. oedoplectron** *Summerh.* in Kew Bull. 1936: 223. *Lissochilus macranthus* Lindl. (1833)—F.T.A. 7: 86. *L. elatus* Rolfe in F.T.A. 87 (1897). Similar in growth to *E. horsfallii*, up to 10 ft. high; flowers pink, rose or purple.
　　**S.Nig.**: Bonny *Shepherd* 6! **F.Po**: Pico S. Isabel, 3,000 ft. (Dec.) *Sanford* 4000! Also in E. Cameroun, Gabon and Congo.

32. **E. latilabris** *Summerh.* l.c. (1936). *Lissochilus schweinfurthii* Rchb. f. (1878)—F.T.A. 7: 88. Habit as in *E. horsfallii*, 3–6 ft. in height; flowers purple, pink and white.
　　**N.Nig.**: Zongon Katab, Zaria (June) *G. V. Summerhayes* 118! Vom, Plateau Prov. *Dent Young* 236a! Gembu, Mambila Plateau, 5,500 ft. (July) *Chapman* 5! Eastwards to Sudan and Kenya, southwards in E. Africa to Zambia and Malawi.

33. **E. juncifolia** *Summerh.* in Kew Bull. 12: 78 (1958). *E. involuta* Summerh. in F.W.T.A., ed. 1, 2: 444 (1936), partly (excl. syn. *Lissochilus smithii*). Leaves 1½–2½ ft. long, rush-like; flowers yellow with red veins on the petals.
　　**Port.G.**: Farim, Cajambarim (Oct.) *Sousa* in Hb. *Esp. Santo* 2317! **Iv.C.**: Mankono (July) *Chev.* 21977! **Ghana**: Anum Plains (Aug.) *Johnson* 1094! Attabubu (Aug.) *Cox* 110! Ejura (Aug., Sept.) *Andoh* FH 5047! *Westwood* 16a! Nungua, Accra Plains (June) *Morton* A911! Kpandu to Golokuati (July) *Westwood* 16!

34. **E. orthoplectra** (*Rchb. f.*) *Summerh.* in Kew Bull. 1939: 499, in obs. *Lissochilus orthoplectrus* Rchb. f. (1878)—F.T.A. 7: 95. *L. milanjianus* Rendle (1894)—F.T.A. 7: 98. *Eulophia involuta* Summerh. in F.W.T.A., ed. 1, 2: 444 (1936), partly. Leaves broad and flat, appearing after the flowers; flowers yellow, petals with red veins or almost entirely red beneath.
　　**Ghana**: Ejura (Aug.) *Hall* CC 329! **Togo Rep.**: Lomé (July) *Davidson* 7! **N.Nig.**: Vom, Plateau Prov. (Feb.) *Dent Young*! *McClintock* 207! Mambila Plateau, 5,000 ft. (Jan., Feb.) *Hepper* 1755! *Wimbush* FHI 48389! **S.Nig.**: Sonkwala, Ogoja Prov. (Dec.) *Keay & Savory* FHI 25120! **W.Cam.**: Jua, Bamenda (Apr.) *Maitland* 1778! Eastwards to Kenya and southwards in E. Africa to Rhodesia and Mozambique.


## 32. PTEROGLOSSASPIS Rchb. f., Otia Bot. Hamburg. 67 (1878); F.T.A. 7: 99.

Terrestrial; stem underground, composed of lobed fleshy tubers; leaves 1 or 2 on very short aerial stem, surrounded by sheaths at base, narrowly lanceolate, long-petiolate, 35–70 cm. long and 6–18 mm. broad; scape erect, up to 85 cm. high, bearing a number of flowers in a short terminal raceme about 8 cm. long; bracts long and slender, over-topping the flowers; tepals all rather similar, oblong- or elliptical-lanceolate, more or less acuminate, about 1 cm. long, greenish; lip nearly as long as tepals but much more fleshy, 3-lobed in lower part, blackish-purple, middle lobe covered with parallel verrucose crests, spur quite absent　.. 　.. 　.. 　.. 　.. 　.. 　*distans*

**P. distans** *Summerh.* in Kew Bull. 12: 82 (1958).
　　**S.L.**: Taigbe, Bonthe Dist. (Oct.) *Adames* 92! *Jordan* 620!

**33. GRAPHORKIS** Thou., Nouv. Bull. Sci. Soc. Philom. Paris 1: 318 (1809). *Eulophiopsis* Pfitz. (1887)—F.W.T.A., ed. 1, 2: 446.

Pseudobulbs cylindrical-fusiform or conical-ovoid, 3–9 cm. long, 1–3 cm. diam., 4–6-leaved; leaves lanceolate, up to 40 cm. long and 3·5 cm. broad; inflorescence appearing before the leaves, 15–50 cm. high, branches spreading; flowers yellow and brown; sepals spathulate-oblong, about 5–6 mm. long; petals elliptical, a little shorter; lip 3-lobed, middle lobe more or less retuse or bifid, disk of lip with 2 keels at the base; spur sharply bent forward, nearly as long as the lip; column with a hairy auricle on each side at the base   ..   ..   ..   ..   ..   ..   ..   ..   *lurida*

G. **lurida** (*Sw.*) *O. Ktze.* Rev. Gen. Pl. 662 (1891). *Limodorum luridum* Sw. in Schrad. Neues Journ. 1: 87 (1805). *Eulophia lurida* (Sw.) Lindl. (1833)—F.T.A. 7: 53; Cat. Talb. 146; Stapf in Johnston, Lib. 654. *Eulophiopsis lurida* (Sw.) Schltr. (1914)—F.W.T.A., ed. 1, 2: 446.
**Sen.:** Oussouye, Casamance (Apr., July) *Adam* 19146! *Berhaut* 6181! **Port.G.:** Acoco, Formosa (fr. Apr.) *Esp. Santo* 1959! Prabis, Bissau (Feb.) *Esp. Santo* 1824! Antula, Bissau *Sousa* 2325! **Guin.:** Timbi-Madina, Pita (Mar.) *Adam* 11634b! Bembaya, Farana (Feb.) *Chev.* 20670! Nzo (Mar.) *Chev.* 21049! Kindia *Jac.-Fél.* 10! **S.L.:** Ninia (Feb.) *Sc. Elliot* 4812! Bumban to Port Loko (fr. Apr.) *Sc. Elliot* 5741! Mt. Aureole, Freetown (Feb.) *Dalz.* 1020! Njala (Jan.) *Deighton* 2460! Fairo, Soro (Jan.) *Deighton* 4566! **Lib.:** Kakatown *Whyte*! Jabrocca, Grand Cape Mt. (Dec.) *Baldwin* 10849! Ziatown (Dec.) *Adam* 16338! Epi (Dec.) *Harley* 1766! **Iv.C.:** Bouroukrou (Jan.) *Chev.* 16916! Adiopodoumé (Dec.) *Giovannetti*! **Ghana:** Dunkwa (Jan.) *Dalz.* 77! Opar Valley, W. Prov. (Jan.) *Irvine*! Aburi (Jan.) *Johnson* 218! Kumasi (Dec.) *Vigne* FH 1522! Amedzofe (Nov.) *Westwood* 10! **N.Nig.:** Assob Escarpment, Plateau Prov. (Apr.) *King* 40a! Wana, Plateau Prov. (Dec.) *Hepburn* 156! Kaciya-Zonkwa, Zaria Prov. (Jan.) *G. V. Summerhayes* 144! Kwoi, Zaria Prov. *King* 40b! **S.Nig.:** 51 miles SE. of Ilorin *Clarke* 28! Nun R. *Barter* 20121! A beokuta to Ijebu-Ode (Jan.) *Burtt* 14! Okomu F.R., Benin (Jan.) *Jones & others* s.n.! Badagry *Thorold* 2021! Oban *Talbot* 772! **W.Cam.:** Ambas Bay (Feb.) *Mann* 782! Kang, Kumba (fr. Mar.) *Thorold* CM14! **F. Po:** S. Isabel to Basilé, 3,000 ft. (Dec.) *Sanford* 4010! Also in E. Cameroun, Gabon, Congo, Uganda and Burundi.

### 34. PODANGIS Schltr. in Beih. Bot. Centralbl. 36, 2: 82 (1918).

Stem up to 11 cm. long, usually much shorter, leafy at apex; leaves 4–16 cm. long, 5–12 mm. broad; inflorescences (including flowers) up to 6 cm. long; sepals and petals more or less elliptical, obtuse, 3·5–5 mm. long; lip more or less orbicular, about 6 mm. long; spur 11–14 mm. long, wide at mouth, constricted in middle, swollen and often shortly lobed at apex   ..   ..   ..   ..   ..   ..   *dactyloceras*

P. **dactyloceras** (*Rchb. f.*) *Schltr.* l.c. (1918). *Listrostachys dactyloceras* Rchb. f. (1865)—F.T.A. 7: 168. *L. forcipata* Kraenzl. (1894)—F.T.A. 7: 168. Flowers white; anther green.
**Guin.:** Kondian, Kissi (Feb.) *Chev.* 20742! Dalaba (Apr.) *Caille* in *Hb. Chev.* 18128! **S.L.:** Shingama, Serabu (June) *Deighton* 1955! **Ghana:** Amedzofe, 2,500 ft. (Feb.–July) *Irvine* 163! *Scholes* 91! *Westwood* 22! **N.Nig.:** Zonkwa-Kaciya, Zaria Prov. (Jan.) *G. V. Summerhayes* 140! Gembu, Mambila Plateau, 4,600 ft. (Jan.) *Hepper* 1824! **W.Cam.:** Buea, 3,200–4,300 ft. (Mar.) *Maitland* 699! Bamenda, 5,000–6,000 ft. (Feb.–Apr.) *Maitland* 1415! *Gregory* 255! Binka, Nkambe, 5,200 ft. (Feb.) *Hepper* 2832! Also in E. Cameroun, Congo, Uganda, Tanzania and Angola.

### 35. CALYPTROCHILUM Kraenzl. in Engl., Bot. Jahrb. 22: 30 (1895).

Inflorescence many-flowered, very dense, up to 5 cm. long, bracts closely imbricate, middle ones 4–7 mm. long; leaves 6–17 cm. long, 2·5–5 cm. broad; lip obscurely 3-lobed in front, side lobes broadly rounded, middle lobe much smaller, acute; sepals about 1 cm. long; lip including spur 2 cm. long   ..   ..   ..   1. *emarginatum*
Inflorescence usually 6–9-, sometimes up to 12-flowered, rather lax, up to 4 cm. long, bracts not imbricate, middle ones 2–4 mm. long, rhachis more or less zigzag; leaves 3–13 cm. long, 0·8–2·5 cm. broad; lip distinctly 3-lobed, lobes rounded, middle lobe much larger and longer than side lobes, emarginate; sepals 6–8 mm. long; lip including spur 1·8 cm. long   ..   ..   ..   ..   ..   2. *christyanum*

1. **C. emarginatum** (*Sw.*) *Schltr.* in Beih. Bot. Centralbl. 36, 2: 84 (1918). *Limodorum emarginatum* Sw. in Schrad. Neues Journ. 1: 86 (1805). *Angraecum imbricatum* Lindl. (1862)—F.T.A. 7: 144; Chev. Bot. 618. Flowers white with yellow or yellow-green lip, very fragrant, especially at night.
**Guin.:** Kissi, Korodou *Chev.* 20727! Nimba Mts. *Schnell* 184! 801! **S.L.:** Makuta (June) *Thomas* 473! Taninahun Bumpe (Apr.) *Deighton* 1728! Roruks (June) *Deighton* 5109! Bubuya (May) *Adams* 175! Bumban to Port Loko (Apr.) *Sc. Elliot* 5734! **Lib.:** Ganta (May) *Harley*! Boporo (fr. Nov.) *Baldwin* 10391! Suacoco, Gbanga (May) *Daniel & Prior* 440! **Iv.C.:** Mt. Goula, Danané (Apr.) *Chev.* 21227! Issia (Aug.) *de Wilde* 426! Tiassalé, Bandama R.(fr. Dec.) *Leeuwenberg* 2147! Brafoуédi (Apr.) *Leeuwenberg* 3339! **Ghana:** Offin Valley, Ashanti (Oct., fr. Dec.) *Miles* 12! Juaso, Akim (Apr.) *Cansdale* FH 4388! Kukurentumi (Apr.) *Johnson* 759! Aburi *Howes* 1172! Ofinso, Ashanti (May) *Vigne* FH 4865! **S.Nig.:** Baba Eko, Ijebu (Apr.) *Ross* 246! Ibadan (May) *Onochie* FHI 20697! Olokemeji F.R. (fr. Feb.) *Onyeachusim* FHI 46997! Onitsha *Barter* 1484! Ubiaja F.R., Ishan Dist. (Apr.) *Daramola* FHI 31267a! **W.Cam.:** Victoria to Bimbia (Apr.) *Preuss* 1240! **F.Po:** San Fernando (Dec.) *Sanford* 4050! Also in E. Cameroun, Gabon, Congo and Angola.

2. **C. christyanum** (*Rchb. f.*) *Summerh.* in F.W.T.A., ed. 1, 2: 450 (1936). *Angraecum christyanum* Rchb. f. (1880)—F.T.A. 7: 142. *A. moloneyi* Rolfe in F.T.A. 7: 145 (1897). *A. ivorense* A. Chev., Bot. 618, name only. Flowers white or cream, the base of the lip often greenish, yellow or orange.
**Mali:** Kita Mt. (Nov.) *Jaeger* 3601! **Port.G.:** Bissau (Jan.) *Esp. Santo* 1635! Chitol, Casselinta (June) *Esp. Santo* 3191! Comura (Nov.) *Raimundo & Guerra* 157! **Guin.:** Kindia (May) *Roberty* 17695! Nzo to Sakomanta (Mar.) *Chev.* 21083! Nimba Mt. (Mar.) *Schnell* 800! **S.L.:** Kambia to Kukuna (May) *Deighton* 5050! Bubuya (May) *Adames* 176! Matiti (May) *Deighton* 5049! Musaia (May) *Deighton* 4306! **Lib.:** (cult.) *Christy*! Jabrocca, Grand Cape Mount Dist. (Dec.) *Baldwin* 10863! Zuie, Boporo Dist. (Nov.) *Baldwin* 10260! Ganta, Sanokwele Dist. (Nov.) *Harley* 1763! **Iv.C.:** Gourémi, Upper Sassandra (May) *Ripert* in *Hb. Chev.* 21640! Béyo (Jan.) *Leeuwenberg* 2556! **Ghana:** Kumasi (Jan.) *Vigne* FH 2736! **N.Nig.:** Beari *Dalz.* 13! Anara F.R., Zaria (May) *Keay & Mutch* FHI 22933! Vogel Peak, Sardauna

W.E.T. DEL.

FIG. 397.—Calyptrochilum emarginatum (*Sw.*) *Schltr.* (Orchidaceae).
A, lip.  B, column.
252

Prov. (Nov.) *Hepper* 1479! **S.Nig.**: Olokemeji F.R. (Feb.) *Keay* FHI 22487! Lagos *Rowland*! Ijaiye F.R., Ibadan Dist. (Feb.) *Keay* FHI 21166! Ubiaja F.R., Ishan Dist. (fl. & fr. Apr.) *Daramola* FHI 31267! Mt. Orosun, Idanre Hills (Feb.) *Brenan & Keay* FHI 22485! Also eastwards to Sudan, Kenya and Tanzania, and south to Angola and Zambia.

## 36. ANGRAECUM Bory, Voy. 1: 359 (1804); F.T.A. 7: 133, partly.

Leaves with sharp points, not closely imbricate at the base; pollinia attached directly to the broad viscidium, stipes absent:

Leaves subulate-terete, more or less falcate, 3–13 cm. long, 1–2·5 mm. diam.; sepals about 4 mm. long; lip boat-shaped, very broad when flattened out, pointed in front, spur 5 mm. long ..  ..  ..  ..  ..  ..  ..     5. *subulatum*

Leaves oblong-lanceolate, flattened, fleshy, 2–4 cm. long, 3–6 mm. broad; sepals 6–7 mm. long; lip much broader than long, spur 4·5 mm. long..       4. *pungens*

Leaves without sharp points, though sometimes acute:

Flowers solitary on short (6 mm. long or less) peduncles; sepals 7 mm. long or less; leaves less than 3 cm. long, closely imbricate at base, lying in the vertical plane of the stem:

Leaves falcately oblong-elliptical, obtuse; spur 5–7 mm. long:

Leaves 5–11 mm. long, 3–7 mm. broad, groove on upper margin narrow, only extending half way to apex; lip indistinctly 3-lobed       ..     1. *distichum*

Leaves 14–25 mm. long, 6–10 mm. broad, groove on upper margin broad, extending nearly to apex; lip distinctly 3-lobed ..  ..  ..  ..     2. *aporoides*

Leaves narrowly lanceolate-oblong, almost straight, subacute, 10–18 mm. long, 2–3 mm. broad; sepals 4·5–5 mm. long; spur 4·5–6 mm. long     3. *podochiloides*

Flowers either solitary on long peduncles or peduncles several-flowered; sepals usually 6 mm. or more long; leaves flat, more or less unequally bilobed at the apex, usually more than 4 cm. long, not lying in the vertical plane of the stem:

Sepals over 3 cm. long:

Spur (measured from apex of ovary) over 10 cm. long, sharply incurved, narrowly conical in basal part; sepals lanceolate-linear, 6–8 cm. long; stem elongated, rooting; leaves oblong or elliptical, unequally lobed, up to 16 cm. long and 4 cm. broad; peduncle 4–7 cm. long, pedicel about equalling the peduncle; lip boat-shaped, broadly elliptical with a long apiculus, equalling the sepals
                                             16. *infundibulare*

Spur less than 5 cm. long, almost straight, broadly conical at the base, constricted in the middle, fusiform in the apical portion; lip broad, nearly orbicular or more or less wedge shaped, nearly as long as sepals, long apiculate:

Sepals 3–5·5 cm. long (usually less than 4·5 cm); lip almost orbicular, 2·5–3·5 cm. broad, side lobes rounded, not projecting forward beyond the base of the central apiculus; leaves 7–14 cm. long, 1·5–3·5 cm. broad  ..    ..    14. *birrimense*

Sepals 3·5–6 cm. long (usually not less than 4·5 cm.); lip broadest in the front, 4–5 cm. broad, side lobes projecting forward beyond the base of the central apiculus so that the lip is more or less cordate; leaves 7–10 cm. long, 1·5–4 cm. broad     ..    ..    ..    ..    ..    ..    ..    ..    15. *eichleranum*

Sepals 2·5 cm. or less long; lip usually with a central low keel in basal part:

Stems short, usually less than 4 cm. long; flowers up to 5, very small, green, sepals less than 4 mm. long; ovary short, twisted round the short obtuse spur; leaves linear or narrowly ligulate, 2–7 cm. long, 2–6 mm. broad; sepals broadly lanceolate; petals linear; lip a little shorter, very concave with a short apiculus.
                                             6. *sacciferum*

Stems elongated; flowers larger, sepals 7 mm. long or more:

Spur about equalling the lip in length; flowers white, flesh-coloured or pale buff; leaves much narrowed in upper part, often with a more or less acute bilobed apex:

Lip triangular-ovate with a broad almost cordate base, 14–21 mm. long; petals thread-like from a broad base, much narrower than the sepals; leaves narrowly or ovate-lanceolate, very unequally and acutely lobed at apex, 4–9 cm. long, 8–23 mm. broad; spur with a swollen base and club-shaped apex, 12–19 mm. long; flowers white or pale buff-coloured  ..    ..    13. *angustipetalum*

Lip narrowly or broadly lanceolate, not much wider at the base and not at all cordate; petals narrower than the sepals but not thread-like:

Leaves very narrowly lanceolate or almost linear, with a long narrow apex, 3–8·5 cm. long, 4–8 mm. broad, unequally and acutely bilobed at the apex; sepals about 2 cm. long, narrowly lanceolate; lip similar, about 15 mm. long; spur a little shorter than the lip, with a narrow conical basal half and a somewhat swollen apical half which is slightly inflexed; inflorescence 1–3-flowered ..  ..  ..  ..  ..  ..  ..  ..  ..     12. *modicum*

Leaves broadly elliptical-lanceolate or elliptical-oblong, with a blunt bilobed apex, 1·5–4·5 cm. long, 7–20 mm. broad; sepals 6–11 mm. long, broadly lanceolate or elliptical; lip lanceolate-ovate, 4–7·5 mm. long; spur about 7 mm. long, swollen at apex with an almost globose or ellipsoid sac  ..     11. *egertonii*

Spur distinctly longer than the lip; flowers green, yellow-green or pale yellow; leaves more or less oblong, usually only narrowed at or near the apex itself:

Lip narrowed just above the column, suddenly widened in the middle, running out at the apex into a long narrow point, about 13 mm. long; spur 2 cm. long, the base wide, middle constricted, apical half rather swollen; leaves very unequally and acutely bilobed, 3–7 cm. long, 6–12 mm. broad; inflorescence 2–4-flowered     ..    ..    ..    ..    ..    9. *angustum*

Lip widest just above the column, lanceolate to ovate, usually less than 10 mm. long:

Leaves narrowly strap-shaped, usually less than 1 cm. broad, often much narrower and never exceeding 1·5 cm., parallel sided almost to apex, apex unequally bilobed with the two lobes rounded, 3–7 cm. long; pollinia attached to a single large strap-shaped pointed viscidium, stipites short and pointed; sepals 10–17 mm. long; lip with a recurved pointed apex, nearly 1 cm. long; spur gently curved, slightly swollen in apical part, 15–20 mm. long    10. *chevalieri*

Leaves oblong or narrowly oblong, usually over 1 cm. broad, narrowing somewhat below the apex, apical lobes more or less connivent; pollinia attached to separate viscidia, stipites wider and rounded at apex:

Spur widest just beyond the middle, tapering in both directions but mouth again widened, apex almost acute, 11–14 mm. long; leaves markedly unequally bilobed at the apex, the shorter lobe very short sometimes scarcely evident, 3–10 cm. long, 1–2 cm. broad; sepals and petals lanceolate, acute, 8–10mm. long; lip ovate, acute, about 6 mm. long      7. *multinominatum*

Spur widest near the apex, where it is swollen and club-shaped, the apex itself rounded, 10–14 mm. long; leaves only slightly unequally bilobed, the shorter lobe quite distinct, 2·5–11 cm. long, 7–22 mm. broad; sepals and petals oblong or oblong-lanceolate, obtuse or acute, 6–11 mm. long; lip ovate, acute, 5·5–7·5 mm. long    ..    ..    ..    8. *pyriforme*

1. **A. distichum** *Lindl.* in Bot. Reg. 21: t. 1781 (1836); Cat. Talb. 146. *Mystacidium distichum* (Lindl.) Pfitz. (1889)—F.T.A. 7: 175. Flowers white.
   **Guin.:** Nzo, foot of Nimba Mts. (Mar.) *Schnell* 829! Tawia to Herimankuna (Jan.) *Sc. Elliot* 4492! **S.L.:** Rokupr (Aug.) *Adames* 116! Njala (Aug.) *Deighton* 1403! Kangama, Gorama Mende (July) *Deighton* 5115! Kruto to Sini-koro, Loma Mts. (Sept.) *Jaeger* 1808! **Lib.:** Suacoco, Gbanga (July) *Daniel & Barker* 219! Bomi Hills (Sept.) *Voorhoeve* 434! Vahun, Kolahun (Nov.) *Baldwin* 10175! Yah R., Sanokwele (Sept.) *Adames* 539! Zwedru, Tchien (Aug.) *Baldwin* 7030! **Iv.C.:** Danané to Goutokouma (Apr.) *Chev.* 21300! 15 km. NE. of Bianovan (Apr.) *Leeuwenberg* 3961! Yapo *Chev.* 16841! Yonkou (Aug.) *Schnell* 1653! **Ghana:** Anjoala F.R., Akim Oda (Apr.) *Andoh* FH 5145! Shiare, Buem-Krachi *Hall* 1358! Bibiani (Nov.) *Darko* 853! Amedzofe (July, Aug.) *Westwood* 30! *Scholes* 85! **Dah.:** Porto Novo *Roberty* 1547! **N.Nig.:** Kagerko, Zaria Prov. *King* 105a! **S.Nig.:** Jamieson R., Sapoba (Nov.) *Meikle* 519! Okomu F.R., Benin (Aug.) *Brenan* 9033! Idanre Hills, Ondo (June) *Keay & Brenan* FHI 22406! Okigwi, Owerri (Dec.) *Jones* FHI 6183! Oban *Talbot* 784! **W.Cam.:** Victoria (Sept.) *Ngongi* FHI 15090! Also in S. Tomé, Principe, E. Cameroun, Gabon, the Congos, Angola and Uganda.

2. **A. aporoides** *Summerh.* in Kew Bull. 17: 560 (1964). *Mystacidium distichum* var. *grandifolium* De Wild. in Ann. Mus. Congo, sér. 5, 2: 240 (1907). *Angraecum distichum* var. *grandifolium* (De Wild.) Summerh. in Kew Bull. 12: 261 (1958). Flowers white.
   **S.Nig.:** Brass R. *Barter* 1854! Ono (Feb.) *Cooper* 82/3! **W.Cam.:** M'bonge *Wright* 58/49! **F.Po:** (Apr.) *Westwood* 92! Also in Principe, E. Cameroun and Congo.

3. **A. podochiloides** *Schltr.* in Engl., Bot. Jahrb. 38: 162 (1906); Cat. Talb. 146. *Monixus aporum* Finet in Mém. Soc. Bot. France 9: 19, t. 3, figs. 9–12 (1907).
   **Lib.:** Tawata, Boporo Dist. (Nov.) *Baldwin* 10299! 10342! Kulo, Sinoe Co. (Mar.) *Baldwin* 11420! Ganta (June) *Harley* 1641! **Iv.C.:** Upper Cavally R. (June) *Pobéguin* 58! Gliké, Tabou (Aug.) *Schnell* 1688! **S.Nig.:** Akampa, Calabar (July) *Latilo* FHI 45831! Oban *Talbot* 894! **W.Cam.:** Bibundi (Oct.) *Schlechter* 15769! Also in Congo.

4. **A. pungens** *Schltr.* in Engl., Bot. Jahrb. 38: 163 (1906); Cat. Talb. 147. Flowers white.
   **S.Nig.:** Calabar *Latilo* FHI 56898! Oban *Talbot* 892! **W.Cam.:** Little Cam. Mt., 3,000 ft. *Cole* 7! Man of War Bay, near Bimbia (Sept.) *Schlechter* 15774! **F.Po:** (May) *Westwood* 91! Pico S. Isabel, 6,000 ft. (Dec.) *Sanford* 4211! Also in E. Cameroun and Congo.

5. **A. subulatum** *Lindl.* in Hook. Comp. Bot. Mag. 2: 206 (1837); Cat. Talb. 147. *Listrostachys subulata* (Lindl.) Rchb. f. (1864)—F.T.A. 7: 168. Flowers white or creamy white, singly or in pairs.
   **S.L.:** Potolo (Dec.) *Sc. Elliot* 4334! Nyandehun, Valunia (Aug.) *Deighton* 6117! Punduru to Pangwama, Gorama Mende (June) *Deighton* 5101! **Lib.:** Wohmen, Vonjama Dist. (Oct.) *Baldwin* 10059! Tawata, Boporo Dist. (fr. Nov.) *Baldwin* 10286! Zolopla, Sanokwele (Sept.) *Baldwin* 9381! Kitoma (Apr.) *Harley* 1799! Ganta (May) *Harley* 1636! **Iv.C.:** Bouroukrou (Dec., Jan.) *Chev.* 16524! **Ghana:** Mim, Goaso Dist. (May) *Andoh* FH 5535! Mampong, Ashanti (Mar.–Apr.) *Westwood* 102! **S.Nig.:** Olokemeji F.R., Abeokuta (Nov.) *Jones et al.* FHI 4916! Akure F.R., Ondo (Aug.) *Jones* FHI 19541! Sapoba, Benin Prov. *Kennedy* 1796! Nun R. *Barter* 2125! Oban *Talbot* 902! **W.Cam.:** S. Bakundu F.R., Kumba (Apr.) *Daramola* FHI 29819! **F.Po:** (June) *Barter*! Also in E. Cameroun and the Congos.

6. **A. sacciferum** *Lindl.* in Hook. Comp. Bot. Mag. 2: 205 (1837). A very small plant with a tuft of narrow leaves and few-flowered inflorescences of pale green flowers.
   **W.Cam.:** Buea, 3,000 ft. (Aug.) *Gregory* 582! Eastwards to Kenya and southwards through E. Africa to Natal and the Cape Province.

7. **A. multinominatum** *Rendle* Cat. Talb. 107 (1913). *Mystacidium clavatum* (Rendle) Rolfe in F.T.A. 7: 172 (1897). Flowers green or yellow-green, often tinged with orange.
   **Guin.:** Kindia *Jac.-Fél.* 89! **S.L.:** Bafodeya (Apr.) *Sc. Elliot* 5555! Musaia (May) *Deighton* 5044! Taninanun, Bumpe (Apr.) *Deighton* 1727! Waka, Kunike Barina (May) *Deighton* 5043! Lunsar, Marampa (June) *Adames* 211! **Ghana:** Aburi (Jan.) *Cox* 88! Jasikan (July) *Westwood* 24! Dukuse, Afram Plains (Apr.) *Cansdale* 86! **Togo Rep.:** Quamikrom (Mar.) *Schlechter* 12952! **S.Nig.:** Otta to Abeokuta *Barter* 3352! Olokemeji F.R. (Mar.) *Keay* FHI 19189! Ibadan South F.R. *Keay* FHI 25305! Onitsha *Barter* 477! Also in Gabon.

8. **A. pyriforme** *Summerh.* in Kew Bull. 1936: 230; F.W.T.A., ed. 1, 2: 452. *A. multinominatum* of Cat. Talb. 147. Flowers green and white.
   **Iv.C.:** Taï Forest (Oct.) *Aké Assi* 6028! **S.Nig.:** Shasha Forest, Ijebu (May) *Richards* 3476! Sapele (June) *Wright* 154! Ajagbodudu, Sapele (July–Sept.) *Wright* 132! Oban *Talbot* 888!

FIG. 398.—ANGRAECUM APOROIDES *Summerh.* (ORCHIDACEAE).

1, flowering stem, × 1. 2, part of stem showing leaves and flower, × 3. 3, intermediate sepal, × 6. 4, lateral sepal, × 6. 5, petal, × 6. 6, lip, three-quarter profile, × 6. 7, ovary with column and anther, pollinaria removed, × 6. 8, apex of column, showing anther and stigma, × 6. 9, pollinarium, × 8. All from *Cooper* 82/3.

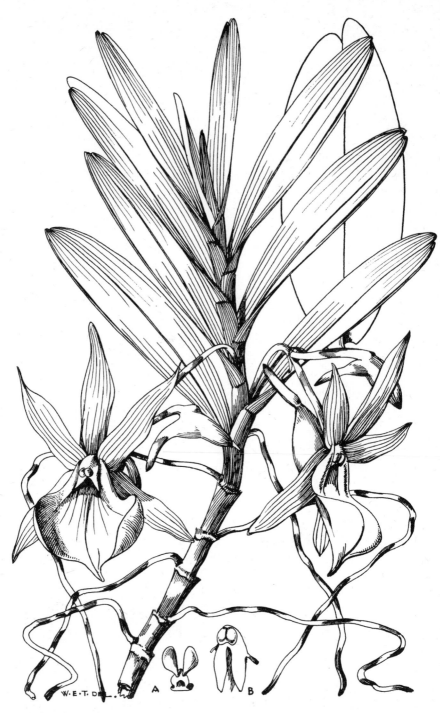

Fig. 399.—Angraecum eichleranum *Kraenzl.* (Orchidaceae).
A, pollinia. B, column.

256

9. **A. angustum** (*Rolfe*) *Summerh.* in F.W.T.A., ed. 1, 2: 452 (1936). *Mystacidium angustum* Rolfe in F.T.A. 7: 570 (1898). Flowers white, tinged brown or yellow.
**S.Nig.:** Itu, Cross R. (Apr.) *Holland* 27!

10. **A. chevalieri** *Summerh.* in F.W.T.A., ed. 1, 2: 452; Kew Bull. 1936: 230. Flowers yellow, yellow-green or greenish white.
**Guin.:** Foredaka, Fouta Djalon (June) *Adames* 281! **Lib.:** Ganta (May) *Harley* 1637! **Iv.C.:** Sogui to Koualé, Toma country (May) *Chev.* 21690! **Ghana:** Aburi (May) *Cox* 95! *Westwood* 13! Efiduase *Cox* 19! **N.Nig.:** Kagoro R., Jemaa, Zaria (June, July) *King* 113! **S.Nig.:** Ibadan South F.R., Oyo Prov. (Apr.) *Mutch* FHI 22806! Akure F.R., Ondo (Aug.) *Jones* FHI 20223! **F.Po:** Lago Louto *Sanford* 4299! Also in Gabon, Congo and Uganda.

11. **A. egertonii** *Rendle* Cat. Talb. 107: t, 15, figs. 1–2 (1913). Flowers white or greenish white.
**S.Nig.:** Sapele, Sapoba (June) *Wright* 157! Oban *Talbot* 889! Also in Gabon.

12. **A. modicum** *Summerh.* in Kew Bull. 12: 84 (1958). Flowers flesh coloured.
**Lib.:** Ganta (May) *Harley* 1923! 2131!

13. **A. angustipetalum** *Rendle* Cat. Talb. 106, t. 14, figs. 10–12 (1913). Flowers white or pale buff-coloured.
**Ghana:** Akim Hills (Dec.) *Johnson* 593! Juaben (July) *Cox* 50! Akatin, E. Prov. (Oct.) *Vigne* FH 4039! Bunsu (Mar.–June) *D. Gillett*! *Westwood* 36! **S.Nig.:** Sapoba *Kennedy* 1918! Ajagbodudu, Sapele (June) *Wright* 153! Oban *Talbot* 890! Also in E. Cameroun, Gabon and Congo.

14. **A. birrimense** *Rolfe* in Kew Bull. 1914: 214. Sepals and petals pale green, lip white with green centre.
**S.L.:** *Bowden*! **Lib.:** Du R. (July, Aug.) *Linder* 77! 212! Gletown, Tchien Dist. (July) *Baldwin* 6985! **Iv.C.:** Niapidou, N. of Sassandra (fr. Jan.) *Leeuwenberg* 2425! **Ghana:** Birrim (Oct.) *Miles*! Mpraeso (Sept.) *Vigne* FH 4252! Kumasi (Oct.) *Vigne* FH 4868! Juaso, Ashanti (Aug.) *Cansdale* 116! Bunsu (July) *D. Gillett*! **S.Nig.:** Onishere F.R., Ondo (May) *Onochie & Latilo* FHI 34699! Sapoba *Kennedy* 280! Mile 20, Owerri to Port Harcourt (June) *Cooper* 25/6! **W.Cam.:** Buea, 3,200 ft. (July, Aug.) *Dundas* FHI 15301! *Gregory* 169! *Keay* FHI 25374! Also in E. Cameroun.

15. **A. eichleranum** *Kraenzl.* in Wittm. Gart. Zeit. 1: 434, fig. 102 (1882); F.T.A. 7: 143; Cat. Talb. 146. Sepals and petals pale green, lip white with green centre.
**S.Nig.:** Calabar (Aug.) *Pierez*! *Binuyo* FHI 41428! Oban *Talbot* 899! **W.Cam.:** Debunscha (July) *Thorold* TN34! Buea (Aug.) *Keay* FHI 25375! Also in E. Cameroun, Gabon and Angola.

16. **A. infundibulare** *Lindl.* in J. Linn. Soc. 6: 136 (1862). *Mystacidium infundibulare* (Lindl.) Rolfe in F.T.A. 7: 170 (1897). Sepals and petals green, lip white, spur green.
**S.Nig.:** Brass R. *Barter* 1860! **W.Cam.:** Victoria *Rosevear* 55/37! Also in Principe, the Congos and Uganda.

### 37. **TAENIOPHYLLUM** Blume, Bijdr. 335, t. 70 (1825).

Plant very small, inflorescence scarcely reaching 2 cm. in height, few-flowered; flowers orange; sepals and petals united in a tube in the lower part, total length 1·5–2 mm.; lip free, very concave with the apex hooded and terminating in an inflexed sharp point, 1·5–2 mm. long; spur almost globose or ellipsoid, 0·6–0·9 mm. long and a little narrower; column very short with two rounded lobes in front; anther 4-lobed; pollinia 4, 2 large and 2 small . .    . .    . .    . .    . .    . .    . .    *coxii*

**T. coxii** (*Summerh.*) *Summerh.* in Kew Bull. 12: 278 (1958). *Ankylocheilos coxii* Summerh. in Bot. Mus. Leafl. Harv. Univ. 11: 169 (1943).
**Ghana:** Aburi (Apr.) *Cox* 92! Also in Congo.

### 38. **LISTROSTACHYS** Rchb. f. in Bot. Zeit. 10: 930 (1852); F.T.A. 7: 150, partly.

Stem usually less than 6 cm., rarely up to 15 cm. long; leaves 8–35 cm. long, 1–2 cm. broad, almost equally or slightly unequally bilobed at the apex, lobes obtuse; inflorescences 10–25 cm. long, peduncle 2–5 cm. long; sepals ovate, 2–3 mm. long; petals a little shorter and narrower; lip obovate or almost quadrate, shortly and broadly apiculate; spur clavate at the apex, 3·5–5 mm. long    . .    . .    . .    *pertusa*

**L. pertusa** (*Lindl.*) *Rchb. f.* in Bot. Zeit. 10: 930 (1852); F.T.A. 7: 61. *Angraecum pertusum* Lindl. (1837)— Cat. Talb. 147. *Listrostachys jenischiana* Rchb. f. in Bot. Zeit. 10: 930 (1852). *L. behnickiana* Kraenzl. in Notizbl. Bot. Gart. Berl. 5: 122 (1909). Flowers white, sometimes with minute red spots towards the base and with a red spur.
**S.L.:** Njala (Nov.) *Deighton* 2552! Messima, Kpukumo Krim (Nov.) *Deighton* 4690! **Lib.:** Fisherman's Lake, Grand Cape Mt. (Dec.) *Baldwin* 10881! Yasono (Oct.) *Harley* 1936! Begwai (Oct.) *Bunting* 26! Peter's Town (Oct.) *Bunting* 97! Nimba (Nov.) *Adames* 739! **Iv.C.:** Mt. Nienokoué *Chev.* 19662! Moossou, Grand Bassam (Nov.) *de Wilde* 807! Niapidou, N. of Sassandra (fr. Jan.) *Leeuwenberg* 2625! **Ghana:** Axim (June) *Cox* 116! **S.Nig.:** Nun R. (fr. Sept.) *Mann* 524! Brass R. *Barter* 1826! Waru (Jan.) *Gregory* 506! Oban *Talbot* 916! Eket *Talbot* 3382! Also in E. Cameroun, Principe, Gabon and Congo.

### 39. **MICROCOELIA** Lindl., Gen. and Sp. Orch. Pl. 60 (1830).

Limb of lip very small, 2–4 mm. long, less than half as long as the straight spur:
  Rostellum very short, arising from near the apex of the short thick column; viscidium reniform-oblong, about as long as broad; limb of lip 2 mm. long with 2 very obscure small lateral lobes at the base; spur about 2 cm. long, narrowed from the mouth to 5 mm. from apex, then suddenly swollen, tapering from there to the end; sepals and petals 2–3·5 mm. long    . .    . .    . .    . .    . .    . .    2. *microglossa*
  Rostellum arising from near the base of the long slender column and projecting upwards, bilobed from the base; stipes long and slender, sharply bent in the middle so that the 2 halves are nearly parallel; viscidium oblong-lanceolate, nearly as long as the rostellum; limb of lip 2–4 mm. long, narrowed towards the base; spur 8–13 mm. long, narrowed just near the mouth, much swollen in the middle and again slightly at the apex; sepals and petals 2·5–4 mm. long    . .    . .    3. *caespitosa*
Limb of lip nearly equalling or longer than the spur which is usually curved:
  Lip obovate, 2·5–3·5 mm. long; column short, truncate at apex; anther transversely placed on column, not drawn out into a point in front; rostellum shortly 2-lobed,

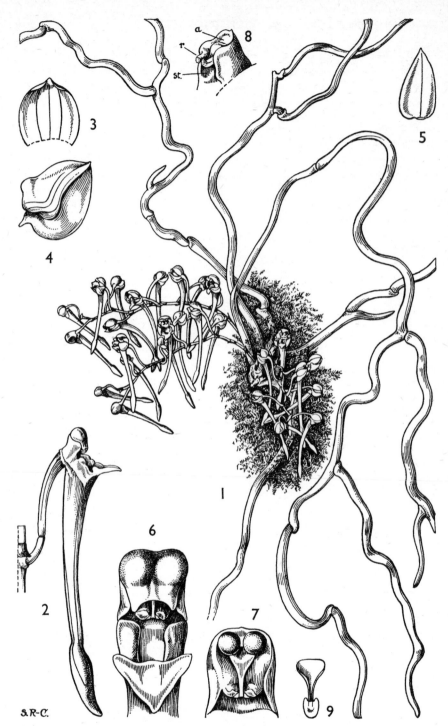

FIG. 400.—MICROCOELIA MICROGLOSSA *Summerh.* (ORCHIDACEAE).

1, flowering plant, × 1.  2, flower with sepals and petals removed, × 4.  3, dorsal sepal, × 8.
4, lateral sepal, from outside, × 8.  5, petal, × 8.  6, column and upper part of lip, × 12.
7, column, with anther removed, × 12.  8, column side view, with anther and pollinarium
removed, × 8:—a, androclinium; r, rostellum lobes; st, stigma. 9, stipes and viscidium,
× 12.  (Reprod. from Hook., Ic. Pl. t. 3466 by permission of Bentham-Moxon Trustees).

258

at apex of column; stipes of pollinium short, viscidium oblong; sepals and petals 2–4 mm. long; spur about the same length as the lip, slightly incurved or recurved, not at all swollen at the apex; inflorescence up to 13 cm. long, many-flowered

1. *guyoniana*

Lip over 5 mm. long, much narrowed at the base, sometimes with a distinct claw; column relatively long, very obliquely truncate; anther obliquely placed on column, drawn out into a long point, covering the pollinarium; rostellum near base of column with 2 projecting fleshy lobes; stipes of pollinium long and slender, viscidium linear; inflorescences up to 9 cm. long, many-flowered:

Spur sharply bent beneath the lip, much swollen and rounded at apex, 5·5–6·5 mm. long; lip ovate or elliptical-ovate narrowed to a short claw at base, total length 6–9 mm. long; lobes of the rostellum narrow and more or less pointed; stipes of pollinia forked in the upper part; sepals 4–7 mm. long, laterals markedly curved

4. *dahomeensis*

Spur more or less S-shaped, incurved with the apex slightly recurved, not much swollen at apex, 6·5–11 mm. long; lip ovate or diamond-shaped, narrowed at base, 5–8·5 mm. long; lobes of rostellum blunt and rounded; stipes of pollinia entire; sepals 4–8·5 mm. long, laterals oblique but scarcely curved  ..        5. *koehleri*

1. **M. guyoniana** (*Rchb. f.*) *Summerh.* in Bot. Mus. Leafl. Harv. Univ. 11: 144 (1943). *Angraecum guyonianum* Rchb. f. (1849)—F.T.A. 7: 148. Flowers white, apical part of spur greenish, brownish or salmon-coloured.
   **N.Nig.:** Gindiri, E. edge of Jos Plateau, 3,000 ft. (Mar.) *King*! Heipang, Jos (Apr.) *Keay et al.* FHI 37609! Also in C. African Republic, Congo, Uganda, Kenya, Ethiopia, Angola and Zambia.
2. **M. microglossa** *Summerh.* in Kew Bull. 1936: 231, and in F.W.T.A., ed. 1, 2: 454. Flowers with a brown line down centre of sepals.
   **S.Nig.:** Oban *Talbot*! Also in Uganda.
3. **M. caespitosa** (*Rolfe*) *Summerh.* in F.W.T.A., ed. 1, 2: 454 (1936). *Angraecum caespitosum* Rolfe in F.T.A.7: 150 (1897); Cat. Talb. 146. *A. andersonii* Rolfe in Kew Bull. 1912: 134. Flowers white, sepals with central green line, lip with central green blotch.
   **S.L.:** Njala (June–Aug.) *Deighton* 717! 5113! Roruks (July) *Deighton* 3250! Jala (Sept.) *Bunting* 77! **Lib.:** Gbanga (Sept.) *Linder* 603! **Iv.C.:** Zago to Gaouloubré (May) *Chev.* 16340! Yapo Forest, N. of Abidjan (fr. July) *de Wilde* 143! Banco Forest, NW. of Abidjan (Sept., fr. Oct.) *de Wilde* 731! *Aké Assi* 6021! **Ghana:** Tarquah (Sept.–Dec.) *Miles*! Kumasi *Cox* 62! Bunsu (fr. Nov., Mar.) *Westwood* 59! 59a! Asuanta (July) *Benton*! **S.Nig.:** Akure F.R., Ondo Prov. (fl. cult. Ibadan, June) *Sanford* WS/744/64! Gambari F.R., Ibadan (Aug.) *Cooper*! Sapele (Aug.) *Wright*! Sapoba, Benin (Aug.) *Keay* FHI 37911! Oban *Talbot* 89! Also in E. Cameroun, Congo and Uganda.
4. **M. dahomeensis** (*Finet*) *Summerh.* in F.W.T.A., ed. 1, 2: 454 (1936). *Dicranotaenia dahomeensis* Finet in Mém. Soc. Bot. Fr. 9: 47, t. 9, figs. 28–38 (1907). Flowers white, fragrant.
   **Iv.C.:** Yapo Forest, N. of Abidjan (Dec.) *de Wilde* 1016! **Ghana:** Bunsu, Akim (May) *D. Gillett*! *Irvine* 3011! Assuantsi (Aug.) *Miles*! **Dah.:** Adja Ouére *Le Testu* 125. **S.Nig.:** Sapoba, Benin Prov. (Dec.) *Sanford* WS/164/66! Gambari F.R., Ibadan Prov. (May) *Sanford* WS/927/925/65! Also in E. Cameroun, Congo and Uganda.
5. **M. koehleri** (*Schltr.*) *Summerh.* in Bot. Mus. Leafl. Harv. Univ. 11: 158 (1943). *Angraecum koehleri* Schltr. in Engl., Bot. Jahrb. 38: 162 (1906). Flowers white, sepals with brownish or salmon-coloured central line, apex of spur orange or salmon-coloured.
   **S.Nig.:** Ibadan South F.R., Oyo Prov. (Aug.) *Keay* FHI 22430! Also in Rwanda, Burundi, Uganda, Kenya and Tanzania.

## 40. ENCHEIRIDION Summerh. in Bot. Mus. Leafl. Harv. Univ. 11: 161 (1943).

Rostellum much longer than the column, arising from near its apex and projecting forward like a bird's beak; stipes of pollinia very narrow, as long as rostellum, suddenly widened just below insertion of pollinia; viscidium linear, much shorter than the stipes; lip 3-lobed from the base, lateral lobes small, erect, triangular, middle lobe broadly transversely elliptical from a narrow claw, retuse with an apiculus, margins lacerate; spur incurved, swollen at apex, 8–11 mm. long (straightened); sepals and petals 4–6 mm. long; stem short; inflorescence 3–15 cm. long

*macrorrhynchium*

**E. macrorrhynchium** (*Schltr.*) *Summerh.* l.c. 162 (1943). *Angraecum macrorrhynchium* Schltr. in Engl., Bot. Jahrb. 38: 22 (1905). *Microcoelia macrorrhynchium* (Schltr.) Summerh. in F.W.T.A., ed. 1, 2: 454 (1936). Flowers white or yellowish, tinged rose or purplish.
**Lib.:** Ganta (May–July) *Harley* 1642! 1819! **Iv.C.:** *Merle* 310! **Ghana:** Pamu Berekum F.R. (Sept.) *Vigne* FH 2490! Juaben, **Ashanti** (July) *Cox* 124! Ejura (Sept.) *Cox* 119! Dodi to Mpreasem (fr. Dec.) *Westwood* 78! Anum, Trans-Volta (Sept.) *Westwood* 35! **N.Nig.:** Kotokerifi, Kabba (Oct.) *Daramola & Adebusuyi* FHI 38404! **S.Nig.:** Olokemeji F.R., Abeokuta (July) *Onochie & Latilo* FHI 38349! *Sanford* 1356/65! Akure F.R., Ondo (Aug.) *Tamajong* FHI 20222! Also in E. Cameroun, Gabon, Congo and Uganda.

## 41. CHAULIODON Summerh. in Bot. Mus. Leafl. Harv. Univ. 11: 163 (1943).

Stem very short; inflorescences slender, up to 60 cm. long, rather laxly many-flowered; bracts very small; sepals and petals 3–5 mm. long, lateral sepals wider in the upper part, curved abruptly in the middle; lip consisting mostly of the large incurved spur; limb very short and acute, with a tooth-like erect callus about 2 mm. tall just in front of the spur opening; spur 12–14 mm. long, 3–3·5 mm. broad at the mouth, narrowing to just beyond the middle where it is abruptly curved forward, the apical portion slightly swollen; column rather short and recurved, anther drawn out into a long point in front; stipes of pollinia much widened above and enveloping the pollinia, viscidium very small ..        ..        ..        ..        ..        ..        ..        *buntingii*

**C. buntingii** *Summerh.* l.c. 164 (1943). Flowers pale or orange-brown.
  **Lib.:** Mt. Barclay (June) *Bunting* 9! **Ghana:** Tarkwa (Jan., July) *Vigne* FH 4838! *Cox* 120! Bunsu (fr. Sept.) *Westwood* 150! 163! **S.Nig.:** Ogba Farm, Benin City (July) *Maggs* 125! Ajagbududu, Benin (Aug.) *Wright* 57! Eket *Talbot* 3287!

## 42. DIAPHANANTHE Schltr., Die Orchid. 593 (1914) (incl. *Rhipidoglossum* Schltr.).

Stipites of pollinia attached to a common viscidium; rostellum pointed, divided into two narrow lobes on removal of the viscidium:
Stem short, bearing the leaves in a relatively dense tuft, sometimes as much as 9 cm. long but with leaves only in the upper half:
Lip quadrate (4-sided) or nearly orbicular, not much narrowed in front, with a short apiculus in the centre of the broad often more or less truncate apex, the margins often toothed or fimbriate:
Sepals 9–11 mm. long; leaves oblanceolate, curved, unequally bilobed or almost entire at the apex, 15–70 cm. long, 2·5–7 cm. broad; inflorescences many-flowered, pendulous, 15–55 cm. long; lip quadrate, 8–11 mm. long, margins shortly fimbriate; spur about as long as the lip .. .. .. .. .. 2. *pellucida*
Sepals less than 7 mm. long; leaves narrowly oblanceolate or ligulate, acutely bilobed or entire at apex, up to 20 cm. long, 7–30 mm. broad; inflorescences slender, less than 15 cm. long:
Spur of lip shorter than lamina, less than 3 mm. long; lip suborbicular, 4 mm. long, 4·5 mm. broad, margins scarcely toothed; leaves linear-oblanceolate, 5–7·5 cm. long, 7–13 mm. broad; inflorescences short, up to 4·5 cm. long, up to 6-flowered; sepals 3·5–4 mm. long .. .. .. .. 5. *suborbicularis*
Spur of lip longer than lamina or rarely equalling it, 5–8 mm. long; lip quadrate or quadrate-ovate, 5–6 mm. long, 4–6 mm. broad, margins usually toothed; leaves narrowly lanceolate or ligulate, up to 20 cm. long, 8–30 mm. broad; inflorescences up to 13 cm. long, many-flowered; sepals 4·5–6·5 mm. long
                                        4. *plehniana*
Lip narrowly ovate, elliptical-ovate, or elliptical, narrowed towards the apex which is obtuse or subacute, the margins almost or quite entire:
Leaves broadly oblanceolate, 2–5 cm. broad, 10–23 cm. long, unequally acute at apex; lip without any callus at mouth of spur, elliptical-ovate or elliptical, 6·5–8 mm. long, 4–5 mm. broad; spur about equalling the lip; inflorescence 10–25 cm. long, many-flowered .. .. .. .. .. .. .. 1. *dorotheae*
Leaves narrowly lanceolate or ligulate, 0·5–2 cm. broad, 4–15 cm. long; lip with callus at mouth of spur, narrowly ovate or elliptical-ovate; spur distinctly longer than the lip:
Sepals 6–8·5 mm. long; spur 12–15 mm. long, apex broadened, truncate, more or less emarginate; stipites of pollinia very broad, adnate along their inner margins, long fimbriate on outer margins; lip narrowly or lanceolate-ovate, 4–8·5 mm. long; inflorescence 6–15 cm. long, 6–10-flowered; leaves lanceolate-ligulate, acute, 8–20 mm. broad .. .. .. .. .. .. .. 3. *bueae*
Sepals 3·5–5·5 mm. long; spur 6·5–13·5 mm. long, swollen in upper half but apex itself narrow, obtuse; stipites of pollinia narrow, linear, quite separate, entire; lip ovate or narrowly ovate, 3–4·5 mm. long; inflorescence 6–28 cm. long, 6–20-flowered; leaves ligulate or oblong-ligulate, acute or more frequently obtuse, 6–15 mm. broad .. .. .. .. .. .. .. 6. *quintasii*
Stems much elongated, reaching over 1 m. in length, bearing the leaves at more or less regular intervals; leaves oblong-lanceolate to narrowly ovate, very distinctly and acutely bilobed at the apex, 5–14 cm. long, 1·5–4·5 cm. broad; inflorescences 5–18 cm. long, many-flowered, pendulous; sepals 3–5 mm. long; lip quadrate, 3·5–5 mm. long; spur a little longer than the lip .. .. .. .. .. .. 7. *bidens*
Stipites each with a separate viscidium; rostellum obtuse or rounded, fleshy, shortly 3-lobed on removal of viscidium:
Stem very short, leaves in a bunch at its apex:
Lip 3-lobed, lateral lobes longly pectinate, front lobe much smaller, tooth-like, entire; petals shortly pectinate; spur filiform, about 2 cm. long; sepals and petals 4–4·5 mm. long; leaves linear-lanceolate, unequally and acutely bilobed at apex, 6–10 cm. long, 7–10 mm. broad; inflorescence 6–11 cm. long, 5–9-flowered
                                        8. *polydactyla*
Lip entire or shortly toothed in front, shortly apiculate at apex; petals entire or very shortly toothed or fimbriate:
Spur not quite so long as the lip; lip obovate from a narrow base, margins shortly toothed in front, 1·7–1·9 cm. long, callus at mouth of spur tooth-like; sepals 1·5–2 cm. long; leaves oblanceolate-linear to oblanceolate, 25–50 cm. long, 2–5 cm. broad; inflorescences 10–30 cm. long .. .. .. 9. *kamerunensis*
Spur more than 3 times as long as the lip, 10·5–16 mm. long; lip ovate or nearly orbicular with a broad base, 2·5–3 mm. long, callus at mouth of spur rim-like;

sepals 3·5–4·5 mm. long; leaves oblanceolate, 4–17 cm. long, 1–4·5 cm. broad; inflorescences 10–35 cm. long, many-flowered .. .. .. **10.** *curvata*
Stem elongated, leaves spaced out along the apical part:
  Spur 2 or more times as long as the lip, not or hardly swollen towards apex; lateral sepals 3–4 mm. long; inflorescence many-flowered:
    Lip longer than broad, 3–3·5 mm. long, quadrate, truncate in front, slightly emarginate with sometimes a small apiculus in the sinus, margins irregular; spur 11–13 mm. long, very slender; viscidia more than half as long as stipites; leaves 9–14 cm. long, 6–16 mm. broad, acute .. .. .. .. **11.** *longicalcar*
    Lip broader than long, about 2–3 mm. long, 2·5–4·5 mm. broad, flabellate-elliptical, obscurely 3–4-lobed in front; spur 5–12 mm. long; viscidia very small, much less than half as long as stipites; leaves 6–15 cm. long, 5–30 mm. broad, unequally and obtusely bilobed .. .. .. .. .. .. .. **12.** *rutila*
  Spur less than twice as long as the lip, distinctly swollen in the apical part; lateral sepals 2–2·5 mm. long; lip transversely elliptical, about 1·5 mm. long and 3 mm. broad:
    Leaves 1·5–2·5 cm. broad, 6–11 cm. long; flowers 1–1·5 mm. apart; inflorescence many-flowered; lip slightly retuse .. .. .. .. **13.** *obanensis*
    Leaves 5–10 mm. broad, 4–11 cm. long; flowers 4–5 mm. apart; inflorescence up to 20-flowered; lip with a broad blunt projecting apex .. .. **14.** *laxiflora*

1. **D. dorotheae** (*Rendle*) *Summerh.* in Kew Bull. 3: 44 (1949). *Angraecum dorotheae* Rendle, Cat. Talb. 107, t. 15, figs. 3–5 (1913). *Rangaeris dorotheae* (Rendle) Summerh. in F.W.T.A., ed. 1, 2: 450 (1936).
   **S.Nig.:** Oban *Talbot* 914! 915!
2. **D. pellucida** (*Lindl.*) *Schltr.* Die Orchid. 593 (1914). *Angraecum pellucidum* Lindl. (1861)—Cat. Talb. 147. *Listrostachys pellucida* (Lindl.) Rchb. f. (1864)—F.T.A. 7: 162. Flowers white, pale yellow or pinkish.
   **Guin.:** Farana (Feb.) *Chev.* 20655! Nimba Mts. (Aug., Sept.) *Schnell* 1868! 3471! 3673! **S.L.:** Kamba (Sept.) *Deighton* 5184! Njala (Nov.) *Deighton* 2554! Sini-Koro (Sept.) *Jaeger* 2040! Kinto to Sini-Koro (Sept.) *Jaeger* 1815! Kordu, Kisi Kama (Nov.) *Glanville* 5911! **Lib.:** Moylakwelli to Tolokwelli (Oct.) *Linder* 1375! Devilbush, Duport (Nov.) *Voorhoeve* 618! Rebbo to Peters Town (Nov.) *Bunting* 30! Sanokwele (Sept.) *Baldwin* 9553! Ganta *Harley* 1829! **Iv.C.:** Massif de Dans (May) *Schnell* 1293! Mt. Tonkoui, SW. of Man (fr. Mar.) *Leeuwenberg* 2962! Oroumba Boka, Toumodi (Oct.) *de Wilde* 650! **Ghana:** Pamu-Berekum F.R. (Sept.) *Vigne* FH 2489! Bon, W. Prov. (Nov.) *Vigne* FH 1586! Akwapim Hills (Oct.) *Johnson* 857! Juaso, Ashanti (Oct.) *Cansdale* 124! Alavanyo, Hohoe (Sept.) *St.-Cl. Thompson* FH 1439! **S.Nig.:** Akure F.R., Ondo (Aug.) *Jones* FHI 19536! E. of Ondo (Sept.) *Keay* FHI 22562! Brass R. *Barter* 37! Ukpon F.R., Ogoja Prov. (July) *Latilo* FHI 36867! Oban *Talbot* 900! **W.Cam.:** Barombi (Aug.) *Preuss* 445! Mahom Nkom-Wum F.R., Bamenda (July) *Ujor* FHI 30482! Also in Annobon, E. Cameroun, Gabon, the Congos and Uganda.
3. **D. bueae** (*Schltr.*) *Schltr.* in Beih. Bot. Centralbl. 36, 2: 96 (1918). *Angraecum bueae* Schltr. (1906). Flowers white and green.
   **W.Cam.:** Buea, 3,300 ft. (June–Aug.) *Deistel!* *Gregory* 153!
4. **D. plehniana** (*Schltr.*) *Schltr.* l.c. 97 (1918). *Angraecum plehnianum* Schltr. (1905). Flowers pink to apricot-coloured.
   **S.Nig.:** Down R., Sapoba (Dec.) *Sanford* 121/66! Okomu F.R., Benin Prov. (May) *Brenan* FHI 25330! *Onochie* FHI 34696! Ajagbodudu *Wright* 140! Afi River F.R., Ogoja Prov. (June) *Jones & Onochie* FHI 17343! **W.Cam.:** M'bonge *Wright* 58/19! Also in E. Cameroun.
5. **D. suborbicularis** *Summerh.* in Kew Bull. 12: 278 (1958). Flowers purplish or pinkish.
   **Ghana:** Kpandu to Wurobong (Oct.) *Westwood* 158!
6. **D. quintasii** (*Rolfe*) *Schltr.* in Beih. Bot. Centralbl. 36, 2: 99 (1918). *Angraecum quintasii* Rolfe (1891)—F.T.A. 7: 141. Flowers greenish white or greenish yellow.
   **Ghana:** Aburi (Oct.) *Cox* 87! Amedzofe (Sept., Oct.) *Westwood* 31! 31a! **F.Po:** Moka, 4,600 ft. (Sept.) *Melville* 436! Also in S. Tomé, Angola, Uganda and Kenya.
7. **D. bidens** (*Sw.*) *Schltr.* Die Orchid. 593 (1914). *Limodorum bidens* Sw. (1800)—F.T.A. 21: 243. *Listrostachys bidens* (Sw.) Rolfe in F.T.A. 7: 160 (1897). *L. ashantensis* (Lindl.) Rchb. f. (1895)—F.T.A. 7: 159. *Mystacidium productum* Kraenzl. (1895)—F.T.A. 7: 174. *Angraecum bidens* (Sw.) Rendle, Cat. Talb. 147 (1913). Flowers salmon-pink, yellowish pink, flesh-coloured or white.
   **Port.G.:** Cabuchangue to Guébu, Catio (June) *Esp. Santo* 2069! Bedanda, Jambarem (fr. Jan.) *Pereira* 2690! **Guin.:** Lola to Nzo (Mar.) *Chev.* 20991! Nimba Mts. (Apr.) *Schnell* 1080! **S.L.:** Moa R. (July, Aug.) *Dawe* 533! Makali (July) *Deighton* 4404! Njala (Aug.) *Deighton* 1210! Rokupr (Aug.) *Adames* 117! Baiima, Gbo (Aug.) *Deighton* 6110! **Lib.:** Du R. (July) *Linder* 79! Mt. Barclay (July) *Bunting* 161! Palilah, Gbanga Dist. (Aug.) *Baldwin* 9209! St. John R., Ganta (June) *Harley* 1816! Putu, Eastern Prov. (Mar.) *de Wilde & Voorhoeve* 3669! **Iv.C.:** Guidéko (May) *Chev.* 16439! Adiopodoumé (fr. Nov.) *Roberty* 15502! Issia Rock, Daloa (fr. Nov.) *de Wilde* 828! Sassandra R. (Apr.) *Leeuwenberg* 4012! 15 km. NE. of Bianovan (Apr.) *Leeuwenberg* 3959! **Ghana:** Juaso, Ashanti (Apr.) *Cansdale* FH 4383! Mankrong, Kwahu (Apr.) *Morton* A566! Aburi (Apr.) *Irvine* 516! Nsawami to Aburi (Apr.) *Morton* A3648! Kumasi (June) *Morton* A3409! Amedzofe (May) *Scholes* 82! **S.Nig.:** Shasha F.R., Ijebu (May) *Ross* 283! Mamu F.R., Ibadan (fr. Sept.) *Onochie* FHI 7892! Udo, Benin Prov. (May) *Umana* FHI 29112! Calabar (May) *Holland* 46! Oban *Talbot* 912! **W.Cam.:** Victoria (May) *Maitland* 725! Victoria to Bimbia (May) *Preuss* 1227! Barombi *Preuss* 335! Kake II, Kumba (fr. July) *Thorold* TN30! **F.Po:** nr. S. Isabel *Sanford* 4011! Also in E. Cameroun, Gabon, Rio Muni, the Congos and Angola.
8. **D. polydactyla** (*Kraenzl.*) *Summerh.* in F.W.T.A., ed. 1, 2: 456 (1936). *Listrostachys polydactyla* Kraenzl. in Engl., Bot. Jahrb. 51: 394 (1914). *Crossangis polydactyla* (Kraenzl.) Schltr. in Beih. Bot. Centralbl. 36, 2: 142 (1918). Flowers greenish white.
   **W.Cam.:** Kufum, Banso Mts., 5,800 ft. (Oct.) *Ledermann* 5716a!
9. **D. kamerunensis** (*Schltr.*) *Schltr.* in Orchis 8: 137 (1914), in obs. *Angraecum kamerunense* Schltr. in Engl., Bot. Jahrb. 38: 161 (1906). Flowers pale green.
   **W.Cam.:** Neu Tegel (July) *Winkler* 209. Ediki (July) *Schlechter* 15768! Buea, 3,300 ft. (July, Aug.) *Deistel!* *Gregory* 556! Also in Uganda and Zambia.
10. **D. curvata** (*Rolfe*) *Summerh.* in F.W.T.A., ed. 1, 2: 456 (1936). *Mystacidium curvatum* Rolfe in F.T.A. 7: 174 (1897). *Angraecum curvatum* (Rolfe) Schltr. in Engl., Bot. Jahrb. 38: 22 (1905), in obs.; Cat. Talb. 146. Flowers white, pale green or yellowish green.
   **S.L.:** Grewahum, Tunkia (Aug.) *Deighton* 4646! **Iv.C.:** Lower Sassandra R. to Cavally R. (June, July) *Chev.* 19248. Boulay I., Ebrié Lagoon (Nov.) *de Wilde* 827! Assakra, Dimbokro (Nov.) *de Wilde* 32171 **Ghana:** Ofin Headwaters F.R. (Nov.) *Vigne* FH 3415! Aburi *Johnson* 285! Konongo, Ashanti (Nov.) *Cansdale* 132! Juaben, Ashanti (June) *Cox* 125! Tepa, Ashanti (Nov.) *Westwood* 120! **S.Nig.:** Mamu

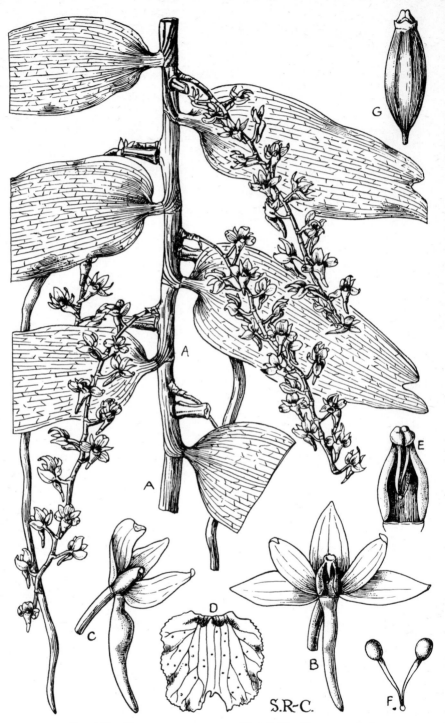

Fig. 401.—Diaphananthe bidens (*Sw.*) *Schltr.* (Orchidaceae).
A, flowering shoot.  B, flower with lip removed, front view.  C, same, side view.  D, lip.  E, column.
F, pollinia.  G, fruit.

F.R., Oyo Prov. (Aug.) *Keay* FHI 16243! Ibadan South F.R. (July) *Ahmed & Chizea* FHI 19778! Gambari F.R., Ibadan (fl. cult. Ibadan, June) *Sanford* 980/65! Akure F.R., Ondo (June) *Charter* FHI 38728! Oban *Talbot* 938! Also in E. Cameroun.

11. **D. longicalcar** (*Summerh.*) *Summerh.* in Kew Bull. 14: 142 (1960). *Rhipidoglossum longicalcar* Summerh. in Kew Bull. 1936: 226; F.W.T.A., ed. 1, 2: 449. Flowers very pale green.
   **S.Nig.:** Ife F.R., Oyo (July) *Latilo & Adebusuyi* FHI 43454! Kwa R., Obutong Beach *Talbot* 939! Oban *Talbot* 937!

12. **D. rutila** (*Rchb. f.*) *Summerh.* in Kew Bull. 14: 143 (1960). *Aëranthus rutilus* Rchb. f. (1885). *Mystacidium rutilum* (Rchb. f.) Dur. & Schinz (1885)—F.T.A. 7: 173. *Rhipidoglossum rutilum* (Rchb. f.) Schltr. (1918)—F.W.T.A., ed. 1, 2: 449. Flowers translucent, pale green or pale yellow, often tinged with rose, brown or purple; in forests.
   **Guin.:** Kouria (Aug.) *Pobéguin* 1678! **S.L.:** Njala (July) *Deighton* 2784! 4639! Musaia (July) *Deighton* 4641! Pujehun, Valunia (July) *Deighton* 5803! **Iv.C.:** Tiapleu Forest (July) *Aké Assi* 8152! Abidjan *Lecoufle* K508! **Ghana:** Wurubong, Kpandu (Oct.) *Westwood* 228! Ofinso (July) *Vigne* FH 4866! Efiduase *Cox* 38! Buasa (May) *D. Gillett*! **N.Nig.:** Pankshin, Plateau Prov. (Aug.) *King* 125! **S.Nig.:** Mt. Orosun, Idanre Dist. (June) *Charter* FHI 38714! Okomu F.R., Benin (June) *Onochie* FHI 31226! Iwo, Oyo (Aug.) *Thorold* TN33! Ibadan (June, July) *Keay* FHI 22433! 25334! **W.Cam.:** Mamfe *Cole* 44b! Also in S. Tomé, E. Cameroun, Gabon and eastwards and southwards to Sudan, Kenya, Tanzania, Angola and Zambia.

13. **D. obanensis** (*Rendle*) *Summerh.* in Kew Bull. 14: 142 (1960). *Angraecum obanense* Rendle, Cat. Talb. 104, t. 14, figs. 1–2 (1913). *Rhipidoglossum obanense* (Rendle) Summerh. in F.W.T.A., ed. 1, 2: 449 (1936). Flowers greenish white.
   **S.Nig.:** Oban *Talbot* 896! Oban F.R., Calabar (Nov.) *Markham* FHI 50084! **W.Cam.:** M'bonge *Wright* 58/66!

14. **D. laxiflora** (*Summerh.*) *Summerh.* in Kew Bull. 14: 141 (1960). *Rhipidoglossum laxiflorum* Summerh. in Kew Bull. 1936: 225; F.W.T.A., ed. 1, 2: 449. Flowers whitish green or yellow-green.
   **Iv.C.:** Mt. Goula, near Danané (Apr.) *Chev.* 21213! **Ghana:** Aburi (Mar.) *Johnson* 858! Amedzofe (Mar.) *Westwood* 52!

## 43. SARCORHYNCHUS Schltr. in Beih. Bot. Centralbl. 36, 2: 104 (1918).

Stem up to 15 cm. long; leaves linear-oblong, unequally and connivently bilobed at the apex, 3–9 cm. long, 5–13 mm. broad; inflorescence 4–22 cm. long, many-flowered; sepals 2·5–4 mm. long; lip 2·7–3·6 mm. long; spur shorter than the lip, swollen in the apical part .. .. .. .. .. .. .. .. *polyanthus*

**S. polyanthus** (*Kraenzl.*) *Schltr.* l.c. 105 (1918). *Mystacidium polyanthum* Kraenzl. (1914). *Sarcorhynchus saccolabioides* Schltr. l.c. 105 (1918). Flowers greenish-white or very pale yellow; in forest.
   **W.Cam.:** Kumbo, 6,700 ft. (Oct.) *Ledermann* 5741! 5763! Bafut-Ngemba F.R., Bamenda (July)! *Daramola* FHI 41573! **F.Po:** Parador to Moka (fr. Jan.) *Sanford* 4304 !

## 44. BOLUSIELLA Schltr. in Beih. Bot. Centralbl. 36, 2: 105 (1918).

Leaves with a deep V-shaped groove along the upper surface, very fleshy, more or less triangular in section, recurved, 1·5–4 cm. long; inflorescences usually overtopping the leaves, up to 5·5 cm. long; bracts shorter than the flowers; sepals 2–3 mm. long; lip oblong, rounded, 2·2–5 mm. long, spur ellipsoid, a little shorter than the lip
   1. *iridifolia*
Leaves not grooved along the upper surface or margin, flattened in a vertical plane, and arranged in a fan-like manner, more or less fleshy:
Bracts nearly as long as, or longer than the flowers, overlapping one another at their base; leaves rather fleshy with convex surfaces, 1–3·5 cm. long, 3–9 mm. broad; inflorescences up to 7 cm. long, densely many-flowered; sepals 3–4 mm. long; lip narrowly oblong, 2–3 mm. long, spur shorter than the lip, not much swollen
   2. *imbricata*
Bracts distinctly shorter than the flowers, scarcely overlapping; leaves only slightly fleshy, the surfaces slightly convex or almost flat, up to 8 cm. long:
Peduncle usually longer than the rhachis; spike dense with flowers almost touching one another; leaves parallel-sided with nearly rounded apices, up to 6 cm. long and 7 mm. broad; inflorescences up to 20 cm. long; sepals and lip about 3 mm. long; spur nearly as long as the lip .. .. .. .. .. .. 3. *batesii*
Peduncle usually shorter than the rhachis but sometimes equalling it in length; spike relatively lax with distinct spaces between the flowers; leaves narrowly lanceolate, distinctly narrowed in the upper half, the apices acute, up to 8 cm. long and 12 mm. broad; inflorescence up to 10 cm. long; sepals 1·7–4 mm. long; lip 1·7–2·5 mm. long, with a broad base and narrow acute apical part, spur conical, a little longer than or equalling the lip .. .. .. .. .. 4. *talbotii*

1. **B. iridifolia** (*Rolfe*) *Schltr.* in Beih. Bot. Centralbl. 36, 2: 106 (1918). *Listrostachys iridifolia* Rolfe in F.T.A. 7: 48 (1897). Flowers white.
   **Iv.C.:** Hiré, Kagbe to Dekadougou (Dec., *Aké Assi* 9342! **Ghana:** Ofinso (Nov.) *Cox* 106! Konongo, Ashanti (fr. Jan.) *Westwood* 79a! Kumasi (fr. Dec.) *Westwood* 79! Prasu (Oct.) *Westwood* 178! Also in E. Cameroun, Congo, Uganda, Kenya and Angola.

2. **B. imbricata** (*Rolfe*) *Schltr.* l.c. 106 (1918). *Listrostachys imbricata* Rolfe in Kew Bull. 1910: 161. Flowers white, spur pale olive green.
   **Ghana:** Akumadan, Ashanti (Aug.) *Vigne* FH 2436! Dunkwa (Oct.) *Blayney*! Ksi to Tamale (fr. Nov.) *Westwood* 98! Tarkwa (July) *Cox* 121! Agu Junction, Anum (Sept.) *Westwood* 42! Also in Uganda and Kenya.

3. **B. batesii** (*Rolfe*) *Schltr.* l.c. 106 (1918). *Listrostachys batesii* Rolfe in F.T.A. 7: 167 (1897). Flowers white.
   **Iv.C.:** Gliké, Tabou (Aug.) *Schnell* 1689! Davo R., E. of Béyo (fr. Jan.) *Leeuwenberg* 2595! Tai Forest (Oct.) *Aké Assi* 6034! Brafouédi rock (Oct.) *Aké Assi* 4437! **Ghana:** Tarkwa (Nov.) *Vigne* FH 4809! Asuansi, Central Prov. *Westwood* 162! Awaso *Blayney* 3! Also in E. Cameroun.

**4. B. talbotii** (*Rendle*) *Summerh.* in F.W.T.A., ed. 1, 2: 456 (1936). *Angraecum talbotii* Rendle, Cat. Talb. 108 t. 15, figs. 6–7 (1913). Flowers white.
**S.L.:** Makump (Aug.) *Deighton* 1436! Makan (Aug.) *Thomas* 1633! Da Oulen Mt., Loma Mts. (Sept.) *Jaeger* 1377! **Lib.:** Kailahun, W. Prov. (fr. Nov.) *Baldwin* 10139! Gowi (July) *Harley* 1818! Nimba Mts., 1,500 ft. (July) *Leeuwenberg & Voorhoeve* 4615! 4686! **Iv.C.:** 56 km. N. of Sassandra, 300 ft. (July) *Leeuwenberg* 4561! **Ghana:** Aburi (June) *Johnson* 4! Tarkwa (fr. Nov.) *Vigne* FH 4809a! Asuansi *Westwood* 162a! **S.Nig.:** Oban *Talbot* 941! Sapoba, Benin Prov. *Sanford* 341/66! 342/66! **F.Po:** Parador to Moka (Jan.) *Sanford* 4310! Also in Annobon.

## 45. CHAMAEANGIS Schltr. in Beih. Bot. Centralbl. 36, 2: 107 (1918).

Flowers always 1 at each node of the inflorescence:
Lip oblong, shortly 3-lobed or 3-toothed at the apex, 5–6 mm. long; spur 13–18 mm. long, swollen part ellipsoid-fusiform, 6–10 mm. long; petals oblong, obtuse; leaves 17–43 cm. long, 2·5–4·5 cm. broad; inflorescences 20–50 cm. long; sepals 5–6 mm. long       .. ..       .. ..       .. ..       .. ..       1. *ichneumonea*
Lip lanceolate or triangular-lanceolate, acute at the apex, 2·5–3·5 mm. long; spur about 12 mm. long, swollen part about 3–4 mm. long; petals lanceolate, acute; leaves 6–28 cm. long, 0·7–2 cm. broad; inflorescences up to 35 cm. long; sepals 3·5–5 mm. long       .. ..       .. ..       .. ..       .. ..       2. *lanceolata*
Most of flowers in pairs, or in whorls of 3–6 at each node of the inflorescence (the lower flowers sometimes occur singly); lip entire; inflorescences 8–35 cm. long:
Leaves oblong to ligulate, with usually rounded unequally bilobed apices, 10–24 cm. long, 1·5–3 cm. broad, leathery but not very fleshy; flowers in whorls of 3–6; sepals obtuse, 1·5–2·5 mm. long; petals orbicular or very broadly ovate; lip elliptical, about 2 mm. long, spur only slightly swollen in apical half, 7–12 mm. long
       3. *odoratissima*
Leaves linear or narrowly lanceolate, with acute apices, often much curved, 15–40 cm. long, 0·7–3 cm. broad, fleshy, often V-shaped in section; flowers single in lower part of inflorescence, in twos or threes in upper part; sepals acute, 1·5–3 mm. long; petals lanceolate; lip lanceolate, 1·5–3 mm. long, spur at apex swollen into an ellipsoid or globular sac, 4·5–10 mm. long       .. ..       .. ..       .. ..       4. *vesicata*

**1. C. ichneumonea** (*Lindl.*) *Schltr.* in Beih. Bot. Centralbl. 36, 2: 108 (1918). *Angraecum ichneumoneum* Lindl. (1862). *Listrostachys ichneumonea* (Lindl.) Rchb. f. (1887)—F.T.A. 7: 163. Flowers white, cream coloured or greenish brown, fragrant, especially at night.
**S.L.:** Juring (Dec.) *Deighton* 318! Massa, Peri (Nov.) *Deighton* 4959! **Lib.:** Mt. Barclay (June, July) *Bunting* 15! 157! Karmadhun, Kolahun (Nov.) *Baldwin* 10201! **Ghana:** Kumasi (Aug.) *Cox* 122! **S.Nig.:** Jamieson R., Sapoba (Nov.) *Keay* FHI 22704! Nun R. (Sept.) *Mann* 520! Degema *Talbot* 3259! Also in Gabon.
**2. C. lanceolata** *Summerh.* in Kew Bull. 12: 279 (1958). Flowers greenish ochraceous, faintly fragrant.
**S.Nig.:** Jamieson R., Sapoba (Nov.) *Meikle* 607! Kennedy 2731! 2732! Oban F.R. (Nov.) *Markham* FHI 50083!
**3. C. odoratissima** (*Rchb. f.*) *Schltr.* in Beih. Bot. Centralbl. 36, 2: 108 (1918). *Angraecum odoratissimum* Rchb. f. in Bonplandia 4: 326 (1856). Flowers pale green, greenish yellow or greenish ochraceous.
**S.L.:** Jau, Gegbwema (Jan.) *Deighton* 3889! Sefadu (Oct., Nov.) *Deighton* 4920! **Lib.:** Soplima, Vonjama Dist. (Oct.) *Baldwin* 10038a! Sanokwele (Sept.) *Baldwin* 9524! **Iv.C.:** Diapé to Adzopé (Oct.) *de Wilde* 746! Davo R., E. of Béyo (fr. Jan.) *Leeuwenberg* 2585! **N.Nig.:** Tibba Plateau, Sardauna Prov. *King* 190! **S.Nig.:** Orakin, Eket Dist. (Oct.) *Daramola* FHI 55294! Ukpon River F.R., Obubra Prov. *Floyer* 6! **W.Cam.:** Buea (May, Oct.) *Gregory* 153a! 299! *Chew* 11! Also in Gabon, C. African Republic, the Congos, Uganda and Kenya.
**4. C. vesicata** (*Lindl.*) *Schltr.* l.c. 109 (1918). *Angraecum vesicatum* Lindl. (1843). *Listrostachys vesicata* (Lindl.) Rchb. f. (1865)—F.T.A. 7: 163. Flowers green to yellow, fragrant at night.
**Guin.:** Socourala (fr. Jan.) *Chev.* 20505! **S.L.:** Bonabie (Aug.) *Dawe* 560! Musaia (Aug., Sept.) *Deighton* 4642! 4904! **Lib.:** Ganta (May, June) *Harley* 1643! 1813! **Iv.C.:** Bouroukrou *Chev.* 16907b! Katiola *Boyco* M7! **Ghana:** Assuantsi *Fishlock* 29! Offin Valley (Oct.–Dec.) *Miles* 10! Apla, E. Prov. (Oct.) *Vigne* FH 4046! Agu, Anum Junction (Sept.) *Westwood* 40! Kumasi (Sept.) *Cox* 102! **S.Nig.:** Olokemeji F.R., Abeokuta *Keay* FHI 26733! Iwo, Oyo (Aug.) *Thorold* TN33a! Idanre Hills, Ondo Dist. *Keay & Brenan* FHI 22428! Akure F.R., Ondo (Aug.) *Jones* FHI 19544! Eket *Talbot* 3758! **F.Po:** Parador to Moka (Jan.) *Sanford* 4315! Also in E. Cameroun, Gabon, Congo, Uganda, Kenya and Tanzania.

## 46. PLECTRELMINTHUS Rafin., Fl. Tellur. 4: 42 (1836).

Stem rather short, rarely up to 15 cm. long; leaves closely imbricate at base, oblong or elliptical-oblong, almost equally bilobed at apex with rounded lobes, 10–30 cm. long, 1·5–3·5 cm. broad; inflorescence 25–60 cm. long, 4–10-flowered; sepals and petals lanceolate-linear, acuminate, the former 3·5–5 cm. long; lip about as long as the sepals, 1–2·5 cm. broad in the middle; spur 17–25 cm. long, spirally twisted
*caudatus*

**P. caudatus** (*Lindl.*) *Summerh.* in Kew Bull. 3: 441 (1949). *Angraecum caudatum* Lindl. (1836)—Cat. Talb. 147. *Listrostachys caudata* (Lindl.) Rchb. f. (1864)—F.T.A. 7: 153; Stapf in Johnston, Lib. 654. *Leptocentrum caudatum* (Lindl.) Schltr. (1914)—F.W.T.A., ed. 1, 2: 457. Flowers with green tepals and white lip.
**Guin.:** Beyla (June) *Jac.-Fél.* 975! **S.L.:** Gbap (June) *Adames* 53! Mongheri to Booma, Lunia (Aug.) *Deighton* 5125! Musaia (June) *Jordan* 2155! Mabum (Aug.) *Thomas* 1523! Bendu (July) *Adames* 67! **Lib.:** Wanau Forest, Ganta (Apr.) *Harley* 2181! Grand Bassa *Bunting* 67! 85! **Iv.C.:** Yapo Forest *Oldeman* 245! Bondoukou (May) *Duffaut* in Hb. *Aké Assi*! Abidjan *Boyco* K223! **Ghana:** Ejura (Oct.) *Vigne* FH4863! Vane to Amedzofe (June) *Westwood* 2! **N.Nig.:** Gwari, Niger Prov. (Aug.) *Onochie* FHI 18265! Kaciya, Zaria Prov. (June–Oct.) *G. V. Summerhayes* 80! 116! Keffi to Jos, Plateau Prov. (July) *Sanford* WS/1116/65! **S.Nig.:** Brass R. *Barter* 1858! Oban *Talbot* 898! Sonkwala Kundeve, Ogoja (June) *Keay & Savory* FHI 25323! **W.Cam.:** Victoria (June) *Nditapah* FHI 52406! Also in Gabon, C. African Republic and Congo.

## 47. AËRANGIS Rchb. f. in Flora 48: 190 (1865).

Spur 13–22 cm. long, spirally twisted; leaves broadly oblanceolate, unequally bilobed at apex, 7–23 cm. long, 1·5–6 cm. broad; inflorescences 15–60 cm. long, up to 15-flowered; sepals and petals 2–3 cm. long; lip about the same length, much wider than the sepals in the middle .. .. .. .. .. .. 1. *kotschyana*
Spur less than 8 cm. long, not much twisted, sometimes curled at apex:
  Leaves linear or narrowly strap-shaped, less than 1 cm. broad; sepals less than 12 mm. long:
    Spur 2·5–4 cm. long, apex usually incurved; leaves 3–8 cm. long, 4–8 mm. broad, shortly unequally bilobed at acutely bilobed at apex, rather thin in texture; sepals 7–12 mm. long, subacute; inflorescence up to 11 cm. long, up to 8-flowered; lip oblong-lanceolate .. .. .. .. .. .. .. .. 5. *calantha*
    Spur 6–7·5 cm. long, not incurved at apex; leaves 10–22 cm. long, 6–9 mm. broad, rounded-bilobed at apex, rather thick and stiff; sepals 7–8 mm. long, rounded at apex; inflorescence 13–40 cm. long, many-flowered; lip ovate or elliptical-ovate
                                                          6. *laurentii*
Leaves more or less oblanceolate, often much wider in the upper half, usually over 1·5 cm. broad; sepals usually over 12 mm. long, often much longer:
  Distance between bracts subtending flowers less than 2·5 cm.; spur 3–5·5 cm. long; sinus between apical lobes of leaf acute or obtuse, shorter lobe with a distinct free end; leaves oblanceolate, 6–22 cm. long, 1·2–6 cm. broad; inflorescences 9–40 cm. long, 5–20-flowered; sepals 1·2–2·5 cm. long; lip very similar to the sepals, usually a little broader .. .. .. .. .. .. .. .. .. 3. *biloba*
  Distance between bracts subtending flowers more than 2·5 cm.; spur 5·5–7 cm. long; leaves obliquely acute or obscurely bilobed at apex, sinus scarcely distinguishable; sepals lanceolate, acuminate; lip almost indistinguishable from the sepals:
    Inflorescence 10–21 cm. long, 2–5-flowered, rhachis almost straight, relatively stout, bracts 7–9 mm. long; sepals 2–3 cm. long; leaves 5–15 cm. long, 1–2·5 cm. broad, no sinus at apex                           2. *gravenreuthii*
    Inflorescence 30–60 cm. long, 5–15-flowered, rhachis zigzag in upper part, very slender, bracts 3–5 mm. long; sepals 2 cm. long or less; leaves 7–18 cm. long, 2–5·5 cm. broad, sinus at apex very shallow; column slender, 5 mm. long
                                                4. *arachnopus*

1. **A. kotschyana** (*Rchb. f.*) *Schltr.* in Beih. Bot. Centralbl. 36, 2: 118 (1918). *Angraecum kotschyanum* Rchb. f. (1864)—F.T.A. 7: 137. Flowers white, sometimes tinged pink, spur pink or salmon coloured, fragrant.
  **Guin.**: Higuelande, Fouta Djalon (June) *Adam* 14627! Timbi Madina, Pita (July) *Adames* 295! **S.L.**: Musaia (June) *Deighton* 5060! **N.Nig.**: Mada Hills *Hepburn* 82! Jos (June, July) *Powis*! Lely P381! Naraguta (June) *Lely* 272! Rukuba *King*! Also in Sudan, Uganda, Kenya, Tanzania, Congo, Zambia and Mozambique.
2. **A. gravenreuthii** (*Kraenzl.*) *Schltr.* l.c. 117 (1918). *Aëranthus gravenreuthii* Kraenzl. (1893). *Mystacidium gravenreuthii* (Kraenzl.) Rolfe in F.T.A. 7: 171 (1897). *Angraecum stella* Schltr. in Engl., Bot. Jahrb. 38: 163 (1906). Flowers white.
  **W.Cam.**: W. of Buea, Cam. Mt., 6,000–6,300 ft. (May, June) *Preuss* 891! *Deistel*! Neu-Tegel (July) *Schlechter* 15794. Bafut-Ngemba F.R. (May) *Keay* FHI 37933! **F.Po**: Pico Boca, 4,500–5,000 ft. *Sanford* 4256! 4257!
3. **A. biloba** (*Lindl.*) *Schltr.* Die Orchid. 598 (1914). *Angraecum bilobum* Lindl. (1840)—F.T.A. 7: 138. Flowers white, sometimes flushed pink, ovary pink, fragrant at night.
  **Sen.**: Emaye, Oussouye (Sept.) *Doumbia* 784! **Guin.**: (Aug.) *Pobéguin*! Kissidougou (June) *Martine* 319! **S.L.**: Karima (Aug.) *Deighton* 4403! Musaia (July) *Deighton* 5108! Hangha (Aug.) *Deighton* 4336! Rokupr (Aug.) *Adames* 119! **Lib.**: Zwedru, Tchien Dist. (Aug.) *Baldwin* 7031! Palilah, Gbanga Dist. (Aug.) *Baldwin* 9208! Ganta (Aug.) *Harley* 1648! **Iv.C.**: Dyola Country, Upper Sassandra (May) *Ripert* in *Hb. Chev.* 21538! Mt. Goula *Chev.* 21221! Danané (Aug.) *Schnell* 1586! **Ghana**: Trawa, E. Prov. (June) *Vigne* FH 4390! 4391! Kintampo, Ashanti (June) *Westwood* 1a! Achimota (May) *Akpabla*! Kpandu (June) *Andoh* FH 5326! Amedzofe (Apr.–June) *Vohringer* in *Scholes* 92! *Morton* A3421! Togo Rep.: *Baumann* 592! **N.Nig.**: Bornu, Gwari, Niger Prov. (Aug.) *Onochie* FHI 35947! Ose stream, Kabba, *Westwood*! Dogon Kurmi, Jemaa Div. (Aug.) *Killick* 41! **S.Nig.**: Ibadan South F.R. (July) *Keay* FHI 22432! Ibadan (fr. Sept.) *Ujor* FHI 42077! Ogbesse F.R., Ondo Prov. (July) *Onochie* FHI 33360! Mt. Orosun, Idanre, Ondo (Aug.) *Jones* FHI 20711! Obubra, Ogoja Prov. (Aug.) *Adebusuyi* FHI 43951! **W.Cam.**: M'bonge *Wright* 58/12!
4. **A. arachnopus** (*Rchb. f.*) *Schltr.* in Beih. Bot. Centralbl. 36, 2: 113 (1918). *Angraecum arachnopus* Rchb. f. (1854)—F.T.A. 7: 140. Flowers white or pink.
  **Ghana**: *cult. specimens*! Also in E. Cameroun, Gabon and Congo. [It is possible that the original specimens were introduced via Ghana and that the species is not a native of that country.—V.S.S.]
5. **A. calantha** (*Schltr.*) *Schltr.* l.c. 115 (1918). *Angraecum calanthum* Schltr. in Engl., Bot. Jahrb. 38: 20 (1905). Flowers white, not fragrant, spur pale greenish.
  **Ghana**: Kumasi (Apr.) *Cox* 123! 123a! Bobiri (Apr.) *Westwood* 135! Jiyeti, Bunsu (Apr.) *Westwood* 216! Also in E. Cameroun, Congo, Uganda, Tanzania and Angola.
6. **A. laurentii** (*De Wild.*) *Schltr.* l.c. 118 (1918). *Angraecum laurentii* De Wild., Not. Pl. Util. Congo 1: 322 (1904). Flowers whitish yellow or cream coloured.
  **Lib.**: Ganta (Jan., Feb.) *Harley* 1769! 1850! **Ghana**: Kumasi (Dec.) *Vigne* FH 4090! Efiduase *Cox* 16! Also in Congo.

## 48. BAROMBIA Schltr., Die Orchid. 600 (1914).

Stem short; leaves oblanceolate, unequally bilobed at apex, 10–25 cm. long, 2–5·5 cm. broad; inflorescence 25–80 cm. long, 3–11-flowered; sepals and petals very narrow, 4–5·5 cm. long; spur very slender, swollen at end, 17–22 cm. long        *gracillima*

**B. gracillima** (*Kraenzl.*) *Schltr.* in Beih. Bot. Centralbl. 36, 2: 124 (1918). *Angraecum gracillimum* Kraenzl. (1893)—F.T.A. 7: 140. Flowers white or pale green, ends of tepals reddish, spur pale brown.
  **W.Cam.**: Barombi (Sept.) *Preuss* 459! Also in Gabon.

## 49. RANGAËRIS Summerh. in Kew Bull. 1936: 227.

Lip 3-lobed, 5–9 mm. long, middle lobe lanceolate or linear, obtusely acuminate, lateral lobes much shorter, almost orbicular, crenulate or dentate; stem very long and slender, leaves placed at intervals of 2–5 cm; leaves narrowly oblong-lanceolate, acute, 4–8 cm. long, 1–2 cm. broad; inflorescence few-flowered, 4–12 cm. long; sepals and petals 6·5–8·5 mm. long; spur 3·5–4·5 cm. long   ..     ..     ..        4. *trilobata*
Lip entire, sometimes slightly crenulate along the lower margins:
Spur 15–19 cm. long; stem elongated, about 0·5 cm. diam.; leaves narrowly oblong, obscurely unequally bilobed at apex, 6–12 cm. long, 1–1·5 cm. broad; inflorescences 8–15 cm. long, 3–4-flowered, bracts 10–15 mm. long; sepals and petals lanceolate. acuminate, 2–3·5 cm. long; lip like sepals; column semi-terete, about 6 mm. long; viscidium rectangular, 3 mm. long, stipites much broadened in the upper part
                                                                    3. *longicaudata*
Spur less than 10 cm. long:
Bracts broadly ovate or almost orbicular, obtuse, sheathing at base, 2·5–7 mm. long; lip lanceolate with a thickened callose apex; viscidium single, rectangular, elliptical, or ovate, longer than broad; rostellum 3-lobed, middle lobe much shorter than laterals; leaves in a bunch at the base, 6–17 cm. long; inflorescences as long as or a little shorter than the leaves; sepals 6–9 mm. long:
Spur almost straight, 5–9 cm. long; leaves thin, V-shaped in section, very shortly bilobed or retuse at apex, the lobes rounded, 1–1·9 cm. broad; ovary glabrous or slightly scurfy      ..     ..     ..     ..     ..     ..        1. *muscicola*
Spur incurved at apex, 7–14 mm. long; leaves fleshy, Iris-like, lying in a vertical plane, acute and entire at apex, 5–10 mm. broad; ovary densely hairy
                                                                    2. *rhipsalisocia*
Bracts lanceolate, acute, not much sheathing, 2–5 mm. long; lip without thickened apex, about 4–6 mm. long, 2–4 mm. broad; viscidia 2, orbicular, scarcely longer than broad; middle lobe of rostellum longer than laterals; spur about equalling the lip, straight; leaves ligulate or slightly wider in the upper part, 6–16 cm. long, 1–2 cm. broad; sepals 4–6·5 mm. long   ..     ..     ..        5. *brachyceras*

1. **R. muscicola** (*Rchb. f.*) *Summerh.* in F.W.T.A., ed. 1, 2: 450 (1936). *Aëranthus muscicola* Rchb. f. (1865). *Listrostachys muscicola* (Rchb. f.) Rolfe in F.T.A. 7: 158 (1897). *Angraecum batesii* (Rolfe) Rendle, Cat. Talb. 146 (1913). Flowers white with orange or flesh-coloured spur, fragrant.
**Guin.:** Dalaba *Caille* in *Hb. Chev.* 18128b! Pita (fr. Apr.) *Chillou* 1225! **S.L.:** Mapaki to Mobonto, N. Prov. (Aug.) *Deighton* 1214! Kabala (Aug.) *Deighton* 4898! Bintumane Peak, Loma Mts. (Aug., Jan.) *Jaeger* 1181! 4003! **Lib.:** Nimba (Sept.) *Adames* 491! 538! **Ghana:** Offin Valley, Ashanti (Oct.–Dec.) *Miles* 9! Efiduase (Sept.) *Cox* 30! Juaben (May) *Cox* 126! Awaso *Blayney* 13! **S.Nig.:** Omo saw mills, Ijebu to Benin (Aug.) *Sanford* WS/1743/65! Sonkwala, Ogoja Prov. (Sept.) *Keay* FHI 25380! Oban *Talbot* 897! **W.Cam.:** Kumba (July) *Schlechter* 15770! Also in E. Cameroun, Congo, Uganda, Kenya, Tanzania, Angola, Zambia, Rhodesia, Mozambique, Natal and Cape Province.
2. **R. rhipsalisocia** (*Rchb. f.*) *Summerh.* in F.W.T.A., ed. 1, 2: 450 (1936). *Angraecum rhipsalisocium* Rchb. f. (1865). *Listrostachys rhipsalisocia* (Rchb. f.) Rolfe in F.T.A. 7: 158 (1897). *L. colarum* A. Chev., Bot. 622, name only. Flowers white or cream coloured, with greenish spur, fragrant at night.
**Sen.:** *Gateau*! Soutou, Casamance (Mar.) *Berhaut* 7184! **Guin.:** Sahadougou (Mar.) *Chev.* 20883! Dalaba (Aug.) *Adames* 325! Nimba Mts. (Apr.) *Schnell* 1210! **S.L.:** Makump (Feb.) *Deighton* 2878! Kpema to Male, Nongowa (Jan.) *Deighton* 4568! Mambolo (Feb.) *Adames* 217! Kambaia, Sulima (Feb.) *Small* 6033! Kabala (Feb., Mar.) *Deighton* 4588! **Lib.:** Ganta (Feb., Nov.) *Harley* 1849! 1949! **Iv.C.:** Issia to Daloa (Nov.) *de Wilde* 841! W. of Niapidou (fr. Feb.) *Leeuwenberg* 2780! **Ghana:** Mampong (Dec.) *Adams* 4527! Kumasi (Dec.) *Westwood* 55! Agu Junction, Anum (Nov.) *Westwood* 75! Efiduase *Cox* 12! **S.Nig.:** Umu Alwa (Jan.) *Carpenter* 177! Ibadan to Abeokuta (Mar.) *Schlechter* 13031! N. of Idi Iya, Ibadan (Feb.) *Keay* FHI 21165! Oke Owa, Oyo Prov. (Mar.) *Keay* FHI 21184! Mt. Orosun, Idanre, Ondo (Feb.) *Brenan* 8969! **W.Cam.:** Kumba (fr. Mar.) *Thorold* CM12! Lus, Nkambe Div. (Feb.) *Hepper* 1876! Also in E. Cameroun, Congo and Angola.
3. **R. longicaudata** (*Rolfe*) *Summerh.* in F.W.T.A., ed. 1, 2: 449 (1936). *Mystacidium longicaudatum* Rolfe in F.T.A. 7: 170 (1897).
**S.Nig.:** Lagos *Millen* 188! Sonkwala Dist. (Nov.) *FHI Staff* FHI 42078! Cross River North F.R. (Nov.) *Cooper* 79!
4. **R. trilobata** *Summerh.* in Kew Bull. 1936: 229. Sepals salmon pink, petals and lip white.
**S.Nig.:** Eket (May) *Talbot* 3299! Onochie & Latilo FHI 32937! Also in Gabon.
5. **R. brachyceras** (*Summerh.*) *Summerh.* in Kew Bull. 1936: 228. *Aërangis brachyceras* Summerh. in Kew Bull. 1934: 213. *Rangaëris biglandulosa* Summerh. in Kew Bull. 1936: 228; and in F.W.T.A., ed. 1, 2: 450. Flowers pale yellowish or white, often with a greenish tinge, fragrant; in montane forest.
**Guin.:** Diaguissa-Dalaba Plateau, 3,300–4,300 ft. (Oct.) *Chev.* 18782! Kouba Nangal, Kourou (Sept.) *Adames* 349! Nimba Mts. (Aug.) *Schnell* 3726! **S.L.:** Bintumane Peak, Loma Mts., 6,000 ft. (Aug.) *Jaeger* 1178! 4274! **S.Nig.:** Obudu Plateau, 3,000 ft. *Cooper* 42! Also in S. Tomé, Congo, Uganda, Kenya and Zambia.

## 50. CYRTORCHIS Schltr., Die Orchid. 595 (1914).

Viscidium linear, gradually tapering from apex to base, equal in texture throughout its length; sepals 1–2·5 cm. long; spur 2·5–4·5 cm. long:
Leaves with nearly parallel sides, rigid and stiff in texture, cross veins not visible, almost equally bilobed at the apex with semi-orbicular lobes, 8–20 cm. long, 1·3–3 (usually less than 2·5) cm. broad; inflorescence up to 16 cm. long, flowers closely placed   ..     ..     ..     ..     ..     ..     ..     ..     ..        1. *ringens*
Leaves oblong-elliptical or oblanceolate, tapering distinctly at both ends, not very rigid or stiff, the cross veins usually visible, especially in dried specimens, unequally

bilobed at apex with more or less connivent lobes, 5–20 cm. long, 1·7–5·5 cm. broad; inflorescence up to 30 cm. long, middle flowers at intervals of 1–2 cm.

2. *monteiroae*

Viscidium composed of a broad rather stiff hardened upper portion with recurved edges and a linear, hyaline and very thin lower portion:

Upper hardened part of viscidium over 3 mm. long, longer than the hyaline part; spur 9–15 cm. long, very slender; sepals 3–5 cm. long; leaves oblong-oblanceolate or oblong, unequally bilobed at apex with connivent rounded lobes, 8–25 cm. long, 1·7–3·5 cm. broad; inflorescences up to 24 cm. long, bracts broad and sheathing, 1–2·5 cm. long .. .. .. .. .. .. .. .. 3. *chailluana*

Upper hardened part of viscidium less than 2·5 mm. long, usually distinctly shorter than the hyaline part; spur usually less than 6 cm. long but sometimes up to 10 cm. and then distinctly wider in the basal than in the apical half:

Leaves long and narrow, with almost parallel sides, often very fleshy and apparently terete, almost equally bilobed at the apex, 8–22 cm. long, 2·5–15 mm. broad; anther only shortly extended in front, which is often much toothed; hardened upper part of viscidium short and broad, not tapering towards the hyaline part; sepals 1–2 cm. long; spur recurved or more or less S-shaped, 1·5–4 cm. long; bracts 5–10 mm. long .. .. .. .. .. .. .. 6. *aschersonii*

Leaves relatively broad, 1·3–4 cm. broad, unequally bilobed at the apex; anther in front extended into a long tapering entire or almost entire appendage; hardened upper part of viscidium tapering towards the hyaline part; inflorescences 5–17 cm. long; bracts often over 1 cm. long:

Spur with a thickened hooked or almost rolled-up apex, 3·5–5 cm. long; leaves ligulate, 9–24 cm. long, 1·5–4 cm. broad; sepals 2–4 cm. long .. 5. *hamata*

Spur almost straight, or more or less curved, but apex not hooked or rolled-up; leaves 8–24 cm. long, 1·5–4 cm. broad:

Spur 6–10·5 cm. long; sepals 2·5–5 cm. long; stipites of viscidium broadly oblanceolate .. .. .. .. .. .. .. .. 4a. *arcuata* subsp. *whytei*

Spur less than 6 cm. long:

Stipites of the pollinia narrowly oblanceolate; hyaline part of viscidium twice as long as the hardened part; anther toothed in front; sepals 1·2–2 cm. long; spur 3·5–4·5 cm. long, not much curved .. .. 4b. *arcuata* subsp. *leonensis*

Stipites of the pollinia broadly oblanceolate; hyaline part of viscidium about half as long again as the hardened part; anther entire in front; sepals 1·5–3·5 cm. long; spur 3–5·5 cm. long, almost straight or somewhat S-shaped

4c. *arcuata* subsp. *variabilis*

1. **C. ringens** (*Rchb. f.*) *Summerh.* in Kew Bull. 12: 87 (1958). *Listrostachys ringens* Rchb. f. in Gard. Chron., ser. 2, 10: 266 (1878). *L. bistorta* (Rolfe) Rolfe in F.T.A. 7: 155 (1897). *L. hookeri* Rolfe in F.T.A. 7: 154 (1897). *L. ignoti* Kraenzl. in Engl., Bot. Jahrb. 51: 395 (1914). *Cyrtorchis bistorta* (Rolfe) Schltr. (1918)—F.W.T.A., ed. 1, 2: 460. Flowers white, weakly fragrant.
   **Sen.:** Oussouye, Casamance (Mar.) *Berhaut* 7256! **S.L.:** cult. plant! **Lib.:** Kitoma (Oct.) *Harley* 1834! **Iv.C.:** Taï Forest (Oct.) *Aké Assi* 6032! Mt. Tonkoui (Oct.) *Aké Assi* 6025! W. of Niapidou (fr. Jan.) *Leeuwenberg* 2497! **Ghana:** No locality (cult. Achimota, Oct.) *Westwood* 157! Nkawkaw *Westwood* 207! **S.Nig.:** Lagos *Moloney*! Calabar (Oct.) *Daramola* FHI 56434! **W.Cam.:** Cam. Mt., 3,000–6,000 ft. (May–Nov.) *Mann* 2114! *Maitland* 735! *Gregory* 199! M'bonge *Wright* 58/28! Bafut-Ngemba F.R., Bamenda (May) *Daramola* FHI 41182! Also in E. Cameroun, S. Tomé, Congo, Burundi, Uganda, Tanzania and Rhodesia.

2. **C. monteiroae** (*Rchb. f.*) *Schltr.* Die Orchid. 596 (1914). *Listrostachys monteiroae* Rchb. f. (1877)—F.T.A 7: 156 (as *monteirae*). *Angraecum aschersonii* of Cat. Talb. 147, not of Kraenzl. Flowers white or cream with greenish or orange-tinged spur, fragrant.
   **S.L.:** Gegbwema, Tunkia (Dec.) *Deighton* 5717! Fairo, Soro (Dec.) *Deighton* 5977! **Lib.:** Totokwelli, Medina (Oct.) *Linder* 1295! Begwai (Oct.) *Bunting* 23! Boporo, W. Prov. (Nov.) *Baldwin* 10388! Gbanga, Central Prov. (Dec.) *Baldwin* 10543! Ganta (Dec.) *Harley* 1845! **Ghana:** Ankobra Valley (Nov., Dec.) *Miles* 18! Ejura (Oct.) *Vigne* FH 4872! Juaben, Ashanti (June) *Cox* 128! **S.Nig.:** Idanre F.R. (Sept.) *Keay* FHI 56544! Sapoba *Kennedy* 1911! Calabar *Williams*! Oban *Talbot* 936! **W.Cam.:** Barombi *Preuss* 418! Buea (Aug., Sept.) *Dundas* FHI 15302! *Keay* FHI 25379! *Gregory* 181! Also in E. Cameroun, Congo, Principe, Gabon, Uganda and Angola.

3. **C. chailluana** (*Hook. f.*) *Schltr.* Die Orchid. 596 (1914). *Angraecum chailluanum* Hook. f. (1866)—Cat. Talb. 147. *Listrostachys chailluana* (Hook. f.) Rchb. f. (1885)—F.T.A. 7: 153. Flowers white or cream.
   **S.Nig.:** Nun R. (Sept.) *Mann* 521! Igbessa *Millen* 193! Sapoba *Kennedy* 1913! Okomu F.R., Benin (Oct.) *Keay* FHI 25438! Calabar (Sept.) *Keay* FHI 13348! Oban *Talbot* 135! **W.Cam.:** Victoria (Sept.) *Ngongi* FHI 15082! Buea *Preuss* 372! Also in E. Cameroun, Gabon, the Congos and Uganda.

4. **C. arcuata** (*Lindl.*) *Schltr.* Die Orchid. 596 (1914).

4a. **C. arcuata** subsp. **whytei** (*Rolfe*) *Summerh.* in Kew Bull. 14: 147 (1960). *Listrostachys whytei* Rolfe in F.T.A. 7: 155 (1897). Flowers white, fragrant at night.
   **S.L.:** Musaia (Oct., Nov.) *Deighton* 4955! Kambaia, Sulima (Oct.) *Small* 492! Jagbwema, Faiama (Sept.) *Deighton* 6139! Bonthe (Oct.) *Adames* 89! source of Niger to Masadou (Oct.) *Jaeger* 171! **Lib.:** Cape Palmas (Nov.) *Harley* 2244! Fisherman's L., Grand Cape Mt. (fr. Dec.) *Baldwin* 10903! **Iv.C.:** Brafouédi, NW. of Abidjan (fr. Dec.) *Leeuwenberg* 2311! **Ghana:** Aframso, E. Prov. (Oct.) *Vigne* FH 4035! Dutukpene, Krachi (Oct.) *Morton* A3733! Plateau F.R., Wiawia (Oct.) *St. Cl.-Thompson* 1559! Kpedze to Kpeve (Sept.) *Westwood* 49! 49a! Also in Tanzania and Malawi.

4b. **C. arcuata** subsp. **leonensis** *Summerh.* in Kew Bull. 14: 149 (1960). Flowers white, fragrant.
   **S.L.:** Mateboi, Sanda Tenraran (June) *Adames* 229! Batkanu (May) *Deighton* 4312! Njala (June) *Deighton* 4313! Musaia (May) *Deighton* 5543! Masingbe, Kunika Sanda (May) *Deighton* 5059!

4c. **C. arcuata** subsp. **variabilis** *Summerh.* in Kew Bull. 14: 148 (1960). *Listrostachys sedenii* Rchb. f. (1878)—F.T.A. 7: 154. *Cyrtorchis sedenii* (Rchb. f.) Schltr. (1918)—F.W.T.A., ed. 1, 2: 461, in greater part. Flowers white or cream with greenish or yellowish spur, fragrant.
   **Guin.:** Pita, Fouta Djalon (Sept.) *Adames* 352! **Lib.:** Ganta (Nov.) *Harley* 1842! **Iv.C.:** Disandougou to Niangouépleu (May) *Chev.* 21527! **Ghana:** Gura (July) *Vigne* FH 1257! Kumasi to Tamale (Apr.)

FIG. 402.—CYRTORCHIS CHAILLUANA (*Hook. f.*) *Schltr.* (ORCHIDACEAE).

A, flowering shoot.  B, column and rostellum, side view.  C, same, front view.  D, anther.  E, pollinia.  F, viscidium, from above.  G, same, from below.

*Westwood* 96! **Dah.**: Atacora Mts. (June) *Chev.* 24215! **N.Nig.**: Tukuruwa R., Zaria (June) *Keay* FHI 25886! Ribako F.R., Zaria (May) *Keay* FHI 22899! Vom, Plateau Prov. *Dent Young*! Munchi, Abinsi (June) *Dalz.* 838! **S.Nig.**: Okwoga to Udah (July) *Kitson*! Eastwards to Kenya and southwards in E. Africa to Tanzania and Zambia.

5. **C. hamata** (*Rolfe*) *Schltr.* Die Orchid. 596 (1914). *Listrostachys hamata* Rolfe in Bot. Mag. 132: **t.** 8074 (1906). Flowers white with a green spur.
   **Ghana**: Aburi (May, June) *Johnson* 227! *Irvine* 3034! Kumasi (July) *Andoh* FH 5049! Awaso *Blayney* 23! Mampong, Ashanti (June) *Westwood* 105! Kpeve (Jan.) *Westwood* 80! **N.Nig.**: Kabba (Aug.) *Westwood* 270! **S.Nig.**: Ibadan South F.R. (Aug.) *Keay* FHI 25373! Olokemeje F.R. (Aug.) *J. B. Gillett* 15338! Idanre Hills, Ondo (Sept.) *Keay* FHI 22439! Akure F.R., Ondo (July) *Onochie* FHI 23420! Owo F.R., Ondo (July) *Onochie* FHI 33245!

6. **C. aschersonii** (*Kraenzl.*) *Schltr.* in Beih. Bot. Centralbl. 36, 2: 129 (1918). *Angraecum aschersonii* Kraenzl. (1889). *Listrostachys aschersonii* (Kraenzl.) Dur. & Schinz (1895)—F.T.A. 7: 156. Flowers white, fragrant.
   **S.L.**: Yagoi (July, Aug.) *Deighton* 5107! **Ghana**: Goaso, Ashanti (May) *Westwood* 119! Awaso *Blayney* 9! **S.Nig.**: Okomu F.R., Benin (Dec.) *Brenan & Keay* FHI 22407! Sapoba *Kennedy* 234! Shasha F.R., Ijebu (May) *Ross* 287! **W.Cam.**: Bambui, 5,000 ft. (June) *Brunt* 494! Also in E. Cameroun and Congo.

*Additional Species*

**C. brownii** (*Rolfe*) *Schltr.* in Beih. Bot. Centralbl. 36: 129 (1918). *Listrostachys brownii* Rolfe in Kew. Bull. 1906: 378. *Angraecum latibracteatum* De Wild. (1916). *Cyrtorchis latibracteatum* (De Wild.) Schltr. (1918). Flowers white.
   **Iv.C.**: Hiré (Nov.) *Aké Assi* 9225! Taï (Oct.) *Aké Assi* 6031! Also in Uganda, Tanzania and Congo.

## 51. SOLENANGIS Schltr., in Beih. Bot. Centralbl. 36, 2: 133 (1918).

Sepals and petals 5–7 mm. long; lip entire, spur 2–2·5 cm. long; inflorescences 2–9 cm. long, bracts 1–2·5 mm. long; leaves elliptical-lanceolate, 3–9 cm. long, 0·8–2·5 cm. broad .. .. .. .. .. .. .. .. .. 1. *scandens*
Sepals and petals 1·5–2 mm. long; lip 3-lobed, all lobes very short, laterals larger than middle lobe, spur 5–7 mm. long; inflorescences 5–15 mm. long, bracts less than 1 mm. long; leaves broadly lanceolate or ovate-elliptical, 1·5–4 (rarely to 5·5) cm. long, 0·6–1·5 cm. broad .. .. .. .. .. .. .. .. 2. *clavata*

1. **S. scandens** (*Schltr.*) *Schltr.* in Beih. Bot. Centralbl. 36, 2: 134 (1918). *Angraecum scandens* Schltr. (1900)— Cat. Talb. 146. Flowers white, greenish white, greenish yellow or pinkish; common epiphyte on cacao and other crops.
   **S.L.**: Yonibana (Oct.) *Thomas* 4112! **Lib.**: Farmington R. Bridge (Sept.) *Voorhoeve* 501! Boporo, W. Prov. (Nov.) *Baldwin* 10248! Harrisburg (Sept.) *Barker* 1421! **Iv.C.**: Banco Forest, Abidjan *de Wilde* 729! 730! **Ghana**: Offin Valley, Ashanti (Oct.–Dec.) *Miles* 14! Tiasi (Oct.) *Vigne* FH 1389! Amentia, S. Ashanti (Apr.) *Vigne* FH 2906! Bunsu (Nov.) *Westwood* 57! Juaso, Ashanti (Apr.) *Cansdale* FH 4387! Tafo, E. Prov. (Sept.) *Andoh* FH 4529! **S.Nig.**: Sapoba, Benin (Aug.) *Onochie* FHI 23423! Igbekhue, Benin (Sept.) *Thorold* TN32! Idanre F.R., Ondo (fr. Aug.) *Okafor & Daramola* FHI 35289! Ikayi, Ondo (Sept.) *Thorold* TN35! Oban *Talbot* 893! Also in E. Cameroun, Gabon, S. Tomé and the Congos.

2. **S. clavata** (*Rolfe*) *Schltr.* l.c. 134 (1918). *Angraecum clavatum* Rolfe in F.T.A. 7: 145 (1897). Flowers entirely white or tepals green and lip white; epiphytic or on rocks.
   **S.L.**: Gbinti, N. Prov. (July) *Deighton* 2507! Kambia (fr. Jan.) *Sc. Elliot* 4223! Petifu Creek (Aug.) *Pelly* 143! **Lib.**: Gbi National Forest, nr. Gran Toun (Oct.) *Voorhoeve* 551! Begwai (Oct.) *Bunting* 14! Jaurazon, Sinoe (fr. Mar.) *Baldwin* 11480! Ganta (Aug.) *Harley* 1826! **Iv.C.**: Bouroukrou *Chev.* 16525! Tiassalé to Divo *Chev.*! Brafouédi, Adiopodoumé (fr. Dec.) *de Wit* 473! Banco Forest, Abidjan (Oct.) *de Wilde* 733! 3193! E. of Beyo (fr. Jan.) *Leeuwenberg* 2540! **Ghana**: Oda (Oct.) *Fishlock* 56! Brenase, Akim (fr. Apr.) *Irvine* 560! S. Fomang Su F.R. (Oct.) *Andoh* FH 4248! Amentia, Ashanti (Aug.) *Vigne*

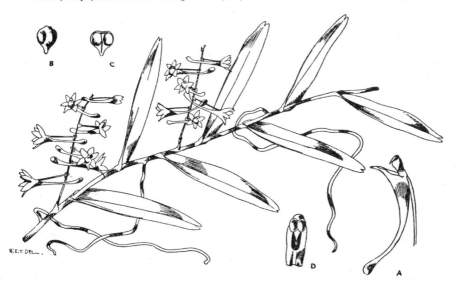

Fig. 403.—Solenangis scandens (*Rolfe*) Schltr. (Orchidaceae).
A, lip and column, side view. B, anther, from above and C, from below. D, column.

FH 3999! Bunsu (Nov.) *Westwood* 58! **S.Nig.**: Lagos *Moloney*! Abeokuta (May) *Killick* 226! Jamieson R., Sapoba (June) *Onochie* FHI 23322! 31237! Usonigbe F.R., Benin (fr. Nov.) *Ejiofor* FHI 24655! Benin (June, July) *Gregory* 552! **W.Cam.**: Kumba to Tombel (May) *Olorunfemi* FHI 30602! Also in E. Cameroun and the Congos.

## 52. DINKLAGEËLLA Mansf. in Fedde, Rep. 36: 63 (1934).

Spur about 2·5 mm. long, much swollen from a narrow base; lip 6·5–7·5 mm. long, tri-lobed in the upper third only, lobes narrowly ovate with narrowed obtuse apices; leaves lanceolate-ligulate, 1·3–2·7 cm. long, 5–9 mm. broad; inflorescences 1·5–2·5 cm. long, 2–3-flowered; pollinia a little larger than the stipes; sepals about 5 mm. long, narrowly oblong  .. .. .. .. .. .. .. .. .. 1. *minor*
Spur 2·5–3 cm. long, gradually narrowed from a wide mouth and only slightly inflated near the apex; lip 16–17 mm. long, trilobed in the upper two-thirds, lobes broadly oblong, rounded at the apex; leaves narrowly elliptical, 1·7–3·5 cm. long, 8–15 mm. broad; inflorescences 5–7 cm. long, about 3–6-flowered; pollinia much smaller than the stipes; sepals 7–8·5 mm. long, elliptical  .. .. .. .. .. 2. *liberica*

1. **D. minor** *Summerh.* in Kew Bull. 14: 156 (1960). Flowers white.
  **Lib.**: Ganta (Aug.) *Harley* 1828! 2158! **Ghana**: Aburi (Oct.) *Cox* 86!
2. **D. liberica** *Mansf.* l.c. 64 (1934). Flowers white tinged with orange, fragrant.
  **Lib.**: Sinoe R. (Nov.) *Dinklage* 2321! **Iv.C.**: Banco Forest (Oct.) *de Wilde* 732! Brafouédi Rocks (Oct.) *Aké Assi* 4435! **S.Nig.**: Sapoba *Kennedy* 1900! Calabar (Oct.) *Lloyd* 9! **W.Cam.**: Buea (Oct.) *Chew* 25!

## 53. EURYCHONE Schltr. in Beih. Bot. Centralbl. 36, 2: 134 (1918).

Stem up to 7 cm. long; leaves broadly oblanceolate, unequally and subacutely bilobed at the apex, 7–17 cm. long, 1·5–5 cm. broad; inflorescences up to 8 cm. long, up to 6-flowered; sepals and petals 2–2·5 cm. long; lip about equalling the sepals, 2–2·5 cm. broad; spur 2–2·5 cm. long  .. .. .. .. .. .. *rothschildiana*

**E. rothschildiana** (*O'Brien*) *Schltr.* l.c. 135 (1918). *Angraecum rothschildianum* O'Brien (1903). Tepals white or greenish; lip white with green centre and chocolate blotch at base.
  **Guin.**: Bossou, Nzérékoré (July) *Schnell* 1428! **S.L.**: Mandu, S. Prov. (July) *Deighton* 2221! **Iv.C.**: Guidéko to Zozoro (June) *Chev.* 19053! **Ghana**: Tano R. Valley (July) *Miles*! Monkey Hill, Bekwai *Cox* 112! Kintampo, Ashanti (June) *Westwood* 115! **N.Nig.**: Amban to Kamwai, Plateau Prov. *King* 196! **S.Nig.**: Afi River F.R., Ogoja Prov. (June) *Jones & Onochie* FHI 18948! Ibadan South F.R. (June) *Keay* FHI 25335! Sapele *Wright* 155! Mamu F.R. *Cooper* 65! Ilaro F.R. (June) *Onochie* FHI 3441! **F.Po**: Parador, 2,500 ft. *Sanford* 4233! Also in Congo and Uganda.

## 54. ANCISTRORHYNCHUS Finet in Mém. Soc. Bot. Fr. 2: 44 (1907). *Cephalangraecum* Schltr. (1918)—F.W.T.A. ed. 1, 2: 461.

Leaves 3 cm. long or less, narrowly oblong, 3–7·5 mm. broad, shortly bilobed with rounded lobes; inflorescences very short, 1–3-flowered; bracts small, much shorter than the flowers; tepals oblong, 2–3 mm. long; lip broadly ovate, 2–3 mm. long; spur 2–2·5 mm. long, slightly swollen above the base; stem elongated, up to 15 cm. long  .. .. .. .. .. .. .. .. .. 8. *schumannii*
Leaves much longer than 3 cm.; inflorescences more than 3-flowered, often many-flowered; bracts small or exceeding the flowers in length;
  Stipes of pollinia single but shortly forked at the apex; leaves usually very unequally bilobed at apex, the shorter lobes sometimes almost or completely suppressed; inflorescence up to 4 cm. long, only moderately dense, bracts short, much shorter than the flowers; lip distinctly 3-lobed, the middle lobe much crisped; spur more or less S-shaped, swollen at apex:
    Apical lobes of leaf very acute and unequal in length, the longer one up to 7 cm., the shorter one sometimes quite lacking so that the apex appears to be entire; leaves stiff and leathery, up to 1·8 m. long and 3·6 cm. broad; sepals 3–5·5 mm. long; lip 5–7·5 mm. long  .. .. .. .. .. .. .. .. 1. *clandestinus*
    Apical lobes of leaf subacute or obtuse, the longer one up to 4 cm., both sometimes shortly lobulate; leaves less stiff, relatively broader, up to 37 cm. long, 0·7–3 cm. broad; sepals 3–4·5 mm. long; lip 4·5–5 mm. long  .. .. 2. *recurvus*
  Stipites of pollinia 2, distinct; leaves with rounded or subacute apical lobes which are sometimes toothed; inflorescences in the form of dense heads, often with large chaffy bracts nearly as long as the flowers; lip entire or only indistinctly 3-lobed; spur straight or slightly curved:
    Apex of leaf serrate or with a few sharp teeth on each side just below:
      Leaves 1·5–3 cm. broad 17–44 cm. long, often tapering in the upper part, with a few sharp teeth on each side below the apex, lobes themselves with a small sharp point; sepals 3·5–6·5 mm. long; lip a little longer than broad, 4·5–5·7 mm. long, 3·5–5 mm. broad; spur 7–11 mm. long  .. .. .. .. .. 3. *capitatus*
      Leaves 7–10 mm. broad, 5–11 cm. long, with parallel sides, with rounded irregularly serrate lobes at the apex, margins otherwise entire; sepals 3–3·5 mm. long; lip distinctly broader than long, 2·5 mm. long, 4–4·5 mm. broad; spur 4–4·5 mm. long  .. .. .. .. .. .. .. .. 6. *serratus*

Fig. 404.—Ancistrorhynchus serratus *Summerh.* (Orchidaceae).

1, habit, × 1. 2, tip of leaf, × 4. 3, flower, side view, × 6. 4, same, front view, × 8. 5, dorsal sepal, × 8. 6, lateral sepal, × 8. 7, petal, × 8. 8, column, × 18. 9, column with anther cap removed, × 18. 10, anther cap, × 18. 11, pollinarium, side view, × 18. 12, pollinarium, front view, × 18. All from *Head* 95.

Apex of leaves bilobed but margins quite entire:
Spur over 5 mm. long, usually distinctly longer than the lip; sepals 4·5–7 mm. long:
Leaves 7–35 cm. long, 8–19 mm. broad; lip as broad as long, or broader, indistinctly 3-lobed with rounded lobes, 4–7 mm. long, 4–9 mm. broad; spur 6–10 mm. long .. .. .. .. .. .. .. .. .. 4. *cephalotes*
Spur 4 mm. long or less, about as long as, or shorter than, the lip; sepals 1·5–5 mm. long:
Sepals 1·5–2·5 mm. long; leaves 3–13 mm. broad, 5–13 cm. long, rather unequally bilobed at the apex, the lobes sometimes divergent; lip 3-lobed at base, about 1·75 mm. long; spur 1·5–3·5 mm. long, equalling or slightly longer than the lip; bracts shorter than the pedicel and ovary .. .. .. .. 7. *straussii*
Sepals 3·5–5 mm. long; leaves 7–24 mm. broad, 5–21 cm. long, slightly unequally bilobed at apex, the lobes straight or connivent; lip entire, 3·5–4·5 mm. long and broad; spur 2·5–4 mm. long, usually shorter than the lip; bracts large scarious, longer than the pedicel and ovary .. .. .. 5. *metteniae*

1. **A. clandestinus** (*Lindl.*) *Schltr.* in Beih. Bot. Centralbl. 36, 2: 138 (1918); F.W.T.A., ed. 1, 2: 462 (excl. syn. *A. recurvus* Finet and *Cummins* 87). *Angraecum clandestinum* Lindl. (1837). *Listrostachys clandestina* (Lindl.) Rolfe in F.T.A. 7: 161 (1897). *Angraecum brunneo-maculatum* Rendle, Cat. Talb. 105, t. 14, figs. 6–9 (1913). *Ancistrorhynchus brunneo-maculatus* (Rendle) Schltr. l.c. 138 (1918). *A. stenophyllus* (Schltr.) Schltr. l.c. 139 (1918). Flowers white, lip with green markings in throat.
   **S.L.:** Kuntaia (June) *Thomas* 430! Kamalu (May) *Thomas* 491! Bubuya (May) *Adames* 174! Karina (Mar.) *Deighton* 5752! Njala (June) *Deighton* 4314! **Lib.:** Ganta (Mar.) *Harley* 1796! **Iv.C.:** Lecoufle K551! Taï Forest (fr. Oct.) *Aké Assi* 6038! **Ghana:** R. Fia, Techiman, Ashanti (fr. Oct.) *Westwood* 201! Akpafu (Apr.) *Scholes* 2074! Doyi bridge, Hohoe (July) *Westwood* 19! **N.Nig.:** Jemaa *King*! Zonkwa, Zaria Prov. *G. V. Summerhayes*! **S.Nig.:** Upper Ogun Estate, Oyo Prov. (May) *Sanford* FHI 56582! Aponmu, Ondo Prov. (July) *Onochie* FHI 33371! Onitsha *Barter* 444! Eket *Talbot* 3744! Ekong to Ndingane, Oban (May) *Talbot* 943! **W.Cam.:** Buea (Jan.) *Schlechter* 12843! Also in E. Cameroun, Gabon and Congo.
2. **A. recurvus** *Finet* in Mém. Soc. Bot. Fr. 2: 46, t. 9, figs. 15–17 (1907). *A. clandestinus* Summerh. in F.W.T.A., ed. 1, 2: 462, partly, not of (Lindl.) Schltr. Flowers white.
   **Guin.:** Timbo to Conakry *Pobéguin* 788. **Ghana:** Nkawkaw (fr. Oct.) *Westwood* 208! Assin-Yan-Kumasi *Cummins* 87! **S.Nig.:** Ajagbodudu *Wright* 158! **W.Cam.:** M'bonge *Wright* 58/50! Also in Gabon and Uganda.
3. **A. capitatus** (*Lindl.*) *Summerh.* in Bot. Mus. Leafl. Harv. Univ. 11: 205 (1944). *Angraecum capitatum* Lindl. (1862). *Listrostachys capitata* (Lindl.) Rchb. f. (1865)—F.T.A. 7: 166. *Cephalangraecum capitatum* (Lindl.) Schltr. (1918)—F.W.T.A., ed. 1, 2: 461. Flowers white or pale rose coloured, yellow patches on inside of lip.
   **S.L.:** Koyeima (Oct.) *Deighton* 3360! **Lib.:** Firestone Plantation No. 3 *Linder* 64! Nimba (Sept.) *Adames* 558! **S.Nig.:** Brass R. *Barter* 1857! Okomu F.R., Benin Dist. (May) *Olorunfemi* FHI 25315! **F.Po:** Pico S. Isabel, 3,000 ft. (fr. Dec.) *Sanford* 4066! Also in Congo and Uganda.
4. **A. cephalotes** (*Rchb. f.*) *Summerh.* l.c. 206 (1944). ·*Listrostachys cephalotes* Rchb. f. (1872)—F.T.A. 7: 166. *L. glomerata* (Ridl.) Rolfe in F.T.A. 7: 166 (1897). *Cephalangraecum glomeratum* (Ridl.) Schltr. (1918)— F.W.T.A., ed. 1, 2: 461. Flowers white, with green or yellow blotches on the lip, fragrant.
   **Guin.:** Farana (fr. Jan.) *Chev.* 20469! **S.L.:** Messima, Kpukumo Krim (June) *Deighton* 5106! Bumpe, Rokupr (July) *Adames* 215! Musaia (June) *Deighton* 4316! **Lib.:** Mt. Barclay (June) *Bunting* 71! **Ganta** (May, Nov.) *Harley* 1811! 1948! **Iv.C.:** Mt. Mafa *Aké Assi* 9658! **Ghana:** Achimota (May) *Akpabla*! Aburi (May) *Westwood* 6! Awaso *Blayney* 21! Breniasi, E. Krachi (fr. Dec.) *Westwood* 159! Amedzofe (June) *Morton* A3416! **N.Nig.:** Zonkwa to Kaciya, Zaria Prov. (Apr.) *G. V. Summerhayes* 85! **S.Nig.:** Ilesha F.R., Ondo (fl. cult. Ibadan, Apr.) *Sanford* WS/1680/65! Okomu F.R., Benin (May) *Keay & Brenan* FHI 25315! Nun R. *Barter* 20106!
5. **A. metteniae** (*Kraenzl.*) *Summerh.* l.c. 209 (1944). *Listrostachys metteniae* Kraenzl. (1893). *L. braunii* Dur. & Schinz—F.T.A. 7: 166. *Cephalangraecum braunii* (Dur. & Schinz) Summerh. in F.W.T.A., ed. 1, 2: 462 (1936). Flowers white with yellow or green blotches on the lip.
   **S.L.:** Roruks (July) *Deighton* 4330! 4833! **N.Nig.:** Jemaa, Zaria Prov. *Cole* 10! **S.Nig.:** Lagos *Moloney*! **W.Cam.:** Cam. Mt., 5,000 ft. *Braun* 19! M'bonge *Wright* 58/29! Bambuluwe to Bamuko, Bamenda (July) *Daramola* FHI 41585! Also in S. Tomé, Congo, Uganda and Tanzania.
6. **A. serratus** *Summerh.* in Kew Bull. 20: 195, fig. 12 (1966). *Listrostachys braunii* Rolfe in F.T.A. 7: 167 (1897), as to *Mann* 2123. Flowers white.
   **S.Nig.:** Obudu Plateau, Ogoja Prov., 5,000 ft. (May) *Head* 95! **W.Cam.:** Cameroon Mt. 5,000 ft. (fr. Nov.) *Mann* 2123! **F.Po:** Ruiché to Caldera San Carlos, 4,000 ft. *Sanford* 4363!
7. **A. straussii** (*Schltr.*) *Schltr.* in Beih. Bot. Centralbl. 36, 2: 139 (1918). *Angraecum straussii* Schltr. (1906)— Cat. Talb. 146. *Cephalangraecum straussii* (Schltr.) Summerh. in F.W.T.A., ed. 1, 2: 462 (1936). Flowers white with green spots on the lip.
   **Iv.C.:** Oroumbo Boka (May) *Aké Assi* 8874! **S.Nig.:** Akampa Rubber Estate, Dunkwe Div. (Mar.) *Latilo* FHI 45447! Calabar *Wright* 130! Oban *Talbot* 940! **W.Cam.:** Moliwe (Aug.) *Schlechter* 15771! M'bonge *Wright* 58/48! Bamenda (June) *Cooper* 13/1! Also in Congo and Uganda.
8. **A. schumannii** (*Kraenzl.*) *Summerh.* in Kew Bull. 2: 281 (1948). *Angraecum schumannii* Kraenzl. (1889)— Cat. Talb. 146. *Mystacidium schumannii* (Kraenzl.) Rolfe in F.T.A. 7: 173 (1897). *Phormangis schumannii* (Kraenzl.) Schltr. in Beih. Bot. Centralbl. 36, 2: 104 (1918). *Tridactyle schumannii* (Kraenzl.) Summerh. in F.W.T.A., ed. 1, 2: 462 (1897). Flowers white.
   **S.Nig.:** Oban *Talbot* 895! **W.Cam.:** (?) near Victoria *Braun* 20! M'bonge *Wright* 58/15! Also in E. Cameroun and Congo.

## 55. ANGRAECOPSIS Kraenzl. in Engl., Bot. Jahrb. 28: 171 (1900).

Spur longer than the lip, 6–38 mm. long, very slightly or not at all swollen in the apical part:
Petals broader than long (measured from apex to base of central vein) broadly triangular, 1·7–2·2 mm. long, 2·3–3 mm. broad; spur 12–38 mm. long; leaves elliptical-oblong, elliptical-ligulate or ligulate, 2–11 cm. long, 4–13 mm. broad, apical lobes unequal, rounded or obtuse; inflorescences 3–9 cm. long .. 1. *ischnopus*
Petals longer than broad, triangular, 1·5–2 mm. long, 1·3–1·8 mm. broad; spur 6–9 mm. long; leaves narrowly ligulate, curved, up to 18 cm. long, 4–12 mm. broad, acute or very unequally lobed with the shorter lobe tooth-like; inflorescences up to 17 cm. long .. .. .. .. .. .. .. .. .. 2. *parviflora*

Spur about the same length as the lip, or shorter, 2–5 mm. long; petals narrowly or
    triangular-ovate, longer than broad:
  Leaves linear or narrowly ligulate, somewhat curved, 2–9·5 cm. long, 4–15 mm. broad;
      side lobes of lip slightly longer than the middle lobe; spur shorter than the lip,
      distinctly swollen at the apex, 2–3 mm. long; lateral sepals scarcely widened in the
      apical half, 3·5–4 mm. long; petals 2–2·7 mm. long, about 1·5 mm. broad; in-
      florescences 4–7 cm. long     ..      ..      ..      ..      ..    3. *tridens*
  Leaves elliptical-oblong or oblanceolate, scarcely curved, 4–10 cm. long, 1–2·5 cm.
      broad; side lobes of lip distinctly shorter than the middle lobe; spur about equalling
      the lip, not much if at all swollen at the apex, 2·7–5·2 mm. long; lateral sepals
      distinctly widened in the apical half, 3–4·7 mm. long; petals 2–2·3 mm. long, 1·5–2
      mm. broad; inflorescences 3·5–18 cm. long      ..      ..      ..      4. *elliptica*

1. **A. ischnopus** (*Schltr.*) *Schltr.* Die Orchid. 601 (1914). *Angraecum ischnopus* Schltr. in Notizbl. Bot. Gart.
    Berl. 4: 170 (1905).
      **S.L.:** Bintumane Peak, Loma Mts. (fr. Oct.) *Jaeger* 289! **Guin.:** Nimba Mts. (Sept., Oct.) *Schnell* 3729!
    3851! **N.Nig.:** Mambila Plateau, 5,500 ft. *Wimbush* in *Hb. King* 132! **W.Cam.:** Buea (May) *Deistel*!
2. **A. parviflora** (*Thou.*) *Schltr.* Die Orchid. 601 (1914). *Angraecum parviflorum* Thou., Orch. Iles Austr.
    Afr. t. 60 (1822). *Mystacidium pedunculatum* Rolfe in F.T.A. 7: 175 (1897). Flowers white or greenish
    white.
      **W.Cam.:** Buea *Gregory* 870a! Also in Tanzania, Malawi, Rhodesia, Mozambique, Madagascar, Mauritius
    and Réunion.
3. **A. tridens** (*Lindl.*) *Schltr.* in Beih. Bot. Centralbl. 36, 2: 141 (1918). *Angraecum tridens* Lindl. (1862).
    *Mystacidium tridens* (Lindl.) Rolfe in F.T.A. 7: 174 (1897). *Angraecum occidentale* (Kraenzl.) Rolfe in
    F.T.A. 7: 142 (1897). Flowers very pale green or yellow-green.
      **W.Cam.:** Buea, Cam. Mt., 4,100 ft. (Sept.) *Preuss* 965! *Gregory*! **F.Po:** 4,000 ft. (Sept.) *Mann* 646!
    Moka, 4,600 ft. (Sept.) *Melville* 418!
4. **A. elliptica** *Summerh.* in Bot. Mus. Leafl. Harv. Univ. 14: 250 (1951). Flowers pale green sometimes with
    orange tinge on lip.
      **N.Nig.:** Kuchamfa, Zaria Prov. (Sept.) *King* 57! Sha, Plateau Prov., 4,300 ft. *King* 57a! **W.Cam.:**
    Wiem L., Bamenda (Aug.) *Savory* 338! Buea, Cam. Mt. *Gregory* 194! 556! (?) Also in Congo.


## 56. EGGELINGIA Summerh. in Bot. Mus. Leafl. Harv. Univ. 14: 235 (1951).

Epiphyte; stems elongated, slender; leaves distichous, ligulate or oblong-ligulate,
2–5 cm. long, 5–9 mm. broad, unequally 2-lobed at apex; inflorescence very short,
2–6-flowered; flowers white or yellowish-white; sepals 3–5 mm. long; lip broadly
lanceolate, 3–3·5 mm. long; spur 2·5–4 mm. long, much swollen..      ..      *clavata*

**E. clavata** *Summerh.* l.c. 238 (1951).
      **Iv.C.:** Lakato to Sassandra (Nov.) *Aké Assi* 9229! **Ghana:** Ofinso, Ashanti (Aug.) *Cox* 99! *Westwood*
    126! Jegeti, Accra to Kumasi (Aug.) *Westwood* 166! Also in Gabon and Congo.


## 57. TRIDACTYLE Schltr., Die Orchid. 601 (1914).

Leaves needle-like, almost terete, with a groove along the upper surface, up to 12 cm.
    long, but only 1–2 mm. diam.; inflorescences very short, up to 7 mm. long, few-
    flowered; sepals 1·5–2·5 mm. long; lip trilobed above the middle, the side lobes
    tooth-like, triangular, much shorter than the middle-lobe; spur slender, 6–8 mm.
    long      ..      ..      ..      ..      ..      ..      ..      ..    5. *tridentata*
Leaves not needle-like, more or less flattened though sometimes fleshy, more than 4 mm.
    broad:
  Spur distinctly shorter than the lip; lip trilobed:
    Inflorescences less than 5 mm. long, few-flowered; spur 4·5–6 mm. long, club-shaped;
      leaves linear-lanceolate or narrowly ligulate, 4–7·5 cm. long, 4–9 mm. broad,
      unequally bilobed at apex; lip 6–7 mm. long, side lobes spreading; sepals 3–4·5 mm.
      long     ..      ..      ..      ..      ..      ..      ..      ..    6. *lagosensis*
    Inflorescences 2–6 cm. long, many-flowered; spur less than 1 mm. long, not at all
      swollen above the base; leaves ligulate, 6–11 cm. long, 7–10 mm. broad, only
      slightly unequally bilobed at apex; lip 3 mm. long, side-lobes projecting forward;
      sepals 2·5–3·5 mm. long      ..      ..      ..      ..      ..    9. *brevicalcarata*
  Spur much longer than the lip, often more than twice as long:
    Spur 3 cm. long or longer:
      Side lobes of lip reduced to teeth; lip 6·5 cm. long with long narrow acute middle
        lobe; spur 3 cm. long, slightly thickened towards apex; leaves linear-oblong,
        broadly and obtusely bilobed at apex, 5–6·5 cm. long, 7–9 mm. broad; sepals
        7–7·5 mm. long; inflorescence unknown, but probably short ..      3. *muriculata*
      Side lobes of lip much longer than middle lobe, divided into short or long fimbriae at
        apex; lip 7–14 mm. long (from apex of ovary), middle lobe triangular-lanceolate,
        2–5 mm. long, side lobes 7–9·5 mm. long; spur 3·5–8·5 cm. long, not at all thick-
        ened; leaves ligulate, 8–20 cm. long, 7–30 mm. broad; sepals 6–9 mm. long;
        inflorescences 3·5–11 cm. long, up to 12-flowered      ..      ..      12. *gentilii*
    Spur less than 2·5 cm. long:
      Side lobes of lip much diverging, as long as or longer than the middle lobe; in-
        florescence up to 8 cm. long, many-flowered; leaves ligulate or lanceolate-ligulate,
        unequally bilobed at apex, 5–20 cm. long:

Side lobes of lip divided in upper part into numerous hair-like portions; leaves
narrowly ligulate, not much narrowed in upper part, 4–10 mm. broad, only rarely
broader; spur slender, 11–23 mm. long; sepals 3·5–5·5 mm. long   11. *bicaudata*
Side lobes of lip entire, or rarely very shortly divided at the apex; leaves ligulate or
lanceolate-ligulate, distinctly but gradually narrowed in the upper part, 1–2 cm.
broad:
   Spur very slender, scarcely thickened in the apical part, 7–11 mm. long; sepals
   3–5·5 mm. long; lip 3·5–6 mm. long, 5·5–8 mm. broad across the side lobes
                                    10. *tridactylites*
   Spur distinctly swollen in the apical two-thirds, 6–8·5 mm. long; sepals 2·5–4·5
   mm. long; lip 2·5–4·5 mm. long, 3·5–5·5 mm. broad across the side lobes
                                      8. *armeniaca*
Side lobes of lip reduced to teeth, much shorter than the middle lobe, or absent:
   Inflorescences up to 4 cm. long; leaves linear-ligulate, unequally bilobed at apex,
   up to 18 cm. long and 13 mm. broad; flowers not lepidote; sepals 2–4 mm. long;
   lip equalling sepals, ovate with a short tooth on each side below the middle;
   spur 5–8 mm. long   ..   ..   ..   ..   ..   ..   .. 7. *fusifera*
   Inflorescences very short, less than 5 mm. long, 1–3-flowered; ovary and sepals
   with numerous black scurfy hairs or scales (lepidote); basal auricles of lip
   prominent:
      Leaves very fleshy, more or less deltoid or V-shaped in section, the upper surface
      concave with a narrow groove along the centre, the lower surface convex, some-
      times with a rounded keel, lanceolate-ligulate, distinctly tapering towards the
      apex, 3–8 cm. long, 6–10 mm. broad, unequally bilobed; sepals 3·5–5 mm. long;
      lip lamina entire or side lobes extremely small; spur slender, 10–15 mm. long
                                      2. *crassifolia*
      Leaves flat or sometimes convex as a result of the recurving of the margins, some-
      times rather fleshy, ligulate, oblong or elliptical-oblong, usually not much
      tapering in the upper part:
         Leaves rather fleshy, broadly ligulate, oblong or elliptical-oblong, broadly and
         obtusely bilobed at the apex, 2–8 cm. long, 7–20 mm. broad; sepals 3·5–5·5 mm.
         long; lip lamina usually entire but basal auricles well developed and acute;
         spur 6–15 mm. long   ..   ..   ..   ..   ..   .. 1. *anthomaniaca*
         Leaves scarcely fleshy, ligulate or narrowly oblong, unequally bilobed at the
         apex, 3–12 cm. long, 7–17 mm. broad; sepals 4–6 mm. long; lip lamina entire,
         basal auricles triangular, acute; spur 10–15 mm. long  ..   4. *oblongifolia*

1. **T. anthomaniaca** (*Rchb. f.*) *Summerh.* in Kew Bull. 2: 284 (1948). *Listrostachys anthomaniaca* Rchb. f.
(1877). *Angraecum lepidotum* Rchb. f. ex Rolfe in F.T.A. 7: 146 (1897). *Tridactyle lepidota* (Rchb. f.
ex Rolfe) Schltr. (1918)—F.W.T.A., ed. 1, 2: 463. Flowers green, yellow-green, pale yellow or white.
**S.L.:** Njala (Oct., Nov.) *Deighton* 2566! 4373! Yagoi, Jong R. (Nov.) *Adames* 102! Massa, Peri (Nov.)
*Deighton* 5974! Roruks (Nov.) *Deighton* 4374! **Lib.:** Sakimpa (Dec.) *Harley* 1843! **Iv.C.:** Moossou
(Nov.) *de Wilde* 979! *Aké Assi* 5450! **S.Nig.:** Jameson R., Sapoba (Nov.) *Ross* 228! Warri *Gregory*
507! Onitsha (May) *Onochie* FHI 35807! Oban *Talbot*! Ibeso, Eket (May) *Onochie* FHI 32090! **W.Cam.:**
Cam. Mt. *King* 123! Bakundu F.R., Kumba (June) *Adebusuyi* FHI 44047! Eastwards to Kenya,
Tanzania, Zambia and Mozambique.
2. **T. crassifolia** *Summerh.* l.c. 285 (1948). Flowers green, greenish yellow or ochre-brown.
**Ghana:** Efiduase *Cox* 33! Awaso (Feb.) *Blayney* 2! Juaso, Ashanti (Jan.) *Westwood* 110! Also in Gabon
and the Congos.
3. **T. muriculata** (*Rendle*) *Schltr.* in Beih. Bot. Centralbl. 36, 2: 146 (1918). *Angraecum muriculatum* Rendle,
Cat. 105, t. 14, figs. 3–5 (1913).
**S.Nig.:** Oban *Talbot* 904!
4. **T. oblongifolia** *Summerh.* in Kew Bull. 2: 286 (1948). Flowers green or whitish green, turning yellow or
yellowish orange.
**S.L.:** Roruks? (cult. Njala, Nov.) *Deighton* 4393! **Iv.C.:** Sassandra R., near Gribo rapids (Nov.)
*de Wilde* 3293! **S.Nig.:** Akure F.R., Ondo Prov. (Oct., Nov.) *F.H.I. Staff* FHI 42061! Also in C. African
Republic, Congo and Uganda.
5. **T. tridentata** (*Harv.*) *Schltr.* in Engl., Bot. Jahrb. 53: 603 (1915), in obs. *Angraecum tridentatum* Harv.,
Thes. Cap. 2: 6 (1863). Flowers whitish, pale ochre-yellow or salmon-pink.
**Guin.:** Dalaba (Aug.) *Adames* 332! Nimba Mts. (Oct.) *Schnell* 3800! Kissidougou (Oct.) *Jaeger* 2153!
Guéckédou (July) *Jac.-Fél.* 1038! **S.L.:** Roruks (Aug.) *Deighton* 4331! **Iv.C.:** Morocro (Feb.)
*Oldeman* 988! **Ghana:** Kwapon (Dec.) *Oldeman* 740! Mlabo, Wurupong, Kpandu (fr. May) *Morton*
A3938! Also in the Congos, Uganda, Kenya, Tanzania and Natal.
6. **T. lagosensis** (*Rolfe*) *Schltr.* in Beih. Bot. Centralbl. 36, 2: 145 (1918). *Angraecum lagosense* Rolfe in F.T.A.
7: 145 (1897).
**S.Nig.:** *Barter*! Lagos *Moloney*! Eket *Talbot* 3298!
7. **T. fusifera** *Mansf.* in Fedde, Repert. 36: 63 (1934). Flowers apricot yellow.
**S.L.:** *Bowden*! **Lib.:** Sinoe R. (Dec.) *Dinklage.* Nimba (Mar.) *Adam* 21110! 21166! Also in E. Cameroun.
8. **T. armeniaca** (*Lindl.*) *Schltr.* l.c. 143 (1918). *Angraecum armeniacum* Lindl. (1839)—F.T.A. 7: 147.
*A. whitfieldii* Rendle (1895)—F.T.A. 7: 148. Flowers orange.
**Guin.:** Socourala *Chev.* 20512! Nimba Mts. (Mar.) *Schnell* 697! Mt. N'zo (Mar.) *Fleury* in Hb. *Chev.*
21033! **S.L.:** *Whitfield*! Boma, Pukumu Krim (Jan.) *Jackson* 17! **Lib.:** Bili (Mar.) *Harley* 1792!
**Ghana:** Kibi Ravine (Dec.) *Morton*! (Nov., cult. Legon) *Morton*!
9. **T. brevicalcarata** *Summerh.* in Kew Bull. 2: 295 (1948). Flowers white, fragrant.
**Ghana:** Kumasi (Aug.) *Cox* 98! Also in Gabon.
10. **T. tridactylites** (*Rolfe*) *Schltr.* in Beih. Bot. Centralbl. 36, 2: 148 (1918). *Angraecum tridactylites* Rolfe
(1888)—F.T.A. 7: 147. *Aëranthus deistelianus* Kraenzl. in Engl., Bot. Jahrb. 33: 75 (1902).
*Mystacidium ledermannianum* Kraenzl. l.c. 51: 393 (1914). Flowers yellow, orange or brownish orange,
sometimes tinged with green, fragrant.
**Guin.:** Nimba Mts. (Feb., fr. Apr.) *Schnell* 4621 1093! **S.L.:** Loma Mts. *Jaeger* 1190! Alikalia (Feb.)
*Bunting* 30! **Iv.C.:** Mt. Dou (Jan.) *Portères* 552! **N.Nig.:** Gurara Falls, Izom (July) *Onochie* FHI 35380!

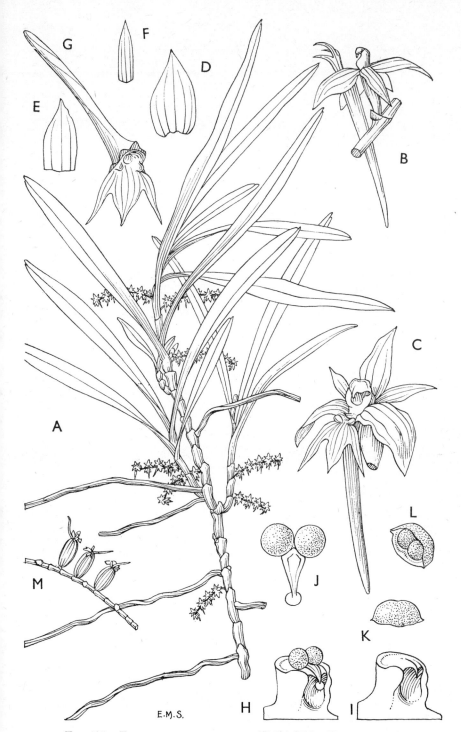

FIG. 405.—TRIDACTYLE TRIDACTYLITES (*Rolfe*) *Schltr*. (ORCHIDACEAE).

A, habit, × ⅓. B, flower, side view, × 5. C, flower, front view, × 6. D, lateral sepal, × 6.
E, dorsal sepal, × 6. F, petal, × 6. G, lip and spur, × 6. H, column with pollinia,
anther cap removed, × 12. I, column with anther cap and pollinia removed, × 12. J,
pollinia, caudicle and viscidium, × 24. K, anther cap, dorsal view, × 12. L, anther cap,
ventral view, × 12. M, fruiting spike, × 1. A—L from *Brenan* 8644. M from *Schnell* 1093.

Zonkwa to Kaciya, Zaria Prov. (Apr.) *G. V. Summerhayes* 86! Assob Falls, Jos (Apr.) *Keay & Jones* FHI 37618! Amban, Plateau Prov. *King*! **S.Nig.:** Lagos *Rowlands*! Mt. Orosun, Idanre, Ondo Prov. (Jan.) *Brenan* 8644! Akure F.R., Ondo Prov. (Mar.) *Jones* FHI 1093! **W.Cam.:** Buea *Deistel* 593! *Schlechter* 12840! R. Katsina bridge, Wum (Feb.) *Hepper & Charter* 1928! **F.Po:** Pobledo alto de Balachá (Jan.) *Guinea* 1447! Also on S. Tomé, Principe, and widely spread on the mainland as far as Mozambique.

11. **T. bicaudata** (*Lindl.*) *Schltr*. Die Orchid. 602 (1914). *Angraecum bicaudatum* Lindl. in Hook., Comp. Bot. Mag. 2: 205 (1837). *A. fimbriatum* Rendle (1916)—F.T.A. 7: 148. Flowers white, pink or yellow, often with a green tinge.
   **S.L.:** Sefadu (July, Aug.) *Deighton* 4834! **Lib.:** Nimba (Oct.) *Adames* 673! **Iv.C.:** Soubré, Sassandra R. (Nov.) *de Wilde* 3323! **Ghana:** Kumawu (Oct.) *Cox* 60! *Cansdale* 135! Atweibo Mt. (Sept.) *Cox* 103! Mampong, Ashanti (Oct.) *Cansdale* 130! Amedzofe (Aug.) *Westwood* 29! **N.Nig.:** Mongu F.R., Plateau Prov. *King* 15! Widely spread in tropical Africa, also in Natal and Cape Province.

12. **T. gentilii** (*De Wild.*) *Schltr*. in Beih. Bot. Centralbl. 36, 2: 145 (1918). *Angraecum gentilii* De Wild. (1903). Flowers white, cream, pale yellow or orange-buff, faintly fragrant.
    **Ghana:** Techiman, Ashanti (May) *Westwood* 101! **N.Nig.:** Amban, Plateau Prov. (June) *King* 192! Wamba to Jos, Plateau Prov. (July) *Sanford* 1118/65! Also in E. Cameroun, Congo and Zambia.

## 58. NEPHRANGIS Summerh. in Kew Bull. 2: 301 (1948).

Stems elongated, slender, usually more or less hanging, up to 35 cm. long, often branched ; leaves needle-like, terete, usually curved, 1·5–9 cm. long, 1–2 mm. diam. ; inflorescences very short, up to 1 cm. long, 1–3-flowered ; sepals broad, obtuse, 1·5–2·7 mm. long ; petals narrowly lanceolate ; lip total length 4–5 cm., bilobed lamina 3·5–5 mm. broad ; spur gradually narrowed to apex, 6–9 mm. long .. .. .. *filiformis*

**N. filiformis** (*Kraenzl.*) *Summerh*. in Kew Bull. 2: 302 (1948). *Listrostachys filiformis* Kraenzl. (1895)—F.T.A. 7: 169. Sepals yellowish, brownish or orange, lip white.
**Lib.:** Ganta (June) *Harley* 1644! 1815! Also in Congo, Uganda and Zambia.